Global Conference on Teaching, Assessment, and Learning in Education (GC-TALE 2017)

SHS Web of Conferences Volume 42 (2018)

Sanur, Bali
28 - 29 August 2017

Editors:

A.G. Abdullah
J. Foley
I.G.N.A. Suryaputra
A. Hellman

ISBN: 978-1-5108-5661-5

Printed from e-media with permission by:

Curran Associates, Inc.
57 Morehouse Lane
Red Hook, NY 12571

Some format issues inherent in the e-media version may also appear in this print version.

Printed by Curran Associates, Inc. (2018)

For additional information, please contact EDP Sciences – Web of Conferences
at the address below.

EDP Sciences – Web of Conferences
17, Avenue du Hoggar
Parc d'Activité de Courtabœuf
BP 112
F-91944 Les Ulis Cedex A
France

Phone: +33 (0) 1 69 18 75 75
Fax: +33 (0) 1 69 28 84 91

contact-edps@webofconferences.org

Additional copies of this publication are available from:

Curran Associates, Inc.
57 Morehouse Lane
Red Hook, NY 12571 USA
Phone: 845-758-0400
Fax: 845-758-2633
Email: curran@proceedings.com
Web: www.proceedings.com

TABLE OF CONTENTS

Statement of Peer review

In submitting conference proceedings to SHS Web of Conferences, I certify to the Publisher that I adhere to the **Policy on Publishing Integrity** of the journal in order to safeguard good scientific practice in publishing.

1. All articles have been subjected to peer review administered by the proceedings editors.
2. Reviews have been conducted by expert referees, who have been requested to provide unbiased and constructive comments aimed, whenever possible, at improving the work.
3. Proceedings editors have taken ail reasonable steps to ensure the quality of the materials they publish and their decision to accept or reject a paper for publication has been based only on the merits of the work and the relevance to the journal.

Title, date and place of the conference.

Global Conference on Teaching, Assessment, and Learning in Education (GC-TALE)
28-29 August 2017
Hotel Grand Inna Beach, Sanur, Bali

Proceedings editor(s)

Ade Gafar Abdullah
Joseph Foley
I Gusti Ngurah Agung Suryaputra
Anette Hellman

Scaffolding in the Indonesian classroom from theory and practise

J. A. Foley

Director of the Language Center

Southeast Bangkok College, Bangkok

Abstract

Too often descriptions of language learning have focused primarily on the language components or the structural features. However, there is an increasing recognition that patterns of discourse realised through text rather than a sentence based approach to language description is turning away from discrete form-based aspects of language learning to focus on the effectiveness of how learners get their messages and ideas across to others. This paper will take as an example the process of writing as it is probably the most difficult of the language skills required of our learners. The cultural context is that of the ASEAN classroom in general and how we can go about *scaffolding* the writing process to help prepare Indonesian students for their potential role in the ASEAN of the future.

Keywords: scaffolding, writing, genre, Indonesian classroom, socio-cultural theories, CLT.

Introduction

Many EFL/ESL textbooks in the past have been largely influenced by psychological views of language inspite of the claim that they are based on Communicative Language Teaching (CLT). Structural behavorists emphasised inductive learning of individual *sentence*s one at a time. What was considered as 'learned'was in fact *form* but not the *function*.

Take the following example from a EFL textbook to illustrate what we mean.

Target Structure: *"If ___ were___, ___ would ___"*

Lesson procedure:

1. Teacher draws the following substitution table on the board:
2. Teacher says the following example of the target structure and asks the whole class to repeat it after him:

   If I were a rich man, I would buy a big house.

3. Teacher asks individual students to repeat the same example after him, correcting mistakes and making each student say the sentence until he gets it 100% correct.

4. Teacher repeats the process with the next example:

 https://doi.org/10.1051/shsconf/20184200001

If he were the principal, he would give us a holiday.

5. Teacher points out the target structure, giving further examples.

6. Teacher gives the following prompts, and asks the whole class or individuals to substitute them into the target structure, correcting mistakes in the responses and making the students repeat the correct response:

a famous movie star	*sign a million autographs*
a doctor	*heal the sick for free*
a magician	*turn rice into chocolate*
an engineer	*build bridges for the village*
a millionaire	*give money to the poor*

The focus here is on the form of the *if clause* in the sentence. But what role does the *if clause* play? What is its function? The *if adverbial clause* is, normally, when placed at the begining of a sentence, to foreground emphasis. In speech this would be done by putting the stress on the *if clause* in the speaker's intonation pattern.

What we have traditionally found in our textbooks can be summarise as follows:

- Modeling of all learning by the teacher
- The mother tongue was not to be used in the classroom, only the target language.
- New language points to be introduced and practiced 'situationally'.

- Minimizing of vocabulary but taught in context until all common grammatical struc tures had been learned
- Graded grammar structures expanded from simple to more complex forms.

However, one central deficiency in such approaches is that: they do not pay sufficient attention to the *social* aspects of language and learning. Both language and learning do not merely have psychological dimensions – they are very much social in nature. More socially and that means more *culturally* oriented explanations of language learning are needed for a complete account of how language learning takes place'.

Socio-cultural theories

In the early 60s saw the publication in English of the works of the Russian psychologist Lev Vygotsky (1896-1934).

For Vygotsky 'cultural development' plays the following roles:

- *Object regulation*: the materials and cultural elements used for learning (textbooks, dictionaries and so on)
- *Other regulated*: teachers, parents, other students and the social environment
- *Self regulation:* to achieve a degree of mastery or control of whatever we are doing and most importantly language as a tool for thinking.

In this context Vygotsky saw *Scaffolding* as essential to attain *Self-regulation.*

Good teaching is about guidance through providing supporting 'structures' to help the students move from one stage of development to the next. For example, the process of *scaffolding* in terms of language teaching would provide structured outlines and lists of possible grammatical structures or vocabulary for students' first attempts at writing. How much scaffolding is needed will depend on the level of the students and the nature of the task.

Within this theory, the concept of *object-regulation* stresses the importance of providing *model texts* so that the student starts learning how language is used to achieve a purpose. On the other hand, *co-operative learning* that is pair-work and group-work, for Vygotsky, was more about opportunities for the more able peer to help the less able. Consequently, tasks and activities should aim to provide for peer guidance in the use of language, rather than just provide for some form of simple oral practise in the language.

The Indonesian classroom

The emphasis in many Indonesian schools and indeed in many classrooms in the region tends to be on mechanical tasks, such as form based grammar drills, memorizing vocabulary and testing. For example, in too many writing programs, the typically writing assignment is a first and final draft, written in class. Students often receive instruction about writing only after the writing tasks have been completed, meaning that writing is more often *assessed than taught.* Teachers' feedback on student writing tends to focus on correction of errors at the sentence level or generalized comments.

The following are examples of first tertiary level students' writing in the first week and the last week of their course (duration 14 weeks) where there was limited *intervention* in terms of the structure of the text or'scaffolding provided in the course.

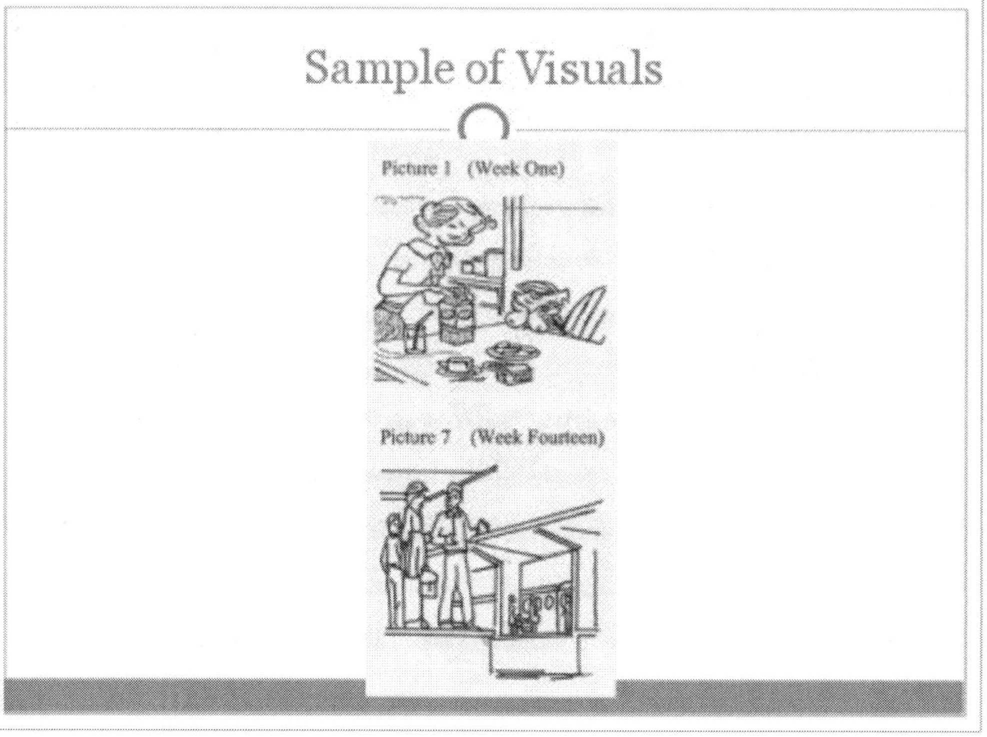
Sample of Visuals
Picture 1 (Week One)
Picture 7 (Week Fourteen)

Student G

Analytical from Writing 1(week 1):

I1 **In the picture**, I see (Me pre) one mother and one son.|| I2 **They** love (Me pre) each other so much.|| I3 **They** have (R pre) a good relationship.|| I4 *And* **Their family** is (R pre) the family [[that be very happy.]]|| I5 *And* **in the picture** may be (R pre) in morning,|| I5.1 **the boy** will go to (M future) school.|| I6 **They** have (M pre) a breakfast.|| I7 **Breakfast** have (R pre) some milk, a cup of coffee, bread and other.|| I8 **The mother** is pouring (M pre) some milk into glass for her son.|| I9 *And* **the boy** feels (R pre) bored|| D9.1 *because* **he** must drinks (M pre) milk everyday.|| I10 **He** wants to eat (M pre) other breakfast [[that not be milk, bread.]]|| I11 *But* **he** must eats (M pre) **it** || D11.1 *because* **it** is (R pre) only one [[that easy to eat.]]|| I12 *And* **the boy** is waiting to go (M pre) to school|| I12.1 *but* **he** must waits (pre) his father send (M) him to school.|| I13 *But* **his father** isn' t getting dressed (M pre).|| I14 **He** bored to wait (Me past) || D14.1 *because* **he** want to go (M pre) to school.|| I15 **He** want to play (M pre) with his friends.|| I16 *And* **he** is thinking to play (M pre) the games with his friends.|| I17 *And* **He** is thinking (Me pre) about his girl friend.|| I18 **He** is thinking (Me pre) about [[what should **he** buy anything [[that the best gift to his girl friend.]]|| D19 *Because* **tomorrow** is (R pre) his girl friend's birthday.|| I20 *And* **he** is thinking (Me pre) [[how to surprise to his girl friend.]]|| I21 **He** is having (R pre) a poppy love.||

Lexical String 1:

1	son			
2		they		
3		They		
4				
5				
5.1	the boy			school
6		They		
7				
8	her son			
9	the boy			
9.1	he			
10	He			
11	he			
11.1				
12	the boy			school
12.1	he – him			school
13				
14	He			
14.1	he			school
15	He			
16	he			
17	He		girl friend	
18	He			
18.1	he		girl friend	
19			girl friend	
20	he			
20.1			girl friend	
21	He			

There are several areas that could be discussed with this student but probably the most important is the use of the nominal group (NG). This is the grammatical unit which would allow the widest range of descriptive meanings to be expressed (Foley 2011). There are three possible parts to the structure of the NG: *Premodifier, Head* and *Postmodifier*. The ability to construct a complex NG is considered an indicator of developing control in a language. EFL learners tend to employ short NGs depending on how much control they have of the target language. In some languages, the NG is constructed differently as found in Thai where there is no pre-modification kao+pat+sorng+jarn (fried+rice+two+plates).

In this example quoted above

The majority of the NGs in theme position are pronouns (*They, He*) or single nouns (*Breakfast*).

* D+H (*Their family, the boy*) are less frequent

* In complement position H+post-m [relative clause] is used but often the structure is incorrect: *(that be very happy*, that not be milk, bread*, that easy to eat*, that the best gift to his girl friend*).*
* The text structure is a description with little structure, very repetitive and few extended nominal groups showing limited control over the target language.

Analytical from Writing 7 (week 14):

[I1] **There** are [E pre] many people in this picture.|| [I2] **Every one** looks [R pre] hurry to go to works.|| [I3] **In this picture** is [R pre] in morning.|| [I4] **They** will go to work [M future] by BTS.|| [I5] **They** are waiting [Me pre] for BTS.|| [I6] **There** are [E pre] two line of BTS.|| [I7] **One of BTS** are going to [M future] final station|| [I7.1] *and* **another BTS** will go back [M future] the first station.|| [I8] **There** is [R pre] one of BTS [[which arrived already.]]|| [I9] **Two man** [[[who are waiting for BTS [[which not arrived yet.]]|| [I10] **A man** [[who is wearing hat,]] holding [M pre] bag|| [I10.1] *and*^^ put [M pre] his suit on his arm,|| [I10.2] **he** just lived [M past] in Bangkok 2 day ago for

M future Me pre
work.|| [I11] **He** is going to work in first day,|| [I11.1] **he** do not know the way to go to
V pre
office [[which he works.]]|| [I12] *So* **He** go to ask with another man for some way [[that
V pre
he can go to work by BTS.]]|| [I13] **Another man** tell him [[how to go to there.]]|| [I13.2]
Me pre
but **a man** [[who is wearing hat]]do not understand.|| [I14] *Therefore* **A man** [[who is
V pre
not wearing hat]] ask him|| [D14.1] *what* **office** [[that he work.]]|| [D15] *When* **a man**
V past Me pre
answered,|| [I15.1] **a man** [[who is not wearing a hat]] know || [D15.2] *that* **a man** [[who is
M pre V pre
talking with him]] work in same office.|| [I16] *So* he tell a man[[who is wearing hat]]||
M future R pre
[D16.1] *that* **he** are going to work in same office,|| [I16.2] **a man** [[who wearing hat]] is
R pre
very happy|| [D16.3] *that* **he** has new friend in new office already.|| [I17] **A man** [[who is
V pre
not wearing hat also]] ask him about his home.|| [D18] *When* **a man** [[who is wearing
Me pre E pre
hat]] know || [D18.1] *that* **their homes** are nearly.|| [I19] **Since that time,** they
M future M pre R pre
will go to work || [I19.1] *and*^^ go back home together|| [I19.2] *and* **they** become close
friend.||

Lexical String 7:

1				
2				
3				
4	BTS			
5	BTS			
6	BTS			
7	BTS			
7.1	BTS			
8	BTS			
9	BTS			
10		wearing a hat	A man	
10.1				
10.2			he	
11			He	
11.1			he	
12	BTS		He-he	another man
13			him	another man
13.1				
13.2		wearing hat	a man	
14		wearing hat	him	a man
14.1			he	
15			a man	
15.1		wearing a hat		a man
15.2			a man	
16		wearing a hat	a man	he
16.1				he
16.2		wearing hat	a man	
16.3			he	
17		wearing hat	him	A man
18		wearing hat	a man	
18.1				
19				
19.1				
19.2				

Again, the majority of NGs are pronouns (*There, They*) or single nouns H or D+H (*another BTS, Another man*) or N+H (*One of BTS*).

* D+H+post-m (relative clause) is used extensively:
A man who is wearing a hat
A man who is not wearing a hat
A man who is talking with him

* The text structure is a description with little structure, very repetitive and again limited use of extended nominal groups. The use of the relative clause as embedded post modifier simply increases the repetitive effect.

What makes this writing unsatisfactory is principally because of the repetitive nature of the writing and the student's inability to develop the use of the descriptive power of the NG.

Theme

Theme is the starting point in English of what the speaker or writer is on about.

Speakers/writers use the first element of the clause to indicate the topic and is the way a text develops cohesion. There are three main types of Theme

Topical theme (TT) [often the subject]

Hamlet is considered Shakespeare's greatest play

Marked Topical theme (MTT) [adverbial]

For 30 years, Shakespeare worked in the theatre

Textual (T)

However, he moved to London later

The students development by Theme

In the picture (MTT)

They (TT)

They (TT)

Their family (TT)

The (TT)

The boy (TT)

They (TT)

They (TT)

Breakfast (TT)

The Mother (TT)

The boy. (TT)

He (TT)

He (TT)

He (TT)

He (TT)

He (TT)

Because (T)

It (TT)

The boy (TT)

He(TT)

He (TT)

He (TT)

He (TT)

His father (TT)

The overall use of either the simple pronoun or a single modified and Head is characterisrtic of immature writing

Scaffolding

Language is scaffolded through guidance and interaction in the context of shared experience. The teacher inducts learners into the linguistic demands of these genres that are important to participation in the school and wider community. Gradually responsibility is shifted to the learners as they gain control of the genre. However, before developing a genre- based programme, the teacher or institution will need to conduct an analysis of the teaching/learning context. At its broadest level, this should take in account the culture, history and tradition of the educational setting, English proficiency, learner expectations and assessment requirements.

To illustrate both the textual and common language features the following are examples of *modeling* as part of the *scaffolding* process (Derwianka, 1996).

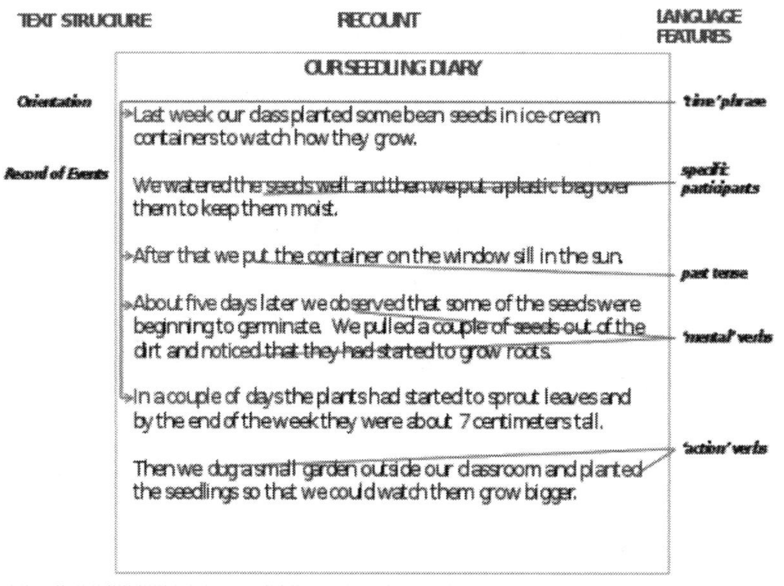

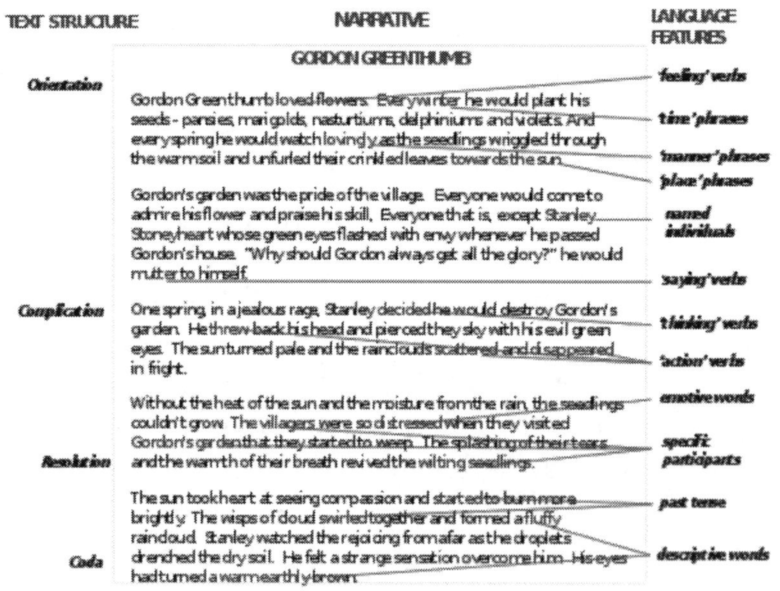

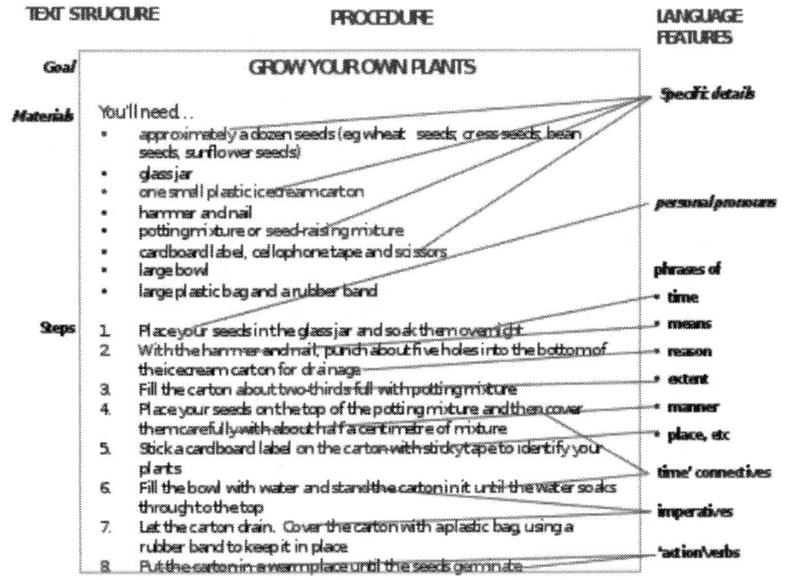

The Volvo Dimension [ad created by the Volvo team to sell the new S90

People see in Volvo cars lots of different things

Some see power and performance

Others strength

Still others advance technology

Each one of them is right

But, in fact, it is all these elements put together that make a Volvo

The S 90 engine illustrates just what we mean

It's not only powerful

It consumes power with precision and economy.

The S 90 has been hailed as a breakthrough in automotive technology.

The computerized v-6 engine includes among its many notable features the highly sophisticated ECCS or Electronic Concentrated Control System, which controls airfuel mix, fuel injection, ignition timing and other vital functions with extraordinary accuracy.

In short, the innovative design of the S90 doesn't come from a mere desire to impress you with power.
It's an innovation that literally brings you the best of everything.
And it succeeds because it has the intangible yet very real quality that makes Volvo what it is. The Volvo Dimension

The scaffolding cycle

In scaffolding the teaching/learning cycle is a process that would be expected to cover several classroom sessions. This is especially true in an EFL/ESL situation where language problems would be an additional focus. The starting point is always *Preparing the Context* and the following explanation of each stage is described below (Shape and Thompson, 1998).

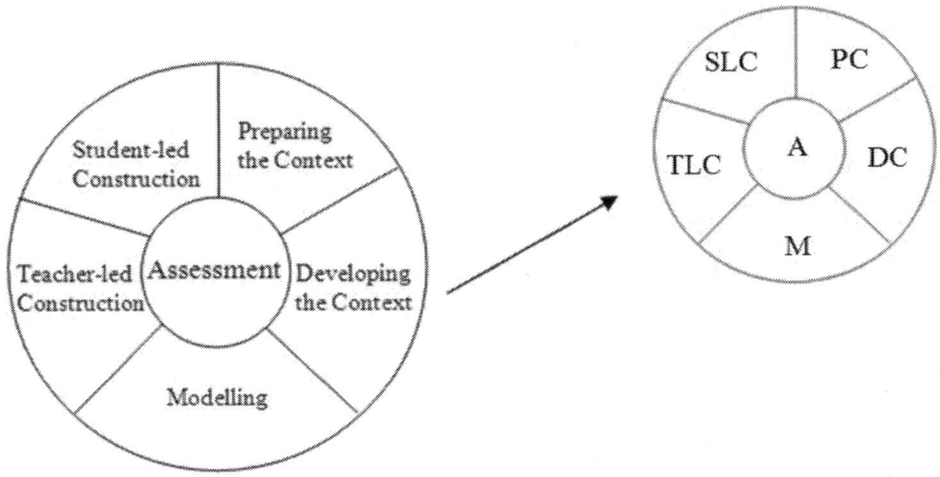

PREPARING THE CONTEXT

In preparing the context of a unit of work teachers need to ask themselves the following questions:

- what is this topic about?
- what do students already know about this topic?
- what concepts and language should be focused on to allow students to comprehend the topic?

The second and third questions are important because students bring both their formal school knowledge (however limited) and their everyday knowledge to the topic

DEVELOPING THE CONTEXT

In developing the context, the teachers builds up knowledge of the topic (the field) through other related reading materials. This moves the students from an everyday understanding to a specialized understanding of the topic.

This part of the scaffolding will vary according to the amount of information students need. For example, students could research information from the library or watch video as part of information gathering

During this phase, students may have to engage with a lot of written language because of the level of technicality and abstraction with which some information is presented when it is written down. Consequently, students will need to be supported as they try to unpack the content.

They will need to explore the ideas orally and build up their understanding through talking about the information and issues presented.

Once students have actively engaged with the content through talking about it, they will need support to express what they have learnt in written language.

MODELING

In this stage in scaffolding the teacher provides model/models of text that approximate what will be required of students. For example, the text using *recount* or *procedure* illustrated above.

TEACHER-LED CONSTRUCTION

At this stage the teacher and the students jointly construct a piece of writing similar to the model previously discussed. The support of the teacher at this stage is critical in reinforcing linguistic features of the text by referring to the language activities as identified in the model.

STUDENT-LED CONSTRUCTION

Here the students independently construct texts of their own similar to the model. It is important that the teacher asks the students to write in the same genre and not ask them to write in a different way.

What often happens in classrooms is that teachers and students spend a considerable amount of time building up the field knowledge of the topic by engaging in a range of activities and looking at a variety of different texts. This culminates in an assignment or task which is a written text of some kind.

While students are working on their text they could be conferencing with other students or the teacher for guidance on both the content and language to be used in their writing.

ASSESSMENT

Assessment lies in the center of the model, with the teacher monitoring the ongoing progress of students at all stages. The teacher determines student understanding of both the content and language demands of the unit of work and focuses on areas of need or passes quickly through areas with which students are already familiar.

Pedagogical issues: scaffolding in the classroom.

As can be seen from the example of scaffolding illustrated here, all the major language skills of listening, speaking reading and writing are involved at the same time.

There is a danger, however, that scaffolding such genres may mean that they are taught as 'items' in a textbook. Rather, we should be asking our students: *what is my purpose in writing/ who are my audience*? Students are often simply taught the name and stages of the genre. But by focusing first on the *social purpose*, students would then be able to understand why the text unfolds in this way, instead of simply following a recipe. Contemporary notions of genre emphasize the flexibility and rhetorical nature of genres rather than formulaic and static rules or conventions.

Teachers need to be aware of and explicitly teach the strategies they expect students to use in the classroom and in real life. This means helping students develop skills for questioning and discussion that enables them to develop the language capacities that display their knowledge. Scaffolding can help learners gain access to educational discourses (the sciences, economics and so on) that they may not otherwise be familiar in their daily lives. At the same time, this can provide them with conceptual tools for reflecting on language as they engage in activities that help them develop their new knowledge.

References

Derewianka, B. 1996. *Exploring the Writing of Genres.* Nottingham: The United Kingdom Reading Association (UKRA)

Foley, J.A. 2011. *Grammar, Meaning and Discourse.* Bangkok: Assumption University Press.

Foley, J.A. 2012. *Unscrambling the Omelette: Second Language Acquisition: Social and Psychological Dimensions.* Bangkok: Assumption University Press.

Johnson, K. and Johnson, H. 1998. Communicative methodology. In K. Johnson, and H. Johnson (eds.), *Encyclopedic Dictionary of Applied linguistics.* Oxford: Blackwell.

Sharpe, T. and Thompson, J. 1998. *Accessing Learning.* Sydney: NSW Catholic Education Office.

Towards a reflective care in Education

The ambivalence of professional caring masculinities in a Swedish Edu-care context

Anette Hellman[1]

Department of education, communication and learning. University of Gothenburg

Abstract: In the past decade, dominant global discourses about individuality and competence have become evident in pedagogy policy and practice [1]. Although traditional ways in Sweden of understanding 'Educare' (A holistic model of education and care) may still exist, Sweden, as several other western countries, have seen a shift towards a more performative culture where the concept of education tends to be related to traditional school subjects – such as the learning of language or mathematics [2,3]. Research has problematized the way care has become blurred and marginalized in relation to learning in educational settings and policy documents. It has also been argued that we need to analyze care as a reflective practice and learn more about how theories of care could be implemented [4]. The purpose of this paper is to explore notions about care and gender in contemporary Sweden and the concept I in this paper suggest for this discussion is reflective educational care [4].

Keywords: Care, reflection, early childhood education, Sweden, gender

Introduction

Discoursers about gender equality are strong in Sweden. Notions on equality are manifested in policy documents and laws, such as 18 months parental leave available to both parents and obligations for teachers in Early Childhood Education (ECE) to counteract gender stereotypes. In recent years, particularly in middle class families, there has been a shift from seeing the ideal man in families as a 'breadwinner' towards being a 'active, present and competent father' who share parental leave and spend more time with their children [5]. The national curriculum for preschool [6] imposes a multifaceted gender equality assignment to preschools, in which one of the preschools' roles is to break down stereotypical gender notions regarding what girls and boys do, how they do things and when they do them. The Government mandate of the Swedish National Agency for Education states that it is 'important even at preschool to show girls and boys that gender will not determine their choice of occupation'. This can be interpreted as an endeavor to also change gender roles regarding in what ways and by whom preschool teaching is done and former traditional gender-coded associations need to be critically analysed. However, idealized norms about caring masculinities within the family sphere, still don't seams to correspond with men's professional work with care in ECE. Sweden still has quite low numbers of men working in professional care for young children (about 4-5%) and several male preschool teachers describe how they have to negotiate quite stereotype understandings of gender in ECE.[2] Hence, un-reflected notions about care seam to make caring for young children ambivalent for men as *professional caregivers*. At the same time, other actions become naturalized for men in ECE, such as teaching, playfulness, leadership and caring for older children.

[1] anette.hellman@ped.gu.se

[2] Brody [7] point out that the general pattern in the world as a whole is that less than 3% of preschool teachers are male. Brody also states that the proportion of male workers is low even in countries such as Norway which has had some very explicit strategies for increasing male workers, but where only 10% of the staff is male in ECE.

To be a professional preschool teacher in Sweden means to work according to the national curriculum. A central model underpinning the curriculum is the concept of Edu-care. Warin [8] describe how the concept of 'educare' (education and care), derived particularly from Nordic approaches to pedagogy and curricula design and has the potential to describe holistic educational purposes where education and care are seen as equally important and inseparable. However, in the past decade, dominant global discourses about individuality and competence have become evident in pedagogy policy and practice [1]. Although traditional ways in Sweden of understanding 'educare' may still exist, Sweden, as several other western countries, have seen a shift towards a more performative culture where the concept of education tends to be related to traditional school subjects – such as the learning of language or mathematics [2,3]. While education is seen as reflected upon and discussed 'on its own' as well as together with care, care is defined in policy documents and in practice more vaguely or in 'combination' with learning. Buiding on Fielding and Moss [9], Warin [8] describes this process as holistic values being obliterated through neo-liberal educational policies to the extent that schooling has become a 'techno-managerial exercise in control and normalization.

In response to these discourses researchers have stressed the need to pay far more attention to practices of care and to the way the concept of care operates in early childhood settings. Haldén [10] are concerned about the way a one sided focus on education/learning and on the individual and competent child, may make children's need of care in the preschool setting invisible. Noddings [2] argues that education should be focused on themes of care rather than traditional subject disciplines. Each participant plays, according to Noddings, an important role in this process. The carer can be seen as attentive and receptive, allowing him or her to understand the feelings the cared-for is trying to express without passing judgments based on a pre-conceived notion. When the behaviour of the carer is centered on the needs of the cared-for, Noddings calls it motivational displacement. She also discusses how motive energy flows towards the needs and wants of the cared-for. However, this does not mean that the carer will always approve of what the other wants and might indeed try to guide the person towards a different set of values, while taking into account the feelings and desires that are there and responding as positively as his or her values and capacities allow. According to Noddings, a caring relationship is one where both the cared-for and the carer contribute. This relationship in the case of the youngest students could be seen to extend to the parents, therefore highlighting the importance of good communication and cooperation between home and preschool settings. It also has greater societal implications because: "To care means to respond to needs, and needs do not stop (or start) at the schoolroom door". . According to her (11), the ethic of caring can create a more effective learning environment and this caring-centered approach to learning allows for active participation and empowerment of all members of the preschool community [12]. This is what Noddings refers to as genuine education: "... we believe that students should be given opportunity to learn how to care for themselves, for other human beings, for the natural and human-made worlds and for the world of ideas. This learning to care requires significant knowledge; it defines genuine education." Also Wrigley et al. [13] discuss ethics of care as a set of values that lie at the heart of intellectually demanding and equitable pedagogies: 'Deep care is central to socially just pedagogies, which understand the need to scaffold from where students are at, in respectful ways' (p. 196). Fielding and Moss [9] have a radical rethink of education based on the concept of 'education in its broadest sense' (EBS) (p. 46) to focus a vision for a more socially just, caring, and relational form of education to draw attention to the increasing androcentrism of Western educational influences alongside a downgrading of caring educational practices. This shift in an understanding of care places care at the centre of education, both as a way of describing teaching and way of conceptualising a key set of values that could be taught. However, as discussed by several gender researchers, discoursers about care and children's needs – even though often discussed as gender neutral concepts – is highly affected by gender stereotypes. Children's needs is hereby often unreflectively understood as boys having particular needs as boys and girls having specific other needs as girls [14, 15]. Therefore, we need to reflect and critically analyze care in relation to norms about gender.

Earlier research shows how male teachers negotiate masculinities when being in and entering a predominantly feminine work area, such as early childhood education [7]. This negotiation can be seen as needed due to gendered ideas of who is seen as suited for which kind of work, which calls for men to find strategies to negotiate masculinities, in order to keep up with a normative locally constructed view on masculinity and care as well as find ways to be an individual and professional teacher. Researchers have also problematized the way un-reflected practices of care might become gender blind at the same time as stereotype norms about gender operates in educational environments [16, 4]. One example of this discussed in this paper, is how the ideal professional position as a preschool teacher is viewed as gender-neutral at the same time as practices of care becomes linked to idealized femininity, naturalized for women and hereby made ambivalent for men (and boys) to perform. These challenges relate to the conflict between how to display norms about Edu-care as a professional Swedish preschool teacher at the same time as care – as a profession - is made ambivalent to perform as a man, a theme further explored in the following section.

The ambivalence of professional caring masculinities

I will in this section present some examples concerning male pre-school teachers and the care of young children by exploring ideas about being professional in relation to conventions about the male pre-school teacher. The data builds on two interview studies with 22 male trainers respectively 8 male preschool teachers and 8 female preschool teachers in Sweden. A common way for all teachers and trainers interviewed was to describe the importance of behaving as a professional and defined this as a 'gender-neutral' position. At the same time, male teachers described themselves as very visible as a minority. Some refereed to this as being positioned as 'the other' in a context supposed to be neutral. As described by Warin [17], to teach often means to put oneself in front of the gaze of different audiences, such as parents, children or colleagues. The way individuals' bodies and actions are exposed to gaze, manifest in different ways depending on context. The male students and preschool teachers discuss that, even if they always where 'visible', particular situations and actions could mean risk of gazing and marginalization. These were often linked to men's performance of care. Johan, a male trainer, discuss this in the following:

> It was when I was working in a group of younger 1-3 years old children. I was then advised by my female colleague, to leave the 'caring parts' of the job to her, like helping children to get dressed. She was my supervisor and she had just had a conversation with one of the parents who thought that it was a little bit strange with men taking care of young children. My supervisor told the parents that all teachers have an obligation to fulfil the goals of the curriculum. Care is a central part of our obligation and something both men and women are supposed to perform and teach. But afterwards, she also told me to play it cool and leave caring activities among the children to female teachers, since some parents had opinions about me as a man doing them. It was not that the parent accused me of sexual activities with the children; it was just that some parents thought that care was an odd thing to do as a man.

Most men interviewed had made an active choice to become a preschool teacher, a choice that might have contributed to the way professionalism, rather than gender, was put in the forefront. Several male and female teachers in the study also describe how they negotiate parents and colleagues gender stereotype expectations through notions in the Swedish curriculum about gender equality. One example is the call for male role models[3]. Underpinning the call for male role models is the influence of sex role theory. The feminine dominance within preschools, together with the moral panic about absent fathers, still serves to support the call for male role models of hegemonic masculinities for boys rather than girls [18, 19]. Care, female teachers, girls and certain spaces labelled as 'feminine' are tied together in this preschool discourse. Care is linked to the institution, female teachers, the building's architecture as well as the design, and informal situations such as meals that operate as a hidden curriculum. Researchers [20] have pointed out the importance of 'loosening the ties' between care, girls, girlishness, women and femininity, and create possibilities to also analyse care in relation to boys, boyishness, men and masculinity. Loosening this tie might also have implications on how jobs that are characterized as caring can open up for men too and how more men are beginning to see their future in such workplaces.

A common theme among men studying to become preschool teachers as well as men working as preschool teachers, was to describe their role as 'breaking' against gender order. As pointed out by Wernersson [22], shifting from a traditional masculine occupation to a traditional feminine one, means breaking the gender order 'downhill' with a potential loss of value, status and salary as well as the risk of becoming 'reduced' in different ways. One example of this in the study, was when traditional norms about men as 'hard working breadwinners providing for their families' clashed with men's choice to become preschool teachers. Some of them – in contrast to the female students in the study - had for instance been labelled as 'lazy' with 'no ambitions'. Tom discusses these norms in the following:

> I get very disturbed by these stereotypes! To work as a preschool teacher is for the first a very important job, I mean, we take care of the future generation! It´s also really hard work and we have a university degree! I´ve been discussing the way this job becomes ridiculed in

[3] This discourse about children's needs and role models may still be found in Swedish preschools, particularly in relation to single mothers. Researchers have showed how common assumptions about 'boys' needs' for male role models and 'girls' needs' for female role models rest on ideas about biological differences between the sexes [18,19,20, 21].

> several ways with my friends here at the university and I think it´s because working as a preschool teacher sometimes is seen as just 'taking care of children'. Something anybody could do without thinking. That's why we have to educate parents and society about what we do in preschool! Maybe it´s also because it´s seen as women's job and obviously it don't seems to be disturbing people if a woman chooses 'a lazy care job'...maybe she is expected to be provided for by her husband.... Ah, I´m so sick of these norms!

In private spaces such as in the family, men in Sweden are expected through discoursers of gender-equality and modern fatherhood, to perform care for young children [5]. However At the same time that professionalism as well as care in national and local policy documents often is understood as 'neutral', practices of professional care becomes ambivalent for men to perform in a public space, particularly in relation to choice of profession. Traditional gendered norms about men as providers and breadwinners becomes visible in the quotation above from Tom, labelling men – but more seldom women - as 'lazy' or without ambitions. The way Tom challenge these norms might be interpreted as breaking with 'old' ideas of gender and a more fluid and flexible way of presenting oneself as male [23]. This can be compared with Jones's [24] study of male teachers which pointed out that conflicting discourses of masculinity operate in early years education settings making different masculinities appear simultaneously. At the same time, working with care and young children as a man and breaking the gender order "downhill [22] means that there is a real risk to be ridiculed in different ways, as discussed earlier. Some male trainers and teachers talked about the possibility of a sexual gaze on their caring actions, a gaze not directed towards their female colleagues. In the following, a male teacher, Nils, and a female teacher, Anna, discuss these issues:

> Nils: The worst thing that could happen is if I take care of a child, say take a child on my knee, and then are accused of sexual harassments by a parent who thinks that caring for children is unnatural for men to do, like, I must have some other intentions... It´s so horrible even to think about that, so I must admit that I sometimes keep a certain distance towards children in order to be more safe or at least I make sure that my colleagues is close to me with some 'extra eyes' of safety to protect me from possible accusations so to speak.
>
> Anna: Yeah, that's so sad; I never have to think like that as a female teacher. I do these things all the time without thinking about it, that is how it should be for all teachers, I mean children need us to care for them despite our gender. If anything we need a man who shows girls as well as boys that care is just as natural for men to do as for women.

Professionalism is thus labelled 'gender neutral' while gender becomes clearly visible in relation to norms about gender, bodies and the gender division of labour. In the heart of this process lays the problem of un-reflected practices of care. In Sweden, one of the most gender equal countries in the world, discoursers also manifested in the curriculum for preschool, where teachers have an obligation to counteract gender stereotypes (Skolverket, 2016), men's caring actions could in practice be ambivalent. I have in this paper discussed that while men are expected to perform care as fathers in the private sphere of the family, men choosing to work with professional care as preschool teachers, are at risk of being labelled as lazy, without ambitions and even as pedophiles.
How can this be understood? The Educare model may still be important, but while there have been a shift towards a more performative culture at risk of favoring education and traditional school subjects also in preschool, the concept of care in national and local policy documents, have become even more blurred and unclear. If care becomes an un-theorized and un-reflected practice, teachers actions in caring practices might pass without critical analyze. As I have shown in this paper, care is hereby at risk of being understood through gender stereotype norms about gender, rather than in line with teachers' gender equal obligations.

In order to promote equality and socially just preschool spaces we need to pay more attention to care. As I have shown, care is in the heart of gender construction and the solution to become aware of how gender operates through care is not to view care and professionalism as gender-neutral, but rather to reflect on and analyze care in order to achieve more knowledge and awareness about the way gender and care manifest locally in preschool settings. A focus on reflective educational care is important in terms of promoting inclusive and socially just preschool spaces for all, adults as well as children. Reflection and critical analyses on care is keywords in order to achieve knowledge and awareness on norms such as gender. But awareness is not enough – we also need action, so let´s start working for change by deconstructing naturalized norms about men as providers and women in need of protection as well as naturalized norms linking care to idealized femininity.

References

1. S.J. Ball, *The education debate: policy and politics in the 21st Century*. (Bristol, Policy Press 8, 2008)
2. N. Noddings *The Challenge to Care in Schools: An Alternative Approach to Education* (2nd ed.). (New York, Teachers College Press, 2005)
3. I. Pramling Samuelsson, P. Williams, S. Sheridan, A. Hellman. Swedish preschool teachers' ideas of the ideal preschool group. Journal of Early Childhood Research,**1**,17. doi: 10.1177/1476718X14559233 (2015)
4. A. Hellman, Teaching reflective care in Japanese early childhood settings. Early Child Development and Care. DOI: 10.1080/03004430.2015.1124869 (2016)
5. T. Johansson, S. Lindgren, & A. Hellman, *Nya uppväxtvillkor*. (Stockholm, Liber, 2013)
6. Skolverket. *Läroplan för förskolan, Lpfö 98* (Rev. uppl.). Stockholm: Skolverket. (2010)
7. Brody, D. L. The construction of masculine identity among men who work with young children, an international perspective. European Early Childhood Education Research Journal, **23**,3 (2015)
8. Warin, J (2014) The status of care: linking gender and 'educare', *Journal of Gender Studies*, 23:1, 93-106, DOI: 10.1080/09589236.2012.754346
9. M. Fielding, P. Moss, *Radical education and the common school*. (London, Routledge, 2011)
10. G. Halldén, Bilden av barnet, I. Engdahl & E. Ärlemalm – Hagsér *Att bli förskollärare*, Stockholm, Liber, 2015)
11. N. Noddings, An Ethic of Caring and its Implications for Instructional, Arrangements, American Journal of Education, **96**, 2 (1988)
12. A. Hellman, K. Lauritsen, Editorial: Nordic perspectives on diversity and social justice in early childhood education. In Anette Hellman & Kirsten Lauritsen (Eds.), *Nordic perspectives on diversity and social justice in early childhood education*. (Nordic Studies on diversity and education. Cambridge: Cambridge Scholars series, 2017)
13. T. Wrigley, P. Thomson, B. Lingard, 2012. *Changing schools. Alternative ways to make a difference*. (London, Routledge,2012)
14. V. Adriany, Gender in pre-school and child-centred ideologies. In S. Brownhill, I. Wernersson, & J. Warin (Eds.), *Men, masculinities and teaching*. (London, Routhledge, 2015)
15. A. Hellman, *Kan Batman vara rosa? Förhandlingar om pojkighet och normalitet på en förskola*. (Göteborg, Acta Universitatis Gothoburgensis, 2010)
16. J. Warin, E. Gannerud, Gender, teaching and care, a compatrative global conversation. Gender & Education, **26**, 3 (2014)
17. J. Warin, Pioneers, professionals, playmates, protectors, 'poofs' and 'paedos': Swedish male pre-school teachers' construction of their identities. In S. Brownhill, I. Wernersson, & J. Warin (Eds.), *Men, masculinities and teaching*. (London, Routhledge, 2015)
18. B. Thorne, *Gender Play: Girls and Boys in School*. (Buckingham, Open University Press, 1993)
19. M. Nordberg, T. Saar, A. Hellman, Deconstructing the normal boy: Heterosexuality and gender constructions in school and preschool. In L. Martinsson & E. Reimer (Eds.), *Norm struggles: Sexualities in contentions*. (Newcastle upon Tyne, Cambridge Scholars Publishing, 2010)
20. S.M. Whitehead, *Men and masculinities: Key themes and new directions*. (Malden, MA, Polity, 2002)
21. R.W. Connell, R. W. *The Men and the Boys* (Cambridge, Polity, 2000)
22. I. Wernersson, More men? Swedish arguments over four decades about 'missing men' in ECE and care. In S. Brownhill, I. Wernersson, & J. Warin (Eds.), *Men, masculinities and teaching*. (London, Routhledge, 2015)
23. J. Warin, A. Vina, Gender flexible pedagogy in early childhood education, journal of gender studies, **26**, 4 (2015)
24. D. Jones, Millennium man: Constructing identities of male teachers in early years contexts. Educational Review, **59**, 2 (2007)

The value of differentiated instruction in the inclusion of students with special needs/ disabilities in mainstream schools

Vasilis Strogilos

Early Childhood and Special Needs Education Academic Group, National Institute of Education, Nanyang Technological University, Singapore

Abstract. The aim of this presentation is to analyse the current international policies on inclusion and to discuss the value and use of differentiated instruction as a means to the inclusion of students with special needs/ disabilities in mainstream settings. The movement to inclusion requires teachers to create inclusive learning environments, which would encourage the use of practices that would benefit all students. In this respect, differentiated instruction has rapidly evolved as a teaching approach to meet the diverse and heterogeneous needs of students with special needs/ disabilities in mainstream classrooms. Modifying and adjusting instruction to allow all students in a classroom to access the general education curriculum is at the heart of inclusive education. In this presentation, I shall discuss the development of differentiated instruction as a means to inclusion for students with special needs/ disabilities in mainstream classrooms. Information will be provided with regard to the main principles of DI as well as the basic criteria in designing individualized adaptations for these students. In addition to this, I will review the factors which influence the development of differentiated instruction based on research findings. Finally, I will argue that differentiated instruction provides a learning environment which takes into consideration the individual characteristics of students and, as such, is a useful approach for the inclusion of students with special needs/ disabilities in mainstream settings.

Introduction

Inclusion presupposes a philosophy of acceptance, where all people are valued and treated on equal terms. It is seen as both a base for future school development and an unending process that fosters participation for all students. Inclusion represents school improvement on many levels. As such, it involves the development of practices which aim to reorganize schools as places of equal opportunity for all students to participate and learn. Thus, inclusive education involves changes from the policy and structural level to the level of schooling through changes to the curriculum and teaching strategies. As Booth and Ainscow (2011, p. 9) argue "Inclusion is about increasing participation for all children and adults. It is about supporting schools to become more responsive to the diversity of children's backgrounds, interests, experience, knowledge and skills". Thus, the focus is not merely on the improvement of the skills of the student with special needs/ disabilities, but mainly on the improvement of teaching for all students.

However, international research has shown that the struggle to become inclusive is still in progress and school and social services reform is a major challenge. The Council of Europe (2009, p.46) states that *'inclusion may be understood not just as adding on to existing structures, but as a process of transforming societies, communities and institutions such as schools to become diversity sensitive.'*. International and national policies and laws (e.g. Every Child

Matters, 2004/UK; IDEA, 2004/USA) have charged schools with developing inclusive practices so as to increase participation and improve the quality of education to all students. Schools are encouraged to design and implement practices which are constantly inclusive since the number of students with diverse learning needs is increasing. Ideally, these schools include all students, eliminate differences, support learning and respond to individual needs. However, as Slee (2013, p.2) argues, in spite of *'financial expenditure and optimistic talk exclusion remains a real and present danger'*. Evidence exists that the placement of pupils with disabilities at mainstream schools has not been followed by meaningful access to learning (Nind & Wearmouth, 2006; Scruggs, Mastropieri, & McDuffie, 2007) even though there is emerging evidence that even students with significant cognitive disabilities can benefit from access to the general education curriculum (Wehmeyer, 2011).

Moving towards inclusive education?

Many countries have adopted the Salamanca Statement on Principles, Policy and Practice in Special Needs Education and a Framework for Action (UNESCO, 1994) and United Nations Convention on the Rights of Persons with Disabilities (2006). As a result, new policy initiatives have recognized the need to work towards the development of inclusive education. Ideally, these schools include all students, eliminate differences, support learning and respond to individual needs. However, as Stangvik (2010) notes, global policies are heavily dependent on local values. Thus, the implementation of international policies on inclusive education is closely related to the different understandings of disability and the meaning of inclusion in different societies.

Internationally there are different understandings of inclusion and the way in which it can be implemented in practice. These understandings of inclusion could be grouped under two broad approaches: a) the radical restructuring of the education system; and b) the implementation of additional special programmes in the existing arrangements in mainstream settings. The proponents of radical restructuring of the education system believe that such a change could provide appropriate education for all students (Ferguson, 2008; Slee, 2010) and see specialized programmes as a form of exclusion, even within a mainstream setting. For example, Slee (2013) argues that the current policies on inclusion employs mechanisms such as the increased enrolment of students in specific disabilities, which then determines the financial support allocated to schools. This process creates more strangers, more surplus children and thus more exclusion. Contrariwise, those who support the continuation of special programmes believe that special education is a service which needs to remain alive in mainstream settings and that not all students' needs are best served in mainstream schools (Forbes, 2007; Zigmond & Kloo, 2011). Zigmond and Kloo (2011, p.170) challenge the idea that special education is so not-special that it can be delivered by mainstream teachers. They argue *'that special education will not survive to serve the special needs of students with disabilities if it loses its identity, its special budget allocation, and its unique special requirements'*.

Even though, internationally, inclusive education is conceived as a broad reform that welcomes diversity among all learners (Unesco, 2001), it is mainly implemented as programmatic regularities which fail to initiate broad school reforms. Thus, I could argue that inclusive education in many countries has not yet provided a shift in approach from serving children with disabilities within mainstream education settings to a more blended practice through a process of eliminating the barriers to participation and learning experienced by students within the school system. To this end, in this presentation, I will argue that differentiated instruction provides a learning environment which takes into consideration the individual characteristics of all students (Janney & Snell, 2006; Strogilos et al., 2017) and, as such, is a useful approach for the inclusion of students with special needs/ disabilities in mainstream settings.

Differentiated instruction

Differentiated Instruction involves responding effectively to the differences that exist among learners in the classroom. According to Tomlinson (2001), teachers differentiate when they reach out to an individual or small group by varying their teaching in order to create the best learning experience possible. Differentiated instruction is considered as one of the essential means to effective education for all students including those with disabilities. As Tomlinson argues (2003), differentiation is not associated with "one size fits all" teaching but mainly with responsive teaching. Thus, when teachers differentiate they 'proactively plan varied approaches to what students need to learn, how they will learn it, and/or how they will show what they have learned in order to increase the likelihood that each student will learn as much as he or she can, as effectively as possible' (Tomlinson 2003, p. 151).

Research findings on differentiated instruction

Despite the perceived importance of differentiated instruction, research has indicated the absence or inconsistent use of this strategy. Limited use of differentiated instruction has been noted for 'typically developing' students (Hertberg-Davis, 2009; Van Tassel-Baska, & Stambaugh, 2005) as well as for students with special needs/ disabilities within mainstream classrooms (Buckley, 2005; Scruggs, Mastropieri, & McDuffie 2007). At the international level, even though the number of students with SEN in mainstream classrooms has increased, the type and the quality of education they receive remains a contentious issue (Morningstar, Shogren, Lee, & Born, 2015). Several authors have urged for the necessary modifications to increase the quality of education provided to students with SEN through differentiated instruction (King-Sears, 2008; Kurth & Keegan, 2014; Lee, Wehmeyer, Soukup, & Palmer, 2010).

The modifications made to the curriculum are considered an essential inclusive strategy for the education of students with SEN in the general classroom. As Janney and Snell (2006) indicate, modifications may be curricular, instructional or alternative. "Curricular" refer to *what* is being taught (i.e. content); "instructional" concern alterations of the *way* instruction takes place (i.e. method); and "alternative" involve *altered goals, instruction and activities.* According to King-Sears (2008), differentiated instruction has the potential to increase the scores on high-test assessments for students with disabilities, students at-risk for school failure, typical students, and students labelled as gifted and talented in comparison to students in schools that promote 'one size fits all' instruction.

Several factors have been identified in the literature with regard to the absence or inconsistent use of differentiated instruction. Among these factors are the lack of content knowledge necessary to extend and differentiate the core curriculum content areas; the lack of time to adjust the curriculum for the students with SEN, or those with gifts and talents; teachers' difficulty to locate and use effectively the appropriate resources; and even their perception that students do not differ in how they learn (Hertberg-Davis, 2009; Van Tassel-Baska, & Stambaugh, 2005). With regard to students with special needs/ disabilities many research studies in differentiated instruction report a lack of curriculum modifications in mainstream classrooms for these students (Buckley, 2005; Strogilos & Stefanidis, 2015; Strogilos, Tragoulia, & Kaila, 2015), and consequently, according to Scruggs, et al.'s (2007) literature review, the education of these students in general classrooms lacks appropriate teaching materials, differentiation in activities and opportunities for individualizing the curriculum. Similarly, Wehmeyer, Lattin, Lapp-Rincker, and Agran (2003) indicated that teachers were adapting their materials in less than 3% of their observations for students with intellectual disabilities. Also, Strogilos, Tragoulia, and Kaila (2015) identified limited curriculum modifications for students with SEN in their descriptive observations in mainstream classrooms. In addition to the limited use of curriculum modifications in co-taught classrooms, teachers indicated a moderate endorsement towards the curriculum modifications for students with special needs/ disabilities (Strogilos & Stefanidis, 2015).

With regard to the type of modifications and their effectiveness in general classrooms, Kurth and Keegan (2014), in their study on the development and use of curriculum modifications for students with special needs/ disabilities, reported that classroom teachers produced modifications of lower quality and clarity than special educators and teaching assistants. In addition, they found that experienced teachers created more simplified modifications in comparison to novice teachers who created more functional alternative modifications. Kern and her colleagues (2001) found that when teachers use curricular modifications, students' on-task behavior and work production improved. Similarly, Lee et al. (2010) found that when curriculum modifications are used, students with special needs/ disabilities increased their engagement in academic-related responses and decreased their problematic behaviors. The authors also indicated that when curriculum modifications are used, teachers performed fewer classroom management activities. In a recent descriptive observational study in 65 inclusive classrooms, Morningstar, et al (2015) identified reductions (51% of the observations) in the cognitive demands (i.e., completing fewer items, picture-based stories rather than written stories) of work for the students with SEN in almost half of these classrooms. Moreover, amongst the most frequent modifications for these students were changes in how materials were presented (31%), environmental adjustments (23%), and response alternations (25%).

Even though many researchers agree that effective inclusion occurs when teachers modify the curriculum to the needs of all students, limited knowledge exists about the types and the quality of modifications understood and used by teachers. In this presentation, I will provide examples of differentiated activities for students with and without disabilities based on the principles of differentiated instruction. In addition, I will present the basic criteria with examples for planning and implementing individual adaptations for students with special needs/ disabilities.

Conclusion

Even though research has identified several challenges in the delivery of differentiated instruction, it is still considered an effective and useful approach in the education of all students. To this end, I will complete my presentation by indicating important prerequisites for the integration of differentiated instruction such as (a) policy makers to include differentiated instruction at the National Curriculum-Policy level; (b) universities to include relevant courses on teachers' pre-service and in-service training; and (c) schools to encourage the development of differentiated instruction as the main approach/ practice for the inclusion of students with disabilities.

References

Booth, T., & Ainscow, M. (2013). *Index for inclusion* (3rd ed). *Bristol: Centre for Studies on Inclusive Education.*

Buckley, C. (2005). Establishing and maintaining collaborative relationships between regular and special education teachers in middle school social studies inclusive classrooms. In Advances in learning and behavioral disabilities: Vol. 18. *Cognition and learning in diverse settings*, edited by
Thomas E. Scruggs, & Margo Mastropieri (pp. 153-198). Oxford, UK: Elsevier.

Department for Education and Skills. (2003). *Every child matters* (Green Paper). London: DfES.

European Commission (2009). *Strategic framework for education and training*. Brussels: European Commission. Electronic source available online at:
http://ec.europa.eu/education/lifelong-learning-policy/doc28_en.htm (accessed 28-01-2013).

Ferguson, D. L. (2008). International trends in inclusive education: The continuing challenge to teach each one and everyone. *European Journal of Special Needs Education* , 23(2), 109 – 120.
Forbes, F. (2007). Towards inclusion: an Australian perspective. *Support for Learning*. 22(2), 66-71.

Hertberg-Davis, H. (2009). Myth7: Differentiation in the Regular Classroom Is Equivalent to Gifted Programs and Is Sufficient: Classroom Teachers Have the Time, the Skill, and the Will to Differentiate Adequately. *Gifted Child Quarterly*, 53: 251-253.

Janney, R. E., & Snell, M. E. (2006). Modifying schoolwork in inclusive classrooms. *Theory into practice*, *45*(3), 215-223.

Kern, L., Delaney, B., Clarke, S., Dunlap, G., & Childs, K. (2001). Improving the classroom behavior of students with emotional and behavioral disorders using individualized curricular modifications. *Journal of Emotional and Behavioral Disorders*, 27, 615-631.

King-Sears, M. E. (2008). Facts and fallacies: Differentiation and the general education curriculum for students with special educational needs. *Support for Learning*, 23(2), 55-62.

Kurth, J. A., & Keegan, L. (2014). Development and use of curricular adaptations for students receiving special education services. *The Journal of Special Education*, 48(3), 191-201

Lee, S., Wehmeyer, M. L., Soukup, J. H., & Palmer, S. B. (2010). Impact of curriculum modifications on access to the general education curriculum for students with disabilities. *Exceptional Children*, 76 (2), 213-233.

Morningstar, M. E., Shogren, K. A., Lee, H, & Born, K. (2015). Preliminary lessons about supporting participation and learning in inclusive classrooms. *Research and Practice for Persons with Severe Disabilities*, 40(3), 192–210.

Nind, M. & Wearmouth, J. (2006). Including children with special educational needs in mainstream classrooms: implications for pedagogy from a systematic review. *Journal of Research in Special Educational Needs*, 6(3), 116-124.

Scruggs, T. E., Mastropieri, M. A., & Kimberly A. McDuffie, K. A. (2007). Co-teaching in inclusive classrooms: A metasynthesis of qualitative research. *Exceptional Children* 73(4), 392-416.

Slee, R. (2013). How do we make inclusive education happen when exclusion is a political predisposition? *International Journal of Inclusive Education, 17*(8), 895-907.

Slee, R. (2010). The irregular school. Exclusion, schooling and inclusive education. London: Routhledge.

Stangvik, G. (2010). Special education in society and culture: comparative and developmental perspectives. *European Journal of Special Needs Education*, 25(4), 349-358.

Strogilos, V., & Stefanidis, A. (2015). Contextual antecedents of co-teaching efficacy: Their influence on students with disabilities' learning progress, social participation and behaviour improvement. *Teaching and Teacher Education,* 47, 218-229.

Strogilos, V., Tragoulia, E., Avramidis, E., Voulagka, A., & Papanikolaou, V. (2017). Understanding the development of differentiated instruction for students with and without disabilities in co-taught classrooms, *Disability & Society. http://www.tandfonline.com/doi/abs/10.1080/09687599.2017.1352488*

Strogilos, V., E. Tragoulia., & Kaila, M. (2015). Curriculum issues and benefits in supportive co-taught classes for students with intellectual disabilities. *International Journal of Developmental Disabilities*, 61(1), 32-40.

Tomlinson, C. A. (2001). *How to differentiate instruction in mixed ability classrooms* (2nd ed.). Alexandria, VA: ASCD.

Tomlinson, C. A. (2003). *Differentiating instruction for academic diversity.* In J. M. Cooper (ed.) *Classroom teaching skills*, pp.149-180. Boston: Houghton Mifflin.

UNESCO. (1994). *The Salamanca Statement and Framework for Action on Special needs Education.* Paris: UNESCO. Electronic source available online at: http://www.unesco.org/education/pdf/SALAMA_E.PDF (accessed 18-08- 2017).

UNESCO.(2001). *The open file of inclusive education*. Paris: UNESCO.

United Nations. (2006). *Convention on the Rights of Persons with Disabilities.* New York: United Nations.

US Department of Education. (2004). *Individuals with Disability Education Act*. Washington, DC: US Department of Education.

Van Tassel-Baska, J., & Stambaugh, T. (2005). Challenges and possibilities for serving gifted learners in the regular classroom. *Theory Into Practice*, 44(3), 211–217.

Wehmeyer, M. L. (2011). Access to general education curriculum for students with significant cognitive disabiltities. In J. M. Kauffman and D. P. Hallahan. (Eds.). *Handbook of Special Education* (pp. 544-556). London: Routledge.

Wehmeyer, M. L., Lattin, D. L., Lapp-Rincker, G., & Agran, M. (2003). Access to the general curriculum of middle school students with mental retardation. *Remedial and Special Education*, 24(5), 262-272.

Zigmond, N., & Kloo, A. (2011). General and special education are (and should be different). In J. M. Kauffman and D. P. Hallahan. (Eds.). *Handbook of Special Education* (pp. 544-556). London: Routledge.

Short bio:

Vasilis Strogilos is an assistant professor in the National Institute of Education at Nanyang Technological University in Singapore. He has earned his PhD from the School of Education at the University of Birmingham in UK. He has experience as a special educator in English and Greek schools. In higher education, he has worked as a researcher at the University of the Aegean/ Greece, as an online tutor at the Roehampton London Online University and at the Greek Open University, and as a lecturer and assistant professor at the University of Thessaly/ Greece. His teaching and research have focused on special and inclusive education. His recent research interests have centered on the development of inclusive pedagogy and curricula with a specific interest on co-teaching and differentiation, and on multidisciplinary collaboration as a means to inclusion for students with severe and multiple disabilities. He has participated in a number of research projects and has published research articles and studies in peer-reviewed journals, books, and conference proceedings.

Using Mobile Phone to Overcome Students' Anxiety in Speaking English

Karmila Machmud[1,*], and *Ridwan* Abdulah[2]

[1]English Department, Faculty of Letters and Culture, Universitas Negeri Gorontalo, Indonesia
[2]SMP Negeri 5 Anggrek Gorontalo Utara, Provinsi Gorontalo, Indonesia

Abstract. The goal of this research is to discover whether the use of mobile phone can minimize students' anxiety in speaking English and in turn will improve their speaking skill. Anxiety has become a major problem for students when asked to speak in a speaking class. Using quasi experimental method, the data of this research was collected through questionnaire of foreign language anxiety classroom scale, and oral test for speaking ability. Factorial design 2 x2 using ANOVA is used to analyze the data of this research. The result has shown: (1) the score of students' learning achievement in speaking English who were taught by using mobile phone application was higher than the students' who were taught by using conventional teaching. It means that the use of mobile phone application results in better influence towards students' speaking ability; (2) The learning achievement of the students with high anxiety in speaking English who were taught by using mobile phone application was higher than the high-anxiety students who were taught by conventional teaching; (3) The learning achievement in speaking English of low anxiety students who were taught by using mobile phone application was higher than low anxiety students' who were taught by conventional teaching.

Key Words: students'anxiety, mobile phone application, mobile technology

1 Introduction

Some problems that appear in teaching speaking are; first, the time allocation is too little, and second, is the anxiety. Such in tenth standard competence at seventh grade, in two meetings or in 4 X 40 minutes of teaching process, the students are asked to describe something/someone with their own words orally in front of the class. Then in the second meeting, the students are asked to tell their friends how to do or to make something.

From the explanation above, it can be concluded that the teaching and learning process is not effective, because in such limited time allocated, the teacher should explain the material and then guide the students to pronounce some words. Thus, it left with almost no time for other important learning activities. Teacher has 80 minutes. In that length of time, for example, 30 minutes is used for explaining the material, 5 minutes for guiding the students to pronounce the difficult words, 5 minutes for preparing the students presentation,

*Correspondingauthor: karmila.machmud@ung.ac.id

and 40 minutes for the presentation. It is obvious that that the times allocated for each activities are not sufficient enough to build students' speaking skill.

In such a short length of time students are pushed to practice speaking in front of the class with a very limited time to prepare. This makes it difficult for them to control their anxiety. The anxiety and worry are associated with three or six symptoms that can be noticed from the students [1]. These problems can be minimized by using Information and Communication Technology (ICT) tool, because studies show that ICT has a positive influence towards students learning process. The ICT tools include computers, radio, television, mobile phone and some other devices. Among the ICT tools that is mostly owned and widely used is mobile phone. The report from the statistic portal shows that the total number of mobile phone users worldwide from 2013 to 2019 shows a significant increase of the mobile phone users from only 4.05 billion in 2013 to 4.77 billion in 2017, and it is predicted that the number will climb to 5.07 billion users in 2019.

Teacher can use this as a potential of improving the teaching and learning processes, because of the availability of this device which is owned by students; thus, this can be considered as cheap technological tool compared to other ICTs, which can be used for teaching and learning [2]. Mobile phone became one of the tools that can be useful in teaching process because most of the students are familiar with the use of mobile phone in their daily live.

Based on the above arguments, we decided to choose SMP Negeri 1 Anggrek as our subject of research. SMP Negeri 1 Anggrek is located at Tolanggo Village, Aggrek subdistrict, Gorontalo Utara regency. The students of SMP Negeri 1 Anggrek come from various villages on the island. The parents'jobs are mostly farmers, fishermen, and civil servants. There are 53 students of seventh grade of SMP Negeri 1 Anggrek, and 29 students out of them usually bring their mobile phones to school. Unfortunately, they do not use it as a learning aid, instead they are busy using it for playing game like face booking, chatting and some other activities that distract them from learning. I think this is happened because teacher does not have a clear design of how to integrate the use of mobile phone in the teaching learning process; thus, in this research I would like to describe how the students use their mobile phones in the instructional process, especially in teaching speaking.

There are some interesting ways of using mobile phone in teaching and learning process, especially in teaching speaking. Some features in mobile phone that can be used in this project are mobile video, voice recorder, a digital storytelling tool, or class video projects. For example, teacher can assign students in different groups. Then each group will create their own video project.

The other advantage of using the features in mobile phone is that the teacher can use the video to assess students' speaking skills, also, can use the video as a document of the students' project, which is in turn can give feedback on their speaking skill.

Many students who have mobile phones have the ability to record video. Then once video sent to you via Bluetooth/cable/wi-fi/other network, you can place on LAN or share to class via YouTube or Facebook. This can then be added to a class blog, added to an intranet/ VLE. When students use video cameras themselves they are given a potential to create something memorable and enjoyable. The task of video-making can provoke genuinely creative and communicative uses of the language, with students finding them doing new things in English [3].

Besides video, the voice recorder's feature can also be used as a means of collecting audio evidence, feedback on work. This can be done by adding the voice recorder to portfolio as evidence of peer feedback, learning, development transfer to machine via Bluetooth, Wi-Fi, email, and link up by embedding, storing on intranet etc. Voice recorder could be used for other purposes too such as: interview, recording instructions, podcast

*Correspondingauthor: karmila.machmud@ung.ac.id

style notes and keywords for revision. Most mobile phone can also record a voice. Children and young learners use their mobile phone to record their voice, watch and exchange videos.

The other activity which can be conducted is Class Video Project. This activity has some instructions firstly, invite students to prepare a minute video in class, then they may prepare, negotiate, discuss, and plan for their project based on the class time, once they are ready, they use one mobile device to record their video. Then, they can share the video file with the teacher using blue-tooth. The whole class can watch those videos using the class projector for feedback and follow-up discussions. Class video project or video creativity can be used to spark students' creativity by encouraging interpretation, provoking thought, asking for language use [3].

2 Methodology

The method of this study is experimental method. In conducting this study, we used quasi experiment. Quasi experiment is a development study of true experiment design [6]. This study was conducted at the seventh grade students of SMP Negeri 1 Anggrek in 2015/2016 academic years. Sixty three participants were given the questionnaire of foreign language anxiety classroom scale to determine their level of anxiety; the results were 20 students who were grouped into high anxiety students, and 20 students as low anxiety students. A number of 13 students who fell into the medium anxiety category were not selected for the experiments. The high and low anxiety groups were then given oral test for measuring their speaking ability. The design of this study was 2 x 2 factorial design. The independent variable was mobile phones application (A_1) and conventional teaching (A_2). While the attribute variable was anxiety involve high anxiety (B_1) and low anxiety (B_2).

3 Results and Discussion

3.1 Results

The findings discuss about the data description of the study, the data analysis of requirement and the homogeneity test.

This section described the data of the students' learning achievement in speaking English after being taught by using mobile phone. The data are presented in eight groups, as follows: (1) The data of students' learning achievement in speaking English who were taught by using mobile phone (A 1); (2) The data of students' learning achievement in speaking English who were taught by using conventional teaching (A 2); (3) The data of students' learning achievement in speaking English who have high anxiety (B 1); (4) The data of students' learning achievement in speaking English who have low anxiety (B 2); (5) The data of students' learning achievement in speaking English who have high anxiety taught by using mobile phone (A1 B1); (6) The data of students' learning achievement in speaking English who have low anxiety taught by using mobile phone (A1 B2); (7) The data of students' learning achievement in speaking English who have high anxiety taught by using conventional teaching (A2 B1) and (8) The data of students' anxiety in speaking English who have low anxiety taught using conventional teaching (A2 B2).

The data description that was obtained in this study can be seen further from the Table 1.

*Correspondingauthor: karmila.machmud@ung.ac.id

3.2 Discussion

Table 1 shows the data findings of students' learning achievement in speaking English. The complete discussions and analysis of the results of this research are as follow.

3.2.1 Students' Learning achievement in speaking English who are taught by using mobile phone application and conventional teaching

The students' learning achievement in speaking English who were taught by using mobile phone application is higher than the students' speaking English who were taught by conventional teaching. It means that using mobile phone application has a better influence towards students' speaking English than using conventional teaching. Using mobile phone application especially video and voice recorder in teaching speaking helps both the teacher and students. The teacher uses fewer times than when using conventional teaching. Then for the students, they enjoy the learning process more than they enjoy the conventional teaching. Students are interested and more motivated to learn.

When students use video cameras themselves they are given potential to create something memorable and enjoyable [3]. Using mobile phone's features, especially video and voice recorder, in teaching and learning process provides opportunity for the students to edit, submit, and re-edit again to make it perfect, the process per se will give more opportunity to the students to improve their speaking skill.

Table 1. The data description of students' learning achievement in speaking English

Data Group	N	Score Min	Score Max	Mean ($\overline{X}$)	Modus (Mo)	Median (Me)	St. dev. (S)	Variants (S^2)
A1	20	9	19	14.20	15.16	14.5	3.06	9.38
A2	20	8	16	12.10	12.84	12.5	2.39	5.73
B1	20	8	18	12.40	14.08	11.5	2.86	8.20
B2	20	9	19	13.80	12.64	12.7	2.99	8.96
A1 B1	10	9	18	14.70	13.5	13.16	2.63	6.93
A1 B2	10	10	19	15.30	16.7	16	2.70	7.29
A2 B1	10	8	15	12.30	13.7	13.5	2.70	7.29
A2 B2	10	9	16	12.30	12.7	12.84	2.57	6.62

3.2.2 Interaction between the model of teaching and students' anxiety level toward students' speaking English

The application of suitable model of teaching by considering the students' anxiety level gave an effect towards students' speaking English. It needs a way to overcome the anxiety level. The model of teaching by using mobile phone application and conventional teaching can reduce the students' anxiety level.

*Correspondingauthor: karmila.machmud@ung.ac.id

Using mobile phone application in English teaching and learning process, especially in teaching speaking enticed the students to perform their competence to communicate their idea.

Using videos into lessons creates enticing visuals and a special interactive environment in the EFL/ESL classroom [5]. Teaching English through videos also allows teachers to be creative when designing language lessons [3]. There are some reasons of using video in teaching and learning process; they are: seeing language-in-use, cross-cultural awareness, the power of creation, and motivation.

The use of mobile phone application gave positive effect, and motivated the students to be more creative. It was shown from the result of the students' scores in speaking English. The students who were taught by using mobile phone application got higher scores than the students who were taught by conventional teaching.

3.2.3 The learning achievement of students who have high anxiety in speaking English, and were taught by using mobile phone application and by conventional teaching

The students' learning achievement in speaking English who have high anxiety, and taught by using mobile phone application, got higher score than the students who were taught by using conventional teaching technique. It means that there is a difference of students' learning achievement in speaking English between those taught by using mobile phone application, and those who were taught by using conventional teaching.

The students who were taught by using mobile phone application were more likely to engage in the learning process, and have more fun than the students who were taught by using conventional teaching. The students who were taught by using conventional teaching tend to show unconcerned to the material. They were panic, sweating, trembling and shaking their body. There are five main types of anxiety disorder; one of them is panic disorder: The focus here is on attacks of panic that appear to come on for no reason ('out of the blue'). Involving racing heartbeat (palpitations), chest pain, sweating, trembling and shaking, many people fear that they are having a heart attack or stroke, dying or going mad [7].

3.2.4 The learning achievement of students who have low anxiety in speaking English, and were taught by using mobile phone application and by conventional teaching

The English speaking achievements of the students who have low anxiety and taught by using mobile phone application were higher than the achievements of the students taught by using conventional teaching technique. It means that there is a difference of students' anxiety in speaking English between students who were taught by using mobile phone application and students who were taught by using conventional teaching technique.

Bailey's in [4] defined that anxiety in second language learning was the key to success. Anxiety causes many potential problems to the students who were learning foreign language. It does not mean that the students' anxiety cannot be reduced. The students' anxiety can be reduced by using a suitable method in teaching and learning process.

*Correspondingauthor: karmila.machmud@ung.ac.id

4 Conclusions and Recommendation

4.1 Conclusion

Based on the findings and discussion presented, it can be concluded that: first, there is a difference of the students' learning achievement in speaking English when they were taught by using mobile phone; second, the students who have high anxiety tend to achieve higher score when they were taught by using mobile phone; and finally, even the students who have low anxiety tend to obtain higher learning achievement when they were taught by using mobile phone.

4.2 Recommendation

Considering that using mobile phone can give a positive effect towards both groups of high and low anxiety students, it is highly recommended that teacher should begin to integrate the use of mobile phone in teaching and learning English as a foreign language in an attempt to minimize students' anxiety, which will eventually increase their learning achievement in speaking English.

We would like to express our gratitude towards all people and institutions that support this research. The first is the government of Gorontalo Utara regency, which has given us the opportunity to conduct this research in one of the schools in the regency. Our gratitude also goes to Universitas Negeri Gorontalo, where we got lots of sources which support our research.

We also want to express our gratitude to the committee and the reviewer who give us the opportunity to present our paper in ***Global Conference on Teaching, Assessment, and Learning in Education*** (GC-TALE) in Bali, 28-29 August 2017

References

1. T. Brown, T. A. O'Leary, D.H. Barlow (ed),Generalized anxiety disorder. *Clinical handbook of psychological disorders, third edition: A step-by-step treatment manual.* (Guilford Publications, 2001)
2. P. W. Mtega, R. Bernard, A. Msungu, Sanare, R., *Using Mobile Phones for Teaching and Learning Purposes in Higher Learning Institutions: the Case of Sokoine University of Agriculture in Tanzania. Morogoro – Tanzania* (Sokoine University, 2012).
3. J. Harmer, *The practice of english language teaching* (Longman: Cambridge, 2004)
4. H. D. Brown, *Principles of language learning and teaching* (Addison Wesley San Fransisco: Longman ,2000)
5. S.Yassaei, *Using original video and sound effects to teach English. English teaching forum* (United Kindom, 2012)
6. Sugiono, *Metode penelitian kuantitatif kualitatif dan R & D.* (Alfabeta:Bandung, 2014)
7. T. Dixon, *Understanding anxiety problem.* (Help-For,2011)

*Correspondingauthor: karmila.machmud@ung.ac.id

Career orientation of senior secondary school students

Bambang Suryadi[1,*], *Dian* Ratna Sawitri[2], *Hamidah*[1], *Farah* Hanifa[1]

[1]Faculty of Psychology, Universitas Islam Negeri (UIN) Jakarta, Indonesia
[2]Faculty of Psychology, Universitas Diponegroro, Semarang, Indonesia

Abstract. The main purpose of this study is to portray career orientation of Senior High School (SMA) and Vocational High School (SMK) students and the role that school counselors should play in providing vocational guidance. This study used a survey method involving 278 students from both SMA and SMK di Jakarta. Career Orientation Inventory (COI) and Counselor Role in Career Guidance Inventory were used as an instrument of data collection. Descriptive statistics and t-test were conducted to analyze the data. The findings of the study reveal that majority of SMA and SMK students wanted to further their study to university and only small number of them wanted to find out jobs or run their own business. The findings also show that school counselors play a very important role in providing vocational guidance services for senior secondary students. Therefore, in order for students to have a bright career in future, school counselors should improve the quality of vocational guidance services and plan comprehensive vocational guidance programs.

Key words: career orientation, vocational guidance, school counselor

1 Introduction

The importance of providing career orientation to students has been studied in numerous researches [1-5]. Savickas found that career orientation explicitly fosters students' awareness of tasks to be faced and decisions to make [1]. Krumboltz and Worthington pointed that learning is essential not only for students to develop the basic employability skills required for the school-to-work transition, but also for the development of work habits, beliefs, interests, and values[2].

Similarly, Lent and Worthington said that responding to national and local initiatives, the career development profession has been engaged in efforts to aid understanding and facilitation of the school-to-work (STW) transition process [3]. Kuijpers, Meijers, and Gundy indicated that career guidance in school, in which a dialogue takes place with the student about concrete experiences and which is focused on the future, contributes to the presence of career competencies among students [4]. A recent study by Loan and Van

* Corresponding author: bambang.suryadi@uinjkt.ac.id

showed that career guidance aimed at assisting students to reflect on their ambitions, interests, qualifications and abilities. It helps them to understand the labor market and education system, and to relate their needs in life and to become responsible global citizen [5].

Previous studies also revealed that there are various factors affecting students' career orientation [6-8]. According to Patton and Creed career orientation is mostly affected by career maturity [6]. Heslin pointed certain factors that affected career orientation which included both contextual and individual factors. The potential role of two contextual factors: whether the career is being pursued in a winner-take-all market as well as the organizational culture in which a person is working. Two individual factors are work orientation and goal orientation [7]. In addition, Sovet and Metz pointed out that toward the end of their high school years, young people face numerous personal decisions that can have wide-ranging and long-lasting effects on their lives. Some of these decisions are career related: joining the military, finding a job, obtaining an apprenticeship, going to college, choosing a major, or gaining skills through volunteer service [8]. Fortunately, adolescents are not alone in their career decision-making. In the context of Indonesia secondary school system, the role of school counselors in career orientation is very important.

The present study is aimed to portray career orientation of Senior Secondary School (SMA) and Secondary Vocational School (SMK) students and the role that school counselors should play in providing vocational guidance services. Based on the findings of the study, it is recommended that for secondary students to have a bright career in future, school counselors should improve the quality of vocational guidance services and well planned vocational guidance programs are required.

2 Methods

The present study applied a descriptive survey design, which is a non-experimental design used to describe students' career orientation and counselor role. The participants were 278 students from two Senior High Schools (SMAN 87 and SMAN 34) and two Vocational High Schools (SMKN 57 and SMKN 20) in Jakarta. The participants of the study were Grade X to XII students in the academic year 2016/2017. The participants consisted of 186 (67%) girls and 92 (33%) boys, aged 14 to 18 years.

The research instruments used for data collection were Career Orientation Inventory (COI) and Counselor Role in Career Guidance Inventory written in the form of a Likert scale. The Counselor Role in Career Guidance Inventory had 9 items and the Career Orientation Inventory had 40 items. Descriptive statistics, was used to analyze the data related to students' career orientation and school counselors' role. T-test analyses were conducted in order to determine differences between demographic subsets of student respondents (school and gender).

3 Findings and Discussion

3.1. Students' career orientation

As mentioned earlier, this study involved students from two senior high schools (SMAN 87 and SMAN 34) and two vocational high schools (SMKN 57 and SMKN 20) in Jakarta. Respondents from SMK took four types of study program, namely catering (16.9%), dance (5%), office administration (12.2%), and accounting (20.9%). As for respondent from SMA, they were from natural sciences (22.3%) and social sciences (22.7%). Analysis was also done to find out career orientation after graduating from schools as shown in Table 1.

Table 1: Student career orientation after graduating from the schools

Career orientation	SMA	SMK
Going to college	114 (91.2%)	105 (68.6%)
Finding jobs	0 (0%)	23 (15%)
Being entrepreneur	0 (0%)	3 (2%)
Others	10 (8%)	17 (11.1%)
Undecided	1 (0.8%)	5 (3.3%)
Total	125 (100%)	153 (100%)

Data on Table 1 show that almost all SMA students wanted to go to college or university after they graduated from schools with percentage of 91.2%. On the other hand, SMK students have various career choices after they graduated from schools. About 68.6% of them decided going to college, followed by 15% decided finding jobs, and 2% decided being entrepreneur, and 11.1% decided other career not mentioned in this study, and about 3.3% undecided. In order to find out whether SMA students differed from SMK students in their career orientation, independent sample T-test was run and the results of analysis is shown in Table 2.

Table 2: Results of T-test analysis for student career orientation based on school type

		Levene's Test for equality of variances		t-test for equality of means		Sig. (2-tailed)
		F	Sig.	t	df	
Career orientation	Equal variances assumed	.008	.928	5.089	276	.000
	Equal variances not assumed			5.112	269.214	.000

Data on Table 2 indicate that there is a significant difference between SMA and SMK students in their career orientation with significance value of 0,000. However, when the analysis was done based on gender, significant difference was not found as shown in Table 3.

Table 3: Results of T-test analysis for student career orientation based on gender

		Levene's Test for equality of variances		t-test for equality of means		Sig. (2-tailed)
		F	Sig.	t	df	
Career orientation	Equal variances assumed	.295	.587	-1.497	276	.136
	Equal variances not assumed			-1.466	172.061	.144

Data on Table 3 show indicate that there is no significant difference between boys and girls in their career orientation with significance value of 0,136. This means that the null hypothesis which said that there is no difference between boys and girls in their career orientation was accepted.

3.1.2 Role of school counselor in providing career guidance

This section is aimed to describe roles that school counselors should play in providing vocational guidance and how important the role is. There were nine statements that described the role of school counselors in vocational guidance and students' responses to each is presented in Table 4.

Table 4: The importance of school counselor role in providing vocational guidance

	Counselor Role	1	2	3
1.	Counselor meets with individual students to discuss career information and/or to help students in career planning.	3 (1%)	25 (9%)	250 (90%)
2.	Counselor conducts a class size of large group workshop for students to help them develop resume writing, interview, and job seeking skills.	8 (2.8%)	48 (17.3%)	222 (79.9%)
3.	Counselor assists student to seek employment (part-time or full-time employment).	17 (6%)	56 (20.1%)	205 (73.9%)
4.	Counselor administers to an individual or a class size or a large group of students a career inventory to assess students' career potential.	11 (4%)	42 (15%)	225 (81%)
5.	Counselor consults with community agencies (education department, social services department) about job vacancies.	13 (4.6%)	35 (12.6%)	230 (82.8%)
6.	Counselor plans activities (discussion, field trip) to stimulate interest in the world of work.	14 (5%)	61 (22%)	203 (73%)
7.	Counselor provides vocational information to students wishing to explore career.	10 (3.5%)	67 (24.1%)	201 (72.4%)
8.	Counselor helps students relate their abilities to future career choices.	5 (1.7%)	28 (10.1%)	245 (88.2%)
9.	Counselor provides information on the qualifications needed for jobs.	14 (5%)	40 (14.4%)	224 (80.6%)

In general, data on Table 4 indicate that majority of SMA and SMK students felt that the role of school counselors in providing vocational guidance services is very important. More specifically, out of nine roles of school counselors, based on respondents' opinions, five roles were perceived to be very high priority with the percentage of more than 80 percent. The other four roles of school counselors were perceived as the second order of importance by the student respondents. These roles include item number 2, 3, 6, and 7.

3.2 Discussion

The findings of this study indicate that both SMA and SMK students were concerned with their future career, however their differed significantly in career choices. Majority of SMA students wanted to go to college and none of them who wanted to find jobs or be an entrepreneur. SMK students, on the other hand, made choices for finding jobs and being entrepreneur. Surprisingly, more than half of SMK students also decided to go to college. In fact, the capacity of university or college for new students is only 30 percent of the total population of graduates from senior secondary schools (SMA and SMK). This means that SMA and SMK graduates who are not accepted in the college will contribute to the increasing number of unemployment in Indonesia.

Data from the Central Statistics Agency (BPS) indicated that the number of unemployment in February 2017 from SMK graduates was higher than SMA with the percentage of 9.27 percent and 7.03 percent respectively [9]. This number of unemployment can be considered as a great challenge for senior secondary school teachers and counsellors to equip and prepare their students with attitudes, knowledge, and skills as expected by work agencies and industrial sectors.

The findings of this study also indicate that both SMA and SMK students felt that the role of school counselors in providing vocational guidance services is very important. The findings support a previous study by Suryadi which found that it is very important for school counselors to provide vocational information to students and meet with students to discuss career information [10]. The only difference is that the present study was conducted in the Jakarta as Capital City of Indonesia, while the previous study was done in the Province of East Java.

This study also supported the regulation made by the Ministry of Education and Culture Number 111 year 2014 which emphasizes on the importance of vocational guidance for secondary students [11]. According to this regulation in relation to the implementation of 2013 Curriculum, every student has to choose his or her academic and vocational interest with assistance from school counselors. Furthermore, the regulation said that the purpose of vocational guidance services is to facilitate students in order to develop, explore, and make career choices.

The present study also gives certain implications for both policy makers and school counsellors. As for policy makers, especially the Minister of Education and Culture as well as Provincial and District Education Officers, they should facilitate secondary schools with professional counselors who meet the academic qualification and competences as required by the regulation. According to the Ministry of Education and Culture Regulation Number 27 year 2008, the academic qualifications for school counselors are having a bachelor degree (S1) from guidance and counseling study program and a certificate of counselor education program. As for counselor competences, they have to fulfil four competencies, namely, pedagogic, professional, personal, and social competences [12]. However, in the implementation process, the regulation compliance was not obediently done as many school counselors were recruited from many other study programs.

As an implication for school counselors, it is strongly recommended to have a grand design and road map for career development programs. According to Tang, Pan, and Newmeyer career development programs should provide meaningful learning experiences that facilitate the development of self-efficacy in students' aspired careers. School counselors also should consider individual variants such as gender, self-efficacy, interests, and outcome expectations, and contextual factors such as socioeconomic background, when developing career intervention plans. School counselors have to be cognizant of the role of self-efficacy and learning experiences in shaping high school students' career interests and choices [13].

4 Conclusion

Senior secondary school students were concerned with their future career orientation. Two important career choices found in this study were going to college and finding jobs. However, due to limited study opportunity at university as well as limited job opportunity at work place, senior secondary student graduates have contributed to the increasing number of unemployment. As a consequence, school counselors are challenged to define their role and function in vocational guidance more clearly.

Obviously, vocational guidance has existed on senior secondary schools for a long time and in this study the roles of school counselors in providing career guidance services were highlighted. The aim of vocational guidance is to help students discover their potentials and make the best of their choices, plans, and adjustment in their vocational life. To meet the above challenges, professional counselors have to conceptualize student career concerns from their perspective and worldview, and provide comprehensive vocational guidance services by considering the 21st century learning skills.

We wish to express our heartfelt gratitude to student respondents involved in this study for their support in data collection process. Our sincerest thanks also go to the reviewers for their invaluable comments and scrutiny, which assisted this study tremendously.

References

1. M. L. Savickas, CDQ. **47**(4), 326-336. (1999).
2. J. D. Krumboltz, R. L. Worthington, CDW. **47**(4), 312-325. (1999).
3. R. W. Lent, R. L. Worthington, CDW. **47**(4), 291-296. (1999).
4. M. A. C. T. Kuijpers, F. Meijers, C. Gundy, JVB. **78**(1), 21-30. (2011).
5. D.T. Loan, N. Van, AIJSC. **4**(5). (2015).
6. W. Patton, P.A. Creed, CDQ. **49**(4), 336-351. (2001).
7. P.A. Heslin, JOB. 26(2), 113-136. (2005).
8. L. Sovet, A. J. Metz, JVB. **84**(3), 345-355. (2014).
9. BPS. Badan Pusat Statistik. *Berita Resmi Statistik: Keadaan Ketenagakerjaan Indonesia Februari 2017*. No. 47/05/Th.XX, 05 Mei 2017. **3**. (2017).
10. B. Suryadi, *The role of public senior secondary school counselors in East Java: Students', teachers', and counselors' perceptions*. Germany: VDM Verlag Dr. Muller Aktiengesellschaft & Co. KG and Licensors. 134. (2010).
11. Kemdikbud. *Peraturan Menteri Pendidikan dan Kebudayaan Nomor 111 Tahun 2014 tentang Bimbingan dan Konseling pada pendidikan dasara dan pendidikan menengah*. Jakarta: Kemdikbud. 1. (2014).
12. Kemdikbud. *Peraturan Menteri Pendidikan dan Kebudayaan Nomor 27 Tahun 2008 tentang kualifikasi akademik dan kompetensi konselor*. Jakarta: Kemdikbud. 4-5. (2008).
13. M. Tang, W. Pan, M. Newmeyer, PSC. **11**(5), 285-295. (2008).

Integrated of Mobile Phone as Interactive Media in Extensive Listening

Abdul Kodir Al-Baekani[1,*], *Iwan* Ridwan[2].

[1]University of Singaperbangsa Karawang (UNSIKA), English Education Department, Karawang-West Java, Indonesia

[2] University of Singaperbangsa Karawang (UNSIKA), English Education Department, Karawang-West Java, Indonesia

Abstract. Learning English is the most difficult to learn by students, especially in learning of listening aspect. This research aims to investigate the process of listening activity in the classroom using mobile phone as interactive media in extensive listening and how the students' responds of learning listening using mobile phone as an interactive media in extensive listening. Methodology of this research is descriptive qualitative. The subject of this research is Private Senior High School Muhammadiyah Karawang with 30 students as the sample of this research. The data analysis of this research uses the result of observation, interview, and documentation. Observation is used to know the learning process in classroom. Interview is used to know the students' respond in learning process. While documentation is used to strengthen the data. The result of observation class shows that the process of teaching and learning listening as follows: (1) the teacher begins learning within 10 minutes, (2) the main activity using mobile phone in learning listening within 25 minutes, and (2) the final activity: the teacher gives a test to measure the students' ability in listening comprehension. Meanwhile, the result of interview with the students shows that students mentioned convenience and interesting using mobile phone (37%) and accessed in anywhere and anytime (30%), easiness (17%), authenticity (10%), and usefulness and fun (7%) to use their mobile phone in English listening.

Keywords : Extensive Listening, Interactive Media, Mobile Phone

1. Introduction

Listening is the most essential in communication. Human can speak verbally if they could listen. Al-Alwan (2013) argued that listening is the fundamental role on communication. In language, there are four basic skill that must be mastered by human to communicate each other. Firstly, listening is the first lesson that must be taught to students. Why is listening aspect should be the first lesson because as logically, human can be able to speak if they could listen something. Secondly is speaking. Thirdly is reading and the last skill is writing. For all, listening skill can be called the hardest skill to learn by the teacher or student because it is needed practice in learning.

Every school especially Senior High School, teacher should teach listening in the classroom and it should be practiced using media. Reinders, H., & Cho, M. Y (2010) stated that Teachers and materials designers are starting to explore the potential of ubiquitous, relatively cheap and increasingly powerful devices as potential supports for learning and teaching. In reality, practicing in the classroom is rare because there is not media availability. Eventually, there is no specific strategy in learning listening in the university (MA, M. A. B., & MA, C. T. M., 2014). Students learn listening section through speech teacher using English language in the classroom. Whereas, learning listening has to practice in learning process.

There are many learning media which can be used teacher to teach listening. One of them is Mobile Phone as a media. Rahmi (2015) clarified that mobile learning did have a significant effect on the listening comprehension ability. Almost all people especially teacher and student was using Mobile Phone which it was called Smart Phone. Every day, they use it to some necessity, such as necessity of communication, looking for information, and also for entertainment. Every human especially teacher and students got entertainment from Mobile Phone like listening music. Beside music, they also got listen the conversation even story and news using English language. In sum, English teacher in teaching listening material can use Mobile Phone as an interactive media in Extensive Listening.

From the problem above, researcher would like to try the research with the title *"Integrated Mobile Phone as an Interactive Media in Extensive Listening.*

The researcher limit this research, that is only learning process of English language listening section which it use Mobile Phone as an Interactive Media in Extensive Listening at Private Senior High School Muhammadiyah Karawang.

Based on limitation of the problem above, the formulation research question are as follows. (1) How does the English teacher use Mobile Phone as Interactive Media in Extensive Listening?; and (2)How are students' responses about using Mobile Phone as Interactive Media in Extensive Listening ?.

There are two aims of this research as follows. (1) to know the process of learning English using Mobile Phone as Interactive Media in Extensive Listening; and (2) to know the students' responses about learning English using Mobile Phone as Interactive Media in Extensive Listening.

1. Review of Related Literature

1.1 Extensive Listening

Many reading researchers argued that extensive reading has an good essential to learn reading comprehension. In article of Willy A. Renandya, reading researchers (for example Adams 1998; Day and Bamford 1998) have now confirmed the importance of extensive reading in the acquisition of reading skills and that "reading is best learned through reading" (Adams op.cit.:73). The evidence for extensive reading is strong: it can improve students' word recognition skills, vocabulary, reading comprehension, fluency, and general language proficiency. In addition, students also develop more positive attitude towards reading and language learning in general. What is amazing is that the students obtain all these benefits by simply doing something that is pleasurable. They just read anything that they find enjoyable. The only condition is that they should choose reading materials that they can understand on their own and that they read a lot of these materials. In other words, they do a lot of practice in comprehension, not practice *incomprehension* as many EFL students often complain about their skills-or strategies-based reading lesson. In sum of them, extensive reading is a good way to improve reading comprehension skill. Wherever they are, whenever they are, they can do practice reading and it was not realized their skill in reading comprehension skill have increased.

Basically, the teachers have to know the principle of language learning and teaching. In handbook of Maya Pentcheva and Todor Shopov (2003:6) said that H. Douglas Brown notes in his *Principle of Language Learning and Teaching* that there are "...best of times and worst of times" in language teaching profession (Brown 1994a), we can safely say that this is the best of times for foreign language teacher. Today, we know much about foreign language acquisition, about child acquisition of language, about cognitive process, etc. it is also very important that we have come to an appreciation of the extreme complexity of this field. This gives us caution optimism to plunge even deeper into a problems.

In Indonesia, all of school especially in Senior High School, English is foreign language. The teachers teach English as a foreign language. They usually innovate to teach students better. Maya Pentcheva and Todor Shopov (2003:6) stated that foreign language teacher and educators are often confronted with the question "What method or what system do you use in teaching a foreign language?" Most often the answer does not come easily or if one gives a straightforward answer, he risks to be subjected to criticism. Teacher always have *to make choices*. These choices are motivated by the fact that they rest on certain *Principles* of language learning and teaching. Now that we know much more about on what we know about language itself. Often, swept by fashionable theories or a desire to sound "scholarly", we forget a simple truth – *we, as human beings, teach a human language to human beings*. "Students and teachers of language", says Osgood, "will discover the principles of their science in the universalities of humanness" (Osgood et al. 1957, 301). A concise but true in definition of man will probably include three major characteristics : (i) one who can reflected and interpret the world around him; (ii) one who can express feelings; and (iii) one who can use language. These characteristics underline three major principles of language teaching and learning.

From the explanation about the principle of language teaching and learning above, it seems to relate the principle of teaching listening in the classroom because many language learners assume that learning listening is difficult even it is the most difficult to learn. To understand better the difficulties in teaching listening, we should begin by identifying the difficulties the student faces when undergoing listening (Natasya Walker, 2014). A lot of learners think the most difficult in learning listening are lack vocabulary and pronunciation. The evidence of difficult in learning listening about pronunciation (Bloomfield et al 2010:3) was stated in article of Natasya Walker (2014) that "the pronunciation of words may also differ greatly from the way appear in print". With the acoustics of the spoken language often varying dramatically to the form of the written language, the identification of words that constitute the oral discourse could prove to be problematic for the students. Likewise, "listeners, unlike readers, do not have the luxury of regular spaces that signal where words begin or end" (Vandergrift, 2007:296). Besides that, lack vocabulary and grammar still become problem to teach listening. Osada (2004:59) in Natsya Walker (2014) stated that vocabulary and grammar also tend to be far more colloquial and much less formal than what the students may be accustomed to, making for potential confusion should their vocabulary knowledge not be familiar with such expressions, as "an obvious factor that can influence comprehension of spoken passage is the overlap between the listener's vocabulary knowledge and the vocabulary of the passage" (Bloomfield et al 2010:12).

In sum, teachers should innovate to teach listening in order to learn listening easier especially to improve students' vocabulary. Pronunciation also should be focused by the teacher with someway like extensive listening in the classroom. It is the way to increase listening comprehension skill. One of this way is using mobile phone to help the process of listening section in the classroom.

1.2. Mobile Phone Media

To make definition of mobile-assisted language learning (MALL) is complex because it should be understood about cultural, disciplinary, and temporal context. It sheds light on the problem of definitions. Which are often circular, insufficient, varying over time, and contradicted by different sources, by helping to distinguish which definition is more accurate, and why. Because, to a great extent, defining is contextualizing, this method can help to define subfield or concept in their fuller dimensionality (Steave McCarty et al. 2017). The example of MALL is technology-enhanced language learning (TELL) to help teacher in teaching using technology (e-learning). Because, each eras are different using technology in learning especially language learning. Steave McCarty et al. (2017) stated that TELL must also listed in chronological order, which helps to gauge the accuracy of many competing or completed definitions of terms: (a) *distance education,* which is still mostly correspondence education in practice around the world, (b) *e-learning,* which is still most widely to cover all forms of digital learning (including CD-ROMs in CALL Labs), and (c) *online education,* which is most recent and relies on the Internet for synchronous and/or asynchronous communication. Online education include blended learning, because use of the Internet is incorporated into such classes. In this era, e-learning always use computer in teaching language learning especially listening material. Listening section of language learning uses computer as a media to help teacher and students in the process of listening activity. In fact, many school do not have a lab or CALL (Computer-Assisted Language Learning). This is a problem that makes teacher confuse to solve it. But, there is a solution: using mobile phone in teaching and learning listening section.

From the explanation above about teaching and learning listening that should be available media to support teacher in teaching listening, it can be concluded that teaching listening has to use media. There are a lot of media which can be used by teacher to deliver material listening. One of them is Mobile Learning. Gautam (2014) stated that mobile learning is an effective way in the process of learning language. Listening skill is an ability to identify and understand what people said. This process involves understanding of phone or speaker's pronunciation, grammar and also vocabularies, and understood of meaning. An active listener will understand the meaning overall (Howat and Dakin, 1997). Listening ability for students using mobile phone media will make easier to understand and to learn listening skill.

The procedures of learning English using mobile phone are as follows : (1) teacher deliver indicator of learning which it was achieved, (2) teacher explain what should students do for the competence, (3) teacher divides students into group with the same skill, (4) teacher allows students to open android mobile phone with relevance topic, (5) teacher allows students to conclude about the content of recording on their mobile phone, (6) teacher concludes students' opinion, and (7) teacher make a concept of learning listening using mobile phone.

2. Research Method

The site of this study was a private school in Karawang, West Java. This school had a populations of more than 500 students with various sociocultural backgrounds. Most of students were bilingual (e.g. Bahasa Indonesia and Sundaness). Most students come from at the low-middle economic background; so, they are relatively highly motivated to learn English. This positive enthusiasm positively affects their engagement in the classroom. They are active and curious in learning, especially with innovative learning strategy, integrated Mobile Phone as an interactive media in extensive listening. We chose the site because of two considerations: (1) learning listening at this school was supported by Mobile Phone as an interactive media in extensive listening, and (2) the authors got an entry access to this school.

In order to accomplish this research, it is needed an appropriate research design. Since this research aims at exploring how Mobile Phone is used in teaching listening and to reveal students' responses toward the used Mobile Phone in teaching listening then basic interpretative & phenomenological study are recommended one. It is in line with Ary, Jacobs, Sorensen & Razavieh (2010) point out mentioned research design aims to describe phenomenon and to reveal subjects' perspective on what they experienced. This method can be called basic interpretative study.

The participants of the present study were 30 students in Private Senior High School Muhhammadiyah Karawang. These students were supported learning listening using Mobile Phone. They were given instruction by teacher to learn listening material using Mobile Phone, thus these students were chosen as participants of the research.

Mostly, the traditional pedagogical procedures to teaching listening is through a process of ; (1) the teacher presents a topic of listening material, (2) the teacher talks about the topic of material, (3) student listens what the teacher said, and (4) student looks for the words and translate them into Indonesian Language. In contrast, in this study, teaching procedure is outlined as follows. Students are working in a groups of two equipped with a unit of smartphone or Mobile Phone with the application which was shared by teacher to students. The apps that was used in this activity is *English Listening Apps.* They have been downloaded those application on their Mobile Phone. And also the topic or material of listening which should be listened by students, it was downloaded by them. Meanwhile, to test their skill in listening comprehension, teacher uses *English Listening Test Apps.* There are some steps done by the teacher to teach listening using mobile phone, as follows. (1) The teacher asked the students to open mobile phone application, then choose play store menu, and next is to download the English listening skill menu. After done, it can be transferred to other students by using share it menu; (2) the teacher taught the students about listening by using mobile phone. The theme about appointment. The teacher asked the students to have aural drill, then aural and oral drill. Next, they repeated it with their seatmates; (3) the teacher asked the students to summarize the content of the dialog they listened to; (4) the teacher concluded the listening subject they learn in mobile phone about appointment; and the final one (5) the teacher provided the English Listening Test on Mobile Phone with multiple choice type.

The data can be analysed through as follows. (1) Class Observation: The researcher observes learning activity in the classroom using Mobile Phone in teaching listening. The researcher observes the teacher from preparation in the class till end, such as, in the first activity with the duration of 10 minutes, the teacher did apperception, motivated the students, expalined the achieved goal of learning, and explained the steps of learning. Next, in the main activity within 25 minutes, the teacher teaches the studnets using mobile phone with the proper steps, such as the students observe the English lsitening material using mobile phone, the students practise aural and oral drill using mobile phone, the students summarize the content of the listening material from mobile phone. And in the final one, teachers tests the students using mobile phone to find out the result of learning process done in 10 minutes test. (2) Interview: it is proof against the information obtained previously. "Interview can be done in a structured and unstructured, and can be done face to face or using the phone"(Sugiyono, 2006; 138-140); and (3) Documentation: Sugiyono (2011:329-330) argued that documentation is a record of events that have already passed. The documentation can be shaped writing, pictures, or the monumental works from someone. Documentation in the from of images such as photos, videos and other sketches. Research result from observation or interviews, it would be more credible or trustworthy if supported by a personal history of childhood life, school, in workplace, in society and autobiography. The data will be needed by researchers is the form of notes non-statistical the profile, goals, lesson plan, syllabus, material, the photos at the time of learning, vision and mission of the school.

3. Findings

Based on the result of analysis by using interviews for 30 participants or students showed that, as follows: (1) the favorite part in using the apps for the surdents on the mobile phone to improve listening skills indicated that most of the students responded positively. To be more specific, the students mentioned convenience and interesting using mobile phone (37%) and accessed in anywhere and anytime (30%), easiness (17%), authenticity (10%), and usefulness and fun (7%) to use their mobile phone in English listening. (2) the responses of the second question, the least favorite aspects of using the mobile phone app, the students' ideas. The students mentioned 'preference using textbooks in class' (50%), 'difficulty to understand' (25%), and 'no scripts' (25%). However, not many of the students showed negative aspects of using the mobile technology. (3) Based on the thirdly, how effective it is to use the mobile phone apps to develop listening skills, the result presents the responses. The detailed responses revealed that most participants'feedback focused on a variety of English expressions and vocabulary (33%). The second most frequent response was 'good to have more chances to access English anytime anywhere' (27%) and 'to be good for self-studying' (23%). Furthermore, the students stated that they were able to have confidence in listening authentic materials (17%) ; (4) From the last result that the points the students considered to be ineffective to use the mobile phone apps, the students' ideas. Not many responses to this question were elicited, as seen in Table 4. The responses of the ineffectiveness of using the apps were 'difficulty to understand' (40%), and 'difficulty of speed' (60%).

It can be concluded from the findings that the mobile phone app would be a more efficient way to expand English listening skills. In addition, mobile device allowed learners to integrate the world outside with the authentic materials and give more positive attitudes for language learning.

4. Conclusion

Based on the analysis of the questionnaire result that the favorite part in using the apps for the students on the mobile phone to improve listening skills indicated that most of the students (N=30) responded positively .To be more specific, the students mentioned convenience and interesting using mobile phone (37%) and accessed in anywhere and anytime (30%), easiness (17%), authenticity (10%), and usefulness and fun (7%) to use their mobile phone in English listening. In return, the least favorite aspects of using the mobile phone app, the students (N=5) mentioned 'preference using textbooks in class' (50%), 'difficulty to understand' (25%), and 'no scripts' (25%). However, not many of the students showed negative aspects of using the mobile technology. And, how effective it is to use the mobile phone apps for the students (N=30) to develop listening skills, the detailed responses revealed that most participants'feedback focused on a variety of English expressions and vocabulary (33%), the second most frequent response was 'good to have more chances to access English anytime anywhere' (27%) and 'to be good for self-studying' (23%), Furthermore, the students stated that they were able to have confidence in listening authentic materials (17%). In reverse, the students (N=5) considered to be ineffective to use the mobile phone apps, not many responses to this question were elicited. The responses of the ineffectiveness of using the apps were 'difficulty to understand' (40%), and 'difficulty of speed' (60%).

In addition, the average result of listening test using mobile phone also showed positive result from the first session to final one. First test of average result was about 60,33. The second one was about 65,33, and the last one was about 73,83. So, it can be concluded that using mobile phone can improve students' listening skill.

References

1. Al-Alwan, A., Asassfeh, S., & Al-Shboul, Y. EFL Learners' Listening Comprehension and Awareness of Metacognitive Strategies: How Are They Related?. International Education Studies, **6(9), 31**. (2013).
2. Archana Gautam. *Mobile Learning; an Effective Way of Teaching and Learning English Language*, http://www.arcjournal.org (2010)
3. Bloomfield, A. et al.. *What makes listening difficult? Factors affecting second language listening comprehension.* (2010)
4. Brown, H. Douglas. Teaching by Principles: *An Interactive Approach to LanguagePedagogy*. 2 Edition.England: Pearson Longman. (2000).
5. Catherine Marhall, Gretchen B Rossman;Designing *Qualitative Research, Second Edition*. Sage Publications, International Educational and Professional Publisher, London, (1995).
6. Creswell, J. W. Qualitative inquiry and research design: Choosing among five designs. Thousand Oaks, CA: Sage. (1998).
7. Howatt, A., and J. Dakin. *Language Laboratory Materials*, ed. J.P.B Allen, and S.P. Corder. (1974)
8. Jack C. Richards. *Communicative Language Teaching Today.* (Cambridge University, Press 2006)
9. Maya Pentcheva and Todor Shopov, *Whole Language, Whole Person: A Hanbook of Language Teaching Methodology,* (Viseu, 2003)
10. Merriam, S. B. *Qualitative research in practice: Examples for discussion and analysis*. San Francisco, CA: Jossey-Bass. (2002).
11. MA, M. A. B., & MA, C. T. M.. Listening comprehension difficulties encountered by students in second language learning class. Journal of Educational and Instructional Studies in the World, **3**. (2014)
12. Natasya Walker. Listening: the most difficult skill to teach, **23** (2014).
13. Steave McCarty, Takehi Sato, and Hiroyuki Obari. *Implementing Mobile Language Learning Technologies in Japan.* (2017)
14. Rahimi, M., & Soleymani, E. The impact of mobile learning on listening anxiety and listening comprehension. English Language Teaching, **8(10), 152**. (2015).
15. Reinders, H., & Cho, M. Y. Extensive listening practice and input enhancement using mobile phones: Encouraging out-of-class learning with mobile phones. (2010).
16. Sugiyono. *Metode Penelitian Pendidikan (Pendekatan Kuantitatif, Kualitatif, danR&D)*. Bandung: Alfabeta. 2011.

TADALURING Microteaching Learning Model (TMLM): Preparing Professional Teacher by Pre-Service Training

Arifmiboy [1*], *Jamaris* Jamna[2],
Mega Iswari[2], and *Agustina*[2]

[1] Institut Agama Islam Negeri Bukittinggi (IAIN Bukittinggi)
[2]Universitas Negeri Padang (UNP)

Abstract: As a certified professional, teacher forms indispensable roles in achieving the learning objectives. The complexity of teacher's profession decidedly requires the teacher to master a number of competencies and skills. One of them is to master the basic teaching skills. Realizing the complexities of the teacher's profession and the requirement of teacher's profession, LPTK –Faculty of Teacher Training and Education– which produces professional teacher, requires enhancing the teacher's quality and creativity through pre-service teaching and training. Microteaching is one of the efforts to prepare the candidate for a professional teacher in higher education. The growth of a number of students being served with the availability of micro teaching laboratories is, however, disproportionate. Consequently, the problems faced are the time management, the laboratory utilization, the availability of sufficient microteaching infrastructures as well as the implementation of micro teaching itself. One of the ways to solve the problems is to develop a microteaching model adapting the force driving sophisticated ICT named TADALURING Microteaching Learning Model (TMLM). The Instructional Systems Design (ISD) made use of developing the model is Borg and Gall procedures. The cyclical phases of conducting the research were preliminary research, development, field-testing, and dissemination. The research finding indicates that the trial-tested model is stated valid, practical, and effective. The TADALURING Microteaching Learning Model (TMLM) is essentially accentuated on a face-to-face classroom, online and offline practices.

Key Words: micoreaching, model, tadaluring

1 Introduction

The International surveys' results reported that quality of Indonesian education still extremely declines. The 1997 to 2007 *World Competitiveness Year Book's* survey reported that of the 47 countries surveyed, Indonesia ranked 39th in 1997 and 46th in 1999. Then in 2002, of 49 countries surveyed, Indonesia ranked 47th and in 2007, of the 55 countries surveyed, it still ranked 53rd [1]. Similarly, in 2005, A Global Monitoring Report released by UN agency, UNESCO, Indonesia ranked 10th among 14 developing Asia Pacific's countries [2]. Another international survey, The United Nations Development Program (UNDP) reported that the Indonesian's Human Resource Quality ranked 109th out of 177 countries in the world [3]. Meanwhile, The Political and Economic Risk (PERC) Consultancy Agency in Hongkong informed that the quality of education in Indonesia still extremely declines among 12 Asian countries studied. This indicated that the *quality of*

* Corresponding Author: arifmiboy@yahoo.co.id

education in Indonesia drew level with Honduras and Nigeria although it is *better* than the *Vietnam*.

At the praxis level, the educational problems designate the obstacles to realistically achieving the goals of national education as legally mandated in the Law No. 20 of 2003 on the National Education System. The declining quality of Indonesian Human Resources (HR) becomes one of the real causes of the obstacles. The decline can be analyzed from several macro indicators, among others, are the 2008 to 2009 Global Competitiveness Report from the World Economic Forum in Martin stating that Indonesia ranked 55th out of 134 countries in terms of the achievement of Competitiveness Index (CI). Subsequently, the United Nations for Development Program in Human Development's 2007/2008 research results reported that Indonesia ranked at the 107th of 155 countries in terms of Human Development Index's (HDI) achievement [4].

The declining quality of Indonesian education had been clearly elucidated by the Study Program for International Student Assessment (PISA) in 2003. PISA reported that, for Science and Mathematics literacy, 15-year-olds ranked 38th out of 40 participating countries and for reading literacy, Indonesian students ranked 39th. In 2006, the reading literacy achievement of Indonesian students ranked 48th out of 56 countries while mathematical literacy ranked 50th out of 57 countries, and science literacy ranked 50th of 57 countries. Furthermore, in 2006, the Progress in International Reading Literacy Study's (PIRLS) research report on the grade 4 elementary school children's reading literacy worldwide under the coordination of The International Association for the Evaluation of Educational Achievement (IEA) in 45 countries/states, both from developed countries and developing countries, found that Indonesian students ranked 41st out of 45 participating countries [5].

Similarly, Balitbang's Data indicated that quality of Indonesian education extremely devastated. The data explained that 8 of 146,052 elementary schools nationwide which has been internationally recognized as The Primary Years Program (PYP). Similarly, 8 of 20,918 Indonesian Junior High Schools which has been internationally recognized as The Middle Years Program (MYP) and lastly, 7 out of 8,036 Senior High Schools which have been internationally recognized as The Diploma Program (DP). The results of the Teacher Competency Test (UKG) held by the Ministry of Education and Culture in November 2015 were still beneath the expected standards. As disclosed by the former Minister of Education and Culture, Mr. Anis Baswedan, PhD, "the average national UKG's score exceptionally declined. The average of national UKG's score was 53.02, whereas the government targeted the average score was at 55. In addition, the average professional competence's score was 54.77 while the pedagogic competence's score was 48.94" [6].

In response to the aforementioned conditions, it requires central government and stakeholders' efforts to improve the quality and competence of Indonesian teachers. These efforts and improvements will automatically assist the achievement and accomplishment of Indonesian education. The process of enhancing Indonesian teachers' professionalism must be early begun when the prospective teachers are studying in university (pre-service). As prospective teachers, the students should possess, experience, and master a variety of courses related to pedagogic competence such as learning planning, learning strategies, instructional media, curriculum development, educational psychology, and learning evaluation. All of these courses will provide pieces of fundamental educational knowledge to prospective teachers in the context of enhancing their pedagogic competence. One of the efforts that can be undertaken to develop the teachers' pedagogic competence is the process of educational training that is packaged in a micro teaching course. Theoretically, Dwight Allen clearly explicated that the primary goal of micro learning is to separately provide the real learning experiences, practices and some indispensable teaching skills to the prospective teachers. They are expected to develop their teaching skills prior to teaching at the real classroom setting [7].

The development of the sophisticated information technology in the field of education opens an access or provides opportunities to undertake the long distance education using the internet media to connect the students with the lecturers, access college schedule, check finances, and send the task files and or documents given by teachers and so on. Everything can be easily done online. Taking into account of the problems faced in microteaching learning along with the development of ICT's facilities, there is a call for developing a Model of Microteaching Learning adopting the today's technological developments as a realistic solution to the problems.

2 Literature Review

Conceptually, the "microteaching" stems from two words, namely micro and teaching. Micro means small, limited, and narrow, while teaching means educating or teaching. Allen and Ryan in Lakshmi explicate that microteaching as a scaled down teaching encounter, scale down in term of class size, lesson, length, and teaching complexity [8]. Whereas Allen point out that "microteaching as a system of controlled practice that make it possible to concentrate on specific teaching skills and to practice teaching under controlled conditions" [9]. Singh states that microteaching is a training technique, which requires pupil teachers to teach a single concept, using specified teaching skills to a small number of pupils in a short period of time [10].

Cooper and Hillary affirm that microteaching forms a teaching situation undertaken in very limited time i.e., for 5-20 minutes with 3 to 10 students [11]. Furthermore, Mc. Laughlin and Moulton describe microteaching is a performance training method designed to isolate the component part of teaching process, so that the trainee can master each component one by one in a simplified teaching situation. Lakshmi adds, microteaching is a scaled-down teaching encounter' in which a teacher teaches a small unit to a group of 5 to 10 pupils [12]. Barmawi highlights that the difference between microteaching and teaching in terms of the number of students. In microteaching class, it comprises of 5 to 10 students whereas in teaching, it includes 30 to 40 students. In conclusion, microteaching is an extremely controlled teaching training to master certain basic teaching skills in a minimised learning condition in terms of time, materials, skills, and a number of students [13].

Dwight Allen explains that the purpose of microteaching for prospective teachers is to: 1) provide real teaching experience and practice a number of basic teaching skills, 2) the prospective teachers can develop their teaching skills ahead of entering the field; and 3) provide a variety of basic teaching skills. For teachers themselves, it refreshes educational programs, and expands teaching experience individually to develop professionalism, and build up an open attitude for teachers to innovation [14].

Wight Allen describes the implementation of microteaching done through seven stages. The six stages of microteaching form a cycle and they can be repeated in accordance with meeting the needs for improvement. The followings are the steps of microteaching learning.

1. Planning a micro-lesson determines the right subject matter that maximizes the practice of teaching skills within 5 to 7 minutes.
2. The teaching session is a lesson plan implemented in front of mentors or classmates. The performance of teachers who practice teaching skills is observed and recorded. The evaluation sheets, tape recorders, and/or video tapes can be used for such purposes.

3. The critique session is the lecturer and the participants discuss the performance of the trainees. Feedback and key points are communicated to the trainees for self-improvement. Evaluation tools provide an opportunity for micro teachers to view their performance objectively. The micro-participants are not given the opportunity to question for self-defence. This is the strength and uniqueness of microteaching
4. The re-planning session is a micro participant develops a teaching plan based on the feedback offered in the critique session. The time provided for this stage is 5 to 7 minutes
5. The re-teaching session is a step to provide the opportunity for micro participants to teach the same unit and skills. Nevertheless, of course, the performance of the micro-teacher, in this session, should have paid attention to the mentors and/or peers' feedbacks. In this session, supervisors and peers evaluate the performance of the participants who perform using the observation sheet.
6. The re-critique session is the same procedure adopted as mentioned in the critique session. The micro participant gets feedback and knows how far the improvement is. This step has the potential to motivate micro participants to improve their appearance in the future [19].

3 Methodology

The Borg and Gall's Instructional Systems Design forms the 4 pieces most important cycles in developing the TADALURING model. In general, in developing the model, TADALURING, it has four main stages. They are 1: Preliminary Research, Stage 2: Model development, stage 3, Field test or model validation and, stage 4: dissemination The stages of developing the model were the combination of a planning along with Preliminary research results, relevant research, and literature study served as the basis for the preparation of microteaching model TADALURING. The stages of learning model development included some of the following activities.

1. Drafting and or composing the design of the model. The basic drafting and or composing of this design used was the preliminary study, the foundation of learning theories, the theory of model development, and the feasibility of its implementation.
2. Experts' validation and practitioners. To validate the design of the developed model was undertaken through experts' judgments. The experts who had been involved were Prof. Dr. Harris Effendi Thahar, M. Pd as a linguist, Dr Jsrial, M. Pd as the design expert, Dr H. Darul Ilmi as a material expert, Dr Asrul Huda, M. Kom as a media and ICT's expert, and Dr Ir. Khairani, M. Pd as an evaluation and instrument's expert. The instruments used in conducting the feasibility test were the validation sheets for the experts. The validated components were language, developed model structure, teaching preparation, basic teaching skills, observation sheets, questionnaires, observation checklists, and interview guides. Every expert had provided insight and input regarding the design of the developed model. The inputs and suggestions had been used as a basis for the draft revisions of the designed model. More importantly, the researchers had used validation sheets which could shape the revision processes of the designed model. The result generated the hypothetical design that has been tested on subsequent activities.
3. Revising the draft of the designed model. The revised draft of the designed model referred to the suggestions and inputs on the desk evaluation comprised of the experts' validation sheets.

The developed TALALURING microteaching-learning model was validated by a number of experts in their respective fields (experts' validation). The expert validity was undertaken in order to obtain the theoretically, substantially, and methodologically experts' recognition. Experts in question are design, linguists, microteaching learning materials, and ICT media experts. Validated objects are instruments, model books, and learning tools. The resutl showed an average score of 3.72 and validly interpreted.

The effectiveness test was undertaken in the form of quasi-experiment research with Non-Equivalent Group Design.. The experiment was given to 24 students in the English Department of teacher training Faculty of IAIN Bukittinggi the seventh semester of 2016/2017 academic year. Adapting to the research design model used, the researcher divided the samples into two groups comprising of 12 participants tailoring to the amount of set by the study programs for each study group in microteaching lectures. The sample used was accidental technique. Based on the pre-test and post, the gained avarage scores were 14. 10. Besides, the hyphothesis which suggested that the mean's score of the mastery of the basic teaching skills is higher then 80, is accepted. The score of $t_{obtain} > t_{table}$ (2,065>1,711).

The practicality's test involved three universities namely IAIN Bukittingi, IAIN Batusangkar, and UIN Imam Bonjol Padang. The number of lecturers involved was 15 people with different areas of expertise. The result showed that an average score of 83,53 and effecftively interpreted.

4 Result

The Tadaluring is the acronym of a face-to-face, inside and outside the network. Tadaluring Microteaching Learning Model (TMLM) is a learning model that combines the forms of teaching practices in an integrated way using communication technology media. These combinations form the face-to-face classroom's practice, face-to-face Online's practice, and face-to-face offline's practice.

Tadaluring microteaching learning model is classified into a family behaviour modification model. In the behaviour modification models are also known a number of model namely; Contingency management model, self-control model, training model, stress reduction model, desensitization model, and assertiveness training model. From a number of various branches of the model, the microteaching Tadaluring learning model includes training model. The construction of the micro teaching course flow is illustrated as follows.

Figur 2. Tadaluring *Microteaching* Learning Model

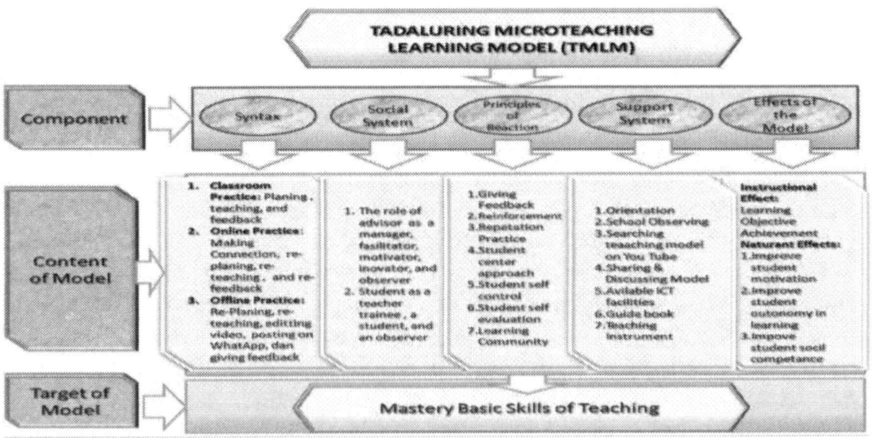

Based on the above diagram, it can be understood that the Tadaluring microteaching-learning model is based on the syntax components, social systems, principles of reaction, support system, and effect of the model. The details are elucidated as follows.

4. 1. Syntax

The *syntax* of Tadaluring microteaching-learning model is described as follows.

4. 1. 1. Classroom Practice

The classroom practice is a teaching practice activity undertaken in the classroom directly attended by supervisors and trainees in microteaching learning. The steps of teaching practice in the classroom are planning, teaching, and providing feedback. The planning activity is an activity in formulating the training strategy, such as determining the type of skills trained, determining the topic of discussion, method, learning approach, and the form of students' engagement and participants.

a. Online Practice

In-class practice activities continued with the online practice. Online practice is a practice activity carried out online with the help of means of communication over the Internet using Skype facility. With Skype facility allows supervisor and all participants can interact directly at the same time and different places. All participants and supervisors meet directly on their computer screens. Each participant and lecturer can see and greet each other.

4. 1. 2. Offline Practice

Offline practice is a follow-up activity of online classroom practice. Offline practice is the practice of teaching activities conducted independently by involving several students or colleagues as a media in the project. Offline practice places emphasis on maximizing opportunities for practice. Each participant recorded his or her training activities independently of both basic and integrated teaching basic teaching training.

4. 2. Social System

In the Tadaluring microteaching learning model, the role of the student is more dominant than the supervisor. The role that students play in microteaching learning is as a practising teacher, as a student in other conditions, and as an observer or evaluator. Students as teachers in microteaching learning when they practice mastering the basic skills of teaching, they will act as real teachers, starting from planning lessons, strategizing, choosing media, methods, implementing learning to carry out the evaluation.

On the other hand, the students will also act as students. Students as microteaching participants will behave and act like a student, ask questions, carry out teacher commands, answer teacher questions, hear explanations, and write various materials presented in accordance with the conditions that other participants who are practising as teachers.

Furthermore, the student as a microteaching participant is an observer as well as an assessment team. As an observer, the student will observe every gesture and learning process conducted by his colleague and then provides an assessment through the observation sheet prepared by participants who perform the practice. Even students will also provide comments in the form of suggestions and criticisms that are constructive in order to improve the appearance for the next exercise.

In Tadaluring microteaching-learning model, supervisors have a very important role in realizing learning goals. The supervisor is the director and the actor in charge. The role of lecturers is a demonstrator, facilitator, motivator, innovator, inspiratory and evaluator/observer.

4. 3. Principels of Reaction

In Tadaluring microteaching-learning, there are a number of reaction of principles namely repetition and improvement, giving immediate feedback either directly or indirectly, giving both verbal and non-verbal reinforcement, providing motivation, student-centred learning, self-evaluation, and learning Independent.

4. 4. Supporting System

Tadaluring microteaching-learning model can be done effectively if it meets a number of activities and the availability of other supporting facilities are orientation, school observing, searching teaching model on You Tube, sharing and discussing, available ICT facilities, guide book, and teaching instrument..

4. 5. Nurturant Effect of the Tadaluring Microteaching Learning Model

Tadaluring Microteaching Learning Model provides two forms of Effects. The first is direct effect. The Tadaluring Microteaching Learning Model is able to achieve the goal of microteaching learning itself. The students of microteaching participants are able to master a mixture of basic teaching skills trained. Second is indirect effect. The indirect effects can improve student's learning motivation, student's learning independence, and students' social competence such as cooperation, mutual respect, mutual help, and reminds well behaved.

5 Conclusion

In respond to the force driving ICT and the difficulty in microteaching learning process, the learning model, which adopts the ICT development into microteaching learning process, is considered as a practical and effective solution to microteaching's problem or difficult situation. This is a way of dealing with it so that the difficulty is removed. Tadaluring Microteaching Learning Model (TMLM) emphasizes on learning process optimizing classroom, online and offline practice by each participant using some kinds of ICT facilities. Tadaluring Microteaching Learning Model (TMLM) was stated valid, effective and practice in using.

6. Reference

1. Martin, Xavier, Sala. *The Competitiveness Index: Measuring the Productive Potential of Nations.* In *The Global Competitiveness Report 2008-2009,* **15** (2008)
2. Martin, Michael O., Mullis. *TIMSS 2007:International Science Report*. Chestnut Hill, MA: Boston College, **5** (2008)
3. HDI. http://en.wikipedia.org/wiki/List_of_countries_by_Human_Development_Index. "List of countries by Human Development Index",**15** (2008)
4. Martin, Michael O., Mullis. *TIMSS 2007: International Science Report*. Chestnut Hill, MA: Boston College. (2008)
5. Program for International Student Assessment (PISA), (2003)
6. www.okezone.com (04/01/16).
7. Allen, Dwight., Kevin Ryan. *Microteaching*. Addison-Wesley Publishing Company Inc. Reading, Massachusetts Menlo Park, California, **15** (1969)
8. Lakshmi, Majeti Jaya. *Microteaching and Prospective Teachers.* Discovery Publishing House Pvt. Ltd. Sachin Printers New Delhi, **4** (2009)

9. Allen, Dwight., Kevin Ryan. *Microteaching*. Addison-Wesley Publishing Company Inc. Reading, Massachusetts Menlo Park, California, **29** (1969)
10. Singh, L. C. *Microteaching: An Innovation in Teacher Education.* New Delhi, NCERT, **25** (1979)
11. Cooper, Hillary. *The Teaching of History Implementing The National Curriculum.* London: Davis Fulton Publishers, **85** (1992).
12. Lakshmi, Majeti Jaya. *Microteaching and Prospective Teachers.* Discovery Publishing House Pvt. Ltd. Sachin Printers New Delhi, **12** (2009)
13. Barmawi &M. Arifin. *Microteaching: Teori Praktek Pengajaran yang Efektif & Kreatif.* Jogjakarta: Ar-Ruzz Media, **46** (2015)
14. Allen, Dwight., Kevin Ryan. *Microteaching*. Addison-Wesley Publishing Company Inc. Reading, Massachusetts Menlo Park, California, **26** (1969)
15. Lakshmi, Majeti Jaya. *Microteaching and Prospective Teachers.* Discovery Publishing House Pvt. Ltd. Sachin Printers New Delhi, **54** (2009)
16. Sukirman, Dadang. *Pembelajaran Microteaching*, Jakarta: Direktorat Pendidikan Agama Islam, Kementerian Agama RI, **27** (2012)
17. Lakshmi, Majeti Jaya. *Microteaching and Prospective Teachers.* Discovery Publishing House Pvt. Ltd. Sachin Printers New Delhi, **55** (2009)
18. Singh, Shivpal. *Teacing Competency Through Microteaching Aproach,* **75** (2011)
19. Allen, Dwight., Kevin Ryan. *Microteaching*. Addison-Wesley Publishing Company Inc. Reading, Massachusetts Menlo Park, California, **86** (1969)
20. Gall, J., Borg. W., & Gall, M. *Educational research: An introduction* (7th ed.). Boston: Pearson Education, **187** (2003).
21. Emzir, *Metodologi Penelitian Pendidikan,*PT. Raja Grafindo, **105** (2008)

In-Service Professional Development on Supporting Elementary Teachers' Pedagogical Content Knowledge and Efficacy through Inquiry-Based Teacher Training

Muhammad Randy Fananta[1,*], *Tria* Umbara[2], and *Saktiana Dwi* Hastuti[1]

[1]PT. KUARK Internasional, Jakarta, Indonesia
[2]Elemetary Teacher Education, President University, Cikarang, Indonesia

Abstract. This study was conducted to ascertain the role of inquiry in supporting teachers' pedagogical content knowledge (PCK) and efficacy based on the In-Service Profesional Development (INSEP) findings. INSEP program has been conducted by Kelas Lentera Kuark in East Sumba Regency, East Nusa Tenggara Province. This program was conducted from January to July of 2016 within three stages such as: (1) Preliminary assessment (2) Teacher training (consist of two sessions: Motivational and leadership training, and Inquiry-Based Teacher Training); (3) Monitoring and Evaluation. This study was conducted using the qualitative approach of multiple cases study. The data were obtained from multiple sources and analyzed through the approach of constant comparative method. The findings show that inquiry plays an important role in constructing teachers' PCK and efficacy. Inquiry does not only serves as an instructional teaching but also it develops their paradigm to understand more about Nature of Science thus they have the ability to develop PCK that can bring a direct impact to their efficacy.
Keywords: PCK, efficacy, inquiry, teacher training, professional development science, nature of science

1 Introduction

Based on the initial surveys' result towards elementary teachers in East Sumba Regency of East Nusa Tenggara Province in Indonesia, it was found the fact that teachers are still the center of science learning process where they deliver it through lecturing and recitation. Teachers express that they have lack of confidence in delivering science materials. They find it difficult to gain the students' interest to study science subject. The reason behind it is because they have not enough capacity of pedagogical content knowledge (PCK).

PCK correlates with the teachers' knowledge based on experience time to time regarding the (1) content knowledge needed; (2) strategy of pedagogy and teaching to gain

* Corresponding author: randy.fananta@komikuark.net

the students' attention; (3) the knowledge about students intelligence background and experience, therefore they can facilitate the students to learn and develop the understanding about the conceptual knowledge from the particular matter [1, 2, 3, 4]. To develop PCK, science teachers need the knowledge about science material, general pedagogy and context (students and school environment) where the particular teachers work [5]. In this case, teachers may not be able to integrate the nature of science as a whole when delivering the teaching. Regarding its integration, teachers need to believe on their professional capacities (known as teacher efficacy). Teacher efficacy in teaching can be used as a powerful indicator from teachers' professional behavior in class [6, 7]. Teachers with low efficacy will struggle more in teaching, satisfaction under the standard, high level of stress at work [7]. The improvement of teachers' professionalism at teaching has a strong impact to teacher efficacy [8]. Therefore, the improvement of teacher efficacy can be done by increasing teachers' PCK through In-Service Professional Development (INSEP) program

Ministry of Education and Culture of the Republic of Indonesia [9] has established the policy in implementing the national curriculum which is expected to strengthen the scientific approach through the application of learning based on inquiry process. This process is an activity for students to improve their knowledge and understanding of scientific ideas and how scientists learn about the nature [10, 11]. Inquiry has been regarded as an approach to bridge the gap between a scientist's way of doing science and a students' learning of science [12].

However, yet there is no teacher training model that integrates inquiry as an approach in East Sumba. We (Kelas Lentera Kuark) have designed inquiry-based teacher training that was proper in supporting teachers' PCK in Tabalong, South Kalimantan [13]. This program was conducted into three stages i.e. (1) Preliminary assessment (2) Teacher training (consist of two sessions: Motivational and leadership training, and Inquiry-Based Teacher Training); (3) Monitoring and Evaluation. Furthermore, we found that the training of delivering science material through inquiry-based learning has given a direct experience to understand the application of inquiry for teaching in class. In this case, we tried to apply the model to another setting and circumstance in order to strengthen the impact of the model on teachers' PCK and efficacy.

2 Methods

This study was conducted using a qualitative approach through a multiple case study. The multiple-case study was conducted to four participants of elementary school teachers in East Sumba, East Nusa Tenggara that participated in INSEP program with inquiry-based teacher training model within the period of January to July of 2016. The data of the participants were presented on Table 1.

Table 1. Background information of the participants of INSEP program.

Participant Name	Age (years)	Education	Teaching Experience
Magdalena Kana	54	Sekolah Pendidikan Guru (equal to Senior High School)	33 years
Mardiana Radjah	40	Diploma 1 English	12 years
Sri Mimiasti	30	Diploma 2 Elementary Education	6 years
Yuliana Mehang	40	Bachelor of Elementary Education	7 years

The case study focuses on two things which are: PCK and teachers' efficacy. Teachers' efficacy can be revealed using the instrument of a list of questions which can uncover the certainty and development to organize the concept from how that certainty represents the daily habit [14]. As the triangulation, we conducted a semi-structured interview with teachers.

PCK is the combination of teaching, learning, and content [2]. Teachers' expertise in science content was measured through pre-test and post-test. Teaching and learning method were disclosed from multiple sources including classroom observation, lesson plan, teacher written reflection, and semi-structured interview. Non-participant observation method was also conducted during the INSEP process inside and outside the class.

All the data of interview and observation were recorded and transcribed verbatim. The collected data were then coded word per word. Then they were analyzed using the constant comparative method approach. Within this research, the coded data of transcribed interview and observation were compared day by day to identify the pattern that appeared during the process of training and monitoring. The result from each data was connected to one another in order to obtain the whole conclusion regarding the efficacy and PCK of each teacher.

3 Results and Discussion

We conducted the inquiry-based teacher training where teachers are not only trained to deepen their science competency but also to create the subject specific pedagogy in teaching science and also doing the simulation for the implementation. There are three stages of the inquiry-based teacher training which can be seen in Fig. 2.

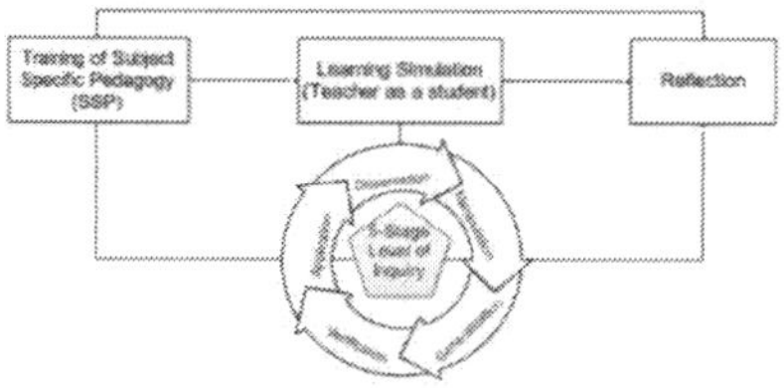

Fig.2. Inquiry-based sub-training [13]

The first stage is introducing the model of instructional inquiry-based learning and SSP of inquiry, then conducting the inquiry learning simulation with teachers where they position themselves as students who perform the inquiry-based learning. In this stage the teachers conducted the inquiry-based learning as observation, manipulation, generalization, verification, and application. Lastly, the teachers conducted a reflection from the learning process to comprehend more when implementing the inquiry-based learning within the national curriculum. Within this stage, we conducted the pre- and post-test to see the increase of competency based on the gain score in each stage. The result of the pre-and post-test can be seen on Table 2.

Table 2. Result of Pre-and Post-Test about Science Competency

Teacher Name	First Session			Second Session		
	Pre-test	**Post-test**	**Gain (category)**	**Pre-test**	**Post-test**	**Gain (Category)**
Magdalena Kana	14,71	55,64	0,48 (Mediocre)	28,95	80,15	0,72 (High)
Mardiana Radja	38.24	46.57	0,13 (Low)	42,87	86,84	0,77 (High)
Sri Mimiasti	29.41	67.89	0,55 (Mediocre)	53,50	92,11	0,83 (High)
Yuliana Mehang	17,65	36,03	0,22 (Low)	34,65	78,51	0,67 (Mediocre)

During the process, the better improvement shows after the second training was conducted. This means they have absorbed a good understanding of science knowledge. This suggests that the inquiry approach is particularly suitable in developing teachers' content knowledge. This is in line with the previous study which revealed that inquiry-based learning conducted by applying a systematic scientific method can build the logical and analytical thinking skills [15, 16, 13] with the result that teacher was easier to understand the content which had been delivered.

The components of the teachers' PCK in teaching science is related with five things such as: (1) orientation in science teaching; (2) knowledge of science curriculum; (3) knowledge of students understanding of science; (4) knowledge of instructional strategy; and (5) knowledge of assessment of scientific literacy [17, 18, 19]. Based on the result of interview and reflection, we can see that there is a shift on their point of views regarding the orientation of science teaching and some of those findings can be seen on Table 3.

Table 3. The Impact of INSEP on Teachers' Science Teaching Orientation.

Magdalena Kana	"... Before we used to learn strictly from the book. But now we can use any material around our school to learn" (Post-Interview)
Mardiana Radja	"...I used to teach what's on the book, that's it. Then after we applied (inquiry base learning), the impact has increase our students' interest..." (Post-Interview) "...Students are more enthusiastic and enjoy... Seeing their behavior and activity sometimes have given me new ideas to be applied for the next class." (Teachers' reflection)
Sri Mimiasti	"... Before I used to teach by preaching and giving more material. But after I attended Kelas Lentera Kuark, the way I teach is more interesting and fun..." (Teachers' reflection)
Yuliana Mehang	"... Creative science material is very compatible with curriculum of 2013(Indonesians' National Curriculum) ..." (Teachers' reflection) "...The impact of this training is not only in science but can also bring into other course such as social science in market session or in theater..." (Post-Interview)

Understanding PCK in teaching science is easier to shape by giving a direct experience to teachers as students learn in class and by exploring and analyzing the relationship between content, teaching and learning [2]. Through the simulation inquiry-based learning in giving the concept of science and reflection, teachers have come to a different point of view in teaching science. From the post-interview and teacher reflection, it was shown that teachers had a new orientation to teaching science. Besides that, the teachers had a good understanding about the implementation of science curriculum in their class.

Teachers' efficacy has a strong correlation in understanding PCK [19]. This can be seen from the questionnaire result on the belief in becoming a professional educator in teaching science, as shown in Table 4. It shows that the teachers have a good efficacy in teaching science and a strong belief in self-improvement. This result indicates that the teachers have the willingness to be professional learners after INSEP. This also shows that having such belief is not related to the age nor educational background which can be seen in Kana case. In one hand, this finding strengthens the theory of teachers' efficacy which is very crucial to their understanding of PCK. On the other hand, however, the teachers have less confident in improving the assessment. After all, this finding indicates their lack in constructing the assessment instrument especially based on process skills.

Table 4. Teachers' Efficacy after INSET.

Aspect	Teacher Name			
	Magdalena Kana	**Mardiana Radja**	**Sri Mimiasti**	**Yuliana Mehang**
Implementing inquiry base learning	84% (Very confident)	66% (confident)	83% (very confident)	80% (confident)
Being a facilitator	88% (very confident)	75% (confident)	85% (very confident)	75% (confident)
Being a Motivator	85% (very confident)	75% (confident)	90% (very confident)	85% (very confident)
Thinking skill training	85% (very confident)	70% (confident)	85% (very confident)	75% (confident)
Becoming expert on science	93% (very confident)	70% (confident)	85% (very confident)	78% (confident)
Class management	90% (very confident)	75% (confident)	85% (very confident)	85% (confident)
Assessment development	73% (confident)	60% (less confident)	80% (confident)	70% (confident)
Self-improvement	90% (very confident)	90% (very confident)	90% (very confident)	80% (confident)

4 Conclusion

The INSEP program can bring a positive impact for science content, teachers' PCK and efficacy as well as its correlation with the inquiry. The internalization of inquiry plays an important role in connecting teachers' understanding of nature of science and the instructional strategy in their class, and also crucial for teachers in developing their paradigm in improving their professionalism [20, 21]. INSEP integrated with the inquiry process through learning simulation enables teachers to build the understanding of science as well as gaining the knowledge of instructional strategy of inquiry-based learning.

Inquiry improved teachers' self-efficacy indirectly by developing their science knowledge and pedagogy [22]. The development of inquiry in improving teachers' PCK and efficacy can be seen in Figure 3. The correlation between inquiry-based learning training for teachers with the development of their PCK and efficacy can be seen where inquiry is the ultimate key in INSEP program. Inquiry has opened a new point of view for teachers where it enables them to understand more about science curriculum, students' ability in understanding science, instructional strategy and valuing science literacy where these lead to the change of science teaching orientation.

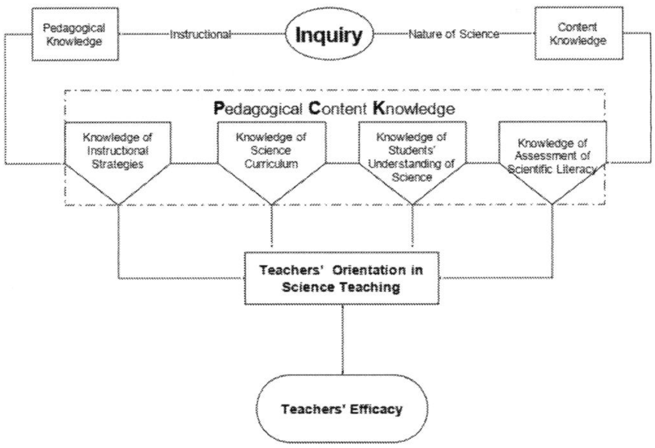

Fig.3. Framework of Inquiry on Supporting Teachers' PCK and Efficacy

There are several implications for further research and science teacher education. This result shows that inquiry plays an important role in shaping and building teachers' PCK and efficacy specifically in teaching science. This finding can help teachers, educators, researcher, and the policy makers in developing the most suitable trainings for elementary teachers to focus on the professionalism. Elementary teachers were always given a very general science training where it raises the possibility of them having a lower PCK than high school teachers [23]. Therefore, INSEP should be set as a sustainable program.

There are still some improvements that need to be added to the INSEP that has been given such as assessment skills based on science literacy. Even though teachers have noted it down within their lesson plan, they are still having lack of confidence in assessment development. Interactive assessment [24] is used during the interaction between teachers and students using the concept of map in science learning [25] which can be applicable for the INSEP program and fascinating to be researched for its easiness and effectiveness for short term training. With the high scores in assessment development of teachers' efficacy, thus we hope the entire PCK components can be built within their mind.

The authors would like to thank PT Kuark Internasional for the support in making this research possible. Many thanks to IOA that conducted the in-service teacher professional development in East Sumba Regency, East Nusa Tenggara, Indonesia.

References

1. D. C. Smith, JSTE, **11(1)**, 27-46, (2000)
2. J. Loughran, A. Berry and P. Mulhall, *Understanding and Developing Science Teachers' Pedagogical Content Knowledge* (Rotterdam, 2012)
3. A. L. Gardner and J. Gess-Newsome, *A PCK rubric to Measure Teachers' Knowledge of Inquiry-Based Instruction Using Three Data Source* (NARST, 2011)
4. J. Gess-Newsome, *A Model of Teacher Professional Knowledge and Skill Including PCK Results of The Thinking From The PCK Summit, in Re-Examining Pedagogical Content Knowledge,* A. Berry, P. Friedrichsen and J. Loughran, Eds., 28-42, (2015)
5. R. M. Schneider and K. Plasman, RER, **81(4)**, 530-565, (2011)
6. C. Velthuis, P. Fisser and J. Pieters, JSTE, 445-464, (2014)
7. R. M. Klassen, V. M. C. Tze, S. M. Betts and K. A. Gordon, Educ Psychol, 21-43, (2011)
8. L. Beauchamp, R. Klassen, J. Parsons, T. Durksen and L. Taylor, *Exploring the Development of Teacher Efficacy Through Professional Learning Experiences* (Alberta Teachers' Association,2014)
9. Ministry of Education and Culture of the Republic of Indonesia, Lampiran Permendikbud Nomor 22 Tahun 2016, Retrieved from http://bit.ly/2ikM1Hq
10. R. D. Anderson, JSTE, **13(1)**, 1-12, (2002)
11. Alberta Learning, *Focus on Inquiry*, (Edmonto, 2004)
12. M. Kim, K. Juuti, J. Holbrook and M. Rannikmäe, IJSME, **11**, 359-383, (2013)
13. T. Umbara, M. R. Fananta and S. D. Hastuti, ASSEHR, **57**, 189-194, (2017)
14. I. M. Riggs and L. G.Sci. Ed., **74(6)**, 625 - 637, (1990)
15. C. J. Wenning, JPTEO, **6(2)**, 9-16, (2011)
16. C. Rooney, EJOLTS, **5(2)**, 99-127, (2012)
17. L. S. Shulman, Harvard Educational Review, **57(1)**, 1-22, (1987)
18. S. Magnusson, J. Krajcik and H. Borko, *Nature, Sources, and Development of Pedagogical Content Knowledge for Science Teaching, in PCK and Science Education*, (J. Gess-Newsome and N. G. Lederman, 1999)
19. S. Park and S. J. Oliver, Res Sci Educ, 261-284, (2008)
20. V. L. Akerson and D. L. Hanuscin, JRST, **44(5)**, 653-680, (2007)
21. P. Nuangchalerm, Higher Education Studies, **2(2)**, 66-71, (2012)
22. S. Choi and Y. H. Lee, 초등과학교육 제34권 제4호, 406-418, (2015)
23. M. M. Thomson, D. DiFrancesca, S. Carrier and C. Lee, Teacher Development, 1-18, (2016)
24. B. Cowie and B. Bell, Assessment in Education, **6(1)**, 101-116, (1999)
25. T. Stoddart, R. Abrams, G. Erika and C. IJSE, **22(12)**, 1221-1246, (2000)

Mathematics Connection Ability and Students Mathematics Learning Achievement at Elementary School

Sabina Ndiung[1], *Fransiskus* Nendi[2]

[1] Lecturer of Primary School Teacher Education Department of STKIP St. Paulus Ruteng, Indonesia
[2] Lecturer of Mathematic Education Department of STKIP St. Paulus Ruteng, Indonesia

Abstract

The orientation of education nowadays should be in line with the fulfillment of 21st century demand which encourage learners to perform the originality of human resources quality. This research aims to know whether there is effect connection mathematics ability toward student's mathematics learning achievement and to know the contribution of connection mathematics ability toward student's mathematics learning achievement at the elementary school. Respondent in this research were 35 students at Watu Weri state elementary school. Proportionate stratified random sampling technique was applied in collecting sample. The research design was ex-post facto. Data were collected through test and documentation. Data were analyzed by linear regression. The research result shows that there was significant effect mathematics connection ability toward student's mathematics learning achievement. The contribution of mathematics connection ability toward students mathematics learning achievement was 21,9% . This result indicates that high or law student's mathematics learning achievement significantly affected by their mathematics connection ability.

Introduction

Recently, education matter in Indonesia continuously becomes a major issue because from year to year there was no significant development in achieving the achievement in international level since from primary school until secondary school. For mathematic learning achievement always fluctuated and occupied at last level [1]. Whereas, on the other hand, Indonesia must be able to compete with 21st education demand which highlights the quality of human resources that be able to be competent in the global society. Instruction process was blamed in unsucessful of student's learning achievement where more stressed much on result than the process.

Educational demands and skills of the new century require new ways in thinking, learning and teaching. In today's society, there is a need for individuals, who could research and structure the information, with critical thinking skills, could resolve the problems with own knowledge and experience, are creative, open to innovations, and at the same time, with effective communication and cooperation abilities within a group. P21 (2009) called these skills, which the individuals should possess, 21st Century skills and classified these as "Learning and Innovation Skills," "Life and Career Skills," and "Information Media and Technology Skills." According to Prensky (2001) and Rotherdam and Willingham (2010), 21st Century skills are not a different set of skills. Every individual is expected to have

---

[1] Punyaku79@gmail.com
[2] fransiskusnendisa@gmail.com

creative thinking and problem solving skills that are required for active participation of the individual in the society and the workplace. For students" achievements in educational life, they should experience these skills at a very early age. [2].

Theoretical perspectives in mathematics education evolved independently in different regions of the world and different cultural circumstances, including traditions of typical classroom cultures, values, but also varying institutional settings (like the location of mathematics educators in the mathematics or education department, their involvement in pre-service or in-service training, the real and the intended curriculum, the books, etc.)[3]. Since, in mathematics, one way to improve teaching and maximize instructional time is to find and utilize connections among mathematical concepts and ideas. Making these connections takes advantage of the fact that knowledge about one topic comprises part of the knowledge of another. That common knowledge does not have to be taught again, but the teacher does have to recognize and make the connection. Recognizing the connections depends on deep understanding of math concepts and processes [4].

There are six mathematics ability in the process of mathematics instruction according to the NCTM (2014) such as problem solving, reasoning and proof, communication, connections, and representation. Furthermore, in NCTM (*National Council of Teacher of Mathematics*) affirms that mathematics connection ability is the important factor in understanding mathematics concept [5]. From pre-kindergarten through grade 12, this standard provides two separate components. Firstly, it helps in connection within and among mathematical ideas. Students learn to make connections among mathematical ideas to build a web of connected ideas. Secondly, the standard refers to connections between the real world and other disciplines. Hence, children can see that mathematics have a great role in science, art, and social studies. Briefly, it offers that mathematics should be integrated with other disciplines and the applications related with the real world should be explored. By doing the connection, the mathematics concepts which have been learned can contribute as strong foundation in comprehending a new concept [5],[6].

The fact shows that Indonesian mathematics learning achievement is still far from the expectation. This has been indicated by PISA (Program for International Student Assessment) result in 2015 which focused reading literacy, mathematics, and science shows that Indonesia was at the 69th from 76 nations [1]. The result from TIMSS (Trends in International Mathematics and Science Study) in 2015 also shows that Indonesian students were at the 45th from 48 nations for science and mathematics in which students had low ability for (1) understanding the complicated information, (2) theory, analysis, and problem solving, (3) media usage, procedure and problem solving, and (4) doing the investigation [7].

The notion of 'connections' in mathematics relates both to those that exist: (i) within and between different content areas in mathematics (e.g., within number or between number and measurement); (ii) between mathematics learning and learning in other areas; and (iii) between mathematics and the context within which a child lives, works or plays (Perry & Dockett, 2008) [8]. Johanning (2010) proposes that, in order to build a coherent curriculum and to foster connections, the big ideas from one topic must be built on in others so that children are given the opportunity to use familiar concepts in new settings [8] .

There were relevant researches conducted by Hendriana, Slamet, & Sumarmo (2014). The study found that students' prior mathematics ability affected the attainment of mathematical connection ability and its N-Gainas well. Therefore, teacher has to pay more attention on comprehending the prerequisite materialsof mathematics before teaching the more complex materials. The grades of self-confidence of students were still average. Improving students' self-confidence is a continuous process, thus the process of acquiring this trait takes time [9], [10] . Then research conducted by Boyaci & Atalay (2016). They have found out as a result of the literature review of studies conducted on 21st Century skills, it was determined that there was not a full consensus on the skills that learning individuals should achieve, and there were differences in certain dimensions. Sub-dimensions of 21st Century learning and innovation skills could be classified as creativity and innovation, critical thinking and problem solving, cognitive skills or skills related to individual's self, and cooperation and communication or inter-personal skills [2].

Kinniburgh & Byrd (2008) have found and advocated connecting the teaching of mathematics and social studies; they suggested that using such connections makes deeper understanding for students as they learn. Today we can even use images and photographs to make connections where students can see both the math and social studies concepts within the photos. Newcombe (2013) says, "Science, mathematics, and social studies are deeply spatial subjects. Currently, students who come to class with higher levels of spatial ability take more easily to learning in these areas, but this fact does not mean we cannot teach in a way that maximizes learning for all. Spatial ability can be improved inside and outside the classroom, as well as by instruction in other subject areas, notably the visual arts. Specializing the curriculum by including and explicitly teaching the spatial symbol systems that lie at the heart of science, mathematics, and social studies is an achievable and worthwhile goal [10].

Method

This is ex-post facto research. Respondent in this research were 35 students who taken from Watu Weri state elementary school. Data were collected through test and documentation. Instrument test was applied in measuring students' mathematics connection ability and documentation was the fifth grade students of their mathematics score in Watu Weri state elementary school in the odd semester of the academic year of 2014/2015. For validity testing, the researcher used the correlation pattern of *Pearson product moment.* Based on trial result toward 10 item test, it found out that 8 items test were valid. The validity of items test should find the reliability. The reliability of items test, the researchers used Alpha Cronbach's pattern. Data were analyzed by using linear regression with SPSS version of 16.00 for windows. By applying assumption testing was normality testing for distribution and linearity data.

Result and Discussion

Before finding hypothesis testing which used linear regression, the researchers did the assumption testing like distribution of normality and linearity data. Referenced to research hypotheis analysis result, it found out that coficient b2 = 1,538 and b1 = 53,250. By means of data, equality regression sample model which formed in this research was $\widehat{Y} = 1{,}538X + 53{,}250$. Furthermore, linear regression coefficient above was tested by using F testing. In line with this, the following are the data of F_{ob} testing with significance grade (α) 0,05. Obtained that $F_{ob} > F_{cv}$ value was 7,40 > 4,125 shows that there was effect mathematics connection ability toward mathematics learning achievement for the fifth grade students at Watu Weri state elementary school. In other word, mathematics connection ability determines students' mathematics learning achievement.

Further analysis is determining the contribution mathematics connection ability toward mathematics learning achievement at the fifth grade students of elementary school. Found determination coefficient (r^2) was 0,219 means the contribution toward mathematics connection ability toward students' mathematics learning achievement was 21,9%, meanwhile 79,1% contributed by another variable which was analyzed in this research. If it combines with regression analysis testing, it can be concluded that the variety of students' mathematics learning achievement was 21,9% can be explained by mathematics connection ability through regression equality

$\hat{Y} = 1{,}538X + 53{,}250$. This result indicates that high or law students mathematics learning achievement significantly affected by their mathematics connection ability.

The hypothesis conclusion above indicates empirical proof that there is effect between mathematics connection ability toward students' mathematics learning achievement because mathematics inter-dependence between concept and material and other subjects as well for man' s life in everyday such as reasoning, problem solving and creativity development as well. Therefore, in mathematics instruction needs the inter-dependence between concept and material, mathematics and other subjects, and daily experience, so instruction process can be meaningful. The implication of it, the students is more creative and having better development of mathematics, particularly connection mathematics which support students to achieve better learning achievement.

Individually, mathematics connection ability contributes significant effect toward mathematics learning achievement. Develop the students' ability to connect among concept, material, mathematics usage for other subjects (non mathematics), and definitely support students' learning achievement. Low or high mathematics learning achievement relies on their mathematics connection ability. That is why students are given the opportunities for mathematics exercising mathematics because by developing students' mathematics connection ability may support their deeper mathematics' understanding. NCTM (2000:274) affirms that mathematics connection ability is one of the important point in comprehending mathematics concepts [6]. By comprehending connection, the mathematics concepts which ever learnt may encourage students to understand new concept. The result of this research has been supported by two previous relevant researches such as research conducted by Hendriana, Slamet, & Sumarmo (2014) where their result shows that students' prior mathematics ability affected the attainment of mathematical connection ability. Then, research conducted by Dixon (2012) which found out that higher scores were possible if students were given time to prepare for the test, as is usually the norm in schools. The intent, however, was to examine students' ability to make connections under impromptu test conditions. 50th percentile were able to connect mathematics and science concepts (16% and 17% respectively) learned in the Project Lead the Way (PLTW) curriculum to the problems they that they believed that these concepts were present in the PLTW courses that they had taken. Their recognition of the concepts may have allowed for greater comprehension of the problem, which likely led to more accurate solutions [11]. Thus, mathematics learning achievement must be supported by their mathematics connection ability.

Conclusion

Reference to the result of the research, it might be concluded that 1) there was significant effect between mathematics connection ability toward mathematics learning achievement of the fifth grade students in Watu Weri state elementary school in the odd semester of academic year of 2014/2015. This conclusion in line with the data of regression significance ($F_{ob} = 7{,}40 > F_{cv} = 4{,}125$). The comparison result indicates that mathematics connection ability gives the significant effect toward students' mathematics learning achievement and 2) there was contribution mathematics connection ability toward mathematics learning achievement in the fifth grade students in Watu Weri state elementary school in the odd semester of academic year of 2014/2015. The contribution of mathematics connection ability was 21,9% toward students' mathematics learning achievement.

This result indicates that mathematics connection ability is an urgent to be done by teacher in instruction process in developing students mathematics learning achivement at primary school by connecting any concepts which are inter-dependence so students are also be able to solve the problem. Content of mathematics knowledge is important to be possessed by teacher in order to run mathematics instruction in creating joyful learning atmosphere. By doing such notion, it can be concluded that more and more mathematics connection ability so much more also student's mathematics learning achievement.

Acknowledgments. The researchers would like to gratefully acknowledge a number of people for the help given to them during the process of this research. Primarily, Maksimus Sehadun as the principal of Watu Weri state elementary school for his warm cooperative, Saint Paul Foundation Ruteng in financial support, Prof. A.A.I.N. Marhaeni, for her encouragement to be involved in the journal publication, and students who have involved in this research.

References

[1] A.Schleicher, "PISA," 2016.
[2] D. Boyaci and N. Atalay, "IJI," vol. 9, no. 1, 2016.
[3] S. Prediger, A. Bikner-ahsbahs, and F. Arzarello, "IJME," vol. 39, no. 2, 2008.
[4] C. C. Molina, "SEDL insights," vol. 1, no. 4, 2014.
[5] NCTM, *Six Principles for School Mathematics*. 2014.
[6] D. Mehmetlioglu and Y. Ozdem, "IJEMST," vol. 2, no. 1, 2014.
[7] A. Mullis, Ina.V.S. Martin, Michael O. Foy, Pierre Arora, "TIMSS," 2015.
[8] W. D. Butler, D. Corcoran, T. Farrell, S. Nicmhuirí, M. O. Connor, and J. Travers, *Mathematics in Early Childhood and Primary Education*, no. 18. NCCA, 2014.
[9] H. Hendriana, "IJE," vol. 8, no. 1, 2014.
[10] J. Graziose and S. L. Ave, "ICTCM," 2013.
[11] R. A. Dixon and R. A. Brown, "JTE," vol. 24, no. 1, pp. 2–17, 2012.

The Rise and Fall of Curriculum 2013: Insights on the Attitude Assessment from Practicing Teachers

Yuliana Natsir, *Yunisrina Qismullah* Yusuf*, *Ulva Fiolina* Nasution

Department of English Education, Faculty of Teacher Training and Education, Syiah Kuala University, Banda Aceh 23111, Indonesia

Abstract. The change of Indonesia's national curriculum from School-Based Curriculum 2006 (hereafter SBC 2006) to Curriculum 2013 (hereafter CR 2013) has caused modifications in the assessment system which requires teachers to use authentic assessment. Many teachers are facing setbacks in the implementation of this assessment, particularly in the attitude aspect. This qualitative research mainly focused on investigating the problems in conducting the attitude assessment faced by teachers in CR 2013. Three English teachers from three junior high schools in Banda Aceh were selected and interviewed with questions adapted from Handayani (2014) and Èliza (2015). The document analysis of attitude assessment was also done to see the reports of students' attitudes. The findings showed that the teachers knew the importance of authentic assessment as a necessity for assessing students' learning outcomes in CR 2013, but they admitted that it was complicated since many categories need to be assessed in detail. They find it impossible to assess all of the students in every school day. When parents do not want to cooperate if their children are in need of attention from both parents and teachers, it is also an obstacle for the teachers to do the assessments. Furthermore, the teachers deemed that self- and peer-assessment were ineffective and the paper further discusses why it is so.

Keywords: Curriculum 2013, assessment, attitude aspect, challenges, teachers

1 Introduction

Along with demands from the era of development, science and technology, the changes of curriculum from time to time in Indonesia is a phenomenon that cannot be avoided in education. A curriculum can only be good for a particular community at a particular time [1]. The occurrence of the curriculum change in Indonesia has begun since 1984 until now. The continuous change requires Indonesia to create future generations with strong national

* Corresponding author: yunisrina.q.yusuf@unsyiah.ac.id

identity, being able to compete and adjust to the global changes. Education should be capable of producing human resources that have the competence of being competitive, innovative, creative, collaborative, and with good characters [2].

Curriculum 2013 (hereafter, CR 2013) is the recent curriculum Indonesia which was created to complement the previous curriculum, which is School-Based Curriculum in 2006 (hereafter, SBC 2006). CR 2013 has the aim to encourage learners to become individuals who are able to conduct observation, questioning, reasoning, and communication that they have acquired during the learning processes in schools [3]. In relation to the implementation of CR 2013, there are two fundamental changes found, namely the change in the learning system and the change in the assessment system [2]. Since this paper focuses on the assessment system, therefore, the discussion revolves around this assessment.

CR 2013 emphasizes on authentic assessments which includes the self-assessment, project assessment, portfolio assessment, process assessment, and performance assessment. The assessment is changed from the knowledge-based to the change-based and is done through process, portfolio, and full and complete output assessment [3]. Hence, the authentic assessment can measure and evaluate the learning outcomes of a student as a whole, both from the aspect of attitude, knowledge, and skill [4]. Moreover, this assessment was also proposed by Wiggins [5] in which it is a form of task which requires the learner to show the meaningful real-world performance.

Nevertheless, the application of this curriculum still rises with various controversies. Alawiyah finds the preparation of CR 2013 was considered too hasty and did not refer to the result of studies that have been done based on the evaluation of SBC 2006, as well as less attention to the readiness of the education unit and the teachers [6]. Maghfirah also finds that the time available for the teachers to conduct the assessment is not sufficient considering other workloads that they need to do in school [7].

In the previous curriculum, the attitude assessment received less attention so it was deemed ineffective by the Ministry of Education. There were scores given by teachers that had gaps between students' report cards and their daily behaviors. Thus, in CR 2013, teachers who were typically accustomed to cultivate the value only in the aspects of knowledge and skill are now required to also pay attention to the aspect of attitude. The attitude competence in CR 2013 is divided into two, namely the spiritual attitude (KI-1) and social attitude (KI-2). Teachers are to assess the students' attitudes through observation, self-assessment, peer-assessment (by using check lists or rating scales of a rubric), and journals (in the form of teachers' notes). Many teachers are still not ready with this system since it is different from the previous curriculum [7] [8] [9].

From our preliminary discussion with some English teachers in a couple of junior high schools in Banda Aceh, it was also found that the teacher guideline book (*buku petunjuk guru*) did not provide sufficient information for them to apply the authentic assessment of the attitude aspect. This is also agreed by Bhakti, Kusairi, and Muhardjito [10] who also found that the guidance for teachers was not sufficient to provide understanding of the application. And so, we are interested to further investigate the problems faced by English teachers in implementing authentic assessment of the attitude aspect in CR 2013. Thus, the research question for this study is: "What are the problems faced by English teachers in the process of attitude assessment based on Curriculum 2013?"

The research result is expected to give more input, knowledge, and reference in improving the quality of education in Indonesia in implementing Curriculum 2013, particularly in the process of authentic assessment. It is also expected to contribute positively to English teachers to be better and conscious in assessing the students' attitude, so that the process of teaching and learning can be carried out optimally.

2 Methodology

This research is qualitative with a case study approach. It was conducted in three public junior high schools in Banda Aceh, namely SMPN 1, SMPN 2, and SMPN 6 since the teachers at those schools have been implementing CR 2013 in teaching English. One English teacher was chosen from each school based on their consent, and thus, making a total of three respondents for this research.

Two instruments were utilized in collecting data, namely interview and document analysis. The interview questions are adopted and modified from the interview guide by Handayani [9] and Eliza [8]. There were nine main questions, with one question that consisted of 3 sub-questions. The questions revolved around the implementation and the problems teachers' faced in conducting the authentic assessment on the attitude aspects. Each teacher was interviewed and audio recorded with a cellphone for about half an hour. This was done after their school time was over. The interview was conducted in Indonesian language and the results were transcribed and translated into English. Moreover, the data from documentation is the students' attitude assessments compiled by the English teachers during the process of teaching and learning. The data from these documents were studied and used to obtain information related to the attitude assessment process.

Data analysis in this research followed the model suggested by Miles and Huberman [11], namely data reduction, data display, and conclusion drawing/verification.

3 Results and Discussion

The first question that we asked was about the suitability between authentic assessment in the attitude aspect and the learning process. The opinion of the teacher can be seen below (E refers to excerpt and T refer to teacher):

E1 *Yes, it is. It depends on the learning process that has been done by each teacher. For example, if the teacher uses the scientific learning approach as suggested by the curriculum, then she must also use authentic assessment for the students' attitudes.* (T2)

T1, T2, and T3 considered that the application of teaching and learning process in schools was in accordance with authentic assessment in the attitude aspect because the assessment must be related to the students' reality, or in other words, everyday behavior and attitude.

The second question consisted of three sub-questions that inquired about the obstacles that the teachers found in the assessment. There are three aspects in the authentic assessment, namely attitude, skill, and knowledge that have to be done by the teachers. The first and second sub-question was about the difficulties they faced while implementing the three aspects.

E2 *There are too many things to be assessed for every single student on every school day. Besides this assessment on attitude, I also have to continue on the evaluation of the skill and knowledge aspects for every single student. Along the way, the learning process has to be done, too. Sometimes I assess each student, but not at the same time, not in one day. We cannot do it all; the workload is a bit overwhelming. Indeed, we are expected to do them all, but in reality, we cannot achieve everything in one day.* (T3)

Similar to T3, T1 also experienced the difficulties when conducting the assessment. Her class consisted of 35 students in one class; due to this large classroom, the time was insufficient to assess each of them while teaching at the same time. There were other activities that also needed to be dome besides making assessments, such as making lesson plans, preparing materials, and so on.

For the last second sub-question, we inquired about the most difficult aspect in the implementation of authentic assessment. T1 elaborated her answer as:

E3 *Of course, the attitude aspect! We know there are two kinds of attitude assessments; they are self-assessment and peer-assessment. The problems in these two assessments are honesty and trust. Sometimes, I find that when the students do the self-assessment, they are not honest. For example, a student said that he never cheated in a test, but I have found him doing so a few times. While in the peer-assessment, they may be worried if they give bad assessment for their friends, they will lose friends. So they write all nice things about each other. However, it can be a lesson for teachers to see whether a student is honest or not. We can judge it from their assessments.* (T1)

Based on E3, T1 as well as the other teachers claimed that attitude assessment was the most challenging one to conduct. Apart from that, T2 believed that the assessments on the skill and knowledge aspects are not as difficult as the attitude one because they are evidently reflected through scores from the students' test results and school work. But the solid challenge is facing the parents; T2 realized that when it comes to students' discipline, some parents often interfered and denied their children's faults. Thus, to gain this kind of parents' cooperation in fostering the children's discipline could be wearisome.

The third question asked teachers' opinion about the process and rule of authentic assessment in CR 2013, whether they agreed to it or not. T3 enlightened:

E4 *I agree but it is too much. I mean, I hope it can be more simplified so that teachers are not burdened with all of the detailed assessments for every student. With our work load, it is impossible! I hope the policy makers can make consideration of this situation so that we can teach more effectively. Otherwise, the assessments will be ineffective if we cannot teach to our fullest capacity.* (T3)

All of the teachers actually agree about the process and rule of authentic assessment in CR 2013, nevertheless, T1 and T3 saw it as too much of a burden for teachers to do everything with the work load that they also have in school besides making assessments. The explanation by T3 can be implied that the process is time consuming, whereas in the previous curriculum the teacher supposed to focus more on teaching materials.

The fourth question was related to the teachers' views in assessing students' learning outcomes through the attitude aspect, whether it is necessary or not. The following statements described their views.

E5 *Even though it is too much, but, yes, it is necessary. This attitude assessment is good, because it can improve the students' characters. Sometimes the students will improve their attitude when they know I am assessing them in that aspect. Although they may feel forced to do good things at first, but I believe, sooner or later they will be accustomed to them, right?* (T1)

All of teachers say that attitude assessment is important because as T1 said, the students would behave well when they knew that their attitudes were being assessed by the teachers. Perhaps they may felt forced at first but T1 believed that the driven attitudes would later become habitual to the students. T2 and T3 further added that the role of teachers is not only to teach, but also to educate and to guide students to become virtuous individuals.

The fifth question probed about the things that must be considered in the process of attitude assessment. One of the answers from the teachers can be seen as follows.

E6 *In the attitude assessment process, I am most concerned about the observation and journal since the peer-assessment is not effective. So, to me, the most important thing is when I observe them personally. This is where I directly see how the students act. If after I access them and their scores are still low, I will remind them. It means that they should try to change to be better. If I do not care, the students will continue making mistakes.* (T3)

Regardless of being overwhelmed with the attitude assessment in CR 2013, the teachers agreed that observation and journals were essentials because they provide direct assessment to students' attitudes. The observation was also considered more effective than self-assessment or peer-assessment. Yet, they also respect the self- and peer-assessments because they could become benchmarks for the teachers' assessments.

For the sixth question, it revolved around the way the teachers assessed students' learning outcomes based on the attitude aspect and time of assessment.

E7 *When I enter the classroom, I must be prepared with various assessment formats, especially the attitude aspect. I do the attitude assessment while I am teaching, I give the materials, and when I am finished, I monitor and assess the students directly as I walk around the classroom. For example, there was a student who had a problem in learning, so I wrote it all down in in my note. It is overwhelming, especially when we have a large class.* (T1)

Here, the teacher already knew how to assess students' learning outcomes in the attitude aspect. It was done for every student in every meeting and overwhelming to the teachers. Thus, T1 suggested should this kind of assessment is done to a small class; the efficiency would be more optimal.

The seventh question was related to the teachers' obstacles in assessing the attitude aspect. The writer asked the teachers about the problem that might happened during the implementation of attitude assessment. The following statements are their opinions.

E8 *Yes, of course there are obstacles. In the authentic assessment, I get directly involved with the students. They may misbehave one or two or three times. Sometimes I do not know what to do. Sometimes when I called their parents, they did not come. It is a little bit difficult for me.* (T2)

The teacher's elucidation in E8 revealed her difficulties in conducting the attitude assessment. It is when there were parents who did not want to cooperate with them when they needed assistance in improving the students' unpleasant attitudes. T1 and T3 added that when these attitudes were rooted from other factors outside of school; the assessment then become more difficult to grade.

Aside from asking questions about the attitude assessment, investigating about their participation in training or seminar concerning CR 2013 was also queried. T2 explained:

E9 *I have attended workshops on CR 13. I was also given a module to assess students' learning outcomes. Although it was complicated, I have studied it all, how to assess the knowledge aspect, and the skill aspect with the portfolio, as well as the attitude aspect, with each instrument.* (T2)

All teachers informed that they had attended workshops on CR 13 and knew the differences between this curriculum and the previous one, SBC 2006. They also understood on how to do the assessment and learning process based on the outcomes they achieved from the training.

Finally, the last questioned probed about the guideline of assessment process, whether they received it or not while they took their trainings on CR 13. T3 revealed that:

E10 *Yes, I have it. But, as far as I know, this assessment is said to be changed again, because it is too complicated for many teachers. Perhaps, they will conduct training again about the new assessment. Personally, I suggest the policy makers to test all rules and policies really well first before instructing all teachers to implement them. If not, we teachers will be the victims of rules that changes all the time.* (T3)

Based on E10, all teachers confirmed they were given the guideline. Thus, T3 still found it complicated. She also added that even though trial and error is common, but to seriously test out a policy before directing all teachers to do it should be done in the future to avoid them become targets of frequent rule change.

4 Conclusions

Based on the findings of this research, it can be concluded that even though the process of attitude assessment had been implemented in the schools under study, the teachers still found the difficulty to deal with it. They were overwhelmed by the many instruments in the attitude aspect. Consequently, they preferred to use more of the observation and journal in the attitude assessment compared to self- or peer-assessments. In addition, they also did not have sufficient time to assess each student in one day. Thus, with the various and numerous instruments, they were hampered with the increasing efforts and time that are needed just to process the students' attitude assessment. And this goes without saying about the other additional changes and workloads in the learning system. Parents, who were not cooperative in assisting their children's attitude progress, were also an obstacle for these teachers in implementing the attitude assessment as required by CR 2013. In spite of the fact that the teachers were obliged to keep on executing the authentic assessments, they expect that the assessment processes of CR 2013 can be simplified in the future to ease its implementation for teachers.

References

1. S. Nasution, *Asas-asas kurikulum* (Bumi Aksara, Jakarta, 2011)
2. Y. Abidin, *Sistem pembelajaran dalam konteks Kurikulum 2013* (Refika Aditama, Bandung, 2014)
3. H. E. Mulyasa, *Implementasi Kurikulum Tingkat Satuan Pendidikan: kemandirian guru dan kepala sekolah* (Bumi Aksara, Jakarta, 2013)
4. Kunandar, *Penilaian autentik (penilaian hasil belajar peserta didik berdasarkan kurikulum 2013) suatu pendekatan praktis edisi revisi* (PT Raja Grafindo Persada, Jakarta, 2014)
5. G. Wiggins, The case for authentic assessment. *Practical Assessment, Research & Evaluation*, **2**(2), 1-3, 1990. Retrieved from http://pareonline.net/getvn.asp?v=2&n=2
6. F. Alawiyah, Kesiapan guru dalam implementasi Kurikulum 2013. *Jurnal Info Singkat Kesejahteraan Sosial DPR RI*, **6**(15), 9-12, 2014.
7. S. Maghfirah, *Kendala guru pada penerapan penilaian autentik dalam pembelajaran Eksponen dan Logaritma di Kelas X SMA Negeri 1 Banda Aceh tahun pembelajaran 2014/2015* (Thesis, Universitas Syiah Kuala, Banda Aceh, 2015)
8. F. Eliza, *Kendala guru Matematika terhadap penilaian otentik pada Kurikulum 2013 di SMA Negeri Kota Meulaboh Tahun Pelajaran 2014/2015* (Thesis, Universitas Syiah Kuala, Banda Aceh, 2015)
9. R. Handayani, *Kendala guru Matematika dalam menyusun Rencana Pelaksanaan Pembelajaran (RPP) di SMP Kota Banda Aceh* (Thesis, Universitas Syiah Kuala, Banda Aceh, 2014).
10. A. S. Bhakti, S. Kusairi, Muhardjito, *Pengembangan model penilaian autentik berbasis Kurikulum 2013*. Retrieved from http://jurnal-online.um.ac.id/data/artikel/artikel2B5157FBA171A8046A 8FBCAD73267BA6.pdf 2014
11. M. B. Miles, A. M. Huberman, *Qualitative data analysis : an expanded source book (2nd edition)* (Sage, Thousand Oaks, 1994)
12. M. Muslich, *Authentic assessment: penilaian berbasis kelas dan kompetensi* (Refika Aditama, Bandung, 2011)

Effectiveness of Problem Based Learning Model toward Biology Learning Outcomes

Nadya Treesna Wulansari*, *I Putu Gede* Sutrisna, *Ni Wayan Kesari* Dharmapatni

Sekolah Tinggi Ilmu Kesehatan Bali, Renon, Denpasar, Bali

Abstract. This study aimed to analyze the effectiveness Problem Based Learning toward Biology learning outcomes in the subject of Basic Nursing Science I. The research design was true experimental design. The population in this study was the first-grade students in Nursing Sciences Study Program in Sekolah Tinggi Ilmu Kesehatan Bali. Purposive sampling technique was applied to select the sample size in experimental and control group. There were 162 respondents included in this study. The data were analyzed using descriptive statistic and independent sample T-Test. The result showed that the average of post-test in control group was 67.01 ± 7.53 and average of post-test in the experimental group was 80.16 ± 5.92 ($p < 0.05$). It could be concluded that Problem Based Learning Model is more effective than a conventional model. It can be used for the alternative model in learning which can improve student's ability in critical thinking and the learning outcomes of Biology in subject Basic Nursing Science I.

Keywords: Biology, Learning Outcomes, Problem Based Learning

INTRODUCTION

Professionalism in nursing requires a good understanding of nursing science. According to Nursalam [1] in 2015 the purpose of nursing science is as a basis for nursing practice, commitment in nursing practice toward the development and basic problem solving of nursing science also the acceptance of nursing intervention according to scientific and rational by other health professions and community. A good understanding of nursing science is gained from the educational system and academic teaching. Based on Ners Profession Program of Education Curriculum (2015), Basic Nursing Science I is one of the subjects that must be understood by nursing students to complete the understanding of nursing problems. One of the learning achievements in this subject is to apply the concept of Biology cell and genetics as an approach in solving nursing problems, supporting the provision of nursing care during nursing practice, which is called Basic of Nursing Science [1,2].

However, in fact, some nurses still have a lack of Biology knowledge [3]. A lack of knowledge leads to faulty action that will have an impact on the patient's condition [4]. This can be caused by failure to record patient problems, communicate nursing plans that have

* Corresponding author : nadyatreesna@gmail.com

been prepared effectively; it is a failure to provide continuous nursing care, and failure to provide patient understandable instructions. Handling errors in the patient during practice are closely related to the application of the classroom learning model.

Based on preliminary observations which were conducted on the first semester students of Nursing Science Program, the learning process was still dominated by conventional model and it caused the learning is one-way learning. It means that the students tend to only listen to the explanation given by the lecturer. This learning process makes the students less able to be active in class. There are some causes of students problem through one way learning, (1) the lecturer role is still dominant and tend to be the main determinant in learning process, (2) most of the lecturers relatively use similar learning strategy and rarely use the approach which involves student activities, so it less motivating students to study [5]. In addition, based on preliminary data that researchers obtained, the quiz score given during the learning process was insufficient, most of the student's score was below 68. This indicated that the average score was still insufficient. Based on these data, an action is needed that can improve student learning outcomes.

Based on the research problem, the study about the effect of Problem Based Learning (PBL) in improving Biology learning outcomes in the study of Basic Nursing Science I in the faculty of Nursing Study Program at Sekolah Tinggi Ilmu Kesehatan (STIKES) Bali. PBL is one of the learning models based on current issues with problem-solving based on theories and concepts that have been studied.

In addition, implementation of PBL on medical teaching concept can improve student learning outcomes [6]. Moreover, learning process with PBL obtained higher learning achievement compared with direct learning strategy on Nursing Clinical learning [5]. Yulianti [7] in 2016, observed the application of PBL model used to scientific approach based worksheet for physics can increase cognitive and psychomotor learning outcomes of students. Besides, PBL model can improve student learning outcomes in science lessons especially in chemistry [8]. However, the study on the effect of the implementation of PBL on Biology on the subject of Basic Nursing Science I in nursing student has not been conducted. The implementation of PBL is appropriate as a teaching strategy in applied subjects to facilitate students in making clinical decisions and ready to work, especially in nursing practice [9].

RESEARCH METHOD

The design of this study was true experiments with post-test of control group design [10]. In addition, purposive sampling technique was used in this study in which class A was a PBL group and class B was a control group that the lectures were given by the conventional model. The sample of this research is first semester students in the faculty of Nursing Study Program that was consisted of 81 students which every group. Furthermore, the data were analyzed by using Independent Sample T-Test [11].

RESULT AND DISCUSSION

The nursing students of class A and B were involved in this study in which every group consisted of 81 students. In addition, control class was taught by using conventional model. Independent Sample T-Test analysis found there is a significant difference between learning outcomes of Biology using PBL and conventional model ($p < 0.05$).

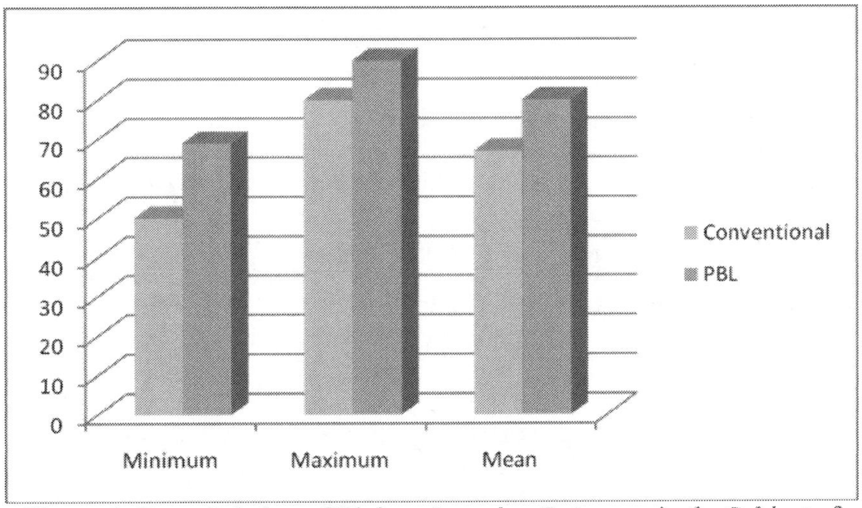

Figure 1. Recapitulation of Biology Learning Outcomes in the Subject of Basic Nursing Science I

Based on Figure 1, the minimum and maximum score of the PBL group were higher than the conventional group. The result of the mean of post-test in conventional group was 67.01 ± 7.53 and the mean of post-test in PBL group was 80.16 ± 5.92 ($p < 0.05$). It means that there was a significant difference in the implementation of PBL model compared with conventional model to the learning outcomes of Biology in the subject of Basic Science of Nursing I. Based on that results, so it concluded that PBL model is more effective than conventional toward Biology learning outcomes in the subject of Basic Nursing Science I. The results of this study supported by Mergendoller [12] in 2006, observed that the learning outcomes of students who were taught by using PBL model is higher than the improvement of students' learning outcomes who were taught by conventional learning model.

PBL is a learning model that requires the students to learn by solving the problems. The problem is given as a stimulus in creating two-way learning. A problem provision is also as a simulation which is used to activate the curiosity of students in learning the subject and as a trigger in improving the knowledge and understanding of a teaching material. Problem-based learning (PBL) process consists of the following stages: (1) the problem; (2) problem analysis and generation of learning issues; (3) discovery and reporting; (4) solution presentation and reflection; and (5) overview, integration, and evaluation [13]. The students will collaborate with their group to hold discussions and negotiation to achieve the solution of the problems.

The implementation of PBL in learning can make the students think critically as an effort to solve the problems given [14,15]. In addition, the student's ability in analyzing becomes very good and positive to the problem in the learning process. The results of this study supported by Graaff & Kolmos [16] in 2003 who conveyed that PBL is a problem-oriented learning model that stimulates the occurrence of an analytically complex level of comprehension through the project work, which would not be possible in conventional classes. The focus of learning is a real problem that is given to students and they should find for solutions by a series of assessments and investigations based on theories and concepts that they have learned [17].

During the learning process of PBL there was an improvement in student activity because the students were not only listening to the lecturers, but students were also actively solving problems, working together in groups to express their opinions so that the students were not only trained to be independent but also active in solving the contextual problem that they were facing critically. Implementation of PBL can improve the student's confidence to solving the problems and students self-efficacy [18]. PBL activities help the students grasp more than just a concept but they will gain an understanding of how, and why, to use that concept [19].

The similar study was conducted by Forbes [20] in 2000 who stated that the average of nursing student achievement is higher by using PBL model. Another study was also conducted by Halla and Neamat [21] in 2011, who stated that by the implementation of PBL the ability of nursing students become higher than the implementation of the conventional model. Celik [22] in 2011 also compare between those models that the implementation of PBL more increased the learning outcome of the students in a physics course.

There are some implications of this study; First, it gives problems to the Biological concepts related to nursing in supporting the exploration of student's knowledge. Second, the purpose of learning should not only understand how the student's concept, but it must consider the attitude of thinking owned by students. Third, PBL can be used as an alternative learning model in order to improve the ability of critical thinking and student learning outcomes in the learning of Biology in the study of Basic Nursing Science I.

Based on the result and discussion on the implementation of PBL and conventional models to the learning outcomes of Biology in the study of Basic Nursing Sciences I, it found that PBL is more effective and the student learning outcomes were better than a conventional model. This is due to PBL based on the philosophy of constructivism, this learning model centered on the learner. In addition, the process of discussion is in the problem solving, PBL requires students not only concerned with the activity in private but more emphasized in groups. It can train and teach students to be able to think critically and responsibly to be able to solve the problems given. A critical thinking and responsibly are the things that students need and they can use in working place in the future.

CONCLUSION

Problem Based Learning model is more effective than conventional model toward learning outcomes of Biology in the subject of Basic Nursing Science I and the implementation of problem-based learning model need to be considered by educators as one of the innovative learning models if the lecturer wants to improve Biology learning outcomes.

ACKNOWLEDGEMENT

The authors are grateful to the research institute and community service in STIKES Bali.

REFERENCES

1. Nursalam. *Metodologi Penelitian Ilmu Keperawatan Pendekatan Praktis* (2005)
2. AIPNI. *Kurikulum Pendidikan Ners Indonesia* (2015)
3. Saryono. Peran Biologi Molekuler dalam Pengembangan Praktik Keperawatan. Evidence based practice in nursing science; Unique, diversity, and innovation. *Proseding Seminar Nasional Keperawatan PPNI Jawa Tengah* (2011)
4. Aziz, A. Tinjauan Kriminologi Mengenai Malpraktik Medik Yang Dilakukan Oleh Perawat. *Jurnal Ilmu Hukum Legal Opinion*, **2**, 1-10 (2014)
5. Yudiernawati, A., Punaji, S., Degeng., Rudianto. 2015. Pengaruh Strategi Pembelajaran dan Gaya Kognitif terhadap Prestasi Belajar Aplikasi Proses Keperawatan pada Pembelajaran Klinik Keperawatan. *Jurnal Pendidikan Humaniora*, **3**, 31-40 (2015)

6. Preeti, B., Ashish, A., Shriram, G. Problem Based Learning (PBL) – An Effective Approach to Improve Learning Outcomes in Medical Teaching. *Journal of Clinical & Diagnostic Research*, **7**, 2868-2897 (2013)
7. Yulianti, D. Problem-Based Learning Model Used to scientific Approach Based Worksheet for Physics to Develop Senior High School Students Characters. The 3rd International Conference on Mathematics, Science and Education. *Journal of Physics* (2016)
8. Suharta., Lynna., Luthan. Application of Cooperative Problem-Based Learning Model to Develop Creativity and Foster Democracy, and Improve Student Learning Outcomes in Chemistry in High School. *Journal of Education and Practice*, **4**, 55-60 (2013)
9. Kushartanti, B.M.W. Pendekatan *Problem-Based Learning* Dalam Pembelajaran Praktik Kerja Lapangan Terapi Fisik. *Cakrawala Pendidikan*, **1** (2010)
10. Arnyana, I. B. P. *Buku Ajar : Dasar-dasar Metodologi Penelitian* (2007)
11. Pallant, Jullie. *SPSS Survival Manual 4th Edition* (2010)
12. Mergendoller, J. R., Maxwell, N. L., Bellisimo, Y. The Effectiveness of Problem Based Instruction: A Comparative Study of Instructional Methods and Student Characteristics. *Interdisciplinary Journal of Problem Based Learning*, **1**, 49-59 (2006)
13. Tan, O.S. *Problem based learning innovation* (2003)
14. Rahman, M.A., Azmi, M.N.L., Wahab, Z. B., Abdullah, A., Azmi, N. The Impacts of 'Problem-Based Learning' Approach in Enhancing Critical Thinking Skills to Teaching Literature. *International Journal of Applied Linguistics & English Literature*, **5**, 249-258 (2016)
15. Sendag, S., Odabasi, H. F. Effects of an online problem based learning course on content knowledge acquisition and critical thinking skills. *Elsevier Journal Computers & Education*, **53**, 132–141 (2009)
16. Graaff, E., Kolmos, A. Characteristic of Problem Based Learning. *International Journal Engineering*, **19**, 657-662 (2003)
17. Bilgin, I., Senocak, E., Sozbilir, M. The effects of Problem-Based Learning Instruction on University Students Performance of Conceptual and Quantitative Problems in Gas Concepts. *Eurasia Journal of Mathematics, Science & Technology Education*, **5**, 153-164 (2009)
18. Rochmawati, J., Djatmika, E.T., Wardana, L. Implementation of Problem Based Learning Model to Improve Students' Problem Solving Skill and Self-Efficacy (A Study on Ix Class Students of SMP Muhammadiyah). *IOSR Journal of Research & Method in Education*, **6**, 51-55 (2016)
19. Hirca, N. Impact of Problem Based Learning to Students and Teacher. *Asia Pacific Forum on Science Learning and Teaching*, **12** (2011)
20. Forbes, H. Beliefs and Learning Approaches of Undergraduate Nursing Students in a Problem- Based Learning (PBL) Environment. *The Australian Electronic Journal of Nursing Education*, **5** (2000)
21. Halla, G. & Neamat, Muhamed. Effect of Problem Base-Learning on Undergraduate Nursing Students Enrolled in Nursing Administration Course. *International Journal of Academic Research*, **3**, 154-162 (2011)
22. Celik, P., Onder, F., Silay, I. The effects of problem-based learning on the students' success in physics course. *Elsevier Procedia - Social and Behavioral Sciences*, **28**, 656 – 660 (2011)

English for Tour Guide: A Need Analysis of a Contextual-Based Language Teaching

Ni Made Ratminingsih[1*], *Made* Suardana[2], and *Anak Agung Ngurah Yudha* Martin[3]

[1]English Education Department, Universitas Pendidikan Ganesha, 81116 Singaraja-Bali, Indonesia
[2]English Department, Universitas Pendidikan Ganesha, 81116 Singaraja-Bali, Indonesia
[3]Hospitality Department, Universitas Pendidikan Ganesha, 81116 Singaraja-Bali, Indonesia

Abstract. This paper reports the results of need analysis of English for Specific Purposes (ESP) for local tour guides in two villages in Buleleng regency. This research is a descriptive qualitative study involving 56 tour guides, in which 30 were based in Ambengan village and 26 others were posted in Sambangan village. There were three instruments applied to obtain the research data such as observation sheet, questionnaire, and interview guide. Findings from the observation showed that the two villages had waterfalls and terraced rice fields as the main potencies. Additionally, the questionnaire specifying the target need showed that both groups had low speaking skill (56.05%) even the rest of them (43.95%) understood English, but was not able to speak English at all. Furthermore, they all required oral English, speaking (69%) and listening (24.5%). Seen from learning need, they demanded conversation (90.5%) as the most important learning experience. The results from interview proved that 92.5% local guides had no academic background in English. From those findings, it can be summed up that they need contextual-based English learning to accomplish their jobs in a more professional way through non formal education.
Keywords: contextual-based language learning, English for Specific Purposes, need analysis, tour guide

1 Introduction

Tourism cannot be separated from the role of tour guides. They are members of the community who have working opportunity in tourism sector becoming members of tourism group or *Kelompok Sadar Wisata*. This group is formed in order to develop the natural potency of the place. Thus, they have a very significant role to facilitate what travelers need and to guide them while traveling in an area. In order to serve them well, language plays an important role in order to conduct communication in general and hospitality service in particular. It is believed that communication skill is greatly an important element of hospitality industry, both oral and written skill [1]. Good tour guides have to be knowledgeable and resourceful, and they have to know how to communicate efficiently and effectively [2]. In addition, they should be skillful because they are the most valuable assets in tourism companies. In many ways, a guide is the 'facade' of a business or company.

* Corresponding author: made.ratminingsih@undiksha.ac.id

Many foreign visitors tend to view them as representatives of the region or country. Besides that, tour guides have tasks to recognize the social environment, arrange transportation, interpret, handle problems, insulate travellers from difficulties and make the environment safe for tourists because they act as 'intermediaries' between tourists and the 'unknown' environment [3]. In addition to this, tour guides have important roles in tourism. They are responsible for leading and interpreting in the context of environment [4].

However, the reality often shows that many tour guides have inadequate proficiency in English, but they are brave enough to conduct guiding. This phenomenon commonly happens in developing countries like in Indonesia. Thus, a need analysis regarding the target situation and present situation should be conducted. It is considered as a factor that drives the development of ESP curriculum [5]. Furthermore, needs analysis can be clarified in two ways, a target situation analysis (TSA) and present situation analysis (PSA) [6]. TSA is a need analysis which focuses on students' needs at the end of the language course. The TSA is the stage at which 'good enough' competence for the job is reached. Present situation analysis (PSA), on the other hands, seeks to establish what the students are like at the start of their language course, investigating their strengths and weaknesses. Henceforth, there are two kinds of needs, needs which come from outside as the target needs derived from the target situation analysis, while needs which come from inside as the felt needs derived from present situation analysis.

There are three basic sources of information, namely the students themselves, the language teaching establishment, and the 'user-institution', for example the students' place of work [6]. Some information that can be traced from the learners are ranging from their levels of ability, resources (financial and technical), and their views on language teaching and learning, attitude toward English and learning and use of a foreign language. As well, information can also be obtained from the surrounding society and culture.

Needs analysis has significant role in developing English [7]. It is considered as the key essence of ESP [8]. Hence, it will help the teachers to make a number of planning decisions, to group the students, to decide the strategies and learning activities, and to capitalize on students' strength and motivation in ESP course [9]. Moreover, the role of needs analysis for ESP is concerning on the use of vocational or specific language needs [10]. ESP as the English teaching and learning in the context of ESL or EFL aims to be used in particular purposes [11]. Therefore, the purpose of an ESP course is to enable learners to function adequately in a target situation, that is the situation in which the learners will learn language based on their needs.

Being able to conduct their tasks successfully, tour guides should possess good competence in English to communicate with their guests, especially related to the context where they use English with them. Contextual teaching and learning is able to enhance the learners to understand concept while practicing to speak with one another [12]. Additionally, if people learn something based on context, the context will give meaning to the content. Hence, learning is meaningful when what is learned is able to be used in the real life context [13].

Previous research on the use of contextual-based learning found that contextual teaching and learning had a significant effect on learners reading skill, besides, it could motivate them in learning and help them to understand the materials [14]. Furthermore, it was proved there was a significant effect toward students' writing competency after getting the treatment [15]. Specifically in ESP, a research discovered that there was a significant effect of using contextual expressions on pre-text learning, oral and writing interaction and word-meaning inferring [16]. Another research also revealed that students in hotel required special English needs and interests in order to motivate them as well as build their self-confidence [17]. In addition, it was found that both verbal and non-verbal communication skills used by tour guides were good in handling the customer [18]. Despite several research conducted previously on contextual-based language teaching, a similar research is still needed in order to analyze its needs for local tour guide who use English as a foreign language. This research is different from the previous ones, which focuses on needs analysis of local tour guides in Ambengan and Sambangan village in Sukasada District before developing training materials which are relevant to their needs. Therefore, this study is beneficial to be conducted in order to help local tour guides in two villages to improve their English skills.

2 Method

This study was a descriptive qualitative research, conducted within one month in the beginning of year 2017. It involved 56 subjects of tour guides who were divided into two tourism groups, named *Kelompok Sadar Wisata (short for Pokdarwis) Banten Sari*, in Ambengan Village consisting of 30 members, and *Pokdarwis Tunjung Mekar* in Sambangan Village comprising 26 members. The instruments utilized to collect data were observation, questionnaire, and interview guide. Observation was held to see the existing potencies of the two tourism villages and the ways the tour guides perform their duties in guiding the guests. Questionnaire was delivered to find out the target needs and learning needs, while the interview was held to the leaders of two groups and several members to identify their educational background, tasks that they have to hold, and problems they faced when handling and communicating to the guests. The data were collected through several steps, such as (1) observing the tourism objects and availability of facilities and activities in the two tourism villages, (2) interviewing the leaders and several members of their present situation in terms of strengths and weaknesses, and (3) distributing the questionnaire to all of 56 tour guides. Then, the collected data were analyzed descriptively by counting the percentage of the responses.

3 Results and Discussion

The result of observation shows that the existing potencies in the two tourism villages are rather similar. In Ambengan there are 6 waterfalls (Jembong, Puncak Sari, Pucuk, Kroya, Kembar, and Aling-aling) while in Sambangan there are 7 waterfalls (Pucuk, Kroya, Kembar, Aling-aling, Canging, Dedari, and Cemara). Among those waterfalls, there are 4 of them under the possession of two villages namely Kroya, Kembar, Pucuk, and Aling-Aling waterfalls. The activities normally conducted in these places are trekking, sightseeing, swimming, cliff jumping, sliding, and taking pictures. Besides that, there are 3 other tourism destinations, terraced rice field, guardian temple, and plantation (clove, durian, mangosteen, coconut). Some activities that are usually done are trekking, sightseeing, having lunch, taking pictures in terraced rice fields, while offering Balinese *canangsari,* praying in guardian temple, and sightseeing, taking pictures, tasting iced-coconut water and flesh in plantation.

The data show that among 30 tour guides of tourism group in Ambengan village, the majority of them (73.2%) only have primary education and only very few of them (2%) have educational background in tourism. Meanwhile, among 26 tour guides of tourism group in Sambangan village, the data show that their educational background tends to be better since some of them (26.9%) are graduates of the tertiary level of institution. From 7 of them, 3 are graduates of English Education Department, and 2 of them is a graduate of hospitality and tourism department respectively. Some others (26.9%) are only graduates of primary education. In terms of tasks, both groups of local tour guides mainly do activities such as handling the guests' arrival in the office, telling tour itinerary, guiding the guests to the tourism objects, conducting tour activities in the tourism object, handling complaints during and after trekking tour. Dealing with the problems, in Ambengan there has not been given English training by the government while in Sambangan the local tour guides have been trained by government and social community service from a university. But, both trainings were really relevant with tour guides' needs.

Table 1. Target and learning needs

No	Statements	Percentage
1	Why do you need English?	
	a. To communicate with foreigner	82.45%
2	What level of English do you have?	
	a. Speak a little bit of English	51.05%
	b. Understand English but cannot speak well	34.00%
	c. Understand English and is able to give response minimally	5%
	d. Able to speak English fluently	5%
3	What kind of English skills do you need to be developed?	
	a. Listening	24.5%
	b. Speaking	69.00%
4	What kinds of learning experience do you need? (the respondents may give more than one answer)	
	a. Discussion	18%
	b. Conversation	90.5%

The result of questionnaire shows that most of the local tour guides need English for communicating with foreigner, it is proved by the percentage 82.45% from both groups' respondents. In terms of English level, 51.05% of the respondents in both groups were able to speak a little bit of English and only 5% of local tour guides were able to give response minimally. Moreover, the majority of respondents from both groups required oral English (listening and speaking) to help them to communicate with their guests since 69% of them chose speaking skill while 24.5% in listening. The fourth item, mostly, both groups demanded conversation as their learning experience while they were learning English.

Based on the result of observation, it is obvious that the two villages have potentials such as waterfalls, terraced rice fields, guardian temples, and plantations. Regarding this, they are classified as ecotourism areas. Besides the objects, the facilities provided are suitable for the activities, such as safety guards and swimming equipment for trekking, swimming, cliff jumping, and sliding. Other activities normally done by the tourists are sightseeing and taking pictures. Additionally, it was observed that there are medication plantations e.g. turmeric which has function to cure mild illness and organic and inorganic dustbins which purposes to preserve the nature. This echoes the definition of ecotourism that it is a kind of sustainable tourism which has conservative function towards the environment by developing the indigenous community [19]. Ecotourism has beneficial impacts such as preserving the nature [20, 21, 22] increasing economy [23] foreign exchange and jobs [20], and encouraging and develop local communities in terms of social aspects [20].

Considering the findings interview and questionnaire, it is clearly shown that the local tour guides of both *Pokdarwis* have lack of ability to speak English. It is proven by the result of questionnaire that only 10% of them could speak English fluently. This problem is crucial since as a tour guide, they need to describe a lot about place and tell the history of a place. This is related to the statement which emphasize that a guide has to be capable in delivering information about tourism destination as well as initiating good conversation with the guests [2]. Besides that, the result of interview showed that there was an English course held in both groups, Ambengan and Sambangan. The English course was about general English which tended to be less relevant since 100% of respondents confirmed that they needed English materials which are related to their jobs. Meanwhile, based on questionnaire, the local tour guides wanted materials that make them to learn. Therefore, the result of needs analysis plays an important role as the foundation of material development. It is in line with statements that needs analysis will help the teachers to make appropriate materials, to decide the strategies and learning activities, and to capitalize on students' strength and motivation [9]. In addition to the finding, the local tour guides also need to learn English contextually in order to accomplish their tasks. This supports arguments [13] that people need to learn a language based on the context since it promotes meaningful learning. In relation to this, contextual teaching and learning helps the language learners to understand the language concept [12]. This means that contextual-based materials are very important to conducted to the local tour guides because they are meaningful when what is learned is able to be used in the real life context.

4 Conclusion

There are three things that can be inferred based on the findings of needs analysis. Firstly, waterfalls and terraced rice fields are the main potencies of the ecotourism which have benefits to increase working opportunities, social and economic development of local communities, national income, and preserve the natural resources. Secondly, the target needs indicates low English speaking skill, very small number of local tour guides who have English education background, limited training programs and resources which is relevant with their tasks as tour guides. Lastly, the local tour guides require contextual-based training materials to improve their English speaking skill.

The authors would like to thank Universitas Pendidikan Ganesha Singaraja for the support toward this research. Special appreciation is also forwarded to the management of the tourism groups in the two villages for the cooperation.

References

1. M. K. Bobanovic, AlmaTourism, **4** (2011)
2. T. O'Brien, H. S. Ham, *Toward professionalism in tour guiding– a manual for trainer* (US Agency, University of Idaho, 2012)
3. H.Q. Zhang, I. Chow, I, Tourism Management, **25**, 1 (2004)
4. J. M. Chilembwe, V. Mweiwa, IJER, **2**, 9 (2014)
5. M.J. Hossain, J.PU, **2**, 2 (2002)
6. R. Richterich, J.L. Chancerel, *Identifying needs of adults learning a foreign language*, (Pergamon Press, Oxford, 1980)
7. N. Prachanant, Procedia Soc. Behav. Sci., **66** (2012)
8. M. Rahman, UJER, **3**, 1 (2015)
9. J. C. Richards, *Communicative language teaching today* (CUP, Cambridge, 2001)
10. N. Rahmani, T. Pushpanathan, IJAMR, **2**, 10 (2015)
11. S.M. Otilia, ES, **2**, 1 (2015)
12. S. Annisa, EEJ, **6**, 4 (2015)
13. E.B. Johnson, *Contextual teaching and learning; What it is and why it's here to stay* (Corwin Press Inc, Thousand Oak California, 2002)

14. K. Khaefiatunnisa, JEE, **3**, 1 (2015)
15. L. Suparman, A.A.I.N. Marhaeni, N. Dantes, e-Journal Program Pascasarjana Undiksha Program Studi Pendidikan Bahasa Inggris, **1** (2013)
16. H. Tian, JLTR, **6**, 6 (2015)
17. H.T.M. Albakrawi, IJELT, **1**, 1 (2013)
18. N. Suardhana, Nitiasih, A.J. Putra, e-Journal Program Pascasarjana Undiksha Program Studi Pendidikan Bahasa Inggris, **3**, 1 (2013)
19. M. Das, B. Chatterjee, IIM Kozhikode Society & Management Review, **4**, 2 (2015)
20. J. F. Koens, C. Dieperink, M. Miranda, Environ. Dev. Sustain., **11** (2009)
21. K.C. Anup, R. Kedar, P.S. Ramesh, Int. J Sust. Dev. World, **22**, 3 (2015)
22. R. Butarbutar, Soemarno, J. Ind. Tour. Dev. Std., **1**, 3 (2013)
23. R. Staiff, R. Bushell, P. Kennedy, JOST, **10**, 2 (2002)

LOCAL WISDOM-BASED CHARACTER EDUCATION IN TEACHING BALINESE TO ACHIEVE NATIONAL INTEGRATION OF A NATION

by
I Kadek Mustika
Universitas Pendidikan Ganesha
Jalan Udayana 11 Singaraj-Bali Telp. (0362) 22570
ikadek_mustika@yahoo.com

Abstract

Indonesia, nowadays, still deals with complex problems, such as national disintegration and social disharmony of a nation. Many regions experience disunity which is started from smallest unit into larger unit. This problem will threaten the unity of Indonesia. It becomes a serious problem so that it is important to optimize the role of education which leads to the insertion of character education in early learning. Teacher is expected to be able to implement the character education, including the preparation, implementation, and evaluation. The implemented character education should be synergized to the local wisdom of each region because it has its own cultural values which could be guidances toward the attitude. Character education which is synergized with local wisdom aims at achieving the vision of integration and social harmony of a nation.

Keywords: local wisdom, character education, integration

INTRODUCTION

Indonesia has a philosophy which becomes the foundation to engage the Indonesian with all of its aspects named *Pancasila*. It also explains that Indonesian value is unity. The fourth paragraph of the opening of Indonesian basis regulation which was passed in 1945 (UUD 1945) also describes the value of peace and social judicature. The point of this concept is creating the integration and the social harmony of Indonesia. However, there are some conflicts which threaten the integration and become the reflection of disharmony in the society.

The issue of disintegration and social disharmony commonly happens together with the existence of the individualism in the human life. In Bali, the disunity also happens in a *banjar* or among *pakraman*. In the national level, some people on behalf of some certain organizations want to separate themselves from Indonesia. It is in line with Sumantri (in Budimansyah, 2011:1) who said that horizontal and vertical conflict which happen in social life is one of the effects of all crisis, which certainly lead to the national disintegration. Because of that, mental therapy is really needed as the effort to fix the divergence of physical attitude and national psychology and also improve the people's awareness to build our spirit as God's creatures, individual, and Indonesian.

Disintegration is a disunity condition; broken condition; loss of unity; disunion (Indonesian dictionary). What is meant by disunity is the threatened of Indonesia's unity which consists of thousand islands and 34 provinces in Indonesia. Whereas, social disharmony could be defined as an imbalance condition in life. Nowadays, the disintegration and social disharmony commonly happens in Indonesia, in the form of big into small scale. Until now, there are some regions which are threatened by national integration, such as Aceh, Papua, Poso and many other regions which might not identified yet. Especially in Bali, social disharmony phenomenon is easily found. For example, in family, big family, *banjar*, until *pakraman* village. The conflict among *banjar* or villages often happens. This kind of conflict has ever been happened on Saturday, 17th of September 2011 between Kemoning village and

Budaga village in Semarapura sub district of Klungkung regency. It happened because there was different perception of the existence of temple and ceremony in *Pura Dalem* which then cause serious conflict and loss of material and victims (Kumar, 2013).

This kind of disunity could threaten the national integration or national unity. The solution which has been arranged are fixing the physical damage and conducting national reconciliation among groups which have been quarreled or got discrimination (Sumantri in Budimansyah, 2011: 2). Education still becomes the key to solve this problem. Education should put the character education forward. Characters also cannot be separated from education, whether it is in formal education, non-formal education or informal education because the point of education is inserting meaningful values in human life through learning. Formal education at school also directs the students to build and improve the positive values. Like what has been stated by Dantes (2014: 69), many countries, including the advance country, believe that formal education is the significant process of nation and character building. Therefore, character education should be implemented in education.

DISCUSSION

Like what has been stated in the previous explanation that character education is very important to be implemented in education field. Talking about the problem in education must be related to learning. Because of that, a teacher must be able to conduct three phases of learning well, such as designing lesson plan, conducting learning process, and evaluating the learning process. A teacher is also expected to be able to integrate the character education into those three phases of learning process. The implementation of character education is integrated with learning. It is supported by Zubaedi (2011: 355) that the format of learning through character education with guidance is relevant as the effort to create the national character which is depreciating.

The first teacher's duty which is related with learning is designing lesson plan. Based on *Permendikbud No. 65 Tahun 2013* which is about the standard of elementary and high school process, the lesson plan designed by the teacher should reflect the character education, which consists of (a) school identity, (b) lesson identity or theme/subtheme, (c) class/semester, (d) topic, (e) time allotment, (f) learning objectives, (g) basic competencies and indicator, (h) learning material, (i) teaching method, (j) learning media, (k) learning sources, (l) teaching scenario, (m) assessment.

According to those components of lesson plan, in implementing character education, ta teacher should modify or adapt the lesson plan that has been arranged. Daryanto and Darmiatun (2013: 189) stated that lesson plan should be able to guide the teacher in creating the learning process which could improve the character, so that lesson plan should be adapted. This adaptation covers (1) the improvement or modification on the components of learning objectives which refer to core competency and basic competency, (2) the improvement or modification on the components of learning activities, (3) on the components of the topic, and (4) on the components of assessment. These four components of lesson plan must show the character education, whether it is explicitly shown through learning objectives or it is implicitly shown through the arranged activities.

Besides those four components, the selection of teaching method also should be considered. According to Curriculum 2013, there are some teaching methods that can be implemented, such as *Problem-Based Learning, Project-Based Learning, Cooperative Learning* and many other methods. Those learning methods teach the students to work cooperatively in a small or even big group. Slavin (2005: 4-8) proposed various teaching methods in which the students could work in small group whose members come from different level of achievement, gender, and ethnic background to help each other in understanding the material. This method is effective to insert cooperativeness in order to avoid social disharmony and national disintegration.

Actually, learning process is the implementation of the plan that has been arranged in the form of lesson plan, but the teacher should conduct the learning process based on the plan. However, teacher could match the plan with the situation and condition. According to Sardiman (2014: 9), no wonder that the unpredicted phenomenon or interaction in the field could create new experience which becomes knowledge.

In implementing the character education into the learning process, the teacher applies the activities which have been arranged in the lesson plan, started from pre-activity, whilst activity and post-activity. If those activities have been done, it means that the teacher has implemented the character education in learning process implicitly. However, the teacher needs to explain verbally the character values which need to be implemented. It is in accordance with Mulyasa (2014: 105), who stated that there are five levels of thinking towards the attitude aspect, namely (1) receiving or attending, (2) responding, (3) valuing, (4) organizing, (5) characterizing. In other words, it is important for the students to receive, whether through the observation or listening to the verbal information about the values that have to be developed by the students.

Besides those activities in each phase, the teacher should understand the material well, implement appropriate teaching method and use media, learning tools and learning sources. In the learning process, there should be good communication between teacher and students. This is the part of cooperation to solve the learning obstacle. According to Lickona (2012: 140), the teacher-student relationship is the foundation of effective teaching. Thus, the success in implementing character education is really determined by the good communication between teacher and student.

In character education, there are some models or methods which could be implemented by the teacher, such as providing a model, giving verbal feedback, motivating, and also giving assignment. In learning, it is important for the teacher to provide good example in the form of good attitude to the students. Mulyasa (2013: 169) stated that teacher's model influences the students' individual development and improvement. This model has significant role and function in building the students' character in order to prepare and develop the human resources, and also bring welfare to society and country's development. In giving feedback, Klausmeier (in Sani, 2014: 31) said that receiving feedback is one of the teaching method in building character beside observing and receiving verbal information. Building the students' character using this method is conducted through giving feedback toward the behavior, such as by receiving or not receiving the behavior that is performed by the students.

After the teacher conducted learning process, the next duty is evaluating. Evaluating the learning result aims at recognizing whether the basic competency has been achieved (Kunandar, 2007: 356). The evaluation should include three aspects, such as cognitive, affective and psychomotor. Cognitive aspect is related to the thinking ability including understanding, recalling, applying, analyzing synthesizing and evaluating. Affective aspect is related to the attitude or character and psychomotor aspect is related to the physical movement and skill.

The evaluation which is conducted by the teacher should be used as the tools to improve the learning process. Arikunto (2005: 25) stated that the conducted evaluation should be based on the plan. It means that the evaluation has been arranged to identify the learning achievement. The information got through the evaluation can be used as the foundation to decide if the learning program, in this case it is related to values and attitudes, needs to be fixed.

The evaluation plan that has been made by the teacher must be implemented to assess the cognitive, affective and psychomotor aspect and it should be based on the technique that has been chosen. For example, in assessing the affective aspect, the teacher can observe by noting the active students, filling self-assessment, and having assessment among students. For cognitive aspect, the teacher can hold spoken or written test. For psychomotor aspect, the teacher can conduct performance test.

The evaluation which is conducted by the teacher is for getting the formative, summative and confirmative score. The confirmative and formative evaluation are very important in improving the students' affective. When the teacher finds the students who perform bad attitude in the classroom, the teacher must admonish and ask them to build good attitude. Summative evaluation is conducted in order to have final evaluation whether the advices and confirmation that have been conducted by the teacher could change the students' attitude.

Character education which is implemented in teaching and learning process must be synergized with the culture or local wisdom. The values of the local wisdom become the basic fact of the education. It is also stated by Yunus (2014: 5) that the development of national character through local wisdom is really needed. The local wisdom should not only be used as a symbol but should also be used to solve the problems and support the improvement of a nation. The revitalization of the local wisdom in solving the problem is necessary to be conducted as an approach (Mustika, 2016). In teaching Balinese, the character education which is implemented should cover the local wisdom of Balinese, especially the values which are related to social awareness and spirit of nationalism in order to achieve the vision of national integration and social harmony of a nation.

The concept which has been believed by Balinese for long time is *Tri Hita Karana*. *Tri Hita Karana* is three sources of happiness which consist of *parhyanagn* (spiritual aspect), *pawongan* (social aspect), and *palemahan* (physical aspect) (Pujaastawa. 2014). The implementation of this concept is important to create the harmony of the relationship among people, especially the *pawongan* aspect. In Bali, for example, there are some terms, such as *angawe sukaning won glen* which teaches the people to make others happy; *sagilik-saguluk sarpanaya paras-paros salunglung sabayantaka* and *asah, asih, asuh* which reflect the togetherness among people in happiness and sadness. Besides those traditional terms, the concepts of harmonious relationship among people also could be seen in the religious concept. That concept is also known as *menyama beraya* which becomes the belief of Balinese society. All of the local wisdom is the social foundation which is very important to improve the solidarity of Balinese society.

In Balinese culture, which mostly contains Hindus belief, the pattern of the harmonious relationship among humans is stated in the concept of *Tat Twam Asi* which means "you are me and I am you". This concept contains the meaning of love and solidarity in having harmonious relationship among humans. Besides, there is also the concept of *ahimsa* (do not hurt others) which persuades the people to avoid violence. The Balinese local wisdom has reflected the values of character education and the teaching methods which are implemented in Curriculum 2013. It is based on *Permendikbud No. 81 A*. cooperative learning concerns on the value of respecting others, living together in a group, caring and keeping the unity. It is in line with the values of education.

The Synergy of Character Education, Local Wisdom, and Teaching Balinese in the Vision of National Integration and Social harmony of a Nation

Character Education	Balinese Local Wisdom	Teaching Balinese (Curriculum 2013)

GENERAL PERSPECTIVE		
Government proposes 18 character values, namely religious, honest, tolerant, discipline, hard worker, creative, independent, democratic, curious, spirit of nationalism, love nation, respect the achievement, communicative, love peace, love reading, environmentally care, socially care, and responsible. (Kemendiknas Balitbang Puskur, 2010)	Bali has many local wisdom which are taken from Balinese life perspective, whether it is religion or culture, such as *Tri Hita Karana, Tat Twam Asi, Karmapala, Ahimsa,* etc.	Teaching Balinese by implementing Curriculum 2013 which based on *Permendikbud No. 81 A* through scientific approach (observing, questioning, collecting information, associating, communicating). In relation with the attitude, there is religious attitude (in core competency no 1) and social attitude 9in core competency no 2).
SYNERGY		
Based on those 18 character values, the characters which are related with the vision of integration and harmony are tolerant, democratic, spirit of nationalism, love nation, communicative, love peace and socially care.	Based on the local wisdom, the vision of integration and harmony is closely related to the concept of *Tri Hita Karana*, especially the *pawongan*. Understanding the terms *manyama braya, asah asih asuh, sagilik saguluk salunglung sabayantaka, paras-paros sarpanaya.* Moreover, the other concepts are *Tat Twam Asi, Ahimsa* and *Karmapala.*	1) In planning, the teacher design lesson plan with character education by inserting the character values and local wisdom in the learning objectives, material, teaching scenario and learning evaluation. It is also used the cooperative learning to perform the character values and local wisdom. 2) In implementing, teacher conducts learning activities which are humanistic based on the plan that has been designed. Teacher builds the close relationship and harmony with the students. 3) In evaluating, teacher evaluates based on the character values which have been planned and conducted through the evaluation principals.

In achieving the vision of the national integration and social harmony of a nation, there are various challenges in internal and external sector. Internal challenges come from the individual itself, such as redundant egoistic and individualistic. In this context, Balinese propose some terms in the society, such as *"mati iba hidup kai"* which means let other people die but I'm still alive. Through this life principal, Balinese are busy to manage their life but they do not care with others, so that the principal of *menyama beraya* is reducing. The external challenges come from external parties which want to trigger disunity or disharmony. In this era, Bali becomes the tourism area that could be the target to be disunited.

In educational context, especially learning Balinese, there are some challenges in achieving the vision which could be started from teacher professionalism context. Based on the government regulation No. 14 in 2015 about teacher and lecturer professionalism, the teacher should have academic qualification in which diploma or bachelor as the minimum degree and the department should be accredited. Besides, the teachers must have competencies which are related to their duty as teacher. The teachers should optimize the four competencies, such as pedagogical, professional, personal and social.

Based on the regulation made by the Minister of Country's Apparatus Development and Bureaucracy's Reformation No.16/2009 about teacher's functional position and the credit number, the improvement of teacher's profession could be conducted by self-development, such as attending functional training, workshop, seminar and conducting classroom action research.

Not only internal challenges, the external challenges such as the minimum facilities in conducting teaching and learning process also influences learning process. Furthermore, the implementation of Curriculum 2013 asks the students to use various media to support the learning process.

CONCLUSION

Indonesia, nowadays, faces some problems in national disintegration and social harmony of a nation. These two phenomena threaten the unity of Indonesia which has been built. It could be seen from the national scale into the smallest unit, like *banjar*. It is also happened in Bali in which these problems are seriously considered and need to optimize the character education. In solving the problem of national disintegration and social disharmony of a nation, the way that could be conducted is by deepening the character values, especially in formal environment, such as school. Teacher plays an important role in planning, implementing and evaluating. Teacher must be able to implement the character education in those three phases of learning by concerning on the regulation and learning theory. The character education which is implemented in learning should be in accordance with the local wisdom. Based on 18 character values, the characters which are related to the vision of integration and harmony are tolerant, democratic, spirit of nationalism, love nation, communicative, love peace, and socially care. These values are in line with *asah-asih asuh, sagilik saguluk salunglung sabayantaka, paras paros sarpanaya* which are included as the Balinese local wisdom. In learning, teacher should implement the values in planning, implementing, and evaluating phases besides the other values, such as social and religious.

In achieving the vision of national integration and social harmony of a nation, there are some internal and external challenges. It is also happened in the implementation at school. The challenges which come from the teacher are the teachers' qualification and competencies. The external challenge is the school facilities which also plays important role in learning process.

The authors would like thank to lecturers at Universitas Pendidikan Ganesha especially I Nengah Suandi for thought supported of this study.

REFERENCES

Arikunto, S. (2005). *Dasar-dasar evaluasi pendidikan*. Jakarta: PT Bumi Aksara

Budimansyah, Dasim and Kokom K. (2011). *Pendidikan karakter: Nilai inti bagi upaya pembinaan kepribadian bangsa. Bandung*: Widya Aksara Press

Dantes, N. (2014). *Landasan pendidikan: Tinjauan dari dimensi makropedagogis*. Singaraja: Undiksha

Daryanto and Suryati D. (2013). *Implementasi pendidikan karakter di sekolah*. Yogyakarta: Gava Media

Kumar, S. (2013). *Peran kepemimpinan majelis madya Desa Pakraman Klungkung dalam resolusi konflik sosial masyarakat (Studi kasus bentrok antara warga Kemoning dengan warga Budaga Semarapura Klungkung Bali*. lib.ui.ac.id/file?file=pdf/abstrak-20350669.pdf. Downloaded on 28th of August, 2016

Kunandar. (2007). *Guru profesional: Implementasi kurikulum tingkat satuan pendidik (KTSP) dan sukses dalam sertifikasi guru (Revision Edition)*. Jakarta: PT Raja Grafindo Persada

Lickona, T. (2012). *Character matters (Persoalan karakter): Bagaimana membantu anak mengembangkan penilaian yang baik, integritas, dan kebajikan penting lainnya*. Jakarta: PT Bumi Aksara

Mulyasa, E. (2014). *Pengembangan dan implementasi Kurikulum 2013*.Bandung: PT Remaja Rosdakarya

Mustika, I K. (2016). Kearifan lokal, jati diri bangsa. Daily opinion on Bali Post on Tuesday, 12th of July, 2016

Permendikbud RI Nomor 65 Tahun 2013 which is about the standard of elementary and high school process. *http://akhmadsudrajat.files.wordpress.com/2013/06/03-a-salinan-permendikbud-no-65-th-2013-ttg-standar-proses.pdf*. Downloaded on 28th of August, 2016

Pujaastawa, IBG. (2014). *Kebudayaan Bali*. Presented on Workshop for Police in Klapa Bali Pecatu Resort on 22nd of August, 2014

Sani, R A. (2014). *Pembelajaran saintifik untuk implementasi Kurikulum 2013*. Jakarta: PT Bumi Aksara

Sanjaya, W. (2008). *Perencanaan dan desain sistem pembelajaran*. Jakarta; Kencana Prenada Media Group

Sardiman. (2014). *Interaksi dan motivasi belajar mengajar*. Jakarta: PT RajaGrafindo Persada

Slavin, R. E. (2005). *Cooperative Learning: Theory, research and practice* (Translated by N. Yusron). London: Allymand Bacon.

Suyanto and Djihad, A. (2013). *Bagaimana menjadi calon guru dan guru profesional*. Yogyakarta: Multi Pressindo.

Yunus, R. (2014). *Nilai-nilai kearifan lokal (local genius) sebagai penguat karakter bangsa: Studi empiris tentang Huyula*. Yogyakarta: Deepublish

Zubaedi. (2011). *Desain pendidikan karakter*. Jakarta: Kencana Penada Media Group

The Effect of REAP Strategy on Reading Comprehension

Andarini Permata Cahyaningtyas[1,*], *Ali* Mustadi[2]

[1]Elementary School Teacher Education Department, Faculty of Teacher Training and Education, Sultan Agung Islamic University, Semarang, Indonesia
[2]Primary Education Department, Post Graduate Program, Yogyakarta State University, Yogyakarta, Indonesia

Abstract. The purposes of this study are to find out the effect of REAP strategy on reading comprehension of the fifth grade elementary students of third cluster of Kotagede Yogyakarta. The research method in this study was quasi-experimental with Pretest-Posttest Control Group Design. The population was all of fifth grade elementary students of third cluster of Kotagede Yogyakarta. The sample was fifth grade students of SD Baluwarti, SD Kotagede 1, and SD Kotagede 5 Yogyakarta that were establised by cluster random sampling technique. The data were collected using tests. The validity of the instrument was measured in terms of content and construct validity. The reliability of the instrument was stated good with the score of Cronbach's Alpha = 0,734. The data of this study were analized by independent sample t-test and Manova with Hotteling's Trace formula. The result showed that REAP strategy gave a positive and significant effect on reading comprehension that is proved by the score of t-test = 4,914 and 3,286 with sig = 0,000 and 0,002. Based on the result, it can be concluded that REAP strategy gives positive and significant effect on reading comprehension of the fifth grade elementary students of third cluster of Kotagede Yogyakarta.

1 Introduction

Reading process basicly has some steps, such as preparing, reading, giving responses, exploring text, and expanding interpretation [1]. This means students will get message by reading text and translate it into something they understand, so there are many understanding activity in their mind. This is confirmed by cognitive scientists who find facts in these last three decades that readers experience mental understanding process while reading text [2]. This shows that there are many cognitive components which involved in reading comprehension. Understanding process in reading is really entangled with the text students read. They need to connect the message they get from the text with prior knowledge they have had. Students can produce a new knowledge by this process, whereas

*Corresponding author: andarinipermata@unissula.ac.id

their prior knowledge will be corrected because there is some mistake information, or expanded because of some new informations.

Reading comprehension basically is an activity to help students to comprehend text. Comprehension of text depends on 2 main things, reader and what they read[3]. Students are reffered to have comprehended the text if they can explain the main points of the text they have read without reading it again, and also they can make connections between the new information they get after reading and their prior knowledge. This is not easy for readers, especially children, so if they failed to connect their prior knowledge with the text, reading activity will become meaningless and boring[4].

Steps of reading comprehension are: 1) literal comprehension, readers need to process written information in the text; 2) interpretation comprehension, readers process information based on text, they need to understand what writer wants to say; 3) critical comprehension, students have to integrate their thought with facts in the text, and they need to evaluate their prior knowledge and experience so they are able to create new information; 4) creative comprehension, students develop their original idea based on the text and be creative to find any conclution or sollution writer presents on it[5].

Reading for understanding in elementary school, especially in fifth grade, sometimes is done in Indonesian Lesson only. In fact, comprehension through text is needed in every activity involved reading, eventhough the way to practice it is in language subject. It is like what teachers of fifth grade of third cluster of Kotagede said, that students still have difficulties in comprehending the text. They only read the text and when they are given some questions about the text, they will read it again from the start to look for the answers. The text is also sourced from Indonesian lesson book only, so their knowledge is less developed. Teachers were only using question and answer method in reading comprehension lesson because they haven't found a method or strategy for reading comprehension yet, encompassed REAP.

REAP (Read, Encode, Annotate, and Ponder) is one of reading comprehension strategy proven to have positive effect to increase students' comprehension while reading text[6]. REAP is defined as a strategy which helps reading process to create uncomplete meanings into coheren conclusion with students knowledge, guide students during and after reading process, and help students to be active readers who are able to understand and internalize the writer's idea they get while reading the text with or without the teacher's guidance [7-10].

In the first step, they read the text in silent. With silent reading, students will be independent readers who are able to reach higher level of literacy capability and academic success, which become prime standard on linguistic [11]. While reading, they will find some important points, they will encode it on their own language. This activity will help students to make some codes about information they get from the text in their long-term memory with some useful ways in every person [12]. After that, annotate it by writing the message, and then ponder what's the meaning of the text [13]. In fact, writing is an activity involves deep toughts, which improve students' ability to firm up their understanding [14]. REAP is expected to help students understand the text. REAP has some advantages such as students become active readers, help students to decipher writer's idea to become reader's by synthesizing into their own language, interpret writer's idea to be reader's, and find out student's thinking developing[15-17]. Based on this theory, it is regretted that the fifth grade teachers of the third cluster of Kotagede have not applied REAP yet, and did not know whether REAP has possitive effect on reading comprehension ability of their students or not. Because of this problem, this study aimed at finding out the effect of REAP strategy on reading comprehension to the fifth grade elementary students of third cluster of Kotagede.

2 Method

2.1 Research Design

This study used quantitative approach, quasi-experiment method with pretest-postest control group design. There were two exprimental groups in this research such as the first and second exprimental group which were chosen by ramdom sampling.

2.2 Time and Place

This reseach was held on second semester of 2014/2015 school year, at March 5-24, 2015 on elementary schools of third cluster of Kotagede, Yogyakarta. Each of exprimental group had 8 times meeting with 1 times for pretest, 6 times for giving treatment, and 1 times for posttest.

2.3 Participants

Population in this study were 224 students of fifth grade of elementary schools of third cluster of Kotagede. Samples were decided by using cluster random sampling technique which appointed 3 classes, fifth grade of SDN Baluwarti as first exprimental group, SDN Kotagede I as the second exprimental group, and SDN Kotagede V as control group. The REAP strategy was implemented in the experimental group, and the control group was taught using underlining method. At the beginning of course, students have been given pretest, and at the end they were given posttest using the same instrument measurement.

2.4 The Phases of the Research

1. Teachers give some instructions about REAP strategy for students in experimental groups.
2. The learning process using REAP:
   a. Use the text for course book or given by the teachers as reading material;
   b. Read and then identify the main idea, key points, facts, orders, and sequences;
   c. Take a short note based on the text;
   d. Ponder the note they write whereas it was right or not.
3. Students were given exercises to find out how their skill on reading comprehension after using REAP strategy.

2.5 Data Analysis Technique

Data for this research were collected by using essay test of reading comprehension. The questions were arranged using Barret Taxonomy because it is trusted to be able to represent questions related to reading comprehension, and more detailed compared to the other taxonomies. The test contained 15 questions that consisted of stages of reading comprehension. Investigation was conducted to measure validity and reliability of the test tools. There were 40 students who were tested and 13 questions were analyzed by using SPSS program, and it was found to have score above 0.312 (t_{table}) for the validity test and

0.785 for reliability test, so these results showed that the questions were valid and reliable. This data was interpreted by using 0.5 significance level.

3 Results and Discussion

Post-data research was divided into pretest and postest data. In the pretest, there were 26 students who had been tested on the first experimental group, 27 on the second experimental groups, and 22 students on the control group. This is the result of pretest of reading comprehension:

Table 1. The Result of Pretest of Reading Comprehension

Description	Exprimental group I	Exprimental group II	Control Group
Total Score	170	175.13	140.51
Average score	6.54	6.48	6.38
The Highest Score	9.23	8.21	7.95
The Lowest Score	6.55	5.13	4.87

Pretest result on the first experimental group showed that the the highest score was 9.23 and the lowest was 6.55, with the average was 6.54. This was categorized as low score compared to the minimum score criteria of Indonesian subject at SDN Baluwarti which was 7.00. Meanwhile, the results of the pretest of the second experimental group showed that the highest score was 8.21 and the lowest was 5.13, with the average was 6.48. This was categorized as low score compared to the minimum score criteria of Indonesian subjects at SDN Kotagede I which was 7.00. The control group also showed similar result. The highest score of the control group was 9.23 and the lowest was 6.55, with the average was 6.54. This was categorized as low scores compared to the minimum score criteria of Indonesian subject at SDN Kotagede 5 which was 7.00. These pretest results from three groups showed that there were no significant differences. This means that students' reading comprehension from experiment and control groups are no difference in the beginning.
However, posttest showed different result.

Table 2. Posttest of Reading Comprehension Result

Description	Exprimental group I	Exprimental group II	Control Group
Total Score	231.79	222.05	171.46
Average score	8.58	8.22	7.47
The Highest Score	10	9.74	8.97
The Lowest Score	6.67	6.15	6.15

Posttest result on the first experimental group showed that the highest score was 10 and the lowest was 6.67, with the average was 8.58. It indicated that their test results increased significantly compared to pretest results as much as 2.02. Meanwhile, the result of the pre-test of the second experimental group showed that the highest score was 9.74, and the lowest was 6.15, with the average was 8.22 that also increased from the pretest score which was as much as 1.74. The control group also showed good result. The highest score of the control group was 8.97 and the lowest was 6.15, with the average was 8.97. This was 0.78 higher than the pretest average score. But this indicated that the posttest scores of control group also did not show any significant progress from pretest scores.

Table 3. Result of Hypothesis with Independent Sample T-test on Reading Comprehension

Groups	T-test	Df	Sig	Note
First Experimental	4.914	48	0.000	There were significant differences
Second Experimental	3.286	48	0.002	There were significant differences

The data showed t-test result on the first experimental group was 4.914 and 3.286 on the second experimental group. Those results are more than $t_{table} = 1.678$. The significant scores < 0.05. These indicated that there were different effect between REAP strategy on experimental groups and underlining strategy on reading comprehension. So that, it could be concluded that REAP strategy had given more positive and significant effect on reading comprehension of the fifth grade students on third cluster of Kotagede Yogyakarta.

REAP strategy steps were proven to be able to help students to comprehend the text. Starting with reading, students absorbed the hidden meaning of the text. Silent reading can guide students to be independent readers who are able to increase their comprehension of the text. Then, students will encode the main points of the text, and then they reform some main point into their language. Students would involve actively in mental activity while they absorbed knowledge and information from what they read. This activity also sharpen their thinking to see something from the other perspective. The next was annotating which allow them to annotate some difficult words and those main points they thought will help them to understand. Annotating helped students to think straight to the points of the text. The last was pondering about the notes they had been written whether they are right or wrong. These steps were really helpful for them to comprehend the whole meaning of the text, go along with the text, and avoid to memorize words by words. These were different from the underlining strategy which made students repeating what words they had underlined to identify the main points of the text. It was attested that REAP was more effective than underlining strategy on reading comprehension.

4 Conclusion

REAP strategy had been proven to have effect on reading comprehension, so it can be applied on reading comprehension activities. REAP strategy can be one of learning strategy alternatives not only can be implemented for Indonesian subjects, but every subjects who need to comprehend some texts. The note the students make can also be used as material for independent-learning so they can be easier to understand the text-sourced subject such as Civic, Science, and Social Studies.

We acknowledge and sincerly thank the lecturers and the Head of of Primary Education Department of Postgraduate Program of Yogyakarta State University for their advice, which help us to complete this research. We are also very grateful for the support, cooperation and useful comments from our collagues. Their supports always encourage us to continue with the study. Finally we would like to thank to the Head of Elementary School Teacher Department in Sultan Agung Islamic University who has given us permission to present this research in GC-Tale of Undiksha.

References

1. G.E. Tompkins & K. Hoskisson, *Language arts: content and teaching startegies (third ed.)*, (Englewood Cliffs: A Simon & Schuster Company, 1995, 202-206).
2. M. Ness, JRCE, **25**, 98.
3. J. Hayon, *Membaca dan Menulis Wacana,* (Jakarta: Gramedia, 2007, 55).
4. A.M. Glenberg, IEJEE, **4**, 9, (2011).
5. D. Rubin, *A Practical Approach to Teaching and Writing (second ed),* (Danvers: John Wiley & Sons, 1993, 196-197).
6. M. Tasdemir, J Soc Behav Pers, **38**, 557 (2010).
7. M.G. Eannet & A.V. Manzo, Journal of Reading, **19**, 648.
8. Manzo & Manzo, *Content Area Reading*, (Ohio: Merrill Publishing Company, 1990, 221).
9. M.R. Ruddel, *Teaching Content Reading and Writing (4th ed)*, (Needham Height: Ally and Bacon, 2005, 261).
10. Y. Abidin, *Strategi Membaca,* (Bandung: Rizqy Press, 2010, 152).
11. K. P. Walker, Journal of Reading, **57**, 185 (2013)
12. S. Schellenberg, M. Negishi, & P. Eggen. Journal of Teaching Educational Psycology, **7**, 18 (2011).
13. K.S. McKnight, *The Teacher's Big Book Graphic Organizers,* (San Fransisco: Jossey-Bass, 2010, 140).
14. J. Coeerey & G. C. Rinnert, Journal of Visible Language, **47**, 3 (2015).
15. M. Tasdemir, J Soc Behav Pers, **38**, 554 (2010).
16. S.L. Strode, Journal of Reading, **36**, 568 (1993).
17. Y.Abidin, *Strategi Membaca,* (Bandung: Rizqy Press, 2010, 153).

Bilingual instruction in early childhood education, can it better develop children?

[1)]*Santri* **E. P. Djahimo**, [2)]*Yulia* **Indahri**

[1)]English lecturer, Universitas Nusa Cendana (UNDANA) Kupang, NTT Indonesia
[2)]Senior researcher, Secretary General of DPR-RI, Jakarta, Indonesia

Abstract. This is a case study of teaching and learning using bilingual instruction in two schools of Early Childhood Education in Kupang-NTT, Indonesia. The aims of this study are to find out whether or not bilingual instruction in Early Childhood Education can better develop children (the outcomes) and if the issue of 'the younger, the better" in children's language acquisition in bilingual setting is acceptable and true. 4 students from one bilingual and one monolingual schools have been observed. In addition, parents and teachers of these students have also been interviewed. The data has then been analyzed qualitatively to come to the answers of the proposed questions, and the results reveal that bilingual instruction alone does not necessarily create better children unless it is supported by other factors, such as encouraging parents, professional teachers, and supporting environment. This is in a line with the findings of the issue about 'the younger, the better' in children's language acquisition, that the younger does not always mean the better. Other aforementioned factors are needed in the development of children's language acquisition. It's recommended that all parties involved in children's development can take part in developing their skills, knowledge as well as characters.
Keywords: Bilingual instruction, Early Childhood Education, Children's development, Children's Language Acquisition

1. INTRODUCTION

Schools have changed over time; not only in relation to the *where* but also to the *how* they are operated. Indonesian schools in particular have gone through the changings: many obsolete buildings and facilities have already been replaced by the modern ones with great management and highly qualified teachers, various textbooks and other supplementary materials, well-designed curriculum as well as other teaching tools with bilingual or even trilingual instruction, well-prepared students, and so on. Without trying to generalize, it can be said that most schools in Indonesia, especially those which are located in urban areas have already had these kinds of changings. The obvious one relates to English as a Foreign Language (EFL) which has been dominantly used in schools. Several schools have even declared themselves as "international" schools which have now been changed to Educational Cooperation Units *(Satuan Pendidikan Kerja Sama),* based on Indonesia regulation (Permendikbud No.31 2014).

Many studies have been conducted about the use of bilingual instruction in school levels but not many of them have looked at how it works to students' achievement, especially on the level of Early Childhood Education program. There are still gaps relate to the issue of language of instruction applied to this level of Education. Actually, we have to put more concern on looking at the answer of the question about whether or not the use of bilingual instruction can really better develop children as well as how it can develop children's language acquisition. In order to bridge the existing gaps, this study has been designed and conducted to focus on one particular issue about medium of instruction. Bilingual instruction in Early Childhood Education will be discussed in particular as the findings of this study. However, this study has been conducted only for a short period of time, therefore there are several psychological issues relate to children's development cannot be assured as some aspects of the longer term issues are beyond the scope of the study.

2. LITERATURE REVIEW

There are three main issues raised in this part, they are the concept of Early Childhood Education; the definition and concept of bilingual instruction, and the use of bilingual instruction in Early Childhood Education; Pros and Cons.

Many authors and researchers see Early Childhood Education with different concepts and perspectives. According to Ojala (1978a), Early Childhood Education is not only a theory but also it can be seen as a practice to be applied before the school age. Later in 1993, Professor Ojala came up with additional issue of Early Childhood Education by saying that it is the process of activity either at home or elsewhere to develop the personality of children between 0 –

6 years of age. Another point of view states that preschool belongs to Early Childhood Education (Hujala et al., 1998) and that Early Childhood Education refers to the activities to educate under school-aged children (Karila, et al., 2001). This study has then adopted to be adapted all these concepts and created its own concept to match with its specific focus, that Early Childhood Education relates to educational activities (teaching and learning) that involve both teachers and students (children of 0-6 years of age) in school environment.

When we are talking about bilingual learners, we will definitely know that it refers to those who use their first language at home or in the community, for example, Bahasa Indonesia, and are learning various subjects through a second and/or foreign language, for instance, English, at schools (Cambridge, 2015). In this study, bilingual instruction is defined as two languages (Bahasa Indonesia and English) which are used as medium of instruction in the teaching and learning process. The two languages are not equally used but in this case, English is more dominantly used than Bahasa Indonesia. There are some positive and negative sides about bilingual education or the use of bilingual instruction in Early Childhood Education program. Mehisto (2012) is one of those who look at the positive side of using bilingual instruction to this level of students by stating that bilingual learners will gain positive benefits from this bilingual program, such as having flexible mental, gaining better understanding about different cultures, and having more opportunity for international trade. Whereas Ball (2011) argues that based on the research result, mother tongue is the basic for children to be able to learn best. This study will look into the issue of whether or not bilingual instruction in Early Childhood Education can better develop children (the outcomes).

The discussion about using bilingual instruction cannot be separated from the diversity of hypotheses in language acquisition. One of the most common belief is from the Critical Period Hypothesis (CPH) about "the younger, the better" in which children are assumed to be easier in acquiring language than adults (Lenneberg, 1967). This point has become the main ground for the use of bilingual instruction in Early Childhood Education. However, there are several critics on this CPH. One of the critics is from Singleton (1995). He agrees that the younger, the better, but conditions apply. So, the younger can only be the better only if the process occurs in the long run. He also adds that adults can also be great language acquirers even though they don't start in their critical period. Krashen (2013) also argues that language acquisition occurs when we naturally pick up a language in a sub-conscious way and this can happen not only to children but also to adults. Another issue to be raised in this study is to prove whether or not "the younger, the better" in children's language acquisition in bilingual setting is acceptable and true.

3. METHODS

This study is about teaching and learning using bilingual instruction in two schools of Early Childhood Education in Kupang, East Nusa Tenggara (NTT) Province and it's a case study. Generally, this study aims at answering the question of whether or not bilingual instruction in Early Childhood Education can better develop children (the outcomes). Particularly, its aim is to find out whether or not the issue of 'the younger, the better" in children's language acquisition in bilingual setting is acceptable and true.

The findings of this study will redound to the benefit of teachers, education practitioners, policy makers and other parties involved in this specific field considering that bilingual instruction in Early Childhood Education tends to be dominantly applied today. The increasing number of schools using bilingual instruction justifies the need for a deeper look into its effectiveness and advantages. Thus, the above mentioned parties that apply bilingual instruction will be able to reflect and assess the process and product they have been going through so far. Furthermore, this study will also help uncover critical issue of "the younger, the better" in children's language acquisition and so a new theory on language acquisition may be arrived at.

Four students from two schools of Early Childhood Education in Kota Kupang – NTT Province have been involved in this study which have then been grouped into two groups and observed. Group 1 consists of 2 students (student 1, PD is a 4.8 – year – old boy and student 2, RR is a 5 – year – old girl) from Early Childhood Education uses bilingual instruction. Group 2 consists of 2 students (student 3, GL is a 5 year – old – girl and student 4, RD is a 5.1 – year – old girl) from Early Childhood Education uses monolingual instruction. Additionally, parents (with different professions) and 2 teachers (with similar qualification) of these four students have also been included in this study to be interviewed. These students have been studying in those two schools for one year.

In collecting the data, interview and observation have been used. Both have been used to validate and strengthen the collected data (methodology triangulation). The data has been collected for eight weeks. The first four weeks have been spent to collect data from group 1 (school 1) and the other 4 have been used for data collection in group 2 (school 2).There have been 40 meetings altogether (20 meetings in each school). During those meetings, the four students have been observed; the observation has been focused on the way they learned, participated, and interacted both verbally and nonverbally in the teaching and learning process in the classroom. Note taking has been made and the results of observation have been documented to be analyzed. Several informal oral tasks have also been conducted to be able to see their language development (i.e English). Parents of these four children have also been involved in this study through interview session in order to find out what and how their children do at home to develop their English language. Teachers of both schools have also been asked about students' achievement and progress in schools. The data has then been analyzed qualitatively to come to the answers of the proposed questions. The main interpretation in this study is qualitatively explained, in order to identify and discuss the findings about whether or not bilingual instruction in Early Childhood Education can better develop children.

4. RESULTS AND DISCUSSION

Each group of the students has been observed for four weeks and their parents and teachers have also been interviewed, and the results show that bilingual instruction alone does not necessarily create better children unless it is supported by other factors, such as encouraging parents, professional teachers, and supporting environment. The detailed findings are as follow:

PD, student 1 of school 1_that applies bilingual instruction, develops better within four weeks. He tends to discover and explore things better than other children in his class. He understands the teacher's instruction well and can communicate in English fluently. However, his Bahasa is not as fluent as his English. According to his teacher, he's the smartest and the most active student in the classroom (which can obviously be observed easily). His mother says that he was born in Australia and spent time in there until 3 years old before moving in to Indonesia (Kupang) and directly enrolled in this school because of its "international" label. Furthermore, she informs that both of she and her husband use English in daily life communication with him because they want to maintain his English. His habits of watching English cartoon movies and playing games using English have shaped his great English ability. Another positive habit, according to his mom is listening to either his mom or his dad reading English stories for him. His oral performance in English is good and his overall knowledge of the lessons is great.

RR, student 2 of school 2_that applies bilingual instruction, doesn't develop herself within four weeks. She tends to be quiet and passive in the classroom. She even rejects to join any kind of sports because doesn't want to mingle around with her classmates. She doesn't speak much during the class and so it's not really clear whether or not she learns something from her teacher. When her teacher explains the lessons and shows how to do it, she prefers to be busy with her own activity with her objects (it can be pen, color pencils, bag, etc.). Her teacher admits that it's not easy to face her because she tends to ignore not only the lessons but also the teacher. She doesn't want to be left alone during school time so her babysitter has to accompany her, even she wants her babysitter to come to the classroom with her. Her language doesn't develop well, that's why she tends not to talk to others but her babysitter. Her mom says that she can't be treated in a hard way because she won't be able to take it. She cries easily to things she doesn't want to. Preventing her from crying, her parents almost always do whatever she wants. Her mom explains that as the only child, she needs to be spoiled. By sending her to this "international" school with English as the dominant language of instruction, her parents hope that she can have good English skill because her teacher always uses English to teach them every school day but Friday.

GL, student 3 of school 2_that applies monolingual instruction, looks great in her academic performance. Her teacher teaches her simple math and science in Bahasa Indonesia and uses English only in teaching English but not all the time during the lesson. She learns English once a week, every Friday as a local content (Muatan Lokal). In her English class, her teacher is very creative in using many kinds of games to attract the students to engage and participate in the teaching and learning process. Realizing that English can only be taught once a week, she says that it's important for her to motivate and encourage her students to study English by presenting the lessons interestingly. This student's English is well-developed during the 4 weeks and so are her math and science. She is very active in sport and tends to be inquisitive. Her teacher claims that her good English has been achieved from her parents. Both of her parents are English teachers and they insist her speak good English. Her father says that he always encourages her to join any kinds of English competition (story telling, speech delivering, and spelling bee). Other subjects are also important for her so both her mom and dad send her to join extra courses. She is literate in both English and Bahasa and always reads English stories by herself.

RD, student 4 of school 2_that applies monolingual instruction, is more or less similar in academic performance to the previous one. Her parents aren't English teachers but they have sent her to English course since she was 4 years of age. Her mother says that she expects her daughter to have good English because she knows how important English is. Many kids' movies and songs in English have been purchased by her parents to make her get used to the language. According to her teachers, she is a type of auditory learner because she can memorize and retain all information only by listening. It's obvious in the observation that she's playing while listening to her teacher's sample of presentation and she can eventually perform that really well.

The observations show that the use of bilingual instruction in school 1 cannot really improve the students' skills and achievement, because RR is still difficult to keep up with the lesson and cannot acquire English well. This might be because of lack of support from parents and environment. She still needs to improve her first language before being able to acquire her English as a language of instruction. It has been noted that the use of English in this school is more dominant than the use of Bahasa Indonesia. English is used from Monday to Thursday and Bahasa Indonesia takes place as the language of instruction on Friday only. This kind of bilingualism is not really clear. It can't be classified as transitional bilingual education because Bahasa Indonesia is not used here as the instruction to learn English. Classifying it as late-exit development bilingual education is also not acceptable because in this type, students' first language has to be taught first to make them easier to acquire the second and/or foreign language. This is strengthened by Catts et al. (1999); August at al. (2005) and Yesil-Dagli (2011) by arguing that in shaping children's oral language and developing their early literacy, not only second language is used but also first or native language is needed. Furthermore, there's an important point to highlight that the use of bilingual instruction for preschool-age children doesn't focus only on a second and/or foreign language but the first language as well because children still learn their native language by those ages (Ballantyne, 2008). On the other hand, GL and RD from monolingual school have shown that they can perform well in both their achievement in their lessons and their English skill. Teachers of both schools might have performed well in their teaching activities but it's obvious that the teacher of monolingual school is more creative in teaching, especially English lesson.

Based on the findings it can be concluded that being students in schools with bilingual instruction doesn't guarantee that all children will develop themselves better. The previously mentioned factors play important roles in developing students to be better. This is in a line with the findings of the issue about 'the younger, the better' in children's language acquisition, that the younger does not always mean the better. These four students are still in the category of critical period but not all can perform better result. The result of this study supports the statements of Singleton (1995) and Krashen (2013) that age matters but conditions apply. Krashen (2013) makes a good point that supports this result by stating that the younger won't be the better if children have anxiety and don't have high motivation. It can happen that two students get exactly similar input but eventually, one will get better than another. In short, the bilingual instruction itself won't do much to develop children in Early Childhood Education but other supporting factor will help the children to have better development.

5. CONCLUSIONS

This research provides the findings that bilingual instruction can't alone better develop children in Early Childhood Education and also the younger doesn't automatically mean the better in language development. Both have to be strengthened by other supporting factors, they are encouraging parents, professional teachers, and supporting environment. However, the limit of time has also limited the reach of findings and analysis of this study and so it opens a door and leaves a room to more extensive research on many levels, especially the one relates to psychological issues about children's development.

6. ACKNOWLEDGEMENT

This study has been conducted with the help of a number of people. Thanks to the principals and teachers of Leaps and Bounds and Right Start Kupang, and TK Pembina Kupang. The appreciation is also addressed to the children and their parents for their great contribution to complete this study.

7. REFERENCES

August, D., Carlo, M., Dressler, C., & Snow, C. E. 2005. *The critical role of vocabulary development for English language learners*. Learning Disabilities Research & Practice, **20**, 50-57.

Cambridge Professional Development Qualifications for Teaching Bilingual Learners: www.cie.org.uk/pdq retrieved on 10th July 2017.

Catts, H. W., Fey, M. E., Zhang, X., & Tomblin, J. B. 1999. *Language basis of reading and reading disabilities: Evidence from a longitudinal study*. Scientific Studies of Reading, **3**, 331-361.

Hujala, E., Puroila, A-M., Parrila-Haapakoski, S., & Nivala, V. 1998. *Päivähoidosta Varhaiskasvatukseen* (From Day care to Early Childhood Education). Jyväskylä: Gummerus.

Karila, K., Kinos, J., & Virtanen, J. 2001. *Varhaiskasvatuksen Teoriasuuntauksia* (Early Eduvation Theory Directions). Juva: PS-Kustannus.

Krashen, Stephen. 2013. *Second Language Acquisition: Theory, Applications, and Some Conjectures*. Cambridge University Press.

Lenneberg, E.H. 1967. *Biological Foundations of Language*. Wiley. *ISBN 0-89874-700-7.*

Mehisto, P. 2012. *Excellence in Bilingual Education: A Guide for School Principals*. Cambridge: CUP/Cambridge International Examinations.

Ojala, M. 1978a. Varhaiskasvatus Tieteenä (Early Childhood Education). *KASVATUS* **(9), 5/1978,** 308-313.

Ojala, M. 1993. *Varhaiskasvatus. Teoriasta Käytäntöön* (Early Childhood Education. From Theory to Practice). Joensuun Korkeakoulu. Kasvatustieteiden Osaston Julkaisuja 2/1978.

Singleton, David,. & Lengyel, Zsolt, eds. 1995. *The Age Factor in Second Language Acquisition : A Critical Look at the Critical Period Hypothesis*. Clevedon (England): Philadelphia. *ISBN 1-85359-302-8.*

Yesil-Dagli, U. 2011. *Predicting ELL students' beginning first grade English oral reading fluency from initial kindergarten vocabulary, letter naming, and phonological awareness skills.* Early Childhood Research Quarterly, **26**, 15-29.

Local culture-based picture storybooks for teaching English for young learners

Ni Made Ratminingsih[1*] and *I Gede* Budasi[1]

[1]English Education Department, Universitas Pendidikan Ganesha, 81116 Singaraja-Bali, Indonesia

Abstract. This paper reports the theoretical perspectives underlying the development of local culture-based picture storybooks for teaching English for young learners. It is a library research which derives from three main sources, journal articles, reference books, and internet articles. Local culture-based picture storybooks are kinds of printed materials which basically contain stories adapted from local culture stories accompanied with visualization in the forms of pictures telling the characters and the settings to help young learners easily understand the stories. They are developed considering the good criteria of storybooks [1, 2]. The benefits of its development are (1) to improve reading skill of young learners, (2) to build children English literacy through listening to the story read by the teachers and communicating interactively with them through dialogic reading, (3) create a fun learning atmosphere which enhances motivation and participation to take part in understanding and using the language, and (4) to preserve local culture. The books are developed using a combination of model [3, 4] which has 5 main steps namely need analysis, developing product, expert validation, and quality assessment by experts and users (teachers), and main product revision.

Keywords: local culture, storybooks, young learners

1 Introduction

Teaching English for young learners (TEYL) should be given as early as possible because early introduction can enhance acquisition. English competence is beneficial to prepare professional working forces that are able to be competitive to get job vacancies in tourism sector and other fields as well as for science and technology advancement in ASEAN Economic Community. The importance of introducing English early is for better and quicker target language acquisition [5]. However, the reality shows that Indonesian learners possess low interest in reading. A research reports [6] that most of the students in Indonesia do not have good reading habit though they spend approximately 10 to 11 years to learn English formally from primary to tertiary level.

In fact, reading is one of the language skills which are important. Through reading, learners can improve their knowledge. Reading stories, story-based approach, is

* Corresponding author: made.ratminingsih@undiksha.ac.id

emphasized to positively affect toward environment which stimulates a higher level of intrinsic motivation and engagement of young learners in primary school [7].

Empirically, a research found that the use of story could motivate young leaners besides build enjoyment and happiness [8]. Moreover, a study also proved that teaching young learners by using story had a strong effect on students' listening and speaking performance [9]. Another study also revealed that teaching by using pictures could support independent reading, develop reading strategies, as well as foster reading motivation [10]. Henceforth, in order to improve students' interest in reading, it is necessary to develop picture storybooks as the main focus of this study. By using this kind of books, the learners can be more interested in reading and more motivated to understand the massage of the stories. Furthermore, through local culture-based stories, English instruction can be made contextual, in which the contents of the stories used local culture which becomes parts of the learners' life, since the familiarity of the topics could decrease cognitive pressure they face [11]. Thus, they can use the language for communicative purposes. In addition, they can also be directly involved in creating interactive dialogue in the process of story reading through dialogic reading. By conversing, they are able to practice communicating while trying to understand the stories. More importantly, through the insertion of local culture young learners will respect, love, and preserve values of local culture. This is the main focus of this current research since the tradition of reading folklores has been decreasing among Indonesian young generation because of technology influence. They tend to enjoy more of the technological games rather than books. Therefore, the purpose of this research is to develop local culture-based picture storybooks for young learners. There are three research questions which become the focus of this study: (1) what is meant by local culture-based picture storybooks and what criteria are used to develop these books?, (2) why is it beneficial to use local culture-based picture storybooks?, and (3) how is the procedure to develop these local culture-based picture storybooks?

2 Method

This paper is the result of library research which obtains the data from three sources, namely journal articles, reference book, and articles of research journals.

3 Results and discussion

3.1 The nature of TEYL and local culture-based picture storybooks

Young learners do not understand yet abstract concepts which is different from adult learners [12, 13]. They focus on concrete concepts of learning. Based on this fact, the teacher should more emphasize on the concretization in learning a foreign language without giving them the abstract concept, in order to ease young learners in understanding the lesson. Young learners are students who are at pre-primary and primary level, from the age of 3 up to 11 or 12 years old [14]. Specifically, there are some important characteristics of young learners [15, 16], namely: (1) young learners have a short attention, (2) they are active learners, (3) they are happy with appreciation, (4) they do not feel embarrassed, (5) they like to imagine, (6) they love to play [17], and (7) they like to imitate what they hear.

Based on the above characteristics, it can be inferred that young learners get bored quickly, be less motivated to learn with difficult tasks, like to imagine, be active, playful, and never shy. Thus, their reading interest can be increased through the use of colorful picture storybooks. They can imagine the characters and places in the storybooks and the vocabulary can be easily understood through pictures. Young learners are also directed to be active in the sense of the story read by teacher, then they actively imitate the vocabulary and sentence, and involve in conversing interactively in the process of understanding the story, so that the power of reasoning can be developed. Therefore, their reading competence will be improved, especially in the form of reading comprehension.

There are two kinds of reading, intensive and extensive reading. Intensive reading requires a deep understanding, while extensive reading is more to do with gaining general understanding to develop good reading habits and to encourage reading for pleasure [18]. It is further stated that reading for pleasure refers to reading which is started at other's request then we continue since we are getting interested in. Reading for pleasure is defined as a reading in which students choose to do reading based on their own [19]. As been noted that reading has benefits such as learners' literacy competence is developed since they read for pleasure and learners who read and enjoy reading make them to be a literate [20]. It provides a means of everyday life information seeking for young teenagers [21]. It helps learners to improve their skills and learn about themselves, their relationships, values, and all that helps them to make transition to adulthood. Thus, reading for pleasure is a type of reading in which readers like to read based on their willingness for self-satisfaction and get the benefits out of what they read.

Picture storybook is a kind of books that integrates the use of text with image created to present aesthetic value. This type of storybook is a unique genre of text in children's and teenage literature, categorized in terms of format rather than content. The use of images is helpful to bridge the meaning and content, to prompt for discussion, to seek access and complement of written language [22]. There are 4 types of storybooks, namely folktale, fable, legend, and myth which are developed by considering local culture.

Concerning with the local culture, it is a culture that exists in a society in a particular place. Local culture is the most basic culture which is possessed by the local community in terms of ideas, beliefs, rules, and materials [23]. In this study, local culture refers to culture of Balinese society in which it is classified into fourteen elements that are influenced by Hindu or Javanese Hindu beliefs [24]. Some of them are related to this research such as the creation of religious literature (*Tutur*, *Ketattwan*, *Kanda*, *Aji*, etc.) and the habit of using palm leaf in writing stories, fairy tales, dramas, and others. In relation to the effort of inserting local culture into English learning, the integration of local culture into English learning does not only educate learners about indigenous cultures within their own countries and in the world, but also make English language learning more relevant for learners [25]. Therefore, it can be concluded that the insertion of local culture in learning English is something that is crucial done by educators to provide cultural education and make learning more relevant to the learners where they live and raise.

3.2 The benefits of using local culture-based picture storybooks

Local culture-based picture storybooks have a lot of benefits. Students who study local culture are able to improve their reading ability [26]. In addition, story can motivate young learners and create a happy and enjoyable learning environment [8]. In line with that statement, a report also states that teaching stories to EFL young learners are able to give a strong effect on improving listening and speaking performance of them [9]. Additionally, it is believed that picture books can support independent reading, develop reading strategies, and foster reading motivation [10]. Furthermore, the use of stories in English class is appropriated as teaching tools to offer opportunities using English through the repetition of language pattern and vocabularies [27]. Therefore, inserting story of local culture in English learning can improve students' competence specifically reading competence.

Based on preliminary research, there are no local culture-based picture storybooks which are available on sale in the market. The book commonly used by teachers is a student workbook or textbook that is not contextual to the local culture. Thus, developing this kind of storybooks has some benefits such as (1) increase students' reading interest by giving materials of reading for pleasure since books packed with images based on topic familiarity in local culture can increase interest and motivation, (2) introduce contextual English vocabulary, which is a local cultural context that can facilitate the comprehension of reading content, (3) lead children to learn to read through dialogue (dialogic reading) in order to construct their own knowledge and use the language in an interactive communicative manner, (4) enhance students' love and appreciation of local culture and as an effort to preserve the culture, and (5) insert character education represented by characters in the story.

Henceforth, it can be concluded that the use of local culture-based picture storybooks for TEYL plays as important teaching tools to enhance students' reading competence through pleasant learning environment with familiar context of learning.

3.3 The steps of developing local culture-based picture storybooks

There are 5 steps in developing local culture-based story books used in this study [3, 4]. The explanations of the steps are presented as follows.

3.3.1 Needs analysis

Needs analysis is conducted by interviewing 9 teachers of 9 primary schools in 9 districts in Buleleng Regency in order to find out whether or not they use local culture-based picture storybooks in teaching English to young learners. Besides that, observation was also done to see how the teacher commonly teaches English, what strategies and media used, and the tendency of using stories in the teaching English.

3.3.2 Developing product

In developing the product, the researcher follows the criteria of a good picture book such as it contains 32-48 pages, the words are approximately 200-1500, it uses simple plot with one until three main characters, and the book is appropriate for children age around 4-10 years old [2]. Besides that, the researcher also considers the criteria of good story book, namely having a conflict, at least one main character, interesting colored and mood reflecting illustrations, and accuracy and variety of cultural markers in making description (varied skin tones, facial features, body types, hair styles/ textures, clothing, homes, language patterns, and names) [1]. Before developing the prototype of the books, the researcher needs to make a draft in the form of dummy books. It refers to a plan of the product which contains scene, text, and illustration.

3.3.3 Expert Validation

The content validity is carried out by two expert judges who give evaluation on the relevancy of the book based on the combination of good criteria of picture book and story book [1, 2] which covers 8 dimensions. Taking into account, the content validity was measured by using Gregory Formula. Additionally, the expert judges are also provided with comment slot for giving suggestion for revising the product, if any.

3.3.4 Quality assessment by experts and users

The quality assessment by experts and users (teachers) is administered after finishing the content validity of the local culture-based picture storybooks. It is designed using Likert scale containing range 1 to 5 (poor, below average, average, good, and excellent).

3.3.5 Main product revision

This step is conducted after the products, the local culture-based picture storybooks, have been given judgment and comments by experts and users (teachers). Both judgment and comments are used to improve the quality of the contents.

4 Conclusion

Local culture-based picture storybooks are teaching media for young learners which cover local culture stories in the form of printed books. These books consist of four types of folklore, namely myth, fable, folktale, and legend. The use of local culture-based picture storybooks is important since it can enhance young learners' reading skill, create enjoyable learning, and build young learners' English literacy, as well as preserve local culture. In developing local culture-based picture storybooks, there are five steps followed, need analysis, developing, expert validation, quality assessment, and main product revision.

The authors of this paper are very grateful to all teachers who had been very helpful and informative to give information needed for this research. Special appreciation is forwarded to the team of the research who spent his valuable time for the research.

References

1. M. J. Wilcox, K. M. Murphy, C. K. Bacon, S. Thomas, *Improving Language Teaching Practice in Preschool Classroom* (Arizona State University, Arizona, 2001)
2. M. Rahardjo, H. D. Waluyanto, A. Zacky, JTE, **6**, 1 (2006)
3. W. R. Borg, M. Gall, *Educational research: An introduction fourth edition* (Longman, New York & London, 1983)
4. K. Hyland, *Second language writing* (CUP, New York, 2003)
5. J. Harmer, *The practice of English language teaching* (Essex, Pearson Education Ltd., 2007)
6. E. Iftanti, TEFLIN J, **23**, 2 (2012)
7. C. Li, P. Seedhouse, EFL J Quartely, **12**, 2 (2010)
8. C. Mart, ELT, **5**, 5 (2012)
9. S. Wu, The Internet TESL J, **14**, 1 (2008)
10. S. Brunsmeier, A. Kolb, CLELE J, **2** (2017)
11. C. Alwasilah, TEFLIN J, **17,** 1 (2006)
12. Juhana, AJEEL, **2**, 1 (2014)
13. K. Klein, English Teaching Forum, **43**, 1 (2005)
14. F. Copland, S. Garton, ELT Journal, **68**, 3 (2014).

15. N. M. Ratminingsih, JIP, **18**, 1 (2012)
16. N. M. Ratminingsih, JIP, **20**, 1 (2014)
17. E. Gürsoy, S.C. Korkmaz, ELT Research Journal, **1**, 2 (2012)
18. F. Liu, IES, **3**, 3 (2010)
19. B. E. Cullinan, Res. J. AASL, **3** (2000)
20. R. K. Kavi, S. N. Tackie, K. A. Bugyei, PROFILE ITPD, **12**, 1 (2015)
21. A. Medaille, EBLIP, **7**, 3 (2012)
22. L. Moses, Language and Literacy, **17**, 3 (2015)
23. J. A. Balshek, ARECLS, **3** (2006)
24. I. M. S. Dharmayuda, *Kebudayaan Bali: Pra Hindu, masa Hindu, dan pasca Hindu* (CV Kayumas Agung, Denpasar, 1995)
25. S. C. Barfield, J. Uzarski, ETF, **1**, 1 (2009)
26. Y. S. Yembise, IJELT, **6**, 2 (2010)
27. J. R. A. Tonelli, H.O. Lima, IJHSS, **3**, 19 (2013)

Enhancing the EFl learners' speaking skill through folktales based instruction

Ida Bagus Nyoman Mantra[1,*], *Wayan* Maba[1]

[1]English Education Study Program, Mahasaraswati University, Denpasar, Bali, Indonesia

Abstract. This present research was conducted to improve the EFL learners' speaking skill through the implementation of Folktales Based Instruction. Consequently this study was intended to explore the valuable usefulness of folktales based instruction implemented intensively in EFL classroom. The subject of the present study consisted of 30 EFL adult learners. The study was conducted in two consecutive cycles in which two learning sessions were carried out for each cycle. The teaching sessions were strongly focused on the learners' active engagement in various communicative classroom activities to ensure that the learners gain higher level of speaking ability. The research result showed that the learners' speaking skill improved significantly in the second cycle compared to the first cycle. The questionnaire result also showed that the learners strongly enjoyed learning speaking through the integration of folktales based instruction.Therefore it is recommended that teachers should consider utilizing folktales based instruction in teaching speaking skill.
Key words: EFL Learners, Speaking, skill, Folktales, Instruction

1 Introduction

Globalization requires students to overstep the boundaries in communication by utilizing various forms of comunication [2], [24, 33]. Having the ability to communicative productively is essentially important for the students [14], [16,28] to be able to convey their ideas, thought, opniion and other matters that they have in their mind [24], [2,4]. Therefore, the students'communication skill should be continually developed by enggaging the students in more contextual speaking activities. The speaking activities should be designed in order to bridge the students to develop their communication skill in line with the context and the purpose of communication [26], [2, 18].

Many researchers have made a great deal of efforts to find out the magic strategies to improve the students' speaking skill, for examples studies conducted by [19], [10, 16, 29], which had contributed several ways of teaching speaking. However, it seems too difficult to get the best strategy for every learning situation, therefore the search for the best way of teaching is continually carried out by the language education researchers. This is done because the students' speaking skill is very important to be improved in order that they can communicate effectively [14, 17]. The necessity of improving speaking skill also

* Corresponding author: bagusmantra@unmas.ac.id

suggested by [28], that having speaking skill is a success key to effective classroom communication.

Looking at the challenges above, an appropriate strategy need to be applied to involve the learners in the speaking learning activities, it is very important to exercise a study to find out more effective way of teaching speaking. In this study, the researcher proposed the use of folktales to improve the students' speaking skill. Folktales have characteristics that have the potential to facilitate the improvement of students' speaking skill. Moreover, the use of local folktales may also develop the students' knowledge and understandings of local culture related to everyday life, customs and norms. Having a good understanding of local culture is very valuable for the students. They can learn many aspects of life which may become useful for their life in the community [22].

Indonesia archipelago is very rich in folktales as thousands of island in Indonesia have their own folktales [11, 15]. Those folktales of course can be utilized to trigger the students' speaking motivation [22, 23]. Several reports of the use of folktales in English language teaching which have significant impacts towards the improvement of the students' learning achievement were conducted by [6, 9, 21, 23]. Moreover, [3] claims that the use of authentic material in teaching English certainly helps communicative learning activities implementation in classroom practices. [7] revealed that folktales help students to use vocabulary in context and language forms by engaging the students in story retelling during language learning in the classroom. Moreover, [30] claims that folktales can promote the ability in using the grammatical structures. Moreover, Folktales are certainly valuable benefits in their own right as folktales, as literature, as social and cultural expression, and as moral teaching [20, 22, 25].

Although there were many studies that have been conducted in regards to the use of folktales in teaching and other perspectives of folktales from cultural-historical aspects [e.g. see 5, 8, 12], sociological and psychological aspects [e.g. see 13, 27, 34], reflecting the wide-ranging and multidisciplinary interest in them. However, in the field of language teaching, the richness and potential of folktales seems to have been under-utilized in today's language classes [1, 32]. In this paper, therefore the usefulness of folktales in teaching speaking become primarily focus of the discussion. The main reasons of conducting this study was due to the facts that for language learners, speaking is an important language skill and folktale as one local cultural identity is an essential aspect of the students' life which they need to comprehensively understand for their sustainable living.

2 Method

2.1 Research Design

Considering the important usefulness of classroom action research, therefore classroom action research design was chosen as research method for this study. There were two teaching cycles undertaken consisting of four interconnected activities such as: planning, action, observation, and reflection. By employing this research design, the researcher was able to do step by step of scientific investigation.

2.2 Participants

The subject of the study was the EFL students of English Study program. There were 30 students chosen as the participants of this study. They were chosen as the subject of the study because after doing interviews and initial reflection using speaking pre-test, the result showed the students still had difficulty in conveying their ideas, thought, opinion and information to other people. The initial reflection indicates that immediate and appropriate teaching learning processes need to be carried out to solve the students' problem.

2.3 Measures

The data obtained from the pre test and post tests were analyzed using Mean score formula in which the total score of the students was divided by the total number of the students, meanwhile, the questioanire was analyzed using rating scale 1 to 4 to figure out the students'changes behaviour and motivation during the learning cycle took place.

2.4 Procedure and Data Analysis

The data were collected by administering pre test, post test and questionnaire. Pre test was conducted before the treatment was given, while two post tests were conducted after treatments were given in every end of each cycle. Menwhile, a questionare was given at the end of teaching cycle. In every teaching cycle, there were three main indicators as the main concern related to the learners' speaking skill, namely fluency, understandability, and accuracy. Those three indicators were measured using scoring rubrics. The students were asked to perform their speaking through retelling a folktale, while the teacher immediately after the presentation asked questions concerning the students' opinion, ideas, feeling, or experiences related to the folktales.

3 Findings

The findings of the study can be tabulated as the following.

Table 1. Pre test and Post Test Result

	Test	Mean score
Initiaal reflection	Pre Test	45.22
Reflection one in cycle one	Post Test One	78.26
Reflection two in cycle two	Post Test Two	85.22

Refering to the table, the mean score of post test one and post test two were much higher than the mean score of the pre test.

Table 2. Result of the Questioanire

Option	Percentage	Qualification
Option A	50 %	Strongly agreed
Option B	40 %	Agreed
Option C	10 %	Undecided
Option D	0 %	Disagreed

Refering to the table, the prencentage of the students who were strongly agreed and agreed were much higher than those who were undecided and diagreed.

4 Results and Discussion

All of data needed for this study were collected in the classroom by using three instruments administered to the students of the study, namely: pre-test, post-test, and questionnaire. The pre-test was administered to the students to obtain their pre-existing ability in speaking skill. The test was in the form of foktales retelling where the students were asked to retell the folktales they have chosen to present the folktale interactively [1, 10, 21]. Post-test was administered once in each cycle to measure the students' progress in speaking skill. There were three sets of raw scores obtained for the present study, those were pre-test, post test and questionire scores.

Pre-cycle was the first step in collecting the data. The researcher made an initial reflection in which the researcher interviewed the lecturers who taught Speaking skill in order to know the problems faced by the students in learning speaking. The result in initial reflection showed that the students had low motivation in learning speaking skill. Additionally, pre test mean score showing 45.22 which indicated that the students had very low ability in speaking. The result indicated that the students' speaking skill urgently needed improving [14], [15, 18, 21]. Therefore, the first cycle was conducted based on the previously made lesson plans.

While doing the teaching action in the first cycle, the researcher also observed the students' active involvement in the learning acivities. The observation was mainly focused on identifying the students' responses toward the application of the folktales as strategy in teaching speaking skill [14, 18, 21]. To collect the data needed for the present study, the post-test was administered to the students. The students were asked to retell the folktales they have chosen in front of the class [14, 16]. The students' speaking ability improved significantly which indicated by the mean figure of the first cycle which reached 78.26.

Considering the significant improvement of the students' speaking skill, the second cycle was started with rivising the lesson plans. The learning activities for the second cycles were completed with the use of pictures and multi media images as well more challeging speaking exercises [14], [17, 18]. The learning activities were conducted based on the learning activities using folktales as proposed by [32]. He suggests that the teaching of speaking through folktales in the simple form can be carried out using the procedures as the following, such as: (1) asking the students to tell a folktale, (2) asking the the students write it down, correcting the grammatical mistakes, (3) returning the written folktales back to the students and asking the students to read it, (4) asking the students revise and expanding their written folktales, (5) practicing the students to comprehend the vocabulary used in the written folktales, (6) asking the students retell the folktales in front of the class, while the others are listening, (7) asking the students to share their correction, (8) discussing the moral value of the folktales. This procedure helps the students involve in learning activity which also develop their other language skills [18], [22, 32].

In cycle two, it was observed that the students were more active in learning speaking. They found that learning speaking through folktales were interesting and challeging. The students' poitive attitude towards the use of folktales based instruction made their learning motivation improved [1, 18, 22, 32], consequently, their speaking ability was also improved. This can be clearly seen fromthe result of the post test two which showed mean figure of 85.22. This result showed that a better improvement was achieved by the students in comparison to this first cycle. Additionally, the questionnaire result showed that there were 50 % of the subjects who strongly agreed, 40 % of the subjects agreed, 10 % of the subjects neither agreed nor disagreed agreed and none of the subjects disagreed on the use of folktales in learning speaking. This questiionaire score indicated that the students really enjoyed in learning speaking through folktales based instruction [20, 21].

In conclusion, Folktales based instruction can be used to improve the students' speaking skill and the students' motivation in learning speaking. Folktales can be used to educate, inspire and enggage the students in various speaking activities and additionally folktales enrich the students' local cultural awareness. This research is expected to contribute to widen the horizon of the English teachers in developing more effective learning activities for the sake of the students' learning progress and to give ideas for the teachers to be more creative in teaching speaking.

References

1. Bean, M., The role of traditional stories in language teaching and learning. In M. R. MacDonald (Ed.), *Traditional storytelling today: An international sourcebook* (London: Fitzroy Dearborn,1999).
2. Cameron, L., *Teaching Language to Young Learners, (*New York: Cambridge University,2001).
3. Carter, R. The new grammar teaching. In R Carter (Ed.), *Knowledge about language and the curriculum:* (The LINC reader. London: Hodder & Stoughton, 1990).
4. Chen, Y.-M. Asian EFL Journal, **8(4)**, 210-232, (2006).
5. Cubitt, C. Folklore and historiography: Oral stories and the writing of Anglo-Saxon history. In E. M. Tyler, & R. Balzaretti (Eds.), *Narrative and history in the Early Medieval West* (Turnhout: Brepols Publishers, 2006).
6. Dardjowidjojo, S. Journal of Southeast Asian Education, **2(2)**, 309-322, (2001).
7. Deacon, B. & Murphey, T, *English teaching forum*, **39 (4)**:10-15, 23, (2001).
8. Dundes, A. Worldview in folk narrative. In S. J. Bronner (Ed.), *The meaning offolklore: The analytical essays of Alan Dundes*, (Logan: Utah State University Press, 2007).
9. Elley, W.B., New zealand Journal of Educational Studies, **15**, 39-53, (1980).
10. Farida, L. A., & Sofwan, Lembaran Ilmu Pendidikan, **41(1)**, (2012).
11. Ghosn, I. K, ELT Journal, **6(2)**, (2002).
12. Gottschall, J., Martin, J., Quish, H., & Rea, J. Evolution and Human Behavior, **25(2)**, 102-112, (2004).
13. Haase, D., Haase,Reception, 230-249, (1993).
14. Harmer, J. *The Practice of English Language Teaching*. (Cambridge: Longman, 2007).
15. Hawanti, S. IJPL, **9(2),** (2014).
16. Hayriye, K., The Internet TESL Journal, **(12)11**, (2006).
17. Koşar, G., & Bedir, H. IJLA, **2(3)**, (2014).
18. Kramsch, C. e-FLT, **11(2)**, 249-254, (2014).
19. Krish, P. J. The Internet TESL Journal, **(7)7**, (2001).
20. Lamb, M. TESOL Quarterly, **41(4)**, 757-780, (2007).
21. Lwin, S.M. *The English Teacher,* **44(2)**, 74-83, (2015).
22. Mantra, et al, IJLLC, **2(2)**, (2016).
23. Muslim, A. B., Nafisah, N., & Damayanti, I. L. Locality and self identity. *Indonesia University of Education (UPI,* 2009).
24. Natasa, I K, The Internet TESL Journal, **12(11)**, (2006).

25. Norton, B., TESL Canada Journal, **28(1)**, (2010).
26. Pollard. L., *Teaching English: A book to Help You through Your First TwoYears in Teaching*., (London: Cambridge University Press, 2008).
27. Ragan, K. Marvels &Tales, **23(2),** 227-247, (2009).
28. Richard, J. C., & Renandya, W., *Methodology in Language Teaching: An Anthology of Current Practice*, (Cambridge: Cambridge University Press, 2002).
29. Shahini, G., & Riazi, A. M. A PBLT Approach to Teaching ESL Speaking, Writing, and Thinking Skills. Retrieved April 15, 2017, from http://eltj.oxfordjournals.org/content/65/2/170.
30. Spiro, J., IATEFL literature and cultural studies, **20**, (2000).
31. Sri Widiastuti, The ASIAN EFL Journal, **3**, (2016).
32. Taylor, Eric K. *Using Folktales*, (Cambridge: Cambridge University Press, 2006).
33. Thornbury, S. *How to Teach Speaking*, (USA: Pearson Longman, 2005).
34. Westland, E. Gender and Education, **5(3)**, 237-249, (1993).

Literacy and reward: teachers' effort to build children reading habit

Komang Widia Helena Arisandi[1,*], *Ni Nyoman* Padmadewi[1], *Luh Putu* Artini[1]

[1]English Education Department, Universitas Pendidikan Ganesha, 81116 Singaraja-Bali, Indonesia

Abstract. Children at early levels of primary school require appropriate guidance in their initial reading skill. They need to be trained on how reading becomes an enjoyable routine activity. This study aimed at describing teachers' effort to build children reading habit. This study employed a qualitative descriptive study and conducted at North Bali Bilingual School Bali. The data were collected through observations and interview. The findings of the study showed that there were several activities conducted by the teacher as efforts to build children reading habit. In terms of building students' reading habit, the teacher used (1) Point-written in Reading Rocket Chart (PRRC), (2) Chip (white, yellow, green) for appreciating good behaviour in reading and using English, (3) Certificate, (4) Class Reward, and (5) Free Play Time. With these efforts, it is evident that the students' literacy improves and they exhibited great enthusiasm in their reading and studying literacy in the classroom.
Keywords: Literacy, Reward, Reading Habit

1 Introduction

Literacy is the foundation which is very important in children's life. Literacy is said as the key to enhance human capabilities which involved the ability in reading and writing [1, 2]. The intention of people write and read is to construct meaning [3-5]. According to [6] in English literacy, students are not only expected to be able to read, but also afford to communicate fluently, independently, and effectively in the society.

National phenomenon today shows that the literacy skills of Indonesian children are very low compared with children in other countries. It will be related to the role of parent as well as teacher. Based on the result of study conducted by [7], English teachers need to increase their knowledge and experience in understanding the concept for using the strategies to teach reading. They should have awareness to the significance in choosing the appropriate strategies for teaching reading. Besides, the results of PISA (the Programme for International Student Assessment) are less satisfactory. In 2015, the average value of reading is 397 in the rank of 62nd of 70 countries that took part in the PISA 2015 [8].

Reading allows children to learn and gain knowledge which is an important gateway to personal development [9]. It is capable in increasing and improving children insight in their early progress. Accordingly, the real efforts how to introduce literacy especially in reading

* Corresponding author: elen_arisandi@rocketmail.com

skill is very necessary. Several attempts have been made nationally, such as: building a library in every school and even the government always provide assistance in the form of books. However these solutions do not have the maximum impact. Reading is still at the level of carrying out the instruction. Most of the students do not have a character in case of loving to read. Most of them perform reading activities for the teacher's instructions, because the demands of the task, the repetition or exams, and have not shown their initiative intrinsically or because of his own desire to go to the library to read. In this regard, it extremely needs a teacher role. This research aimed at describing teachers' effort to build children reading habit. For this purpose, whether there is a significant effort that has to be done by the teacher in terms of students' reading habit.

2 Method

2.1 Subject of Study

The subject of this study was the students of North Bali Bilingual School. The students in grade 1, 2 and 3 were selected as the subject of this study. Besides, the teacher and headmaster were also as the subjects who were observed by the researcher. They were also as informants in gaining the data through interviewing.

2.2 Instruments

In this study, there were several types of instruments used by the researcher in collecting data. They are human instrument, observation checklist, and interview guide. Human instrument was used as the key instrument of qualitative research. The researcher used checklist during the observation to write down the interesting and important information during the process of collecting data. After conducting classroom observation, the interview was conducted by the researcher toward the learning process. Besides, the interview was used to investigate kinds of rewards used in which may influence students' reading habit.

2.3 Procedures of data collection

In eliciting the objectives of this research, the procedures of data collection were designed. The first step was done by observing the students in grade 1, 2 and 3 in North Bali Bilingual School. To explore back the result of observation, the interview was carried out as well. It aimed at eliciting related information from the teacher and also the headmaster about how reading activity was done by the students properly. Besides, the teacher's teaching method and guiding the students were much needed which may bring influences on their reading habit. Then, the data regarding to the teacher effort by giving rewards to build students' reading habit was gained from the observation and interview.

2.4 Data analysis

The data were analyzed to discuss the teacher's effort to build children reading habit through rewards. This study employed a qualitative descriptive study and conducted at North Bali Bilingual School Bali. [10] suggest that qualitative data analysis consists of three procedures; Data Reduction, Data Display and Conclusion Drawing/Verification.

3 Results and Discussion

The findings of the study showed that there were several activities conducted by the teacher as efforts to build children reading habit. In terms of building students' reading habit, the teacher used (1) Point-written in Reading Rocket Chart (PRRC), (2) Chip (white, yellow, green) for appreciating good behaviour in reading and using English, (3) Certificate, (4) Class Reward, and (5) Free Play Time.

Table 1. Reward Application and Students' Response toward Reading Rewards

Name of Rewards	When Reward is Given	The Reward	Students' Response
Point-written in Reading Rocket Chart (PRRC)	- 10 minutes reading = 1 point - 21 minutes reading = 3 points	- Enjoying in the reading corner	- Students were excited in achieving points by reading every day
Chip (white, yellow, green)	- White = using English - Yellow = good attitude in society - Green = enthusiastic in reading and writing	- Chip collection	- Students were interested in collecting the chips by doing a positive behaviours in school
Certificate	- 15 minutes reading routinely = super reader certificate	- Super Reader Certificate	- Students had passion in possessing the certificate by striving to read routinely and make reading as their habit
Class Reward	- 500 points in a week = class reward	- 10 minutes free time for playing	- All students in a class showed their cooperation in collecting and achieving the points by reading every day
Free Play Time	- 500 points or more in a week (accordance to the teacher students agreement) = free play time	- Playing in classroom, playground, and creativity room. - Free time for watching film	- Students were very enthusiastic in following the agreement, and making reading as their hobby

3.1 Point-written in Reading Rocket Chart (PRRC)

The Point-written in Reading Rocket Chart is one of the rewards given in North Bali Bilingual School in the early level of primary students. This reward is given when the students have read. Every 10 minutes reading is appreciated one point. The students who have not read, they give no point. Awarding the point depends on each teacher in NBBS. Every class can determine reading point differently. According to one of the teachers in that school, she checks students' point in every week. There are 5 rules in reading that she and her students mutually agreed. 1 – 10 minutes reading is counted one point. 11 – 20 minutes is counted two points. For 21 minutes of reading are counted three points. The students, who gain 3 points in every day, will have 5 points in a week. Every point will be collected and placed in the *house* owned by the student. The student who is able to be a super reader of the month (full reading by gaining 3 points for a month) is given gift by the teacher. Usually, the gift is enjoying in the reading corner.

3.2 Chip (white, blue, and green) for Appreciating Good behaviour in Reading and Using English

For the students in the early level, Chip is given by the teacher as a reward. The use of chip is used for appreciating the students' behaviours in reading. There are three colours of chip that provided by the teacher, such as: white, blue, and green. The white chip is used for motivating the students in using English. The blue chip is used for students' attitude in their society inside or outside classroom. Meanwhile, the green chip is used for the students' learning attitude. The students who are enthusiastic in reading and writing practice will be given the green chip.

3.3 Certificate

The other reward is certificate. According to[11], the use of tangible reward like certificate is often utilized to increase reading. In NBBS, certificate is usually given for the super reader. Super reader is the name for the students who reading more than 15 minutes in every day throughout holiday. They will obtain 3 points in every reading. If the students read more than 15 minutes routinely, they will be given the Super Reader Certificate. It is given after long holiday in the middle of semester or the end of semester.

A study was conducted by [12] about rewards for reading in the level of elementary students. Based on the result on their research, after they control for pre-reward reading motivation, sex, and grade, and combined with the reward variables (intangible and tangible rewards (prizes of token, certificate, food), reward expectancy) significantly improved the prediction of post-reward intrinsic reading motivation, extrinsic reading motivation, and global reading motivation. From the result, it can be said that the use of certificate as a reward gives positive influence for students' reading habit.

3.4 Class Reward

The Class Reward is the reward given by the teacher who in charge in the class. It depends on the teacher what kind of rewards would be given to the students. Usually, the teacher gives *free time* as the reward. Every class has to have 500 points in a week. If the class successes in achieving the point, the students will obtain the *free time* as the reward from the teacher. The *free time* is usually given for 10 minutes for playing in the playground.

This reward has been applied by the teacher in NBBS. Teaching reading needs a big teacher's effort which cannot be separated from the technique used in the teaching process. A study was conducted by [6] in terms of technique of teaching reading for boosting English literacy for those who in the first grade of elementary school. The result has showed that there are two major reading techniques implemented such as phonic approach and the whole language approach in order to improve the students' English literacy. The two techniques are complimenting each other and found to be effective to improve the students' literacy not only in reading but also in speaking. The students are always happy and cheerful in the classroom because the approaches implemented bring joyful learning in the classroom.

3.5 Free Play Time

Free Play Time is a reward given by teacher for the students who gain the point of reading which accordance with teacher-students agreement. This reward can be the additional time for playing in the classroom, playground or in the creativity room. It also can be a time for watching film. All of the free time are given depend on the teacher creativity.

4 Conclusion

Motivating the students to read leads to success outside of the classroom. The rewards above is given depend on the teacher who in charge in the class. The role of teacher can support students' reading habit by creating positive efforts, such as using point-written in reading rocket chart for appreciating the students in reading, using chip, certificate, class reward, and free play time. With these efforts, it is evident that the students' literacy improves and they exhibited great enthusiasm in their reading.

References

1. N. Burnet. *Education for All: Literacy for Life*. (Unescho Publishing, Paris, 2005)
2. T. Haryanti. Jambore Taman Bacaan Masayarakat. (2004)
3. G. Owocki. *Make Way for Literacy*. (Heinemann Porthsmouth, NH. Washington DC, 2001)
4. L. Morrow. *Developing Literacy in Preschool*. (The Guilford Prss, London, 2007)
5. SB. Neuman, KA. Roskos. *Language and Literacy Learning in the Early Years. An Integrated Approach*. (Harcout Brace Jovanovich College Publishers, United States America, 1993)
6. NN. Padmadewi, PNP. Sari, DAE. Agustini. JPBI. **1**, 1 (2013)
7. LP. Artini, K. Seken, K. Ahmad. JPBI. **1**, 1 (2013)
8. Indonesia PISA Centre. *Hasil PISA 2012*.
9. C. Clark, K. Rumbold. Nat. Li.t Trst. 5 (2006)
10. MB. Milles, AM. Huberman. *Qualitative Data Analysis: Second Edition*. (SAGE, London, 1994)
11. M. Anthuis. The Effect of Extrinsic Rewards and Intrinsic Motivation on Student Reading and Learning. **11**. (2014)
12. PH. Chen, JR. Wu. JIP. **4** (2010)

Designing task-based syllabus for writing class

Hanna Sundari[1,*], *Rina* Husnaini Febriyanti[1], *Gustaman* Saragih[1]

[1]English Education Department, UniversitasIndraprasta PGRI, Jagakarsa-Jakarta Selatan, Indonesia

Abstract. Writing is viewed as the most complex skill to learn and to teach. Beside learner factors, teacher, materials and syllabus may also affect the process of learning language as foreign language. Syllabus, in general, can be defined as a set of what is taught (content) and the way it is taught (procedure. This current research aims to design a task-based syllabus for writing class at university level. This study was conducted by qualitative descriptive design with 92 students and 4 lecturers as respondents. As part of research and development project in one private university in Jakarta, a developed task-based syllabus was based on need analysis and the principles of task-based language teaching. Students' proficiency levels are fair with sentence patterns and grammar as the most difficult aspects. Academic writing is more preferable orientation with the small portions of creative writing. Then, the developed task-based syllabus has been proposed for writing class which covers the components of goal (learning outcome), course description and objectives, a set of writing tasks, features of content focus and language focus and course evaluation. The developed syllabus, then, can guide the lecturers in designing lesson plan and selecting materials for writing class.

Keywords: syllabus design, task-based language, writing skill

1 Introduction

Writing, as one of productive language skills, is communicating messages in written symbols involving other micro skills [1]. Writing activity purposes to express ideas to the readers. In regards to writing in second/foreign language, Hedge formulated the competences achieved by language learners, including the use of rhetorical features in organizing information, the use of technical aspects and accuracy to avoid ambiguity, the use of accurate grammar and vocabulary, and competence in styles to meet the audiences context [2]. From a set of competences mentioned, teaching writing class, particularly in foreign language context, seems challenging and effortful. For these reasons, writing class should be well-planned and prepared by course designers.

Related to preparation, course planning has seemingly become an essential factor in teaching process. Designing course includes several items to do, such as developing syllabus, planning lesson, selecting materials, and preparing tests. Concerning to syllabus,

*Correspondingauthor:hanna.sundari@gmail.com/hanna.sundari@unindra.ac.id

Nation [3] wrote that syllabus, as a part of curriculum, includes learning goals, content and sequence of materials, layout, and evaluation. Moreover, it describes systematic objectives and materials to be taught and achieved in a certain format. Syllabus also focuses on selected and specific contents [4]. For more practical definition, Ur [5] said that the characteristics of syllabus are a set of public document containing a list of systematic contents and process, with explicit learning goals, time, methods, and materials. According to Candlin, syllabus, different from curriculum, is more narrowly localized and based on what actually happens at the classroom level as teachers and learners apply [6]. Syllabus has function as "navigation mode" for both teacher and learners. In short, syllabus can be defined as a set of what is taught (content) and the way it is taught (procedure). Therefore, for the sake of effectiveness and success of teaching and learning process, designing syllabus is inevitable and unquestioned.

In designing language course syllabus,Krahnke has proposed several types of syllabi, such as structural-based, notional/functional-based, situational-based, skill-based, content-based, and task-based syllabus [6]. The latter refers to a syllabus based on tasks given to the learners in order to perform tasks in target language [7]. Task-based syllabus is developed based on the principles of Task-based language teaching (TBLT). According to Feez, TBLT focuses on process rather than products in which learners engage and interact communicatively in activities and tasks [8]. Then, the tasks are gradually sequenced based on difficulty level. The given tasks facilitate learners to use language and promote meaningful communication.

Furthermore, in the notion of TBLT, task is variously defined among the experts. In very general term, Longsaid that a task is a piece of work undertaken for oneself or others, freely or for some rewards, such as painting a fence and dressing a child. This definition implies that task is seemingly a set of activity [9] in everyday life. From pedagogical view, Skehan wrote that tasks are activities in which meaning is as primary focus and resemblance to real-life language use [8]. In linguistic perspective, Ellisdefines task as a workplan requiring learners' pragmatic language process and linguistic resources in achieving an outcome either productive or receptive language activities [10].Then, Nunan summed up that task is a piece of classroom work involving comprehending, manipulating, producing or interacting in target language. In the other words, tasks refer to activities in the classroom involving communicative language use in which focuses on meaning by using learner's own linguistic resources to achieve language outcome related to real-life language use. Concerning to the syllabus, a TBLT syllabus designs a set of the tasks performed by learners within program. Moreover, Nunan suggested the syllabus includes two types of tasks: real-world tasks and pedagogical tasks [8]. The real-world tasks refer to an unrestricted, holistic everyday activities found everywhere; meanwhile, pedagogical tasks are restricted assigned tasks to achieve specific goals of learning [11]. Accordingly, real-world tasks can be as the source and model for pedagogical tasks as long as they are eligible and purposeful.

Related to research and effectiveness, the TBLT has attained extensive attention in the field of language teaching. In reading skill, studies by Mao [9] and Kumara, Padmadewi, and Suarnajaya [12] have proved the effective application of TBLT in reading comprehension. Moreover, Widayanti's [13] research and Kusnawati's [14] study found that writing ability in understanding writing process and writing skill has been increased through action research using TBLT. From Malaysia, study by Ahmed, Jamilah and Bidin [15] has also proved that TBLT gives significant effect on writing skills for university students. Then, comparative study by Tabar and Alavi [16] showed that Task-based writing provides learners with different opportunities for planning prior to performing written tasks of different levels of cognitive complexity. However, the study related to develop a task-based syllabus as part of course program particularly for writing class in university context may be still limited. Indeed, in the matter of English language teaching system at university level, institutions have large opportunity and autonomy to design the system [17]. For these reasons, the current study aims at developing a task-based syllabus for writing class at university level.

2 Methode

The purpose of the current study is to design task-based syllabus for writing class at university particularly in English education major. The research was conducted by qualitative descriptive research design as a part of research and development for writing course design at one private university in Jakarta. The used instruments were questionnaire, worksheets and interviews by taking 92 university students and 4 lecturers in the stage of need assessment as well as expert reviews for the first draft developed-syllabus. The collected data in interviews were analysed qualitatively using themes and categories; meanwhile, descriptive data using percentage were taken from questionnaire. The writing worksheets were evaluated and scored based on the scoring rubric for writing essay by Alice Oshima[18].

In designing syllabus, the researchers followed the steps in design and development of task-based syllabus by Long and Crookes[16], as follow: 1) a comprehensive need analysis, 2) diagnosis of learner need, 3) identification of target tasks learners are preparing to carry out, 4) classification of target tasks into task types, 5) development/selection of pedagogical tasks for classroom use, 6) sequencing of pedagogical tasks to form a task-based syllabus. Besides, the need analysis applied the Graves's stages, including: 1) deciding what information to gather, when, from whom, and how, 2) collecting information, 3) interpreting the information and making decisions[19].

3 Results and discussion

This present research purposes to design a task-based syllabus for writing class at university level. After analysing the collected data and conducting the steps of designing syllabus, the findings cover two main points: the results of need analysis and a developed task-based syllabus.

From the rating of self-assessment, the students' levels of writing are categorized as fair. It means that they believe that they are able to write in English; even thought, they also realize they may fall in to errors, ineffectiveness and inaccuracy of writing. This result support the survey analysis by Floris in which revealed that 51.4% of the respondents thought they had fair in language level as intermediate level learners [20]. Moreover, developing sentences in correct patterns and using grammar seem to be the most difficult writing practices. Composing sentences in the forms of simple, compound and complex looks quite challenging. This may be consistent to what Murray and Christison [2] said that "there is no evidence that simple sentences are easier to understand than compound and complex ones". In contrary, writing mechanism, using conjunctions, and developing topics and ideas are considered as the easiest practices. This indicates that the students have a lot of ideas to write; however, they feel hard to express them in appropriate grammar patterns.

Some interesting results come up from students' writing worksheet. They were assigned to write one essay in descriptive or narrative types. Most of them preferred descriptive writing. The topics they chose were various and attractive, such as my favourite food, my future husband, and why I like Gado-gado. Then, the ideas are fairly organized in the form of 5-paragraph essay. Nonetheless, the thesis statements are quite unclear and ineffective to depict the whole essay. Still, some lacks of S-V agreement, sentence patterns, tenses, conjunction, word selection and writing mechanism appear here and there. This fact may support the study by Hidayahin evaluating writing essay by TESL college students whose errors were mostly on S-V agreement, tenses, part of speech and vocabulary [21]. Moreover, concerning to activities and practices, individual work is more preferable than pair and group work. This result may be different from the study by Tutyandari [17]. She found that pair or group work was useful to encourage passive learners in writing class.

From the perspectives of importance and orientation of writing, students and lecturers agree that all micro skills mentioned are important in writing class. Nevertheless, the lecturers think that creative writing, such as writing poem, poetry and drama scripts is less important. Meanwhile, the writing orientation is addressedto academic writing with the small portion of other orientations to support academic achievement.Besides, the targeted competences in the course should be relevant and adaptive to the real-world language use. Related to the topic, local contents and values can also be included for instructional materials. This opinions support the study by Gunantar which showed that local cultural content has become the purpose of EFL teaching, combined with foreign culture [22].Moreover, materials and activities are gradually classified from simple to complex. Based on these gathered data of need analysis, the researchers then made some decisions dealt with the learner needs and identification and classification of tasks set on the developed task-based syllabus.

The components on the developed syllabus covers several items, such as goal of course (learning outcomes), course description and objectives, both general and specific, a set of writing tasks, features of content focus and language focus and course evaluation. These components are partially derived from the task components by Nunan [23][9]. Tasks requires specific components: goals, input data, activities and settings (classroom modes).The outcome of the course is to develop students' writing skills in the terms of producing various written texts with clear, appropriate, and effective logical structure in the forms of sentences, paragraphs, and essays creatively and confidently.

After that, writing course is then levelled into three categories: basic, independent, and proficient users.Basic level focuses on writing a series of sentences with simple connectors about personal/social life and around. Meanwhile, independent level practices to write a clear detailed text in a linear sequence related to subject field of interest in the form of paragraphs. At proficient level, students can write clear well-structured texts of complex subjects in the form of essays to support academic achievement. The levelling and its objectivesare quite determined by curriculum plan, learning outcomes and graduate profile constructed by university.

Afterward,tasks are identified, classified and sequenced according to the pedagogical objectives that have been formulated and the real-world tasksexist. The tasks are stated in "-ing" form to express activities the students will actually perform in the class room as well as skills they have to achieve in the real-world language use.For example, giving comments and captions for photos/pictures/situations sounds task that people do in real world. Then, the pedagogical task in the classroom become writing caption and comment for photos/pictures/situations, with specific language goal. Therefore, the tasks students perform in the classroom are formulated to use the language both inside and outside the class.This may be consistent to what Skehan stated that one of the key characteristics of a task is there is some sort of relationship to comparable real-world activities [10].

The other components of syllabus are content focus and language focus. Content focus are materials discussed during the course in order to achieve the desired goals. Furthermore, the language focus are linguistic features discussed and practised to support students completing the tasks. Putting the grammatical linguistic features on the syllabus may support of what Nunan said. He wrote that, despite the fact that tasks involve communicative language use focusing on meaning rather than grammar, the form is still important. Meaning and form are highly interrelated. He also added that the grammatical knowledge is to express meaning and to enable the language user to express different communicative meanings [10].

The classroom setting is one of task specific components[239]. In the developed syllabus, the classroom settings are not explicitly stated as the component of the syllabus. However, the modes can bevariously applied in the settings of individual work, pair work, group work, and the whole class mode. The classroom mode is selected by considering several factors, such as levels, class size, the task difficulty and time allocation. This purposes to give a plenty room for syllabus user to make adjustment and adaptation to the class situation.

At the course evaluation, process assessment is adopted both during the lesson and at the end of it. Nunan wrote that task should have a sense of completeness and stand alone as communicative act [10]. The completeness can be seen and measured when the students perform the assigned task completely in the class in the terms of fluency, accuracy and meanings mostly for written work. The evaluation modes consist of task performance, participation (attendance), portfolio, and examinations.

4 Conclusion

The present study aims at designing a task-based syllabus for writing class at university level. After getting the learner needs, orientation and importance of writing course, the developed syllabus was designed based on the principles of task-based language teaching. It covers several components, such as goal of course (learning outcomes), course description and objectives, a set of writing tasks, features of content focus and language focus and course evaluation. Then, the teachers/lectures can use this syllabus for their writing courses. Moreover, it can guide them to design lesson plan and to find suitable and appropriate materials. Just like a map or navigation mode, syllabus gives direction and leads the user to the destination. However, not everything is explicitly stated in it. Adjustment and adaptation are still needed.

This research is supported by Ministry of Research, Technology and Higher Education of Indonesia through the schema of PenelitianProdukTerapan (PPT) 2017 No. 0428/K3/KM/2017. Moreover, we would like to say our gratitude to our colleagues in LPPM Indraprasta PGRI University Jakarta for their assistance and supports.

References

1. Spratt, M., Pulverness, A. & Williams, M., *The TKT Course* (Cambridge University Press, 2011).
2. Murray, D. E. & Christison, M., *What English Language Teachers Need to Know Volume II: Facilitating Learning* (Taylor & Fancis e-Libaray, 2011). doi:10.4324/9780203846292
3. Nation, P. & Macalister, J. *Language Curriculum Design* (Taylor & Fancis e-Libaray, 2010).
4. Brown, J. D., *The elements of language curriculum: A Systematic Approach to Program Development* (Heinle & Heinle Publisher, 1995).
5. Ur, P., *A Course in Language Teaching: Practice and Theory* (Cambridge University Press, 2009)

6. Rahimpour, M. Current trends on syllabus design in foreign language instruction. **2,** 1660–1664 (2010).
7. Richards, J. C., *Curriculum Development in Language Teaching* (Cambridge University Press, 2001)
8. Richards, J. . & Rodgers, T., *Approaches and Methods in Language Teaching. Cambridge University Press* (Cambridge University Press, 2001). doi:10.1016/0346-251X(87)90017-0
9. Mao, Z., Theory Pract. Lang. Stud., **2**, 2430–2438 (2012).
10. Nunan, D., Asian EFL J. Quaterly, **8**, 12–18 (2006).
11. Sanchez, A., IJES, **4**, 39–71 (2004).
12. Kumara, G., Padmadewi, N. & Suarnajaya, J. Pendidik. Bhs. Ingg., **1**, (2013).
13. Widayanti, M. J. Lang. Circ. J. Lang. *Lit.* **6**, 27–39 (2011).
14. Kusnawati,. *Bhs. Sastra,* **14,** 93–108 (2014).
15. Ahmed, R. Z., Jamilah, S. & Bidin, B., Open J. Mod. Linguist.**, 6,** 207–218 (2016).
16. Tabar, N. A. & Alavi, S. M. A Comparative Study of the Effects of Task-Based Writing under Different Pre-Task Planning Conditions on Intermediate EFL Learners ' Written Performance in Personal and Decision-Making Tasks. **5,** 970–978 (2013).
17. Cahyono, B. Y. & Widiati, U., *The Teaching of English as a Foreign Language in Indonesia* (State University of Malang Press, 2011).
18. Oshima, A. & Hogue, A. *Introduction to Academic Writing:Third Edition* (Pearson Education, Inc, 2007).
19. Graves, K., TeacherSource, 308 (2000).
20. Floris, F. D, . *K@Ta,* **10,** 53–62 (2008).
21. Hidayah, A. T., Int. J. Educ. Res., **1,** 1–12 (2013).
22. Gunantar, D. A., Lang. Circ. J. Lang. Lit. **10,** 141–151 (2016).
23. Jeon, I., English Educ, **60**, 87–109 (2005).

Implications of Nursing Clinical Practice to The Student's Spiritual Health

Asthadi Mahendra Bhandesa*, *I G A Puja* Astuti, *Ni Putu* Kartiningsih

Institute of Health Sciences Bali

Abstract. This study aimed to clarify the implications of Nursing Clinical Practice (PLKK) to the spiritual health of STIKES Bali students. This study employed purposive sampling method to determine the number of respondents. To conduct this study, the fourth grade of nursing students were recruited as the sample with total number 136 respondents. A questionnaire about spirituality from World Health Organization (WHO) was used in this study as the instrument. In addition, the data were analysed by using quantitative descriptive technique. The result showed that 50.0% of students had a very good spiritual health, 42.6% had good spiritual health, 6.6% had moderate spiritual health, and 0.7 % had poor spiritual health. It can be interpreted that spiritual health of nursing students of STIKES Bali is good after conducting Nursing Clinical Practice. Thus, this study can be concluded that Nursing Clinical Practice has implication to the ability of students to love themselves and others meaningfully as the evidence of students' spiritual health.

Keywords: Spiritual Health, Nursing Clinical Practices, Nursing Students.

1 INTRODUCTION

Spiritual health in health education is an integrated part to form professional health workers in accordance with international standards. Nurses have an important role in improving the quality of public health. Nurses are required to appear professionally while providing nursing care and able to establish cooperation with various parties to provide services provided comprehensive and can meet basic needs, including the needs of bio, psycho, socio, and spiritual clients [14].

Nursing services provided to the public must meet international quality standards, which can be ensure the safety and the comfort of clients and their families. Nurses are required to perform professionally when providing nursing care and able to establish cooperation with various parties to provide services comprehensively and able meet basic needs, including the needs of bio, psycho, socio and spiritual clients [15].

The health is justly as a pyramid which could divide into three parts: physical, mental and spiritual. A complete health comprised the satisfactory state of physiological, psychological and spiritual needs. The current students enjoy the high quality of material life with the noble clothes and the delicacy, but their spiritual life are vanity [3].

* Corresponding author: Bhandesa_asthadi@stikes-bali.ac.id

As an effort to achieve academic ability, in accordance with the expected competence, the students of STIKES Bali are given the opportunity to Nursing Clinical Practice (PLKK). PLKK is a process by which students practice to improve nursing skills. Through the implementation of the nursing process, students are expected to be able to conduct assessment, planning, implementation and evaluate and do documentation in meeting with bio-psycho-social-spiritual health problems.

Empirically the students make contact with the patient so that in both personality and skills able to develop. As the real cases in practice areas that concern with various aspects of nursing will have implications for students, one of them is in the aspect of spiritual health, whereas students as a prospective nurse as a professional health worker has greater opportunity to provide comprehensive nursing care by helping clients meet their basic holistic needs. Students also have the ability to view clients as bio-psycho-sociocultural and spiritual beings, so indirectly the interaction also affects the personality and skills of students, one of them is in the aspect of spiritual health.

From the philosophical, theoretical, and empirical backgrounds of this study, this research becomes very important. This is because PLKK is one of the processes in nursing education so that students are able to apply knowledge and skills in the form of nursing care as well as one of the factors that can build the spiritual health of the students.

2 METHODS

This study employed descriptive quantitative design. The sample were determined through purposive sampling technique. The sample in this study were the fourth grade of nursing students of bachelor program at Institute of Health Sciences Bali who have done PLKK IV with 136 respondents. The data were collected by using the instruments of spiritual health questionnaires from WHO. Data were then analysed by using quantitative descriptive technique. The category of spiritual health of students after PLKK IV was grouped into four categories, those categories were very good, good, enough and less

3 RESULTS AND DISCUSSION

Nursing Clinical Service and its implication for student's spiritual health was carried out in STIKES Bali and conducted on the students of bachelor of nursing program academic year of 2016/ 2017 who had implemented PLKK IV. From the total population of 206 students, the sample of the study was taken using the sampling formula to obtain a sample of 136 students. Implications of nursing clinical service on the spiritual health of STIKES Bali students were presented with 32 statements.

Table 1. Implications of nursing clinical practice on spiritual health Student Grade IV in Bachelor of Nursing Program

No.	Statements	Frequency(s)					Percentage (%)				
		TA	S	B	SB	SBS	TA	S	B	SB	SBS
1	How much does your spiritual connection help you to overcome your hardest moments ?		1	16	64	55		0,7	11,8	47,1	40,4
2	How much does the influence of your spiritual relationship help you to tolerate stress?		3	31	66	38		2,2	22,8	47,1	27,9

3	How much does the influence of your spiritual relationships help you to understand about the circumstances of those around you?	2	3	47	59	25	1,5	2,2	34,6	43,4	18,4
4	How much does your spiritual connection give you comfort/ tranquility within you?			27	55	54			19,9	40,4	39,7
5	How much have you discovered the meaning of life in your life?		30	57	34	15		22,1	41,9	25,0	11,0
6	How much does someone else give meaning to in your life?		3	51	34	48		2,2	37,5	25,0	35,3
7	How much do you feel the life you live on has a specific purpose?		9	42	53	32		6,6	30,9	39,0	23,5
8	How much do you have reason to survive in this life?	1	2	23	49	61	0,7	1,5	16,9	36,0	44,9
9	How much do you feel the spiritual power within you?		5	38	59	34		3,7	27,9	43,4	25,0
10	How much do you feel in getting spiritual power in your hardest moments?		3	32	55	45		2,2	23,7	40,7	33,3
11	How much trust do you have to contribute to the welfareness of your life?		5	31	65	35		3,7	22,8	47,8	25,7
12	How much confidence do you have in your daily life?		7	28	68	33		5,1	20,6	50,0	24,3
13	How much trust do you have to give strength in everyday life?		6	28	68	34		4,4	20,6	50,0	25,0
14	How much do you feel your spirituality in every beauty within you?		6	35	70	25		4,4	25,7	51,5	18,4
15	How much inspiration do you feel able to help you in your life?		8	36	55	37		5,9	26,5	40,4	27,2
16	How much you be grateful for everything you can enjoy from nature?		5	22	34	75		3,7	16,2	25,0	55,1
17	How is the life expectancy that you feel?	2	1	29	52	52	1,5	0,7	21,3	38,2	38,2
18	How much hope do you have about the life you are living today?			20	64	52			14,7	47,1	38,2
19	How much do you express your admiration for something around you? (i.e. nature, art, music)		2	18	55	61		1,5	13,2	40,4	44,9
20	How much do you feel about connection between mind, body and soul?		3	23	61	49		2,2	16,9	44,9	36,0
21	How much do you feel about your way of life is in line with what you feel and think about?		10	40	66	20		7,4	29,4	48,5	14,7

22	How much trust do you have to help you align what you do, think and feel?		4	38	63	31		2,9	27,9	46,3	22,8
23	How much spiritual power helps you to experience a better life?			19	71	46			14,0	52,2	33,8
24	How much spiritual power helps you to experience happiness in this life?		2	16	75	43		1,5	11,8	55,1	31,6
25	How much do you feel the peace in your heart?		5	30	67	34		3,7	22,1	49,3	25,0
26	How much do you have peace in your heart?		6	30	65	35		4,4	22,1	47,8	25,7
27	How much can you feel for peace in your heart if you need it?		5	35	71	25		3,7	25,7	52,2	18,4
28	How much can you feel the touch of harmony within yourself?		7	34	57	38		5,1	25,0	41,9	27,9
29	How much optimistic are you to develop the quality of life you have today?		2	27	67	40		1,5	19,9	49,3	29,4
30	How capable can you maintain your optimism in the uncertain circumstances you experience?	2	5	31	74	24	1,5	3,7	22,8	54,4	17,6
31	How much trust do you have to help you enjoy life?			26	64	46			19,1	47,1	33,8
32	How satisfied are you, when you find a balance between your mind, body and soul?		1	28	68	39		0,7	20,6	50,0	28,7

Information:
TA : None
S : A little
B : Big
SB : Very Big
SBS : Very Great

The category of spiritual health of the students after the PLKK is categorized into four, that is very good, good, enough and less. The categories are shown in Table 2.

Table 2. Category of spiritual health of fourth grade students of bachelor of nursing after PLKK IV

No.	Category	Frequency(s)	Percentage (%)
1	Very Good	68	50,0
2	Good	58	42,6
3	Enough	9	6,6
4	Less	1	0,7
Total		**136**	**100**

Based on the results of the research above, it could be shown that 50.0% of nursing students of grade IV had excellent spiritual health and 42,6% had good spiritual health after PLKK IV. Kozier et al [3] states that because spirituality was a reflection of internal experience (inner experience) that was expressed individually, then spirituality presents from many aspects in human beings such as religion, belief, hope, transcendence and forgiveness. Spirituality is one that influenced by culture, development, life experiences of trust and value of life. Spirituality could bring love, trust, and hope, see the meaning of life and nurture relationships [12]. The ability to love others and self was meaningfully the evidence of spiritual health. Establishing a relationship with the great, life, or value was one way of developing spirituality. Health of healthy spirituality was something that gave peace and acceptance about self and it was often based on a lasting relationship with the Supreme God. Spiritual health was achieved when one found a balance between the value of life, the purpose of life, the belief system, and one's relationship with oneself or others.

4 CONCLUSION

Based on the discussion above, it could be concluded that there are 50.0% of students have excellent spiritual health, 42.6% have good spiritual health, 6.6% have adequate spiritual health, and 0.7% have inadequate spiritual health after the implementation of PLKK IV. This research is expected to be the basis for Bali STIKES to evaluate the implementation and improve the implementation process of Nursing Clinical Practice. PLKK could be received well and receive good perception among students therefore it could be improved in accordance with the development of the world of health and education. In addition, research on Nursing Clinical Practice in next research should be developed with input and review from other variables so that the improvements in the implementation of PLKK could be improved.

ACKNOWLEDGEMENT

I would like to express my deepest gratitude to I Gede Putu Darma Suyasa, S.Kp.,M.Ng.,Ph.D. as the director of Institute of Health and Sciences (STIKES) Bali for the support in conducting this study. In addition, I would like to thank to the Centre for Research and Community Service (P3M) STIKES Bali; as well as gratitude to the family, and colleagues who have provided input in the preparation of this manuscript.

References

1. Alyssa N. Bryant. (2008). *The Spiritual Struggles of College Students: Illuminating a Critical Developmental Phenomenon.* Vol. 4. Issue 4. Spirituality in Higher Education Newsletter.
2. Anand, Varun; Jones, June; Gill, Paramjit (2015). *The Relationship Between Spirituality, Health and Life Satisfaction of Undergraduate Students in the UK: An Online Questionnaire Study.* 54:160–172. https://doi.org/10.1007/s10943-013-9792-0.
3. Chin-Moucheng,Yao-Yuanliu,& Rong-Lainho. (2006). The *Impact of Internet Technology on Spiritual Health of College Students.* April 16-18. pp33-38. Proceedings of the 5th WSEAS International Conference on Applied Computer Science, Hangzhou, China.
4. Donia Baldacchino. (2015). *Spiritual Care Education of Health Care Professionals.* 6, 594–613. http://doi.org/10.3390/rel6020594
5. Hamid, A.Y.S. 2009. *Bunga Rampai Asuhan Keperawatan Kesehatan Jiwa.* Jakarta: EGC.
6. Jalaludin Rakhmat 2011. Psikologi Komunikasi. Bandung: PT.Remaja Rosdakarya.
7. Kozier, B et al. 1997. *Fundamental of Nursing Practice in Canada (1th Canadian ed).* New Jersey: Prentice Hall Inc.

8. Kozier, B et al. 2004. *Fundamental of Nursing Consepts, Process, and Practice (7th Canadian ed)*. New Jersey: Prentice Hall Inc.
9. Lee-Ann O'brien, Simon Denny, Terryann Clark, Theresa Fleming, Tasileta Teevale & Elizabeth Robinson (2013). *The impact of religion and spirituality on the risk behaviours of young people in Aotearoa, New Zealand*. Vol. 32 Number 4. Youth Studies Australia.
10. Maryam Azizi, Abolfath Azizi, Ahmad Abedi, Mohamad Baqher Kajbaf, Reza Ali Fallahzadeh. (2016) *The Effect of Spiritual Therapy on Hope of Life in Female Students Who Referred to Isfahan University Counseling Center*. 7. 126-131. http://www.scirp.org/journal/psych http://dx.doi.org/10.4236/psych.2016.71015.
11. Norwati Mansora, Nur Syahidah Khalidb. (2012). *Spiritual well-being of INSTED, IIUM Students' and Its Relationship with College Adjustment.* Vol. 69. 1314 – 1323. Procedia - Social and Behavioural Sciences. Elsevier.
12. Perry&Potter. 2005. *Buku Ajar Fundamental Keperawatan Konsep, Proses, dan Praktik.* Ed.4.Vol. 1.Jakarta : EGC.
13. *Park, J. J., & Millora, M. (2010). Psychological well-being for White, Black, Latino/a, and Asian American students: Considering spirituality and religion. Journal of Student Affairs Research and Practice,* 47(4), 1–18. https://doi.org/10.2202/1949-6605.6143 Available at http://journals.naspa.org/jsarp/vol47/iss4/art3/. Innovations in Research and Scholarship Feature.
14. Ristekdikti dan PPNI. 2015. Kurikulum Pendidikan Ners Indonesia. Jakarta: AIPNI.
15. Ristekdikti dan PPNI. 2015. Kurikulum Inti Pendidikan Ners 2015. Jakarta: AIPNI.
16. Ya-Chu Hsiao, Hui-Ying Chiang, Hsiang-Chun Lee & Su-Hui Chen. (2012). *The Effects of a Spiritual Learning Program on Improving Spiritual Health and Clinical Practice Stress Among Nursing Students*. VOL. 20, NO. 4. The Journal of Nursing Research.

The Influence of Social Support, Self Confidence, and Motivation to Economic Learning Results Class XI SMK Negeri 2 Jambi City

Suratno dan Zuhri Saputra Hutabarat

Causal associative research uses a quantitative approach. The research was conducted at SMK Negeri 2 Jambi City. The method used. The population and samples numbered 124 people, using a saturated or census sample technique. Result of research: 1) Positive positive influence between social support to learning result sebeesar 0,455, t count> t table that is 5,640> 1.67. 2) Positive and significant influence between self-confidence to the learning result of 0.402, t arithmetic> t table is 4.849> 1.675. 3) Positive and significant influence between the motivation of learning outcomes 0.419, t arithmetic> t table is 1.675> 0.05. 4) Positive and significant influence between social strengthening to motivation of 0.272, t arithmetic> t table ie 3.119> 1.655. 5) Positive and significant influence between work environment on work motivation of 0.844, t arithmetic> t table that is 10.177> 2.015. 6) Positive and significant indirect effects between support to economic learning outcomes through motivation of 0,559, F hit> F table or 518,919> 2,698. 7) Positive and significant indirect effects between self-confidence in learning outcomes through motivation of 0,504, F hit> F table or 18,919> 2,68.

Keywords: Social Support, Self Confidence, Motivation, and Learning Outcomes.

CHAPTER I INTRODUCTION

The purpose of education is said to be achieved if the learning outcomes of learners experience development and improvement. What is meant by learning is a process of changing the behavior of individuals through the interaction of learners with the environment, Hamalik (2003: 36). While the results of learning is the result of learning efforts undertaken learners. In formal education, learning outcomes are always accompanied by measurement and assessment activities, as well as in the process of teaching and learning activities is intended to determine whether or not the learning objectives have been achieved, as well as to know the position of students who are clever, moderate or later, which later can be made As a valuable input for educators as a reference to improve the quality of teaching that has been done through reports of learners' learning outcomes.

Reports of student learning outcomes can be seen from the results of replication and submitted in a certain period in the form of report cards. In an effort to achieve an optimal learning outcome of the learning process a student is connected by internal and external factors. Internal factors are factors that arise from within the learners themselves include physical circumstances, intelligence, talent, interests and attention, emotional state and discipline.

The form of successful learning in schools can be seen from the results of student learning, because the learning results show the results of efforts achieved by students during their learning activities at school. In accordance with the provisions of the applicable curriculum to measure the success of learning is the stipulation of the minimum limit of learning completeness for economic subjects that is 70.

CHAPTER II THEORY BASIS

1. Learning Outcomes

Learning is a familiar word to all levels of society. For students or students the word "learning" is a familiar word. In fact it is an integral part of all their activities in studying in formal education institutions (Djamarah, 2008: 12). In this case, Slameto (2010: 2) states that learning is a process of doing a business to gain a whole new behavioral change, as a result of its own experience in interaction with its environment. Meanwhile, according to Shah (2010: 90) learning is the stages of change in the overall behavior of individuals who are relatively settled as a result of experience and interaction with the environment that involves cognitive processes.

2. Social Support

Surafino (2007: 192) states that social support refers to providing comfort to others, caring for, or appreciating. The same opinion is also expressed by Sarason (in Smet 1994: 135) which states that social support is the interpersonal interaction shown by providing assistance to other individuals, where the aid is generally obtained from the courageous person for the individual concerned. Social support can be the provision of information, behavioral assistance, or material derived from familiar social relationships that can make individuals feel cared for, valued and loved. Social support is information or feedback from others indicating that a person is loved and cared for, valued and respected, and involved in mutual communication and mutual obligation (King, 2012: 226).

3. Confidence

Confidence is included in the positive self-recipient. This self-acceptance is a result of the satisfaction / frustration experience of the positive needs of others (Rogers, 1960 in Alwisol, 2009). Confidence is also included in the need for self-esteem to obtain satisfaction from this need allows the individual possess a sense of confidence in his ability and appearance to be more competent, productive in all aspects of life, otherwise if someone fails to gain kepuasn or experiencing lack of self-esteem he Will experience inferiority, powerlessness, lack of enthusiasm, and lack of confidence in his ability to overcome the problems of life he faces (Maslow, 1960 in Hamdali & Jaenudin, 2013). Siri-esteem is the overall evaluative dimension of the self. Confidence is also referred to as self-esteem or self-image (Santrock, 2003: 215). Confidence can also be interpreted as a mental attitude optimism of the ability of children to self-ability to solve everything self-ability to make adjustments to the situation faced (Surya, 2007: 67).

4. Self Confidence

Confidence is included in the positive self-recipient. This self-acceptance is a result of the satisfaction / frustration experience of the positive needs of others (Rogers, 1960 in Alwisol, 2009). Confidence is also included in the need for self-esteem to obtain satisfaction from this need allows the individual possess a sense of confidence in his ability and appearance to be more competent, productive in all aspects of life, otherwise if someone fails to gain kepuasn or experiencing lack of self-esteem he Will experience inferiority, powerlessness, lack of enthusiasm, and lack of confidence in his ability to overcome the problems of life he faces (Maslow, 1960 in Hamdali & Jaenudin, 2013). Siri-esteem is the overall evaluative dimension of the self. Confidence is also referred to as self-esteem or self-image (Santrock, 2003: 215). Confidence can also be interpreted as a mental attitude optimism of the ability of children to self-ability to solve everything self-ability to make adjustments to the situation faced (Surya, 2007: 67).

Confidence is an individual's beliefs to determine how to behave in accordance with what is needed (Susanti, 2008). Someone who has confidence will be more confident to do something or enter in an environment (Shinta, 2011: 55). True self-esteem means having several things that include self-integrity, knowledge insight, courage, vantage point, and positive self-esteem. The main reason someone feels lack of confidence is because they do not know what they can do and have no experience (Lumpkin, 2005: 88).

5. Motivation

Suprijono (2011: 163) says that the essence of learning motivation is internal and external encouragement to learners who are learning to make behavioral changes. Motivation to learn is a process that encourages learning, direction, and persistence of behavior. Motivation, according to Uno (2011: 8) is the impetus and strength in a person to do certain goals that want to achieve. The purpose in question is something that is outside of human so that human activities more focused because someone will try more spirit and enterprising in doing something.

According to Sardiman (2011: 75) motivation is a series of attempts to provide certain conditions, so that someone wants and wants to do something, and if he does not like it, it will try to nullify or avoid feelings of dislike it. Meanwhile, according to Mc. Donal cited by Discourse (2008: 148) motivation is a change of energy in a person (person) characterized by the emergence of feelings and reactions to achieve goals. Learning is an emphasis obtained because of the interaction between individuals and their environment. In this case, according to Shah (2010: 90) learning is a stages of change throughout the relatively resident individual tingah as a result of experience and interaction with the environment that involves cognitive processes.

RESEARCH METHODS

1. Research Design

Judging from the formulation of this research is explanatory research (explanatory research) is to explain the state of each variable. This study also belongs to the associative class of causality. Associative classified because of this research to determine whether or not there is influence between the variables in the perusal. Categorized causality, because this study aims to determine whether the observed variables have a certain causal effect as theoretically suspected. This research is called explanatory research because the purpose is to explain the effect of causation between variables through hypothesis testing. The description of the influence between these variables is as follows:

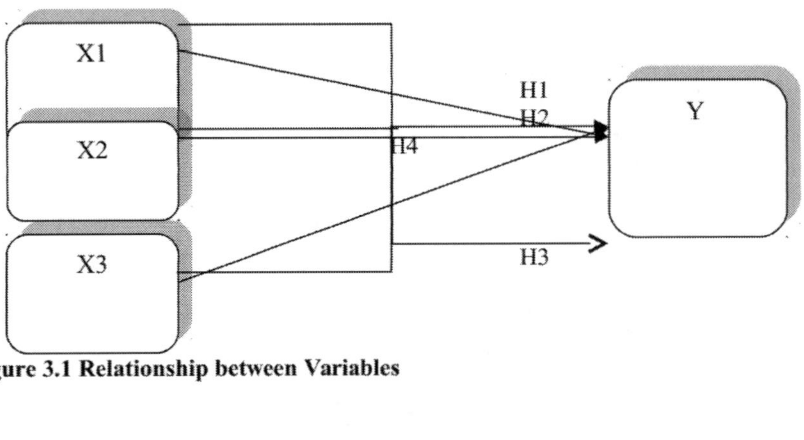

Figure 3.1 Relationship between Variables

Information :
X1 = Social Support
X2 = Self Confidence
M = Motivation
Y = Learning Outcomes

2. Place and Time of Research

The place of research is SMK NEGERI 2 Jambi City which is located at Jalan Gelatik Pasir Putih Kecamatan Jambi Selatan. The study was conducted in March. This study I did in less than a month, because I use a questionnaire.

3. Sample Research

Arikunto (2002: 109) states that the sample is part or representative of the population under study. The sample is part of the population that has the same properties as the population, a good sample is a sample that can represent or reflect to the maximum state of the population.

Sampling technique in this research use saturated sampling technique. Saturated Sampling is a sample determination technique when all members of the population are used as a sample (Sugiyono, 2012: 96). This is often done when the population is relatively small, less than 30 people, or research that wants to generalize with a very small error in which all members of the population are sampled. The technique for sampling used in this research is proportional sampling. Determination of the sample size was taken based on the formula of Slovin in Umar (2011: 67) about the size of the sample determined according to the following formula::

$$n = \frac{179}{1 + (179x0{,}05x0{,}05)}$$

$$= \frac{179}{1{,}4475} = 123{,}66$$

$$= 124$$

4. Research Methods

This research used survey research method. This method is appropriate because in addition to describe the current condition, can also answer questions about the characteristics of variables, relationships and influence between variables.

To analyze the pattern of relationship between variables with the aim to determine the influence of direct or indirect set of independent variables (exogenous) to the dependent variable (endogen) used path analysis model (path analysis).

RESEARCH RESULT AND DISCUSSION

1. Learning Outcomes

The results of economic learning in the intention in this study is the mastery of students on learning materials and the success rate of students in this study is the cognitive aspects, namely aspects related to knowledge or intellectual. Economic learning results obtained from test results. From the result of the research, the students score as follows:

Table 4.1 Economic Value Statistics Students Class XI Accounting SMK N 2 City Jambi odd semester of 2016-2017

No	Score	Category	Frequency	Percentage
1.	< 49,99	Very less	0	0
2.	50-59,99	Less	12	9.7
3.	60 – 69,99	Enough	54	43.5
4.	70 – 79,99	Good	47	37.9
5.	> 80	Very good	11	8.9
Total			124	100

Source: Results of SPSS 2017 data

From the table above shows that the students who classified the test results are very good 11 people or 8.9% of the overall sample. For students who have good category criteria that is 47 people or 37.9%. For good enough kateogri amounted to 54 students or 43.5% of the sample. While students who have poor test results amounted to 12 students or 9.7% of the sample. For more details can be seen in table 4.2 below. Table Statistics of students' economic learning outcomes Statistics Learning outcomes

Statistics

Learning outcomes

N	Valid	124
	Missing	0
Mean		67.26
Median		65.00
Mode		65
Std. Deviation		7.553
Range		30
Minimum		50
Maximum		80
Sum		8340

Source: Results of SPSS 2017 data

From the overall value of learning results can be concluded that the highest value with the value of 80 while the lowest with a value of 50. Reach from the highest value to the lowest value or the so-called range of 30. While for the middle value is 65.00. For the value that often appears or mode is the value of 65. From 124 respondents then got the mean or average value of the report card as a whole is 67.26 or including good enough category. So it is stated that the variables of economic learning outcomes of students of class XI Accounting SMK Negeri 2 Kota Jambi including good enough.

2. Social Support

Data on social support were collected through a questionnaire distributed to 124 respondents. The questionnaire value was obtained based on 30 questionnaire statements consisting of 5 answer choices. Each answer option has different value weights. For the social support questionnaire that has the weight of value 5 is the answer S (Always), the weight of value 4 is Sr (often), the weight of value 3 is Pr (ever), the weight of value 2 is Jr (Rarely), the weight of value 1 is Tp (No ever). Based on the results if the data from the 3 indicators are obtained information as presented in the following table.

No	Score	Category	Frequency	Percentage
1.	< 45	Very less	0	0
2.	46 – 75	Less	13	10.5
3.	76 – 105	Enough	110	88.7
4.	106 – 135	Good	1	0.8
5.	> 136	Very good	0	0
Total			124	100

Source: Results of SPSS 2017 data

Dari tabel di atas terlihat bahwa siswa yang tergolong dukungan sosialnya baik hanya 1 orang atau 0.8% dari keseluruhan sampel. Untuk siswa yang memiliki kriteria kategori cukup dukungan sosialnya berjumlah 110 siswa atau 88.7 % dari sampel. Sedangkan siswa yang memiliki dukungan sosial yang kurang berjumlah 13 siswa atau 10.5% dari sampel. Namun secara keseluruhan dapat dilihat pada tabel dibawah ini.

4.3 Tabel Statistik hasil angket dukungan sosial

Statistics

Social Support

N	Valid	124
	Missing	0
Mean		86.21
Median		86.00
Mode		85
Std. Deviation		7.955
Range		34
Minimum		72
Maximum		106
Sum		10690

Source: Results of SPSS Year 2017 data

From the total results of the social support questionnaire can be concluded that the highest score with a value of 106 while the lowest with a value of 72. Range from the highest value to the lowest value or the so-called range of 34. As for the middle value is 86.00. For the value that often appears or mode is the value 85. Of the 124 respondents then got the mean or average social support as a whole is 86.21 including enough supportive categories. So it is stated that the variables of socio-economic support of students of class XI Accounting SMK Negeri 2 Kota Jambi is quite supportive category.

3. Self Confidence

Data on the confidence of Daddy was collected through a questionnaire distributed to 124 respondents. The questionnaire value was obtained based on 34 questionnaire statements consisting of 5 answer choices. Each answer option has different value weights. For the social support questionnaire that has the weight of value 5 is the answer S (Always), the weight of value 4 is Sr (often), the weight of value 3 is Pr (ever), the weight of value 2 is Jr (Rarely), the weight of value 1 is Tp (No ever). Based on the results if the data from the 4 indicators are obtained information as presented in the following table.

No	Score	Category	Frequency	Percentage
1.	< 45	Very less	0	0
2.	46 – 75	Less	3	2.4
3.	76 – 105	Enough	105	84.7
4.	106 – 135	Good	16	12.9
5.	> 136	Very good	0	0
Total			124	100

Source: Results of SPSS 2017 data

From the table above shows that students who belong to self-confidence either only 16 people or 12.9% of the overall sample. For students who have sufficiently self-confident category criteria of 105 students or 84.7% of the sample. While students who have less trust amounted to 3 students or 2.4% of the sample. But overall can be seen in table below.

Table 4.4 Description of confidence questionnaire statistics

Statistics

confidence

N	Valid	124
	Missing	0
Mean		96.08
Median		96.00
Mode		89[a]
Std. Deviation		8.700
Range		49
Minimum		70
Maximum		119
Sum		11914

Source: Results of SPSS Year 2017 data

a. Multiple modes exist. The smallest value is shown

Sumber: Data diolah Penulis

From the overall confidence questionnaire results can be concluded that the highest score with a value of 119 while the lowest with a value of 70. Range from the highest value to the lowest value or the so-called range of 49. While for the middle value is 96.00. For the value that often appears or the mode is the value 89. Of the 124 respondents then got the mean or the average confidence as a whole is 96.08 or including the category is quite confident. So it is stated that the economic confidence variable of students of class XI Accounting SMK Negeri 2 Kota Jambi is quite confident.

4. Learning Motivation

Data on learning motivation were collected through a questionnaire distributed to 124 respondents. The questionnaire value was obtained based on 30 questionnaire statements consisting of 5 answer choices. Each answer option has different value weights. For the social support questionnaire that has the weight of value 5 is the answer S (Always), the weight of value 4 is Sr (often), the weight of value 3 is Pr (ever), the weight of value 2 is Jr

(Rarely), the weight of value 1 is Tp (No ever). Based on the results if the data from the 4 indicators are obtained information as presented in the following table.

No	Score	Category	Frequency	Percentage
1.	< 45	Very less	0	0
2.	46 – 75	Less	14	11.3
3.	76 – 105	Enough	110	88.7
4.	106 – 135	Good	0	0
5.	> 136	Very good	0	0
Total			124	100

Source: Results of SPSS 2017 data

From the table above shows that students who classified as learning motivation is good enough only 110 people or 88.7% of the overall sample. For students who have less motivation category criteria amounted to 14 students or 11.3% of the sample. But overall can be seen in table below. Description of learning motivation statistics

Statistics

Motivation

N	Valid	124
	Missing	0
Mean		83.18
Median		83.50
Mode		88
Std. Deviation		6.464
Range		32
Minimum		66
Maximum		98
Sum		10314

Source: Results of SPSS Year 2017 data

From the overall questionnaire motivation to learn can be concluded that the highest value with a value of 98 while the lowest with a value of 66. Reach from the highest value to the lowest value or the so-called range of 32. While for the middle value is 83.50. For the value that often appears or the mode is the value 88. Of the 124 respondents then got the mean or overall confidence as a whole is 83.18 including the category is quite motivated. So it is stated that the variables of student's economic motivation of class XI Accounting SMK Negeri 2 Kota Jambi including enough motivated category. From the results of research and discussion presented in Chapter IV, the conclusions of the study are as follows: There is a strong and significant influence between social support on learning outcomes. This implies that improved social support will lead to increased learning outcomes.: a) There is a strong and significant influence between self-confidence in learning outcomes. This implies that a stronger self-confidence will lead to increased learning outcomes. b) There is a strong and significant influence between learning motivation on learning outcomes. This implies that increasing motivation will lead to increased learning outcomes.c) There is a weak and significant influence between social support on learning motivation. This implies that improved social support will lead to a slight increase in learning motivation d) There is a weak and significant influence between self-confidence in learning motivation. This implies that an increasingly strong confidence will lead to a slight increase in learning motivation.

CONCLUSIONS AND RECOMMENDATIONS

1. Conclusion

From the results of research and discussion presented in Chapter IV, the conclusions of the study are as follows: a) There is a strong and significant influence between social support on learning outcomes. This implies that improved social support will lead to increased learning outcomes.b) There is a strong and significant influence between self-confidence in learning outcomes. This implies that a stronger self-confidence will lead to increased learning outcomes.c) There is a strong and significant influence between learning motivation on learning outcomes. This implies that increasing motivation will lead to increased learning outcomes. D) There is a weak and significant influence between social support for learning motivation. This implies that improved social support will lead to a slight increase in learning motivation. F) There is a weak and significant influence between self-confidence in learning motivation. This implies that an increasingly strong confidence will lead to a slight increase in learning motivation.

2. Suggestions

From the above conclusions, the author gives some suggestions as follows: a) Social support has a strong and significant influence on learning outcomes. Based on this matter, it is important for SMK Negeri 2 Kota Jambi to be able to increase Social support by way of teacher pay more attention to student learning difficulties. For parents are expected to give advice and hints about problems faced by students. b) Confidence has a strong and significant influence on learning outcomes. Based on this it is important for students to increase their confidence by way of improving speech ability can help to feel more comfortable and confident in larger social situations will result in overall belief.

BIBLIOGRAPHY

A.M, Sardiman. (2011). Interaksi dan Motivasi Belajar Mengajar. Jakarta: Rajawali Press.

Apollo dan Adi Cahyadi. 2012. Konflik Peran Ganda Perempuan Menikah yang Bekerja Ditinjau dari Dukungan Sosial Keluarga dan Penyesuaian Diri. Widya Wart

Crowford,J.C dan Henry, J.D. 2003. The depression anxiety stress scale (DASS): normative data and late structure in a-large non-clinical sample. British journal of clinical psycology, 42, 111-131

Dimyati dan Mudjiono,1994.Belajar Dan Pembelajara.Jakarta:Depdikbud.

Djamarah, Syaiful Bahri, Psikologi Belajar, Jakarta : Rineka Cipta, 2008

Govaerst, S dan Gregoire, J. 2004. Stressfull academic situations: study on appraisal variables in adolescence. British journal of clinical psychology, 54, 261-271

Gronlund., dan Linn.1985. Measurement and Evaluation in Teaching. New York: Macmillan Publishing Company

Hamdali, A. dan Jaenudin, U. 2013. Psikologi kepribadian lanjutan: studi atas teori dan tokoh psikologi kepribadian. Bandung: pustaka setia

Keeves, J.P.1994. *National Examinations: Design, Procedures, and Reporting.* Paris: UNESCO, International Institute for ducational Planning

Krisdalina, A. E. 2009. Hubungan pemahaman akreditasi rumah sakit dan karateristik dengan kinerja perawat manajer di rumah sakit epolision pusat raden said sukanto Jakarta. Tesis FIK UI Depok. Tidak dipublikasikan.

Yulianto, F. dan Nashori, H.F. 2006. Kepercayaan Diri dengan Prestasi Atlet Tae Kwon Do Daerah Istimewa Yogyakarta. Jurnal Psikologi Universitas Diponegoro vol. 3(1), 55-62

Putri, R.L & Hadi,Ch, (2007) Bagaimana memahami seorang diri remaja ?, http://www.fpsi.unair.ac.id, diperoleh tanggal 17 februari 2016).

Rini, J.F (2002) . Memupuk Rasa Percaya Diri, http://digilib.itb.ac.id, diperoleh tanggal 17 Februari 2016.

The Equating of Battery Test Packages of Mathematics National Examination 2013-2016

Badrun Kartowagiran[1] , *Sudji* Munadi[1], *Heri* Retnawati[2,*], and *Ezi* Apino[2]

[1]Educational Research and Evaluation, Graduate School, Universitas Negeri Yogyakarta, Indonesia
[2]Mathematics and Natural Science Faculty, Universitas Negeri Yogyakarta, Indonesia

Abstract. When a test implements several test instruments, there is an assurance that the test instruments that will be implemented are equal and this equality is an urgent matter. Therefore, this study aimed at confirming the equality of the test instruments that had been implemented in Mathematics National Examinations (MNE) 2013-2016. This study was conducted using quantitative approach. The data were gathered using students' response documentation technique and MNE test instruments; the test instruments were drawn from the packages that had been administered to the national examinations for junior high school students in the Province of Yogyakarta Special Region. The data were analyzed by using the stages of item parameter estimation, designing the equating equation, equating through concordance model, drawing the test characteristics curve, and interpretation. The results of the analysis showed that the instruments that have been administered are almost equal, both from one package to another and from one year to another, with the standards of test instrument 2013.

1 Introduction

In order to improve the National Examination (NE) quality and to decrease the fraud, the Republic of Indonesia Government implements a policy that the administration of NE should make use of several test packages. This policy has been implemented since 2010; in that year, the national examination only made use of 2 test packages. One year later, the number increased into 5 test packages. In 2013 and 2014, the NE was administered using 20 test packages. Then, in 2015 and 2016 the NE was administered using 5 packages again. The test packages that had been administered were developed based on the same guidelines so that these packages were expected to measure the same skills. Thereby, the test packages that had been administered in these NE are assumed as the parallel ones.

The use of several parallel test packages in the NE can decrease the potentials of fraud during the test administration [1]. In the same time, the use of those packages has another advantage namely that the secrecy of the test items are more assured [2]. With the abundant number of test packages that will be administered in the NE, the potentials of leakage can be minimized [3, 4]. However, the main challenge in administering several test packages during the NE is related to the assurance that these packages are equal and measure the

* Corresponding author: heri_retnawati@uny.ac.id

same indicator. In other words, there should be a guarantee that the test packages that will be implemented in the NE are the equal ones and this is an urgent matter.

The guarantee that the test packages are equal can be confirmed both theoretically and empirically. This confirmation is related to the concept equating or concordance of test score [5]. The equating can be performed using the classical approach (classical test theory approach) and the modern approach (item response theory approach) [6, 7]. The equating using the modern approach basically calculates the students' ablities and level of difficulty into certain scores with a linear equation [2, 8]. In order to perform this calculation, parameter of index discriminant (*a*), level of difficulty (*b*), and pseudo-guessing (*c*) should be estimated first. The estimation that involves these parameters in the item response theory is known as the 3PL Model. On the other hand, several methods that can be implemented in order to perform equating are namely mean and mean method, mean and sigma method, and item characteristics curve method which includes the Stocking and Lord method [9, 10].

In the mean and mean method, the equating constant α and β only involves the mean of the item difficulty index (*b*) and the mean of item discriminant index (*a*) after the parameters have been estimated first [2]. For example, if the score of test 1 is equated to that of test 2 using the 3PL Model using the mean and mean method according to [9], then the linear equation model will be as follows:

$$b_2 = \alpha b_1 + \beta \tag{1}$$

So that the researchers attain the following equations:

$$\overline{b_2} = \alpha \overline{b_1} + \beta \Leftrightarrow \beta = \overline{b_2} - \alpha \overline{b_1} \tag{2}$$

and

$$a_2 = \frac{a_1}{\alpha} \tag{3}$$

then

$$\overline{a_2} = \frac{\overline{a_1}}{\alpha} \Leftrightarrow \alpha = \frac{\overline{a_1}}{\overline{a_2}} \tag{4}$$

$\overline{b_1}$ and $\overline{b_2}$ are mean of common item difficulty index from test 1 and test 2. $\overline{a_1}$ and $\overline{a_2}$ are mean of common item discriminant index from test 1 and test 2. While, α, β is equating constants.

Then, in the mean and sigma method the determination of equating constant between α and β involves the mean and the standard deviation of the item parameters. For example, if the score of test 1 is equated to that of test 2 using the 3PL Model, according to [9], the equating model will be as follows :

$$b_2 = \alpha b_1 + \beta \tag{5}$$

As a result, the researchers attain the following equation :

$$\overline{b_2} = \alpha \overline{b_1} + \beta \Leftrightarrow \beta = \overline{b_2} - \alpha \overline{b_1} \tag{6}$$

and

$$S_2 = \alpha S_1 \Leftrightarrow \alpha = \frac{S_2}{S_1} \tag{7}$$

with S_1 and S_2 are standard deviation of item difficulty index from test 1 and test 2

Furthermore, in the item response theory, if the item response model is compatible to a data set then any linear transformation from the measurement scale will also be compatible for the data. This means that there has been relationship between the measurement scales from both tests. Thereby, if the scale of test 1 is equated to that of test 2 for the 3PL Model then the relationship between the item parameter and the participants' abilities for both scales can be stated as follows [10].

$$q_{1i}^{*} = aq_{1i} + b \tag{8}$$

$$a_{1j}^{*} = \frac{a_{1j}}{\alpha} \tag{9}$$

$$b_{1j}^{*} = \alpha b_{1j} + \beta \tag{10}$$

$$c_{1j}^{*} = c_{1j} \tag{11}$$

a_{1j}, b_{1j}, c_{1j} are item parameter for item j on the test 1 scale. a_{1j}^{*}, b_{1j}^{*}, c_{1j}^{*} are item parameter for item j on the test 1 scale after having been equated to test 2. q_{1i} is the ability of i participant on the test 1 scale and q_{1i}^{*} is the abilty of i participant on the test 1 scale after having been equated to test 2. The c parameter is not transformed because its value does not depend on the metric q or, in other words, the parameter c is free from the scale transformation [10]. Moreover, the constant that links the score of test 1 and that of test 2 can be calculated using multiple methods known as test score linking method.

In relation to the use of several equating methods, the results of a study by [11] showed that the Stocking and Lord-type characteristics curve equating and the mean and sigma method have equally been good. On the other hand, [12] through the results of their study stated that there is not any best single method for equating the test scores. As a result, the researchers can select one of the equating methods in order to confirm the quality among the test packages according to the needs. Based on the above problems and the theoretical review, this study then aimed at confirming the equality of the test packages that had been administered in the MNE 2013-2016

2 Method

This study was a descriptive explorative which aimed at describing the equality of the test packages that had been administered in the MNE from 2013-2016. The data were gathered using the students' response documentation technique and the MNE test instruments from the test packages that had been administered in the NE for junior high school students in the Province of Yogyakarta Special Region. There were 9 test instruments that had been selected as the object of this study from the MNE 2013, 8 instruments from MNE 2014, and 5 instruments from MNE 2015 and 2016. Then, the test package 1 from MNE 2013 had been defined as the standard of reference for equating the inter-year test packages. The data analysis was performed through the stages of item parameter estimation, design of equating equation, equating using the concordance model, drawing of test curve characteristics (TCC), and interpretation.

3 Results and Discussions

After the item parameter estimation that included the index of difficulty (b), the index discriminant (a), and the pseudo-guessing (c) had been performed, the researcher equated the model without common items (concordance) using the mean and sigma method. The results of concordance analysis yielded the transformation equations for each item parameter. These parameters were the equal ones and then these parameters were manipulated in order to illustrate the characteristics curve from each test instruments. From the test characteristics curve (TCC), the researchers attained the information regarding the test instruments equality both the inter-package one and the inter-year one. Then, the TCC from each instrument was presented in Figure 1, Figure 2, Figure 3, and Figure 4.

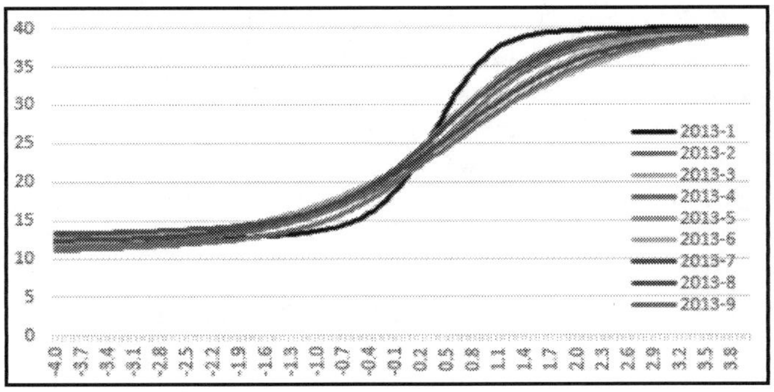

Fig. 1. The TCC of the MNE 2013 Test Instruments

By looking at Figure 1 the researchers found information that the equating that had been conducted resulted in mutually approaching curves. In a closer look, there was one curve that had been inclined to less coinciding namely test package 1 (2013-1). On the contrary, the curves of the other test packages (2013-2 to 2013-9) were inclined to more coinciding from one to another. This finding implied that the test instruments that had been administered in MNE 2013 had been equal from one to another.

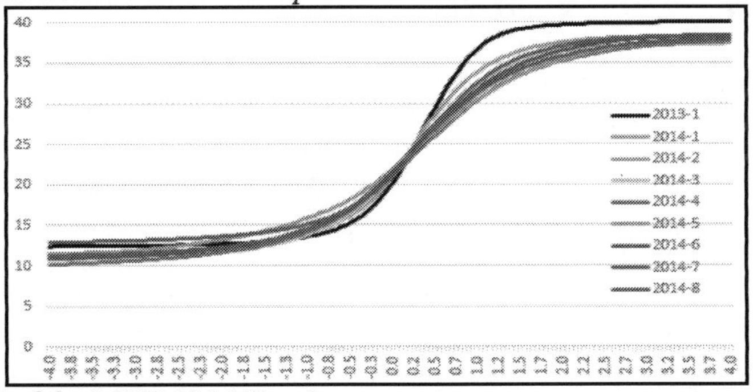

Fig. 2. The TCC of the MNE 2014 Test Instruments

Looking at Figure 2, it was apparent that the curves that had been resulted were also inclined to coincide one to another. In a closer look, it was also apparent that the inter-test package curves (2014-1 to 2014-8) were also coinciding from one to another. These findings implied that the test packages that had been administered in the MNE 2014 for Junior High School had been equal. Then, in comparison to the characteristics curve of test package 1 from the MNE 2013 (2013-1) it was apparent that these curves had been less coinciding although they had been close to each other. Such curves indicated that the test instruments that had been administered in the MNE 2014 had almost been equal to those that had been administered in the MNE 2013.

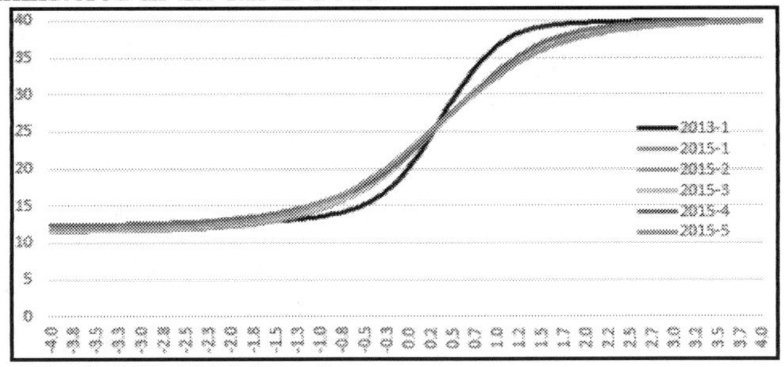

Fig. 3. The TCC of the MNE 2015 Test Instruments

Looking at Figure 3 it was apparent that the curves that had been yielded were also mutually closer to each other. In a closer look, it was also apparent that the inter-test package curve in MNE 2015 (2015-1 to 2015-5) had been very coinciding. Such coincidence implied that the test packages that had been administered in the MNE 2015 had been equal. In comparison to the test package 1 from the MNE 2013, it was apparent that the curve of the MNE 2015 had been inclined to coincide only on certain ability and to move away on the other ability. This condition also indicated that the test instruments that had been administered in the MNE 2015 had been equal to those that had been administered in the MNE 2013.

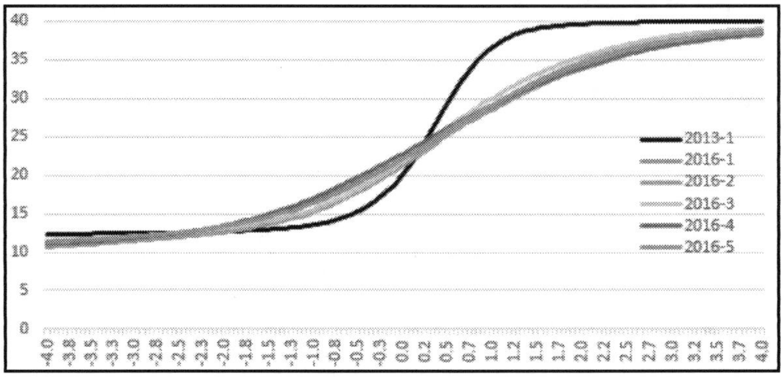

Fig 4. The TCC of the MNE 2016 Test Instruments

By looking at the TCC in Figure 4 the researcher found that the yielded curves were closer to each other as well. In a closer look, it was apparent that the inter-test package curve from the MNE 2016 test packages (2016-1 to 2016-5) had been mutually coinciding. This coincidence implied that the test packages that had been administered in the MNE 2016 had been equal. Next, in comparison to the item characteristics curve package 1 from the MNE 2013 (2013-1), the researchers could see that the curves had been less coinciding on certain ability scales; however, these curves still have similar tendency. Thereby, based on the information of the TCC there was an indication that the test instruments that had been administered in the MNE 2016 had been equal to those that had been administered in the MNE 2013.

The results of the analysis showed that the test instruments that had been administered were almost equal, both in terms of package and in terms of year, according to the standards of test instruments 2013. One of these results were in concordance to a study by [2] which showed that one of the test instruments that had been administered in the MNE, namely the 2014 one, which consisted of 20 test packages, had a tendency to be equal. Moreover, the presence of these results clearly indicated that the test instruments that had been administered, especially the test packages, in the NE had been considered fair. As a result, the test participants did not consider themselves being in disadvantages. This matter certainly could be confirmed empirically using the equating procedures that had been performed. [13] proposed the importance of test equating procedures in order that the NE test participants would not be in disadvantages and the scores that would be awarded to the NE test participants became fair although each participant completed different test package and experienced different difficulty level.

Looking at inter-year equating results, it appears that TCC of test instruments of the MNE 2015 and 2016 are less closely aligned with test instruments of MNE 2013 (see figure 3 and 4). This is because the NE 2015 and 2016 began to accommodate Higher Order Thinking Skills (HOTS) items which refers to PISA and TIMSS items. This is inseparable from the Indonesian government's policy to implement the Curriculum 2013, where one of the main focus in this curriculum is the development of students' HOTS [14, 15]. In addition, in the NE 2015 and 2016 has been implemented Computer Based National Examination (CBNE), where the CBNE score is more accurate than Paper and Pencil Test score [16].

4 Conclusions

Based on the results of the analysis, the researchers conclude that the instruments that have been administered in the MNE (2013-2016) are almost equal, both from one package to another and from one year to another, with the standards of test instrument 2013. On the other hand, the researcher propose the following suggestions namely (1) for the future studies, it is necessary to carry out the equating procedure of test instruments of the NE that employ wider-scale object, both in terms of coverage (province) and of sample, so that the information that can be attained from the equating procedures will be more accurate ; (2) the equating procedures are very important to perform in each test instruments especially for the implementation of Final Examination and National Examination so that information regarding the quality of the test instruments can be attained. This information can serve as the foundation for the government to make policies regarding the design of Final Examination and National Examination test items.

References

1 H. Retnawati, B. Kartowagiran, J. Arlinwibowo, E. Sulistyaningsih, Int. J. Instr. **10(3)**, 257–276, (2017)

2 H. Retnawati, J. Kependidikan, **46(2)**, 164–178, (2016)

3 T. Rijanto, J. Penelit. dan Eval. Pendidik. **16(1)**, 365–383, (2012)

4 A. P. Herkusumo, J. Pendidik. dan Kebud. **17(1)**, 455–471, (2011)

5 N. J. Dorans, J. Appl. Psichol. Meas, **28(4)**, 219–226, (2004)

6 J. Ryan, F. Brockmann, *A practitioner's introduction to equating with primers on classical theory and item respons theory*. (Council of Chief State School Officers, 2009)

7 İ. Uysal, S. Kilmen, Int. Online J. Educ. Sci. **8(2)**, pp. 1–11, (2016)

8 H. Retnawati, *Teori respons butir dan penerapannya* (Parama, Yogyakarta, 2014)

9 R. K. Hambleton, H. Swaminathan, H. J. Rogers, *Fundamental of item response theory* (Sage Publication, Newbury Park, CA, 1991)

10 M. J. Kolen, R. L. Brennan, *Test equating: Methods and practices* (Springer, New York, NY, 2004)

11 X. Pang, E. Madera, N. Radwan, and S. Zhang, *A comparison of four test equating* (Methods Research Report, 2010)

12 C. H. Yu, S. E. O. Popp, Pract. Assessment, Res. Eval. **10(4)**, 1–19, (2005)

13 D. S. Sukirno, Cakrawala Pendidik. **26(3)**, 305–321, (2007)

14 H. Retnawati, S. Hadi, A. C. Nugraha, Int. J. Instr. **9(1)**, 33–48, (2016)

15 E. Apino, H. Retnawati, J. Phys. Conf. Ser. **812**, pp. 1–7, (2017)

16 H. Retnawati, Turkish Online J. Educ. Technol. **14(4)**, 135–142, (2015)

Contribution Of Infrastructure Facilities Towards The Enhancement Of Students' Learning Outcomes

Ade Fitri Rahmadani
Pendidikan Teknik Informatika dan Komputer, Universitas Bung Hatta, Indonesia
e-mail : adefitri.rahmadani@bunghatta.ac.id

Abstract. This research intends to show the contribution of infrastructure facilities towards the outcomes from "Mediagnosis Permasalahan Pengoprasian PC dan Peripheral" class on students with specification. The method of this research is descriptive correlational. The data comes from the questionnaire that were spread, with scale model of likert, which comes with five answer scales. With total of 105 population, which all from X grader students majoring in Computer Network Engineering; using the random sampling technique, 53 students were picked. The questionnaire was analyzed using the SPSS (Statistic Product and Service Solution) 16.0 program. The outcome of data processing was that the analysis uji-t of thitung was 6,233, which there was a correlation of contribution of facilities towards learning outcomes. Meanwhile, ttabel was 2,008 with df = n – 2 at a significance level of 0,05 with the provisions of thitung > ttabel. This means that the correlation of the facilities and learning outcomes is as big as 43,2%, while the other factors are affected from other variables.

1 Introduction

Education is a system which includes input, process and output. As Arikunto (2009) said, input is a raw ingredient which inserted to transformation. In school system, 'raw ingredient' means the student candidates who are just about to enter school. Before these students enter an institution or school, their abilities would be assessed first. The scores would indicate whether or not the student would do well in school.

These students' scores could be measured by their abilities and process of learning. According to Arikunto (2012), learning outcomes are the results that the students get from their processes of learning, so that their abilities could be measured by their own learning outcomes.

Students' abilities that could be measured by their process of learning could also influenced by two factors, which are internal and external (Slameto, 2010). Internal is a factor which is in someone's body. In this case, internal factor includes interest, talent, motivation, memory, intelligence, and creativity; meanwhile, the external factor includes the environment, like family, friends, and school environment.

Educational facilities include in the external factor that would influence the outcomes of learning. These educational facilities could be the tools and equipment that students use at school; however, all these equipment could be the reason why the learning outcomes are not optimal (Bafadal, 2008: 2).

From the observation that was held in the vocational high school, there was an indicator that shows how low the learning outcomes were, which caused by lack of facilities. For example, the amount pf computers in the computer room were not enough for 35 students with only 2 hours period a week. This situation makes the graduating students have not been able to compete in a workplace setting and also have not been able to create their own business because of how limited the lab equipment is; therefore, this situation makes the students have not been able to grow in their own field of interest and have not been able to reach the 'Kriteria Ketuntasan Minimum' or KKM from the subjects that have been set.

2. RESEARCH METHOD

This research is a descriptive correlation research. The population includes 105 students which were picked with Proportional Random Sampling method; this method is suitable with the sampling technique, which then obtained a sample of 52 students. Then, each sample got separated by 3 classes using the formula from Sugiyono that was shown by Ridwan (2010):

$$n_i = \frac{N_i}{N}.n$$

So, the total sample that was used in this research is 53 students, while the rest of the sample would be used in the trial instrument. The sample respondents in this instrument trial were students who were not included in the sampling and in the actual data retrieval.

Questionnaire is the instrument that was picked to get the data for this research; every question in the questionnaire has a value that was set in the likert scale model, which has five scales in each question, which are 'Selalu' or SL, 'Sering' or SR, 'Kadang-kadang' or KD, 'Jarang' or JR, and 'Tidak Pernah' or TP. Each question has its value from 1 to 5; if the question either tend to be positive or negative , each question has the value of either 1, 2, 3, 4 or 5.

3. RESULTS AND DISCUSSION

The variable data of the facilities was picked by the questionnaire that had 25 questions and have been checked each one of their validity and reliability. The valid questionnaire will be tested to those 53 sample respondents. The description of the distribution of infrastructure infrastructure scores can be seen in table 1 below:

Table 1. Distribution of Infrastructure Score Variable Score

Kelas Interval	Frekwensi	Persentase (%)
69 – 74	3	5.66%
75 – 80	6	11.32%
81 – 86	9	16.98%
87 – 92	11	20.75%
93 – 98	8	15.09%
99 – 104	13	24.52%
105 – 110	3	6.66%
Total	53	100%

From the data obtained, the average level of achievement score of vocational student infrastructure facilities amounted to 83.07%, and entered into the category very strong.

The result of product moment correlation from pearson can be seen in table 2 below:

Table 2. Product Moment Correlation with SPSS

Variable	Pearson Correlation	Significance
X1 – Y	.658	.000

According to the correlation of the simple analysis in table 2, there is a correlation between facilities and the students' learning outcomes (r) which is 0,658. This shows that there is a strong relation between infrastructure facilities with the learning outcomes, because it falls between 0,60 – 0,799, with positive direction because of the r value is positive, therefore, the complete the facilities, the more positive the outcome.

Then, the analysis of significance test to see the significance of the correlation or significance relationship of two variables. According to Ridwan (2009), to test the significance of one independent variable with dependent variable used t calculation formula that can be seen in table 3 below:

Table 3. Uji -t

Variabel	t hitung	Signifikan
X1 – Y	6.233	.000

The result of uji-t is thitung = 6,233 and ttable = 2,008, with df = n – 2, which can also be said that thitung (6,233) > ttable (2,008), or significance < alpha (0.000 < 0,05); H0 got rejected, while H1 got accepted, meaning that there is a correlation infrastructure facilities (X1) towards the outcome of student learning (Y).

After the t-test is done, then the last thing that needs to be done is the determination analysis to find the percentage of the correlation between facilities and the outcome of learning, which can be seen in table 4 below:

Table 4. Analisis Determinasi

Variabel	r Square
X1 – Y	.432

According to the table number 4, the r2 (r square) is as big as 0,432, therefore the total contribution of the facilities towards the outcome of learning could be seen in the Koefisien Determinan (KP) score, which 0,432 x 100% = 43,2%. This shows the percentage contributes the facilities towards the outcome is 43,2%, while the rest is from other variables.

4. SUMMARY AND RECOMMENDATION

4.1 Summary

Infrastructure facilities have a significance contribution towards the learning outcomes of vocational student, which is 43,2% with 95% of confidence level. This shows that the better the facilities, the better the learning outcomes for these students.

4.2 Recommendation

The researchers are expected to study more deeply about other factors that have an influence on learning outcomes. Thus, efforts to improve student learning outcomes can be implemented in a real way. Other than that, the school is expected to add more facilities in school to boost up the outcome of student learning, that would automatically boost up the percentage of graduating students. And to the further researchers, to further expand the study of the factors that affect learning outcomes in terms of aspects studied, the number of respondents, as well as the area of research, because there are still many factors that affect the student learning outcomes that have not been revealed in this study.

5. Acknowledgement

This article was included in thesis paper when the writer was only a college student strata 1 in Engineering Department of Padang State University. This article has already made it through the seminar and also tested by the professors in Padang State University.

References

1. Arikunto, Suharsimi. 2002. Prosedur Penelitian Suatu Pendekatan Praktek. Jakarta: Rieneka Cipta.
2. Arikonto, Suharsimi . 2009. Dasar-Dasar Evaluasi Pendidikan, Jakarta: Bumi Aksara. Edisi Revisi.
3. Ibrahim, Bafadal. 2008. Manajemen Perlengkapan Sekolah Teori dan Aplikasinya. Jakarta: Bumi Aksara.
4. Riduwan. 2010. Dasar-dasar Statistika. Bandung: Alfabeta.
5. Slameto. 2010. Belajar dan Faktor-Faktor yang Mempengaruhinya. Jakarta: Rieneka Cipta.
6. Sukardi. 2009. Metodologi Penelitian Pendidikan. Jakarta: PT. Bumi Aksara.

Computer-based peer assessment as a part of learning process

Komang Setemen[1,*], *Luh Joni Erawati* Dewi[1] , and *I Ketut* Purnamawan[1]

[1]Management Informatics Department, Universitas Pendidikan Ganesha, Singaraja, Indonesia

Abstract. This study aimed to develop prototype of computer based peer assessment application. This application is used for learning assessment which involves the students in learning process, either individually or in group. This study was research and development (R&D) model which adopted 4D model (Define, Design, Develop, and Disseminate). Prototype of this application was developed by utilizing internet media (online). The users of this system are lectures and students. In this application, the lecturers has roles to design the scoring rubrics, give the tasks, and assess the students' tasks. The students had roles to submit the tasks, and assess their pairs. Every kind of scoring rubrics can be implemented in some subject matters which have same characteristics (sharing rubrics). This application is expected to produce transparent, quick, correct, and accurate scoring. It is also efficient for lecturers in students assessment. The online-based application design will make this application become an electronic assessment which can be accessed whenever and wherever.

1 Introduction

The competence-based education paradigm covers curriculum, pedagogy, and assessment which stress the importance of standard or result [1] the implication is that syllabus and assessment development have to enable the students to demonstrate knowledge, and skill. The educational system applied to the students in competence based education largely determines students' success in learning achievement. This assessment process is often neglected in teaching. The assessment performed should not only in the form of final evaluation as to know the final learning achievement, but it should put more emphasis on learning process.

The birth of Curriculum 2013 (often its called K13) brings a reasonably great impact on the change in education in Indonesia. The impact on the change that occurs is not only in learning process but also in assessment process of the students. Assessment in K13 has shifted toward a better one, in which formerly it only measured competences in terms of learning achievement only (which used to use a test) toward authentic assessment which measures competences in attitude, skill and knowledge based on process and product. This assessment process also encourages the use of students' portfolios as a major assessment instrument. In relation to portfolio assessment, there is one thing which is very

* Corresponding author: k.setemen@undiksha.ac.id

interesting, that is the existence of assessment and feedback given by peers and this is called peer assessment. Peer assessment is a process in which an individual assesses the number, level, value, or success of his or her peer [2]. The characteristics of peer assessment includes a process in which a members of a team assesses each other's work, thus, the goal setting has to be fully understood by the students. Peer assessment as an alternative assessment gives freedom to the students to express opinions. The assessment is based on performance (performance-based assessment) so that in the peer assessment learning also occurs, the students develop cognitive skill and social skill [3].

Research on assessment and evaluation have been done by many researchers in education sector. Setemen [4] did a research on developing online-teaching evaluation development. The functions that have been accommodated are to manage students' data, to manage test item data, and some system settings to regulate the user's accessing right. This on-line system has also been tried out with the students in small groups, and can be accepted well by the students as an assessment system which is computer-based.

Improvements in learning achievements of the students, in this case university students, cannot be separated from the types of assessment used in learning process. This was proven by the study conducted by Setemen [5], in which the study done was an experiment comparing the use of the portfolio assessment and project assessment in computer programming. The result of portfolio assessment was better than that of the use of project assessment in computer programming. It is pointed out strongly that in portfolio assessment process the assessment is done continuously.

From the result of assessment that has been shown it can be seen that assessment process plays an important role in learning, in which the assessment done is oriented toward the students. This agrees with the change in teaching process from teacher-oriented to student-oriented. This means that not only in learning the orientation is toward the students, but also as long as it is possible, in the assessment process the students can also be involved. Thus, recently, the terms self-evaluation and peer assessment started to be used. Actually, self-evaluation and peer assessment are elements in portfolio assessment. Regarding peer assessment, some researchers have also published research results. An example is the study conducted by Hämäläinen, et al. [6] in their study they applied peer review in the programming tasks. The benefits gained from the application of peer-review are: 1) the students can learn from other students and receive a feedback, and 2) the teacher is helped with the assessments that have been done by other students. Another study was done by Wing-Shui [7] on the effect of peer assessment and feedback strategy on computer programming course. The finding shows that the students felt satisfied with peer assessment and feedback strategy in Computer Programming Course. Besides, their performance was also better than those with the use of traditional teaching method.

From some studies that have been mentioned above, peer assessment or peer review has been conducted by using computer as an aid. However, the assessment process done was limited to the course material used by the researcher. In the study, the features in the application made has not expressed the followings: 1) whether the system can be used for materials for other teaching practices, 2) whether the assessment rubric can be used by other teachers and 3) whether the peer group formed is individual or a peer group.

2 Method

This research is research and development (R&D) model. This study was a teaching assessment development done in two phases by adopting the 4D model (Define, Design, Develop and Disseminate). The first phase was determined as Define and Design phases, aiming to design a web (on-line) peer assessment prototype. In this first phase, the study started with a descriptive study which aimed to determine and to decide the conditions, empirical and theoretical prerequisites of Peer Assessment model which was going to be developed. The main activity was doing a series of need assessment. This need assessment activity would be done through library study and field survey. The next phase was Develop and Disseminate stage which was basically aimed to develop a model and its valid and reliable materials. The first activity is a limited model and its materials try-out. This was done through two activities. The first activity took the form of a limited try-out of the model and its materials. The second activity was a try-out in the form of an experimental research to test the strengths of the model and to see the effectiveness of its model.

3 Results and Discussion

3.1 Define Phase

In the define phase, a need assessment had been done. The need assessment took the form of a library research related to peer assessment and field survey which targeted SMA and SMK teachers from some schools in Buleleng Regency. The library study conducted aimed to compare some results of studies done by researchers in the computer and education sector which took the data from some scientific articles. This library study is important to see the position of the researcher in the study that would be done. After the study of some articles, it turns out that the related studies to the development of computer-based evaluation application that would only be used by the researcher. This means that if the application developed would be implemented in the material of other teaching materials, then modifications are needed in the application to be developed. Besides, in the developed application there has not been available a feature to share an assessment rubric. Thus, any if there is a topic that needs the same rubric, the teacher has to rebuild the rubric. Based on the analysis, it is clear that at this define phase, the most competitive feature in the prototype of assessment developed is that it can be used by anyone as a teacher who has to enlist in the system through the administrator. Another feature is the facility to share rubrics which enable the use of the same rubric among teachers for different teaching materials.

The second activity in the need assessment is field survey which targets teachers at SMA's and SMK's in various schools in Buleleng Regency, Bali. The number of the respondents in the field survey was 103 teachers. The survey was conducted by distributing questionnaires that are related to assessing students. There were 3 (three) important categories asked in the survey, namely: 1) planning the assessment, 2) assessment strategy, and 3) implementation of assessment. The description of each category and the result of the survey can be seen in Table 1, Table 2 and Table 3.

Table 1. Results of the survey about assessment planning by the teacher.

No	Statement	Percentage of answers	
		Yes	No
1	Always plans an assessment in the teaching process	98%	2%
2	Involves teacher/ colleage in planning an assessment	35%	65%
3	Uses standard guide in assessing	94%	6%
4	Uses the existing assessment instrument/rubric in assessing	90%	10%
5	Determines by oneself type and criteria of assessment	26%	74%
6	Assessment criteria are open (communicating with the students)	91%	9%

Table 1 shows that the teacher has planned assessments well. However, in planning the assessment, the involvement of a teacher or colleague in planning is below 50%. This can be seen from point 2 in Table 1, that the involvement of other teachers or colleagues in planning assessments is only 35%. Besides, based on Table 1 point 5, the teacher tends to determine by him/herself the type and criteria of assessment only 26%. This occurred since the teacher in general uses the existing standard guide.

Table 2. Results of survey on assessment strategy used by the teacher.

No	Statement	Percentage of answers	
		Yes	No
1	Assessment only involves the teacher	58%	42%
2	Assessment involves other teachers in its process	42%	58%
3	Assessment involves students in its process	22%	78%
4	Gives the chance to students to assess him/herself	21%	79%
5	Give the change to a student to assess other student individually and in group (peer group)	26%	74%
6	Assessment is done through the colllaboration of teacher, another teacher, and students	45%	55%

Table 2 shows that in terms of strategy used by the teacher, the involvement of another teacher, and students in assessing is below 50% on average. This means that the teacher still plays the central role in his or her students' assessment process. Thus, the teacher has started to do an assessment by involving another teacher and students so that the assessment process is more objective. With the presence the students' assessment, then indirectly, the process of knowledge creation occurs in the students.

Table 3. Results of the implementation of scores done by the teacher.

No	Statement	Percentage of answers	
		Yes	No
1	Assessment is done manually	83%	17%
2	Assessment is done with the aid of computer (for example, using Excel application or the equivalent)	88%	12%
3	Assessment is computerized (using the existing assessment system)	24 %	76%
4	Assessment is done through local network or wide network (on line)	25%	75%
5	Assessment is just in a day (has to be complete at the time only)	28%	72%
6	Assessment is done in a range of time	91%	9%
7	Assessment is done manually	83%	17%

Table 3 shows that in implementing assessments of the students, many teachers have not used computer-based assessment system. This can be seen in Table point 3, only 24% of teachers used computer-based assessment system.

3.2 Design Phase

After a library research was done and after analyzing needs related to peer assessment with self assessment features, the strong features in this computer-based peer assessment are: 1) a teacher can create the course; 2) The teacher can create the class to be taught;

3) the teacher can create the group of assessments; 4) the teacher can create or choose rubrics to be used in assessing ; and 5) the teacher can create or start assessment. Broadly, the design of the prototype of this computer-assisted peer assessment looks like the one shown in Figure 1.

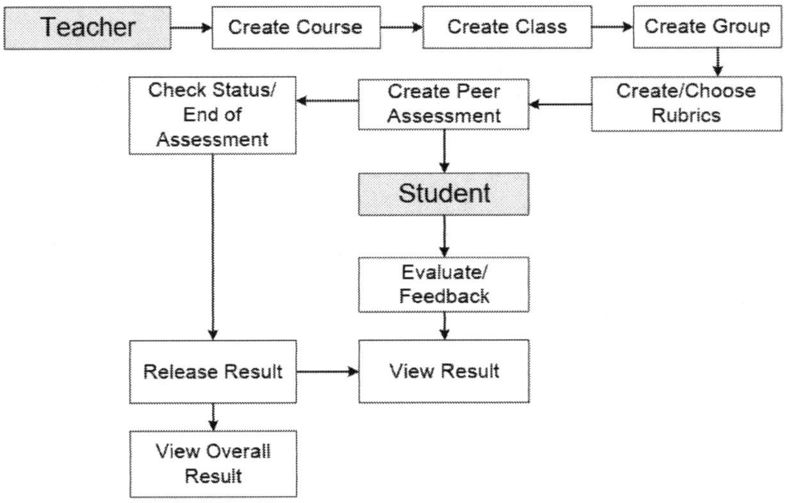

Fig. 1. The Flowchart of Peer Assessment.

Figure 1 shows the steps that have to be taken by a teacher in doing peer assessment. Before the teacher does peer assessment, he /she has to create the course first. The teacher can create more than one course. The next step is to create a class at this step, the teacher has to enter the class and class members who take the course which was created before. Every course created has to have a class and class members who take the course. After creating the course and class members, the next step is to create a group of assessors for peer assessment. The group of assessors created are both individuals and some individuals in a group. This group is automatically created by the system, according to the number of members per group who are entered by the teacher. The next step is to create an assessment rubric or to choose from the already existing ones. The creation of a rubric starts from the determination of components, weight of each component and its descriptor and its score range. In addition to creating a rubric, the teacher also has to choose or use a rubric that already existed before if it is still relevant. Then the teacher creates or maps out the assessment to be given to the group of assessors.

The task of the student is to do the assessment according to the map given by the teacher. In addition, he or she can also give comments on the work that he/she assesses. If the time for assessment is up, then the teacher will announce the results of all of the groups. Thus, the students can see the comments given on the works that they did. The comments given by the group of assessors will become inputs for every student to do correction to the materials that he/she has learned.

4 Conclusion

The process of assessment by using peer assessment needs much time and energy. This is caused by the great variation in the results produced in an assessment in which an individual can be assessed by a group of students so that the total number of the results become the square of the total number of the groups. Therefore, by developing the prototype of computer-based peer assessment it will help a teacher to correct and give a final assessment to a learning process. Computer in this process does not only serve as an aid but it has become part of the process of assessment. If this assessment application can facilitate the teacher in assessing the students in assessing the students, the assessment will become more transparent, quick and accurate. The peer assessment application is designed to be on-line based, so that it becomes an electronic assessment and can be accessed from anywhere and whenever as long as there is internet connection.

This research was funded by Ministry of Research, Technology, and Higher Education of the Republic of Indonesia.

References

1. H. B. Uno, *Perencanaan pembelajaran* (Bumi Aksara, Jakarta, 2011)
2. K. Topping, AERA **68**, 249 (1998)
3. Rochmiyati, Jurnal Penelitian dan Evaluasi Pendidikan **17**, 333 (2013)
4. K. Setemen, Jurnal Pendidikan dan Pengajaran **43**, 207 (2010)
5. K. Setemen, *Seminar nasional forum pimpinan pascasarjana LPTK indonesia* (Undiksha Press, Singaraja, 2014)
6. H. Hämäläinen, V. Hyyrynen, J. Ikonen, J. Porras, Int. Jour. on I. T. & S. **1**, 3 (2011)
7. W. S. Ng, Informing Science and Information Technology **9**, 17 (2012)

An Analysis on Communication strategies Employed by Indonesian English as Foreign Language Adult Students

Mohamed Nizar, *Putu Kerti* Nitiasih, and *Wayan* Suarnajaya

English Education, Ganesha University of Education, Singaraja, Bali, Indonesia

Abstract. As an EFL country, English Language Teaching (ELT) in Indonesia should be more focused on developing students' strategy, particularly communication strategy (CS). The reason is because these students are triggered by instrumental motivation in learning English. However, awareness toward CS should also be possessed by teachers so they can develop students' CS through certain way. Unfortunately, it is rarely found English teacher who is aware about this notion. Thus, this study was conducted to provide data about initial CS of adult students in private university in Bali. According to experts, this age level has already got fully developed cognitive and linguistic capabilities. Thus, they are considered to be able to employ and use CS well. Through this study it is found that they possess 20 initial CSs out of 38 CSs from the referred taxonomy. The findings of this study could be beneficial for the development and novelty of future research concerning with CS and to enhance teachers' awareness toward CS, particularly the ones who establish private course offering intensive speaking program as well as for the lecturers in the concerned university.

Keywords: Communication Strategies, EFL, Adult Students, English Language Teaching.

1 Introduction

Strategy in language learning can be defined as actions or thoughts that a person uses in solving problems he/she encounters while learning or speaking in foreign language [1]. This strategy is divided into two divisions, 'Learning Strategy' and 'Communication Strategy' (CS) [1]. In relation to English language teaching, these strategies are equally important. However, considering the trends of English teaching nowadays is to develop communicative skills more [1], CS is the strategy that should be well mastered by students and those who are learning English, especially the ones living in EFL countries.

In educational context, awareness of CS, should not solely be possessed by students/ language learners. Teacher should also have a good mastery of it. By being mindful to students' CS, teacher could develop students' initial CS by implementing what is called Strategies-Based Instruction (SBI) in language learning. Experts states that strategy mastery, especially CS, could lead to autonomous learning [2]. Therefore, understanding and knowledge toward students' CS is really essential to be possessed by foreign language teachers so that they can their students' CS.

However, it is seldom found teachers that really pay well attention to this strategy. This normally occurs in countries in which English is taught as foreign language; for example Indonesia. Since English is taught as foreign language, students in Indonesia rarely speak English on regular basis in their daily activities. Teachers tend only to focus on developing students' competency according to the syllabus and curriculum. More specifically, in a private university in Indonesia, it is known that there are still found lectures who have insufficient knowledge about strategies used by language learners, including CS [3]. The main reason is because these teachers graduated from linguistics or literature department instead of teacher training and education program. Corresponding to the observation, similar matter also occurred in the university where the study took place. Based on the researcher's experience while doing teaching practice program in 2017, some of lecturers would prefer to develop students' speaking skill through other teaching methods instead of concerning on developing the students' initial CS. This fact is also supported by the statement of the students obtained from personal interview. Hence, some students of this private university were chosen as the subject of this study.

Besides, the reason why adult students were observed is because experts' descriptions on adult students imply that they are considered to be more able to use CS well. They are also considered to have a mature cognitive and linguistic ability and tend to use their prior knowledge to overcome problems while speaking in L2 [4]. Moreover, the reason why adult students learn L2 in Indonesia as EFL country is mostly based on instrumental motivation (i.e. to be able to study or to work abroad or at least to get a job in English fields to get considerable salary). Pertaining to the instrumental motivation, good communication skill is needed for every job provider certainly needs employees who have good communication skills. Thus, CS of these adult students is really important to develop.

In order to develop students' CS, teachers should definitely identify them first. This study was conducted to provide data regarding to students' initial CS to develop and recommendation for the concerned teachers to implement method or teaching technique to develop students' CS. It is expected that the findings could be beneficial and could enhance teachers' awareness toward students' CS especially for teachers who are not majored from teacher training and education.

2 Methods

The research was conducted in the form of qualitative design. Judgment sampling was applied in determining the sample of this study. There were 16 (8 males & 8 females) 6th semester students majoring in teacher training and English education in Mahasaraswati university, Bali, Indonesia. These students age 20 to 21 years old. In addition, in this study data were collected through Documents or artifacts analysis, Observation and Interview. This study also involve some instruments as data collection tools such as: the researcher himself as the main instrument, observation sheets, video recorder, interview guide and an interactive oral production task: role play instruction. Pertaining to role play instruction, students were basically required to perform speaking performance in which they were told to pretend as a tour guide or tourists visiting a certain place for approximately three and half minutes. Students, pretending as tour guide and tourist, were then told to explain and to have conversation related to things representing Hinduism in Bali and Balinese cultures. Their performances were recorded and were later analyzed.

Technically the recordings were transcribed and every CS found was coded and classified based on CS taxonomy proposed by Dörnyei & Scott [5]. There are 41 types of CS based on the taxonomy, they are as follows: Message Abandonment (1), Message

Reduction (Topic Avoidance) (2), Message replacement (3), Circumlocution (paraphrase) (4), Approximation (5), Use all of purpose words (6), Word coinage (7), Restructuring (8), Literal translation (transfer) (9), Foreignizing (10), Code switching (language switching) (11), Use of similar sounding words (12), Mumbling (13), Omission (14), Retrieval (15), Self-repair (16), Other-repair (17), Self rephrasing (18), Over-explicitness (waffling) (19), Mime (nonlinguistic/ paralinguistic strategies) (20), Use of fillers (21), Self-repetition (22), Other-repetition (23), Feigning understanding (24), Verbal strategy makers (25), Direct appeal for help (26), Indirect appeal for help (27), Asking for repetition (28), Asking for clarification (29), Asking for confirmation (30), Guessing (31), Expressing non-understanding (32), Interpretive summary (33), Comprehension check (34), Own-accuracy check (35), Response: repeat (36), Response: repair (37), Response: rephrase (38), Response: expand (39), Response: confirm (40), Response: reject (41). Once classification is done, conclusions were then drawn by previously matching the obtained data with the related books (theoretically) as well as with the related findings from similar researches.

3 Findings and Discussions

Based on the observation, adult students are known to consistently employ 20 of 41 types of communication strategies referred. They are Use of fillers (21), Self-repetition (22), Retrieval (15), Mime [the use of non-linguistic act] (20), Self-repair (16), Indirect appeal for help (27), Code switching (11), Mumbling (13), Circumlocution (paraphrase) (4), Message abandonment (1), Restructuring (8), Asking for confirmation (30), Approximation (5), Expressing non-understanding (32), Asking for repetition (28), Self-rephrasing (18), Other-Repetition (23), Own-accuracy check (35), Message reduction [topic avoidance] (2) and Omission (14). For more detailed display as well as to ease understanding the explanation, Table 1 is provided below.

Table. 1. Communication strategies consistently employed by adult students

CS consistently employed	Frequency of the employed CS in each performance		Number of students employing the CS in each performance	
	1st performance	2nd performance	1st performance	2nd performance
(21)	567	571	16	16
(22)	43	45	12	14
(15)	61	48	13	13
(20)	9	21	8	9
(16)	16	17	10	12
(27)	25	17	12	7
(11)	3	10	3	5
(13)	7	9	5	7
(4)	3	7	2	3
(1)	3	7	3	6
(8)	4	6	4	5
(30)	10	6	8	6
(5)	1	5	1	3
(32)	5	4	4	4
(28)	1	3	1	3
(18)	1	2	1	2
(23)	4	2	3	2
(35)	1	2	1	2
(2)	1	1	1	1
(14)	1	1	1	1

Pertaining to the use of CS no. 21 (use of filler), students tend to say "aah", "umm" or extend certain words as the filler to gain time for thinking what to say next; for example, student no. 9 said,"...before aah(21) ceremony *nyepi nyepi* and we doo(21) tooo(21) goo(21) aah(21) to go the beach ..." (Selected from the transcription of adult student no. 9).

In employing CS no. 22 (Self-repetition), students tend to repeat words they previously pronounced to gain time for thinking what to say next; for example, student no. 1 said, "...before aah(21) *nyepi* day we we we do(22) a parade..." (Selected from the transcription of adult student no. 1).

The use of CS no 15 (Retrieval) is dominated by series of incomplete or wrong forms or structures (sentences or words) that are pronounced before reaching the optimal form; for example "...will be held for the tena' teenager(15) so the teenager will..." (Selected from the transcription of adult student no. 2).

Furthermore, the form of CS no. 20 (Mime [the use of non-linguistic act]) employed by these students was kind of visual description using hands; for example "...do you know drum? Drum? (*exemplifying drum with his hands)(20)..." (Selected from the transcription of adult student no. 6).

Next, the use of self-repair as CS no. 16 is dominated by self-initiated corrections of errors made by the student when they spoke; for example, "...popular name *nyepi* is aah(21) silent way eh silent day(16)." (Selected from the transcription of adult student no. 12).

In relation to the use of CS no. 27 (Indirect appeal for help), adult students are more likely overcome their L2 inaptitude by asking for help from the interlocutor indirectly through face expression and eye contact; for example, student no. 14 said, "...and a' red for brahma and purple for aah(21) (*appeal for help by making an eye contact)(27) aah(21) *siwa* (Shiva)..." (Selected from the transcription of adult student no. 14).

The following strategy employed by the students is CS no. 11, Code switching. These students simply use either L1 or L3 words or phrase in their L2 speech once they had no idea what to say; for example, student no. 2, "...will cut your point(5) (*pointing to his front teeth)(20) or *gigi seri* (incisor)(11)..." (Selected from the transcription of adult student no. 2).

Another CS employed by the students to gain time for thinking what to say next is CS no. 13, Mumbling. Once they stuck in the middle of their speech, some students occasionally construct what they want to say correctly by muttering inaudibly; for example, student no. 15 said, "...aand(21) aand(21) rice paddies okay so (*paused for some seconds mumbling)(13) usually wee(21)..." (Selected from the transcription of adult student no. 15).

Some of adult students were also known to employ Circumlocution (paraphrase) or CS no. 4 in which they give illustration, description or example of certain objects in overcoming their L2 inaptitude; for example, student no. 16 said, "...how we make it? we use *semat*. *Semat* is very small thing (*describing it with his hands) and that is made from bamboo(4)..." (Selected from the transcription of adult student no. 16).

On the other hand, the form of CS no. 1 (Message abandonment) is dominated with messages which is intentionally left unfinished due to L2 inaptitude, for example, student no. 13 said, "...we will burn our body aah until mmm and our body will back into the nature(1)..." (Selected from the transcription of adult student no. 13).

Furthermore, Restructuring (CS no. 8) is a kind of CS in which the speaker abandons the executions of verbal plan and substitutes it with other alternative plan instead due to L2 inadequacy. This is one of many examples of this CS employed by student no. 13, "That's all aah about *ngaben* and thank you for, thank you very much and have a nice day(8)" (Selected from the transcription of adult student no. 13).

Additionally, in employing Asking for confirmation (CS no. 30) adult students tend to be kind of confirmation statement or question to assure that one heard or understood something correctly; for example, (Student 4 asked a question, "Aah(21) okay I would like to ask about aah(21) how about if we don't a' don't do it?(15)) (*Student 8 confirmed the question by whispering)(30) (Selected from the transcription of adult student no. 8).

Other strategy used by adult students is Approximation (CS no. 5). Its form is a kind of single alternative lexical items (e.g. superordinate or related terms) which share semantic features with the target word in L2 speech; for example, student no. 9 said, "...function in *melasti* is aah(21) aah(21) foor(21) for the Balinese people(15) aah(21) a' clear body (*laughing feeling embarrassed) clear your body(5)..." (Selected from the transcription of adult student no. 9).

Furthermore, the form of Expressing non-understanding (CS no. 32) that mostly employed by the students is a kind of re-explanation request through face expression; for example, (*a female student asked a question, "*speaking in Japanese* aah(21) could you tell me umm(21) aah(21) only woman that can do the baleganjur or not? *student 6 was frowning*(32)..." (Selected from the transcription of adult student no. 6).

Referring to the use CS no. 28, Asking for repetition, adult students tend to use the regular phrase such as "pardon" or simple repetition request; for example, (*Student 16 asked a question, "okay my name is made *the mic. is screaming* check my name is Made and I'm froom(21) California, aand(21) is that just something like traditional drum or anyone else umm(21) to go with baleganjur?) *"pardon me?", student 6 asked for repetition*(28)..." (Selected from the transcription of adult student no. 6).

In addition, Self-rephrasing (CS no. 18) can be described as a paraphrase of either sentence or word previously said to make clear the intended meaning. Here is one of the examples of this strategy employed by student no. 11, "...I think *lawar* aah spicy or not its depend aah of what consume it I think I think its depend on who want to eat(18)..." (Selected from the transcription of adult student no. 11).

Other-Repetition (CS no. 23) is other CS which is also employed to gain time to think. Its form can be a repetition of what the interlocutor said; for example, (*Student 10 asked a question, "mmm(21) my name is Ogek from Singapore aah(21) I wanna ask you a' how many times aah(21) do the melasti?") "okay how many times(23), the first is..., Student 4 answered" (Selected from the transcription of adult student no. 4).

Own-accuracy check (CS no. 35), as the other initial CS possessed by adult students, can be described as a strategy in which the speaker check that what him/herself said was correct by asking a concrete question to the interlocutor or repeating a word with a question intonation. Here is example of it selected from the transcription of adult student no. 16, "Student 16 said, "...we don't have pen or pencil yet in the past so we use kind of very aah(21) shave (*his voice raising when saying 'shave' as if he is checking his accuracy(35)..."

The last, there are 2 CSs which were equally employed for 1 time in both performance. They are namely Message reduction [topic avoidance] (CS no. 2) & Omission (CS no. 14). The form of the former one is kind reduced message due to L2 inaptitude, here is the example as selected from the transcription of adult student no. 11, student no. 11 said, "...I think lawar just can aah(21) consume just for aah(21) one day (*paused for a second) one day(2)..." (*She was actually about to proceed but canceled it because it will lead to aspecific topic she couldn't tell well). on the other hand, the form of the latter one, omission, is exactly the same as described in the taxonomy; here is the example as selected from the transcription of adult student no. 6, student no. 6 said "...That's aah(21) that's aah(21)(22) speci' aah(21) name for Balinese people(14)...".

Before making a conclusion, these findings were compared to findings of similar researches [6, 7, 8, 9, 10] which are conducted in other EFL countries. It is found that adult students in EFL countries tend to employ similar CSs, especially the use of Mime [the use of non-linguistic act], code switching, Approximation, Asking for confirmation, circumlocution, and appeal for help.

4 Conclusions

There are two conclusions in relation to the findings of this study. Firstly, adult students in EFL countries are known to employ similar CSs to cope with their L2 inadequacy. Secondly, CSs identified in this study attest the maturity of cognitive and linguistic ability of adult students in overcoming problems while speaking in L2 as stated by referred expert [4]. These CSs, especially use of fillers (21) and Message reduction [topic avoidance] (2), are students' good assets to develop as it reveals capability to provide sense of security for themselves so that they can do sort of maneuver to help them deal with difficulties in delivering message just like what expert says [11]. It is expected that these findings could be beneficial for the development of future research concerning with CS and to enhance teachers' awareness toward CS, particularly the ones who establish private course with intensive speaking program as well as for the lecturers in the concerned university so that they can develop students CS through SBI or other ways.

I would like to express my full gratitude to The Almighty God "*Ar-rahmaan Allah Subhanahu Wa Ta'ala*" for His blessings to fininsh this article. Furthermore I would like to address my sincere appreciation to my beloved family, Abah, Mama, Faisal and Ira, for their total support. At last, I would like to address my gratitude and appreciation to my principal supervisor Prof. Dr. Putu Kerti Nitiasih, M.A. and my second supervisor Drs. I Wayan Suarnajaya, M.A., Ph.D., whose patience, expertise, motivation and gentle guidance have helped me to finish this article well.

References

1. H.D. Brown, *Principles of Language Learning and Teaching (fifth Edition), (*New York: Pearson Education, 2007)
2. A.L. Wenden, TESOL Newsletter, **19**, 1-7 (1985)
3. S. Wahyuni, *L2 Speaking Strategies Employed By Indonesian EFL Tertiary Students across Proficiency And Gender*, (Canberra: University of Canberra, 2013).
4. A.F.V. Smith, G. Strong, *Adult Language Learners: Context and Innovation* (Maryland: TESOL Publications, 2009)
5. Z. Dornyei, M.L. Scott, Communication Strategies in a Second Language: Definition and Taxonomies. LL., **47,** 173-210 (1997)
6. C.A.R. Cervantes, R.R. Rodriguez, Gist: Edu. n. L. Rsrch. J, **6**, 111-128 (2012)
7. Ş. Yaman, P. Irgin, M. Kavasoğlu, JESR. **3**, 255-268 (2013)
8. R.A.A. Alawi, PJEL, **2**, 1-11 (2016)
9. N.T. Thu, N.T.K. Thu, AERJ, **4**, 283-287 (2016)
10. R.L. Ugla, N.I. Adnan, M.J.Z. Abidin, IJERE, **2**, 44-50 (2013)
11. Z. Dornyei, On the Teachability of Communication Strategies. TQ, **29,** 55-85 (1997)

Text Attributed with Language Attitude in Indonesian Language Instruction as The Efforts to Knit Nationalism in a Frame of Diversity

I Putu Mas Dewantara

Universitas Pendidikan Ganesha

Jl. Udayana No. 11, Singaraja, Bali. 81116, Indonesia

Abstract. This research aims to 1) describe the use of text attributed with language attitude in Indonesian language instruction and 2) describe student's response to the instruction. It was a descriptive research which data were collected by using observation method and questionnaire. The subjects were 4 lecturers of Indonesian language and 258 students of Unversitas Pendidikan Ganesha. The results of this study indicate that 1) text attributed with language attitude was applied in 4 main steps, namely (1) preparation of conditions, (2) reading of text in groups, (3) intergroup discussions, (4) inferences; 2) Student response in terms of interest, 55.91% of students were very interested, 38.76% of students were interested, and 5.33% were less interested. In terms of textual ability to inspire the sense of nationalism, 63.70% of students felt the text was very capable to do it, 32.43% of students felt the text was capable enough to do it, and 3.88% of students felt the text was less inspiring to raise nationalism within the frame of diversity. Thus, learning was able to attract students and the text used was able to awaken nationalism within the frame of diversity. The implication is the existence of teaching materials to nurture nationalism.
Keyword: text, language attitude, Indonesian language, nationalism, diversity

1 Introduction

Recently, the discussions of nationalism in the framework of diversity are widely practiced. This can not be separated from various problems related to the spirit of nationalism that occurred. There has been a degradation of the spirit of unity[1,2]. Therefore, it is necessary to build multicultural awareness to foster the spirit of nationalism in a diversity frame. Multiculturalism is a spirit that appreciates diversity, whether cultural diversity, ethnicity, religion, or other forms of diversity [3,4,1,5]. Multicultural awareness is the instrument of national therapy [6]. In addition, an understanding of nationalism, identity politics, and a sense of solidarity also needs to be enhanced to foster a spirit of nationality in order to overcome this nation's growing problems [7,8]. Diversity should have been viewed as a destiny of God's grace from a long process of the nation's journey that must be cherished [9].

Permendikbud No.21 of 2015 on Character Building Movement at School is a form of understanding of the awareness of national identity and nationalism. In addition, the slogan "Aku Indonesia, Aku Pancasila" is also a sign of the necessity of strategic efforts in fostering the spirit of nationalism within the framework of diversity in the present. One of various strategic efforts such as by applying the project model of citizens in the learning of Citizenship [10]. The result of the implementation of this model is able to develop students' nationalism attitude. The study of the movie "Land of Heaven Said" as a media for the growth of students' nationalism attitude has also been done [11]. The results of this study indicate the influence of film media as a medium for growing nationalism attitudes. It seems that fostering the nationalism spirit needs to be done through the review of nationalism history itself [12,8].

The growth of nationalism spirit is also done in learning Indonesian and Literature. The study of the novel and its implications for Indonesian Language and Literature learning has been done in an effort to foster nationalism [13,14]. Another step taken is the development of anti-corruption education materials integrated with Indonesian language learning which one of its goals is to foster nationalism [15]. Until now there has been no research on learning Indonesian in Higher Education in an effort to grow nationalism. Therefore, this research tries to offer an attempt to overcome the problem of nationalism in the framework of diversity. The effort is through a text charged with language attitudes. The content of language attitudes can be used as a means to love the Indonesian language in particular and the Indonesian nation in general. Indonesian is the foundation of the nation [16]. This is the adhesive language of the nation.

2 Method

This research is descriptive research. The data of this research were 1) the steps of using the text of the attitude of the language in Indonesian language learning and 2) the student's response toward the learning. The subjects were 4 lecturers of Indonesian language subjects and 258 students of Universitas Pendidikan Ganesha. There were two methods of data retrieval used in this research, those were observation and questionnaire. The observation was done to know the steps of learning applied by the lecturer and the questionnaire was done to know the student's response toward the learning process. Data analysis was done through three stages, namely data reduction, data presentation, and inference and verification.

3 Results and Discussion

3.1 The steps of Indonesian Language Learning by using the Text of Language Attitudes were applied in 4 steps

1) Preparation

This step aimed to stimulate interest and motivation of students in learning. At this stage the lecturer explains the various problems facing by this nation related to nationalism in the frame of diversity. This stage is an important step considering that motivation is the overall driving force in the learners that leads to learning activities, ensures learning activities, and gives direction for learning objectives can be achieved [17,18]. The results showed that motivation factor contributed 67% to reading culture [19]. Therefore, it is important to generate students' motivation before learning Indonesian by using phenomenon text of language attitudes. Besides storytelling, the other activities that was done by the lecturer in this step was showing the picture or video related to nationalism problem. The use of historical stories and problems in everyday life was aimed to make the problems more easily understood by the students.

2) Reading text in groups

When the students were motivated to learn, the next stage was grouping. The groups were heterogeneous group consists of 3-5 students. The heterogeneous group formations are very appropriate for the college level and can improve learning outcomes [20,21,22]. The next stage was distributing the phenomenon text with language attitude. The students were assigned to read the texts and discuss it in groups and trace information related to the reading material. What students do at this stage was to find the source and choose the best source. A variety of sources of information can be used in the form of photographs, drawings, texts, diagrams, videos, etc. so it is necessary to sort out related sources that can be used to better understand the text of phenomena distributed by professors [23]. The investigation of the information wasa divided into 5 stages, namely (1) determining the information needs, (2) accessing the information needed effectively and efficiently; (3) providing critical evaluation and sorting information; (4) understanding the information; and (5) using it ethically And legaly [24]. The result of the research showed that the steps had been done by the students with the instruction of the lecturer.

3) Intergroup discussion

At this stage, the students did intergroup discussion by conveying their understanding of the text that had been read and enriched by other information. Other groups should listen and give comment on the other groups' understanding. This intergroup discussion aims to re-evaluate the understanding like what had been done by other groups at the previous step.

4) Conclusion

This step is a step that was done together by students and lecturers. The steps were done by the lecturers, where the lecturers gave students the opportunity to make conclusions regarding to the texts that have been distributed. The lecturers seen to help the students to conclude the text in accordance with the results of the discussions that had been done.

3.2 Student Response

The student's response toward Indonesian Language Learning by using the Text of Language Attitudes could be seen from the students' interest. It can be seen in Figure 1.

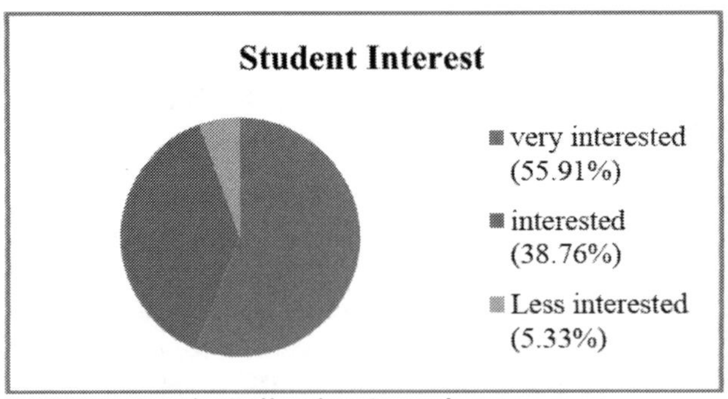

Fig. 1 Student Interest Response in Following Learning

Figure 1. shows that students are interested in learning Indonesian by using text of language attitudes. Four questions are used to determine the interest of students in following the learning process, such as (1) the attractiveness of learning, (2) the feelings of learning, (3) motivation, and (4) the desire to learn by using text of language attitudes. During this Indonesian language learning has been running well, but cannot attract the students' interest due to the limited resources or learning media used by the lecturers [25,26]. The att ondance of language-laden text is expected to overcome the weaknesses of previous Indonesian language learning.

The ability of the Text of Language Attitudes in arousing the spirit of nationalism in the frame of diversity can be seen in Figure 2.

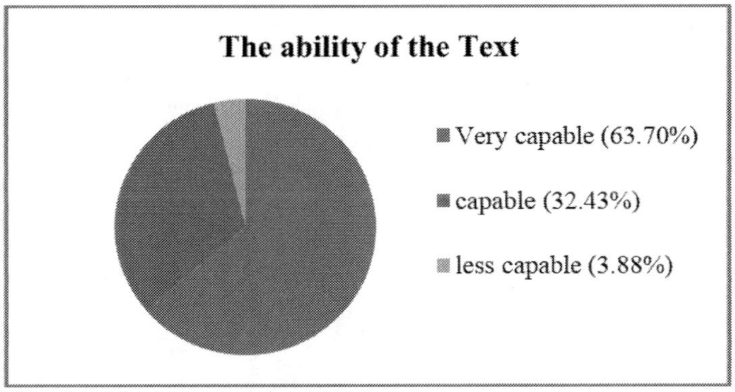

Fig. 2 Text Ability Awakens the Spirit of Nationalism

Figure 2 shows that the text of the attitudes of language is capable in arousing the spirit of nationalism in a diversity frame. The six indicators used to measure nationalism are (1) the position and role of the Indonesian language, (2) awareness of threats and challenges, (3) awareness of diversity, (4) awareness of the role of self in maintaining nationalism, (5) the love of the motherland, and (6) positive attitude toward language and nation. The results of this study were emphasizing on the premise that education, language, and culture are important factors in fostering nationalism [27]. In addition, the results of this study also reinforce opinions that emphasize the relationship of nationalism and language learning by saying that the formation of nationalist attitudes can be done through language learning [28]. This is also done by Malaysia through 'Enforcing Bahasa Malaysia and Strengthening English [29] as well as by the Taiwanese community through the curriculum of' Mother Language Education [30]. Therefore, language learning is packed using text of language attitudes with the goal of developing positive language attitudes that can arousing nationalism within a diversity frame. A strong and prominent national identity can overcome ethnic barriers to believe in diverse communities [31].

4 Conclusion

Indonesian Language Learning by using the Text of Language Attitudes is conducted through four main stages: preparing conditions, reading in groups, intergroup discussions, and inferences. Students interested and feel that the text of language attitudes used in learning process can foster the spirit of nationalism. The implication of this research is that the text can be used as a medium of growing nationalism to unite nationalism in the frame of diversity.

References

1. A. M. A. Shofa, J. Pancasila dan Kewarganegaraan, **1(1)**, 34–41 (2016)
2. L. Agung, Int. J. Hist. Educ., **12(2)**, 392–403 (2011)
3. D. A. Tanudirjo, *Membangun Pemahaman Multikulturalisme: Perspektif Arkeologi* (Solo, 2011).
4. G. Lestari, *J. Pendidik. Pancasila dan Kewarganegaraan*, **28(1)**, 31–37 (2015)
5. R. B. Pratama and S. D. Kharisma, "Insight oriented multicultural education nationality for resilience in moral era mea," in *Proceedings International Seminar FoE (Faculty of Education)*, 2016, pp. 71–83.
6. A. Abdullah and Hannati, "Nasionalisme Dipertanyakan Kembali (Renungan Perjalanan Sejarah Bangsa 71 Tahun Indonesia Merdeka)," in *Pendidikan Ilmu-Ilmu Sosial Membentuk Karakter Bangsa Dalam Rangka Daya Saing Global*, 2016, pp. 377–390.
7. M. Z. Alfaqi, J. Pancasila dan Kewarganegaraan, **28(2)**, 111–116 (2015)
8. R. Wodak and S. Boukala, J. Lang. Polit., **14(1)**, 87–109 (2015)
9. G. Kasnowihardjo, J. Berk. Arkeol., **36(2)**, 161–172 (2016)
10. H. Maksum, PIONIR J. Pendidik., **3(1)**, 1–11 (2015)
11. R. R. Wijaya, *E-CIVICS*, 5(2), 1–13 **(2016)**
12. M. Z. Alfaqi, *J. Civ.*, 13(2), 209–216 **(2016)**
13. A. Hirata, "Nasionalisme Dalam Novel Sebelas Patriot Karya Andrea Hirata dan Implikasinya Terhadap Pembelajaran Bahasa dan Sastra Indonesia di Sekolah," 2016.
14. S. Kirom, "Novel Tak Ada Lain Karya Suparto Brata (Kajian Psikologi Sastra, Nilai Karakter, dan Relevansinya dengan Pembelajaran Bahasa Indonesia di SMA)," 2015.
15. M. W. Nugraheni, "Pendidikan Antikorupsi dalam Model Pembelajaran Bahasa Indonesia Terintegrasi Siswa Kelas VII Semester I SMP Negeri 1 Tembarak Tahun Ajaran 2010/2011," vol. 12, no. 1, pp. 14–27, 2016.
16. Mahsun, "Merajut Kebinekaan Bahasa sebagai Pemerkukuh Ikatan Kebangsaan." [Online]. Available: http://fkip.untidar.ac.id/wp-content/uploads/2016/10/Mahsun-merajut-kebinekaan-bahasa-sebagai-pemerkukuh-ikatan-kebangsaan.pdf. .
17. M. U. Usman, *Menjadi Guru Profesional*. (Bandung: Remaja Rosdakarya, 2002)
18. P. Buckley and E. Doyle, *I*nteract. Learn. Enviroments, **24(6)**, 1162–1175 (2016)
19. Pranowo and A. Herujiyanto, Masy. Linguist. Indones., **33(2)**, 153–171 (2015)
20. V. D. Tran, A. Giang, and A. Giang, "The Effects of Cooperative Learning on the Academic Achievement and Knowledge Retention," vol. 3, no. 2, pp. 131–140, 2014.
21. R. M. Gillies and R. M. Gillies, Int. J. Educ. Psychol., vol. 3, no. 2, pp. 125–140, 2014.
22. A. Macpherson, *Cooperative Learning Group Activities for College Courses*. Surrey, BC(Canada: Kwantlen Polytechnic University, 2015)
23. A. Nurohman, *J. Kependidikan*, **2(1)**, 1–25, (2014)

24. B. Johnston, S. Webber, and B. Johnston, Stud. High. Educ., **28(3)**, 335–352 (2003)
25. I. K. Dibia, I. P. M. Dewantara, and M. S. Indriani, "Evaluasi Pelaksanaan Pembelajaran Bahasa Indonesia MPK Undiksha," Singaraja, 2015.
26. I. P. M. Dewantara, Int. J. Lang. Lit., 1(1), 19–29 (2017)
27. H. A. . Tilaar, *Mengindonesia Etnisitas dan Identitas Bangsa Indonesia*. Jakarta: Rineka Cipta, 2007.
28. B. Lestyarini, J. Pendidik. Karakter, 2(3), 340–354 (2012)
29. P. Le Ha, J. Kho, and B. Chng, J. Int. Comp. Educ., 2(2), 58–71 (2013)
30. J. Dupré, *Curr. Issues Lang. Plan.*, 15(4), 393–408 (2014)
31. A. L. Robinson, *Comp. Polit. Stud.*, 49(14), 1819–1854 (2016)

The effect of blended learning model on senior high school students' achievement

Iga Setia Utami[1,*]

[1]Department of Informatics and Computer Engineering Education, Universitas Bung Hatta, 25133 Padang, Indonesia

Abstract. The purpose of this research was to determine the effect of blended learning model on senior high school students' achievement. This study used experimental research method with randomized control group pretest-posttest design. The study was carried out with 63 students attending information and communication technology course, where 31 of whom were in the experimental group and 32 of whom were in the control group. In the experimental group, teacher used blended learning as instructional model, while in the control group, the course was taught based on traditional teaching model. Data collected from the result of learning objective test with 35 questions. The research showed that the learning result of experimental group is higher than the learning result of control group. Based on the result of this research, it can be concluded that the blended learning model contributed more to the students' achievement.
Keywords: blended learning, instructional model, students' achievement

1 Introduction

Instructional model is an important component that can affect the effectiveness and efficiency of teaching and learning process. An instructional model serves a good master plan for teaching. Besides, it is also a conceptual framework which describes a systematic procedure in organizing learning experiences to achieve certain learning goals and serves as a guide for the teachers in planning and implementing learning activities [1]. To implement an instructional model, the teachers provide various learning aspects to achieve learning objectives [2,3]. Furthermore, instructional model encourages students to get involve in teaching and learning process. By doing that, student centered learning can be realized.

The rapid development of Information and Communication Technology (ICT) gives different views in the learning process. Conventional or traditional learning is no longer used by the teacher. The teachers are required to use a variety of methods that provide more opportunities to learn by utilizing various sources. The implementation of ICT in learning may change the teacher paradigm of learning from teacher centered learning into student centered learning [4].

The use of technology in education has an important role and it should be applied in the learning system. On the other hand, there is still a learning process which is done by

* Corresponding author: igautami@bunghatta.ac.id

emphasizing the method of lecturing and memorizing. By using this old method, the students have less experience in getting information in the learning activities. Information technology-based learning is one of technology utilization in education and teaching systems. One of the technologies that can be implemented is internet. In this globalization era, it is easy for the teacher and the students to find the internet connection because of spread of computer and the existences of internet connection [5].

Internet technology supports the learning process through online learning (e-learning). E-learning is distant learning environments in which internet and network technologies are used for presenting and receiving the content used [6]. E-learning ensures the flexibility and efficiency which cannot be found in classroom environment. It is allowed the student to learn everywhere and every time.

Moreover, learning is not only based on technology but also it is a process of interaction between teachers, students and learning resources. That is not all learning can be done by online environment [7]. Although e-learning has several advantages, there is also several limitation of online learning environment. It cannot replace learning in the classroom. Face to face learning provides the social interaction which is needed for learning. In other words, face-to-face processes are important and it should not be left behind in learning [5,8,9].

The two instructional models, online and traditional (face to face) model have several advantages and disadvantages. It is better to the teacher to combine the two teaching models. The combination of e-learning and traditional face to face learning may integrate in teaching and learning process [10,11]. It is known as blended learning.

Blended learning is the combination of traditional (face-to-face) and online learning so that instruction occurs both in the classroom and online [11,12,13]. Blended learning integrates the advantages of online learning with some advantages of face to face learning. Then, it maximizes the efficiency of classroom training and provides better reinforcement in the web-based elements. Through blended learning, the students had prepared for the course in the classroom and it can be more efficient because the students may discuss with their teachers and their friend that they could not do during classroom interaction [5].

Because it combines the two advantages of instructional models, blended learning has positive impact to the learning process. Some researchers stated that blended learning can enhance students' learning outcomes, improve students' motivation, and it is effective way for achieving learning objectives [4,5,14,15]. Blended learning also spends lower cost for training and it may enhance the students' learning experience [16,17].

On the other hand, the formula for organizing a well blended learning course is not easy. The teachers encourage to re-think and redesign courses that afford students more, and it requires to the teacher to practice how to teach by using blended learning in the classroom to get the good result [18,19,20]. In short, applying the blended learning as instructional model will take the teachers' serious effort in order to get maximum benefit.

Recently, the use of blended learning as a learning model has increased especially in colleges. However, this learning model also can be used in high schools. This study tried to determine the effect of blended learning model on senior high school students' achievement.

2 Method

In this study applied experimental method. The randomized control group pretest-posttest design was used. Students were divided into two groups, experimental group and control group. Both of the groups were given different treatment. In experimental group teacher used blended learning model (part online, part face to face), while in control group traditional learning model was used.

In the experimental group, students were informed about blended learning and they were introduced how to use online learning. Online learning (available at http://elearningsman1lintau.com) was designed by using Moodle, a Learning Management System (LMS). In this web, before login to the system, students have to register themselves by creating an account on e-learning web as in Figure 1 below

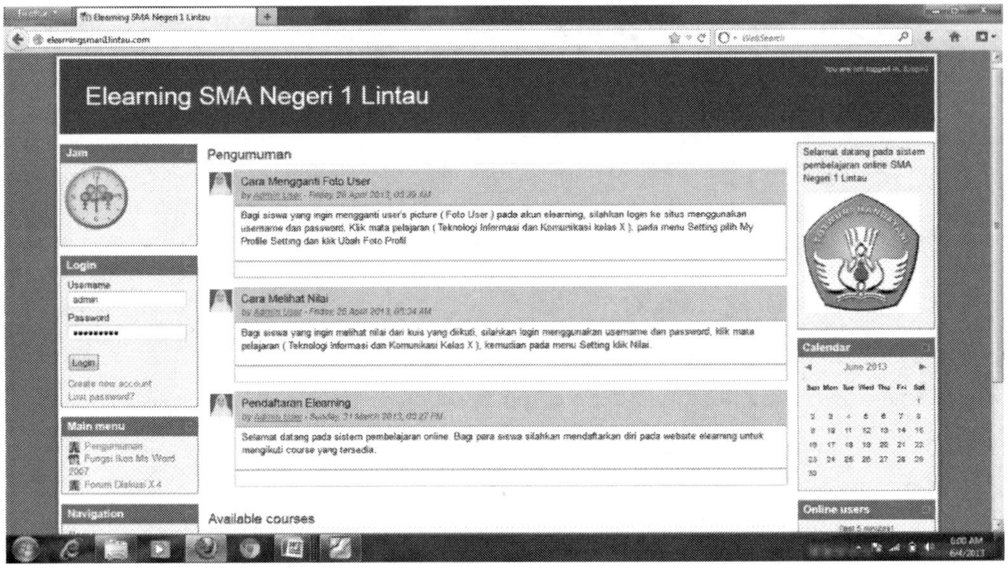

Fig. 1. Homepage screen of online learning

To follow up learning activities, students were asked to login by using their own account. Online learning was design to support face to face learning. In this online learning teacher may give lesson materials, quizzes, assignments and discussion forum. Then, students can follow all the learning activities provided by teacher on the web page.

The population in this study consisted of all of students enrolled in the information and communication technology course at the first grade of SMA Negeri 1 Lintau Buo 2012/2013 academic year. Selecting of sample used random sampling technique. There were 31 students in the experimental group and 32 students in the control group.

Objective test as research instrument in form of multiple choices was conducted to collect data. Before the tests were given, a trial test including 50 questions was established. After analyzing the question items which include the validity, difficulty index, discrimination index and reliability, there were 35 questions were used to achievement test. Then, both groups were given the same pretest and data obtained were analyzed to look at the students' initial ability. Furthermore, each group was treated differently. At the end of the learning both two groups will equally given posttest to see the results of the learning model that has been given. The data obtained were analyzed to know the normality and homogeneity. Finally, a t-test for independent samples was conducted.

3 Result and discussion

Based on the test were given to both of the sample groups, the average of learning outcomes in experimental group was 57.8 for pretest and 82.5 for posttest. While in control group the average scores was 58.3 for pretest and 72.9 for posttest. Those result can be seen in Figure 2.

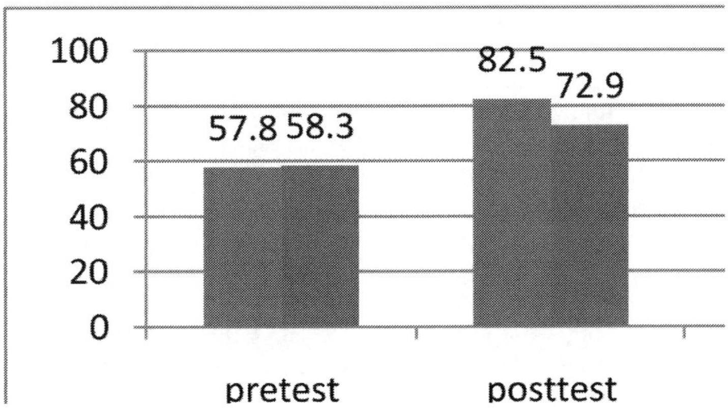

Fig. 2. Student learning outcomes

To investigate the effect of blended learning model in terms of students' achievement, in comparison to traditional instruction, this study conducted a t-test analysis for independent sample.

Hypothesis test was established to scores of pretest. The findings obtained are presented in Table 1 below

Table 1. The result of independent t test for pretest

Groups	N	$\bar{x}$	Sd	df	t	sig
Experimental	31	57.8	6.340	61	-301	0.764
Control	32	58.3	7.215			

Based on the result on the table above, it can be interpreted that there was no significant difference ($t_{(61)}$=-301, sig >0.05) between the experimental group and control group scores. It states that the students in the experimental group and control group had similar levels of knowledge.

At the end of application, the posttest score of the two groups were analyzed. The finding was obtained and it is presented in Table 2 below:

Table 2. The result of independent t test for posttest

Groups	N	$\bar{x}$	Sd	df	T	sig
Experimental	31	82.5	6.117	61	5.657	0.000
Control	32	72.9	7.328			

From the table above, it shows that there was a significant difference ($t_{(61)}$=-5.657, sig <0.05) between score of the experimental group and control group. It can be concluded that the students in the experimental group had higher levels of learning achievement than control group.

The present study was to determine the effect of blended learning on senior high school students' achievement. In this instructional model, learning activities can be done by online and face to face learning, it takes advantages of both two teaching models. By combining these two models, efficient learning can be achieved [21,22,23]. In this study, the submission of material was done through online learning so that face-to-face learning can run effectively and efficiently. Learning can be done anytime and anywhere. It also reduces instructors' workloads and they have more time to work with other material. Furthermore, blended learning helps the students in achieving learning objectives. This result is similar with the resembles studies [19,20].

Blended learning model encourages students to learn more actively. They have opportunity to learn on their own pace. They can prepare themselves for the course before coming to the class and it can realize student centered learning and reflect the value of 21 century education [5,23,24,25]. The research indicated that student who attended blended learning course had higher achievement scores than those who attended traditional teaching. The survey results are in line with other researchers [4,14], who also argued that blended learning effective for increasing students' performance. Most of those studies also found that blended learning improves the academic achievement averages [15,16,26].

The present study was not only give a concern about how to balance the online learning and face to face learning, but also give attention to content to be provided. Blended learning can help teachers to provide several formats of learning material [21]. Then, it can increase student motivation and they can get new experience on learning process. The other relevant study also stated that blended learning can help for enhancing students' motivation [19,24,27]. In this research, content was presented in various formats such as text and also video format. Each learning topic was provided with a quiz as an evaluation. Last, discussion forums were also available as a communication and sharing media between teachers and students and among fellow students.

4 Conclusion

The aim of recent study was to determine the effect of blended learning model on students' achievement. The scores obtained from the learning objective test that applied to the experimental and control group were compared. There is a difference to the learning outcomes of students, where the average learning results of experimental class is higher than control class. The other finding was found that there is a significant increment in student learning outcomes on information and communication technology subjects after using blended learning. Based on the research finding, revealed that the blended learning model contributed more to students' achievement than traditional learning model did. Blended learning can be used as an alternative learning model for teachers in order to support students' learning achievement. To implement this model, teachers should develop their computer and internet literacy rate. School has the important rule to provide training or workshop for the teachers in applying this model. This research is still limited to information and communication subject, it is hoped that further research will be conducted for other courses and also it is suggested to give a concern about how to blend professionally between face to face and online instruction.

I am pleased to acknowledge the assistance for this research received from the principal of SMAN 1 Lintau for allowing me to conduct research there and also from Drs.Sukaya, Muhammad Adri, S.Pd, M.T, Drs.Ahmad Jufri,M.Pd., Drs.Hanesman, M.M and Ahmaddul Hadi, M.Kom.

References

1. U. Winataputra. *Model-model pembelajaran inovatif* (UT, Jakarta, 2005)
2. P.Eggen, D.Kauchak, *Educational psychology : windows on classrooms* (Pearson/Merrill/Prentice Hall, Upper Saddle River, N.J, 2007)
3. R.M. Branch, T.J.Kopcha, *Instructional design models* (Springer, New York, 2014)
4. V. K. Ceylan, A. E. Kesici, Journal of Human Sciences, **14**, 1 (2017)
5. I.U. Yapici, H. Akbayin, TOJET, **11**, 2 (2012)
6. W. Horton, *Designing web-based training* (John Wiley, New York, 2002)

7. J.Bersin, *The blended learning book; best practices, proven methodologies and lesson learned* (Pfeiffer Publishing , San Francisco, 2004)
8. N. S.Sukmadinata, *Landasan psikologi proses pendidikan* (Remaja Rosdakarya, Bandung, 2007)
9. C.J.Bonk, C.R.Graham, *Handbook of blended learning: Global Perspectives, local designs* (Pfeiffer Publishing, San Fransisco, 2004)
10. C.C.Chen, K.T.Jones, The Journal of Educators Online, **4**, 1 (2007)
11. B.Akkoyunlu, M.Y.Soylu, J. Educational Technology & Society, **11**, 1 (2008)
12. A.P. Rovai, H.M. Jordan, International Review of Research in Open and Distance Learning, **5**, 2 (2004)
13. D.R.Garrison, N. Vaughan, *Blended learning in higher education* (Jossey-Bass, San Fransisco, 2008)
14. S.B.Sjukur, Jurnal Pendidikan Vokasi, **2**, 3, (2012)
15. N.Vernadakis, M.Giannousi, V.Derri, M.Michalopoulus, E.Kioumourtzoglou, *Procedia technology*, **1**, 439-443 (Elsevier, The Netherlands, 2012)
16. C. Akarawang, Edulearn, **10**, 1 (2016)
17. J. Poon, JOLT, **9**, 2 (2013)
18. L.M. Jeffrey, J. Milne and G. Suddaby, JITE :Research, **13**, 121-140 (2014)
19. D.R.Garrison, H.Kanuka, The Internet and Higher Education, **7**,2 (2004)
20. B. Kistow, Caribbean Teaching Scholar, **1**, 2 (2011)
21. Husamah, *Blended learning* (Prestasi Pustaka, Jakarta, 2014)
22. A.M. Bliuc, P. Goodyear, R.A.Ellis, Internet and Higher Education, **10**,231-244 (2007)
23. P.Sharma, ELT Journal, **64**, 4 (2010)
24. P.Zaka, Journal of Open, Flexible, and Distance Learning, **17**, 1 (2013)
25. T.Aytac, UbiCC Journal, **4**, 3 (2009)
26. Y.C.Lee, K. Lau, V.W.Y. Yip , AAOUJ, **11**, 2 (2016)
27. A.Buran, A. Evseeva, *Procedia - Social and behavioral sciences*, **206**,177-182 (Elsevier, The Netherlands, 2015)

CONSTRUCTING SELF AWARENESS USING EDUCATION HUMAN VALUE IN SCHOOL CULTURE

Dewa Nyoman Wija Astawa
IKIP Saraswati Tabanan
dw.wija@gmail.com

ABSTRACT

The big number of poverty in Indonesia impact to the hope of having a free learning service, especially education for early age and elementary school students from the less fortunate families. Many people usually ask the quality of such kinds of free of charge schools. The low price makes a low standard for the students. Sathya Sai School in Denpasar has proven that the free learning service does not mean the standard quality of the school is low. This study will explain how the teachers and the members of the foundation build the awareness of the students and parents by socializing and internalizing the value that empower their collective awareness to help the school achievement. By using local cultural approach, the school builds a program that involves the parents especially the woman.

Key words: human value education, free-education, self-awareness

A. Introduction

The education for children in their early ages is fundamental to the children's development in the next years. The period is the golden time in life but comes only once. During the time, education, stimulation, counseling, maintaining, caring, will all help the potency of the children.

In those early ages there should start a process of self-awareness. Thus, the primary level of formal education, which includes the pre-school stage, is not only about understanding important formal lessons but also about implementing this self-awareness in language, cognitive, psychomotor, and creativity class acitivities.

The self-awareness is marked with physical claims and particular activities that challenge morals.[1]. Their ideals come from the material world around. The children identities are carefully develop from their social and psychological environment.

Modern society believes that self-awareness is one of the domains in the children development that demands attention. It is believed that the process of transferring things would be contributive. The awareness is a concept with which the children see themselves through different perspectives and know how to behave in accordance with their beliefs and the social norms. According to Catron[2], children in their early ages need stable emotion as a form of self-awareness along with their physical growth. The self-awareness concept may be formed by internalizing self-reliance, sex-awareness, gender role, tolerance, and personal health care.

1 Darmadi, H. 2006. *Dasar Konsep Pendidikan Moral: Landasan Konsep Dasar dan Implementasi.* (Bandung: Alfabeta, 2006) p. 191

2 Catron, C.E. 2006. *Early Childhood Curriculum: A Creative Play Model.* Upper Saddle River, New Jersey: Pearson, 2006) p. 199-201

B. Research Methods

This article discusses how education for children in their early ages is done with some construct that adjusts the background and environment of the children. With qualitative approach, the study was carried out at the Sathya Sai School in Denpasar, Bali, a school with its own characteristics. The data were collected through observation and FGD *(focus group discussion)* involving the teachers, the headmaster, the foundation members, and the students'parents. The analysis was done to the text resulting from the observation and FGD.

C. Results and Discussion

1. The Profile of Sathya Sai School Denpasar

The Sathya Sai School of Denpasar is located at jalan Kemuda III No. 9 Peguyangan Kangin, Northern Denpasar. It started with the establishment of kindergarten on the 12th July 2002 and was inspired by the Sathya Sai School at India and around the world. The school applies EHV that centres on character building to prepare the students with discipline, social care, religious teachings, and hardworking.

The entrance test for the students includes interview with their parents, regarded as a kind of "contract" as so-called by Mr. I Ketut Wiana, the foundation chairman, between the school and the students'parents, in which the mission and vision of the school are introduced and to be understood by the parents. The interview is also the time when the commitment of the parents for their children education is emphasized as a bridge to strengthen the basis of EHV. This has been seen important as most of the students'time is spent with their parents. The interview is also a time for the foundation to show that the school program is based on truth, rightenousness, peace, love, and non-violence principles. The five become the characteristics of the school curriculum. Then, after five years, on the 12th July 2007, the school established its elementary school accredited A.

2. The Implementation of EHV

EHV comes from the combination of 3-H, i.e. *head*, *heart*, and *hand*. It suggests that any words delivered by the head and then spoken should be first screened by the heart, before finally being executed as action. EHV has a strategic role in developing quality human resources. The program of EHV at the elementary school purposed to provide the children with knowledge, attitude, moral, social skills, and services so that they can continue to study at the higher levels. The program expects individuals with good social attitudes and be able to solve problems found in their environment[3].

The EHV at the Sathya Sai School in Denpasar was derived from the model developed by the Institute of Sathya Sai Education in India, which has been an adaptation of the one promoted by Dr. Art-Ong Jumsai Na Ayudhya of ISSE Thailand. The chairman of the Sathya Sai Indonesia Education, Mr. Anuraga Duarsa said that in Indonesia the development of EHV should consider the local cultures which include the religions and faiths that belong to the local people.

3 Wortham, S.C. *Early Childhood Curriculum: Developmental Bases for Learning and Teaching*. (Upper Sadle River, New Jersey: Pearson, 2006) p. 7

EHV always involves social skills that pedagogically are not possible implemented through the conventional teaching. It requires optimal growth of the students, especially their service attitudes, which can never be taught without concrete practice. This means the program should facilitate the students'potencies optimally so that their studying becomes fully meaningful.

EHV at the Sathya Sai School is not merely aimed at the teachers and their students. The parents and the school assistants are also targeted. The end is to create a conducive surrounding for the whole community. Thus, a parenting program is scheduled for once a month. In addition, the teachers and the school employees are given a *Personal Development* (PD) program of which the material is given by invited speakers.

EHV, according to Na-Ayudhya, is developing as a stimulus for the 5 senses though it is not always understood properly since most information is stored in the subconscious mind. For children this condition might be bad or good. Consequently, the teachers should be able to integrate the important values to stimulate what are kept subconsciously[4].

The most important part of EHV is inspiring the students for some transformation that emphasizes the balance of mind, words, and action. Thoughts should be controlled by the heart before becoming speech and action. The human values are not to be taught in words yet delivered through *afirmation, habit,* and *inspiration* at schools and homes. The values are truth, right conduct, peace, love, and non-violence and love is the core of all because it creates closeness. Action with love will change into mild attitudes which in turn become right conducts. When the five values are maintained the *self realization* or *self awareness* can be attained.

3. Parenting: Developing the Parents' Care

Parenting is an obliged monthly program for the students'parents. It is purposed to keep an effective communication about the school agenda, the needs of the parents and the community. With the program the parents are hoped to stimulate and begin the change. This is based on the fact that most of the time the children spend is with their parents.

With the material that is based on the need of the children, parenting is done to exercise critical mind towards *collective consciousness*, which is based on good education for the generation. Parents and teachers are like the two wheels of a bike that support to each other. When the two wheels are good, the bike will run well. Yet, when one of them is flat, the bike cannot move properly. This means that parents and teachers should work hand in hand for the children to reach the goal[5].

4. EHV: Developing the Self-Awareness and the School

The construction of self-awareness at the Sathya Sai School Denpasar is aimed at establishing the foundation for developing humanity. It needs the cooperation of parents at homes and teachers at schools.

There are three strategies to be used in teaching EHV. First, it can be done by integrating the values into the curriculum. This means any lesson in the classrooms should teach the values and the values should ensoul all of them. When a teacher fails to teach them,

4 Jumsai Na Ayudhya, Art Ong, *A Development of The Human Values Integrated Instructional Model Based on Intuitive Learning Concept* .(Chulalongkon University, Bangkok, Thailand, 2003) p. 12-20

5 Marilyn Price-Mitchell, PhD, *Self-Awareness: How Parents Foster Meaning and Purpose in Kids. This article is republished and updated with many new articles each year at RootsofAction.com. Copyright 2012 – 2016*

he (she) is not a teacher but a cheater. Second, EHV is also possible by inserting them in the extracurriculum program, e.g. drama, discussion, musics, etc. Thirdly, EHV may be provided to the children at schools by teaching them directly, through some class activities, i.e.: (1) silent sitting or meditating. Yet before having the actvities at school all teachers should do them at homes with their own families; (2) praying and reciting holy verses; (3) group chanting; (4) story telling; and (5) role playing, etc. There should be remembered here that learning is the same as entering the uncomfort zone that is full of risk. A teacher is teaching because he (she) is being allowed by the students and not because he (she) is obtaining an authority by the institution. These strategies are such a construction that moves learning as an uncomfortable zone to be a comfortable zone.

At Sathya Sai School the EHV is not taught in a traditional way. When the students are doing silent sitting, for example, methaporical words in simple EHV philosophy are spoken and implanted. They can be also delivered through listening to particular songs, and let the children think deeply about the meanings. These are all expected to raise the spirit of the children to be positive. Another activity is named the morning circle. In this activity, the teacher is reminding the children to do good things as habit, either at homes, schools, or other places. And the last activity is having breakfast together. In it, the children are conditioned for learning to be patient, to keep clean, to share, and to be thankful.

Such constructive activities are not effective at all if they are done only at the school because of the limited school hours. The role of the parents at homes, especially the mothers, is wholly needed. This is the reason why the parenting program is considered important. It is supposed to build the parents'senses of belonging so that they are contributive to the EHV program and involved actively in concrete activities, e.g. preparing the meal and clean around the school area. When the parents understand correctly the EHV the self-awareness within the children will be definitely realized. Then, the children will become honest, independent, tough, and studious. In the end, it creates a good school atmosphere.

D. Conclusion

The self-awareness of the students of Sathya Sai School has been developed through EHV program. The values, i.e. truth, right-conduct, peace, love, and non-violence are the main content of the program. The concepts are transformed into affirmative, habitual, and inspiring concrete activities that are supposed to be easily understood. Then they are programmed and modeled by the school teachers and the parents. The self-awareness provides the children with the competence for controlling their emotion, having the spirit to fight and be independent, either physically or mentally.

E. Acknowledgements

Great thanks are addressed to management of Denpasar Sathya Sai School and all of the organizers of GC-TALE 2017 due to the publication of this article. Many thanks for all of the important people who involved in the completion of this article. Without their guidance, supports, and advices, this article would never have been completed. Hopefully, it can give benefit for the readers.

References

Budiningsih, C. A. 2004. *Pembelajaran Moral: Berpijak pada Karakteristik Siswa dan Budayanya*. Jakarta: PT. Rineka Cipta.

Bruce, Rita, 2003. *Sathya Sai Parenting*, Sri Sathya Sai Books and Publication Trust, Prasanthi Nilayam (India).
Bruce, Rita, 2005. *Love of Conscience, Four Aspects of Human Natur*, Sri Sathya Sai Books and Publication Trust, Prasanthi Nilayam (India).
Calhoun, C. 1993. *Habitus, Field and Capital: The Question of Historical Specificity in Bordieu: Critical Perspectives.* Cambridge: Polity Press
Catron, C.E. 2006. *Early Childhood Curriculum: A Creative Play Model.* Upper Saddle River, New Jersey: Pearson
Darmadi, H. 2006. *Dasar Konsep Pendidikan Moral: Landasan Konsep Dasar dan Implementasi.* Bandung: Alfabeta.
Jumsai Na Ayudhya, Art Ong, 2003. *A Development of The Human Values Integrated Instructional Model Based on Intuitive Learning Concept* . Chulalongkon University, Bangkok, Thailand.
Marilyn Price-Mitchell, PhD, *Self-Awareness: How Parents Foster Meaning and Purpose in Kids. This article is republished and updated with many new articles each year at RootsofAction.com. Copyright 2012 – 2016*
Maton. K. 2008. *Habitus in Pierre Bordieu: Key Concepts*. North Yorkshire: Acumen
Michal Einav , 2014. *Perceptions about parents' relationship and parenting quality, attachment styles, and young adults' intimate expectations: A cluster analytic approach, The Journal of Psychology: Interdisciplinary and Applied*, 148:4, 413-434.
Reed-Dannahay, D. 2005. *Locating Bordieu.* Indianapolis: Indiana University Press
Seetharam, Lakshmi and Kallidaikurichi Easwaran Seetharam, 2005, *VIDYA: Values Integration for Developing Young Adults*, Institute of Sathya Sai Education, San Pedro, Philippines.
Thomson. P. 2008. Field in Pierre Bordieu: Key Concepts. North Yorkshire: Acumen
Wardhani, W. D. L 2011. *Konstruksi Identitas Kebangsaan dalam Pendidikan Kewarganegaraan di SD Kelas Awal* . Thesis. Tidak Dipublikasikan.
Wortham, S.C. 2006. *Early Childhood Curriculum: Developmental Bases for Learning and Teaching*. Upper Sadle River, New Jersey: Pearson

RECONSTRUCTION OF CHARACTER VALUES BASED ON MANGGARAIAN CULTURE

Yuliana Wahyu[1], dan *Ambros* Leonangung Edu[2]
[1]*Program Studi PGSD STKIP Santo Paulus Ruteng, Indonesia*
[2]*Program Studi PGSD STKIP Santo Paulus Ruteng, Indonesia*

ABSTRACT

This study aims at exploring and finding the character values in penti ceremony, Manggarai. Penti means the harvest thanksgiving ceremony of the Manggarai people. Penti is a sign of gratitude to God the Creator (*Mori Jari Dedek*) and to the ancestral spirits over all the crops. This research is conducted in East Manggarai district, NTT Province. There are several stages of ceremony penti, they are *teinghang, barong lodok, barong wae, libur kilo*, dan *tiba meka landang*. Each of these preparations with ritualistic rituals have their oun meanings. Technique of collecting data is done through interview and literature study. The research shows that *penti* emphasizes customs, imitation, exemplary, command, reward, and authority in character formation. Character formation is inseparable from the cultural environment in which children grow and develop.

Keywords: Character Values, Penti Ceremony, Manggarai

1. INTRODUCTION

1.1. Background Problems

One source of character formation is local cultural values. So far, local culture has been isolated into the periphery due to the effects of globalization that uphold Western civilizations. Along with the rise of grassroots movements as the antithesis to globalization, where locality is now more emphasized than globalism, the learning context is shifting from the domain of macro theories to the more practical and implicative local domains. In Manggarai, Nusa Tenggara Timur Province, there are many local assets that have been buried, and now education practitioners continue to explore more deeply about the local wealth as a source of value for the personal formation of learners of character. Among the rows of cultural assets, one of them is a pent ceremony. Penti is a thanksgiving ceremony. This ceremony is firmly attached to the life of the Manggarai people because the majority of Manggarai people (70%) are farmers. The ceremony is always celebrated every year and successfully unites every people from various backgrounds. Based on this brief description can be seen how the ceremony *penti* is a cultural and aesthetic expression of the character of Manggarai people who packed in a cultural ceremony. To further examine the values of characters in the distribution of ceremonial lodges, the research team tried to make a study entitled *"Reconstructive Values of Character in Manggarai Culture in NTT Province"*.

1.2. Problem Formulation

The problem becames the basis of this writing is: what are the character values in *penti* ceremony in Manggarai, NTT Province?

1.3. Purpose of the Study

This study aims at finding the character values in penti ceremony in Manggarai, NTT Province.

1.4. Benefit of Study

Firstly, the Manggarai teritory, which is currently divided into three districts (East Manggarai, Manggarai and West Manggarai) must maintain the unity of penti ceremony unaffected by geographical sectoral interpretations and segregations that may obscure the original Manggarai heritage. *Second,* the research team try to explore more about the values of the characters in penti ceremony.

2. Literature Review

2.1. Character Values

Professional education process will certainly form the character of the students. The characters that are meant are honesty, integrity, justice, courage, virtue (Fajarini, 2014: 6). The ASPEN Declaration presents six main ethical values that form the basis of character building: (1) *trustworthy* which includes honesty and integrity, (2) *treating others with respect,* (3) *responsible*, (4) *fair*, (5) *affection* , and (6) *good citizens* (Mundiri, 2012: 42). Strengthening the character is not only the role of the school but also the family, religion, culture, and community or environment. Indonesia has a diverse cultural diversity. In that culture there is a dimension of philosophy, psychology, sociology, and religion that contains values of character that can be developed along with character education at schools. This paper raises the local wisdom of Manggarai region, NTT Province, about *penti* as a cultural asset forming character values. There are some stages of the rituals performed.

2.2. The Nature of Penti Ceremony

2.2.1. Meaning and Purpose of Penti

Penti is the harvest thanksgiving ceremony of the Manggarai people. Penti is a sign of gratitude to God the Creator *(Mori Jari Dedek)* and to the ancestral spirits over all the crops and the turn of the season as well. At the turn of the season between July to September every year is also done thanksgiving, penti. There is a belief of Manggarai people, if this ceremony is not implemented then God will give anger in the form of disasters that befall the Manggarai people.

2.2.2. Order of Penti

2.2.2.1. Barong Lodok

Lodok is a system of division of land (rice field or garden) in Manggarai in the form of cyber crebs. Lodok consists of three parts: (1) *Lingko*. *Lingko* is a *lodok* center planted by a tree called *haju teno*. *Haju teno* symbolizes fertility. *Lingko* is all the land area owned by a descendant *(wa'u)* in a village *(golo)*. *Lingko* is not private but communal land belongs to all villagers. (2) *Cicing.Cicing* is the outer boundary of a circular garden. (3) *Langang.Langang* is the line of radius or barrier between the garden.

Lodok in a *lingko* is divided by the authority *(tu'a teno)*, chosen from the clan leaders *(tu'a-tu'a panga)*. The division of the circle begins with planting a *teno* wood by placing it into a hole that has been excavated earlier and located in the center of the circle. From the center where we planted the wood drawn the radius of the circle which then becomes the boundary among the gardens. The wood planting *(tente teno)* ceremony begins with a custom ritual by offering a chicken egg or a goat. The goal of the ceremony is to take a prayer and hope to God and the ancestors to give them sustenance for the land to be divided later. After prayer, the egg is placed in the place of *haju teno* then plugged. After the hole is covered by soil, around the *teno* wood will be plugged in small woods called *lance koe*. In this lance is tied a rope as a center to draw a boundary line or a circle of nets.

Barong means invitation, summon the spirit in the *lodok* center or "circle" to follow the *penti*. So, *barong lodok* is a customary rite to invite the spirit of the gardener in the center of the circle to attend the *penti* ceremony. Customary leader *(tu'a gendang)* starts the rite by doing betel meal *(pandeng cepa)*. *Cepa* is a typical meal welcoming guests with a mixture of betel leaves, areca nut, and lime *(saung kala, raci agu tahang)*. The next stage is *"pa'u tuak"* or watering a drink of wine *tuak* stored in bamboo to the ground. After that the event continued by slaughtering a pig to be dedicated to the spirits of the ancestors. Before the pig is insulted, the customary leader says *"tudak ela"* means to deliver the words of prayer. *Cepa, pa'u tuak,* and the pig slaughter ceremony are performed to make the land as a place of cultivation, endowed with abundant income, and far away from catastrophe. Participants who joined the circle continued with the song called *"sanda lima"*. After they return to the traditional house while singing some poetries telling the joy and honor of the rice that has given life.

2.2.2.2. *"Barong Wae".*

Wae means water. The participants and customary leader go to water spring called *"ulung mata wae"* around the village. Arriving at the place the leader say about the purpose of the arrival by hogging a rooster. This rite is done to invite the ancestral spirits as the guardians of the spring to follow the penti ceremony, so that the spring never subsided. The goal is to express gratitude to God who has created water for the lives of the villagers. After they return to the village, then they head to the *compang* to do *barong compang*.

2.2.2.3. "Barong Compang".

According to Deky (2016), compang is a special place composing by selected stones, on which there is a flat stone *(watu lempe)* and is located in the center of the village to offer offerings. After the entourage arrived at compang, the customary leader express greeting by conveying to the *naga golo (naga:* dragon), ancestral spirits *(wura agu ceki)*, God the owner the creator *(Mori Agu Ngaran, Jari Agu Dedek)*. The rite is to invite the spirits to follow the ceremony at the traditional house *(mbaru niang)*. *Naga golo* existence is very important for Manggarai people, because the Spirit protects the villagers from fire disasters, hurricanes, or other hazards. After that, the participants return to the *custom house* to perform *wisi loce* (span mats) ceremony. The purpose is to invite the ancestral spirits to sit together and join the *penti*.

2.2.2.4*Libur Kilo.*

Libur means to gather, *kilo* means family. *Libur kilo* is a family gathering to renew life of the whole family. The offering in rite is a chicken and a small pig. This starts with *"renggas"* as the opening ceremony. *Renggas* is singing with "sanda lima" song. The "sanda lima" has a meaning about the basic five human needs: (1) *mbaru tara kaeng* (residential house). (2) *natas bate labar* (playground). (3) *wae bate teku* (water springs). (4) *uma bate duat* (garden as a place to planting). (5) *compang* as a place to pray. After the ceremony teing hang each kilo or thanksgiving offerings from each family.

2.2.2.5. The peak of the ceremony is marked by the gathering of customary heads of the village, the authority who divide the *lingko*, the head of the family, and other villagers who are invited in the ceremony penti. After that, each family do *teing hang* (each family's thanksgiving offerings.

3. RESEARCH METHODS

3.1. Type of Research

The authors combine literature and field research. In the literature study, investigation of secondary data has been related to the issues studied. To collect primary data from the field, the authors use qualitative methods in which the data collected is described in the form of narrative descriptions.

3.2. Research Subjects

The main subjects of this research are indigenous leaders and community leaders. Selected consideration is the broad and deep knowledge contribution of the ceremony.

3.3. Data Collection Procedures

3.3.1. Literature review

Library study is done by deepening the writings of previous experts to then explore whether there are values of characters in ceremony penti and lodok division.

3.3.2. Interview

Interviews were conducted to explore qualitative data about penti. Open-ended questions are created to get data about character values in the ceremony.

3.4. Data Analysis Techniques

Techniques of analyzing research data through the following stages. 1) Grouping the data, researchers select the data from literature studies and interviews. 2) The data are described qualitatively to find the meaning of *penti* and the character values in it. 3) Analyzing the data. The data analyzed by reading and interpretation technique. 4) Drawing conclusions, the data has been processed can be used as a stepping stone to answer the formulation of research problems. 5) Presentation of data.

4. RESULTS AND DISCUSSION

4.1. Research Findings

Meanings of "*TUDAK PENTI*" OR EXPRESSION OF *PENTI*RITE				
No	**Expression**	**Meaning**	**Character values**	**Meaning**
1.	***Barong Wae:*** *Denge le meu empo, ho'o de manuk kudut barong wae. Wali di'a kamping ite Morin agu Ngaran, ai ite poli teing ami wae bate tekugm ho'o. Tegi kali dami, lami agu riang koe wae teku ho'o kudut mboas kin wae teku, kembus kin wae bate tekug'm ho'o. Kudut neka koe do'ong le roho agu rowes le ronco wae teku ho'o. Porong inung wae ho'o te guna laing latang te weki agu wakar dami. Porong mese bekek kali dami, gerak ranga.*	Hear the ancestors, this chicken is offered in this spring. Thanks the Lord, for God has given us water for our needs. We beg, keep this drinking water! Hopefully this drinking water is always sufficient for our needs. Maybe this water is kept away from any disturbance. May it is useful for the soul and body. Hopefully, it always gives us freshness.	Religious	Attitudes and behaviors that obedient in implementing religious teachings someone embraced.
2.	***Barong Compang:*** *Denge di'a lite wura, ho'o manuk barong compang, ai to'ong wie penti. Tegi dami, kudut lengkang koe pa'ang, nggalu koe ngaung, kudut tadang koe darap de tanah, agu kolang de leso. Tadang koes raja leso, deu koes tai wie, sika koes ringan, wur koes rucuk, agu kando koes dango.*	Dear the spirits, the village guardians, this chicken we dedicate to you all in this place, because tonight will be held *penti* ceremony. Please, protect the whole villagers and all the circumstances. Maybe we are kept away from disease outbreaks. Maybe we are kept away from nasty maneuvers and evil interference. Keep away from health problems.	Religious and togetherne ss	Attitudes and behaviors that obedient in implementing religious teachings someone embraced.
3.	***Libur Kilo:*** *Ho'o manuk agu ela kudut libur kilo, tegi dami kali ga, neka manga tuka bana kut caka salang duat, neka caka salang we'e, kut beka agu buar, kete api one, tela galang peang, kut raci kali lebo kala, kudut mbaun eta mose dami lemekn wa, kut malir di'a tumbu wai lau, kut mese bekek*	Dear the ancestors, a chicken and a pig are offered to this "libur kilo" rite. Please, we hope that the meal will not disturb our health. Keep away from interruptions when we leave and leave work. May our families multiply. Our life needs are met and our pigs are growing health. May be planted	Religious, peace-loving and socially concerned	Attitude and obedience to follow the religious teachings someone embraced, with respect for each other and protect the environment

	gerang ranga.	betel nut and betel leafy. May experience good life and God's grace abound.		

4.2. Discussion

Penti implies a great variety of meaning for Manggarai people. *Penti* is a form of expression or symbol of communication to God Almighty. In addition, there is a good interaction and communication between members of the community that can foster strong bonds of brotherhood. The functions of the rite for the community are as follows.

1) Ritual function, in the form of a phrase or symbol of communication to God the Almighty. The function is seen from the recitation of prayers or mantras and various offerings that must be met at the time of the event stages take place.
2) Social function, *penti* attract social interaction. Good interaction and communication between community members will foster strong bonds. The social contact in holding the ceremony is seen during the cooking together, until the entire preparation of the penti is held.
3) The function of aesthetics, it can be seen from the way of dress, song or music that is playing, and the beauty of the language in the prayer (*tudak*).

5. CONCLUSION

Penti is a thanksgiving ceremony that shows the character values for the Manggarai community. In addition, the ceremony shows the unique identity of Manggarai culture. Penti teaches many things: *courage, sportsmanship, self-esteem, beauty, brotherhood, and religious, hard work, tough,responsibility, or discipline values.* It is a form of idiom or symbol of communication to God. In addition, there is a good interaction and communication between community members in fostering strong fraternal ties. The ceremony also teaches the value of discipline and responsibility and togetherness in running a life of community. It has a religious function, aesthetic function and ritual function., and so forth. Indirectly penti implementation includes nature, humans, and God. Maybe not so many people in Indonesia who know or recognize the rite. It is not only a ceremony, but also a thanksgiving to God Almighty. Therefore, this ceremony needs to be preserved and introduced not only in Indonesia but to many people around the world.

References

1. Abut, E. Y. & Kaleb S. *Identitas Budaya Dan Maknanya Dalam Upacara Penti Orang Manggarai*. Artikel. Mahasiswa Pascasarjana jurusan PBI UNS Semarang, (2013).
2. Adisti.. *Penti Sebuah permainan Rakat Manggarai Flores NTT*. Artikel. Di download tanggal 6 Januari 2017 dari alamat website http://sarkomkar.blogspot.co.id/2009/03/penti-sebuah-permainan-rakyat-yang-sarat.html, (2015).

3. Chyntia.. *Goet Manggarai Nilai Religius Mori Jari*. Artikel di download tanggal 15 Januari 2017 dari alamat website http://chyntia-abbo.blogspot.co.id/2010/09/goet-manggarai-nilai-religius-mori-jari.html. (2010)
4. Ditindb. *Permainan Penti*. Artikel. Di Download tanggan 13 Januari 2017 dari alamat website http://kebudayaan.kemdikbud.go.id/ditwdb/2015/12/17/permainan-penti/. (2015).
5. Deky, N. *Compang Orang Manggarai (Menggulung Riuh Compang Pantai Pede)*. Artikel. di download tanggal 6 Januari 2017 dari alamat website http://www.nusalale.com/detailpost/compang-orang-manggarai. (2016).
6. Fajarini, U. *Peranan Kearifan Lokal dalam Pendidikan Karakter*. Sosio Didaktika: vol. 1. No.2. (2014).
7. Lickona, T. *Pendidikan Karakter Panduan Lengkap Mendidik Siswa Menjadi Pintar Dan Baik*. Bandung: Nusamedia. (2014).
8. Nucci, L. P. & Narvaez, D. *Handbook Pendidikan Moral dan Karakter*. Bandung: Nusamedia. (2016).
9. Pantar, F. H. *Penti Sebuah Permainan Rakyat Yang Sarat Akan Nilai-nilai Budaya*. Artikel. Didownload tanggal 5 Januari 2017 dari alamat website http://sarkomkar.blogspot.co.id/2009/03/penti-sebuah-permainan-rakyat-yang-sarat.html. (2009).
10. Ruyadi, Y. Model *Pendidikan Karakter Berbasis Kearifan Budaya Lokal. Proceedings of the 4th International Conference On Teacher Education; Join Conference UPI & UPSI*. Bandung. (2010).
11. Salves, G. *Upacara Penti Sebagai Ekspresi Budaya*. Artikel. Di download tanggal 6 Januari 2017 dari alamat website http://baimsalves.blogspot.co.id/2013/09/upacara-penti-sebagai-ekspresi-budaya.html.. (2013).

Teacher Pedagogic Competency and National Examination Result at Elementary School

Putu Eka Suarmika
University Abdurachman Saleh Situbondo, Jawa Timur

Abstract: This study aims to determine the relationship of teacher pedagogic competence to the national science exam results in elementary schools in the district Gerokgak. The population in this research is all the sixth grade teachers in gerokgak sub district and the sample in this research is 29 teachers. Instrument in this research using pedagogic competence questionnaire and the value of national examination result of academic year 2016/2017. The research method used is *ex post facto* while the data analysis using correlation analysis with the help of SPSS 16. Based on data analysis, it is found that the pedagogic competence relationship of teacher to the result of national science test shows the correlation of being with kofesien rho= 0.47. While the pedagogical competence of teachers contributes as much as 22.09% to the national science exam results, and 67,91% is influenced by other variables not implemented in this study. Pedagogical competence is primarily concerned with the level of understanding of learners, instructional design, and implementation of learning the diagnosis, evaluation of learning, the development of learners has provided significant support in the form of professional pedagogical teaching
Key words : *elementary school, national examination. natural science, and teacher pedagogic competency*

1. Introduction

The imperative in recent years about improving student outcomes is also about improving the quality of the teaching workforce. In recent years, however, recruiting and retaining quality teachers has become a challenge among some OECD countries. In addition to the ageing of the teaching workforce, some countries experience high rates of attrition among new teachers and a shortage of quality teachers in high-demand subject areas and disadvantaged schools. There is also concern about attracting high-achieving and motivated candidates into teacher education programmes and the lowering of qualification requirements in the certification and licensing of new teachers [1].

One of vital factors and generally discussed in relation to the quality of education is teacher. A quality teacher is one who has a positive effect on student learning and development through a combination of content mastery, command of a broad set of pedagogic skills, and communications/ interpersonal skills (Hightower et al.,2011) [2]. [3] It is very important because teachers have a crucial factor to determine the quality of education system. Teachers should have the competencies affecting the students' achievements and performances. To meet the demand, teachers should also be able to answer the students' need in their learning. According to the Minister of National Education Regulation No.16 of 2007, there are four competencies of teachers' academic qualifications through teacher education; pedagogical, personal, professional, and social competencies. Those competencies are aimed to actualize the need in improving the quality of education in widely scope

The nature of teachers' pedagogical knowledge and how new knowledge is incorporated into the profession is an especially relevant issue for policy makers. The learning environments of today's classrooms are becoming more diverse and teachers are expected to teach "21st century skills" that are the new priority among OECD countries and partner economies. In our increasingly complex, knowledge-based and interconnected digital societies, the labour market requires a new set of skills previously of little significance for seeking employment. Today, skills such as creativity, critical thinking, problem-solving, collaboration, communication, among others, are necessary to succeed in the labour market. These new demands require teachers to deviate from traditional teaching methods and to employ innovative, evidence-based teaching practices. For some countries, this might entail a re-skilling of the current teaching workforce and upgrading of the profession's knowledge base within teacher education institutions and through professional communities. Therefore, understanding what the current knowledge base looks like will help determine whether and to what extent re-skilling is required [1].

Science is a way of asking and answering questions about the physical aspects of the universe. Science is not just a collection of facts or a collection of answers to questions, but rather a process of continuing dialogue with the surrounding physical environment. Saintis with special expertise, generally have language, methods and habits of mind to construct explanations of nature. This knowledge is sometimes separate and even contrary to the usual way of finding out. Science has a role to make choices. Scientific knowledge as a discipline knowledge, constructed identically and symbolically in nature. Scientific reasoning is characterized by explicit theoretical formulations that can be communicated and tested by supporting evidence [4].

Science learning is independent of the science itself, that science is regarded as a body of knowledge, a way of thinking, a way of thinking, and a way of investigating (science as a means of inquiry) . This shows that science learning involves the use of a number of senses, hands-on and tools or manipulative minds-on. That is, in the process of learning science there is a form of interaction both in thinking and working. Chiappetta & Kobala (2010) [5].

To measure the achievement of elementary science students' competence using the National Standard School Exam. The National Examinations held at Elementary School are the activities of measuring and evaluating the students' competence to gain recognition of learning achievement, in this case the achievement of science learning. [6] The changes in science and educational system have also affected teachers' competencies. There is strong relationship among the other sciences and system of society and they affect educational system and teachers' competencies. Educational system can be based on scientific results of educational sciences, psychology, economy, technology, sociology and so on.

The results of the study Wardoyo (2015) found that pedagogic competence had a significant effect on teacher performance. The better the teacher's pedagogic competency is accompanied by improved performance, and vice versa[7]. In the research literature on teaching and teacher education, there is a shared understanding that teachers' professional knowledge is an important determinant of instructional quality that impacts students' achievement gains (Baumert & Kunter, 2006; Bromme, 1997; Munby, Russell, & Martin, 2001) [8]. Yet few empirical studies have assessed the different components of teachers' knowledge directly and separately to predict instructional quality or student outcome. The main goal of this study was to explore whether teachers' pedagogical content knowledge as a crucial component of teachers' professional knowledge makes a contribution to explaining differences in students' learning outcomes in elementary science education.

Based on the backdrop above then the purpose of this study find out descriptively the pedagogic competence of sixth grade teachers in elementary school and aims to determine the relationship of teacher pedagogic competence to the national science exam results in elementary schools in the district Gerokgak.

2. Research method

Research method used is ex post facto correlation research. Ex post facto that research in which the independent variable orvariables have already occurred and in which theresearcher starts with the observation of a dependentvariable or variables.He then studies the inde-pendent variables in retrospect for their possiblerelations to, and effects on, the dependent variableor variables (Kerlinger, 1964) [9].The sample was taken by 29 sixth grade teachers from 45 elementary schools located in sub district gerokgak district Buleleng Bali. Random sampling with eastern, middle and western divisions and samples are all civil servant teachers with an average of 10 years teaching experience

Pedagogical competence was taken using a questionnaire with 30 questions with three choices of answers that are often, occasionally and never, consisting of ten dimensions : (1) controls the characteristics of leas from the physical aspect, the spiritual moral, social, cultural, emotional and intellectual; (2) control of learning theories and principles of learning that educates ; (3) develop curriculum related to the subject matter; (4) conducting educational learning ; (5) utilize information and communication technology for the sake of learning; (6) facilitate the development of potential learners to actualize their potential; (7) communicate effectively, empathetic, and manner with the students; (8) conducting the assessment and evaluation processes and learning outcomes ; (9) utilize the assessment and evaluation for the sake of learning ; (10) take action to improve the quality of reflective learning.. While student exam data is taken from data of science exam result 2017 issued by Buleleng District Education Office.

Descriptive data analysis of teachers' pedagogic competence and national exam results were analyzed by calculating the mean. Furthermore, to determine the high or low quality of research variables, the mean scores for each variable are converted using the Mean Ideal and deviation Ideal criteria. While to analyze the relationship of pedagogic competence with national exam result in this research uses correlation Spearman's rho technique with the help of SPSS 16.

3. Results and Discussion

Pedagogic competence is one of four competencies that must be possessed by professional teachers. Maclellan (2008) 10] discusses pedagogical competence using the term "pedagogical literacy" and argues that it is essential that teachers develop both their knowledge and skills in order to be able to adapt to different situations, and that writing and documentation is a suitable method to support their professional development.

Table 1. Summary Table on the Level of Pedagogic Competence of Elementary Teachers and National exam Result

Variable	Mean	Interpretation
Pedagogic competency	77,970	Very high
National exam result	74, 855	Very high

Based on the analysis of competency data of elementary school teachers in Gerokgak sub-district, it was found that 77,97 teachers were competent with very high category. Teacher pedagogical content knowledge mainly influences student learning whereas teacher motivation mainly influences students'interest [11]. [12] Pedagogical competence is primarily concerned with the level of understanding of learners, instructional design, and implementation of learning the diagnosis, evaluation of learning, the development of learners has provided significant support in the form of professional pedagogical teaching.

The teacher understands, Competency and Attitude towards the teaching profession directly impacts the student's performance in all spheres. Teacher's beliefs and their Attitude towards the teaching profession affect the behavior of the students. Teacher is an activator of social change. He / She are directly linked with the future of the students and today's students are responsible citizens of tomorrow [13].

Table 2 Correlation between pedagogic competence and national exam results

			comp ped	nat exam
Spearman's rho	komp ped	Correlation Coefficient	1.000	.140
		Sig. (2-tailed)	.	.470
		N	29	29
	un ipa	Correlation Coefficient	.140	1.000
		Sig. (2-tailed)	.470	.
		N	29	29

Based on data analysis with the help of SPSS 16, in correlation coefficient rho roughly 0.47 which means having a moderate level of relationship or having a significant relationship. To see the significance of the relationship continued with the significance test with t test. Based on t test results in t count can be 2.765 and t table at 5% significance level of 2.052. Meaning there is a significant relationship because t count greater than t table (2.765> 2.052). To see how much influence pedagogic competence of science exam results, followed by calculating the coefficient of determination. The calculation of coefficient of determination is 0.2209 (22.09%) That is, teachers' pedagogic competence gives 22.09% influence to the science exam results and 78.91% is influenced by other factors not implemented in this study. Pedagogic competence of teachers give influence to the results of national science examinations, it is proved that the results of student learning is influenced by external factors that is pedagogic competence.

[13] It is science education that prepares fusion talent required by modern society and countries that lead the world economy innovatively reform the science education. It has predicted that the country would be a positive and creative brain. The science education in the early childhood education field is either demonstrating the experiment plan and process of experiment or choosing to experience activities follow the instructions of the teacher rather than selecting child's ability and intelligence. Learning effi-ciency is increased when infants recognize the individual and the individual teaching and learning in a suitable way. Real intellectual power is the power that the amount of knowledge, but knowledge can operate properly

In order to ensure meaningful and consistent science learning opportunities at the primary level, primary teachers must be supported to confront their existing ideas about science and science teaching and learning if they are to begin to articulate the problematic nature of the perceptions they use to define their practice. Getting at the ways that teachers think about science and science education is critical if we are to understand what matters in a primary school context and why. With support, primary teachers may begin to think differently not only about their role as science teachers but also the type of learning they value for their students and the role they want to see their students play as active, critical thinkers and learners in science. Documenting stories of primary school teachers drawing on aspects of quality teaching and learning in primary schools is useful in showcasing to others what is possible and encouraging that shift in thinking. Providing alternative ways of thinking about science teaching may enable primary teachers to relinquish their personal feelings of inadequacy, build on their existing pedagogical strengths and provide a consistent science learning experience for all primary students. In this way primary teachers may come to see new possibilities and opportunities for science learning as well as realizing the potential science learning which exists in the experiences that they presently provide for their students [14].

Conclusion

Science education need to implement innovative classroom activities, such as research activities and real problem solving, which enables students to mobilize and integrate different competences in order to construct explanations and/or coherent arguments, related with everyday life socio-scientific issues. For that teacher must have a good pedagogic competence in science learning. Pedagogical competence is primarily concerned with the level of understanding of learners, instructional design, and implementation of learning the diagnosis, evaluation of learning, the development of learners has provided significant support in the form of professional pedagogical teaching. In addition to the role of teachers there are several factors that affect the learning outcomes of science include: motivation, interest, early knowledge of students about science, environment and parents factor.

Acknowledgement

A thank you to Prof. Dr. I Wayan Suastra. M.Pd as the Director of Graduate University of Education Ganesha Singaraja which gives a lot of encouragement so that this research can be completed. Do not forget to the friends of Undiksha basic education doctoral program which gives many suggestions related to this research.

References

1. S. Guerriero. (ed.) *Pedagogical Knowledge and the Changing Nature of the Teaching Profession*, OECD Publishing, Paris. http://dx.doi.org/10.1787/9789264270695-en. (2017)
2. D. Sulisworo, R. Nasir, I. Maryani. *Identification of teachers' problems in Indonesia on facing global community*. https://www.researchgate.net. (2016)
3. S. Aimah, M. Ifadah, D.A. Bharati, D. A. *Building Teacher's Pedagogical Competence and Teaching Improvement through Lesson Study. Arab World English Journal,* **DOI:** https://dx.doi.org/10.24093/awej/vol8no1.6
4. N. Y. Rustaman. *Education and Science in Developing High Level Thinking Skill for Character Development*. Proceedings of a national biology seminar. Vol **8. No 1**. http://jurnal.fkip.uns.ac.id. (2011).

5. Rahmawati. *Integration of Culture Value Siri Pese / Pacce of Bugis-Makassar Society in Science Lesson*. www.fkip.unismuh.ac.id. (2015)
6. K. Selvi. *Theacers' Competencies*. https:www.researchgate.net/publication/. (2010)
7. C. Wardoyo. *The Measurement of Teacher's Personality Competence and Performance Using Embedded Model*. files.eric.ed.gov/fulltext/EJ1077384.pdf (2015).
8. K. Lange, T. Kleickmann. K. Möller. *Elementary Teachers' Pedagogical Content Knowledge and Student Achievement in Science Education*. https://www.uni-muenster.de. (2012).
9. L. Harold G. *Ex Post Facto Studies as a Research Method. Special Report*. https://eric.ed.gov/?id=ED090962. (1973)
10. J. From. *Pedagogical Digital Competence - Between Values, Knowledge and Skills*. https://www.researchgate.net. (2017)
11. M.M. Keller, K. Neumann, H.E. Fischer. *The Impact of PhysicsTeachers' Pedagogical Content Knowledge and Motivation on Students' Achievement and Interest*. http://onlinelibrary.wiley.com/doi/10.1002/tea.21378/full. (2016).
12. A. Hakim. *Contribution of Competence Teacher (Pedagogical, Personality, Professional Competence and Social) On the Performance of Learning*. www.theijes.com/papers/v4-i2/Version-3/A42301012.pdf. (2016).
13. A. Shokeen. *Relationship among Pedagogical Understanding, Teaching Competencies and Attitude towards the Teaching Profession of B.Ed. Student Teachers: An Exploratory Study*. http://www.confabjournals.com. (2016)
14. E.G. Yun, S.H. Park. *The Research of Early Child Scientific Activity according to the Strength Intelligence*. http://onlinepresent.org/proceedings/vol131_2016/32.pdf. (2016)
15. Fitzgerald, A., & Smith, K. *Science that Matters: Exploring Science Learning and Teaching in Primary Schools*. Australian Journal of Teacher Education, *41*(4). http://dx.doi.org/10.14221/ajte.2016v41n4.4. (2016).

The Role of Public Relations as A Management Function in Higher Education

Putri Anggreni
(University of Mahendradatta)
E-mail: gekcay@gmail.com

Abstract

Nowadays, public relations in higher education institution become an integral part of the management team. It must contribute to achieve organizational goals and demonstrate an organizational accountability identified by measurable result. Placement public relations at certain positions in management has consequences in the process of implementation of the public relations function. It happened because the management, workflow and hierarchy greatly affects work patterns as well as patterns of decision making in carrying out communication tasks performed by public relations.The goal of research to find out the role and position in public relations in management education institution (university) in achieving its objectives. After identification of roles and positions performance public relations firm, the study aims to evaluate the performance of public relations as part of management.This study used a qualitative descriptive research method with the study subjects were Mahendradatta University Public Relations and internal stakeholders. Data drawn from interviews with the help of the interview guide and then processed into information that can answer the research objectives.The result of the research shows that not all public relations officers in Mahendradatta University, do their role as real public relations officers. They also do not have the same understanding about the role of a public relations officer will just place public relations in same position as marketing.

Key words: public relation, private university, role, reputation.

INTRODUCTION

The communication function conducted by universities has no difference which with other institutions both government and private institutions. That distinguishing the communication function undertaken by these institutions is public communication component. The public then expanded his understanding Become stakeholders who are individuals and groups within or outside the college institution that has a role in determining Success of the company. College is interesting to be examined from the side of communication conducted by Public Relations because there are specific challenges.

According to Cutlip (1985) colleges face four problems in communication are: (1) Inadequate funding support for institutional roles public in the wider community; (2) Competition to get qualified students needs passion and expensive; (3) Constraints and regulations make the college administration difficult and costs; (4) Academic and ownership freedom is a challenge for internal stakeholders or external.

The issue of accountability is a matter that must be maintained by the university. One indicator of accountability is good communication. Other than that Universities have a significant role in educational development Community. Not just as an educational provider but a college is expected to contribute academically and practically in development. Accountability becomes a more crucial issue in state universities because of the cost the education used for raising one of them is obtained from taxes paid by the people. Form of social responsibility of state universities become heavier than related to finance is also related to that output resulting from graduates.

By making Public Relations (PR) the main channel of communication is a big challenge as it relates to media perceptions of relationship work society. Parents of students and prospective students are also the ones who it is important to manage the communication for the benefit of the college. Another interesting aspect is the independence of public relations as a tool management, status and position determine the performance of Public Relations as a tool communication. Position of Public Relations in a company can be a tool identification of work objectives and may also reflect a communication policy outlined by management.

FORMULATION OF PROBLEMS

The Role of Public Relations in an institution of higher education (university and polytechnics) is very important how management positions relationships society as a management tool greatly determines the output generated by Public Relations. Work program is a sign of Public Relations and management to perform the task. Position and role of Public Relations as well can be identified with the implementation of the work program, what is Public Relations have a work program integrated with management policies or just to be a tool complement the company's overall work program. To what extent flexibility or work autonomy related to structure and work can also be affect the performance of Public Relations as a means of communication between companies and public.

The formulation of the problem in this research are: (1) What is the role and position of Public Relations in management?; (2) How is the performance of Public Relations related to roles and positions provided by university management?

OBJECTIVES OF RESEARCH

To know the role and position of Public Relations in management at educational institutions in this case universities in achieving goals. After identifying the role and position of Public Relations in the company, with this research aiming to evaluate the performance of Public Relations as part of management.

RESEARCH METHOD

1. Type of Research

Descriptive with qualitative approach. The basic concept of qualitative research defined by Bogdan and Taylor (1975) is as a research procedure resulted descriptive data in the form of written or oral words of people and behavior which can be observed. This approach is directed to the background and the individual as a whole.

Further qualitative research is a particular tradition in the social sciences that are fundamentally dependent on observation in humans within their own territory and in contact with such persons in its language and in its terminology. (Moleong, 2002). Qualitative research at the organizational level in this case can complement research with information to estimate and design a strategy deals with organizations that are difficult to break through. Interpretation methods can be used to generalize the findings in depth to some complexity the course of an organization.

2. Research Subject

This research will be conducted at Mehendradatta University with research subjects are:

Public Relations Officer of Mahendradatta University (Unmar) along with 2 internal stakeholder informants.

3. Data Source

The main sources in qualitative research are the words and actions, this is the case is the result of observations or interviews by conducting activities to see, hear and ask questions while the types of data support in other qualitative research is as follows: (a) Written source: Related to this research then the written sources are become a source of data, among others, are corporate documents and agencies related can also be a source of data that can add input from research this; (b) Statistics: In qualitative research can also use statistical data which has been available as an additional source of data for research purposes.

But in qualitative research is not too much basing themselves on the data statistics, but make use of statistical data it is just as a way deliver and direct to the events and events found and sought for themselves in accordance with research objectives (Moleong, 2004).

4. Data Collection Technique

The data needed in the researcher is collected by doing interview indepth interview. In digging information using the method interviews, Interviews conducted with unstructured but focused on depth (indepth interview). It aims to find information that is not raw (rigid) or single and more free information, questions and statements can be flowing like a casual conversation. Strengthening informants through the process Interview. Because all communications have consequences, the field of research interviews with forms and content that are not "neutral". The interviewer is a powerful tool which researchers can see about humanity and create opportunities for develops and changes in the person being interviewed. It must be fixed anyway is maintained so that research activities are an opportunity to reveal anything which goes beyond the story into something useful.

ROLE AND PERFORMANCE OF UNIVERSITY

In this section will be described an analysis of the role and position by the PR University of Mahendradatta. In addition, at the end will be presented the results of performance assessment of PR based on the assessment of stakeholders.

1. The Role of University

Before understanding the role of Public Relations in carrying out management functions it is necessary to begin by identifying the PR working model. There are four the Public Relations model formulated by Grunig and Hunt in his Managing Public Relations 1984: press agentry/publicity, public information, two way asymentric and two-way symentric. In detail the models can be described as follows: (a) Press agentry/publicity: intended for the benefit of propaganda, one-way communication, so the message is often incomplete; (b) Public information: The main purpose is dissemination or dissemination of information. One-way communication, not necessarily in a persuasive way; (c) Two-way asymentric: This model is intended for scientific persuasion. Communication two-way, with unbalanced effects; (d) Two-way symentric: This model aims to gain mutual understanding (mutual Understanding), while the communication is two-way with the effects balanced (Gozali, 2010).

Public Relations of Mahendradatta University is working with media/journalists to report objectively to the stakeholders. Two-way communication with stakeholders can not be identified in this research. If analyzed from the results of interviews about the tasks that PR do tend to doing media relations activities. Media relations activity is one of the activities PR is important, but in creating two-way communication required for more direct communication (communications) where the PR is active facilitate management to conduct communication process between management and stakeholders. By directional communication is dynamic, in which information is exchanged in two direction between university management and stakeholders. In that communication two-way is the most important thing is the occurrence of feedback. Feedback is information which is important and useful to be input management in carrying out its duties to achieve organizational/university goals.

After discussing about the communication model conducted by Public Relation of Mahendradatta University then discuss about the role. The dominant role of Public Relations in management is identified to be four levels: (a) Expert prescriber, PR officer is considered as an expert who can give advice to the boss. This kind of role is needed at the institution which many have threats and many/often experience change; (b) Problem solving facilitator, acting as a facilitator in the process of solving problems, PR is involved in any management, being a team member or become a leader in crisis management. This kind of role is required on institutions that have many threats but little change; (c) Communication facilitator, acting as facilitator of communication between institutions with the public or as a communication bridge and as a medium or mediator in case of miscommunication. This kind of role is required for the company has few threats but many/often changes; (d) Communication technician, acting as technical communications implementer including providing technical services. This kind of role is required for the company has few threats and changes slightly (Cutlip, 2005).

2. Position In The Management

When viewed the structure of the organization PR in the management of the University, they are not in a dominant position. Structurally Public Relations university Mahendradatta is a technical implementation unit. Technical implementation unit in organizational and system of documents work is a necessary supporting element for the organization of education within the University. Public Relations of Mahendradatta University is directly responsible to the rector and coaching is done by Vice Rector II. If seen from the documents of organization and working procedure of Vice Rector II is the only petition Rector in the field of Public Relations in addition to carrying out activities in the field of general administration, planning and finance. The process of bridging management with the public. The Public Relations position has influence in the process of establishing relationships between management and the public. Such that delivered, a public relations practitioner who successfully practices public relations effectively those who use the new model of symmetry as two-way practice. Model Placing the organization with its public on two opposite poles.

Effective public relations practice according to this model is the result of accumulation of the practice of two-way asymetrical and two-way symmetrical models. Each party will persuade to get agreement. In this condition, it could happen that is pure asymetrical model, ie the situation of dominant coalition trying to press the public to accept the position of the organization, or it could happen coorporation model, that is where the situation public use communications to convince dominant coalition to accept positions public.

Furthermore, if the purpose of public relations is incorporated into an organizational goal, public relations executives sit as a dominant coalition. Target public program relations will be more adaptable if the public relations executive is a coalition dominant in the organization. Conversely, if the public relations executive does not sit in a coalition which is dominant, it will be difficult to determine the function of public relations, no matter how the public executive relations tries to fulfill its communication objectives. (Ananto, 2004).

3. Performance Of Person

In this research performance of Public Relations seen from stakeholder perception. understanding above contribution of PR in management between internal stakeholders Mahendradatta University. The first informant at the University of Mahendradatta gave an answer to the authority of the publicist that is documentation and publication for activities that scale university and other university leaders informant gave additional that the PR is structurally directly under the rector. Informants at the University of Mahendradatta give an opinion that preach and then if there are outsiders who need information about Mahendradatta University that is public he can issue.

Other informants argue that the responsibility and authority of PR is imaging campus. So how to imaged this campus well so that it can be known all over community layer. Opinions given by each informant have not been able provide an appropriate and detailed perception of responsibility and authority of PR in the university. That function PR is a multi-perception. Confusion about the role and function of PR is still very high, even to the level of decision makers. In Khasali (2010), Management PR, stated that there are several viewpoints in viewing Public Relations. Viewing PR in the viewpoint: (a) Personality; (b) In terms of communication; (c) Publications; (d) Marketing; (e) Management; (f) Advertising.

The view of the perception of public relation is an advantage. Being a excellence when made a However tasks and functions of Public Relations as in the document organization and working procedures is not something which is suitable for university characteristics that have challenges in solving problems as disclosed by Cutlip (1985), universities face four problems in communication that is: (a) Insufficient funding support to run the role of a public institution in the wider community; (b) Competition to get a qualified student needs to be passionate and expensive; (c) Constraints and regulations make the college administration difficult and requires cost; (d) Academic and ownership freedom is a challenge for internal stakeholders or external.

Especially for Public Relations Mahendaratta University, stakeholders stated that still need to be improved coordination between Public Relations with other parts. Head of Public Relations Mahendradatta University held by a lecturer is considered as a separate obstacle in coordinating internally PR and coordination with other parts. According to informant job informants are responsive and need high coordination speed.

CONCLUSION

Answering the purpose of this study then concludes research on roles and position Public Relations as a Management Function in Higher Education is:

1. The role of Public Relations at Private University tends to act as Communication technician and little role as Communication facilitator. This matter due to the organization and working arrangements of private universities do not provide many opportunities to play a role.
2. The position of Public Relations at the State University is a marginal position with the proven still many levels of bureaucracy that must be passed in carrying out its functions.
3. Performance of Public Relations related to the role and position in this research is considered positive and in accordance with the portion of work.

IMPLICATION OF STUDY RESULT

Practically, this research provides an explanation that the role and position of PR in private universities can not play a role and gain a position good. The role and position of the PR is influenced by the rigid rules relating to organization and work procedures endorsed by the government. Limited resources are obstacles are quite heavy faced by PR. Humas run public relations activities not to the strategic level. Academically, the concept of PR excelence and the dominant coalition of Public Relations in an order organizations at state universities can not be realized optimally. Roles and positions the strategic is not obtained by PR. Public Relations should have a role up on level of expert presciber by conducting two way communication (two-way communications).

The ideal position is where the PR is in a central position, has a coalition which is dominant in organizational communication. Socially, this research can provide an explanation that the role problem and the strategic position of PR in an organization is ideal. Roles and positions ideal has implications PR can work optimally in performing functions communication as an organization.

BIBLIOGRAPHY

Ananto, Elizabeth Goenawan, 2004, *Public Realationa As Dominant Coalition, Is it Possible ?*, in Ridwan Nyak Baik & Irmulan Sati T (eds), *Dominant Coalition, Critical Reflections on the Role of Public Relations in Management*, Jakarta, BPP PERHUMAS.

Bogdan and Taylor, 1975, in J. Moleong, Lexy (1989), *Qualitative Research Methodology*, Bandung, Remadja Karya.

Cutlip, Scott M., et al., 1985, *Effective Public Relations, Fifth Edition*, New Jersey, Prentice Hall International Inc.

Gozali, Dodi M, 2010, *Communication Measurement: Concepts and Applications of Public Relations Performance Measurement*, KITA.com Books.

Grunig, James E., and Todd Hunt, 1984, *Managing Public Relations*, New York, Holt, Rinehart and Winston.

Khasali, Rhenald, 2010, *Public Relations Management: Concepts and Applications in Indonesia*, Graffiti Publisher.

Moleong, 2004, *Qualitative Research Methods*, Bandung, Youth Rosda Karya.

Jolly phonics: effective strategy for enhancing children english literacy

Ni Putu Pebri Ariati, *Ni Nyoman* Padmadewi, *and I Wayan* Suarnajaya

English Education Department, Ganesha University of Education, Singaraja Bali Indonesia.

Abstract. Having strong foundation in English literacy nowadays is believed to be a very crucial requirement for the children to survive and success both in educational and social lives. One of the most effective strategies for enhancing children's early reading and literacy skill is through Jolly Phonics. Despite the popularity and its effectiveness, there is still factor to the absence of the strategy, namely lack of teacher knowledge. This paper is intended to share the techniques of how to implement Jolly Phonics strategy for teaching and developing children's English literacy. This study was conducted in one of innovative bilingual schools in Denpasar, Bali. The study employed descriptive qualitative method as a research design and used observation, documentation and interview to collect the data from the teacher. The findings showed that the teacher had successfully implement the 5 skills in Jolly Phonics, namely (1) learning the letter sounds, (2) learning letter formation, (3) blending-for reading, (4) identifying sounds in words-for writing and (5) tricky words, through variety of enjoyable techniques involving children's sight, sound and kinesthetic means. Thus, implementing Jolly Phonics through techniques explained will improve children's overall English literacy skill.

Keywords: Jolly Phonics, English literacy

1 Introduction

The term literacy has become a very crucial terms in education during the last few decades. As states by Roskos, Christie & Richgels, extensive research in the field of literacy, particularly research on literacy development of young children has been conducted and becoming a basis for everyday practice [5]. This is because having a strong foundation in literacy, specifically English literacy is very beneficial for the continuity of children's life, as today English is used as lingua franca and international language of communications.

The fact that having strong literacy foundation is fundamental for children is supported by Alberta Education, who claims that literacy has always been a basis of learning [4]. Moreover, Morrow also states that children's futures are partly determined by how well the children are guided to literacy [9]. Regarding to the urgency of literacy, the children therefore should be guided to literacy right from the very early age. The earlier they get into literacy, the more experiences they will gain to survive both in educational and social lives.

However, finding an appropriate yet innovative strategy to develop and enhance children's Engish literacy is not an easy thing to be done. Udosen & Ukpak (in Ekpo et al) states that most pupils in preschool come without a slight reading readiness [3], and they have no idea of how to sound out the letters in English. Consequently, Mullins points out that many children struggle to decode words [2]. In this case, one of the most effective strategies that could be used to enhance children's early reading and literacy skill is Jolly Phonics.

Jolly Phonics is a fun- systematic program designed for young or beginner learner to develop their reading and literacy skill. According to Lloyd, Jolly Phonics teaches the 5 main skills, where the children are firstly taught the 42 sounds in English, and then continued into blending and reading skill, at the same time they are taught to write by identifying the sounds in words [13]. In addition, Farokhbakht & Nejadansari also state that Jolly Phonics incorporates multisensory approach, where the information is delivered through sight, sound and kinesthetic means [6], which is really suitable for young children's characteristics.

Dixon et al, as well as Farokhbakht & Nejadansari have conducted a study to see the effectiveness of Jolly Phonics in improving the children's early English literacy. The findings of both study shows that children who are taught by using Jolly Phonics has better performance on reading, spelling and literacy compared to those who are conventionally taught [12,6]. Other study conducted by Ruhaena has also proved that Jolly phonics implementation affects not only the children's English literacy ability, but also their Indonesian literacy [7]. As a result, Ogbemudia & Alasa claims that Jolly Phonics instructional strategy is a very effective panacea to reading difficulty faced by children [10].

Despite the popularity and its effectiveness, there are still factors to the absence of the strategy. In this case, Mullins have listed few factors to the absence of Jolly Phonics at school, such as lack of materials and lack of teacher knowledge [2]. For that reason, this study was carried out to assist teachers on how to develop and enhance children's English literacy skill through Jolly Phonics strategy. In addition, the problem in this study were directed to answer questions about thet kinds of technique used for implementing Jolly Phonics and how those techniques implemented in the classroom.

The remaining sections of this paper will discuss about method of study, findings and discussion of jolly phonics implementation in the classroom, which are ended with closure. Therefore, this paper is expected to be able to inspire other teachers on how to develop and enhance children's early English literacy through Jolly Phonics strategy.

2 Methods

This study was a descriptive qualitative study which was conducted at Dyatmika Kindergarten, Denpasar Bali Indonesia. The subject of this study was a qualified and experienced native teacher who taught Jolly Phonics at Dyatmika Kindergarten. She graduated from one of university in Australia and had been been teaching children through Jolly Phonics for about 3 years. The data were collected through a series of observation and documentation in Dyatmika kindergarten class, and at the same time through interview with the teacher, which was then analyzed descriptively.

3 Results and Discussion

Jolly Phonics at Dyatmika Kindergarten was implemented following the principles of effective literacy instructions states by Roskos, Christie & Richgels, in which the teacher had succesfully engage in a rich conversation with children through story reading, shared reading activities, and etc [5]. As have been stated, English literacy has become the core important of all learning, because nowadays English is used as a universal language to explain most knowledge and technology. It is not only used for the sake of schooling but the children can use it for life. Realizing the fact stated, Dyatmika Kindergarten then used Jolly Phonics to help the children master the English literacy skill. In this case, the skills in Jolly Phonics are attractively taught and creatively developed by the teachers based on children's need and situation, so children could easily engage with it. It was also taught and introduced systematically following the letters groups and skills sugessted, starting from the easiest to the complex one. However, the findings of this study was not in the line with Lloyd, where it was suggessted to introduce 1 sound per day [13]. Here, the teacher adapted the program into longer one, due to the children's multicultural language background, so the section for the sounds introduction were maximized by introducing only 1-3 sounds per week. This action was in accordance with Ehri, et al and Boyadzhyan who propose that phonics should be systematically taught as a part of literacy program and, it should be adapted based on children's ability [8, 1]. Therefore, the following are the techniques used by the teacher for teaching English literacy through the five skills in Jolly Phonics.

3.1 Learning the Letter Sound

Knowledge of letter sounds was the first skill needed to be mastered by children in learning English literacy through Jolly Phonics. Actually, the 42 sounds in English were divided into 7 groups of letters, where in a group consisted of 6 letters, such as *s, a t, i, p, n* and so forth. Based on the findings, the teacher introduced 1-3 sounds in a week, and it would always be reviewed and reinforced. There were some techniques used for teaching letter sound skill to children, namely story reading, song and action as well as games, as can be seen below.

a) Story Reading

Story reading was done everyday for the children either as introductory activity or main activity for learning letter sound. The story read for the children could be any story, story with rhyming or storyline from Jolly Phonics book itself. Through story reading the children was not only able to enjoy the story read by the teacher, but also listen to the words and sounds in English as well as develop their comprehension skill and critical thinking.

b) Song & Action

The second one was song and action, where the children were asked to sing the Jolly Phonics song, at the same time they did the action for the sound. For instance, the sound */s/* represented by <u>s</u>nake and the children needed to sing and weave their hand in an 's' shape like a snake and say *ssssss*. And as a follow up acivity, the teacher asked them to find words or things in the classroom which was started with */s/* sound, like *sun, ship, starfish*, etc.

c) Games

The teacher provided Jolly Phonics games, which were software games to be played through computer by the children. Here, the games designed interactively where the children could practice all the five skills of Jolly Phonics independently. For instance, they could learn as they played with the letter sound and the picture represented by the sound, such as 'i' for 'inky' the mouse, 's' for 'snake', etc. Thus, with a clear instruction provided by the game that could guide them through the activities, they would be able to train their self-independency and determine what to learn and what to play.

3.2 Learning Letter Formation

When children knew the letter sound, at the same time they needed to know the letter formation. Firstly, teacher showed the letter example through song or flashcard and wrote it on the board. After that, the teacher showed the letter formation in the air as she gave instruction to be followed by children. For instance, in the letter /l/, the teacher said '*start at the top, down and cross*' and it should be noted that teacher needed to do the formation in a mirroring position, so children could immitate it correctly. And as a follow up activity, the teacher provided a kind of sound sheet, consisted of picture represented the letter sound, space for the children to trace and do the letter formation, as well as some pictures where children needed to determine which picture started with the letter sound being practiced. At the end, they were asked to circle the best letter they formed to boost their confidence.

3.3 Blending – For Reading

Blending can be started when the first six letter sounds have been taught by teacher. The skill needed to be frequently practiced by looking at the letters, saying the sound and hearing the words. In this case, teacher firstly planned lists of words needed to be blended by children, so it would be easier in teaching it. Flashcard was the main media used by the teacher, where at the first time teacher always did a kind of review through the letter sound. Variety of techniques was employed, such as having children in a small group and giving one flashcard for each of them to be blended. This was done in order to make them took responsibility toward themselves firstly, and when they found any obstacles in blending the word, the teacher would fix it together with the group. Another way used by teacher was having small whiteboard and some single letter flashcards, as can be seen below.

Pict. 1. Whiteboard Blend

Picture 1 above shows the example of whiteboard blend with CVC word. The example of word *'yab'* above actually has no meaning, but when the children knew most of letter sounds they would be able to blend the word correctly. The teacher put 1 by 1 letter to be sounded out and then blended by the children. This technique was found attractive, because it was taught slowly but sure, and required the children to look at the letter, sound out the letter 1 by 1, and hear the word before they finally could say the word. As a follow up activity, the teacher run a home reading program, where the children needed to practice their blending and reading skill at home with their parents and noted their progress in their reading records. Actually, those program was in the line with techniques of of promoting autonomous learning proposed by Padmadewi where providing lots of book and facilitate children's early reading skills through reading log or reading records could motivate and lead them to be an autonomous learners, which was very beneficial not only in present, but also future life [11].

3.3 Identifying Sounds in Words – For Writing

In learning to write, the children were required to have *phonological awareness,* where they could listen for the spoken words, identify, and write the sound in that word. There were variety of techniques used by the teacher, starting from copying, modeling, in which the teacher could slowly rub the words written on the board and asked them to try to write on their own. And finally, gave chance for the children to write independently, either through news writing, story writing, or free writing as can be seen below.

Pict. 2. Free Writing Guide

Pict. 3. Free Writing Example: Friends

Picture 2 above shows the guide for children's free writing. In free writing, they could write about friends, toys, food etc. Here, teacher always provided alpabeth and words for writing, like *a, and, an, brother*, etc, just in case they forgot how to write it because those words were a kind of tricky words. Picture 3 is the example of one child's writing, where she tried so hard to write about her friend independently. Firstly, the teacher let the children drew while thinking about what they needed to write. When she finished drawing, she started writing by hearing the sounds in words. Based on the example above, it could be said that this child was in the preparation step to become independent writer. She put the message she wanted to say into her writing, which could be understood by others. She knew why she loved her friend and stated the reason clearly. She only needed to try to hear sounds in words more carefully, and remember some more tricky words, like *friend, to, me*, etc. By doing this activity frequently, the children would be used to writing and ready to be independent writer.

3.4 Tricky Words

Tricky words can be defined as irregular words which are difficult to spell, but frequently happened in many text. The study found that the teacher introduced 2-3 new tricky words per week for the children through variety of techniques, such as flashcard activity where teacher introduced tricky words through flashcards, and they needed to create sentence with the tricky word stated. For instance, the word 'live' where the /e/ sound was the tricky part, and beyond expected the children could say the tricky word and create sentence like '*let's live in another place*', '*I live all the time*', etc. The other way was through snapping game, where teacher spreaded tricky words flashcard on the floor and children needed to snap and say the word. Moreover, the teacher also used outdoor activity, as can be seen below.

Pict. 4. Look, Copy, Write & Say the Words

Picture 4 above shows the example of look, copy write & say the words activity for tricky word skill. This was done outside the classroom to boost children's learning motivation. Firstly, the teacher gave 1 flashcard randomly to each child, and had them to look, copy and write the word on the ground by using chalk. Surprisingly, the children could write correctly, with neat and perfectly formed letter, even they wrote by using tool other than pencil. And then, they required to swap with another children's card and try the same steps, so they would be used to the written tricky words. Finally, all children were asked to line up and each of them needed to snap, read and say at least 3 tricky words written on the ground by using snapping tools. Therefore, through those activities, they simultaneously tried to remember, read and say their tricky words, at the same time they learned to write it correctly, which was very beneficial for them and could affect their English literacy skill.

The techniques explained in this paper were the techniques used by the teacher in teachig English literacy through Jolly Phonics. Even though the succes of the techniques could be seen through the children's performance, it should be admitted that the limitation of this study was not investigating the strategy quantitatively. Despite of the limitation, children's response and performance in learning the 5 skills in Jolly Phonics as described above indicated that they were in the middle of developing and enhancing their English literacy skill, where they knew most of letter sounds, they remember most tricky words, they were able to blend, read and at the same time write independently on their own.

5 Closure

English literacy skill should be introduced earlier to children as today English took the prime position in communication. Moreover, it was not an instant process to shape the children into literate individual due to the complexity of literacy itself. The 5 skills in Jolly Phonics which were delivered through variety of techniques proposed in this study has shown to be powerful ways to develop and enhance children's English literacy. For the best result, it should be carefully planned by the teacher with regard to the children's need and ability.

6 References

1. A. Boyadzhyan. *A Systematic Approach for Teaching Phonics by Interactive Whiteboard.* http://www.sierranevada.edu/assets/ABoyadzhyanfinalproject.pdf (2012)
2. A.K. Mullins. An Analysis of the Phonics Dance in a Semi-Rural Midwestern Elementary School (doctoral thesis, 2013)
3. C.M. Ekpo et al. JPS & ESL Ppl R. Dev (presented at conference in Ibadan Uni, 2007)
4. E. Alberta. *Literacy first: a plan for action* (Alberta Education, Canada, 2010)
5. K. A. Roskos, J.F. Christie, D.J. Richgels. Y. Ch. 52-60 (2003)
6. L. Fakhrokhbat, D. Nejadansari. Ef. Syn. Mlt. Ph. T.L EFLYL. I. J. Rsc. Stu. In Edu. **4**, 39 (2015)
7. L. Ruhaena. Pgr. Mtd. JP. thd. Bc Tls. B. Ind & B.Ing. J. Pnlt. Hmn. **9**, 192 (2008)
8. L.C. Ehri et al. Systm. Ph. Ins. Hlp. Stdnt. Lrn. R. J. Drct. Ins. **2**, 121 (2001)
9. M. Morrow. *Developing literacy in preschool* (The Guilford Press, New York, 2007)

10. M.I. Ogbemudia, V.M. Alasa. JPIS. Pnc. R. Dfcl. Cdhd. E. J. Ed. Rsc. Bhv. Sc. **3**, 1 (2014)
11. N.N. Padmadewi. Tch. Prmt. A. L. Cl. J. Edu. Soc. Sc. **3**, 45 (2016)
12. P. Dixon et al. Synthetic phonics in low-income schools in Hyderabad, India. http://jolly2.s3.amazoneaws.com/research/jph%20research%20Hyderabad.pdf. (2011)
13. S. Lloyd. *The Phonics Handbook* (Jolly Learning Ltd, England, 2010)

Competency test for selecting majors to produce competitive vocational graduates in industry

Agus Aan Jiwa Permana [1,*], *Ni Ketut* Kertiasih[1] , and *Kadek* Suranata[2]

[1] Department of Information Management, Universitas Pendidikan Ganesha, Indonesia

[2] Department of Counseling Guidance, Universitas Pendidikan Ganesha, Indonesia

Abstract. President of Indonesia, Jokowi reoriented vocational school graduate toward demand driven graduates that is graduates who have certificate and skill required by industry. The initial stage of the new student on vocational school is choosing a major. At first step, students often confuse in choosing the majors they want. The mistake of choosing a department will be a set beck to the motivation of learning and skill, later will impade students future career. Thus competence test is needed to helping them in choosing the majors according to their competence. The solution to this problem is to conduct online competency tests for new students. The Results of research with 60 responden, 78% corresponds stated that the majors they were in match with their interests. Then the remaining 22% did not match. But the result of the competency assessment of students in match with the majors of is 40% and 60% of students need counseling for selection majors.

1 Introduction

The public opinion in Indonesia about education is now shipting. Public in Indonesia is more interested in sending their children to Vocational High School (VHS) than Senior High School (SHS) after finishing study at junior high. Studies shown that parents want their children to be ready for work after finishing study if they can not afford for higher education [1]. Consequently government has set a policy to cover the shortage by increasing the ratio of VHS : SHS To 70:30. The idea behind the increasing ration in favor of VHS is that it is expected to produce skillfull and intellectual class among the work force [2]. Many other countries are now follow the policy of increasing the ration. Among them is Malaysia. It is clearly stated by [4] who emphesized that fruitfull shifting toward Vocational High School is a major transformation in vocational education in Malaysia. President Jokowi stressed the importance of reorientation of education and vocational training in the direction of demand driven by the certification in accordance with the demands of industry. He also mentioned that the Indonesian population is dominated by young people by 60%. With such a large percentage, Indonesia should be able to increase

* Corresponding author: agus.aan@undiksha.ac.id

its workforce without having to import labor from abroad. One of the leading sectors of VHS is the computer majors that entered in 15 majors VHS promising the future in the field of multimedia [5].

In a National Conference of Vocational and Technological (Semnasvoktek) organized by the Faculty of Engineering and Vocational, Undiksha Dr. I Ketut Suarnaya, M.Pd as a keynote speakers mentioned that the slogan VHS is Working, Continuing, and Self-employed abriviated as WCE. The slogan indicates that the vocational school graduates has aplicable competence. However, practically the effort to reduce unemploiment by increasing the ration is not good. The same data was also recorded by the Central Bureau of Statistics which stated that viewed from the educational background, the largest proportion of unemployed are those graduates of VHS 9.84%. This figure is higher than the unemployed high school graduates 6.95%, junior 5.76%, and primary 3.44% [4] Vocational students need to be equipped with risk analysis preventing risk Many feel that they themselves are primarily responsible for performing their duties safely while in practice [6].

It is commonly found that students in choosing education program are baseless. They often choose education program randomly whitout consulting their potention such as the basic common abilities (intelligence), talents, interests and physical and socio-cultural conditions and their career interests. One of the efforts that can be done is to prepare the assessment tool to help students understand the potential and the condition of self including the general basic ability (intelligence), talents, interests and physical condition as well as social culture and career interests of students [7]. After the selection process, it should also be noted that the curriculum is based on a competency approach to train vocational students [8]. The success of those affected by many factors which allow cognitive variables, interest and personality should be evaluated together in predicting success [9].

[10] said that an inventory personal survey tests can be used to screen for potential new students. The test is aimed to place students according to their interests with the right areas of expertise which can increase their motivation. The success of a vocational education according to Erlin Leigh Parker (2008, in [10]) is influanced by many factors, one of them is the recruitment of the students. Good recruitment process and adequate facilities will automatically produce competent graduates. In the near future, the system will be online-based, making it easier for participants who are away from school and has limited time can take the test. it is supported by [8] who stated that technology can be used for steering student in choosing a career in accordance with personality and competence. With a good selection process, and adequate facilities will produce graduates who are competent and ready to compete in the world of work. After the selection process, it should also be noted that the curriculum is based on a competency approach to train vocational students [11]. Meanwhile, the limited number of conselor and inbalance in ratio between the conselor and students is the main constraint. It makes face-to-face counseling service less effective. In addition, there is a tendency of current trend that counselee feel uncomfertable if the conseling is conducted face to face [12]. Based on the above explanation, it is necessary to develop an assessment instrument that can be accessed online through the internet network to assist the selection that can give the students recommendations in choosing the majors, especially in the vocational school.

2. Literature Review

2.1 Vocational Education

Vocational education embraces an open system (multientryexit system) and multimakna (oriented to culture, empowerment, character formation, and personality, and life skill) [13]. Even overseas The Italian Government has been in the context of implementing its Digital Agenda overhauling the education system in the country where all children have the right to attend education and training for at least 12 years, applicable for students aged 15 years or older who can be accepted for internship in the company because it is already Signed an agreement with MIUR (Department of Labor) [14]. Based on the vocational exposure above, in the vocational education system there is a difference with the education system in high school because students are directed to become a workforce ready to compete at the industry level. Industry is in dire need of manpower who has the skills to work in certain fields eg in tourism, computers, fisheries, maritime, shipping, agriculture, plantation, pharmacy, catering, fashion, automotive, hotel, accounting, office administration, banking, travel, And marketing. So that the role of vocational schools and colleges and colleges should cooperate with the government and industry to produce demand driven labor such as the expectations of Mr. President Jokowi. [15] wrote an article that says that there are three features in vocational education, among others : 1. More Practical: If you feel attending college classes are boring, do not like doing assignments and assessed by exam, you may consider vocational education. Vocational education actually trains practical skills, so of course more practice than theory. 2. Many institutional options: Unlike undergraduate degree education and so on, vocational education is offered by more institutions, whether universities, colleges, polytechnics, training centers or other institutions that specialize in vocational education programs. 3. Various programming options: Vocational education is suitable for those who are clear and confident about what they want to pursue as future careers. Lots of fields are available ranging from tourism and hospitality, retail management, software development, interior design, automotive engineering, hair styling to culinary.

2.2 Test Inventory Personal Survey

Before humans recognize measurement tools that are now known as personality inventory, attempts to get personality traits have existed in such a way as to see the hand line, face, date/day of birth, and other endeavors [1]. Personality measurement is important in the knowledge of careers since 75 years ago (1918s), although much centered on abnormal personality, this has also been done in normal personality, especially to help individuals make selection in the choice of job As well as positions [1, 16]. Characterizes people with similarities in six personality types: realistic, investigative, artistic, social, enterprising, and conventional (RIASEC) [17]. A person's suitability and environment can lead a person to predict and understand knowledge about personality types and environmental models. This will be highly beneficial for job options, vocational stability and achievement, choice of education and achievement, individual competence, social behavior, and to affect one's sensitivity. Self Direct Search Inventory (SDS) inspired from Holland's personality and career theory (1975). SDS is a type of developed by Holland through the adaptation and refinement of Vocational Preference Inventory (VPI) made in 1958, has been revised three times and the linventory ast of 2001, has been used and adapted over 20 nations in the world with versions over 15 Language, and has been proven through various studies in measuring personality types and providing career predictions suitable for students and students as well as for those already in the workforce [1, 16]. The developed Personal Inventory Personal Survey (TIPES) has four indicators: (1) interest and involvement in an activity, (2) self-perception of ability to perform a field/skill, (3) interest and interest in a particular occupation or occupation, (4) self-assessment to be able to work on certain areas of expertise.

2.3 Web Based

Terminologically website is a collection of web pages, usually summarized in a domain or subdomain, located within the World Wide Web (WWW) on the Internet. A web page is a document written in HTML (Hyper Text Markup Language) format, which is almost always accessible via HTTP, a protocol that conveys information from a website server to be displayed to users via a web browser. All publications from these websites can form a huge information network. The pages of the website will be accessible through a URL called Homepage. This URL sets the pages of a site to be a hierarchy, though, the hyperlinks on the page govern the readers and tell them the overall susceptibility and how this information flows through.

3 Method

In the early stages of the study, the subject were students of state VHS and Private VHS in Buleleng. Research in phase I (year 2017) focused on identification of personality type model that corresponds to vocational areas (majors) in VHS based on personal inventory survey test and its correlation with student achievement involving sample of student of VHS in Buleleng Regency determined by cluster technique sample area The validity test involves: (1) counselors expert; (2) educational/ICT media specialists, and (3) Counseling teacher practitioners (school counselors) as assessors of the design of online test systems. Small-scale test of the implementation of a online test system will be done with the subject of Vocational High School students in SMKN 3 Singaraja for engineering majors (Engineering), SMKN 1 Sukasada for industrial majors(Industry), State SMKN 1 Seririt for tourism majors (Tourim).

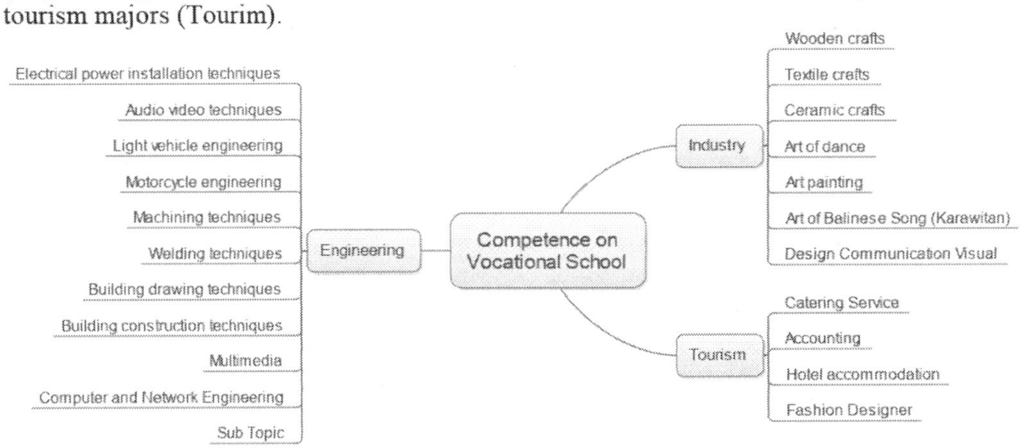

Fig. 1. Mapping majors at vocational schools

$$C = \{ 3(c1) + 2(c2) + 1(c3) \} \tag{1}$$

Information about equation 1 :

C = Conggressive Scores or Conformity Scores
C1 = first digit match score
C2 = second digit match score
C3 = third digit match score

The process of collecting data is done by giving questionnaires to students which contains four parts. Section A from the numbers 1 (one) to 66 (sixty six) were statements related to daily activities. With a response time of 15 minutes. Section B contains items of statements from number 67 (sixty seven) to 132 (one hundred and thirty two) were statements related to areas of expertise or competence. With a response time of 15 minutes. Section C contains

Item statements from number 133 (one hundred and thirty three) to 210 (two hundred and ten) were relatated to various types of work. Those types of works were devided into two parts, one part is the kinds of work the students love and the other part is the kind of work students do not want. With an answer time of 20 minutes. Section D contains The statement items from 211 (two hundred and one) to 216 (two hundred and sixteen) corresponded to different types of talents and areas of expertise. With answering time for 5 minutes.

Table 1. Relation of Majors & Interest Base on School Type

School Type	**Relation of Majors & Interest**		**total**
	Correct	**Incorect**	**Responden**
Engineering	**27**	**3**	**30**
Industry	**10**	**10**	**20**
Hotel dan Tourism	**10**	**0**	**10**
Totally :	**47**	**13**	**60**
Percentage :	**78%**	**22%**	**100%**

4 Result and Discussion

After the test is done then the results will be displayed directly, whether the user's competence in accordance with the profile of the majors are planned to be selected. Based on the results of research from three schools with different skills obtained from 60 responden the result is that 78% corresponds stated that the majors they were in match with student interests. Then, the remaining 22% did not match. There was a difference to the degree of mismatch between the departments and the students' interest, 50% in industrial schools and only 10% occurred in technical schools in Table 1. The result of competency indicated that, 60% of students need counseling and more specifically for primary, secondary, tertiary test results can be seen as Figure 2.

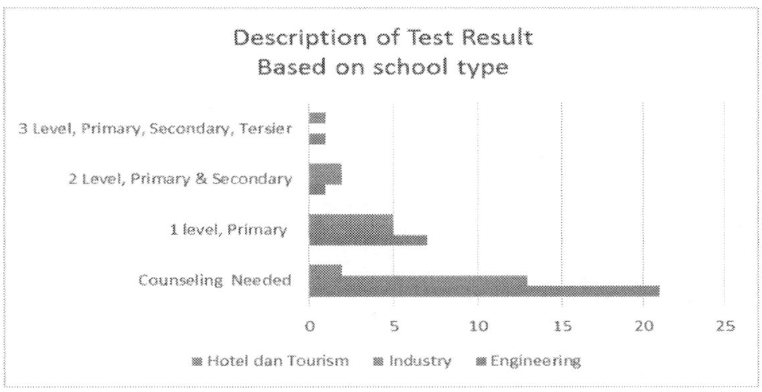

Fig. 2 Description of Result Based on School Type

5 Conclusion

The conclusion of this research is to give the right recommendations to the students in choosing the department. The degree of conformity of majors with student interest is good. But the result of the competency assessment of students in accordance with the department is 40%. With the system developed, students are more confident in choosing the majors in accordance with their interests and desired skills. The right placement will improve students' learning motivation in the classroom. Thus after students finished their education at VHS, they will have a successful career in the industry that is Demand driven graduates and also produce graduates who are equipped with entrepreneurial spirit to create new jobs oportunity to reduce unemployment in Indonesia.

5 Acknowledgment

Thanks to teachers and schools that provide valuable data such as SMKN 3 Singaraja, SMKN 1 Sukasada, and SMKN 1 Seririt. The greatest thanks is given to Universitas Pendidikan Ganesha that provided funding for the early stages of this research through DIPA 2017 funding entitled "Online Assessment Personal Survey (TIPES) as a Selection Recommendations Major On Vocations School".

References

1. Suranata, K., 2009, Relationship Between Conformity of Personality Type and Environmental Model with Maturity of Career Choice Choice (Study on Grade XII Students at State VHS 1 Padang), *Thesis*, Guidance and Counseling Study Program, Graduate Program, State University of Padang, Padang. *(in Indonesian)*
2. Lewis, H. W. and Jorgensen, B. (2015) 'Sweat and Brains : A Look into Intellectual and Vocational Trends , and their Importance'.
3. Rahman, A., Hanafi, M. and Ibrahim, M. (2014) 'Assessment Practices for Competency Based Education and Training in Vocational College , Malaysia', *Procedia - Social and Behavioral Sciences*. Elsevier B.V., 112(Iceepsy 2013), pp. 1070–1076. doi: 10.1016/j.sbspro.2014.01.1271.
4. Public Relations, the Majority of Unemployed Vocational Graduates, President Jokowi Ask Vocational Education System Removed, *http://setkab.go.id/category/berita/*, (accessed: October 28, 2016) *(in Indonesian)*
5. Supriani, Nanda Fitri, Here are 15 Promising Future Vocational Schools in the Future, *http://Papasemar.Com/Author/Nanda-Fitri-Supriani/*, (accessed: October 28, 2016) *(in Indonesian)*
6. Andersson, I., Gunnarsson, K., Rosèn, G. and Åberg, M. M. (2014) 'Knowledge and Experiences of Risks among Pupils in Vocational Education', *Safety and Health at Work*. Elsevier Ltd, 5(3), pp. 140–146. doi: 10.1016/j.shaw.2014.06.002.
7. Kemendikbud. *Specialization Guidelines for Learners. Agency for Human Resource Development Education and Culture and Quality Assurance of Education: Ministry of Education and Culture*, 2013 *(in Indonesian)*
8. Yang, H. H. (2016) 'Have Learned about Students in Educational Technology Major'. doi: 10.1109/EITT.2016.53.
9. Volodina, A., Nagy, G. and Köller, O. (2015) 'Success in the fi rst phase of the vocational career : The role of cognitive and scholastic abilities , personality factors , and vocational interests ☆', *Journal of Vocational Behavior*. Elsevier Inc., 91, pp. 11–22. doi: 10.1016/j.jvb.2015.08.009.
10. Asmara, Andik dan Haryanto. Development of Interest And Talent Test Using Artificial Neural Network Method (ANN) to Predict Potential Student in Robotics Field, *Journal of Vocational Education*, Vol 5, Number 3, Page. 273-286, November 2015 *(in Indonesian)*
11. Kateryna, R. (2015) 'Competency-build approach to the unitizing of vocational schools', (September), pp. 471–476.
12. Hidayah, N & Ramli,M. *Development of an online Behavioral Cognitive Counseling Model based online to manage Junior high school students SRL in Malang Raya. Research Report*. Malang: Research Institute of UM (2013) *(in Indonesian)*
13. Universitas Brawijaya, About Vocational, *http://vokasi.ub.ac.id/tentang-vokasi/*, (Accessed : 28 Oktober 2016) *(in Indonesian)*
14. Nazzaro, Antonio, The Italian national plan for digital school in the framework of the Digital Agenda for Europe, *http://s3.amazonaws.com/academia.edu.documents/33373326/nazzaro_dubai_invio.pdf?awsaccesskeyid=akiaj56tqjrtwsmtnpea&expires=1477648954&signature=lcmvzpvxw%2b6khvxijbg67mrnlhw%3d&response-content-disposition=attachment%3b%20filename %3dthe_italian_national_plan_for_digital_sc.pdf* , 2016, (diakses: 28 Oktober 2016)
15. HC Indonesia, 3 Vocational Education Features (Vocational), *http://www.hotcourses.co.id/studyabroadinfo/subjectinfo/3keistimewapendidikankejuruanvokasi/*, 2014, Accessed (October 28, 2016) *(in Indonesian)*
16. Nauta, M.M. The Development, Evolution, and Status of Holland's Theory of Vocational Personalities: Reflections and Future Directions for Counseling Psychology. *Journal of Counseling Psychology*, 57 (11),pp. 11–22. DOI: 10.1037/a0018213. 2010
17. Holland, J., *Making Uccupation Choise*, Prentice-Hall, New Jersey, 1985
18. Miller, Springer, Tobacyk & Wells. 2004. Congruency Between SDS Personality Type and Environmental Model. College Student Journal. (online). v.8, n.6073197, pg.2. *(http://findarticles.com/p/articles/mi_m0FCR/is_1_38/ai_n6073197/pg_2)*, Diakses 24 Juli 2008.
19. Alfindasari, Dessy. Steps to Conduct Research and Development (R & D), *http: //www.eurekapendidikan.com/2014/12/karakteristikrresearchanddevelopment.html*, 2014, Accessed (October 28, 2016) *(in Indonesian)*

Teachers' Beliefs and Perceptions of Code Switching in English as Foreign Language Classroom

Sondang Pondan Perlindungan Leoanak*, *Bonik* Kurniati Amalo

Politeknik Pertanian Negeri Kupang, Nusa Tenggara Timur, Indonesia

Abstract. In Indonesia, the use of only English as a medium of instruction in English as a Foreign Language (EFL) classroom is highly demanded. The present study investigates how code-switching is perceived by High School teachers in Kupang city, East Nusa Tenggara province, Indonesia, in teaching English. Specifically, it aims to find out the teachers' beliefs and perceptions towards the use of Indonesian language (L1) as their pedagogical instrument in teaching English in EFL classrooms. To achieve that goal, 40 EFL teachers were asked to give their responses to a 24-item of questionnaire which focused on teachers' roles and beliefs of code-switching applied in the EFL classroom. After analyzing the data, it was found that the teachers applied code-switching to serve pedagogical aims and to facilitate the EFL teaching and learning process. As a result, they used code-switching when explaining difficult words, encourage students' participation and managing and organizing the classroom. Another finding was that, the teachers also believed that, the advantages in applying code-switching exaggerated the disadvantages in ELF classroom.
KEYWORDS: *code switching, teachers' belief and perception, Kupang city*

1 Introduction

In English as foreign language (EFL) classroom, Indonesian teachers face several challenges in classroom interaction. Teachers in EFL classroom use more Indonesian language than English as a medium of instruction. Nowadays, Indonesian EFL classroom facts say that, the demanding of using only English as a medium of instruction is highly increased. However, the actual practices are different. The EFL teachers may switch from English to Indonesian (or vice versa) during their teaching practice. This practice is known as code switching and it is done to facilitate teaching and learning interaction among classroom participants.

Code switching (CS) is a natural part of being bilingual and it is a common phenomenon used by those who have developed two or more languages [1]. This phenomenon is also known as code alternation or code-mixing [2]. CS is used globally in communication

* Corresponding author: sonleo2002@yahoo.com

practice. Moghadam *et al.* [3] state that the use of CS in a conversation is a normal practice in global communication for various reasons and it is usually an unconscious process.

However, the use of CS in EFL classroom has been a subject of controversy. It has been regarded as negative and undesirable behavior where there is a failure in using the target language [4, 5]. CS has also been considered as a sign of laziness of language [6]

In contrast to the above controversies, some researchers [7, 8] have argued that in EFL classroom context, CS is useful in assisting teaching and learning process of English. Code switching can no longer be considered as errors interference, but as bilingual resources [9]. In addition, it helps low proficiency students gaining better comprehensions especially when giving classroom procedures [10]. Code switching should not be considered as a sign of defect in the EFL classroom, yet, it is a careful strategy employed by the teachers [11].

The proposed definition of code switching has been defined by numbers of researchers. For Gumperz [13], CS is the juxtaposition within the same speech exchange of passages of speech belonging to two different grammatical systems or subsystems. Code switching is also seen as the alternating use of more than one language [14]. Code switching also defines as using more than one language in the course of a single communicative episode [15], in the same conversation or even within the same sentence [16].

Code switching takes a variety of forms. It can occur within a sentence, at the beginning or at the end of sentences. In the level of syntactic, Poplack [17] identifies three different types of code switching, namely tag, intersentential and intrasentential switching. Gumpers [13] also introduces the concepts of CS into situational (participant switch) and metaphorical (topic switch). Another linguist, Lin [18] categorizes code switching into alternational switching and insertional switching.

Code switching has also been observed in English language teaching (ELT) classrooms. Some of these studies were focused on elementary level [19, 20] and some other were in higher education level [11, 21]. In secondary school level, Selamat [22] studied CS on Malaysian ESL (English as Second Language) classroom. She found that teachers and students generally exhibit positive attitudes towards code switching in the classroom and perceive its pedagogical merits as an effective teaching and learning resource. Canagarajah [23] studied CS in ESL classroom in Jaffna. He summarizes the categorization of CS in classroom into classroom management and content transmission. In addition to CS in English classroom, Ferguson [24] reported that, there are three functional categories of CS in the classroom context as for curriculum access, classroom management and interpersonal relation.

This study adopted the similar stance on the issue, particularly, when English Only classroom would only lead to frustration since the input is incomprehensible to the learners [12]. The present study also believes that the application of CS in EFL classroom by teachers would be able to ensure that the transfer of intended skills to the students is done effectively. Thus, the present study aims to investigate how CS is perceived by High School teachers in Kupang city, East Nusa Tenggara province, Indonesia, in teaching English. Specifically, it aims at finding out the teachers' beliefs and perceptions towards the use of Indonesian language (L1) as their pedagogical instrument in teaching English in EFL classrooms. Hopefully, the present study may provide information to EFL educators regarding the concept of code switching and how it is believed and perceived by EFL teachers in Kupang city High Schools.

2 Methods

To find out the teachers' beliefs and perception of CS applied in EFL classrooms in Kupang city high schools, the present study used a quantitative research design. In gathering the data, a questionnaire was used. The questionnaire was administrated based on Selamat's [22] Canagarajah's [23] and Ferguson's [24] categories of CS. The questionnaire consisted of three parts. The first part collected teachers' personal information. The second part measured the teachers' self-reported frequency in using code switching during EFL classroom interaction. It consisted of 14 items and for each item, the teachers had to choose between 1 (never) to 5 (every time). The third part of the questionnaire consisted of 10 items of five-point (1=strongly agree to 5=strongly disagree) of teachers' beliefs in the practice of CS. There were 40 teachers who took part in this study.

3 Results and discussion

The present study aims to find how CS perceived by high school EFL teachers in Kupang city. Based on the result of the distributed questionnaire, it was found that the teachers' frequency in using CS during EFL classroom interaction was considered as positive. Most of the respondents used it '*in most of their classroom interaction*'. The teachers' response in the frequency of using code switching is presented in Table 1 below.

Table 1. Teachers' frequency in using code switching

No	I switch English to students' first language (L1) (Indonesian):	Never (%)	Almost never (%)	Often (%)	Most of time (%)	Every time (%)
1.	To explain the meaning of new words and sentences	0	0	40	57.5	2.5
2.	To explain difficult concepts	0	0	45	50	5
3.	To explain grammar explicitly	0	0	35	62.5	2.5
4.	To check for students' comprehension	0	0	55	40	5
5.	To introduce unfamiliar materials/topics	0	0	45	52.5	2.5
6.	To explain the differences between the students' L1 and English	0	0	50	45	5
7.	To draw students' attention to the correct pronunciation of sounds in English	0	0	46.1	50	5
8.	To organize classroom tasks	0	0	47.5	45	7.5
9.	To maintain classroom discipline and structure of the lesson	0	0	50	45	5
10.	To provide praise/feedback/personal remarks about students' performance	0	0	55	42.5	2.5
11.	To encourage students' participation in classroom activities	0	0	55	37.5	7.5
12.	To build/strengthen interpersonal relationships between the teacher and students	0	0	52.5	40	7.5

13.	To reduce students' anxiety in learning English	0	0	45	47.5	7.5
14.	To increase students' motivation and confidence in learning English	0	0	40	52.5	7.5

The Table 1 above indicates teachers' response in the frequency of using CS in classroom interaction. This frequency then categorized into: curriculum access (items no. 1-7), classroom management (items no. 8 & 9) and interpersonal relation (items no. 10-14). In term of curriculum access, about 46.07% of the respondents *often used CS*, half of them (50%) were *used it in most of their teaching time*, while only 3.93% *applied it every time in their daily teaching routines*. In terms of classroom management, 50% of the respondents *often applied CS*, 45% *applied it in most of their teaching practice* and 5% *applied it every time*. For interpersonal relation, 49,5% *often applied CS*, 44% were *applied it in most of their teaching activities* and there were only 6.5% *used it every time in the classroom*. The questionnaire also revealed that none of the respondents said that they were never or almost never used CS in their teaching practice.

The data on the Table 1 above reveals that all the teachers applied CS in their EFL classroom interaction. They used CS for curriculum access such as in explaining meaning of words, sentences or grammar. They also applied CS to deal with classroom management such as when organizing classroom task and maintaining classroom discipline. The teachers also applied CS for interpersonal relation, as to provide feedback, strengthen interpersonal relationship and increase students' motivation.

In relation to teachers' belief in the practice of using CS, most of the teachers also showed positive attitude towards the use CS during their EFL classroom interaction. The Table 2 below shows the responses from the teachers.

Table 2. Teachers' beliefs in using code switching in EFL classroom

No	As an EFL teacher, I believe that:	Strongly agree (%)	Agree (%)	Un-decided (%)	Dis-agree (%)	Strongly dis-agree (%)
1.	CS will facilitate the language learning process	75	25	0	0	0
2.	The practice of CS will increase the students' reliance and dependency on the teacher	10	65	25	0	0
3.	CS should be included as an integral part of the EFL lesson	15	75	10	0	0
4.	There should be a strict separation of the mother tongue and English in the EFL classroom	0	0	5	70	25
5.	CS should only be used as a last resort when all other options have been exhausted	20	55	25	0	0
6.	CS is an efficient, time-saving technique	40	60	0	0	0
7.	English is best taught in English-only classrooms	10	35	25	25	5
8.	The use of other languages in the EFL classroom will result in	0	30	25	45	0

	a decline in the standards of English					
9.	The ideal teacher of English is a native speaker	10	20	15	20	35
10.	The more English that is used, the better the results for the learners	70	20	10	0	0

The Table 2 above shows teachers' beliefs in using CS in EFL classroom. The Table 2 tells that all the teachers (100% of repondents) agreed that CS will facilitate the language learning process and is an efficient, time-saving technique. About 90% of the respondents agreed that CS should be included as an integral part of the EFL lesson and whether the more English used, the better the results for the learners. In the concepts of the practice of CS will increase the students' reliance and dependency on the teacher and whether CS should only be used as a last resort when all other options have been exhausted, about 75% of the respondents agreed to the concepts.

However, regarding the concept of having a strict separation of the mother tongue and English in the EFL classroom, whether the use of other languages in the EFL classroom will result in a decline in the standards of English and the ideal teacher if English is a native teacher, most of the respondents disagreed to the statements. The percentage of disagreement of those concepts about 95%, 45% and 55% respectively.

Meanwhile, in the concept of whether English is best taught in English-only classrooms, the respondents'responses were varied. About 45% of the respondents were agreed, 25% were neutral and the last 30% were disagreed.

4 Conclusion

The result of the present study argues that high school teachers in Kupang city applied code switching as a useful strategy in assisting their EFL classroom teaching and learning process. All the respondents agreed that, in the most of their EFL classroom interaction, CS helped them in accessing curriculum (explaining the meaning of new words, phrases, grammar, unfamiliar topics), managing the classroom (maintain classroom discipline) and maintaining interpersonal relation (increase students' motivation and confidence). They also believed that CS was efficient and able to facilitate them in EFL classroom teaching and learning process.

References

1. D.K. Palmer. Bilingual Research J., **32**(1), 42-59 (2009)
2. D.O. Fakeye. Asian Social Science, **8**(8), 149-154 (2012)
3. S.H. Moghadam, A.A. Samad, E.R. Shahraki. Theory and Practice in Language Studies, **2**(11), 2219-2225 (2012)
4. J. Eldridge. ELT J., **50**(4), 303-311 (1996)
5. M.K. Bailey, D. Nunan. *Voices from the Language Classroom*, (1996)
6. S.L. McKay, N.H. Hornberger. *Sociolinguistics and Language Teaching*, (1996)
7. S. Greggio, G. Gil. Linguagem and Ensino, **10**(2), 371-393 (2007)
8. D.M. Kang. System, **36**, 214-226 (2008)

9. J. Jenkins. International J. of Applied Linguistics,**16** (2), 137-162 (2006)
10. C. Tien, K. Liu. In Azirah Hashim & Norizah Hassan. (Eds). *English in Southeast Asia: prospects, perspectives and possibilities*, (2006)
11. B.H. Ahmad, K. Jusoff. English Language Teaching, **2**(2), 49-55 (2009)
12. H.G. Widdowson. *Defining issues in English language teaching*, (2003)
13. J. Gumperz. *Discourse Strategies*, (1982)
14. P. Auer. *Bilingual conversation*, (1984)
15. M. Heller (Ed.). *Code-switching: Anthropological and sociolinguistic Perspectives*, (1988)
16. C. Myers-Scotton. *Dueling languages: Grammatical structure in code switching*, (1993)
17. S. Poplack. Linguistics, **18**, 581-616 (1980)
18. A. Lin. *Teaching in two tongues: language alternation in foreign language classrooms*, (1990)
19. G. Anselmo, M. Williams. *Proceedings of The National Conference On Undergraduate Research (NCUR) Weber State University*, 1-6 (2012)
20. E. Rezvani, A.E. Rasekh. English Language Teaching, **4**(1), 18-25 (2011)
21. K. Bista. World Issue, **29**(9), 1-19 (2010)
22. J.T. Selamat. *Code Switching in the Malaysian ESL classroom.* Unpublished Master Thesis, (2014)
23. A.S. Canagarajah. J. of Multilingual & Multicultural Development, **16**, 173-195 (1995)
24. G. Ferguson. AILA Review, **16**, 38-51 (2003)

The primary school teachers' competence in implementing the 2013 curriculum

Wayan Maba*, *Ida Bagus Nyoman* Mantra

English Education Study Program, Mahasaraswati University, Denpasar, Bali, Indonesia

Abstract. This study was conducted to analyze the competence of primary school teachers in implementing the 2013 curriculum. The 2013 curriculum has been implemented in almost all schools and there are still many unsuccessful implementations in several Indonesian schools. Therefore it is important to study the teachers' competence in implementing the 2013 curriculum. A qualitative research design was carried out in this study by utilizing argumentative descriptive analysis. The data was collected by carrying out in depth interviews to the primary schools teachers who were selected by random sampling techniques. The results of this study indicated that primary school teachers have insufficient competence in implementing the 2013 curriculum especially in designing lesson plan, lesson plan implementation and assessment practices. Consequently, it is recommended that further intensive training and focus group discussion should be held to improve the teachers' competence in implementing the 2013 curriculum. **Keywords:** Primary School, Curriculum 2013, Teachers, Competence

1 Introduction

Recently, the Indonesian government has introduced a new educational curriculum which is called '2013 Curriculum'. This curriculum is implemented in an effort to make Indonesian students to be more productive, creative, and innovative through strengthening the attitude, skills, and knowledge of the students. The implementation of 2013 curriculum was also intended to provide answers to some of the problems inherent to the curriculum in 2006 and in the same time also to motivate students to be more scientifically minded people [17], [13, 11].

The 2013 curriculum put more emphasis on exercising the students to their three core competences, covering congnitive, affective, and psychomotoric competence. These competences are strongly immersed into several set of plans and arrangements consisting of the learning objectives, learning materials, teaching methods, learning media, learning experiences, and source of learning. These then are used to guide the organization in conducting the learning activities to achieve the main educational goals [14], [2, 7, 19]. Moreover, there are other things that should be considered in educational reforms that

* Corresponding author: baguspapers@gmail.com

happened today, among which are the internal challenges in the form of national standards [5], [14, 18].

Additionally, in the implementation of the 2013 curriculum in the classroom, several obstacles were found in many fields, including related to the unpreparedness of teachers as implementers in the classroom. The problems are mainly in the form of lesson planning, implementation of the lesson plan and assessment [17], [9, 12, 19]. The teachers are required to be really dedicated to their profession to be able to implement the lesson plan and the assessment properly. The assessment in particular as the curriculum suggested it should be more towards assessment for learning [20], [1, 21]. To able to implement the curriculum, teachers are required to have pedagogic competence, personality competence, social competence and professional competence [2], [3, 10, 11].

As part of the main competencies, the professional competence of the teacher becomes an important spotlight especially for the target teacher as an implementer of the curriculum [15]. Teachers who are not competent in their profession, they are clearly going to face obstacles in the implementation of 2013 curriculum. To anticipate this government has attempted to provide continuous training to the teachers. The trainings that are conducted by the government are still not enough to provide adequate knowledge and understanding to the teachers [9], [5, 7, 15].

Many studies have been conducted related to the general implementation of 2013 curriculum, therefore this study was specifically conducted with the aim mainly to investigate the competence of elementary school teachers in implementing 2013 curriculum viewed from their professional competence in carrying out the lesson in a real classroom setting. The primary school teachers became the primary choice in this study because successfull implementation of 2013 curriculum in primary school strongly affects the implementation of the curriculum at the higher level of education institutions.

2 Method

2.1 Research Design

This study employed qualitative research method and the data are described descriptively and argumentatively. The data were collected by conducting in-depth interviews and classroom observations to selected elementary school teachers participated this study. The schools were sampled by means of purposive sampling technique. The consideration was mainly based on School location limitations for the area of Bali province, as well as the limitations of elementary schools which have been implementing 2013 curriculum in Bali province.

2.2 Participants

The participants were randomly selected from the total population of primary high school teachers in Denpasar, and they were categorized based on edcuation background, teaching experience, age and the lenght of training on curriculum during the first term of 2017-2018 education year. Finally, there were 10 teachers were selected as the participants of the study.

2.3 Measures

The data of the primary school teachers' competence in implementing the 2013curriculum were transcribed in a written form. All data were arranged and coded according their category. The data then were triagulated to establish reliable research findings. Cross verification across all data within its category ensures all research findings can be generalized to the whole population of the study.

2.4 Procedure and Data Analysis

The teachers were intensively interviewed in relax situation to ensure that teachers provide a valid and reliable information regarding the implementation of 2013 curriculum in their teaching practices. Direct classroom observations were conducted to collect data related to the implemenation ofthe curriculum. The data from the interviews and the observations were then coded and analyzed critically and urgumentatively to establish the reliable findings.

3 Findings

The main finding of the study can be tabulated as the following.

Table 1. Lesson Planning

Teachers' competence in lesson planning										
	T1	T2	T3	T4	T5	T6	T7	T8	T9	T10
Basic competence	R	R	R	R	R	R	R	R	R	R
Learning objectives	R	Ur	Ur	R	Ur	Ur	R	Ur	R	R
Learning materials	Ur	Ur	Ur	R	Ur	Ur	R	Ur	R	R
Learning Media	Ur	Ur	Ur	R	Ur	Ur	Ur	Ur	R	R
Learning method	Ur	Ur	Ur	R	Ur	Ur	Ur	Ur	R	R
Assessment	Ur	Ur	Ur	R	Ur	Ur	Ur	Ur	R	R
Learning resources	Ur	Ur	Ur	R	Ur	Ur	Ur	Ur	R	R

'R'refers to relevant, meanwhile 'Ur' referes to unrelevant

According to the table, most of the primary school teachers had insufficient competence of lesson planing

Table 2. Lesson planning implementation and assessment

Teachers' competence in implementation and assessment										
Techers'code	T1	T2	T3	T4	T5	T6	T7	T8	T9	T10
Pre activity	A	Ina	Ina	Ina	Ina	Ina	A	Ina	Ina	A
Whilst activity	Ina	Ina	Ina	A	Ina	Ina	A	Ina	A	A
Post activity	Ina	Ina	Ina	A	Ina	Ina	A	Ina	A	A
Assessment	Ina	Ina	Ina	A	Ina	Ina	A	Ina	A	A

'A'refers to appropiate, meanwhile 'Ina' referes to inappropiate

According to the table, most of primary school teachers had insufficient competence in implementing the 2013 curriculum and most of teachers conducted inappropiate assessment.

4 Results and Discussions

Based on the analysis of the result of in depth interviews and classroom observations which focused on three aspects of teaching and learning processes, namely lesson planning, lesson plan implementation, and assessment of 2013 curriculum, several findings about teachers' competence in those three aspects can be describes as the following.

In implementing a curriculum in a real classroom setting, having a good lesson plan is a highly vital component of the teaching-learning activity. Appropriately designed lesson plan certainly keeps teachers manage the classroom learning activities correctly, thus it makes easier for the teachers to help the students to reach objectives [2], [3, 7, 8, 21]. A lesson plan at least contains competence standards, learning objectives, learning material, learning experience, learning methods, assessment and source of learning.

Teachers in this study were found to have insufficient competence in determining the suitable learning materials and learning media. They did not know exactly the appropriate learning materials, this is due to the learning materials provided by the government are considered too difficult for the students and are not contextual to the students' learning environment. Ideally, teachers should strongly comprehend the learning materials to be taught to their students [17], [5, 6, 11].

In addition, mot teachers were found to have insufficient competence in determining the teaching methods to be used in teaching. They found it difficult to figure out what appropriate teaching method in line with the principle of scientific approach as suggested by the 2013 curriculum. They still confused whether the scientific approach should be done merely using scientific teaching method or other methods which lead to more social activities, such as communicative learning methods, cooperative learning and community learning. A good understanding on the use of teaching methods is certainly highly essential for the teachers to be successful in teaching [7], [5, 12, 11, 14].

In addition, some teachers mentioned that they did know how select and design suitable teaching media. Appropriate use of teaching media obsulutely help the students in understanding the learning material [8], [12, 15]. Moreover, the teachers' competence in determining the correct assessment is also important [20]. Moreover, in regard to 2013 curriculum, the teachers are required to carry out more productive assessment in the form of authentic assessment and formative assessment should be conducted regulary accompanied with corrected feedback and follow up actions [11], [1, 18]

Most teachers had insufficient competence in conducting the lesson plans. During the classroom observation, it was found that the teachers did not properly implemented the lesson plan based on the lesson plans previously made. This discrepancy might be due to low competency in classroom activities implementation which consists of five scientific steps. The success of classroom implementation of the 2013 curriculum is measured by how successfull the teachers in carrying out the scientific stages in class. Implementation stages of scientific still considered difficult for the school teachers [14]. Although, it was found that the teachers efforts to create maximum students' creativity and higher thinking process, but it was not yet completely in maximum level. Their effort can be considered as a positive sign that the teachers are willing to create a better learning atmosphere [12], [7, 18].

The teachers' competence in carrying out the assessment was also considered to insufficient. Teachers' ability in conducting authentic assessment, performance and product based assessment certainly needs to be improved [1], [4, 11]. The 2013 curriculum requires teachers' competence in assessing the students' ability and conduct follow up actions based on the results of the assessment. The follow up actions are absolutely needed to be done by the teachers to ensure the students' learning progress [21]. .

As mentioned in the regulation of Minister of Education and Culture number 104 year 2014, learning assessment is conducted through authentic and non-authentic assessment [1], [20, 21], but authentic assessment is the main assessment [4], [17, 18]. It was found that most teachers still conducted assessment to measure the students' cognitive rather than the attitude and psychomotoric of the students. This is clearly can be seen from the types of the assessment the teachers used are mainly non-authentic. Some teachers who used authentic assessment explained that they could not use the scoring rubric properly due to lack of understanding and a big number of the students in a class.

This study also found that teachers' insufficient competence in implementing the 2013 curriculum due to some difficulties they faced, such as (1) the subject matters are complicated and mostly inapplicable in supporting of the students' productive skill. Consequently, the teachers have to find materials to meet the students' need and ability, (2) the content of the material in each chapter are not clearly defined, especially for materials that should be repeated at the next level, (3) the implementation procedures are rigidly designed to suit the scientific principles, (4) the assessment processes require the teachers' competence in using scoring rubrics, (5) feedback and follow up actions should continually given to maintain and improve the students' learning progress.

References

1. Ariev, P.R. A theoretical model for the authentic assessment of teaching. *Practical Assessment, Research & Evaluation*, 10(2), 1-11, (2005).
2. Chairani, N. Curriculum Reflected in an International Oriented Senior High School, Yogyakarta. Sino-US English Teaching, August 2015, Vol. 12, No. 8, 568-574, (2015).
3. Eraslan, A. Teachers' reflections on the implementation of the new elementary school mathematics curriculum in turkey. *HU journal of of Education*, 28(2), 152-165, (2013).
4. Gulikers, J.T.M, Bastiaens, T.J., & Kirschner, P.A. A five-dimensional framework for authentic assessment. *Educational Technology Research and Development*, 52(3), 67-86, (2004).
5. Hasan, S. H . History Education In Curriculum 2013:A New Approach To Teaching History. *HISTORIA:*International Journal of History Education, Vol. XIV, No. 2, (2013).
6. Hussain, Z., Adeeb, A., Aslam, H.D. Curriculum Implementation and Feedback Mechanism at Secondary School Level in Punjab Pakistan. *International Journal of Learning & Development*, 1(2), 92-98, (2011).
7. Lewis, Anne C. *Teacher Development.* Teach Directions, ProQuest Education Journals, Vol 67, No.6, (2008).
8. Lumadi, M.W. Challenges besetting teachers in classroom assessment: an exploratory perspective. *Journal of Social Science*, 34(3), 211-221, (2013).
9. Maba, Wayan. 2017. The Implementation of Education National Standard in the Instrument of School Accreditation of Bali Province Education Authority. International Research Journal of Engineering, IT & Scientific Research (IRJEIS). Vol. 3 Issue 4, July 2017, pages: 1-6, (2017).
10. Maba, Wayan. Teachers' sustainable professional Development Through Classroom Action Research. International Journal of Research in Social Sciences.Vol. 7 Issue 8, August 2017, (2017).

11. Maba, Wayan. International Journal of Social Sciences and Humanities. Teachers' Perception on the Implementation of the Assessment Process in 2013 Curriculum. Vol. 1 No. 2, August 2017, pages: 1-9, (2017).
12. Mantra, I.B.N. Promoting Primary School Teachers' Competence through Dynamic Interactive Workshop and Partnership *International Journal of Linguistics, Language and Culture (IJLLC)*Vol. 3, No. 1, (2017).
13. Mulyasa, E. *Pengembangan Implementasi Kurikulum 2013*. Bandung:PT. Remaja Rosdakarya, (2013).
14. Nugraheni, A. S. Controversy a Policy Change in the Curriculum in Indonesia in Terms of the Point of View of Indonesian Language Subject. Journal of Education and Practice. (Online), Vol.6, No.2, (2015).
15. Oliva, Peter F. *Developing the Curriculum*. Boston: Little, Brown, and Co. (1982).
16. Ornstein, A.C. & Hunkins, F.P. *Curriculum, Foundations, Principles, and Issues*. Fifth Ed. Singapore: Pearson, (2009).
17. Retnawati, H., Hadi, S., Nugraha, A. C.Vocational High School Teachers' Difficulties in Implementing theAssessment in Curriculum 2013 in Yogyakarta Province of Indonesia. *International Journal of Instruction. Vol.9, No.1, (2016)*
18. Rusman. Curriculum Implementation at Elementary Schools A Study on "Best Practices" Done by Elementary School Teachers in Planning, Implementing, and Evaluating the Curriculum. Journal of Education and Practice . (Online)Vol.6, No.21, (2015).
19. Shilling, T. Opportunities and Challenges of Curriculum Mapping Implementation in One School Setting: Considerations for School Leaders. *Journal of Curriculum and Instruction (JoCI),* , Vol. 7, No. 2, Pp. 20-37, (2013).
20. Sri Widiastuti, I.A.Md. EFL Teachers' Beliefs and Practices of Formative Assessment to Promote Active Learning. The ASIAN EFL Journal. Volume 3, (2016).
21. Sri Widiastuti, I.A.Md. *Teachers' Understanding of Formative Assessment. Jurnal Bahasa dan Seni.* Vol 45, No 1, (2017).

THE EFFECT OF PROBLEM BASED LEARNING AND SELF-ASSESSMENT ON STUDENTS' WRITING COMPETENCY AND SELF-REGULATED LEARNING

I Putu Suyoga Dharma[1], Pande Agus Adiwijaya[2]

[1]English Education Department, STKIP Suar Bangli, 80613, Bali, Indonesia
[2]English Education Department, STKIP Suar Bangli, 80613, Bali, Indonesia

Abstract. This experimental study aimed at investigating the effect of Problem Based Learning (PBL) and self-assessment (SA) on students' writing competency and self-regulated learning in Tabanan Regency. This research applied 2x2 factorial design. 96 students were selected as sample through random sampling. Data were collected by test (writing competency) and questionnaire (self-regulation). Students' writings were scored by analytical scoring rubric. The obtained data were analyzed statistically by MANOVA at 5% significance level. This research discovers: 1) there is a significant effect of PBL which occurs simultaneously and separately on students' writing competency and self-regulated learning, 2) there is a significant effect of SA which ocurs simultaneously and separately on students' writing competency and self-regulated learning, 3) there is a significant interaction between teaching model and assessment type on students' writing competency and self-regulated learning which occurs simultaneously, 4) there is no significant interaction between teaching model and assessment type on students' writing competency, and 5) there is a significant interaction between teaching model and assessment type on students' self-regulated learning. This research results implies that PBL and SA should be applied in instruction process as a way to improve the quality of students' writing competency and self-regulated learning.

Keywords: Problem Based Learning, self-assessment, writing competency, self-regulated learning

1. Introduction

Until this era, Indonesian is still faced by low students' competency. Competency is actually the estuary of instruction process on certain range of time. It consists of three domains, namely: cognitive, affective, and psichomotor [1]. Operationally, it is knowledge, skills, and norms. To achieve it, good planning, implementation, and assessment should be applied [2]. It must be applied to all subjects including in English Foreign Language. Writing is one of the skill which should be mastered. Writing itself is a cognitive process involving complex mental activities [3]. In writing, the students should be able to produce and organize ideas into a peice of paper by using quality language. Through writing, the students' creative, critical, and analytical thinking can be trained.

Based on the observation in senior high schools in Tabanan regency, teaching and assessing writing are still dominated by teacher. The teacher explained writing materials and then asked the students to write it down. Assessment was just done at the end of learning with certain scratch and score. There was no report on the students' learning progress. This situation made the students feel boring and some of them just copy their friends work. As the result, the students' writing competency was still low. Furthermore, they seemed to have low effort in searching their learning materials and they just waited to be "fed up" by their teacher. This condition is absolutely out of the modern learning paradigm.

To find solution on the problems, it was experimented Problem Based Learning (PBL) and self-assessment (SA). PBL is learning model which is based on constructivist perspective [4]. It uses problems as the trigger for learning [5]. Learning is ill-structured since the students contruct it through social interaction in which one student may differently acquire knowldge from others [6]. In PBL, there is no rote learning of facts and figures, students brainstorm problems, arrange possible solutions, decide collectively on their learning objectives, do individual work to seek out necessary information, then report back to synthesize and apply their new knowledge collectively to the problem at hand [7]. These features parallel to the important factors that are necessary for authentic learning environments and for creating authentic real world tasks used in assessment [8].

To support the PBL which demands real world task, authentic assessment is needed [9]. Self-assessment is parralel to PBL since it gives chance for students to do reflection on their learning. In self-assessment, the students judge their work (writing) by using evidence and clear criteria by themselves [10-11]. This conditions makes the students have critical and analytical thinking on their learning [12]. It leads the students to produce quality writing. Self-assessment can be guidance for the students to write which means it can also lead the students to be self-regulated [13].

2. Methods

This research applied posttest only control group design. Four classes in two different schools were randomly selected to be sample (class X IPA 4 and X IPA 5 of SMAN 1 Tabanan and Class X IPA 1 and X IPA 2 of SMAN 1 Baturiti). After they were proven to be equal in nature, treatment were given. Class X IPA 4 were treated by using conventional model and conventional assessment; class X IPA 5 were treated by PBL and self-assessment, class X IPA 1 were treated by PBL and conventional assessment; class X IPA 2 were treated by conventional model and self-assessment. They were treated for 12 times, then posttest was done to collect students' writing competency data and questioniare was for self-regulated data. then, the obtained data were analyzed by using MANOVA formula.

3. Results and Discussion

Finding of this research is divided into two parts, descriptive and hypothesis testing. The result of the descriptive analysis can be presented in the Table 1.

Table 1 The Result of Descriptive Analysis

Descriptve Analysis	A1				A2			
	B1Y1	B1Y2	B2Y1	B2Y2	B1Y1	B1Y2	B2Y1	B2Y2
Mean	85.46	91.00	81.83	81.00	77.83	75.96	74.96	76.67
Std. Deviation	7.989	5.125	8.781	8.016	6.585	7.992	8.073	5.917

Note : A1 = PBL, A2 = Conventional model, B1 = self-assessment, B2 = conventional assessment, Y1 = writing competency, Y2 = self-regulated learning

Based on the Table 1, it is known that students' mean score writing competency and self-regulated learning instructed by PBL is higher than conventional model. It means that PBL gives better effect than conventional model. Furthermore, the standard deviation of students' writing competency and self-regulated learning instructed by PBL is higher than conventional model. It means that PBL students are more hetergonous than conventional students.

Next, it was continued to hypothesis testing done by MANOVA formula and the result can be presented in the Table 2.

Table 2 The Result of Hypothesis Testing

No	Compared Groups	Probability	Decision
1	PBL vs CM	0.000	Significant
2	SA vs CT	0.001	Significant
3	Simulatnuous Interaction	0.001	Significant
4	PBL vs CM Y1	0.000	Significant
5	PBL vs CM Y2	0.000	Significant
6	SA vs CT Y1	0.047	Significant
7	SA vs CT Y2	0.001	Significant
8	Interaction on Y1	0.317	Insignificant
9	Interaction on Y2	0.000	Significant

Based on the Table 2, it is known that 1) there is a significant difference in students' writing competency and self-regulated learning occured simulatnously between the students instructed by using PBL and conventional model, 2) there is a significant difference in students' writing competency and self-regulated occured simulatnously between the students assessed by using SA and conventional assessment, 3) there is a significant interaction between teaching model and assessment type on students' writing competency and self-regulated learning occured simultanously, 4) there is a significant difference in students' writing competency between the students instructed by using PBL and conventional model, 5) there is a significant difference in students' self-regulated learning between the students instructed by using PBL and conventional model, 6) there is a significant difference in students' writing competency between the students assessed by using SA and conventional assessment, 7) there is a significant difference in students' self-regulated learning between the students assessed by using SA and conventional assessment, 8) there is no significant interaction between teaching model and assessment type on students' writing competency, and 9) there is a significant interaction between teaching model and assessment type on students' self-regulated learning.

Those results prove that PBL combined with SA gives positive effects on students' writing competency and self-regulated learning. In the implementation of PBL, SA is implemented in discussion phase. In writing, the students were firstly oriented into writing problems. They were given some examples of writing to be analyzed and decided the best writing in group. To do it, each group was assigned to find out the criteria of good writing both from ideas and linguistics aspects. In this time, opened communication is done by the teacher to help the students make hypothesis of good writing. Then, presentation was done and other groups could complete their friends presentation. Here, the students interacted and discussed each other in group so they could mutually complete. They could also make strategic planning to finish their assigenement. It trained the students the students' self-regulated in learning. Finally through interaction, discussion, and guidance, the good writing criteria was established.

Based on the explanation, PBL creates four positive atmospheres on students' learning facilitating the improvement of writing competency and self-regulated learning. *First*, it provides scaffolding in writing. The students learn to identify and understanding posed problems writing. Here, writing concept is constructed step by step. By having the concept, good writing can be produced.

Second, productive and critical thinking is trained. Writing demands products consisting on certain elements. The habit of problem solving exercises to think productively and critically. It facilitates creative and deep writing product. *Third*, interaction is opened. During writing, problems is discussed in group firstly. If it is stuck, the students can ask to the teacher. It is only guidance and clues are given, not fixed answers. *Fourth*, it leads the students to make strategic planning. By realizing the learning, the awareness of planning or making decision arise on students themselves. It leads into self-regulated learning.

This finding is also supported by other researchers discovering PBL is effective for writing competency [14,15,16] and self-regulated [17,18]. All of the researchers agree that PBL creates opened ended, problem solving, interactive, and discovery learning. By posing the problems, the students do efforts to solve the problems and formulate hypothesis about good writing. It directs the students to think actively, critically, and creatively so that the students are accustoming to be productive thinker. In addition, interaction and discussion while the PBL is implemented make the students become self-regulated. It is because they have peer to work together so if they have problems they can consult them.

After the students have good writing concept, the students start to write certain genre based on the criteria established. During writing, the students were also introduced into self-assessment and how to do it. Here, open communication was also held by the teacher. The students could freely ask questions and guidance if they had difficulties in assessing their writing. Self-assessment was in the form of checklist so the students judge their writing by just focusing on the criteria. This process led the students to think about their thinking in writing or doing reflection. When the students did reflection, it facilitated their critical and analytical thinking which directing into productive learning. In addition, by understanding the criteria, the students could know what they should do. It was actually a process to be self-regulated.

From the implementation, it is known that SA provides guidance on criteria of good writing and writing strategic planning. It makes the students know what should be done in writing. At least, it can keep the students in tract. Furthermore, self-reflection is constructed. By seeing the checklist, reflection will be done on the work/writing; whether or not it has been matched to their works. Gradually, the awareness increases leading into taking decision independently and confidently. Lastly, the availablity of feedback along learning process helps the students to monitor their learning. SA is on going assessment which make feedback is available. It can be from the students or the teachers. From it, the students are aware on their strengths and weakness so improvement can be done as fast as possible.

The effectiveness of SA on students' writing competency [19,20,21] and self-regulated learning [22,23] are also proven by others. They discover that SA gives chance for the students to do self-reflection and on going process which enable the students to revise their writing. In addition, the use of checklist can be used as guidance for the students to write so they can keep in tract while writing. In terms of self-regulation, SA has strong correlation on it since the students' awareness increases. When they are aware of their writing, the students can do certain efforts to produce it in good quality. Finally, this research and previous researches agree that SA is very hard to be implemented at the beginning.

Besides of their effectiveness, there are several challenges which might be faced in implementing them. Both PBL and SA are very students centered so it will be very hard for the students who are previously treated by lecturing models. The students should be accostumed to do it firstly. This research proves that the students can apply them well after eighth meeting. Cultural background should also be taken into account. Indonesian culture which is thick of east culture impedes the PBL and SA implementation. Indonesian is usually keep silent or affraid in telling their problems or ideas in public. Some students might have problems or solution, but they do not share it. Lastly, it is harmful for less motivated students. Increasing learners' motivation can become first thing and should be done before class to keep them follow the learning process.

4. Conclusion

This research has proven that PBL and SA have better effect on students' writing competency and self-regulated learning than conventional learning model and assessment. Learning experiences created by PBL and SA, such as : scaffolding, productive thinking, students centered, process oriented, reflective learning, problem solving, and democratic grow the students' writing competency and self-regulated learning. Those conditions actually make the students active in constructing knowledge of writing and attitudes of self-reguation.

This result implies that EFL learning, especially writing have to be instructed by PBL and assessed by SA. Implementing both of them brings positive effects on cognitive, psychomotor, and affective domains. Cognitive and psychomotor domains are trained in students' writing, while the affective domain is self-regulated learning. In addition, it implies that combining PBL and SA strengthen that attitude is constructed through learning process, not in just a particular subject.

By seeing the research result, some recommendations are given for three parties. For the EFL teachers, implementing PBL and SA should be implemented step by step since they are hard at the beginning. The teachers have to be opened for the students' problems and give strategic solution. For the Indonesian government, PBL and SA should be used as the main instruction and assessment approach in teaching EFL for secondary school. In addition, the government can make a team to develop self-assessment checklist for each topic in English, such as: writing, speaking, listening, reading, learning strategy, etc in each grade so the EFL teachers can just use it. Finally, the other researchers is recommended to research the effectiveness of PBL and SA in different language skill, grade, and attitudes (motivation, democratic, hard working, etc). In addition, it is also important to find out how and why the different results in separated are occured.

Through this paper, special thank is given to some parties, namely: headmaster of SMAN 1 Tabanan and SMAN 1 Baturiti for the permission in doing this research, EFL teachers in SMAN 1 Tabanan and SMAN 1 Baturiti for becoming judges and experiment executors, and leacturer in STKIP SUAR Bangli for the supports for this research.

5. References

[1] Depdiknas.(2003). *Kurikulum 2004 Standar Kompetensi Mata Pelajaran Bahasa Inggris*, Jakarta, Departemen Pendidikan Nasional.

[2] D. Brown. *Principle of Language Learning and Teaching*, New York, Pearson Education, (2007).

[3] G.D. Boric. *Effective Teaching Methods: Research-based Practice*, New Jersey, Pearson Merril Prentice Hall, (2007).

[4] J.R. Savery. Overview of Problem Based Learning: Definitions and Distinctions. In Ertmer, P.A., Macklin, A.S., Alper, A., Bodner, G., Duffy,T., Hmelo-Silver,

C., Jonassen, D., O'Rourke, K., Savery,J.Watson, G., Woods, D., and Mong, C. (Eds). TIJPBL, **1** 9-20. Accessed on December 5th,2013. It is available at http://www.isetl..org/conference/presentation.cfm+pid=600-23k. (2006)
[5] J.Wilson. Problem Based Learning Revolution: Revolution or an Empty Promise? A review of the Constructivist Approach to Problem Based Learning in Undergraduate Medical Education. *JEL* **1**,1. Accessed on December 5th, 2013. It is available at http://www.mdx.ac.uk/schools/arts/research/docs/ (2007)
[6] G. Arazt, S. Sungurs. Effectiveness of Problem-Based Learning on Academic Performance in Genetics. *Bambed*, **35**, 448-451, (2007)
[7] M. Savin-Baden. "Challenging Models and Perspectives of Problem-based Learning." Management of Change: Implementation of Problem-based and Project-based Learning in Engineering, 9-30 (2007)
[8] W. Barber., S. King., Sylvia Buchanan. Problem Based Learning and Authentic Assessment in Digital Pedagogy: Embracing the Role of Collaborative Communitie. *EJEL*, ***13***, 59-67 (2015)
[9] V.Bozalek., D. Gachago., L. Alexander., K. Watter., D. Wood., E. Ivala., J. Herrington, J. (2013). The Use of Emerging Technologies for Authentic Learning: a South African Study in Higher Education. *BEJT*, **44**, 629-638 (2013)
[10] S. Bloxham., P. Boyd. *Developing Effective Assessment in Higher Education: A Practical Guide, New York,* Open University Press, (2007)
[11] K.H.K. Tan. Qualitatively different ways of experiencing student self-assessment. *HERD,* **27**, 15-29 (2007).
[12] C. M. Chen., C.C. Chen. Problem-based Learning Supported by Digital Archives; Case study of Taiwan Libraries' History Digital Library. *TEL*. **28**, 5-28 (2010).
[13] D. Little. Language Learner Autonomy: Some Fundamental Considerations Revisited. *ILLT*. **1**, 14-29 (2007).
[14] K.D. Witte., N. Rogge. Problem-based Learning in Secondary Education: Evaluation by a Randomized Experiment, HUB RESEARCH PAPERS, 1-23 (2012)
[15] G. Dastgeer, M.T. Afzal, Improving English Writing Skill: A Case of Problem Based Learning, *AJER*, **3**, 1315-1319 (2015)
[16] N. Othman, M.I.A. Shah, Problem-Based Learning in the English Language Classroom, *ELT*, **6**, 125-134 (2013)
[17] Y. Tas, S. Sungur. The Effect of Problem-Based Learning on Self-Regulated Learning: A Review of Literature, *CJE*, **14**, 533-560 (2012)
[18] M. C. English, A. Kitsantas, Supporting Student Self-Regulated Learning in Problem- and Project Based Learning. *IJPBL*, **7**, 128–150 (2013).
[19] T.T. Purwanti, The Implementation of Self-assessment in Writing Class: A Case Study at STBA LIA Jakarta, *TEFLIN Journal*, **26**, 97-116 (2015)
[20] H. Zheng, J. Huang, Y. Chen, Effects of Self-assessment Training on Chinese Students' Performance on College English Writing Tests, *Polyglossia*, **23**, 33-42 (2012).
[21] M.R. Javaherbakhsh, The Impact of Self-Assessment on Iranian EFL Learners' Writing Skill, *ELT*, **3**, 213-218 (2010)
[22] E. Khodadady, H. Khodabakhshzade, The Effect of Portfolio and Self-assessment on Writing Ability and Autonomy, *JLTR*, **3**, 518-524 (2012)
[23] Y.G. Butler, J. Lee. The Effect of Self-assessment among Young Learners of English, *LT*, **27**, 531 (2010)

Effectiveness of Contextual Science Practicum Module to Improve Science Process Skills and Environmental Caring Character

Ni Luh Putu Rusmana Dewi*, *I Wayan* Suastra , and *Ni Made* Pujani

Science Education, Post-Graduate Program, Universitas Pendidikan Ganesha, Singaraja, Bali, Indonesia

Abstract. This study aims to examine the effectiveness of contextual science practicum module to improve science process skills and environmental caring character. The type of this research is research and development refers to the Thiagarajan development model which carried out until development. Before effectiveness test, the validity and practicality of the module were tested. The effectiveness data of practicum module was obtained through a limited field trial conducted with 30 sample of student. The effectiveness test of the module were analyzed using pre experimental method with One Shot Case Study design. The results are: (1) the validity of the practicum module is 3.40 belong to valid category; (2) the practicality of practicum module is 3,50 belong to very good category; and (3) the effectiveness of the practicum module viewed from the skills of the science process is 85,86 belong to good category and the score of students environmental cares at moral knowing is 3,53 belong to entrust category and the score at moral behavior is 3,22 belong to developing category, so the module can be declared effective.The implication of this research is the teacher should understand the component of contextual approach and provide the supporting equipment of practicum activity. **Keyword**: *science practicum module, contextual, science process skills, environmental caring character*

1 Introduction

The quality of science learning, especially the science process skills and the environmental cares Indonesian students are still low. This fact is supported by the PISA 2015 value index, indicating that the scores of Indonesian students' science is only 403 are still below the average score is 493 [1]. This problem caused by the practicum system is not applied as a source of student learning, whereas lot of IPA topics, Particularly pollution topic will find a lot of application in the daily life of students in the environment. Currently, the assessment used to assess student learning outcomes 100% from cognitive aspect by using quiz, test in ending of topics, final test, and home work without assessing student performance so the students' science skills become less developed [2].

*Corresponding author: rusmanadewi@rocketmail.com

Meanwhile, the emergence of crisis in character of the nation, especially the character of the environment caring caused by the current education failed to develop the education value in school [3]. There are lot of learning in schools that only transfer knowledge to student without inserting noble values or character education so the impact are moral decline and behavior not responsible.

The solution of this problems is developing the contextual science module and doing test of effectiveness. The characteristics of the new developed practicum modules be adapted with student life (contextual) and modified by a simple set of practicum and following the contextual learning component. With this concept, the practicum will be more interesting and fun for students because it brings the science problem into everyday life[3].

Contextual learning is an alternative to a teacher to teach by connecting the lessons with the real-world situations of the students or the students' daily experiences [4]. This is as a stock of students to solve problems in their life as a member of society[5] . CTL approach is very relevant if raised and used in science practicum activities because it considers the contextual aspects and components of the CTL approach, such as: contructivisme, modelling, learning community, inquiry, questioning, refleksi, dan autentic assesment.

Science process skills is process approaches used by the teachers in discussing content that refers to the process of science [6]. Scientific process skills are also defined as skills possessed by scientists [7]. Scientific process skills are needed as a basis for students to solve a problem. Therefore, curriculum in the schools for the science topics are advised to apply science process skills to teach science. Scientific process skills can be trained through the laboratory activities (practicum) contextually [8] . When practicum activities were done with contextual approach, the interaction of knowledge of the students with learning environment will happened. Therefore, the student interest of IPA topics will be increasing so that students' science process skills will also increase. The indicators of science process skills studied include; 1) formulate hypotheses, 2) doing experiments, 3) observations, 4) implementing concepts, and 5) communicate.

The environment caring character is an attitude and action that always strives to prevent the destruction of the surrounding natural environment and always try to repair the existing natural damage. The achievement indicators of environmental caring values are: 1) hard work to protect nature, 2) appreciate health and hygiene, 3) wise in using natural resources, and 4) responsibility to the environment [9]. The process of internalizing characters through three important stages [10]: First, students have a knowledge of the good (moral knowing). Secondly, from the knowledge of goodness, the commitment of the students to the good (moral feeling) was arises. Third, after students have a commitment to goodness, they end up actually doing good (moral behavior). The process of applying the characters in the daily life of learners can be done through contextual learning, because learning is done naturally, so that student can practice the characters has been learned [11].

The contextual science practicum module in this study is practical module adapted with the daily life of the students (contextual) with a simple set of practicum and follows the components of contextual learning. Given the essence of practicum in science learning, the effort to make the practicum as a simple and easy activity is very important. One way for practicum to be done easily is to relate it to the real situation in the student environment (contextual). In this way, the practicum will be more interesting and fun for students because it brings the science problem into everyday life[12].

2 Method

This research is aimed to a model testing through the development of a product of teaching moduls of science SMP. The method of this research is research and development. The developed product is contextual science practicum module for junior high school second semester class VII on environmental pollution topic. The development model for teaching modul used in this study is a model of development suggested by Thiagarajan et al. (1974) is a 4-D model. This model consists of 4 stages of development, such as define, design, develop, and disseminate. In this study only done until the stage of develop because of limitations of researchers. The design of the effectiveness test of the module is done by using pre experimental method with One Shot Case Study design.

X	0

Figure 1. One Shot Case Study Research Design [13].

The subjects in this study were the people involved to obtain information about the practicum module consisting of the experts, the practitioners and the students. The experts are from the lectures and focus for obtaining data on the validity of the practicum modules. Practitioners come from the science teachers of junior high school and focus in obtaining data about the practicality. The students focus in obtaining data about the practicality and effectiveness of teaching modul. Students as subject are taken as one class that is class VII SMP.

The practicum modul that were tested for effectiveness is the contextual science practicum module for Junior High School developed on environmental pollution topic. The practicum activities developed include: 1) polluted water features, 2) water purification methods, 3) air pollution in the surrounding environment, 4) contamination of tobacco smoke, 5) contamination of soil in surrounding environment, and 6) detergent effect on plant growth. The data collection technique used in this research is by using several instruments such as: (1) teaching material validation sheet, (2) observation sheet of learning implementation, (3) teacher response questionnaire, (4) student response questionnaire, (5) observation sheet Science process skills, (6) observation sheets of environmental cares, and (7) environmental character questionnaires. This modul will have effectiveness in improving students 'science process skill when it reaches a minimum score of 80 with good category and will have the effectiveness of improving students' environmental cares character when minimum average score 2.5 with category start to develop. To test the effectiveness of the module is done by testing the hypothesis through t test one party with the help of software SPSS 16 for windows.

3 Result and Discussion

The validity of module is 3,40 with valid category, practicality result of the module are 3,50 in implementation module, 3,54 in teacher response, and 3,53 in student response belong to very practical category. Furthermore, the effectiveness of the module is derived from the value of the science process skills and students environmental caring character.

Table 1. Result of Student Science Process Skills Observation

No.	Indicator	Value	Category
1	Formulate Hypotheses	90,13	Very good
2	Doing Experiments	84,86	Good
3	Observation	85,27	Good
4	Implementing Concepts	85,97	Good
5	Communicate	83,05	Good
$\bar{x}$		**85,86**	**Good**

Table 2. Result of Student Environmental Caring Character Observation

No.	Indicator	Average Score	Category
1	Hard work to protect nature	3,51	Entrust
2	Appreciate health and hygiene	2,99	Developing
3	Wise in using Natural Resources	3,29	Developing
4	Responsibility to the environment	3,09	Developing
$\bar{x}$		**3,22**	**Developing**

Table 3. Result of Student Environmental Caring Character Questionnaire

No.	Indicator	Average Score	Category
1	Hard work to protect nature	3,60	Entrust
2	Appreciate health and hygiene	3,49	Developing
3	Wise in using Natural Resources	3,57	Entrust
4	Responsibility to the environment	3,46	Developing
$\bar{x}$		**3,53**	Entrust

The result of science process skill of student obtained the average value of 85.86, that's mean that students' science process skills are in good category. This result is reinforced by hypothesis testing through left side t test which got result that t = 4,007 bigger than t table, that's mean, the average score of student science process skill is in good category. So it can be concluded that practicum activities using the contextual module is effective to improve students' science process skills. The results of this study are in line with the results of research conducted by Ting & Siew [6] which shows that contextual learning can improve the skills of the science process. This is because contextual learning can help students to build more complex cognitive structures with real activities. Through the interaction between environments, students will be able to apply the skills of the science process to solve better everyday problems. In line with Ting & Siew, Crawford [15] reveals that realistic, authentic, challenging, and contextualistic exercises or tasks were more effective for students and can build new knowledge or strengthen existing knowledge.

The success of the contextual practicum module in improving the science process skills is also due to the relationship between each component of contextual learning with science process skill indicators, such as: The 'contructivisme' component facilitates the process skills for formulating the hypotheses and applying concepts, the 'modeling' component facilitates the skills of the experimental process, the 'questioning' component facilitates the process of formulating the hypotheses, the 'inquiry' component facilitates experimental process skills and makes observations, the components of the 'learning societies' facilitating experimental process skills and applying concepts, the' reflection 'component facilitates the communication process skills, the' authentic assessment 'component facilitates the communication process skill.

The average score of environmental caring characters obtained based on the results of character observations of 3.22 with the developing category, while based on the questionnaire obtained an average score of 3.53 with the entrust category. This difference indicates that the process of internalization of the environmentally caring character in the understanding phase (moral understanding / moral knowing) occurs faster than the moral action. Students are easier to understand knowledge about environmental cares than to implement environmental awareness indicators in the form of concrete actions. Even the result of students 'environmental caring show different categories between observations and questionnaires, both of them show that the contextual science module is effective for improving students' environmental caring character. This result is reinforced by hypothesis testing that obtained the result that the value of $t = 16{,}089$ bigger than t table, which shows the average score of environmental cares are in the developing and entrust category. This is because the practicum module was presented in a contextual manner and gives students the opportunity to practice the character of environmental care in a real way in accordance with the condition of the students' daily life (contextual). In addition, the components of the contextual practicum module also increment the indicator of environmental caring character, ie; (1) 'modeling' component, facilitating wise characters using natural resources, (2) 'inquiry' components, facilitating the character of environmental responsibility, (3) the 'learning community' component, facilitating the character of health ,hygiene appreciation, and character of hard work to protecting nature, (4) the 'reflection' component, facilitating the character of environmental responsibility. The results of this study are in line with the research conducted by Yustina & Febrita[16], states that environment-based learning (contextual) has a positive impact on students 'attitudes and participation, involving students in learning with the environment can increase students awareness of the environment.

4 Conclusion

Based on the findings obtained in the research and discussion that has been done, it can be concluded several things as follows. (1) The validity of practicum module which developed is 3.40. (2) The practicality of practicum modules which developed is 3.50. (3) the effectiveness of the practicum module viewed from the skills of the science process is 85,86 belong to good category and the score of students environmental cares at moral knowing is 3,53 belong to entrust category and the score at moral behavior is 3,22 belong to developing category.

Referring to the research findings that have been done, then the suggestions that can be submitted as follows. First, teachers should be associate students with the real world problem in the practicum process. Second, teachers should prepare the completeness of the tools for support the practicum activities. Third, the schools and all components must be commitment to develop the expected character in the learning process.

Acknowledgement

The authors would like to thank Prof. Dr . I Wayan Suastra, M.Pd for technical assistance and Dr. Ni Made Pujani, M.Si as supervisor. We also thank to all the parties involved in this research.

Reference

1. OECD, *PISA 2015 Result and Focus* (2015).
2. I.W.Suastra, I.B.P. Mardana, I.N.P. Suwindra, *Research Grant* (2006).
3. I.W.Suastra, *Penelitian dan Pengembangan Pendidikan*, **5**, 258 (2011).
4. M. Crawford, Contextual Teaching and Learning:Strategies for Creating Contructivist Classrooms, NTPN, **11**, 2 (2001).
5. C.C.Hudson, V.R. Whilsler, Contextual Teaching and Learning for Practitioners, SCI, **6,** 55 (2012).
6. K.L. Ting, M.NY.Siew, *Education and Learning,* **3**, 99 (2014).
7. I. W. Suastra, *Pembelajaran Sains Terkini:Mendekatkan Siswa dengan Lingkungan Alamiah dan Sosial Budayanya*, 5 (2013).
8. P.C.Buntod, P.Suksringan, A. Singsreevo, Effects of Learning environmental Education on Science Process Skills and Critical Thinking of Mathayomsuksa 3 Student with Different Learning Achievement, JSS, **6**, 61(2010).
9. Kemdiknas, *Panduan Pelaksanaan Pendidikan Karakter,* 2001.
10. T. Lichona, *Educating For Character*, 105 (1943).
11. K.Komalasari, The Living Values-Based Contextual Learning to Develop the Student'Character, JSS, **8**. 246-251 (2012).
12. R.J. Rezba, C.S. Sparaque, R.L. Fiel, H.J. Funk, J.R Okey, H.H. Jaus. Learning and Assessing Science Process Skills, Hunt Publishing Company (1995).
13. W.R. Borg, M.D. Gall, *Educational Research: An Introduction. Fourth Edition* (1983).
14. S.Thiagarajan, D.S. Semmel, M.I. Semmel, M.I., *Instructional Development for Training Teachers of Exceptional Children: A Sourcebook* (1974).
15. M. Crawford, Contextual Teaching and Learning: Strategies for Creating Constructivist Classrooms (Conclusion), JEL, **11**, 4 (2013).
16. Yustina, E.Febrita, Increase Motivation and Knowledge the Environment through Contextual Model, ASS, **9**, 241 (2013).

TEACHER'S INSTRUCTIONAL BEHAVIOUR IN INSTRUCTIONAL MANAGEMENT AT ELEMENTARY SCHOOL REVIEWED FROM PIAGET'S COGNITIVE DEVELOPMENT THEORY

Ni Putu Eni Astuti
STKIP Suar Bangli
Bangli, Bali
putu.eniastuti@gmail.com

ABSTRACT

This writing aimed at (1) describing the importance of teacher to review instructional management at elementary school based on Piaget's cognitive development theory; and (2) describing teacher's instructional behavior in managing instructional at elementary school reviewed from Piaget's cognitive development theory. In general, Piaget' cognitive development theory divides children' cognitive development into four stages. In the elementary school ages of 7 to 11 or 12 years old, Piaget classified the cognitive development into concrete operational marked by the use of clear and logic rules. The children implement logic thinking on concrete object, yet abstract or hypothetical. Although the intelligence of this step has been advanced, the way of thinking of the children is still limited because still based on concrete object. Therefore, teacher should comprehend the importance of concrete operational instructional at elementary school so the children could maximally achieve the learning goal in accordance to their thinking level that they acquire. The suggestion that can be proposed is that the teacher should synergize the characteristics of elementary school students in concrete operational stage with the instructional readiness in the steps of planning, implementation and evaluation.

Key words: *teacher's instructional behavior, concrete operational stage.*

I. INTRODUCTION

Indonesia's Regulation at UU No. 20 in 2003 stated that elementary school is an educational level which become a base for the students to enter secondary level. It means elementary school is an education level which underlies high level education (high school). This level is intended for seven years old children by the assumption of these children have comprehension level and education needs in accordance to themselves. Elementary education is indeed held for giving basic knowledge, attitude and skills for the children. This elementary education is further developed to improve their quality. We should understand the definition of elementary school so we can follow every activity in accordance to the students' development in this level.

According to Piaget, 7 to 11 or 12 years old children in Elementary School are in the stage of concrete operational [1]. It is stated as concrete operational because in this stage, the students are able to think logically, but still limited on concrete things, which means solving problem that they directly face.

Considering this condition, the teacher should understand how the students' thinking ability in this phase. Teacher's instructional behavior should concern students' thinking development stages on the concrete operational stage and implementing personal and social constructivism learning approach, by concretizing learning close to students' environment, so the learning would be real for the students and indeed can be understood in their development stages.

Concrete media is anything that real and can be used to deliver a message from the sender to the receiver so it would stimulate the thinking, feeling, and students' interest which finally the learning process can be more effective and efficient to achieve the learning goals. There are three concepts of concrete learning media such as (1) concrete learning media functions to attract students' interest to the delivered learning material, (2) concrete learning media is useful in improving students' understanding on the given learning material, (3) concrete learning media can present valid and reliable data [2]. Generally, the function of concrete learning media can be described as (a) a tool to create effective learning condition, (b) an integral of the entire learning situation, (c) setting concrete basis and abstract concept so it could reduce verbal understanding, (d) developing students' learning motivation, (e) elevating learning quality [3].

Piaget's point of view highly suggests that the teacher to use concrete media or learning material so the students are able to accept the material in concrete operational cognitive stage. Instructional which is based on limited instructional behavior in the class, cognitively will orientate on students' ability to comprehend as many as possible learning material rather than empowering and optimizing students' cognitive structure ability for the sake of adapting with their environment. As the consequences, the students may acquire many separated knowledge, but these are not systemic and less powerful (meaningful, integrated, value-based, challenging, and making the students to be active). Based on the above background, this research aimed at describing the importance of teacher in reviewing instructional management at elementary school from Piaget's cognitive development theory and describing teacher's instructional behavior in managing instructional at elementary school reviewed from Piaget's cognitive development theory.

II. METHODS

This research used qualitative methods with concurrent embedded design. It combines qualitative and quantitative methods simultaneously, but the method weight was different. Quantitative method was used as primary method and qualitative method was the secondary method to find primary and secondary data. Subjects of this research were teachers of elementary school in Sukasada District, Buleleng Regency, Bali-Indonesia. They were chosen because teachers in this district were having problems in motivating students to study, which influenced students' learning achievement. Both primary and secondary data were collected through interview, observation, and documentation. The collected data were analyzed inductively and descriptively.

III. RESULT AND DISCUSSION

A. The Nature of Children at Elementary School

Indonesia's Regulation at UU No. 20 in 2003 concerning National Education System article 6 stated that every citizen aged 7 to 15 years old are obliged to follow elementary education. Moreover, Indonesia's Government Regulation No 17 in 2010 concerning Management and Implementation of Education article 69 paragraph 4 stated that all schools at elementary education level are obliged to accept all citizen aged 7 to 12 years old as their students up to the limit of its capacity.

Development era in this stage is called as elementary school era because in this era, a child is expected to get an essential elementary knowledge for self-preparation and adaptation in adultery life. A child is expected to learn certain skills, both for curricular or extracurricular.

There are four skills that are expected to be mastered by elementary school students namely: independence, social, school and motoric cooperation skills [5].

There are some elementary school children characteristics that should be acknowledged by the teacher, in order to comprehend more the condition of the students especially in elementary school. Teacher should implement teaching method that is appropriate to the students' condition, so it is importance for the teachers to know their students' characteristics. First characteristic is the elementary school children like to play, second characteristic is happy to move, third characteristic is happy to work in group, and fourth characteristic is happy to feel or do something directly.

There are some development assignments at school [4], as follows: (1) learn to acquire physical skill to play, (2) learn to form positive attitude for themselves as a biological creature, (3) learn to socialize with their friends, (4) learn to play their roles based on gender, (5) learn basic skills in reading, writing and counting, (6) learn to develop daily concepts, (7) develop consciousness, (8) learn to have personal freedom, and (9) develop positive attitude upon institutions and social groups.

In relation to dimension of elementary school students' development, there are some development dimensions experienced by the elementary school students [5]. Those dimensions are physical development, emotional development, social development, mental-intellectual development, moral development, interest development, and personal development.

B. Teacher's Instructional Behavior

Behavior is a response or action that is conducted by an individual as a reaction of condition in his/her life. In *KBBI* (Indonesian Big Dictionary) it is stated that instructional is teaching or learning (guidance or enlightenment). Instructional derived from instruction [6]. Further, Saylor and Alexander stated that instructional is defined as curriculum implementation or specifically refers to teaching learning process [6]. Therefore, it can be defined that instructional is the implementation of instructional as the implementation of curriculum.

Instructional development is a systematic way in identifying, developing and evaluating series of material and strategy which are directed to achieve a certain educational goal. Final result of the instructional development is an instructional system, that is an empirical developed teaching learning strategy and material which is consistently achieving certain instructional goal. This instructional development consists of a set of activities including planning, development, and evaluation upon the developed instructional system so after experiencing some revisions, the instructional system would satisfy the developer.

Based on Indonesia's Government Regulation No 41 of 2007, standard of learning process consists of planning instructional process, implementation of instructional process, evaluation of instructional process, and supervision of instructional process. It is intended to implement effective and efficient learning process. Related to this writing, it is only limited to the first three points.

Based on the analysis, teacher's instructional behavior was categorized into three steps, namely planning, implementation, and evaluation. In planning instructional, teacher's instructional behavior was categorized as very low (33.3%), in implementing the instructional it was categorized as very vlow (33.3%) and in evaluation step it was categorized as very low (54.2%).

C. Concrete Operational Learning at Elementary School

Learning process is an educative interaction process related to its goal, directed to its goal and specifically conducted to reach the goal [7]. Instructional is a conducted activity to initiate, facilitate and improve the intensity and quality learning of the students [8]. Since instructional is a systematic and systemic effort to initiate, facilitate, and improve learning process, so instructional activity is closely related to the kinds of principal, learning, and the learning achievement.

In general, Piaget's cognitive development stages can be described in the specific characteristics in the scheme of the following table:

Table 1

Development stages with specific characteristics

Stages	Sensor-motoric	Pre-operational	Concrete Operational	Formal Operational
Ages	0-2 years old	2-7 years old	7-12 years old	Above 12 years old
Rationale	Action and imitating	Symbolic/ language and intuitive, imaginative	Reversible transformation and permanency, still concrete	Deductive and inductive, abstract

Moment of thought	Now	Start to "not now"	Limited concreteness	Leaving the present and start the future
Other characteristics	Reflect, habit, differentiation of material and result	Egocentric	Decentering, series, classification, concept, number, time, probability, causality	Combination, proportion, multi reference, two reversible, flexible

Further, the stages of instructional process consisting of planning, implementing and evaluating the learning process which have adapted concrete operational stage of learning from Piaget's theory.

1) Stage of Planning learning process

1. Teacher formulates learning goals which are customized with students' ability to follow concrete learning.
2. Teacher arranges learning goals which adopt indicator from basic competency by customizing students' ability by modifying teaching method to ease the learning process
3. Teacher arranges learning material from the simpler to complex level, from easy to difficult and/or concrete to abstract based on the learning goals and arranges flexibility and difficulty of learning material by considering students' potential (including the fast and slow learner, high and low motivated students).
4. Teacher arranges learning material based on concrete life context and based on the science and technology development.
5. Teacher arranges relevant strategy, approach, and learning method to achieve learning goals and competency that should be mastered by the students so it will ease students' understanding to state the excepted learning.
6. Teacher allocates time proportionally by considering level of material complexity and/or needs of the students in every learning stages.
7. Teacher is able to choose learning sources/ learning media that help the students so it can be used to achieve learning goals or competency that want to achieve.

8. Teacher can optimize more than one learning source by arranging learning so it can be attractive for the students and can be easily understood.

2) Stage of Learning Process Implementation

1. Teacher connects learning material with the students' prior knowledge by proposing problems close to the students' world.
2. Teacher creates freedom to the students to find their own answer of the given problems and does not give them the answer.
3. Teacher forms learning group and gives assignment or problem to the students in order to arise questions from the students.
4. Teacher involves students in experiment at the class to find a concept.
5. Teacher uses learning sources from concrete things and close to the students' environment in order to be easily understood.
6. Teacher uses tracker question to measure level of students' understanding.
7. Teacher provides help as empowerment from the learning of the students.
8. Teacher creates opportunity for the students to conclude their learning with their own language.

3) Stage of Learning Evaluation

1. Planning test that is appropriate with learning indicator to measure students' learning improvement from cognitive aspect.
2. Variety test so maximally able to measure every students' development from cognitive aspect.
3. Using authentic assessment (quiz, verbal question, assignment, and so on) to supervise students' learning development.
4. Implementing many portfolios in form of structured assignment
5. Using appropriate test based on the learning goals and material as arranged in lesson plan.
6. Using test analysis to identify easy, moderate or difficult topic/basic competency so it can be found the strength and weakness of every student for the remedial.

7. Teacher uses many tests to provide feedback for the student about their learning development and becomes the basis of arranging further lesson plan.

IV. CONCLUSION

Based on the finding of this research, it can be concluded that teachers in Sukasada District, Buleleng Regency, Bali-Indonesia had implemented their teaching without optimally considering the psychological developmental stages of Piaget. It could be seen from their achievements that were categorized as very low, in term of planning, implementation, and evaluation in their instructional behavior. Therefore, it is suggested that more intensive in-service training and seminar which emphasizing the importance of considering Piaget's cognitive development stages in teaching elementary school students should be conducted by the authorities in Buleleng such as the Principal of the schools, head of Education Office of Buleleng, as well as the teachers who need to improve their own understanding of this issue.

Acknowledgements

The author thanks the Principal of STKIP Suar Bangli for supporting and giving permission to conduct this research.

REFERENCES

[1] Nurkanca, Wayan. (2001). *Perkembangan Jasmani dan Kejiwaan.* Surabaya. Usaha Nasional.

[2] Hamdani (2005). *Media Pembelajaran di Sekolah Dasar*. Bandung: Alfabeta.

[3] Sumantri, Mulyani dkk. (2004). *Media Pembelajaran*. Jakarta: Rineka Cipta.

[4] LN, Syamsu Yusuf. (2004). *Psikologi Perkembangan Anak dan Remaja*. Bandung. Remaja Rosdakarya.

[5] Munandar, Utami. (1992). *Mengembangkan Bakat dan Kreatifitas Anak Sekolah.* Jakarta: PT. Gramedia Widiasarana Indonesia.

[6] Harjanto (2006). *Perencanaan Pembelajaran.* Rineka Cipta, Jakarta.

[7] Suastra, I.W. 2009. *Pembelajaran Sains Terkini*: Mendekatkan Siswa dengan Lingkungan Alamiah dan Sosial Budayanya. Universitas Pendidikan Ganesha. Singaraja.

[8] Udin S. Winataputra, et al. *Teori Belajar dan Pembelajaran.* (Universitas Terbuka, Jakarta, 2007).

Inclusive Education for Students with Disability

Ikhfi Imaniah[1*], *Nurul* Fitria[2]
[1] English Education Department, Faculty of Teachers Training and Education, Muhammadiyah University of Tangerang, Indonesia
[2] Young Learners Education Department, Faculty of Teachers Training and Education, Muhammadiyah University of Tangerang, Indonesia

Abstract. This paper identifies and discusses major issues and trends in special education in Indonesia, including implications of trends for the future developments. Trends are discussed for the following areas: (1) *inclusion and integration*, issues will remain unresolved in the near future; (2) *early childhood and postsecondary education with disability students*, special education will be viewed as lifespan schooling; (3) *transitions and life skills*, these will receive greater emphasis; and (4) *consultation and collaboration*, more emphasis but problems remain. Moreover, the participant of the study in this paper was an autism student of twelve years old who lived at Maguwoharjo, Yogyakarta. This study was qualitative with case study as an approach of the research. The researchers conclude the autism that has good academic, communication and emotional skill are able to go to integrated school accompanied by guidance teacher. But in practice, inclusive education in Indonesia is inseparable from stakeholders ranging from government and institutions such as schools, educators, school environment, community and parents to support the goal of inclusive education itself. Adequate infrastructure also needs to be given to the school that organizes inclusive education for an efficient and effective students understanding learning-oriented of inclusive education. In short, every child has the same opportunity in education, yet for special education which is aimed at student with special educational needs.

Keywords: *inclusive education, early childhood, integrated school.*

1.1. Introduction

In recent years, the concept and practice of inclusive education have gained importance. Internationally, the term is increasingly understood more broadly as a reform that supports and welcomes diversity amongst all learners [1]. Inclusive education can be seen as a process of strengthening the capacity of an education system to reach out to all learners. It is, therefore, an overall principle that should guide all educational policies and practices, starting from the belief that education is a fundamental human right and the foundation for a more just society. Education takes place in many contexts, both formal and non-formal, and within families and the wider community. Consequently, inclusive education is not a marginal issue but is central to the achievement of high quality education for all learners and the development of more inclusive societies [2]. Inclusive education is essential to achieve social equity and is a constituent element of lifelong learning. In short, inclusive education is a process that involves the transformation of schools and other centres of learning to cater for all children – including boys and girls, students from ethnic and linguistic minorities, rural populations, those with disabilities and difficulties in learning and to provide learning opportunities for all youth and adults as well. Its aim is to eliminate exclusion that is a consequence of negative attitudes and a lack of response to diversity in race, economic status, social class, ethnicity, language, religion, gender, sexual orientation and ability.

Most countries adopt the principles of Education for All (EFA) in their policies and laws, but in practice, education is for "almost all" or for "most people" and the people excluded are precisely those who need it most, in order to compensate their disadvantaged social and educational situation. According to the EFA Global Monitoring Report 2008, there are still 72 million of children in the world who have no access to primary education and 774 million of young people and adults who are illiterate, 64% of which are women.

Access to early childhood care and education, necessary to guarantee equal opportunities, is quite limited in the three first years, and children from the most disadvantaged socio-economic contexts and from rural areas hardly benefit from these services. Access to secondary education has increased by 5%, from 2000 to 2005, reaching 66%. Nevertheless, a greater effort should be made since nowadays primary education is not enough for being included into the knowledge society and emerging from poverty.

Education quality and equality are unresolved matters even in those countries with high schooling rates. There are large disparities among and within countries as regards the access to the different educational levels and the distribution of knowledge. Only 63% of countries with available data have reached gender parity in primary education, decreasing to 37% at the secondary level. Students from economically disadvantaged sectors, rural areas or minorities are those who, due to a structural situation, show higher repetition and drop-out rates and achieve lower learning outcomes [3].

* lqvee.sorrow@gmail.com

Moreover, segregated schools and programs exist, which are aimed at people with special educational needs, of different ethnic origin, or migrant families. Many of these students are not treated with human dignity; their culture is not respected and they are victims of physical or psychological violence.

Therefore, educational exclusion is an important phenomenon which is not restricted to those who do not attend school because of a denial of access, or because they dropped out of school due to repetition, the lack of relevance of the education, economic obstacles or the students' life circumstances. Exclusion also affects those who attend school but are segregated or discriminated against because of their ethnic origin, gender, social condition, of other individual characteristics or capacities, as well as those who cannot learn due to the low quality of the education they receive.

Based on the observation at Mugowoharjo, Yogyakarta, Indonesia; there is an eight year student who has disability which study at one of formal school at Yogjakarta. He looks normal like other students, but he is not; he is autism. He needs guide from the teacher during the learning process. So far, inclusion has been presented through a disability lens. Inclusion should, however, not be seen as an issue solely about students with disability. Many, however, now view inclusive education as being concerned with diversity more generally. According to Shaddock and colleagues (2009), inclusion implies that if participation becomes an issue for any student, whether arising from disability, gender, behaviour, poverty, culture, refugee status or any other reason, the desirable approach is not to establish special programs for the newly identified individual or group need, but to expand mainstream thinking, structures, and practices so that all students are accommodated.

2.1. *Result and Discussion*

2.1.1. Inclusion and Integration

Inclusive education is a process of strengthening the capacity of the education system to reach out to all learners and can thus be understood as a key strategy to achieve EFA. As an overall principle, it should guide all education policies and practices, starting from the fact that education is a basic human right and the foundation for a more just and equal society. The major impetus for inclusive education was given at the World Conference on Special Needs Education: Access and Quality, held in Salamanca, Spain, June 1994. More than 300 participants representing 92 governments and 25 international organizations considered the fundamental policy shifts required to promote the approach of inclusive education, thereby enabling schools to serve all children, particularly those with special educational needs [4].

Although the immediate focus of the Salamanca Conference was on special needs education, its conclusion was that: Special needs education – an issue of equal concern to countries of the North and of the South – cannot advance in isolation. It has to form part of an overall educational strategy and, indeed, of new social and economic policies. It calls for major reform of the ordinary school.

An 'inclusive' education system can only be created if ordinary schools become more inclusive – in other words, if they become better at educating all children in their communities. Inclusion is thus seen as a process of addressing and responding to the diversity of needs of all children, youth and adults through increasing participation in learning, cultures and communities, and reducing and eliminating exclusion within and from education. It involves changes and modifications in content, approaches, structures and strategies, with a common vision that covers all children of the appropriate age range and a conviction that it is the responsibility of the regular system to educate all children.

The term inclusion has different interpretations in various countries. It is sometimes associated with those students living in marginalized or poor contexts, but frequently it is related to the participation of the disabled or those with special educational needs in mainstream schools. In this way inclusion is considered to be almost the same as integration, when they are in fact two different approaches with different visions and perspectives. As a consequence of this misconception, inclusive policies are regarded as a responsibility of special education, restricting the analysis of all the common forms of exclusion and discrimination that take place within education systems.

In the case of inclusion, on the contrary, the focus is on the transformation of education systems and schools so that they can cater for the diversity of students' learning needs resulting from their social and cultural background and their individual characteristics as regards learning motivations, abilities, styles and rhythm. According to this perspective, it is not the students enrolled in school that must adapt to the existing educational provision, but rather the school that should be adapted to the needs of every student, since all students are different. Corps, Ceralli, and Boisseau (2012) explain that special education system means children with disabilities receiving an education in a segregated learning environment such as a special school or centre that is often isolated from the community, from other children, or from the mainstream education schools. In many countries this type of special education system has no link to the Ministry of Education. [5]

The key element of inclusion is not individualization but the diversification of the educational provision and the personalization of common learning experiences in order to achieve the highest degree of participation of all students, taking into account their individual needs. This implies advancing towards universal design, where the teaching learning process and the curriculum consider from the very beginning the diversity of needs of all students, instead of planning on the basis of an "average" student and then carry out individualized actions to respond to the needs of specific students or groups who were not taken into consideration by an education proposal based on a logic of homogeneity instead of diversity. The response to diversity—an essential condition to achieve high quality education— is probably the main challenge currently faced by schools and teachers, as it involves substantive changes in the existing conceptions, attitudes, curricula, pedagogical practices, teacher training, evaluation systems and school organization.

In short, the inclusive education system takes a systematic approach to change rather than a school by school approach. [6]. Integrated education can also mean that a child with a disability is placed into a mainstream school and class, but that the school makes only minimal attempts to address any specific academic or social needs the child might have, and accordingly the child must adapt his/herself to the environment. It means classes for children with disabilities that are located in mainstream schools but in a separate classroom with other disabled learners and with a dedicated teacher. These children will often have little or no contact with their non-disabled peers. Moreover, Education system considers the measures it must take to be able to provide an appropriate education with all children learning together. Links are made with support services both special and mainstream.

2.1.2. Early Childhood and Postsecondary Education with Disability

The World Declaration on Education for All, adopted in Jomtien, Thailand (1990), sets out an overall vision: universalizing access to education for all children, youth and adults, and promoting equity. This means being proactive in identifying the barriers that many encounter in accessing educational opportunities and identifying the resources needed to overcome those barriers [7].

Early childhood is a quickly expanding focus in special education. Research has supported the belief that early interventions are critical for significant improvements in exceptional individuals. There have been new laws, funding, and training established in the field to assist in supporting early childhood special education. Unfortunately, there is still a much higher demand for these services than what is currently being provided. The current emphasis in legislation has been toward expanding special education from birth to adulthood. This newer perspective is based on a view of lifespan schooling.

School-going age for children both with and without disabilities may vary from region to region and country to country. National governments have policies and laws stipulating the age limits for primary and secondary school: 1) average preschool age 3-6 years, 2) average primary school age 6-15 years, 3) average secondary school age 14-25 years. In this case, the subject of student is at twelve years old which has disability on his communication, socialization and emotional. Autism spectrum disorder (ASD) and autism are both general terms for a group of complex disorders of brain development. These disorders are characterized, in varying degrees, by difficulties in social interaction, verbal and nonverbal communication and repetitive behaviours [8]. For children with disabilities, who often begin school later than their non-disabled peers and who sometimes require additional time, the process of setting an age limit law can prevent the completion of a basic education.

Education for children with disabilities in most low and middle-income countries largely consists of segregated special schools in urban areas where infrastructure is more developed. However, evidence is beginning to show that this type of educational system is not a viable solution over the long term and invariably excludes the poorest individuals, especially in rural settings.

The past ten years have witnessed a significant global drive on the issue of access to education. Whilst much progress has been made, as demonstrated by a global decrease in the number of children out of school, there remain serious challenges in terms of quality of teaching, learning, retention and success in education. For children with disabilities, the situation is critical. Education systems and services worldwide are simply failing to meet their needs, both in terms of access and quality of education.

2.1.3. Transition and Life Skills of Disability Students

Most individuals learn a majority of their life skills from their families or peer groups. For individuals with special needs, these skills often become a priority for their education. Individuals who have sensory or physical disorders such as visual impairment, hearing loss, cerebral palsy or spinal bifida need to learn how to use the specialized materials and equipment that will assist them. Transition has become a growing concern of special education. The results of research on vocational and postsecondary outcomes spurred special educators to develop programs to help students with disabilities make the transition from school to work and independent living.

In line with the transition and life skills, the curriculum and teachers methods in teaching and learning process are connected to the success of students' life skills in the environment.

An inclusive curriculum addresses the child's cognitive, emotional, social and creative development. It is based on the four pillars of education for the twenty-first century – learning to know, to do, to be and to live together [9].

It has an instrumental role to play in fostering tolerance and promoting human rights, and is a powerful tool for transcending cultural, religious, gender and other differences. An inclusive curriculum takes gender, cultural identity and language background into consideration. It involves breaking negative stereotypes not only in textbooks but also, and more importantly, in teacher's attitudes and expectations. Multilingual approaches in education, in which language is recognized as an integral part of a student's cultural identity, can act as a source of inclusion. Furthermore, mother tongue instruction in the initial years of school has a positive impact on learning outcomes.

An inclusive approach to curriculum policy has built-in flexibility and can be adjusted to different needs so that everyone benefits from a commonly accepted basic level of quality education. This ranges from varying the time that students devote to particular subjects, to giving teachers greater freedom to choose their working methods, and to allowing more time for guided classroom-based work.

According to the *EFA Global Monitoring Report 2005*, one way to move towards a relevant, balanced set of aims is to analyse the curriculum in terms of inclusion. An inclusive approach to curriculum policy recognizes that while each learner has multiple needs – even more so in situations of vulnerability and disadvantage – everyone should benefit from a commonly accepted basic level of quality education. This underlines the need for a common core curriculum that is relevant for the learner while being taught according to flexible methods [10].

Accessible and flexible curricula, textbooks and learning materials can serve as the key to creating schools for all. Many curricula expect all students to learn the same things, at the same time and by the same means and methods. But students are different and have different abilities and needs. It is important, therefore, that the curriculum be flexible enough to provide possibilities for adjustment to individual needs and to stimulate teachers to seek solutions that can be matched with the needs, abilities and learning styles of each and every student [11]. This is particularly important in the development and practice of learning activities for youth and adults.

2.1.5. Consultation and Collaboration

For students that are involved with integrated programs, it is essential to implement a consultation or collaborative model of service delivery [12]. Consultant models were originally thought to be "expert models" where special education specialists would work with general educators and instruct them on how the teachers were to meet the needs of their exceptional students. This model was not appropriate in some states, where the special education teachers receive no significant additional training compared to their regular education counterparts. This model also had difficulties when experienced teachers did not want or appreciate advice about their own classrooms. Current models of consultation are more reciprocal and give the specialist either equal, or less authority than the regular education teacher. The collaboration model is a more popular method being advocated by several researchers and practitioners. Collaboration models (like consultation models) assume that the regular teacher is the primary provider of instruction and has the ultimate responsibility. The special education teacher's role is to act as a support and a resource to the regular teacher including working in the general classroom with students. The General Education Collaboration Model proposed by Simpson and Myles (1993) is composed of five essential components: 1) Flexible departmentalization, 2) Program ownership, 3) Identification and development of supportive attitudes, 4) Student assessment as a measure of program effectiveness, and 5) Classroom modifications that support mainstreaming. (p. 65)

Although collaboration models may sound ideal, there are several problems that have not yet been resolved. One problem is communication, for teachers do not always possess the skills to problem solve or work well together. Another concern is the lack of time to work with other professionals. There is also sometimes little common ground between regular and special education, with educators holding many misconceptions of the other field. Teacher education is behind in training special educators and especially regular educators to deal with these changes. The field is also at a loss as to how to adequately train teachers working in the schools. Although in-services are the current method of training working educators, there needs to be additional support, school adaptations, and classroom modifications for collaboration to be successful. The future of special education seems to be swinging toward increased collaboration. It appears that special education will be able to adapt to these changes, but it remains to be seen if the general education programs will be as receptive.

4.1. Conclusion

In sum up, an inclusive education system can only be created if ordinary schools become more inclusive – in other words, if they become better at educating all children in their communities. Inclusion is thus seen as a process of addressing and responding to the diversity of needs of all children, youth and adults through increasing participation in learning, cultures and communities, and reducing and eliminating exclusion within and from education. It involves changes and modifications in content, approaches, structures and strategies, with a common vision that covers all children of the appropriate age range and a conviction that it is the responsibility of the regular system to educate all children. Moreover, an inclusive curriculum addresses the child's cognitive, emotional, social and creative development. It is based on the four pillars of education for the twenty-first century – learning to know, to do, to be and to live together. But, there is common ground between regular and special education, with educators holding many misconceptions of the other field. Teacher education is behind in training special educators and especially regular educators to deal with these changes. The field is also at a loss as to how to adequately train teachers working in the schools. Although in-services are the current method of training working educators, there needs to be additional support, school adaptations, and classroom modifications for collaboration to be successful. Accessible and flexible curricula, textbooks and learning materials can serve as the key to creating schools for all. Many curricula expect all students to learn the same things, at the same time and by the same means and methods. But students are different and have different abilities and needs. It is important, therefore, that the curriculum be flexible enough to provide possibilities for adjustment to individual needs and to stimulate teachers to seek solutions that can be matched with the needs, abilities and learning styles of each and every student. But in practice, inclusive education is inseparable from stakeholders ranging from government and institutions such as schools, educators, school environment, community and parents to support the goal of inclusive education itself. Adequate infrastructure also needs to be given to the school that organizes inclusive education for an efficient and effective students understanding learning-oriented of inclusive education.

References

1. UNESCO. *Policy Guideline on Inclusion in Education*. Paris (2009)
2. S. Miles and N. Singal. The Education for All and Inclusive Education debate: Conflict, contradiction or opportunity?. *International Journal of Inclusive Education*. (2008)
3. UNESCO. *Defining an Inclusive Education Agenda: Reflections around the 48th session of the International Conference on Education*. Geneva, Switzerland. (2009)
4. UNESCO. *The Salamanca Statement and Framework for Action on Special Needs Education*. Paris, UNESCO/Ministry of Education, Spain. (1994)
5. H. Corps, G. Ceralli, and S. Boisseau. *Inclusive Education*. (2012)
6. H. Corps, G. Ceralli, and S. Boisseau. *Inclusive Education*. (2012)
7. UNESCO. *Policy Guideline on Inclusion in Education*. Paris (2009)
8. Autism Speaks Inc. *Autism Speaks and Autism Speaks It's Time to Listen & Design are trademarks owned by Autism Speaks Inc*. All rights reserved. (2012)
9. J. Delors, et al. *Learning: the Treasure Within. Report to UNESCO of the International Commission on Education for the Twenty-first Century*. Paris, UNESCO. (1996)
10. UNESCO. *EFA Global Monitoring Report 2005. The Quality Imperative*. Paris, UNESCO. (2004*b*)
11. UIS. I*nvesting in the Future: Financing the Expansion of Educational Opportunity in Latin America and the Caribbean*, Montreal, Que.,UIS. (2004*c*)
12. J. Siegel. *Special Education Issues, Trends, and Future Prediction*.

The Use of Bilingual Methods through Scientific Approach at Integrated Subject to Improve the Student's English Skills

Cholifah Tur Rosidah[1,*], *Suryaman*[2]

[1]Teacher and Education Science, Primary Teacher Education Program, PGRI Adi Buana University of Surabaya, Indonesia
[2]Post-Graduate Study, Educational Technology Program, PGRI Adi Buana University of Surabaya, Indonesia

Abstract. This study was aimed to know the PGRI Adi Buana Surabaya primary teacher education program student's English skill after being applied through scientific approach at integrated subject. Therefore, this study was conducted to support the student's readiness in facing globalization era and international challenge, also created competent, superior, and high character. This study was an action research which conducted in two cycles. There were four steps namely planning, action, observation or evaluation and reflection. The subject of this study was 40 students of C class 2015. The data obtained was analyzed through descriptive analytic static. The result showed that there was 32, 74% enhancement of the student's English skills after applied bilingual method through scientific approach. That enhancement could be seen on the average result of the cycle I which showed 47, 58% as a low category and increased to 80, 32% at cycle II as good category.
Keywords: Bilingual method, scientific approach, English skills

Introduction

Teaching learning process is meant to build the student's knowledge, attitudes, and skills. By having those three elements, they are intended to be able facing global challenge. The global challenge itself couldn't be avoided especially when WCU (World Class University) comes. As cited by www.antaranews.com in 2007, The Minister of National Education at that time, Prof. Dr. Bambang Sudibyo, MBA was preparing 50 universities for WCU. Since then, both state and private universities has competed in preparing great skilled graduates in their fields and also in international arena.

Educational institutions especially universities are required to preparing the student's skill in facing the globalization such as developing the science and language fields. Foreign language proficiency must be applied in order to improve the education quality. By doing so, English subject must be put on the requirement of graduation. However, the students of PGSD Unipa only get once English subject for eight semesters. As a result, it hasn't given a great result yet. This problem can be solved by applying bilingual program for all subjects; one of them is integrated learning course.

Literally, bilingualism is also called as two languages (KBBI: 2016). According to that literary meaning, it can be drawn that bilingualism is related to the use of two languages or two language codes. Sociolinguistically in general, bilingualism can be interpreted as the use of two languages by a speaker in association with others. Therefore, Bilingual method is a method of learning by using a language other than mother tongue (Chreese: 2010). In this study, the writer both uses Indonesian and English language. The result of the previous study on bilingual methods can be several reasons why bilingual methods are important to apply in a lesson. As presented by Turnbull (2016) which concludes that bilingual model learning may be applied to train foreign language abilities of children other than the original language. But, in the learning process, bilingual methods become less effective if only done by the teacher, so it is also required the active participation of students. Herdia (2016) explains on his research that the application of bilingual methods is a challenging learning in any country, in addition to preparing students in global challenges also requires the ability of teachers who master foreign languages.

* rosidahcholifah@gmail.com

To support the application of bilingual methods in order to run effectively and obtain optimal results, it can be supported by a scientific approach that centralizes learning to students. The scientific approach is a learning approach designed in such a way that learners actively construct concepts, laws, or principles through the stages of observing, questioning, gathering information, associating, and communicating (Wieman: 2010).

From some research results indicated that the lecturer is obliged to equip his students with other skills, such as proficiency in language to compete in the international world. Based on this study, the application of bilingual method should be done by using a scientific approach that centralizes learning to students so as to optimize students' English skills.

Methods

Research method is a tool, procedure and technique selected in conducting research (Djajasudarma: 2010). The research approach is concerned with some analysis steps in processing scientific information. It is the basic concept used as the framework of thinking in a study. This research includes a classroom action research, a reflective form of research based on specific actions to improve the competence of learning process in classroom (Agung: 2012). This research was conducted during the process of integrated learning and lecturing on the students of primary teacher education program at C class 2015 at PGRI University Adi Buana Surabaya.

There are some steps in conducting this study, namely planning, acting, observing, and reflecting. At the planning step, the researcher prepares teaching materials, design of the action in form of a lesson plan, an evaluation plan that includes assignment and performance, and the media to be used in teaching learning process.

At the acting stage, implementation is done in accordance with the prepared planning. During lecturing process, lecturers will deliver integrated learning materials through bilingual method by using scientific approach. The lecture process is packed by prioritizing student activities in observing, questioning, gathering information, reasoning, and communicating in 2 languages, Indonesian and English. Furthermore, at the observing step, observation is conducted during the learning process. Taking observation data aims to observe directly the behavior of students, especially speaking, listening, writing and reading skills. The reflecting step will be done to examine the results of actions in cycle I which relates to bilingual method through scientific approach. The results of this study then will be used to look for and set some alternative and new effective action to be made as an action research cycle II. The cycle can be seen in chart 1 below.

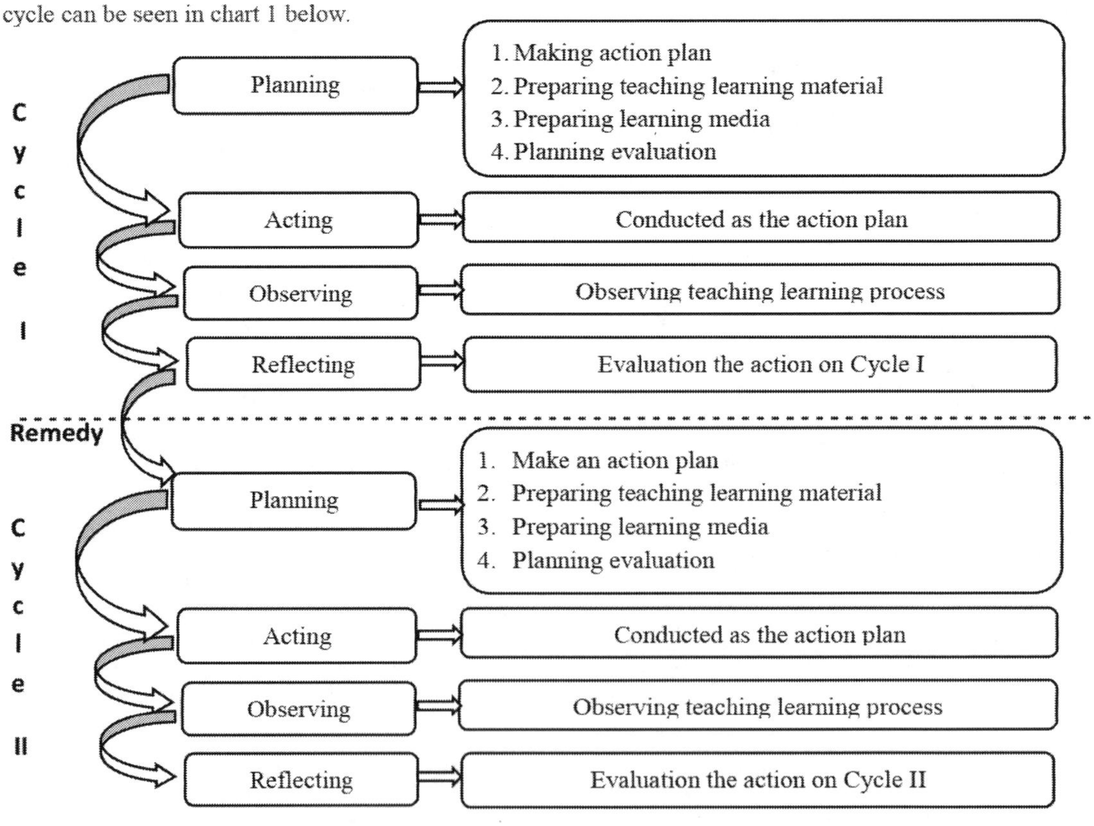

Figure. 1 Research cycle

The cycle model used in the study refers to the model developed by Arikunto (2014) which mentions the class action consisting of four stages: planning, acting, observing, and reflecting. The methods used to collect data are observation, documentation, and tests. Observations are made using an observation sheet during the learning process. Documentation includes all recording materials and photographs during the study. Tests conducted in research in the form of oral and written tests to determine the level of English language skills students' development.

Data analysis of this study used descriptive statistical analysis and quantitative descriptive. Agung (2012), descriptive statistic analysis method is a way of data management by applying descriptive statistic techniques and formulas such as frequency distribution, graph, mean number (mean, median, mode) to describe the state of an object to obtain general conclusion. While the method of quantitative descriptive analysis is used to determine the students' speaking abilities that are converted in the assessment of the five-point benchmark (PAP).

Results and Discussions

Here are the results of the study applied to the primary teacher education program students of PGRI Adi Buana Surabaya C Class 2015 which held in 2 cycles starting on March, 08 to July, 21, 2017. Cycle I consisted eight meetings which starts from March 8, 2017 to April 25, 2017. Seven times meetings used for teaching learning, and the rest were used for evaluation at the end of the cycle. Furthermore, cycle II also consisted of eight meetings. Teaching learning process was held during seven meeting and the last meeting was used as the final evaluation. The evaluation was done starting on May 12, to July 21, 2017. The description of statistical analysis result is presented in Table 1.

Statistik	**Cycle I**	**Cycle II**
Mean	5,15	8,05
Median	8	15
Modus	10	15
M %	47,58 %	80,32 %

Table 1. Statistical Analysis Result

Based on the result of descriptive statistic analysis of cycle I, the mean was 5,15. While the median showed the 50% upper limit and upper 50% frequency which containing the cumulative frequency of $1/_2$ N was 8.00 and the visible mode of the score indicating the highest frequency in cycle I was 10.00.

Based on the calculation of polygon graphs can be known M, Md, Mo indicated Mo> Md> M (10.00> 8.00> 5.15), so it can be concluded that the distribution of English language proficiency data on the first cycle was a negative crossover curve and most scores were low. Furthermore, to determine the level of students' English proficiency can be seen from the percentage ratio (M%) with the five-level PAP criteria obtained by M% 47.58%, which converted into five-level PAP is at 0-50% mastery level which means that Level of students' English proficiency in cycle I on low criteria. Then this action research class needs to be continued into cycle II for further improvement and refinement.

Furthermore, descriptive analysis of cycle II obtained the mean of 8.05 while median was 15 and mode was 15. This means M, Md, Mo seen Mo> M (15.00> 8.05), so it can be concluded that the distribution of English language ability data in cycle II was a negative crossover curve and score Obtained tend to be high. The level of English language ability can be seen from the average percent (M%) with the five-point PAP criterion that shows the number 80.32%. After converted into five-level PAP shows the level of students 'English proficiency reaches 70-85% which means that the level of students' English proficiency in cycle II was in good criteria. This shows an enhancement in the average percentage of the development of students' English skills from cycle I to cycle II of 32.74%.

From the observation result, we know that bilingual method with scientific approach can make the learning process become more varied, interactive, and fun. Bilingual method with scientific approach done in practicing the language skills, among others, 1) listening skill, by giving questions based on material exposure; 2) reading skills, students read materials related to integrated learning courses using various English literature; 3) speaking skills, exposing or presenting the material of integrated learning course using English language; 4) writing skills, making an English resume at the end of each lecture. Based on the above description it appears that bilingual method with scientific approach in integrated learning course has tried to provide one of the services to students by using more varied learning method which oriented to the learners.

From the data obtained based on the results of evaluation and observation, it was known that there was a very significant increase in the ability of English speaking students from cycle I and cycle II. The application of bilingual method with scientific approach can give a good influence on the students' English skills. The success in this research showed that the application of bilingual method with scientific approach to improve the English language ability was very effective. To achieve optimal capability hence required also optimal learning process. The optimal learning process was not only a lecturer's job, but also the active participation of students. In addition the availability of facilities and infrastructure were also being a supporter in the success of learning. Based on the results and the description above, application of bilingual method with scientific approach in the integrated learning course can improve student's skill of PGRI University of primary teacher educational program.

Conclusion

Based on the result of data analysis, it can be concluded that after implemented the application of bilingual method with scientific approach in integrated learning course, there was improvement of English language skill of PGSD student of class 2015 C equal to 32,74%. The enhancement was seen from the average increase and the percentage of students' English proficiency in the first cycle of 47.58% to 80.32% in cycle II. Thus the application of bilingual method with scientific approach in integrated learning course can improve the English speaking skill of elementary school teacher education program at PGRI Adi Buana Surabaya.

Based on these conclusions, the following suggestions may be proposed: 1) for lecturers, it is advisable to increase creativity and ability in selecting and developing more innovative learning methods and in accordance with the needs of students, 2) for researchers, it is advisable to make further research as a refinement From bilingual method with scientific approach in improving students' English proficiency.

Acknowledgements

Education running dynamic followed the development of the world globally. Therefore we must also prepare our students to become part of globalization. One of them with unprepared to have English language skills with use of bilingual methods through scientific approach in learning process.

References

1. A. Creese, A. Blackledge. MLJ, **94**, 103 (2010)
2. Agung, Iskandar. (2012). *Panduan Penelitian Tindakan Kelas Bagi Guru.* Jakarta: Bestari Buana Murni.
3. Anderson, L.W., Karthwohl, D.R. (2010). *A Taxonomy for Learning, Teaching, and Asessing.* New York: Addison Wesley Longman.
4. Araromi, Akinsola. JLLL, **13** (2015)
5. Arikunto, Suharsimi. 2014. *Penelitian Tindakan Kelas.* Jakarta: PT Bumi Aksara.
6. Bambang. 2007. *50 Universitas Disiapkan Jadi Wordl Class University.* http://www.antaranews.com/berita/76042/50-universitas-disiapkan-jadi-world-class-university. Downloaded at January 10th 2017.
7. B. Turnbull. IJBEB, 1-8 (2016)
8. C. Wieman. Change, **39**, 9-15 (2010)
9. Djajasudarma. (2010). *Metode Linguistik: Rancangan Metode Penelitian dan Kajian.* Bandung: Reflika Aditama.
10. Heong, et al. Procedia, **59**, 197-203 (2012)
11. Kamus Besar Bahasa Indonesia (KBBI) Luar Jaringan (offline), Pusat Bahasa Kementrian Pendidikan Nasional (2016).
12. Limbach, Waugh. Instructional Pedagogies, **3**, 1-9 (2010)
13. R.R. Herdia, J. Altarriba, A.B. Cieslicka. MBRCR, **18**, 383-395 (2016)
14. Turiman, et al. Procedia, **59**, 110-116 (2012)

Third-Party Intervention in Asymmetric Conflict Resolution Pasar Raya Padang Post-Earthquake in 2009

Ikhsan Muharma Putra (Ikhsan@stkip-pgri-sumbar.go.id)[1], Rinel Fitlayeni[2], Marleni[3], Afrizal[4], and Indraddin[5]

[1,2,3]Sociology Education Department of STKIP PGRI Sumatera Barat, Indonesia
[4,5] Sociology, Faculty of Social and Political Sciences, Universitas Andalas, Indonesia

Abstract. After earthquake 30 September 2009 Raya Market Padang began to be disturbed. Padang City Government responded quickly with construction of an emergency store, but has Opposed by some traders because decreased profits and low visitors. This condition triggered a long conflict and has been going on for five years (2009-2014). Many efforts have been made by stakeholders to find a conflict resolution, so it is important to further elaborate who actor has become a third party facilitating asymmetric conflict resolution. Based on the findings, the parties who really are in the position of a third party is the DPRD of Padang City and KOMNASHAM RI. The DPRD of Padang is structurally in the position of the government, but its function is as a representative of the people in the government. This position makes the slices between the DPRD and the traders as the people (constituents) and slices with Padang City Government as executive of the city government. The intervention undertaken by the DPRD was Hearing to Raya Market traders, and involving their representatives at the hearing, the DPRD working meeting discussing the issues related in 2nd Commission. Other interventions are controlling the rehabilitation and reconstruction process, and the realization of budgeting in APBD based on the needs of traders.

Keywords: Conflict Resolution, Market, Third Party.

1. Introduction

Market is one of an important part in the community institutions. The market can be seen not only as place and can also be seen as institutions that have value and norms. Some interesting things from a market is the presence of time and space [1]. It means the aspect of the space is a market place. A market is rendezvous the seller and buyer, there are transactions, exchanges, reciprocities, and various economic activities. Referring to the time when there is need and no, which are would affect in the market. Traditional markets also have two aspects, which mainly is a space for reconcilable of seller and buyer. Market facilities having a different value, depending on their function in economic activity. The market can also be seen point of view of conflict, where market is a place of conflict [2][3]. When a market is space and time are full of competitions or conflicts then market must be special attention from various parties. Many parties are involved, there are too many real and hidden interests. However as far as conflict is a conflict latent hence, the market will still be running as where it should be. But if the conflicts have shaped manifest such as the existence of protest, resistance, sabotage, destruction and others, then market function will did not run as intended. Raya Market Padang is one of the largest traditional markets in West Sumatra. Peak of the development was held during 1931-1970. At the time trade rapidly growing in the city, the construction of city hall in Raya market and stimulate the buildings shops and building market to the 1990s. The building of Raya Market is local asset which managed by city government. Management and orderliness facilitated by the market office for all problems related to the market. All of the social system in the market running by itself after all the process of market development and deals in every actor the market kept running.

The journey of Raya market deeply disturbed when earthquake 30 September 2009 was broken down and destructive building facilities in the market[4]. Almost all building is did not passed an eligibility for use as market facilities, so the recommendation for a number of buildings market has to rehabbed. After the earthquake, the city government issued a number policy related to the act of response disaster as rehabilitation and reconstruction of Raya Market. Essentially the constitution no 24 / 2007 has set up how the action stakeholders related in time of disasters, including the act of and policies which should be made by local governments. According to research conducted by Firdaus, there were clashes prolonged when rebuild Raya market [4]. This conflict occurs between the Padang government as holders of authority with Raya market traders who are victims of the earthquake victims the right owner the stores, kiosks and shops [4][5].

In the case of the conflict after earthquake 30 September 2009 have occurred in a very long time where happened since 2009 up to peaked in 2010 and 2011 until 2014 is still taking place despite experiencing a decrease in scale conflict[5]. The protest done by the merchants become the foundation of the contention so that there must be a settlement effort. An effort to resolve and looking for a solution of the conflict have involved various parties especially the existence of the third side of the conflict. But which party should be said to be third-party? This question is essential to be answered in this study. So that it becomes important to do more depth study about intervention third parties in conflict resolution asymmetrical Padang Raya market post-earthquake in 2009. This paper presenting analysis asymmetric conflict between the government of the field after raya market traders, where any formal resolution done by various parties to look how the role of a third party in the resolution.

2. Methods

This writing resulting from research titled resolution conflict asymmetrical; case study conflict market raya Padang City after earthquake 2009.The research was done with a qualitative approach through collect the data in-depth interviews, documents and observation. The data in gather since February 2017 until June 2017 later in analysis with a model analysis adjustment Miles and Huberman (1992) .Data of the results of interviews, a document already in learn and the evidence field observations then analysed and reduced until then can be presented in the form of writings.

3. Result and Discussion

3.1. Literatur review

The conflict in the case of Raya market of Padang City after the quake 30 September 2009 have been examined in some study conducted by students, off researcher or lecturer. However, they are looked for different sides in the studies. Studies have initially still in the study organization, political and social movements as research conducted by Fidel in 2011. This study attempts to see contribution FWK (*Forum Warga Kota*) as social movement within the democracy in the city. He found that FWK is a great society organization where is a combination of some groups such as; victim traders earthquake in the city the highway, victims year 2007 that aid has not been given by the city government. Public mobilization is large enough joined FWK. But FWK is not contribute to the democracy in the city[6]. Other research was in 2011, which having a resemblance to research conducted by Fidel, however have a difference at the focus of his studies. Suryanto was about more deepen on strategic done by FWK in order to concern for municipal the policy in the development of stall emergency Raya market, Padang City. The finding was about is that there are some strategies done by FWK enforcing the interests of the formal; strategy and strategies demonstrations and violence. Two strategies was carried out by FWK and reflect where their demands granted by the city council city the field to form a team to solving problems and the redevelopment in the market town [7].

Similar far from previous research, Firdaus in 2011 also examine deeper related with conflict Raya market Padang City. This study focuses attention on protest disaster victims; study conflict disaster management in Raya market Padang City. In this study, Firdaus found that there are some important event that brings protests Raya market traders municipalities, especially related to the issued by City Government. These policies are; the construction of emergency kiosks that are considered to be injurious to traders. Development policy repeated inpres market I, II, III and IV considered traders would not in accordance with their needs traders and out of context disaster management, as well as the construction shops phase VII the second floor and iii requiring emptying on the 1st floor. The form of protest done traders is through APPR (*Asosiasi Pedagang Pasar Raya*) Padang, forming joint strategy with FWK (*Forum Warga Kota*) and protest strategy together PBHI West Sumatra [4].

Still related to the conflict in Raya Market, Padang City that is the assessment conducted by Ira Ariesta in 2014. This study more focus on the discussion the role of women to the process of conflict resolution rehabilitation and reconstruction the market. In the analysis on the acts, there were several parties of women who owned a role in efforts conflict resolution of them are women from government agencies, women from NGO, some women activists from colleges and traders. However, those roles still quite low in the process of resolution. Of several action performed apparently women have been a victim of violence and a very in spare is when the negotiation process is in opposing sides when that moment women not involved [5].

Next study also involved in the study at conflict in Raya market Padang City which is assessment was conducted by Elvawati in 2015. Elvawati's , Padang City, study focused on model resolution conflict in local institutional with a case study Raya market. In this study Elvawati identify the causes of conflict with the findings the following first; the notion of emergency stall, secondly kiosks purchased , thirdly the number of kiosks built, fourthly the market building block 1 , the fifth, reconstruction inpres market II, third and fourth and fifth, the phase VII the floor 2 and 3[8]. The study also relevant to this review are review come to the scene and different contexts the role of the when conflict resolution. This study has been done by afrizal in his writings titled "*Third-party Intervention in terminating Oil Palm Plantation Conflicts in Indonesia; A Structural Analysis*". In this study Afrizal formulate in the findings of her research the role of third parties in conflict resolution company palm with local communities. In his innovation there were several parties who contributed to mediate conflict. The parties are the local government and NGO. Both sides was having different results on efforts to resolution done. The local government is considered to have failed as the party being a mediator or third party to conflict resolution. Parties who are able to mediate NGO's success is to make an agreement between the company[9].

Several studies have been reviewed before it shows the conflict in the Raya market during the post-earthquake 30 September 2009. This study will see from different points of view where efforts to resolve conflicts that has been done in document and analysed it by the relationship structure between the government and the community. There are two layers of the different structures of the conflict, where the government as up structure and down for the community. For type conflict like this review of the actually potentially good for conflict resolution requires analysis to learning conflict resolution on to others context.

3.2. The identified role in mediation Asymmetrical conflicts

Parties identified has a role in conflict Raya market the post-earthquake 2009 is the first; warring parties are Raya market traders with municipal the City. Both; party role as an assistant. Third party; function as the working on ways the common ground on efforts to resolve conflicts. The warring parties on conflict Raya market is the party who is arguing the market traders oy the of Raya market disaster victims. A suit in regard to the municipal the next. There is disquiet, concerns the disaster victims with government policy not pro traders and community. Concerns would be the seeds of contention which the government issued policies are incompatible with the condition of the traders and alleged would be harmed traders. Relation a merchant by the city government at the time of this conflict being broken until making a difference of opinion. The Position does not balance in the power of the side of relationships between the government and traders herding traders to raise a force with involves several interest groups. Any of a large group the ones involved are Forum Warga Kota (FWK) [7]. The group also involved to strengthen traders is the representation of traders who joined to organizations traders, Aliansi Pedagang Pasar Raya (APPR). The alliance is being representation of traders to continue sound traders to the city government.

The second is the companion. Parties here is PBHI (Pusat Bantuan Hukum Indonesia) of west Sumatra. PBHI is legal aid organisations who work with his client to provide legal assistance on clients[11]. PBHI is non-profit organization that works on social care in the community involved in legal problem.The principle which is embraced in the process of mentoring is improving human resources who became his client. As for some of the approaches made by PBHI for conflict resolution the padang city was first; make improvements to human resources, second; with the strengthening of the Organization[11]. Two things are done by PBHI increasingly strengthening position PBHI as escort agency. PBHI not involved directly in the conflicts, but only provide protection, awareness, and strengthening organizations that perform resistance. On its activities have indeed PBHI alignments i.e. favours the client namely market Kingdom. And cannot be said to be a neutral party. In this case the Government of the city of Padang also cannot assume the PBHI as third-party neutral. Could have been in this position is a NGO's, the government, or Dewan Perwakilan Rakyat (DPR) [9][10]. Then a third party will conduct the mediation in an effort to accelerate the completion of the conflict. The deal that awakened from the results of the mediation will have the consent of the warring parties to a resolution of the conflict. In the case of a conflict the Padang earthquake 2009 identified there are two parties who acted as a third party. First is the KomnasHam and second is the DPRD Kota Padang.

3.3. The DPRD Padang city as one of the third party

The involvement of DPRD as one of the parties involved in the conflict resolution asymmetric the Padang earthquake was over the demands of the society in DPR as a representative of the people in the Government. The position of DPRD in principle is a balanced relationship on the position of this Asymmetrical conflict [12]. City council has main task of performing the surveillance on any policy conducted by the city. The council also has the right to the ratification of the proposed by the government of the . Is known in the individual in parliament loaded with interests, but interest in this context is also related to public interest as costituent council members. The council on the view the government is as a partner in running governance function. In the organization of the council and municipal in the same government structure. Those roles also the result of encouragement parties a companion namely west Sumatra PBHI to DPRD the keep running its function as an institution which oversees the administration city[12].

The city council the this , into a position assessed balanced by the warring parties .There are some activity in relation to the resolution involving city the council .The involvement of this could have been involvement council members as individuals or as representing institutional affairs council .Facilitating early done by central KomNasHam that unite representation of the warring parties by involving members of the council .Next to the hearing council and an audience with market traders highway by members of the council on 8 November 2011 until the results devoted through KomNasHam. In the case of burning occurring in the shelter in April 22, 2012 by the personnel of city, vice chairman directly the scene to perform investigation as well as silencer conflict that occurred on Raya market conflict. To the process of tension council budgeting process is struggling rehabilitation and reconstruction Raya market stay in the city [12]. Padang council also submitted discussion market cases conflict Raya market on council II commission who performs various functions including seek conflict resolution[12].

Mediation efforts to the process resolution was carried out by the field city council meetings and also council members individually or institution do investigation on various the case with Raya market. Government meeting town with traders in facilitation city council the at Padang city council to the point of a successful. Until in 2015, 2016 year and in 2017 rebuilding Raya market with the local and national budget.

4. Conclusion

Conflict in the Raya market Padang City post-earthquake in 2009 is one form of conflict asymmetrical. The warring parties basically structurally and power flaw. The government as parties who have power or power fight with traders who in structurally have no power to attempt conflict resolution of the role city Padang council is important in effort to resolution. The council has relation by the community as its constituents or the council is the representation of the community. And that city council has representatives function of council of the society sued to do their job as an institution control and escort the government. Here the supervisory function of DPRD to municipal the policy in line legislation and regulations. With the implementation of this parliament interests market traders of victims earhquake 2009 Raya market well accommodated.

5. Acknowladgements

We are very grateful because we can complete write this paper by a title "Third-Party Intervention in Asymmetric Conflict Resolution Pasar Raya Padang Post-Earthquake in 2009". This paper results from our study entitled asymmetric conflict resolution model (Case Study of Conflict Resolution of Pasar Raya Post-Earthquake in 2009) in 2017. This paper cannot be completed without the effort and co-operation from our group members, Group members' Ikhsan Muharma Putra, M.Si, Rinel Fitlayeni, S.Sos, M.A, Marleni, M.Pd. And Patners Groups members' Prof.Dr. Afrizal M.A and Dr. Indraddin, M.Si. we also sincerely thank our informan, our lecturer in STKIP PGRI Sumatera Barat for completed this research. The possibility of this article is still a lot less, and we beg in forgiveness. We welcome readers' criticism and suggestions for better paper.

6. Reference

[1] Damsar, *Sosiologi Pasar*, 1st ed. Padang: Laboratorium sosiologi, FISIP Universitas Andalas, 2005.
[2] Damsar, *Sosiologi Ekonomi*, 2nd ed. Jakarta: Raja Grafindo Persada, 2002.
[3] Sudarmo, "The Implications of Taditional Market Development for Governance of Conflic : an Experience of Klewer Market of Surakarta, Indonesia," *J. Gov. Polit.*, vol. 7, no. 4, pp. 619–639, 2016.
[4] Firdaus, "Protes Korban Bencana; Studi Konflik Penanggulangan Bencana di Pasar Raya Kota Padang," *J. Ilmu Sos. Mamangan*, vol. 1, no. 2, pp. 27–39, 2014.
[5] I. Ariesta, "Peran Perempuan Dalam Resolusi Konflik Rehabilitasi dan Rekonstruksi Pasar Raya Padang," *J. Ilmu Sos. Mamangan*, vol. 1, no. 2, pp. 41–51, 2014.
[6] Fidel, "Kontribusi FWK Sebagai Gerakan Sosial dalam Demokratisasi di Kota Padang," Andalas, 2011.
[7] Suryanto, "Strategi Kelompok Kepentingan Memperjuangkan Kepentingannya Dalam Konflik Forum Warga Kota (FWK) Terhadap Kebijakan Pemerintah Kota Padang," Andalas, 2011.
[8] Elvawati, F. Rinel, Marleni, Afrizal, and Indraddin, "Model Resolusi Konflik dalam Institusi Lokal, Studi Kasus Pasar Raya Padang Pasca Gempa 2009," Padang, 2015.
[9] Afrizal, "Third-Party Intervention in Terminating Oil Palm Plantation Conflicts in Indonesia: A Structural Analysis," *SOJOURN*, vol. 30, no. 1, pp. 141–172, 2015.
[10] H. Mail, O. Ramsbotham, and T. Woodhouse, *Contemporary Conflict Resolution*. Malden: Polity Press, 1999.
[11] Firdaus, "Dari Aksi Demonstrasi ke Negosiasi; Strategi Penguatan Gerakan Pedagang Korban Bencana Pasar Raya Padang Oleh PBHI Sumbar," in *Konferensi Nasional Sosiologi V*, 2016, pp. 2289–2296.
[12] I. M. Putra, F. Rinel, Marleni, Afrizal, and Indraddin, "Model Resolusi Konflik Asimetris (Studi Kasus Resolusi Konflik Pasar Raya Pasca Gempa tahun 2009)," Padang, 2017.

Research Based Learning (RBL): How to Improve Critical Thinking Skills?

Tri Saptuti Susiani*, *Moh* Salimi, and *Ratna* Hidayah

Primary Teacher Education, Universitas Sebelas Maret, 57126 Surakarta, Indonesia

Abstract. Critical thinking is a part of the 21th century necessary skills trained to future teachers. Research-based learning (RBL) presents as an alternative learning model that can develop the critical thinking skills. This study aims at describing the effect of the implementation of the research-based Learning (RBL) to develop future teachers' critical thinking skills. Qualitative approach was used in this classroom research. The participants in this study consisted of 106 future teachers. Data were collected using observation and test. The results of the research indicate that the implementation research-based learning (RBL) can develop critical thinking skills. It is proved by the fact that the students were able to show their interpretation, analysis, evaluation, inference and explanation abilities. Meanwhile, aspect of self- regulation ability has not yet been able to be found. This research were used to make more sure previous research that implementation of the RBL can improve the quality learning process and make positive changes for the students in the intellectual and emotional. This research provides a different learning experience which can develop critical thinking skill.

Keywords : Research Based Learning (RBL), Critical Thinking Skill

* Corresponding author: trisaptuti@staff.uns.ac.id

1. Introduction

Educational Institution for Education Workforce (Lembaga Pendidikan Tenaga Kependidikan, or LPTK) should be able to provide qualified teachers in ccordance with the 21st century demands. According to National Research Counccil, individuals need to possess competencies in 5 main skills, such as: adaptability; complex communication skills; problem-solving skills; self-management and self-development; and systems thinking. [1] In addition, other skills required to be improved in education to cope with the 21st century challenge involve work ethic, collaboration, good communication, social responsibility, critical thinking and problem solving. Frydenberg & Andone, point out that to deal with learning in the 21st century, every individual should have critical thinking skills, digital literacy knowledge and skills, information literacy, media literacy, as well as information and communication technology skills. [2] To create human resources with critical thinking, problem solving, communication skills and technology literacy requires adequate education and learning through the 21st century education. [3] The improvement of critical thinking skills have not yet been the main focus in lectures. To train students to improve their critical thinking skills, a lecturer should apply a teaching method which allows them to respond to a phenomenon.

Research-based learning is a learning model which is associated with such activities as analyzing, synthesizing, and evaluating, and enables learners and lecturers to improve their assimilation and application of knowledge. The research-based learning is conducted under constructivism which covers four aspects: learning which constructs students' understanding, learning through developing prior knowledge, learning which involves social interaction process, and meaningful learning which is achieved through real-world experience. Research is an important means of enhancing learning quality. The research components include background, procedures, implementation, research results and discussion, and publication of research results. [4] Research- based learning is a system of instruction which used an authentic – learning, problem – solving, cooperative learning, hands on, and inquiry discovery approach, guided by a constructivist philosophy. Its usefulness had been recognized for many decades but "research in classroom" had not been adopted as a teaching method by many. [5] Research-based learning involves interdisciplinary collaborative research on university students in the learning, meaning that there is a relationship between their learning experience at class and new invention in research. RBL encourages students' active participation in the learning. RBL some learners' competencies include: 1) strong understanding of basic concept and methodology, 2) problem solving in creative, logical, and systematical manners, and 3) scientific attitude which covers respect for evidence, honesty, open-mindedness. [6] Learners are expected to have communication skills, competent technique and analysis to adapt, collaborative work, competition. In reference to such explanation, it follows that RBL is alearning model which integrates research in learning process to construct knowledge by formulating hypotheses, collecting data, analyzing making conclusion, and writing a report. The RBL model gives university students opportunities to learn and construct knowledge from such research procedures as finding information, formulating hypotheses, collecting data, analyzing, making conclusion, and writing a report. The procedures of RBL in a learning: (1) formulating a general question; (2) overview of research-literature; (3) defining the question; (4) planning research activities, clarifying methods/ methodologies; (5) undertaking investigation, analyzing data; (6) interpretation and consideration of results; (7) report and presentation of results. [7]

Critical thinking is defined as the ability to think logically, reflectively, systematically, and productively, applied in judging a situation to make appropriate consideration and decision. "Critical thinking is a process of making reasoned judgments based on the consideration

2. Method

The participants in this study consisted of 106 students at Pre- Service Teacher Training and Education.Qualitative method was employed in the present research. The research belongs to classroom research due to its orientation of developing quality of educational practices, or of improving educators' skills at class by examining theory-practice in education, evaluating, and applying in educational institutions). Three phases in observation cycle in classroom research according to are: planning meeting, classroom observation, and feedback discussion. [15] Data in the research were collected using observation and test. Instruments used in observation were observation sheets and question sheets. Meanwhile, instruments to assess critical thinking were formulated based on main aspects including interpretation, analysis, evaluation, inference, explanation, and self-regulation. Data analysis techniques used in this research that is done continuously for data collection continued until the end of the study. Data analysis in this research include; data collection, data reduction, the presentation of the data and make conclusions.

3. Result and Discussion

RBL was implemented in the course lecture of Integrated Thematic Curriculum Development. The procedures of RBL implemented in the learning were adopted from Peter Tremp's theory, including: formulating a general question; overview of research-literature; defining the question; planning research activities, clarifying methods/ methodologies; undertaking investigation, analyzing data; interpretation and consideration of results; report and presentation of results. First, the stage of formulating a general question, a topic, or a problem on the nature of learning and integrated learning, and the implementation in primary schools was given to the students. The students at this stage had not recognized the concepts of learning and integrated learning. Next, at the stage of overview of research-literature, students were asked to discuss in groups to examine materials of integrated learning from any references and literatures. After they discussed the concepts as well as various types of integrated learning, questions related to the implementation of integrated learning in primary school arose. Then, at the stage of defining the question, students made temporary hypotheses on the implementation of integrated learning in primary schools. in order to reveal the hypotheses, they were required to conduct a research, and therefore they moved on to the stage of planning research activities, clarifying methods/ methodologies. At the stage, they planned and prepared an investigation to analyze the implementation of integrated learning in primary schools, and focused on the Lesson Plans and the implementation itself by primary school teachers. They discussed in groups and constructed instruments used to observe both Lesson Plans and the implementation. They developed instruments from aspects of integrated learning. Then, at the stage of undertaking investigation and analyzing data, students did investigations by making observation in a primary school. The first investigation was done by making observation on the Lesson Plans of integrated learning constructed by a primary school teacher, while the second one on the implementation of integrated learning in the primary school. At the stage of interpretation and consideration of results, they were asked to analyze their observation results on the Lesson Plans and the implementation of integrated learning. Finally, at the stage of report and presentation of results, students reported their research results and analysis in a presentation.

Aspects of critical thinking presented in the research included: interpretation, analysis, evaluation, inference, explanation, and self-regulation. Test results, such aspects of critical thinking as interpretation, analysis, evaluation, inference, and explanation were identified, while aspect of self-regulation was not.

Table 1. Test Results of Critical Thinking

No	Aspect of Critical Thinking	Total	Percentage (%)
1	Interpretation	106	100
2	Analysis	75	70.75
3	Evaluation	100	94.34
4	Inference	32	30.19
5	Explanation	71	66.98
6	Self- regulation	0	0

The observation showed similar results to the test results. Such aspects of critical thinking as interpretation, analysis, evaluation, inference, and explanation were seen, while aspect of self-regulation was not.

The observation showed similar results to the test results. Such aspects of critical thinking as interpretation, analysis, evaluation, inference, and explanation were seen, while aspect of self-regulation was not. To find out aspect of interpretation ability through test, students were asked to answer a question related to the given case: "What did the teacher do before starting the lesson?" All students made correct answer. Based on observation, the aspect of interpretation ability was found in students in each stage of the implemented research-based learning. To find out the aspect of analysis ability, students were asked to answer a question related to the case of learning, "Why does a teacher need to prepare students' worksheets?" 75 students answered correctly. Based on observation, the aspect of analysis skill was found in students in each stage of the implemented research-based learning. To find out the aspect of evaluation ability, students were asked to answer a question related to the case of learning, "What should a teacher do when playing video on environment?" 100 students were able to make correct answer. Based on observations, the aspect of evaluation ability was also seen at the stage of planning research activities, clarifying methods/ methodologie, undertaking. Find out aspect of inference ability, students were asked to answer a question "What should the teacher and students do after break time?" 32 students made correct answer. Phase of report and presentation of results, the inference ability was seen when students presented their research report and gave solutions each other. Students' explanation ability was seen when students were asked to narrate media and their use by teacher in the presented case of learning. The observation results indicate that the explanation ability emerged at the stage of interpretation and consideration of results when students had second discussion about the implementation of integrated learning. At the stage of report and presentation of results, the explanation ability was seen when students explained their written research report. Self-regulation refers to an ability to control and manage individual self both when thinking and acting.

Based on observation and test results, the implementation of RBL model was proved to be able to improv students' critical thinking skills, particularly on the aspects of interpretation, analysis, evaluation, inference, and explanation. Implementation of the RBL is a model that integrates research with learning. Learning that make students active scientific measures. By implementing the scientific method are able to develop critical thinking abilities of students. Research-based learning, interesting discussion and cause gain insight in learning how students can make a design research. [16] Research-based learning have pointed to the need to develop an enthusiasm for critical enquiry, resourcefulness and creative solutions in undergraduate students.[17] The RBL model gives opportunities for students to actively participate in the learning. Through research experience, they are expected to implement collaborative learning during the research and the obtained knowledge. Therefore, the RBL model is useful to develop inquiry skills. Students are given chance to learn not only from lecturer's materials, but also from practice of research involving searching for literature, constructing hypotheses, collecting data, analyzing/ testing data, and drawing conclusion. RBL used for the exercise provides students with an opportunity to practice their metacognitive abilities and foster critical thinking, abilities to make predictions, propose causative factors, and present constructive arguments.[18] This is done/measured via/through oral presentations and final reports which all are the core components of any scientific research. Critical thinking nowadays is required to be mastered by university students. The development on science and technology will rapidly lead to the provision of information, either the source or the essence. In order to be able to use the information well, evaluation on data and source of information is required. The evaluation ability and decision making to use accurate information requires critical thinking. Procedures of RBL enable to individuals to equip with deductive, inductive, reflective, critical, and creative thinking abilities. With such abilities, one can deal with more complex life. Students can have critical thinking and problem solving skills in critical manners. In addition, critical thinking is one of initial factors of emergence of creativity, which develops due to observation of a phenomena or problem. The implementation of RBL, students not only receive information from the lecturer alone. This RBL allow students to participate in the plot and investigated a problem. Characteristics of the rbl more refers to the flow of education kontruktivistik, where learning is an active process of learning to find and build knowledge. Active processes is not only mentally but also physically. Kontruktivistik instruction can help the development of skills to think through problem-based teaching style openeded used teacher. Step-by-step problem-solving activities there are RBL.[19] Activity breakdown not only can develop critical thinking skills, but can also deepen understanding of students against the concepts studied.[20] When students are involved in solving problems, students felt having a role in the effort to find a solution and need to take real action.

4. Conclusion

4.1 Acknowledgement

The research results conclude that RBL model can be implemented in the learning of future primary school teachers in such stages as formulating a general question; overview of research-literature; defining the question; planning research activities, clarifying methods/methodologies; undertaking investigation, analyzing data; interpretation and consideration of results; report and presentation of results. Its implementation enables to develop their critical thinking skills. It is revealed from the emergence of abilities of such aspects as interpretation, analysis, evaluation, inference, and explanation. The self-regulation ability, however, was not found. Test and observation were employed in the research. The future research is expected to continue the present research with technique of clinical interview.

References

1. J.A. Koenig, Rapporteur, *Assessing 21st century skills: summary of a workshop* (The National Academies Press, Washington DC, 2011)
2. M. Frydenberg, D. Andone. IEEE. (2011)
3. M. Salimi, *Guru yang efektif dan pengajaran yang efektif dalam rangka menyongsong generasi emas 2045: Sebuah Kajian Awal.* (Universitas PGRI Semarang Press, Semarang, 2016)
4. D. T.Widayati, D. Luknanto, E. Rahayuningsih. *Pedoman umum pembelajaran berbasis riset* (UGM. Yogyakarta, 2010).
5. S. Poonpan, S. Siriphan, *Indicators of research-based learning instructional process: a case study of best practice in a primary school*, (Bangkok, Faculty of Education, Chulalongkorn University Phaya Thai, 2001
6. P. Arifin, *Makalah seminar nasional research based learning,* (ITB, Bandung, 2010)
7. P. Tremp, *Research-based teaching and learning A LERU project* (University of Zurich, Munich, 2010)
8. P. A. Facione, CA Sánchez, NC Facione, J Gainen, JGE, **44** (2010)
9. P. C. Seifert, AORN, **91** (2010).
10. P. Facione, *Critical thinking: what it is and why it counts* (California Academic Press, Millbrae, 2013)
11. D. Rosyada, *Pembelajaran berbasis penelitian.* [http://www.uinjkt.ac.id/pembelajaran-berbasis-penelitian/] (2016)
12. M. Fikri, *Research based learning, asah kemampuan praktis mahasiswa ITB.* [https://www.itb.ac.id/news/read/3211/home/research-based-learning-asah-kemampuan-praktis-mahasiswa-itb] (2011)
13. K. C. Suryandari, Pedagogia, **16**, 2 (2013)
14. Buschor, C. Bieri, Kamm, Esther, ERPP, **14**, 3 (2015)
15. D. Hopkins, *A teacher's guide to classroom research fourth edition* (Open University Press, New York, 2008)
16. Ellen Bastiaens, *Three Educational Models for Positioning the Maastricht Research-Based Learning Programme (*Springer, Maastricht, 2017)
17. P. Guinness, *JGHE,* **36**, 3 (2012)
18. A. Al- Maktoumi. et.al, JGHE, **40** (2016)
19. D. Mujis, D. Reynolds, *Effective Teaching* (Sage Publications, London, 2008).
20. R. Killen, *Effective Teaching Strategies: Lessons From Research and Practice* (Cencage Learning, Victoria, 2009)

"GENIT" AS EFFECTIVE DESIGN OF LEARNING MEDIA

Mardiki Supriadi

Abstract

There are two main approaches to using media in schools: students can learn "from" Media and technology, and they can learn "with" media and technology. The basis for The use of media and technology as a "tutor" in schools is "educational communication," that is the deliberate act of communicating content (teaching content) for students by assuming that they will learn something "from" this communication, so communication is not again free but controlled and conditioned for educational purposes. Findings on the impact of technology-based instruction in education can be concluded that technology as a tutor has a positive effect on learning, one of them is the student can complete a set of educational goals in less time than necessary in a traditional approach. However, these two approaches only make the task become easier but do not activate and facilitate them to think critical and higher learning. "Media Genit" uses a new approach as an approach that can optimize perspective-based cognitive processes, constructivism, which constitutes an environment in which the student as a designer thinks creative about content combined with real-world tasks, student learning content, enjoy the learning process, and recognize that they have created something that is valuable.

Keywords: Constructivism, "Media Genit", Learning

Introduction

One of the main reasons for widespread attention is focused on media and technology in education today is a huge financial investment made in the media and technology in education around the world. For example, in Indonesia, education budget in the 2016 APBN has made history. Because, the education budget in APBN 2016 became the largest compared to previous years, where the education budget in APBN 2016 reached Rp 419.2 trillion or 20 percent of the total state expenditure of Rp 2,095.7 trillion. (Source: APBN 2016, processed by Sapulidi Research Center (SRC) NGO Sapulidi).

Why is so much attention paid to media and technology within education? First, in correlation with the media, there are many problems that become the attention of students, parents, educators, government, and society in general. For example, questions about the effects of different media towards the cognitive and moral development of children. With regards to technology, people want to know if the various new technologies are more effective to teach and learn compared to a more traditional classroom approach, whether few technologies are more motivating than others, or at least, what technology could be used to improve access or reduce costs in education.

Despite so much attention paid to media and technology in education, unfortunately there are still many teachers who are not comfortable with moving from the learning orientation shifting from teacher-centered teaching to student-centered learning, and this kind of transformation requires considerable time and support (Fisher, Dwyer, & Yocam, 1996), plus many media and technological innovations of false learning promises to make learning fun and easy teaching. Cognitive tools do not make these promises, either for students or teachers. Instead, cognitive tools activate complex cognitive learning strategies and critical thinking.

The purpose of writing this paper is the integration of Genit models in the 2013 curriculum could contribute in the formation of cognitive constructivism of students, one of the applicable application is using LEGO / logo.

Genit Media

There are two main approaches to using media in schools: students can learn "from" media and technology, and they can learn "with" media and technology (Jonassen & Reeves, 1996). Learning "from" media and technology is often referred to terms such as television learning, computer-based instruction, or an integrated learning system (Hannafin, Hannafin, Hooper, Rieber, & Now, 1996; Seels, Berry, Fullerton, & Horn, 1996).

Apart from this approach, media have been introduced to schools because it is believed to have a positive effect on teaching and learning. The purpose of writing this paper is to design the learning media of Genit on fundamental education. For the purpose of this report, the media is defined as "all means of communication, in all formats." In this case, the media includes such diverse symbols such as print, graphics, animation, audio, and motion pictures. Technology is defined as "every objects or processes from humans that can be used to deliver the media." In this, technology includes a variety of phenomena such as books, movies, and the internet.

In correlation to education, the media is a symbol system, used by teachers and students to represent knowledge; Technology is a tool that allows them to share knowledge representation with others. Unfortunately, the meaning of media and technology in education is defined synonymously. The difference between media and technology should be clarified because, the impact must be understood. Here is an excerpt from the Sixth Edition of The Encyclopedia of Educational Research (Alkin, 1992) explains these differences:

Computer-based technology could not be considered a "media," because of the variety of programs, tools, and devices that can be used with them are not limited to certain symbol systems, or to certain classes of activities ...... in this case, the "computer" is actually "a few stagse of discovery" of many uses, media to create symbols, explore, and think in different domains. This matter is used to represent and manipulate symbol systems - language, math, music - and to create symbolic products - poetry, mathematical proofs, compositions. (Salomon, 1992, p. 892)

The basis for the use of media and technology as a "tutor" in school is "educational communication," which is the deliberate act of communicating content (teaching content) to students assuming that they will learn something "from" this communication, thus communication is no longer free but controlled and conditioned for the purpose of education (Krendl, Ware, Reid, & Warren, 1996).

Findings on the impact of technology-based instruction in education conclude that technology as a tutor has a positive effect on learning, one of them being the student can complete a set of educational goals in less time than necessary with a more traditional approach. Limited study and evaluation indicates that integrated learning systems are effective forms of the technology-based instructions that are very likely to play a bigger role in classrooms in the future, some experts suggest that "... the corresponding role for technology is not a teacher / expert, but rather, that of the extension's mind 'cognitive tool' "(Derry & Lajoie 1993, p.5)

The media to optimize cognitive processes have been deliberately adjusted or developed to function as an intelectual helper, to activate and facilitate critical thinking and higher learning. In the media approach to optimize the cognitive processes, media and technologies provided directly to learners, is used to represent and express what they know. The student functions as a designer using media and technology as a tool for analyzing the world, accessing and interpreting information, organizing private knowledge, and represent what they know to others.

Ideally, a task or problem for a media application for optimizing the cognitive processes will lie in a realistic context with that result that is meaningful personally to the learner. And there are things that at least need to be considered about media and technology in education, which are:

Bringing media and technology to school can take advantage of the quality of strong motivation for children. The education system that utilizes this motivation for having the chance for success is much greater ...... Each media has a cognitive excess profile, and any media can be used to enhance other impacts. (Greenfield, 1984, p. 178)

In a media approach to optimize cognitive processes, information is not encoded in predefined educational communications which are then used to transmit knowledge to students. Indeed, with media to build cognitive understanding, the need for formalized learning system design process is reduced. Instead of specialists like designers learners shaping student learning through determined communication and interaction, media and technology provided directly to learners to use to represent and express what they know. Students themselves serve as designers using media and technology as tools for analyzing the world, accessing and interpreting information, organizing their personal knowledge, and represent what they know to others.

In recent years, learning theory has gone through what can be called "Paradigm shift." Constructivist learning theory gradually gets a sense of the same respect and long attention given to the theory of behavioral learning (Duffy & Jonassen, 1992). Constructivism concerns the process of how students are creating meaning and knowledge in the world as well as the result of constructive processes.

Media to optimize cognitive processes based on constructivism perspective offers a new paradigm in the world of learning. As the cornerstone of the learning paradigm, constructivism calls for the need for active participation of students in the learning process, the need for independent student development, and the need for students to have the ability to develop his own knowledge. The point is that with forming an environment where creative thinking about content is combined with real-world tasks, students learn content, enjoy learning, and recognize that they have created something of value.

To change students can learn "from" the media, to learn "with" Media based on the perspective of constructivism, it is necessary to find and improve the media design that exist today, so as to contribute in optimizing students' cognitive processes. As for the characteristic of the media is as follows:

1. GOOD, which is a good learning medium is to create an environment where creative thinking about content is combined with real-world work, students learn content, enjoy learning, and recognize that they have created something of value.
2. EFFECTIVE, which is learning media will have the greatest effectiveness when applied in a constructivist learning environment. The media to optimize cognitive processes based on constructivism perspectives can empower teachers to design their own knowledge.
3. NATURAL, namely: Designing learning materials that tell the natural phenomenon or environment.
4. INNOVATION, an innovative strategy of project-based learning.
5. TECHNOLOGY, which is a technology-based learning media.

To easily remember it, the author gave the Media Genit name which is the acronym of Good - Effective - Natural - Innovation - Technology.

Conclusion

The students can learn "from" the media, and they can learn "with" media, but these two approaches only make the task easier but do not activate and facilitate them for critical thinking and higher learning. Media Genit offers a new approach to the world of learning, which is transforming students from learning "from" the media, to learning "with" media-based perspective constructivism, that is, shaping an environment, where students as creative content designers are combined with real-world tasks, student learning contents, enjoying the learning process, and recognize that they have created something of value.

References

Arsyad, A. (2013). *Media Pembelajaran*. Jakarta : PT Raja Grafindo.

Munadi, Y. (2008). *Media Pembelajaran: sebuah pendekatan baru*. Jakarta: Gaung Persada Press.

Smaldino, S. E. et al (2011). *Instructional Technology & Media For Learning Teknologi Pembelajaran dan Media Untuk Belajar. Edisi Kesembilan*. Jakarta : Kencana Prenada Media Group.

Rusman. (2012). *Belajar dan Pembelajaran Berbasis Komputer, Mengembangkan Profesionalisme Guru Abad 21*. Bandung : Alafabeta

Rusman, dkk. 2012. *Pembelajaran Berbasis Teknologi Informasi dan Komunikasi: mengembangkan profesionalitas guru*. Jakarta: Raja Grafindo Persada.

Johnson, E. B. (2007). *Contextual Teaching and Learning*. Bandung: MLC.

Schunk, D. H. (2012). *Learning Theories and Educational Perspective*, Yogyakarta: Pustaka Pelajar

David, J. L. (1996). Developing and spreading accomplished teaching: Policy lessons from a unique partnership. In C. Fisher, D. C. Dwyer, & K. Yocam (Eds.), *Education and technology: Reflections on computing in the classroom* (pp. 237-250). San Francisco, CA: Jossey-Bass.

Jonassen, D. H. et al (1996). Learning with technology: Using computers as cognitive tools. In D. H. Jonassen (Ed.), *Handbook of research for educational communications and technology* (pp. 693-719). New York: Macmillan.

Hannafin, M. J. et al (1996). Research on and research with emerging technologies. In D. H. Jonassen (Ed.), *Handbook of research for educational communications and technology* (pp. 378-402). New York: Macmillan.

Alkin, M. C. (Ed.). (1992). *Encyclopedia of educational research (6th Ed.)*. New York: Macmillan.

Salomon, G. (1992). New information technologies in education. In M. C. Alkin (Ed.), *Encyclopedia of educational research (6th Ed.)* (pp. 892-903). New York: Macmillan.

Krendl, K. A. et al (1996). Learning by any other name: Communication research traditions in learning and media. In D. H. Jonassen (Ed.), *Handbook of research for educational communications and technology* (pp. 93-111). New York: Macmillan.

Derry, S. J. et al (1993). A middle camp for (un) intelligent instructional computing: An introduction. In S. P. Lajoie & S. J. Derry (Eds.), *Computers as cognitive tools* (pp. 1-11). Hillsdale, NJ: Lawrence Erlbaum.

Greenfield, P. M. (1984). *Mind and media: The effects of television, video games, and computers*. Cambridge, MA: Harvard University Press.

Vocational Aptitude Test

I Made Candiasa[1,*], *Nyoman* Natajaya[2] , and *Ketut* Widiartini[3]

[1]Universitas Pendidikan Ganesha, Singaraja, Indonesia
[2]Universitas Pendidikan Ganesha, Singaraja, Indonesia
[3]Universitas Pendidikan Ganesha, Singaraja, Indonesia

Abstract. Test for measuring vocational aptitude has been formulated and validated. There are three main constructs involved in vocational aptitude test, which are individual characteristics, activities that are likely to be selected, and professions that tend to be idolized. Individual characteristics indicate the individuals talents, whereas the activity that tends to be chosen leads to student interest in the activity, and the intended profession gives clues about the capability of themselves to pursue the profession. Content validity test with Lawse technique yields content validity ratio (CVR) for all items are in the range 0.82-0.94 and content validity index (CVI) = 0.88. The construct validity test yields comparative fit index (CFI) = 0.918 and chi square coefficient ($\chi 2$) = 5.85 with significance (p) = 0.002. These findings indicate that the test is valid either by content or construct. Furthermore, the reliability test with Alpha Cronbach found the alpha coefficient (α) = 0.82. Finally, it can be concluded that vocational aptitude test can be utilized for early identification of student vocational aptitude. The hope, the test can help students to choose the appropriate vocational school, in order to obtain the better learning outcomes. Keywords: vocational aptitude test, individual characteristics, activities, professions

1 Introduction

Vocational Schools are being encouraged to produce graduates ready to work in various sectors. When viewed from the angle of a graduate, then all parties hope that the resulting school graduates have an optimum competence in their respective fields. On the other hand, when viewed from resources, all parties hope that the resources owned can be used effectively and efficiently to support the learning process. This means that prospective students who entered a vocational school have the capacity to learn and choose the program that suits their aptitude each. Scientific aptitude is a major determiner of science learning [1]. By taking analogy in science, prospective students of vocational schools need to know their respective aptitude in order to choose the right program. Parents are also very important to know the aptitude of their son or daughter in order to give the consideration of correct program will be selected. Experience in the field shows that not many prospective students considering vocational aptitude in selecting a program. Prospective students choose programs that crossed his mind or simply following the advice of friends or other parties without consideration. When this is the case then the prospective students are likely

* Corresponding author: mailto:candiasaimade@yahoo.co.id

to have problems in their learning process. Even if the aptitude is not the only determinants of success of learning in vocational schools, aptitude still is the dominant factor. Motivation, creativity, and curiosity students will increase further when they choose program in accordance with his aptitude.

The aptitude is innate ability to learn specific areas quickly and easily [1, 2]. In this context does not mean successful people on a field solely because of his aptitude, because in reality all successful people always make the effort. It's just for people who are his works in accordance with his aptitude, work done would be nice, so that their work does not feel as forced work. Scarred scientists will always be interested in finding the problems need to be solved accordingly. Scarred musicians will enjoy in learning music. Meanwhile, scarred computer engineers will be working on computer repair in the laboratory with a zealous and feeling happy. Individual aptitude refers to how fast or how easy that individual can learn in the future [3, 4, 5]. Aptitude is measured with a test of aptitude. The Aptitude tests is more specific and more limited compared to the intelligence test [6]. Intelligence tests concentrate on abstract functions, include the use of verbal symbols or numeric, therefore more specific interests, relating to the ability of a more concrete or practical are forgotten. The gap is resolved by a test of aptitude. Single aptitude tests only measure a single domain capabilities, such as the mechanical aptitude test, aptitude test music, or a test of numerical aptitude. Multiple aptitude tests generate a score for some area of different capabilities, such as the Scholastics Aptitude Test (SAT), Differential Aptitude Test (DAT), and Multiple Intelligences Test-based on Howard Gardner Theory.

The importance of vocational school selection and counseling service spurred the development of aptitude test to measure mechanical aptitude, administration aptitude, musical aptitude, as well as aptitude in other fields. Even for the selection of employees in a particular field still requires an aptitude test more specifically, such as tests of auditory, observation, and agility. For instance, the selection of employees for the industry or the selection of members of the military requires aptitude of observation, hearing, and agility. Map of the program and aptitude pattern of each individual has given a strong impetus to the development of the aptitude test. Similarly, the map between job vacancies with the aptitude pattern of each individual has provided a powerful impetus to the development of the aptitude test. Individuals who have mechanical aptitude are directed to proceed to the field of mechanical vocational school. Individuals who have musical talent is advised to study in the field of music. Similarly, individuals who have a strong clerical talent placed on matters that much to do with the field of administration. The aptitude in this study is restricted to the vocational aptitude expected to help prospective students choose the program in vocational school.

Vocational aptitude tests developed into a multiple aptitude tests will be classifying vocational aptitude into four fields, namely engineering, administration, information, and health. Theoritical study identified three dimensions of vocational aptitude, i.e. individual characteristics, activities that are likely to follow, and the profession that tends to idolized. Individual characteristics indicates the basic potential belonging to that individual. Activities that are likely to follow activity that tends to be chosen leads to student interest in the activity, and the intended profession gives clues about the capability of themselves to pursue the profession. The classification of aptitude in this research are determined based on multiple intelligence theory from Gardner and cognitive style theory from Witkin.

Gardner [7] sort aptitude into linguistic, logical-mathematical, musical, bodily kinesthetic, visual-spatial, interpersonal, and intrapersonal. The main indicator of linguistic talent was words and language. Activities that are able to do interpretation and explanation of ideas and information by using language, and understands relationship between communication and meaning. Professions that tends to be choosen include publict relation, presenter, media consultants. Main indicator of the logic aptitude involve number and logic.

Competence that tend to be owned are detecting patterns, perform mathematical calculations, scientific reasoning, and problems analyses. As an instance, strong numerical knowledge helped the students to achieve higher scores in programming [8]. The profession tends to be chosen include engineers, accountants, computer, bookmakers, analyser. Music, sounds, and rhythm is a music aptitude indicators. Activities that are likely to be able to do such as use of sound, understands relationship between sound and feeling. A profession that tends to be chosen are singer, composer, entertainers, party-planners.

Bodily-Kinesthetic aptitude main indicators there are in physical movement. Activities that are likely to be able to do is to involve the physical agility, dexterity, and balance. Pictures, shapes, images, plane trees, and space is the primary indicator of visual-spatial talent. Activities that are likely to be able to do is interpretation pictures and images, imagination and expression of space. A profession that tends to selected designers, architects, photographers, sculptors, engineers, beauty consultants. Communications and teamwork or cooperation is the main indicator of interpersonal talents. Activities that are able to be done includes the perception of pepople behavior, ability to relate to others. Professions that tends to be done include human relation, therapists, counsellors, politicians, doctors, healers. Self-reflection, self-discovery is the primary indicator of intrapersonal aptitude. Activities that are likely to be followed is self-awareness, the capability to understand oneself, one's relationship to others and the world. The work tends to be chosen is self-actualization.

Cognitive style is characteristic of the individual in feel, remember, think, solve problems, and make informed decisions [9]. Witkin et.al. [10] sorting out cognitive style into two polar inclination, i.e. the field independent and field dependent. Some individual characteristics of field independent cognitive style are: 1) has the ability to analyze to separate the object from the surrounding environment, so his perception is not affected when the environment changes; 2) had the ability to organize objects that have not been organized and reorganized the objects that already organized; 3) has the impersonal orientation, so that the less sensitive, cold, keep the distance with the other individual, and individualistic; 4) tend to define their own goals, and 6) tend to work concerned with intrinsic motivation [10, 13]. Job title tends to be favoured by field independent individual is work that prioritizes theory and analysis, as well as allowing to do independently, among others, include the fields of mathematics, natural sciences, medicine, or technology.

Some characteristics of field dependent individuals are: 1) tend to think global, looking at objects as a single unit with its surroundings, so that its perception is easily affected by changes in the environment; 2) tend to accept the existing structures due to lacking capability restructuring; 3) has a social orientation, so it looks more friendly, kind, thoughtful, considerate, and affectionate towards another individual; 4) tend to follow the existing goals, 5) tend to work with emphasis on external motivation [10, 13]. The profession tends to be selected is a profession that places emphasis on social skills, such as social workers, publict relations, extension officers, trademark, advertising, administration, or politics.

Many researches have been done related to cognitive style and it's impact in individual activity, especially in learning. Hansen found that the selection of a major was influenced by cognitive styles [11]. Not differ so far Hansen, Salameh found that there were statistically significant differences in the field-independent cognitive style among academic specialization [12]. Research by Davis resulted that subjects with academic backgrounds in the physical sciences were more field independent [13]. In addition, engineering students tend to field independent. It proved by research of Zimmerman et.al. that found where three groups engineering students indicated a strong preference for the field-independent learning style [14]. In the area of computer programming, Mancy and Reid come to conclusion that in introductory programming courses, field dependency is an important factor in determining success [15]. Futhermore, it is also found that cognitive style of language learners may affect the language performance of individuals [16]. In searching information, Santos Jr. et.al. found that analytic group tends to focus on documents with significantly more specific information than the wholist group [17]. In science process skills field independence were more successful than field dependent students [18].

Individual characteristics, activities that are likely to follow, as well as a profession that tends to favoured are used as dimensions of vocational aptitude. There are four vocational program identified, namely, techniques, administration, relationships, and health. Techiques includes computer and electrical engineering, automotive, and building. The program of administration includes, management, accounting, and cooperation. The program of relation includes tourism services, such front office or guide, production of food, and fashion. Health program covers health care services.

Program of production technique supported by logical-mathematical, visual-spatial, bodily-kinesthetic, and strong independent cognitive style. The field of administration supported by logical-mathematical, linguistic, interpersonal, relatively balanced field independent and field dependent cognitive style. The program of relation supported by linguistic, music, interpersonal, and strong filed dependent cognitive style. Health program supported by visual-spatial, interpersonal, bodily-kinestetic, and relatively balanced field independent and field dependent cognitive style.

2 Method

Vocational aptitude indicators that have been identified from the review of the theory developed into a test comprising 35 items each with five options. The test has been validated either by content or construct validity. The validity of the content tested qualitatively and quantitatively using Lawse technique. The construct validity is done by using factor analises. Finally the reliability of test is counted by using Alpha Cronbach reliability technique. Validated vocational aptitude test has been tried out its effectiveness by trying to apply it on the vocational school students with various program. The test results are analyzed descriptively.

3 Results and Discussion

The developed of vocational aptitude test consisted of 35 items. The content validity of the test with qualitative approach done by seven experts to get 29 items worth using without revision and 6 items worth using with revision. After the revision, content validity is tested with quantitative approach by using Lawse technique to get content validity ratio (CVR) for all items are in the range 0.82-0.94 and content validity index (CVI) = 0.88. The results of content validity test show that the test developed feasible to be tested further. The vocational aptitude test then tested the validity of the construct. The tests were tested empirically by involving the vocational school students as respondents. The test results were analyzed by factor analysis. The first iteration yields the Kaiser-Meyer-Olkin coefficient (KMO) shown on the KMO and Bartlett's Test table of 0.512, with a significance of 0.08. Since the KMO is less than 0.5 and the significance of more than 0.05 means that the analysis can not be continueed. Therefore, item that has the smallest Measure of Sample Adequacy (MSA) coefficient in the Anti-image Matrices table is elemited. Furthermore factor analysis was repeated and resulted in KMO coefficient of 0.628 with a significance of 0.01. The KMO coefficient has exceeded 0.5 with significance less than 0.05, so factor analysis can be proceed further.

The MSA coefficient in the Anti-image Matrices Table for each variable indicates that there are two items having MSA coefficients less than 0.5. In order for further analysis to be carried out, the MSA coefficients for each variable should be more than 0.5. Therefore, the two items must be eliminated, and re-analyzed. Factor analysis of 32 items is done because three items had been eliminated. After re-analysis, there was an increase of KMO's efficiency to 0.918 with a significance of 0.00. The MSA coefficients for each variable are all more than 0.5. Thus, the analysis can be continued.

Table Total Variance Explained shows there are seven factors that are formed. Varimax rotation results show that the first factor consists of five items, the second factor consists of three items, the third factor consists of five items, the fourth factor consists of four items, the fifth factor consists of five items, the sixth factor consists of six items, and the seventh factor consists of eight items. The first, second, fourth and fifth factors are indicators of individual characteristics. Meanwhile, the third sixth factors become an indicator of activities that tend to be followed. The seventh factor is a indicator of the preferred profession.

The above results show that the developed vocational aptitude test consists of three dimensions, namely individual characteristics, activities that tend to be selected, and the idolized profession. The dimensions of individual characteristics consist of three indicators, namely numerical ability, analytical skills, visual ability, and communication skills (language). The dimensions of activity that tend to follow consist of two indicators, namely formal activities and social activities. The dimensions of the idolized profession consist of only one indicator, the profession itself.

The constructs already found are tested by factor analysis. The test results obtained comparative fit index (CFI) = 0.918 and chi square coefficient ($\chi 2$) = 5.85 with significance (p) = 0.002. These findings indicate that the construct of the developed aptitude test is valid. Finally, reliability of test was counted by using Alpha Cronbach technique where Alpha coefficient (α) equal to 0.82. This means that vocational aptitude tests developed are reliable, so as to provide constant results.

Vocational aptitude test was also tried out on vocational school students' samples from various programs. The results show that the fit of the aptitude test result with the program chosen by the students is only 68%. Of the 32% of students whose aptitude did not match the chosen program, 78% were aware of the incompatibility, while the remaining 22% were unaware of the incompatibility. The results of the limited interview to 32% of students who aware a mismatch of aptitude test results with the selected program stated that 58% of them chose the program because of following friends or suggestions of others, including parents; 26% of them chose the program because it was easier to get a job; and the remaining about 16% did not say anything.

4 Conclusion

Vocational aptitude test has been tried to develop in order to give consideration to prospective vocational school students about their aptitude, so they can choose programs that match their aptitude. The vocational aptitude test developed consisted of 32 items representing three dimensions, namely individual characteristics, activities that tend to be selected, and idolized professions. The dimensions of individual characteristics consist of three indicators, namely numerical ability, analytical skills, visual ability, and communication skills (language). The dimensions of activity that tend to follow consist of two indicators, namely formal activities and social activities. The dimensions of the idolized profession consist of only one indicator, the profession itself.

The developed vocational aptitude test is sufficiently feasible to use, both content and construct. Content validity test with Lawse technique found content validity ratio (CVR) for all items are in the range 0.82-0.94 and content validity index (CVI) = 0.88. Factor analysis got comparative fit index (CFI) = 0.918 and chi square coefficient ($\chi 2$) = 5.85 with significance (p) = 0.002. In addition, the test is also quite reliable. Test reliability analysis with Alpha Cronbach technique obtained Alpha coefficient (α) of 0.82.

Trial of the test at vocational student samples from several programs indicates that the fit of the test result with the program selected by the student is 68%. There are many arguments against it. Nevertheless, efforts to refine the already developed aptitude test need to be pursued, for example by the addition of indicators that are viewed more carefully giving clues to students' aptitude. It is hoped that the developed vocational aptitude test can really give consideration to the vocational school students' prospectives to choose the right program with the aptitude. Thus the learning process will take place more meaningfully and learning resources can be used more effectively and efficiently.

References

1. Rajib Mukhopadhyay, "Scientific Aptitude – Some Psychometric Considerations with Special Emphasis to Aptitude in Physics", Educationia Confab. Vol. 2, No. 1, January 2013
2. Cindy Rosner, Christine McMullin, Kishma Patnaik, Doug Hastings, Krista Hamrin, Anne Steiner. *Understanding Your Aptitude*. Boston: Johnson O'Connor Research Foundation, Inc., 2012.
3. Philip Carter, *IQ and Aptitude Tests*, London: Kogan Page, 2007.
4. Lewis R. Aiken, *Psychological Testing and Assessment*. Boston: Allyn and Bacon Inc., 1988.
5. Lee J. Cronbach, *Essentials of Psychological Testing*, New York: Harper & Row Publishers, 1984.
6. Robert J. Gregory, *Psychological Testing: History, Principles, and Applications*, Boston: Allyn and Bacon, 2000.
7. Howard Gardner, *Frames of Mind: The Theory of Multiple Intelligences,* New York: Basic Books, 1983.
8. V. Martin, Maria Antoniate, T. Lucia Agnes Beena, "Prediction of Association among Numerical Aptitude, Programming Skills, Trait Emotional Intelligence on Students Performance", *International Journal on Computer Science and Engineering (IJCSE)*, Vol. 4, No. 09, Sep 2012.
9. Ok-choon Park, "Adaptive Instructional Systems," Handbook of Research for Educational Communications and Technology, ed. David H. Jonassen, New York: Simon & Schuster Macmillan, 1996.
10. H. A. Witkin, P. K. Oltman, E. Raskin, dan S.A. Karp, " "Field-Dependent and Field-Independent Cognitive Style and Their Educational Implications", Review of Educational Research, Vol. 47, 1977.
11. John W. Hansen, "Cognitive Styles in Postsecondary Technology Programs", Journal of Technology Education Vol. 6 No. 2, Spring 1995.
12. Emad M. Al-Salameh, "A Study of Al-Balqa' Applied University Students Cognitive Style", International Education Studies Vol. 4, No. 3; August 2011.
13. Gregory A. Davis, "Learning Style and Personality Type Preferences of Community Development Extension Educators", Journal of Agricultural Education, Volume 47, Number 1, 2006.
14. A. P. Zimmerman, R. G. Johnson, T. S. Hoover, J. W. Hilton, P. H. Heinemann, D. R. Buckmaster, "Comparison of Personality Types and Learning Styles of Engineering Students, Agricultutural Systems Management Students, and Faculty in an Agricultural and Biological Engineering Department", American Society of Agricultural and Biological Engineers, Vol. 49(1), 2006.

15. Rebecca Mancy, Norman Reid, "Aspects of Cognitive Style and Programming", 16th Workshop of the Psychology of Programming Interest Group. Carlow, Ireland, April 2004.
16. Kamal Heidari Soureshjani, "Cognitive Styles on C-Test and Cloze-Elide Test: Which Style Acts Better? ", Language Testing in Asia Volume two, Issue two, May 2012.
17. Eugene Santos, Jr., Hien Nguyen, Fei Yu, Deqing Li, John T. Wilkinson, "Impacts of Analysts' Cognitive Styles on the Analytic Process", IEEE/WIC/ACM International Conference on Web Intelligence and Intelligent Agent Technology, 2010.
18. Mehmet Mutlu and Burak Kagan Temiz, "Science process skills of students having field dependent and field independent cognitive styles", Academic Journal, Vol.8(11), 2013.

The Implementation Of Character Education Values In Integrated Physical Education Subject In Elementary School

Ayi Suherman[1*]

Program Studi PSD Penjas, Universitas Pendidikan Indonesia Kampus Sumedang, Jl. Mayor Abdurachman No. 211 Sumedang, Jawa Barat, Indonesia

Abstract. The issue of this research emphasizes on the implementation of character building values through physical education learning in elementary school. The effort in developing this character building practice is essential to be done in order to tackle moral and character crises, which already occur in both individual and collective levels reflected in educational institution from elementary school to higher education. Hence, to form culture and national character, educational program and process are inseparable from environmental factor including the values of society, culture, and humanity. Physical education subject that is based on 2013 Curriculum has significant difference compared to the previous physical education subject. This is due to the fact that integrated physical education has its own uniqueness in terms of planning, systematic implementation, and instructional medium. This research aims at producing guidance in implementing character values integrated in physical education in elementary school. The method used in this research is research and development (R&D) method, which includes preliminary research, model designing, limited trial, and extensive trial, as well as validation and dissemination. The findings of the research show that character values can be implemented in physical education in elementary schools in Sumedang Regency.
Keywords: Character Values Implementation, Integrated Physical Education, and Character Building.

1. Introduction

One of the causes of problems in the educational field is the lack of education that can develop intrinsic values of humanity, since "intellectual education" and "education of values" are considered separated. This dichotomy shows the weakness in planning because moral values are not in a different dimension with intellectual values; in fact, it is the foundation of the intellectual development. As stated by Sapriya, [1] this kind of program planning is the main source of moral and character crisis among students, graduates, teachers, even education management. It is undeniable that there is a degradation of national identity values in our younger generation, which is caused by the lack of character education. Whereas, education character is the transformation vehicle of culture, value, science and technology, art and character education, which has been the center of the development of national character in both formal and informal education.

* Corresponding author: ayisuherman1960@gmail.com

Based on the writer's observation, nowadays, character education is generally executed only in classrooms. This somehow burdens the students to develop, since what they actually need are freedom to move and higher thinking skills, which need to be actualized without pressure. It is believed that character education implementation done in schools will be more effective if it is integrated with various subjects, not excluded into one single subject.

To overcome the problems of character education, a learning strategy is needed in building the students' character, since it has to be positioned as an effort to develop a strong character. Thus, every education unit has to have their own character building programs integrated with all intracurricular and extracurricular activities.

According to Lickona [2], character education is a potent medium to motivate democratic living, which becomes the reflection of human character grown in each individual in the context of collective living. In a democratic society, every person has a moral commitment in terms of living together, such as respecting their own rights and the rights of others, following the norms and rules, participating in togetherness, believing in the benefits of living together, and so on.

In relation to the aforementioned issue, one of the teachers' strategies to implement character education to the students is through physical education that is integrated with various activities in other subjects. The best implementation of character education for the students will have an impact on the expected development of national character values.

This research has developed nine character values from 18 characters developed in the elementary schools, namely creativity, independence, democracy, appreciation of achievement, friendship/communication, peace, reading interest, social care, and responsibility. This character implementation will be carried out in elementary school.

According to Depdiknas [3], character can be interpreted as a good individual quality, meaning that an individual knows virtues, has willingness to do good deeds, and is well behaved, as the result of intellectual development (*olah pikir*), spiritual and emotional development (*olah hati*), physical development (*olah raga),* affective and creativity development (*olah rasa dan karsa*). Thus, research on the implementation of character values in fun physical education in elementary school can be one of the efforts to build national character.

Albertus [4] provides examples on the characters classification. Character values implemented to the students cannot proceed if they are not maintained properly. In order to build a more effective character education, there are three basic approaches that need to be embedded, namely school-based education (class), school culture, and community. In-class character education is integrated with the curriculum that consists of kinds of quality the teachers need to possess, so a special subject for character education is not needed.

To reach the objective, a whole and thorough character education is mandatory. According to Lickona[2], character education is a potent medium to motivate democratic living, which is the reflection of human character grown in each individual in the context of collective living.

Linda [5] explains that those values are 'values of being" and "values of giving". *Values of being* are values embedded within the self that develop into actions and how we treat others, for example: being honest, kind, peaceful, dependable, discipline, and in the limit. *Values of giving* are values that need to be practiced or given; with consideration that what will be received is equal to what will be given, for example: fidelity, trust, respect, love, affection, sensitivity, unselfishness, virtues, friendliness, justice, and generosity.

Nurgiyantoro [6] defines physical education at school as a part of thorough education that includes character education, which strives to implement good behavior so that children will be able to understand (cognitive), feel and make choices (affective), and implement them to their behavior on a daily basis (psychomotor). Similarly, Suryanto [7] believes that through the implementation of character education, children will be more sensitive and reflective to the sense of humanity, to themselves, to their surroundings, to others, and to God.

2. Methods

This research refers to the research and development procedure as explained by Borg and Gall [8]. The systematic steps of research and development in implementing character education values in the physical education subject in elementary school are:

1. Doing an preliminary research, which means gathering information that includes literary review regarding the effort of implementing character values integrated in physical education subject;
2. Composing the first draft of the implementation of character values integrated in physical education subject;
3. Conducting trials that include limited and extensive trials focusing on the implementation of values integrated in physical education subject; and
4. Conducting validation and dissemination tests.

The subjects of this research were all elementary schools in Sumedang Regency. To determine which school would be the sample of this research, cluster sampling technique was employed. Based on the sampling, there are seven elementary schools in 26 districts in Sumedang Regency appointed as samples that represent each district.

The data collection was conducted using the following instruments: observation, interview, questionnaire, and test. Based on the data collection technique, qualitative and quantitative data were gathered. Qualitative data were analyzed using descriptive analysis technique, which employed sustainably based on the achieved data by Maleong [9]. As for quantitative data, the data were analyzed by using statistical analysis approach, or T-test analysis technique, in which the data processing used SPSS version 21 software on a computer.

3. Result and Discussion

The result of introductory research finds out that a majority of teachers are not used to implement character values integrated in physical education subject. In fact, several previous research reveal that form of learning model of Curriculum 2013 has not been implemented yet. This is caused by three factors: First, elementary school teachers have not received training on the implementation of character values integrated in physical education subject. Second, 2013 Curriculum will be implemented in 2016/2017 Academic Year, so that the implementation of character values is still scattered or not yet integrated. Third, the teacher forum has not been able to develop characters in physical education for elementary schools students, since the numbers of qualified trainer are still limited.

The planning stage of designing the implementation of character values integrated in physical education subject in elementary school consists of the following syntaxes. First analyzing students' needs based on the implementation of character values. Second, adding activities filled with character values to several learning activities in an integrated manner. Third, evaluating character-based learning that will be added or reduced based on the students' needs.

The results of implementing character values in integrated physical education subject can be drawn from a limited test conducted at SDN Panyingkiran III in Sumedang Regency. The instrument used in this research was a questionnaire consisting 10 questions, which are related to character values programmed in 2013 Curriculum such as independence, discipline, responsibility, honesty, cooperation, spirituality, communication, mutual care, creativity, and peacefulness. There are 10 questions used as the data collecting instrument, which have five alternative answers: never (1) rarely (2) sometimes (3) often (4) very often (5) regularly.

Based on the above respondents' distribution answers, it can be identified that there were 34 students who are observed by one physical education teacher. From the limited test, ten character behaviors can be seen in the questionnaire's average result on this following table.

Table 1: The Result of Students' Character Questionnaire

Number of question	1	2	3	4	5	6	7	8	9	10
Frequency	66	65	64	66	62	62	68	66	66	59
Percentag e	94.2 9	92.8 6	91.4 3	94.2 9	88.5 7	88.5 7	97.1 4	94.2 9	94.2 9	84.2 9

The table above shows that the average percentage of the questionnaire result was 92% which can be categorized as excellent. The highest percentage of the questionnaire average result was derived from the question number 7 that is about the students' communication skill, and the percentage result was 97.14%. Meanwhile, the lowest percentage was derived from the question numbers 5 and 6 that is regarding the cooperation skill and sincere praying, and the percentage as 88.57%. Furthermore, it can be concluded that character values have been integrated in the subjects, but there is still a gap between subjects and character values.

Based on the aforementioned data, it can be identified that the character values integrated in physical education can be implemented for students in elementary schools in Sumedang Regency. In line with character education model, Kesuma [10] explains that the main principles of implementing character values integrated in physical education subject that need to be put forward are: (1) the foundation of teacher-student interaction in the learning process is affection, (2) teachers' attitude and behavior has to reflect the believed values, (3) teachers see their students as developing subjects in which their development is related to the teachers' role. Hereinafter, coaching and implementation of this character model can be done through designing a lesson plan (*Rencana Pelaksanaan Pembelajaran* (RPP), exploring, elaborating, and confirming the meaning of the materials discussed as well as evaluating the observation which will be reported to the school community for further follow-ups Masitoh [11].

The implementation of character values integrated in physical education subject in elementary school is beneficial for the students to reduce the teachers' domination, which oftentimes imposes their desire in teaching academic-oriented subjects while disregarding psychological aspects that are considered as intrinsic needs of elementary school students. One of the fundamental problems in the educational process in elementary school is that there are some faults done by teachers in treating their students in the learning activity by Agustin[12]. With curriculum implementation training program conducted by Departement of Education for teachers, especially the teachers in elementary schools in Sumedang Regency, it is expected to boost the teachers' encouragement in seeing the importance of learning essence is in elementary school.

The results of the research show that there is a difference in the subjects' implementation in the area of city center, districts, and rural areas. This is quite reasonable because there are variations and demands in each region, especially the presence of teachers. This statement is supported by the research conducted by Haenilah [13] which says that, "teachers as educators whose sole responsibility is to teach, has personality characteristics which are influential to the success of human resources' development". Teachers' personality is an important factor in determining the level of success of the students' learning process. The above research result is in accordance with Syah [14] who asserts that the aforementioned personality will be the one that determines whether a teacher will be a good educator and coach for their students or the one who destroy their future, especially elementary school students and those who are still in pursuit of their sense of identity.

The result of the data analysis shows that the actualization of the implementation of character values in integrated physical education subject varies at each elementary school. However, they all give significant influence towards the elementary school students' personal skills. This statement is supported by the result of a research conducted by Haryani [15] which says that children who have a lower level of quality character will shows low social emotional, so that the children might experience some difficulties in learning, having social interaction, and controlling themselves.

Based on the school's policy in implementing character education, Lickona [2] accurately says that the strategies of the implementation of character education at school has to take an integrated path. First, integrating character education content formulated into all subjects. Second, integrating character education into daily school activities. Third, integrating the haracter education into structured or planned activities. Fourth, building communication and cooperation between schools and parents.

The implementation of character education in elementary school refers to tens character quality pillars by Megawangi [16], they are: love to God and His creations, responsibility, discipline and independence, being honest, trustworthy, and wise, respect and hospitality, mutual cooperation and confidence, creativity and hard work, leadership and justice, being kind, humble, and tolerant, as well as peace and unity. Character education is in line with students' behaviors, particularly their personal skills and the students generally will have positive behavior as the result of studying the characters. According to Surya [17], the characteristic features of behaviors based on the result of studying is intentional, positive, and effective. The change of behavior is intentional, meaning that experience or practice is intentionally and consciously done and it does not happen accidentally. The change of behavior is effective, meaning that the change of behavior remain relative and if it is needed, it can be reproduced and used, like in problem solving, tests, or daily lives in order to sustain its survival.

4. Conclusions

The result of the research on the implementation of character values integrated in physical education subject in elementary school students shows that: First, seen from the teachers as curriculum developers aspect, character education that is currently implemented has only been executed in the forms of direction and advice, and has not yet integrated in the learning scheme. Most teachers are still confused to conduct character education due to the fact that the implementation of 2013 Curriculum that consists of character education will only be enacted in 2016/2017 Academic Year. Second, the students are not used to the character education presented in a learning model, as it is usually, showcased by the teachers. Third, the effectivity of implementing character values into integrated physical education. It can be concluded that character education for elementary school students does not only come in the form of characters that are programmed in the lesson plans. However, the teachers must begin it with good examples done both in the classroom and daily lives.

References

1. Sapriya. "*Perspektif Pemikiran Pakar tentang Pendidikan Kwarganegaraan dalam Pembangunan Bangsa (sebuah Kajian Konseptual Filosofis Pendidikan Kewarganegaraan dalam Konsteks Pendidikan IPS*". Disertasi tidak Dipublikasikan, PPS UPI Bandung.(2007).

2. Lickona, Tom. *Eleven Principles of Effective Character Education.*Journal of moral Education. 25:93-100. (1996).
3. Depdiknas. *Grand Design Program Pendidikan Anak Usia Dini Nonformal* Tahun 2007-2015. Jakarta: Direktorat PAUD Dirjen PLS. (2010)
4. Albertus, D.K. *Mengembangkan Kultur Akademis Bagi Pembentukan Karakter.* Makalah disampaikan dalam Konfrensi Nasional dan Workshop Asosiasi Psikologi Pendidikan Indonesia. Peran Pendidikan dalam Pembangunan Karakter Bangsa. Malang: Program Studi Psikologi UM. (2010)
5. Linda. N. Eyre. *Teaching Your Children Values.* New York: Simon and Chuster. (1995)
6. Nurgiyantoro, Burhan. *"Implementasi Pendidikan Karakter dalam Bahan Ajar".* Makalah Disajikan dalam Stadium Generale di Jurusan FBS FKIP Universitas Sebelas Maret. Surakarta. 27 Nopember 2011. (2011)
7. Suryanto, Edi. *Pendidikan Budi Pekerti Berbasis Cerita Anak Melalui Penanaman Nilai Etis Spritual di SD.* Jurnal Sekolah Dasar. No. 1: (2015).
8. Borg, Wolter R., Gall Meredith Damien. *Educational Research an Introduction (Third Edition).* New York. Longman Inc. (2003).
9. Maleong, Lexy J. *Metodologi Penelitian Kualitatif. Bandung.*Remaja Rosdakarya. (2006)
10. Kesuma, D dan Salimi, M. *Kajian Teori dan Implementasi Pendidikan Karakter di Sekolah (Sebuah Pengalaman Pusat Pengkajian Pedagogik).* Prosiding Seminar Internasional Prodi Pendas SPS UPI Bandung. (2012)
11. Masitoh, dkk. *Strategi Pembelajaran di Taman Kanak-Kanak.* Jakarta: Universitas Terbuka. (2005)
12. Agustin, M. *Deteksi Dini dan Stimulasi Kecerdasan Jamak Anak Usia Taman Kanak-Kanak.* Prosiding Seminar Internasional Penguatan Pendidikan Karakter Pada Jenjang Pendidikan Dasar di Era Global. Prodi Pendas SPS UPI. (2012)
13. Haenilah, E. *Analisis Kebutuhan Profesionalisme Guru Sekolah Dasar Pasca Sertifikasi.* Jurnal Sekolah Dasar. No.2: 99-198. (2015).
14. Syah, Muhibin. *Psikologi Pendidikan dengan Pendekatan Baru.* Bandung: PT. Remaja Rosdakarya. (2000)
15. Haryani. *Pendidikan Karakter Melalui Bimbingan dan Konseling Bagi Siswa Sekolah Dasar.* Prosiding Seminar Internasional PenguatanPendidikan Karakter Pada Jenjang Pendidikan Dasar di Era Global. Prodi Pendas SPS UPI. (2012).
16. Megawangi. *Pendidikan karakter; Solusi yang Tepat untuk Membangun Bangsa,* Depok: Indonesia Heritage Foundation. (2003).
17. Surya, Muhammad. *Psikologi Pembelajaran dan Pengajaran.* Bandung Yayasan Bhakti Winaya. (2000).

Communication strategies of English native speaker teacher in the teaching and learning process in a kindergarten

Ni Made Mestriani*, *I Ketut* Seken, and *I Nyoman* Adi Jaya Putra

Universitas Pendidikan Ganesha, Singaraja, Bali, Indonesia

Abstract. In Dyatmika Kindergarten, the students came from different cultural backgrounds and nations. They were taught by English native speaker teacher in English sessions. The students sometimes might understand the intention of the teacher; however, sometimes they might not understand at all. Therefore, the teacher should know how they communicate with the students to avoid misunderstanding. This study aimed at identifying, describing and explaining communication strategies used by English native speaker teacher at Dyatmika Kindergarten during English sessions of the teaching and learning process. This study used descriptive qualitative design, and the data collected through observation and interview. The data were analyzed using taxonomy of communication strategies by Dornyei and Scott (1995). The result indicated there were thirteen communication strategies used by the English native speaker teacher in teaching and learning process. The use of communication strategies by English native speaker teacher was expected to inspire other teachers inside and outside Dyatmika Kindergarten when they communicate with students and make the students understand the message conveyed by the teachers.

Keywords: communication strategies, English native speaker teacher, Dyatmika Kindergarten

1 Introduction

Teaching and learning are two things that cannot be defined apart. According to Brown [1], teaching is a guiding and facilitating learning, enabling learner to learn and setting condition for learning. [1] learning means acquiring knowledge of a subject or a skill by study, experience or instruction. [2] supported that learning serves a basis for future progress in society. The purpose of teaching and learning could be accomplished when the strategy of teachers in communicating in the class could be understandable by the students.

Dyatmika Kindergarten is a bilingual school which is located in Denpasar, Bali, Indonesia. The school have students who come from different cultural background and nations. The English native speaker (ENS) teacher teaches in English sessions. The students have to communicate using English to be able to interact with other students and the teacher. When the students have less English vocabulary, they would have less

* Corresponding author: made.mestri@gmail.com

comprehension of what the teacher expressed during the teaching and learning process. Furthermore, they would be hard to comprehend the message. Thus, the teacher should know how to communicate the message as clear as possible to the students.

English native speaker teacher in Dyatmika Kindergarten owns communicative competence. The teacher tries to reach communication goals in the class that could promote students' English learning. Canale and Swan in [1] mentioned one of communicative competence is strategic competence which means verbal and nonverbal communication strategies that might be called into action to compensate for breakdowns in communications due to performance variables or insufficient competence. [3] added communication strategy is an effort done by individuals to solve the problem when they tried to communicate and they have limited available linguistic resources. Relating to this study, communication strategies were used to keep the communication going well in teaching learning process and to make the students involve in classroom interaction.

[4] conducted research on communication strategies used by English teachers at junior high school in Sukawati, Gianyar, Bali. The result showed the teachers used fourteen communication strategies. They were used to enhance students' speaking abilities during the process of learning. Moreover, [5] conducted a study on communication strategies used by Mathematics and Science teachers to communicate in English in morning sharing at junior high school. The result showed the use of communication strategies by the teachers aimed at keeping the conversation in the class going smoothly.

Related to this study, a preliminary observation was done to see how the English native speaker teacher communicated with bilingual kindergarten students and made them understand in defining teacher's intentions. Sometimes the students could understand easily; however, sometimes they could not comprehend at all. This situation made the teacher should have certain communication strategies to assist them solve the problem in communicating with the students especially during English sessions.

Therefore, this study was conducted to identify, describe and explain communication strategies used by English native speaker teacher at Dyatmika Kindergarten. A classification of communication strategies by Dornyei and Scott in [6] was used as a basis to analyze types of communication strategies used by the teacher. They classified the strategies into three namely direct strategies (message abandonment, message reduction, message replacement, circumlocution, approximation, use all-purpose word, word-coinage, restructuring, literal translation, foreignizing, code switching, use similar sounding words, mumbling, omission, retrieval, self-repair, other-repair, self-rephrasing, over-explicitness, mime), indirect strategies (use of fillers, self-repetition, other-repetition, feigning understanding, verbal strategy makers), and interactional strategies (direct appeal for help, indirect appeal for help, asking for repetition, asking for clarification, asking for confirmation guessing, expressing non-understanding, interpretative summary, comprehension check, own accuracy check, responses).

The communication strategies used in Dyatmika Kindergarten showed how the teacher attempted to communicate clearly with kindergarten students and make them understand the intentions. Consequently, this could make them involve in the lesson and they have meaningful learning. Further, this study result could give inspiration for other teachers in communicating with kindergarten students. It could let the people who do not teach at Dyatmika Kindergarten knowing the way how communication strategies happen in the school and how the strategies could work well for students' understanding of the message. Thus, the communication strategies used by English native speaker teacher at Dyatmika Kindergarten could be a model in teaching kindergarten students.

2 Methods

This qualitative study focused on the use of communication strategies by English native speaker teacher in English sessions at bilingual school. In addition, the main source of the data was English native speaker teacher at Dyatmika Kindergarten B in academic year 2016/2017 to know the communication strategies used in teaching bilingual students.

The data of this study were collected through observation and interview. The observation was done by videoing the teaching and learning activities spesifically when the teacher sat together with the students and they learnt classically on the carpet. Moreover, interview was another method to collect the data in this qualitative study to gain more information related to the use of communication strategies.

Once the teaching and learning activities in English sessions were recorded and transcribed, communication strategies taxonomy of Dornyei and Scott in [6] was used as basis to analyze types of communication strategies used by the teacher. In analysing the data of observation and interview, this study adapted a number of data analysis by Miles and Huberman [7] namely data reduction, data display and conclusion drawing/verification.

3 Results

As shown in Table 1, the English native speaker teacher used thirteen communication strategies proposed by Dornyei and Scott in [6]. There were four direct strategies, three indirect strategies and six interactional strategies.

Table 1. Types of Communication Strategies used by English Native Speaker Teacher

No.	Types of Communication Strategies	Sub-type of Communication Strategies	Motives
1.	Direct Strategies	Code-switching	To make the students understand the intentions
		Self-repair	To tell the students appropriate words to be used in saying something
		Other-repair	To make the students know the appropriate words to be used in telling their intention
		Self-rephrasing	To simplify the message delivered to the students
2.	Indirect Strategies	Use of fillers	To be more friendly to the students and to gain time to think what to do next in the discussion
		Self-repetition	To tell information to the students and to order students to do something
		Other-repetition	To make sure all students listen and know the appropriate answer in the discussion
3.	Interactional Strategies	Asking for repetition	To have clear information from the students
		Asking for clarification	To gain more information and it would make other students learn from their friends
		Asking for confirmation	To avoid misunderstanding between the students and teacher
		Guessing	To support the students by giving the words that they tried to think of.
		Comprehension check	To ensure students' understanding and involve the students in the lesson
		Responses	To respond what the students shared so they know the answers exactly

In direct strategies, first, the English native speaker teacher used code switching. She switched English to Indonesian language to tell the context of words to the students. She used this strategy because some local students did not understand the words in English. The motive of using this strategy was to make the students understand the teacher's intentions easily and this could prevent misunderstanding of the teacher's intentions especially for students who had low English proficiency. This result was in line with the notion of code switching proposed by Dornyei and Scott in [6]. They mentioned code switching means the speakers use L1/L3 words with L1/L3 pronunciation in L2 speech; this may involve stretches of discourse ranging from single words to whole chunks even complete turns. [8] supported code switching strategy was believed as effective strategy in English language classroom when dealing with low English proficient learners.

Second direct strategy used by the teacher was self-repair. She repaired her previous utterance as she realized she used wrong words when she read stories and explained lesson to students. She directly repaired the wrong words in order to avoid misunderstanding among teacher's intention and students' comprehension. This result was supported by Dornyei and Scott in [6]. They mentioned self-repair means the speakers make self-initiated correction on their own speech. The teacher in this study used this strategy to motivate students to use appropriate words and this could keep the communication running well.

Other-repair strategy was another direct strategy used by the teacher when the students told her about something. Usually, she repaired the structure of the sentence especially when the students told about past events. She used this strategy in order to make the students know the appropriate words that they had to use in telling past events. Dornyei and Scott in [6] mentioned other-repair means the speakers correct something in interlocutor's speech. Thus, related to this study, the other-repair strategy used by the teacher to correct words in students' speech to make the students use correct words later.

The last direct strategy used by English native speaker teacher was self rephrasing. She rephrased her previous words to simpler one as not all of the students understand the teacher's intention. By using self-rephrasing, the teacher could use simple word to explain a term to the students. Dornyei and Scott in [6] stated self-rephrasing means the speaker repeated a term but it is not quite as it is. The speaker might add something or use paraphrase. The teacher in this study paraphrased her previous utterance to make the students understand her intention easily.

Moreover, in indirect strategy, there were three strategies used by English native speaker teacher. First, she applied use of fillers such as OK, actually, you know, and well when she started talking to the students because she wanted her speaking sound friendlier. Moreover, she needed time to think what she could share to students. Dornyei and Scott in [6] mentioned use of fillers means speakers use gambits to fill pause, to stall and to gain time in order to keep communication channel open. This was in line with the result of this study; the teacher used this strategy to gain time to think what to say next [9] supported use of fillers was intended to mark of hesitation, time creating device, editing term and hold control of conversation while the teachers think what to share to the students.

Self-repetition was another indirect strategy used by the English native speaker teacher in order to make sure all students listen and to tell information to the students clearly. When the students understood the intentions, the students could do what the teacher mentioned during the learning time. According to Dornyei and Scott in [6], self-repetition means the speakers repeat words immediately after they were said. This study showed the teacher used this strategy immediately after first utterance was mentioned to explain lesson and give instruction to the students.

Other-repetition was the last indirect strategy used by the English native speaker teacher. She repeated students' utterance when she asked questions to the students. This was done to make sure all students listen and know the appropriate answer in the discussion. Other repetition, according to Dornyei and Scott in [6] means the speaker repeated something the interlocutor said to gain time. In this study, the teacher repeated students' appropriate answer or good opinion to make other students could learn from their friends. By repeating students' words, it could give opportunity for the students who had limited English vocabulary to learn more from their friends and teacher. This is in line with a study done by [10]. They showed the teacher' reason used self-repetition is to increase the amount of input especially when the teacher gives information or correction.

In interactional strategies, the teacher used six strategies during the English sessions. First, asking for repetition strategy was used when the teacher could not hear and get the students' intentions. This was in line with Dornyei and Scott in [6]. Asking for repetition means the speaker asked the interlocutor to repeat as the speakers do not hear and understand properly. In this study, the teacher used this strategy to have clear information from the students. When she asked the students to repeat what they mentioned previously, she also wanted other students listen to the opinion. Thus, they could learn from the students who share opinion during discussion.

Second interactional strategy used by the English native speaker teacher was asking for clarification in order to make students talk more and to gain more information from the students. When the teacher asked the students to clarify their intention, the teacher, indirectly, trained the students to be confident in telling their intention. According to Dornyei and Scott in [5], asking for clarification was same with the result of this study. The speakers request explanation of unfamiliar meaning. Teacher asked the students to clarify their intention to make information become clearer. Moreover, Chin; Nassaji and Well in [11] supported the teacher used asking for clarification of the students' previous contribution to enhance students' existing knowledge.

Asking for confirmation was another interactional strategy used by English native speaker teacher to have clear information and to make sure what she heard was same with the students' intention. This was in line with asking for confirmation mentioned by Dornyei and Scott in [6]. The speakers request confirmation that they heard. In this study, the teacher asked for confirmation of students' opinion during stories and discussions to have clear understanding of the message shared by the students.

The fourth interactional strategy used by the teacher was comprehension check to ensure students' understanding toward the stories and topic in the class. This result was in line with Dornyei and Scott in [6]. They mentioned comprehension check means the speaker ask questions to check that the interlocutor can follow what the speakers want. This strategy was to make students engaged in classroom interaction. Teacher could make sure students were following the lesson and kept focusing on the discussion.

Guessing was another interactional strategy used by the teacher. She used this strategy to help and support the students in telling their intentions in English. Some students had limited vocabularies in English. Dornyei and Scott in [6] mentioned guessing strategy is similar to a confirmation request but latter implies a greater degree of certainty regarding the key word. Relating to this study, the teacher used this strategy as students did not know how to say the words and they only gave code. The teacher tried to guess; thus, students could know how to tell their intentions and this could develop their English vocabularies.

Last interactional strategy used by English native speaker teacher was response. The teacher applied this strategy usually after the students gave opinion during discussion. According to Dornyei and Scott in [6], response strategy consisted of response repeat, repair, rephrase, expand, confirm and reject. However, in this study, the teacher responded only by confirming appropriate answers or rejecting students' inappropriate answers. This was done to acknowledge students' contribution in the class and prevent misunderstanding as students would know whether their answers were appropriate or not.

4 Conclusion

The use of communication strategies is necessary in teaching and learning process. In this study, communication strategies used by English native speaker teacher at Dyatmika Kindergarten showed the teacher attempted to make students understand the teacher's intentions and solve the problem when the teacher communicated with the students. Classification of communication strategies proposed by Dornyei and Scott in [6] was used in analysing types of communication strategies used by the English native speaker teacher. The result showed there were thirteen communication strategies used by the teacher during the carpet time in English sessions. This result could be used as reference to inspire teachers who teach inside or outside Dyatmika Kindergarten in teaching bilingual students.

Greatest gratitude goes to all lecturers of English Education of Postgraduate Program at Ganesha University of Education, Bali, Indonesia especially the principal supervisors, Prof. Dr. I Ketut Seken M.A and Prof. Dr. I Nyoman Adi Jaya Putra, M.A for their invaluable guidance, correction and motivation during this study.

References

1. M.E. Gredler. *Learning and Instruction Theory into Practice (Sixth Edition)*. New Jersey: Pearson Education, Inc. (2009)
2. H.D. Brown. *Principles of Language Learning and Teaching (Fifth Edition)*. New York: Pearson Education. (2007)
3. A. Maleki. *Techniques to Teach Communication Strategies*. JLTR **1** 640-646. (2010)
4. I.G.A.I.T. Juliari. *The Profile of Communication Strategy used by Junior High School English Teachers in Teaching and Learning Process in SMPN 2 Sukawati*. EPPUPG **2** (2014)
5. K.A. Wedananta. *Communication Strategies Used by Mathematics And Science Teachers to Communicate in English in Morning Sharing in SMP Nasional Plus Jembatan Budaya: A Sociolinguistic Analysis*. EPPUPG **3** (2014).
6. Z. Dornyei, M.L. Scott. *Communication Strategies in a Second Language: Definition and Taxonomies*. LL, **47** 173-210 (1997)
7. M.B Miles, A.M. Huberman. *Qualitative Data Analysis (Second Edition)*. California: Sage Publication, Inc (1994)
8. B. H. Ahmad, K. Jussoff. *Teachers' Code Switching in Classroom Instruction for Low English Proficient Learners*. ELTJ **2** 49-55 (2009)
9. V.A. Pamolango. *Types and Functions of Fillers used by the Female Teacher and Lecturer in Surabaya*. P **15** 11-15 (2015)
10. M.D.M. Viano, P.G. Conejos. *Communicative Functions of Repetition in Classroom Interaction*. RESLA **11** 129-141 (1996)
11. M.Y. Nothash, M.A. Karafkan. *Teacher-employed Communication Strategies: Investigating Function Type Occurrence in Iranian EFL Context*. IJALEL **4** 73-80 (2015)

Integrating self-regulated learning and discovery learning into English lesson plan

Ni Kadek Heny Sayukti

English Language Education, School of Post-Graduate Studies, *Universitas Pendidikan Indonesia*

Abstract. The notion of learner-centeredness has been embedded in the National Curriculum of Indonesia, 2013 Curriculum. However, most of the teachers seem to be hardly acquainted with the concept of Self-Regulated Learning (SRL) and discovery learning in the lesson planning. Considering the phenomenon, this study intends to explore the concept of Self-Regulated Learning in the lesson plan of English subject for a tenth-grade level by employing a qualitative design with data obtained from a teacher-made lesson plan and a semi-structured interview. The researcher used content analysis to analyze the lesson plan. Meanwhile, the qualitative data from interview result were preceded through a coding sheet and transcribed modified figure. The findings revealed an integration of SRL cyclical phase and discovery learning in the teacher-made lesson plan. Based on the discussion, the results need to be applied in a considerably large context, in order to see thoroughly dynamic integration between Self-Regulated Learning model, lesson planning and the concept of learner autonomy.

Keywords: self-regulated learning, discovery learning, learner autonomy, lesson plan

INTRODUCTION

Due to the development of learner-centered methods and approaches to teaching English as a foreign language in the current curriculum, the role of teachers and learners have changed. Based on the Indonesian Ministry of Education and Culture (*Permendikbud No. 22*, 2016) [6], 2013 Curriculum enhances active learning, multiple sources, scientific approach, competence based approach, integrated learning, applicative learning, and balance of hard skills and soft skills. Based on the overviews, the learner's role in the learning has given an insight to the concept of learner as a critical thinker which means the learner has the authority to control and to develop their own learning. In fact, most of teachers in Indonesia seem to rarely become acquainted with the notion of learner autonomy, especially in lesson planning. Having a well-designed lesson plan actually determines the teaching success (Cunningham, 2009) [3]. However, most of teachers experience several common difficulties in lesson planning such as the way of formulating indicators, selecting materials, determining learning activities, selecting media and developing assessment procedure (Farid, 2014) [9]. Additionally, based on a study conducted in a Learner Autonomy-based teaching training, it was revealed that most teachers actually desired to integrate Learner Autonomy principles in their teaching-learning contexts, but the principles seem to be not feasible for the implementation (Lengkanawati, 2017) [10]. In line with the phenomenon, Farid (2014) [9] also mentions in his case study that the teacher made lesson plans tend to be not specific in depicting autonomy into the learning procedure. Thus, the Indonesian teachers tend to be uncertain whether this could be integrated into the lesson plan and teaching practice.

There are some research results which present evidences about SRL and school curriculum in the English as foreign language learning setting. In undergraduate level, Hoops et al. (2016) [8] assert it is essential to teach *how* to learn, rather than only *what* to learn, meanwhile all faculty members from all disciplines needs training of curriculum development and effective teaching practices that promote student learning and achievement. However, Borg and Al-Busaidi (2012) [13] report that teachers tend to feel a gap between their desire to implement the concept of autonomy and their practices in particular objectives, assessment and materials. Therefore, teachers play a significant role in motivating student's self-regulation by being a self-regulated teacher (Alsamadani, 2010) [4]. Based on the overview, the concept of SRL in 2013 Curriculum is integrated with Piaget's cognitive theory about cognitive and Vygotsky's social constructivist theory in which learners receive more chances to discover and to interpret ideas by themselves (Asri, 2016) [2]. For teaching procedure, modification is required in the element of learning objective, material development, teaching method and assessment.

Considering the previous elaborations, the present study is intended to extend existing knowledge about Self-Regulated Learning by elaborating the theories in cyclical phase. As past studies dominantly concerned at lesson plan analysis or the theory in teaching practice, this study explores the concept of Self-Regulated Learning in two sections; its integration with discovery learning in the teacher made lesson plan and teacher's perspective about the concept of Self-Regulated Learning (SRL) in lesson planning. The underlying reasons behind this study are there are few researches on Self-Regulated Learning, especially its integration to language learning and lesson planning. Since the current curriculum does not view learners as passive recipients of information, exploring the role of SRL theory in lesson planning contributes an important insight to design a comprehensive lesson plan. It would support the principle of lesson planning based on the content standard (*Permendikbud No. 21*, 2016) [5], namely to empower students' motivation, creativity, initiative, interest, innovation, inspiration and autonomy.

The Concept of Learner Autonomy and Zimmerman's model of Self-Regulated Learning Cyclical Phase

The concepts of learner autonomy, independence, and Self-Regulated Learning have been growing throughout the recent years. Learner autonomy is the principle of learners to be encouraged to assume a maximum amount of responsibility for what they learn and how they learn it (Richards & Schmidt, 2002) [7]. In general, self-regulation refers to the process through which learners systematically direct their thoughts, feelings, and actions toward the attainment of their goals (Rajabi, 2012) [14]. In addition, Lengkanawati (2017) [10] believes that if we teach learners learn how to learn by the use of learning strategies, it is inevitable that we make them become autonomous which in turn will help them have better learning outcomes. As a result, SRL allows learners to learn to control their own behaviour through the social context in which learning occurred and interaction with self-reward.

Zimmerman's model of SRL has been revised several times. In general, it is a cyclical process, where the student plans for a task, monitors their performance, and then reflects on the outcome. Initially, it develops through the integration of cognition, behaviour, emotions and motivation into curriculum and learning. The cyclical phase consists of three major parts, namely *forethought phase, performance phase* and *self-reflection phase* as illustrated by the following figure. The first phase, Forethought phase commonly consists of two major process, task analysis and self-efficacy. Just like brainstorming phase, this phase fragments learner's background knowledge into smaller pieces and the personal strategies for the performance (Winne, 2001) [11]. The second phase concerns about learner's performance which consists of self-control and self-observation. In this phase, *self-control* aims at maintaining concentration and interest during performance. The last phase is when learners judge their work and formulate reasons for their results (Zimmerman and Moylan, 2009) [1]. Self-judgment occurs as the process assessing self-performance. As a result, in context of language learning, social cognitive theory views self-regulation as comprising process of self-observation, self-judgment, and self-reaction. Relevant to its purpose, this study explores only the formulation of learning objectives and learning procedure to the theory of Self-Regulated Learning cyclical phase. Thus, extended parts of metacognitive goals such as self-assessment and reflective thinking are embedded in the learning procedure assignments.

RESEARCH METHODS

This study attempts to explore the integration between a teacher made lesson plan based on 2013 Curriculum and the theory of Self-Regulated Learning cyclical phase through a qualitative method. The design adopted triangulation method in which document analysis was supported by other qualitative methods such as document analysis. The participant in this study was an English teacher who currently teaches at Grade X in a Senior High School level (in this case, vocational school) and her lesson plan. Since the purpose of this study was exploring the integration of SRL cyclical phase to lesson planning, the teacher was expected to be knowledgeable to the concept of learner autonomy and professionally has conducted the concept into language teaching. The main instrument covered the utilization of focused interview to answer the research question about the teacher's perception and a lesson plan checklist to examine the component of the teacher made lesson plan. Focused interview gives an insightful method for gathering information in which the respondent is free to answer the question in her words through open-ended questions (Malik & Hamied, 2016) [12]. Then, content analysis was deliberately optimized to gather detailed information based on the unit of analysis meanwhile coding process attempted to examine the interview results and to eliminate unnecessary parts which were not related to the research. After lesson plan check, coding and transcription, the data was analysed in four central themes, namely the analysis of the learning objectives, approach/teaching method, learning procedure/activities organization and the teacher's perceptive toward Self-Regulated Learning integration.

RESULTS AND DISCUSSION

In particular, findings and discussions sections are divided into three subsections following the central themes.

The Learning Objectives

This lesson plan was made by the teacher for teaching English at tenth grade in a vocational high school. The teachers cover the 90 minutes lesson based on the Core competence and Basic competence written in the national syllabus of English subject for Secondary. The topic of the lesson is about expression of compliment and congratulation which later evolved into eight learning objectives covered in one lesson plan for one meeting:

(2) Learning Objectives

1. Through discussion, learners will be able to mention the social function, language feature and generic structure of the text about giving compliment.
2. Through discussion, learners will be able to define the generic structure of the interactive interpersonal text about giving compliment and giving response autonomously.

3. Through discussion, learners will be able to mention the social function, language feature and generic structure of the text about congratulation.
4. Through discussing and gathering information, learners will be able to make a congratulation card by using appropriate language feature.
5. Through discussing and gathering information, learners will be able to express compliment and react appropriately.
6. Through discussing and gathering information, learners will be able to arrange the jumbled dialogue into a good conversation about giving compliment and expressing response elaborately.
7. Through discussing and gathering information, learners will be able to compose a proper conversation about giving compliment and congratulation appropriately and elaborately.
8. Through discussing and gathering information, learners will be able to perform a role play about giving compliment and congratulation appropriately and elaborately.

Based on the point (2), action verbs used in the learning objectives vary more than the indicators (*to perform, to compose, to arrange, to express, to make, to mention,* and *to define*). The congruence between the indicator and learning objectives are shown in the choice of action verbs and types of objectives written by the teacher. Meanwhile, psychomotor domain appears in the form of making congratulation card, and performing role play. However, affective domain is explicitly integrated through discussion and gathering information session at some point when learners are expected to collaborate together.

The Approach and Teaching Method

The lesson is designed with scientific approach and the model of learning is discovery learning in which students are engaged to discover ideas and interpret the meaning of the lesson.

(3) Approach: Scientific Approach
Methods: Discovery Learning (Small Group Discussion)

Regarding to the form of 2013 Curriculum lesson planning, the point (3) basically shows complete information of approach, method and technic.

The Learning Procedure Organization

Since the approach is scientific, the learning of teaching procedure is organized based on the discovery learning. The procedure is presented in three phases, pre-activity, whilst-activity and post-activity.

(4a) Pre-Activity:

- Orientation
  - The teacher greets the students, pray together and takes an attendant.
  - The teacher asks the students:
    - *When is your birthday, students?*
    - *What do you say when some is having birthday party?*
    - *What do you say if she's wearing a very beautiful dress on the party?*
    - *Well, students, today we are going to study about how to congratulate and give compliment to other people*
- Apperception
  - The teacher gives a prominent perceptive about the topic
- *Today we will study about giving compliment and congratulation*

- Motivation

- The teacher gives a prompt of *giving compliment and congratulation*

- Reference
  - The teacher states the learning objective and scope of learning.

Regarding to the design, point (4a) already covers pre-activity stages which consists of orientation, apperception, motivation and reference. Concerning the written form, the teacher focuses only on writing the teacher's instruction instead of mentioning learner's activities in each phase. At the same time, the lesson starts with the *"when is your birthday?"* which depicts SRL forethought phase of setting task goal and building a self-motivation.

4b) Whilst Activity

- Stimulation
  - The teacher gives an example of congratulation card
  - The learners observe the cards
  - The teacher shows a picture of beautiful lady, a picture of victory, and a picture of graduation day.
  - The learners observe the pictures
- Problem statement
  - The teacher asks the learners about the pictures
  - The learners give their responses
  - The learners ask about the way for expressing congratulation in a situation
  - The learners ask about the way for expressing compliment.
  - The learners give their responses
- Data Collection
  - In a small group consist of four people; learners discuss the structure of the text about expressing compliment, giving response and extended expression.
  - In the group, learners gather information about interactive text, both in written and spoken form, about expressing compliment, giving response and extended expression.
- Verification
  - The teacher asks the learners to identify the social function, generic structure and language feature of the text.
  - The learners arrange a jumbled conversation into a good dialogue.

4c) Post-Activity

- With the teacher, the learners derive a conclusion from the lesson about social function, generic structure and language feature of the text.
- The teacher gives instruction about the upcoming homework about making a congratulation card.

Regarding the suitability of the learning procedures with the learning objectives, this whilst activities seem to affect more cognitive domain. In fact, the lesson plan seems already provide a space for doing task strategies elaboration through the stimulation part. In contrast, in the point (7) and (8) of the learning objectives, the students are expected to be able to compose a congratulation card and to perform a dialogue. Meanwhile, in verification session, learners are expected to only identify social function, generic structure and language feature and to arrange jumbled dialogue. According to point 4c, the instructions tend to be not sufficient enough to complete all learning objectives written in the first part. In line with Alsamadani (2010) [4] who states that self-regulation would happen only if the teacher could model it by being a self-regulated teacher first. In this case, she gives reinforcement and reward to the learners in order to help them become self-regulated. For this reason, external motivations such as reward and types of reinforcement need to be mentioned in the lesson plan in order to comprehensively inform every teacher who could use the lesson plan.

CONCLUSION

Based on the overview, the teacher made lesson plan already contain all component of lesson plan based on the process standard of 2013 Curriculum. However, there are several sessions of the learning procedure in the lesson plan that does not constitute a complete feature of the cyclical phase. As a result, a comprehensive reinforcement and reward need to be attached in the lesson plan in order to give a better insight of metacognitive goals of the lesson plan.

ACKNOWLEDGEMENTS

I would like to express my gratitude to Prof. Dr. Nenden Sri Lengkanawati, M.Pd. for her constructive suggestions during the course and my peers for their supportive advice and my very great appreciation to LPDP (Indonesia Endowment Fund for Education) Scholarship for supporting my postgraduate study.

REFERENCE

1. B. J. Zimmerman, A. R. Moylan, *Handbook of Metacognition in Education*, 299-315 (2009)
2. D. N. Asri, *Counsellia: Jurnal Bimbingan dan Konseling*, 4 (2016).
3. G. Cunningham, *The new teacher's companion: practical wisdom for succeeding in the classroom,* (2009)
4. H. A. Alsamadani, Eur. J So. Sci., **16**, 51 (2010)
5. Indonesian Ministry of Education and Culture, *Permendikbud* No. 21 (2016)
6. Indonesian Ministry of Education and Culture, *Permendikbud* No. 22 (2016)
7. J. C. Richards, R. Schmidt, *Longman dictionary of language teaching and applied linguistics* (2002)
8. L. D. Hoops, S. L. Yu, Q. Wang, V. L. Hollyer, IJTLHE, **28**, 75-93. (2016)
9. M. Farid, *Teachers'difficulties in lesson planning* (Doctoral dissertation, Universitas Pendidikan Indonesia) (2014)
10. N. S. Lengkanawati, IJAL, **6,** 222-231 (2017)
11. P. H. Winne, *Self-regulated learning and academic achievement*, 153-190 (2001)
12. R. S. Malik, F. A. Hamied, *Research methods,* (2016)
13. S. Borg, S. Al-Busaidi, *ELT J*, **12**, 1 (2012)
14. S. Rajabi, Procedia So. Be. Sci., **47**, 344-350 (2012)

Teaching English Pronunciation of Suprasegmental Features on Students of English Education

Yousef Bani Ahmad,
yousef.baniahmad@fkip.unsika.ac.id
Pendidikan Bahasa Inggris, FKIP, Universitas Singaperbangsa Karawang
Jl H.S Ronggowaluyo Telukjambe Karawang

Abstract. The objective of this research is to know the technique and activity in teaching English pronunciation on suprasegmental features (intonation and stress). This research uses qualitative approach with descriptive method. The subject of this research is 6 students from English education department. Technique of collecting data by doing observation, interview and documentation. The results showed In teaching English pronunciation for suprasegmental features is very concerned about the how to teach students, give materials and do exercises. There are some materials that are taught to improve students' ability in stress words of English sentences. Students learnt combination of words adjectives and nouns are generally stressed is in the first, students are taught about the prefix, learnt about words with suffixes and students were also given exercise with compound words. Furthermore, in teaching intonation, students are also given understanding and practicing the reading text, analyzing and pronouncing the English word in accordance with the correct intonation. The impact, students understand how to use rising and falling intonation.

1 Introduction

In communicating English language, learners must have good competencies in speaking. In English conversation there must be something to understand and do not. It often happens that it is difficult to comprehend our English speaking because of English pronunciation what learners used to say differently or wrongly in the native speakers. Therefore pronunciation is very important to be learned so as not to experience the wrong meaning in English conversation. For example, the words *keen* and *kin*, try to see as if the same way to pronounce. However, in pronunciation the word *keen* is pronounced by /ki:n/ with a long vowel sound, while the word kin /kin/ sound is short. The purpose of pronunciation learning is to make better understand the English language and also understand more easily by native speakers or people. Whatever our English surfing without good pronunciation and accent skills, our English sometimes will be hard to understand by native speakers and take different meaning.

According to results of research were obtained by Abbas (2011) with the title of *A Study on the Situation of Pronunciation Instruction in ESL/EFL Classroom study*, he mentioned some of the main problems were faced by ESL and EFL students they are lack of exposure to target language, lack of motivation, less pronunciation skill, and lack of attention, rhythm, intonation and word stress in English.

Error pronunciation of learners in learning English can hamper communication. For example, when a student says, the soap /soup/ in the restaurant where they should put the soup /sup/, inaccurate phonemic production can cause misunderstandings between waiters and students. An incorrect student pronouncing a phoneme may be very difficult to understand by the recipient of the message. This can be very frustrating for learners who may have good grammar and lexis but having problem with the pronunciation.

In pronunciation, learners also need to understand about suprasegmental features in enhancing the quality of pronunciation mastery. It is not only developing the correctness of pronunciation but more about understanding the meaning of speakers. The emphasis on one word in the English conversation sentence certainly has a strong meaning to be conveyed. While intonation will determine whether the sentence has the meaning of the question, command or statement. Stress and intonation in English are part of suprasegmental features. Clark, Yallop and Fletcher (2007) inform that suprasegmentals can be refered as prosodic features or nonsegmental features. They are features of spoken language such as pitch, rhythm, and tempo which are not easily identified as discrete segments. According to Longman Dictionary of Applied Linguistic (Richard, Platt, and Weber, 1985, p.281), suprasegmentals are units which extend over more than one sound in an utterance such as stress and tone. Stress is the emphasis on a particular syllable in one word. Every word that has two or more syllables always has stress when reading it in one of its syllables. Not only in words is stress but also in sentences. It would sound strange if in one sentence there is no stress in one or several words in it. Because the stress on certain words in a sentence can make others more easily understand what is being said.

Intonation in speaking is very essential besides the correct way of pronouncing words. Intonation makes others better to recognize and even differ sentence in statement, question or instruction. The intonation can be divided into two types: (1) rising intonation which is regularly used in the sentence in question with the yes/no question and the expression "indecision" or "amazement", (2) Falling intonation is used in the customary statement and Wh-question.

Previous research on English pronunciation was conducted by Lawrence Jun ZHANG (2004) with the aim of raising students awareness about the importance of intonation as they learn English with their teachers, Zhang's research focused on examining sociocultural and psychological trends in the selection of phonological models. The results showed that students' responses in learning aspects of English pronunciation such as the suprasegmental feature were very important and necessary to understand the meaning in social interaction.

There are several previous studies that discuss about suprasegmental feature. However, the focus and study are different, as Salvador who discussed Segmental and Suprasegmental Mismatch in Lexical Access. He examined the role of suprasegmental and segmental information in the activation of spoken words. Based on the background of the problem and statement are mentioned the research is focused on Teaching English Pronunciation on Suprasegmental features (stress and intonation) on students of English Department that exist in phonetics and phonology class.

3 Method

This research used descriptive qualitative approach which is a way to apply deep understanding aspect to a problem. This study is descriptive and tends to use analysis and see the phenomenon that occurs with the aim of providing a broad and deep understanding a formulation of research problem. The subjects of this study are students who take courses phonetics and phonology. In conducting interviews and data collection, the researcher took 10 samples. In addition, researcher conducted observations, interview and observation to explore sources which are related to teaching suprasegmental (intonation and stress) in the classroom. Furthermore, researcher analyzed the data.

4 Discussion

As teachers, the learning model used in the classroom is frequently close to the language used outside the classroom. Many teachers adjust their accents to be straightforwardly understood by their students, but rare can steadily teach with significantly different accents. English teachers need to be conscious of sound differences in English pronunciation properly, and the more knowledge a teacher has for different accents and variations of English will add knowledge to his students when they engage directly in conversations with native speakers or English users.

In teaching suprasegmental in the classroom, teachers used some efforts are conducted to make students understand in pronouncing words. Providing an understanding of how to master the correct word stress and intonation will help students to speak and deliver messages. In teaching stress in words and sentences, teachers provide understanding related to some formulas that must be known by the students. Understanding is given to provide students easily identify where the stress in the word and sentence.

English learners need to know the right keywords are very important. If someone speaks an English word with the wrong stress then the English listener may have an understanding in the sense of the word. In listening, if English students expect a word to have a certain stress pattern, they may not recognize it when there is a native speaker who says it. In other words, what they hear does not match what they understand all along.

Some ways of making learners mindful of word stress forms in the English. The first set of activities focuses on the pronunciation of proper names (Marry, Richard, Thomson) and place names (London, Jakarta, Bali) Because of the constraint in the vocabulary corpus, these accomplishments are suitable for the lower levels of proficiency. The second set of activities involves more vocabulary and writing and reading skills, and is appropriate and reading skills, and is applicable for the higher levels of proficiency. There are some important things students must understand about stress.

The first rule is that adjectives and nouns are generally stressed in the first syllable, example *green tea*. Here, students are given an exercise by reading the text that the teacher has given and then the students are asked to find and identify which includes the adjective or noun in the text and then pronounce it according to the correct word pressure rule.

Second, students are taught about the prefix-related insights in words that have two or three syllables. Prefixes that are taught to students include *be, in, dis, ex, un,* the stress is almost always on the second or third syllable. Prefixes are not stressed in English words. The important note here is that students understand the majority of these words are verbs. Example *because, increase, distrust, expostulate, unbelieve.*

Third, students also have experience learning about words with suffixes. Any suffixes are never stress. Some of the suffixes are -ive (administrative), -ly (quietly), -al (original), -ic (organic). After given the understanding by the teacher, students are given a text that has a lot of suffixes and asked to find words where there is suffix and then after the words containing suffix students are required to look for stress in the word and mark it.

Fourth, students are also given a sense allied with compound words. There are also some procedures for defining the stress words compound. These are words which are made by joining two nouns, nouns and adjectives, verbs and prepositions. It is very mutual for compound words that have nouns to have stress on the first syllable. Example: *crossword, hotdog.* Students are given an exercise to find any word which is compound word then after it is compound word list compiled and then check every word by using dictionary application in students' smartphone.

In dealing with intonation in the language classroom, students focus on practice based on theories are given by lecturer. Learners study how the choices they make with regard to intonation chose to determine the meaning of utterances. As well as helping to determine meaning. Intonation is the linguistic use of *pitch* in discourse. It is *linguistic*, in the sense that it carries meaning; changing the intonation of an utterance can easily change the meaning of that utterance. For instance, the clause : *you are hungry, aren't you ?* has one meaning if the tag is accompanied by a falling tone, but a different meaning if it is accompanied by a rising tone. The notion of linguistic can be extended to include the paralinguistic use of into- nation, in which something of the mood or attitude of the speaker is conveyed, for instance whether the speaker is angry, bored, insistent, etc. *Linguistic* might also be extended to include *sociolinguistic* variation of the kind that shows where a person comes from.

In intonation learning, students are given knowledge of how the speaker behaves and also feels what is being conveyed. When students listen to the words in low tones or slow tones of course they learn how the meaning is in it. Students also learn and practice to convey messages by using the intonation of each utterance. Student's voices rise and fall in tones like notes in a musical scale, high to low or low to high.

Actually, intonation and stress work together to express meaning. Usually, the last word or the next to last word in an utterance is stressed. Students voices go down at the end of an utterance. Students understanding how to distinguish high or low intonation in saying English words or sentences is helpful in improving the ability to speak English. They know the formula or rule to use falling intonation.

First, statements and commands in sentences are using falling intonation. Example. He goes to college (statement), open the door (command). Example of the sentence above using intonation that descends at the end of the sentence. In this case students practice to read and search for any sentence that is in the text using falling intonation.

Second, students learn to use falling intonation with wh-questions. Wh-questions begin with such words as *"when," "what," "where," "who," "who," "who," and "which." When will you go to bali? What is that? Who is eating rice? Where do you go? How are you? Whose pen is this?* By knowing the rules of falling intonation with wh-questions students are also given some English pronunciation exercises to properly train the sounds are produced.

In the teaching of rising intonation, students are given materials and practice to improve English pronunciation skills. Students are given learning in order to use intonation appropriately and in accordance with the meaning of the message to be conveyed. Students understand when to use rising intonation at the end of an utterence when ask yes-no questions. Yes-no questions require "yes" or "no" for an answer. Example. *Did you drink milk?* In this exercise, students are asked to pair up to ask questions and answer with yes no question pattern. In this practice, students get the learning to use high intonation correctly. Then to train rising intonation, students are given several sentences in which mention the names of fruits or other names and then in each word using rising and end using falling intonation as the end of the sentence. Example: *I like fruits such as banana, watermelon, strawberry, apple, and mango. The words from banana, watermelon, strawberry, apple (rising), mango (falling).* In addition, if there are sentences in the form of choice, then the first option using rising intonation and both falling intonation. *Example: Do you want milk or tea.?* It can be concluded that students can use rising intonation when naming items on a list or when offering choices, except for the last item or choice use falling intonation.

4 Conclusion

The results of research on English pronunciation teaching from suprasegmental features are:

1. In teaching English pronunciation for suprasegmental features is very concerned about the proper procedures, materials and exercises so that students can understand and also master the emphasis of word or intonation in English.
2. In teaching the pronunciation of stress in the English language students are able to know four key ways to determine stress in English and also be able to distinguish which is stress or unstress with some exercises.
3. Pronunciation intonation in English conversation plays an important role in knowing the meaning of the speaker. Teachers provide excellent teaching when providing an easy formula in knowing and practicing what is rising or falling intonation in words or phrases. Thus, students are able to speak English according to the correct intonation.

References

1. A.P. Gilakjani, Studies in Education. J. E, **1**, (2011).
2. Clark, J., Yallop, C. and Fletcher, J. *An Introduction to Phonetics and Phonology 3rd ed,* (Victoria, Blackwell, 2007)
3. Dobrovolsky, M., & Higgins, E.C. *Phonology: The Function and Patterning of Sounds.* In O'Grady, W., & Archibald, J., *Contemporary Linguistics,* (Boston, Bedford/St. Martin's, 2001).
4. Fromkin,V., Rodman, R. and Hyams, N. *An Introduction to Language* (8 th ed.), (Boston, MA, Thomson Wadsworth, 2007).
5. Kelly, Gerald. *How to Teach Pronunciation.* Series Editor Jeremy Harmer, (Charlbury, Oxfordshire, UK, Bluestone Press, 2010).
6. Kenwhorty, Joanne. *Teaching English Pronunciation*, (UK, Longman Group, 1987).
7. L.J.Zhang, Applied Linguistics. J. E, **145**, 219 (2004)
8. Ladefoged, P. *A Course in Phonetics (5th ed),* (Boston, Thomson Wadsworth, 2006).
9. Richard, J., Platt, J., and Weber, H. *Longman Dictionary of Applied Linguistics.* (Essex, Longman House, 1985).
10. S.S-Faraco, N.S-Galles, A-Cutter, Mem.Lang. J. E**, 45,** 412-432 (2001).

The Attitude of socio-harmony and Local Wisdom: an Indicator of the Development of Social Tolerance of High School Students

Deni Adi Wijaya[1], *Djono*[2], dan *Suryo* Ediyono[3]

[1]Graduate Program in History Education, Sebelas Maret University, Surakarta, Indonesia
[2]Faculty of Teacher Training and Education Science, Sebelas Maret University, Surakarta, Indonesia
[3]Faculty of Cultural Science, Sebelas Maret University, Surakarta, Indonesia

Abstract. This article discusses the dilematis aspect about the shift in traditional values-based local wisdom about the ecological aspects. Close to the learning in school emphasized the sides of cognitive ability and do not offset the granting of traditional values education. The purpose of this article is to promote tolerance of students to be more sensitive to the surroundings so as not to become individualists. Research results at the high school level, students are not given the attitude of socio-harmony in the form of tolerance. The purpose of this article is to enhance social tolerance that can deliver value to prevent the behavior concept asocial.

1. Introduction

This past decade, the educational paradigm switching to a pragmatist because of current globalization. Directly the values in traditional education replaced with instant education. Refer to the fact that resulted in terpinggirnya subjects of history because the students assume more importance facing the subjects examined in the national exam. Many cases of student fights, declining manners towards teaching system used due to gurus on cognitive and Psychomotor domains. Affective value no longer prioritized to build student character.

In the 21st century this rapidly changing society accompanied social change in culture. These changes affect to the system of teaching the nuances of pattern that follows a pragmatic period, in this case the history teacher was forced to membelajarkan the material related to the world of work. As a result of learning history to be saturating and lost of charge values. Rapid cultural change caused a cultural globalization's influence can be seen in the development of students in a class. As one example of this is the phenomenon of the rate of development of the students ' positive attitudes formed through education at school is not as fast as the pace of development of the students ' negative attitude that is formed through the influence of modern teaching: hedonism, a materialistic, and idividualistis, of course it is not in accordance with the personality of the nation of Indonesia. It gives rise to a deterioration in moral values learners result from cultural changes leading to modernization. While local cultural values still exist though not commensurate with modernization that emphasize instant results. Values education can answer about the waning value of character and can build attitude of students through local wisdom values. It can be associated with the conclusion of researchers who stated, "Javanese people so protect and care about her children as contained in the proverbial Java *Memayu Hayuning Bawana* means add to the beauty of the world that's been created in such a beautiful". It is a tradition of values education has been used by instilling character values to students.

Positive impact of advances in technology education indeed makes it easy for students to search for information such as on whatsapp, line, blackberry massengger, etc. The information is not all of it leads to positive things and the impact it makes students prefer communicating via digital than socialization. Changing the paradigm of communicating to the world of digital and tend to be more individualistic, American scientist D. Tapscott consider computerization becomes a measure of progress, intellectual capacity, and transforming the learning process (Naydenova & Shaposhnikova, 2016).

2. The issues socio-harmony and possible solution?

One of the things that need to be examined is the problem egaliterianisme and reject the existence of social segregation in a multicultural society. This research highlights the crucial aspects of the traditional education of Javanese people are principled on the concept (1) *memayu hayuning bawana* means add to the beauty of the world with good works, (2) *mangan ora mangan asal kumpul* means everything despite not having anything required on the family group, respectively, and (3) *rukun agawe santosa, crah agawe bubrah* means that the pillars make safe peace and a sense of invidualistis make the split. Etymologically there are three Java language level is *ngoko* or coarse language , *krama madya* or fine language used to equal levels, and *krama inggil* or fine language used for someone who had degrees above. The Java community is therefore very much appreciate and respect each other, therefore socio-harmony so highly in tune with *andap asor* attitude or humble. Summed up each individual is very mindful of manners and culture Java very tolerance.

This way of thinking and behavior of Java against the influx of very elastic culture outside resulted in the creation of residential dwellings giving priority to harmony with elements of socio – harmonization between the implications of the establishment of the Joglo House with harmony with society. These elements created a development strategy to become a village *"Gemah Ripah Loh Jinawi Tata Tentrem Kerta Raharja" Gemah Ripah Loh Jinawi* Term etymologically refers to a region with fertility which brings prosperity, whereas *Tata Tentrem Kerta Raharja* described an area that is orderly, peaceful, and prosperous. The phrase is in tune with the founding of Javanese traditional House who prefer the harmonization as noted in the Serat Dewa Ruci, Haryanto (2007:11) argued, "the overall content of the Serat Dewa Ruci contains very valuable lessons about ethics and mysticism that until now the cultural values held by the people of Java". Therefore, the Javanese traditional house construction not only to stand upright but there are values and symbolic meanings.

The problem of socio-harmony high school level turns out to be in harmony with the local wisdom. The crucial problem, how students can absorb all local wisdom for personal development and social skills. It is the excellence of the students, in addition to being given a cognitive based learning also enriched the knowledge of social and personal knowledge. According to Titov & Cherkashin (2016) knowledge is divided two ability-based Ecology, among others:

1. The ability of the emotional attitude, readiness of students in social activities;
2. Ecological knowledge Capabilities, allowing skills to troubleshoot the situation of ecology to form private student attitudes.

Problem Solving?

In the process of learning at the high school level, to improve the ability of personal and social students, teachers should design the class as there is in a society that is multicultural classes emphasize the similarities egaliterianisme students to reduce the asocial action leads to attitudes of discrimination. Teachers as facilitators of learning as a mediator to bridge their differences in class to avoid social segregation. This can be done by using the humanistic learning. As for the steps among others; 1) arrange the seating Division students to not clumped in one group alone but rather gives rise to a process of socialization, 2) gives the task group discussion by dividing the group into a heterogeneous.

3. Socio-harmoni values and local wisdom

In the context of Javanese culture, the term word often used to describe a harmonious society, harmonious Society (Balchindorzhieva & Tsyrendorzhieva, 2016) means harmony between society and nature, individual and community, individuals and others, as well as individuals with her inner world. Thus giving birth to a social capital, (Ratanakosol, Pathumcharoenwattana, & Kimpee, 2015) classified into six namely;

1) Human capital refers to the people, the example of the richness of local wisdom;
2) Commitment social relationships are formed within the community, such as rules and norms;
3) Natural capital, such as the environment's natural resources in the community;
4) Cultural capital;
5) Financial capital;
6) Capital base that is built as a facilities and infrastructure.

The purpose of the system of education is the development of cognitive, personal, social, and moral students. The hopes of researchers in this study in order to reduce the student's moral degradation due to globalization and to consider local characters such as example *andap asor* attitude or humble, with *andap asor* students can appreciate that life must respect differences. Specifically with the example of the attitude of the local characters who built in the expectation of history learning can improve social harmony students aligned with the principle of harmony of Javanese culture. Therefore the principle of Javanese people who actually is *memayu hayuning bawana* attitude which means where they live consider harmonization. Therefore students should be improved the attitude of social harmony in order to prevent the existence of a religious war raises the issue of tribe, race, religion, and war pose thought-provoking dissent that continuously lead to disintegration of the nation due to the pudarnya local culture as identity formation of character education. So, concluded the social harmony is the concept of respect for differences, mutually tolerant, accepting cultural diversity, Mahejabin (2015) describes indicators socio-harmony there 10 stages, among others;

1. Mutual understanding differences;
2. Developing social Harmony through cooperation;
3. Gender equality to decrease discrimination;
4. Social Tolerance;
5. Emphasize the love of the environment;
6. Social Harmony depends on the attitude of non-violence;
7. Humanism;
8. Respect for all religious, racial, ethnic, and tribal;
9. Emphasize social interaction;
10. The recognition of the values in the form of local wisdom of local communities.

Development of indicators of social tolerance of high school students form social affective values, among other things:

Humanism;

= attitudes to appreciate cultural diversity.

= attitude of responsibility.

= attitude of social skills in the form of socialization and social interaction in the form of courtesy, respect the differences between individuals in the community.

Empathy;

= job skills in the form of mutual. solidarity against a friend

= is experiencing a disaster.

= keep the natural environment as an attitude of maintaining continuity of society. such as preserving local wisdom.

Tolerance;

= acknowledge diversity of religion, tribe, ethnicity, and race.

= attitudes reducing social conflict.

Egalitarian;

= gender equality be sexes form social discrimination not attitude

= equation of culture, religion, race, and ethnicity.

The Value of Local Wisdom;

= character's attitude reflected in the students an appreciation for cultural identity.

= preserve relics of the forms of cultural discernment, wisdom in the form of social culture born in the community with nature, society and the individual, and the community with the creator.

4. An alternative model of social inquiry to promote tolerance in Learning History

The term learning already is best known in the world of education is issued UU RI No. 20 Year 2003 about the system of national education, UU Sisdiknas (2003: Article 1) stated:

> Learning is defined as "... the process of interaction with educators and learners learning resources in an environment of learning". Learning as a concept of pedagogy, technically defined as the systematic and systemic effort to create learning environments that are potential yield learning process boils down to the development of individual potential as learners.

From that understanding, it appears that the learning and learning has a substantive and functional attachment, Winataputra argued, "the interconnectedness of substantive learning and learning is located on the node onset behavior change in individuals. The functional linkages with learning learning is to produce learning "(2007:1.1). About behavior change in learning behaviors can be associated with the conclusion of researchers who stated, "learning is the process of changing the behaviour (in the broad sense) incurred or modified through practice or exercise" (Chauhan, 1979:4). Regarding the substantive attachment in learning such affective side reflected the attitude of socio-harmony that is the attitude of tolerance and will peak later boils down to social tolerance awareness of students in an environment of family, school, and community. It can be associated with the conclusion of experts who declared, "learning is a series of activities of the soul in order to obtain a change in behavior as a result of the experience of individuals in interaction with the environment that concerns the cognitive, affective, and psychomotor" (Mayer, 2000:7).

Learning History have aspects the purpose of skills development students to train a self-reliance and responsibility. On the other hand, the aspect of social values also need to be developed for the construction of the attitude of the character of students. One of the learning model used in this study using Inquiry model. Based on the opinion of Gulo quoted Al Tabany (2015:78) stated:

> Inquiry strategy means a series of learning activities involve maximum throughout the students ' ability to find and investigate the systematic, critical, logical, analytical, so that it can formulate its own discovery. The main objectives of learning activity is Inquiry; 1) maximum student involvement in the process of learning activities, 2) keterarahan activities logically and systematically on the learning objectives, and 3) to develop an attitude of trust in themselves the students of what is found in the Inquiry process.

Based on explanation Inquiry learning in General turns out there is a special purpose. Massylas States Inquiry can improve skills to review the environment in more critical. The implications of learning History for training the critical thinking through the process of discovery. Regarding the advantages of learning in Inquiry model History, Widja (1989:49) argued, "method of Inquiry strongly support the strategy of teaching that emphasizes Local History material". It is when analyzed more deeply because the function Inquiry model could push students actively discuss relate to historical sources directly via the assignment in the group discussion. In addition, students are given an understanding of how the interpretation of the facts obtained using historical research include; 1) collection of historical sources/heuristics, 2) assessing and evaluating the collected facts/criticisms, 3) describes the style of the language's own historical sources obtained/interpretation, and 4) renders the analysis of a collection of facts through structured task/phase of historiography. The steps above when applied in the study of history should innovation with learning based on problems to train students skilled in problem solving. It was when combined with model Inquiry can create a learning situation problematic history, stimulus of students asked, curiosity and encourage students seeking answers through experimentation in the process of discussion. It can be associated with Bruner's opinion stated, "Discovery Learning can improve the cognitive skills of reasoning also trains students how to find and solve problems with knowledge and generate meaningful knowledge" (Winataputra, 2007:3.18).

From the above, the Exposure model learning of Inquiry has discussed the advantages according to experts and writers chose the social Inquiry model of Suchman. Richard Suchman developed a modified Inquiry learning. Suchman argues about the importance of bringing students on the attitude that all knowledge is tentative, an expert on education outlines a theory of Suchman as follows:

First, invite students to imagine in actual conditions;
second, identify the components that are around them;
third, formulate problems and making hypotheses on the condition;
fourth, get data by making the questions with answers "Yes" or "no";
fifth, make inferences from data acquired (Joyce in Al Tabany, 2015:85).

Based on the opinion of Bruce Joyce above, Social Inquiry is the learning strategies of social groups and subgroups of society. Students should be given experience of how to solve the problem, through the experience of each student will be able to build a useful knowledge for themselves and society. There are three characteristics of social Inquiry and strategy development formulated as follows:

First, the existence of social problems in the aspects of the class are considered important and can encourage the creation of class discussions;
second, the formulation of the hypothesis as a focus for Inquiry;

the third use of the fact as a hypothesis testing (Al Tabany, 2015:88)

In this study, learning learning stages using the syntax of Inquiry expressed by Eggen & Kauchak (1996) that has been modified to researchers. Inquiry learning syntax can be seen in the table below.

Table 1. Inquiry Learning Stages

Na	Phase	Teacher Behavior
1	Presenting questions or problems	Teachers guide students to identify a problem and the problem is written on the Board and divide the students in the group discussion. Each group discussion should be evenly men, women, tribal, ethnic origin and describes the multicultural groups.
2	Make a hypothesis	The teacher gives the student the opportunity to argue form the hypothesis. Teachers guide students in determining the hypotheses that are relevant to the issue.
3	Designing experiment	The teacher gives the student the opportunity to specify the steps in accordance with the hypothesis. Teachers guide students to sort steps the process of discovery.
4	Experiment to obtain information	Teachers guide students get information through discussion with other groups. Teachers guide each group to try to solve problems together.
5	Collect and analyze data	The teacher provides opportunities for each group to present the data collected. Teachers invite students to ask and guide all members of the group to argue.
6	Make inferences	Teachers guide students in making conclusions are shared equally between group discussions in class. Teachers invite other groups to make a conclusion beyond a group presentation.

5. Conclusion

Positive impact of advances in technology education indeed makes it easy for students to search for information. The information is not all of it leads to positive things and the impact it makes students prefer communicating via digital than socialization. Changing the paradigm of communicating to the world of digital and tend to be more individualistic. One of the things that need to be examined in the socio-harmony is a matter of egaliterianisme and reject the existence of social segregation in a multicultural society. The purpose of the system of education is the development of cognitive, personal, social, and moral students. Therefore students should be improved the attitude of social harmony in order to prevent the existence of religious wars cause problems the tribe, religion, and race. Specifically with the example of the attitude of the local characters who built in the expectation of history learning can improve social harmony students aligned with the principle of harmony of Javanese culture.

References

1. Al Tabany, T.I.B. *Mendesain Model Pembelajaran Inovatif, Progresif, dan Kontekstual: Konsep, Landasan, dan Implementasinya Pada Kurikulum 2013*, (2015).
2. Balchindorzhieva, O.B & Tsyrendorzhieva, D.Sh. Harmonious Society and Ecological Civilization: The Concepts' Relationship, *SHS Web of Conferences*, 28, p.1-4 (2016).
3. Chauhan, S.S. *Innovations in Teaching- Learning Process*. (1979).
4. Eggen, P.D & Kauchak, D.P. *Strategies for Teachers Teaching Content and Thinking Skills*, (1996).
5. Haryanto, S. *Pembangunan Berwawasan Kultural. Senate Hearing Opens Inaugural Speech Professor*, p.4-19 (2007).
6. Kemendikbud. *Undang-Undang Republik Indonesia Nomor 20 Tahun 2003 Tentang Sistem Pendidikan Nasional*, (2003).
7. Mahejabin. Social Harmony and Prosocial Behavior: Two Aspects of One Coin, *Shaikshik Parisamvad An International Journal of Education*, 5 (1), p.67-72 (2015).
8. Mayer, J.D. Emotion, Intelligence, and Emotional Intelligence. In Forgas, J.P. (Ed). *The Handbook of Affect and Social Cognition*, (2000).
9. Naydenova, N.N & Shaposhnikova, T.D. Student in Information Society: Indicator of socio – moral Development of Individual in Quality Education Evaluation, *SHS Web of Conferences*, 29, p.1-5 (2016).
10. Ratanakosol, K., Pathumcharoenwattana, W, & Kimpee, P. Learning Process for Creating Community Identity, *SHS Web of Conferences*, 26, p.1-5 (2016).
11. Titov, E.V & Cherkashin, E.I. Readiness of Senior Higher School Students for self-determination in Ecology Field as an ability to Form Subjective Attitude, *SHS Web of Conferences*, 29, p.1-3 (2016).
12. Widja, I.G. *Dasar - Dasar Pengembangan Strategi Serta Metode Pengajaran Sejarah*, (1989).
13. Winataputra, U.S. *Teori Belajar dan Pembelajaran*, (2007).

Needs Analysis of the English Writing Skill as the Base to Design the Learning Materials

Andi Tenri Ampa[1,*], and *Hidayah* Quraisy[2]

[1] English Department of Faculty, Teachers Training and Education Unismuh, Jl. St.Alauddin 259, Makassar 92114, Indonesia
[2] Social Sciences Department, Faculty of Teachers Training and Education Unismuh, Jl. St.Alauddin 259, Makassar 92114, Indonesia,

Abstract. This research used a descriptive method. It was aimed at identifying students' learning needs for the English writing skill as the base for designing the learning materials. Writing skill covered the analysis of the types of paragraph, types of text, the components of writing and paragraph development. The subjects of the research were the fourth semester students that consisted of 330 students. The samples were taken 15 % randomly, so the number of samples was 50 students. The research used a questionnaire as the instrument to get responses from the students about their learning needs. The results showed that the learning needs for the writing skills coped with the types of paragraph development, the types of text, and components of writing skill. The types of paragraph development included the ways by definition (79.7%), classification (67.0%), listing (59.3%), cause effect (47.7%), example (47.3%), and comparison (45.7%). The types of text consisted of description (66.0%), news items (59.7%), narration (58.7%), discussion (56.7%), recount (57.0%), and exposition (50.7%). The components of writing skill contained structure (79.6%), vocabulary (79.4%), content (62.0%), organisation (53.6%) and mechanic (34.0%). The implication of the findings would be the base of teaching and learning process, especially in designing the learning materials for the English writing skill.

Key words: Needs Analysis, Writing Skill, Learning Material

1. Introduction

There are two categories of language skills, namely receptive skills and productive skills. Receptive skills are the skills to understand the message being heard and read, while the productive skills are those that produce messages or ideas through spoken and written text.[1] The processes of productive skills are the processes of producing the language codes that begin with the idealization stage of the speaker's thinking. Furthermore, the election stage of language forms as ideas that appear at the stage of idealization, then at this stage the verbal linguistic codes produce the speech.

In other words, the productive skill process begins with a semantic encoding, which is the stage for the preparation of ideas, then the grammatical decode of conceptualization in grammatical units and semantic decode, namely the understanding of the concepts of ideas. Both skills are basic in obtaining language especially in English. This research focuses on

*Corresponding author: atenri59@yahoo.com

the writing skill as one of the productive skills. The results of the researcher's observations indicated that from the four English skills (listening, speaking, writing & reading), speaking and writing skills as the produuctive skills are most difficult to be perceived by the students. They have difficulties expressing ideas. The research results show that fifth semester students still low ability in writing English essays, especially in content, grammar and the use of transitional signals in writing. Similarly, in English speaking skills, the results of the observations showed that the students are still lacking in English. They generally have difficulty on the use of vocabulary and grammar [2]. Therefore, as a lecturer needs to provide solutions to improve English writing skills.

In writing skills, there are ten types of paragraph development: by listing, by examples, by comparison, by contrast, by definition, by classification, by space and time, by process, by cause and effect, and by generalisation [3]. These types of paragraph are different in content, but the components of each paragraph that should be paid attention in writing paractice are similar; those refer to structure, vocabulary, content, organisation and mechanic. Besides, there are several kinds of text, such as: narrative, recount, procedure, report, exposition, descriptive, discussion and news items [4]. These types of text are important to discuss in teaching and learning the English writing skills. However, the types of paragraph development, text and the components of writing should be met by the students' needs.

To design the learning materials, it is necessary to match with the students' needs. Therefore, prior analysis of learning needs for the English writing skills is required. Needs analysis for the English speaking skills had been investigated and the results indicated that the learning needs for speaking skills are English pronunciation (88%), vocabulay (92%), and structure (90%). Furthermore, they also need to learn the function of speaking skills, those are interaction (43%), transaction (92%) and performance (92%)[5]. By identifying the students' learning needs, the materials of learning will match the students' needs, so they can be motivated to learn well.

This research is intended to describe the students' needs to learn the English writing skills as the base to design the learning materials. It concerns with the types of paragraph development, the types of text to be learned, and the components of writing that are needed to be discussed in the teaching and learning process. To achieve the effectiveness of teaching, it is required to match the needs and the ability of the students. The previous research of needs analysis for a developmental reading, writing, and grammar course had been conducted, and the results focused on the appropriate approach for the development of the course [6]. In addition, another previous research of needs analysis to find out the learning style, skills priorities, current deficiency, and alternative methodologies chinese college students. Therefore, this research focused on the material needed by students as the novelty of the other reseachers [7].

3. Literature Review

3.1 Learning Needs

Learning needs deals with the needs of the individuals since they have learned. This need will be motivated for individuals to learn continuously. The basic human needs are from the lowest level to the highest level. This called the human needs level'. In other words, every situation that is less than it should be indicates a "need". If the gap is large or causes consequences, it is necessary to place as a priority that must be overcome [8].

*Corresponding author: atenri59@yahoo.com

The learning needs vary that means everyone tends to have different learning needs [9]. So, the learning needs wanted by someone in the country may be different from the learning needs of people living in urban areas. The perceived need for learning last year may also be different from the perceived the learning needs in the future. If a learning need has been met, and then the other learning needs that must be met through the learning activities must be met. Therefore, the learning needs should be identified through an individual approach.

Needs are the permanent tendencies that exist within a person's motivation to achieve a certain goal. The need for learning is essential to identify as the foundation for the preparation of the learning program. The learning needs that have been identified will provide direction where the program activities are addressed. It is necessary because the need in learning is the basis that describes the distance between the learning objectives desired by the learners or the actual learning conditions. Each learner has different needs that should be identified to determine what needs the learner will have in potential and eventually become his needs.

In addition, the learning need is a current learner's competence compared with the competence of learners who should be mastered. The gap includes knowledge, skill and attitude. In the implementation of the learning needs, involvement of learners is needed because the potentials that exist in each learner will be indicated according to their individual learning needs. The learning needs will be arranged carefully and sequentially, then determined by the priority of learning needs or basic interests to be fulfilled through the learning activities.

3.2 Needs Analysis

Needs Analysis is a procedure for collecting information about learners' needs. Its purpose is to establish key learning outcomes and requirements in the design and delivery of a course or learning activity [10]. The needs relate to the characteristics, concerns and potential constraints of the students. The analysis seeks to match possible techniques and materials to these needs and thus identify whether the design is appropriate to the intended goals [11]. In the other words, needs analysis involves doing some kinds of activity with a learner in order to find out what their learning needs are, because a good understanding of learners' needs can contribute to successful course planning. Furthermore, need analysis means a plan for gathering information to make decisions about priorities [12]. In curriculum development, need analysis is the process by which one defines educational needs and decides what their priorities are [13]. Furthemore, needs analysis is defined as a process of needs while determining priorities. Needs analysis is a method to know the difference between the desired conditions with the existing conditions. The desired conditions are often referred to as ideal conditions, while existing conditions is often called real conditions [14]. So, needs are basically the gap between what is already available and what is expected.

There are several things that attach to the notion of need analysis. First, need assessment is a process that means there is a series of activities in the implementation of need assessment. Need assessment is not a result, but a certain activity in the effort to make certain decisions. Second, the need itself is essentially a gap between hope and reality [14]. Thus, the need assessment is an activity to gather information about the gaps that each student should have with what they already have.

*Corresponding author: atenri59@yahoo.com

4. Method

The research was a descriptive method that aimed to describe the learning needs for the English writing skills. The populations of the research were the fourth semester students that consisted of 330 students. The samples were taken 15 % randomly, so the number of samples was 50 students. The researcher used a simple random sampling. The instrument used was a questionnaire that aimed to get the responses from the students about the learning needs. Data analysis used were: first, identifying and scoring the students' responses; second, doing percentage of students' responses; third, ranking the percentage of students' responses; fourth, deciding the level of students' needs.

5. Results and Discussion

The results of the research cover the students' need analysis in learning the English writing skills. Writing skills refer to the types of paragraph development, the types of text, the components of writing that are needed to be learned by the students.

There are six types of paragraph proposed to be selected by the students. The description of students' responses are shown in the following table:

Table 1. Types of Paragraph Development

No	Types of Paragraph Development	F	%
1	Development by Definition	239	79.7
2	Development by Classification	201	67.0
3	Development by Listing	178	59.3
4	Development by Cause Effect	143	47.7
5	Development by Example	142	47.3
6	Development by Comparison	137	45.7

The students' responses indicate that the ranks of paragraph development types needed to be learned by the students are (1) development by definition, (2) development by classification, (3)) development by listing, (4)) development by cause effect, (5) development by example, and (6) development by comparison. The highest level needed by the students is the paragraph development by definition. From 300 responses, there are 239 responses (79.7%) that have determined the development by definition as the first choice followed by the other types of the paragraph development. It indicates that all types of paragraph development are needed by the students in learning the English writing skills. Recall the previous statements [3] that there are ten types of paragraph development. However, only six types are firstly needed by the students. The four types of paragraph development: by contrast, by space and time, by process, and by generalisation are not responded by the students at that moment.

Furthermore, there are six types of text which are needed by the students in learning the writing skills. The frequency of students' responses are shown in the following table:

Table 2. Types of Text

No	Types of Text	F	%
1	Description	198	66.0
2	News items	179	59.7
3	Narration	176	58.7
4	Recount	171	57.0

*Corresponding author: atenri59@yahoo.com

5	Discussion	170	56.7
6	Exposition	152	50.7

The students' responses in Table 2 indicate that the rank of needs on the types of text is from description to news items, narration, recount, discussion, and exposition. However, the greatest percentage of the type is a descriptive text (66%), while the fewest level of need is an exposition (50.7%). The other types of text are not responded by the students.

In writing skills, the components needed to be learned by the students are structure, vocabulary, content, organisation and mechanic. The percentages of the students' responses are shown in the following table:

Table 3. Components of Writing

No	Components of Writing	F	%
1	Structure	199	79.6
2	Vocabulary	176	79.4
3	Content	155	62.0
4	Organisation	134	53.6
5	Mechanic	85	34.0

Data in Table 3 indicate that the English structure (79.6%) and vocabulary (79.4%) in writing take the first and the second ranks to be discussed in learning the English writing skills, being supported by the other components: content, organisation, and mechanic. Vocabulary is a list of words, and the word is a core of language. However, the words should be put in correct positions or a string of words should refer to the structure of language. Each class of words has its own use and function in any string. Every language has its own system as a rule that should be obeyed in that language.

Conclusion

The English writing skill as a productive skill is normally put as the last sequence among the four language skills. It seems right that writing is the last skill which is acquired by children in getting their first language. For the purpose of achieving the writing skill, a number of components are essential to have for the students; they are structure, vocabulary, content, organisation, and mechanic. Structure and vocabulary are the first two needs to be possessed by the students in learning the English writing skill. In making the English paragraph, a variety of ways for the paragraph development are certainly needed by the students. In fact, there are six important types of paragraph development that the students really need in their writing task. Then, the types of text also frequently make the students confused, so that they also need them. Therefore, the matters being discussed in this research should be taught to the students for the sake of their writing development in the future.

Acknowledgements

The research received a grant from Directoriate General of Higher Education, Ministry of Research and Technology, Indonesia. Thus, it is a must for us to express indebtedness and our thanks.

References

*Corresponding author: atenri59@yahoo.com

1. G. B. Mundhe. *Teaching Receptive And Productive Language Skills With the Help of Techniques*. IJE. **1,** 2:1-6. (2015)
2. A.T. Ampa. *The students'Ability to write the English Productive Skills*. Makassar: Unismuh (2015)
3. M. Imhoof and H. Hudson. *From Paaragraph to Essay: Developing Composition Writing*. England: Longman Group Ltd. (1975)
4. M. Anderson & K. Anderson. *Text Types in English*. Australia: Macmillan Education. (1997)
5. Ampa, et al. *The Students' Needs in Developing Learning Materials for Speaking Skills in Indonesi*. JEP. **4** (17):171-178. (2013)
6. R. C. Cunningham. *Needs Analysis For a Developmental Reading, Writing, and Grammar Course at a Private Language School in Cambodia University of Hawai'i at Mānoa*. (2014). w.hawaii.edu/sls/wp-content/uploads/2014/08/Cunningham.pdf
7. C. Jie. *English Learner Needs Analysis:* A Case Study of Beijing Institute of Petrochemical Technology (BIPT). IJHSS. **3**, 1, (2013).
8. A.H. Maslow. *Toward a psychology of being*. Start Publishing LLC. (2013).
9. R. Frost. *A training course and manual on the learning process*. https://www.quora.com/ What-does-learning-needs-mean-What-are-they-How-are-these-needs-best-me. (2014)
10. J.D. Brown. *Foreign and second language needs analysis*. In Long, M.H.&Doughty,C(Eds.),The handbook of language teaching(pp. 269-293). U.K.: Wiley-Blackwell. (2009)
11. M. Shang, M.Dr. *Needs Analysis*. http://slideplayer.com/slide/5352155/. (2008).
12. Seels and Glasgow. *Making Instructional Design Decisions*. Second Edition. http://ocw.metu.edu.tr/pluginfile.php/3298/course/section/1175/20121127191305441. Pdf. (1990)
13. C. McNeil. *Training Needs Analysis*. https://4good.org/cheryl-mcneil/training-needs-analysis-table. (1985).
14. G. Anderson. *A Proactive Model for Training Needs Analysis*. JEIT. **18**, 3, 23-28. (1994).

*Corresponding author: atenri59@yahoo.com

An Analysis of the Basic Clause Types of the Students' Syntactic Competence

Muhammad Basri D.[1*], Sitti Halijah[2]

[1]Department of English, Faculty of Letters, UMI Makassar, South Sulawesi, Indonesia
[2]Department of English Education, Faculty of Letters, UMI Makassar, South Sulawesi, Indonesia

Abstract. The article was concerned with the results of a descriptive study about the basic clause types that the students had recognized in the objective test. The aims of the study were to know the students' syntactic competence in the case of the seven basic clause types and to describe their problems to make the distinctions between one type and the others. The research samples consisted of 80 persons. The data were analyzed by using a mean percentage. The research findings indicated that the six semester students got the fewest scores of SPOA (3.75%) and SPOC (13.75%); the fewer scores of SPOO (21.25%) and SPO (22.50%); and the few scores of SPA (30.00%), SPC (31.25%), and SP (36.25%). The main problematic issue faced by the students was to distinguish the SPO type from SPC (26.25%), meaning that the difference between object and complement had become the confusion for the students. It was recommended that these basic clause types be discussed in syntax class by giving more examples to the main problems faced by the students in order that they could differentiate among the clause types.

Keywords: Basic Clause Types, Syntactic Competence

1. Introduction

In this world there are about 6,000 languages spoken, and English has become the global language of the world. English has become the first largest language from a number of countries using it as an official language. Today it is still ranked first as the international language and the language of the global world [1]. English has become the third largest if it is seen from a number of native speakers, i.e. 341 million inhabitants, with at least 140 countries, after Mandarin Chinese with 874 million inhabitants of 16 countries, and Hindi (India) with 366 million inhabitants of 17 countries [2].

The influence of global information that flows very rapidly makes the language very important to highlight the role of the social status and function, so that all of them can be applied well within the scope of both linguistic and non-linguistic matters. Besides, the matters of competence and performance are obviously crucial in learning a language [3]. The knowledge of linguistic structure is referred to as competence, whereas the realization of this knowledge in actual event is called performance [4]. In learning a language, both competence and performance should be achieved by the students.

The linguist sees a language as form, namely sounds or letters and their combinations into larger units, such as words, sentences, and so forth. This is the problem of syntax in a language form [5]. It is obvious that syntax is the one dealing with the way in which words are put together to express thoughts or ideas sensibly [6].

The example is that 'The dog frightened the child' is a sentence. At an intuitive level *the dog, frightened* and *the child* appear relevant constituents in a way that strings like *frightened the* or *dog frightened* do not. The word 'string' is supposed to be used as a neutral term to refer to any sequence of constituents. The words *the dog or frightened the child* or indeed the whole of sentence as strings without a commitment as their status as constituents or any identification of the type of constituents. Furthermore, again at an intuitive level the strings *the dog* and *the child* seem to be constituents of the same type [7].

A study of syntax contains analyses. The well-known analyses are bracketed, immediate constituent, and tree diagram analyses. Each of these has a definite use to make the students understand for identifying and classifying the syntactic categories with their functions. The subject requires competence and performance and the linguistic knowledge and language skills. Therefore, to design a model for the instructional materials of the English syntax, we need to know all syntactic matters with these analyses.

The areas of syntax cover the English phrases and clauses. A phrase is a word or a group of words that form a particular meaning grammatically. The phrase may identify a referent, indicate the action of, modify, or qualify grammatical elements in sentences [8]. Words are organized into phrases, groupings of words that are clumped as a unit and a sentence can be modeled as a set of phrases [9]. The previous study found that the phrase is significantly importance to analyze the distribution clause using machine translation [10].

The English phrases have various divisions, i.e.: noun phrase (NP), verb phrase (VP), adjective phrase (Adj P), adverb phrase Adv P), and prepositional phrase (Prep P). The English NP is very important because this type can fill in four functions: subject, object, complement, and adjunct. A predicate function is filled in by a VP. The English clauses include independent or main clauses and dependent or subordinate clauses. This article focuses on discussing the independent or main clauses; they are seven basic clause types: (1) Subject Predicate, (2) Subject Predicate Adjunct, (3) Subject Predicate Complement, (4) Subject Predicate Object, (5) Subject Predicate Object Adjunct, (6) Subject Predicate Object Object, and (7) Subject Predicate Object Complement.

It is stated that the subject function is normally filled in by an NP; the predicate function is always filled in by a VP; the object function may be filled in by either an NP or a Prep P; the complement function may be filled in by either an NP or an Adj P; and the adjunct function is filled in by an Adv P, a Prep P, or an NP [11].

The basic structure of clauses is simply composed of 'S+P' as constant elements, and what follows these elements depends on the kinds of clause types. These types are called main or independent clauses. The other clauses are called subordinate or dependent clauses; they are very complex. Therefore, the English phrases are required to be mastered immediately as the basic constituents in constructing any clause. For example, in a sentence (an independent clause) 'The Indonesian students should understand the English word classes' has a syntactic structure as an SPO (Subject, Predicate, and Object). The function of the phrase 'The Indonesian students' is a subject; the function of 'should understand' is a predicate; and that of 'the English word classes' is an object. These three functions are obligatory constituents in this clause. The meaning of a clause is not clear if one of these constituents is omitted.

It should be remarked that syntax establishes the coordinated system of form and meaning. Any thought in the language can be expressed in different forms. Sometimes two semantic descriptions in a sentence appear: real or concrete and idiomatic or figurative [12]. In syntactic evidence, the sixth semester (level 3) students have got greater accurate performance than their inaccuracy (54.97% > 36.34%). Anyhow, there are still a number of non-productive features of syntactic evidence, i.e.: future tense, perfective aspect, simple past, past perfect, and present perfect voices, and interrogative of indicative and purpose of subjunctive moods. The implication of the above statements suggests that those students be trained intensively to acquire morphological and syntactic categories with their related features, so that they will be able to use these categories with their features in their interactive utterances [13].

2. Method and Materials

The research problems were oriented to syntactic issues of the learning products. The population consisted of the sixth semester students of English department, Faculty of Letters, Indonesia Moslem University in Makassar. The samples consisted of 80 students. The research variables covered the students' competence of the seven basic clause types and the problematic issues shown in deciding the basic type of independent clause.

The research instrument that had been used to gather the data was a competence test. The test consisted of seven items in which the testees had to choose the numbers provided in brackets after each of the clause type. These items were included in the whole parts of the test. The time used for doing the whole parts of the test was 90 minutes. The results of test were classified as the quantitative and primary data. The rating score was ranged from 0 to 10.

The above categories were concerned with the substances of the independent clauses, including the seven basic clause types: (1) Subject Predicate (SP), (2) Subject Predicate Adjunct (SPA), (3) Subject Predicate Complement (SPC), (4) Subject Predicate Object (SPO), (5) Subject Predicate Object Adjunct (SPOA), (6) Subject Predicate Object Object (SPOO), and (7) Subject Predicate Object Complement (SPOC). Based on the results of competence test, the correct and incorrect options were counted for being further computed in percentage rates. Finally, the results of calculations were tabulated, exposed, and interpreted.

3. Results and Discussion

The findings specify the interpretations of the students' competence of the seven basic clause types and the problematic issues shown in deciding the basic type of independent clause by the sixth semester students of English department, Faculty of Letters, UMI Makassar. Table 1 below is concerned with the sixth semester students' competence of the basic clause types that had been recognized in the syntactic test. And, Table 2 refers to the problematic issues faced by the sixth semester students in making the distinction among the seven basic clause types.

Table 1. The students' competence of the seven basic clause types

No	Types	Examples	Q	Scores	%
1	SP	A few people cannot sleep.	29	3.63	36.25
2	SPA	The book is on the table.	24	3.00	30.00
3	SPC	They have been full.	25	3.13	31.25
4	SPO	The Indonesian students should understand the English word classes.	18	2.25	22.50
5	SPOA	The English language has a foreign status in Indonesia.	3	0.38	3.75
6	SPOO	My uncle will give Mary a dictionary.	17	2.13	21.25
7	SPOC	I imagined your aunt much fatter.	11	1.38	13.75
Total			127	15.90	158.75
Mean			18.14	2.27	22.68

Table 1 indicates the results of a descriptive study about the basic clause types that the sixth semester students have recognized in the objective test. The results of analysis can be classified into three rates: the most problematic issue (the fewest percentage of syntactic competence), the more problematic issue (the fewer percentage of syntactic competence), and the problematic issue (the few percentage of syntactic competence).

It has been shown the table above that the SPOA type with the example 'The English language has a foreign status in Indonesia' has been recognized as the most problematic issue in which their competence is 3.75%. It is the fewest percentage of syntactic competence for the SPOA type. The example of this type can be analyzed by using brackets as follows: The English language (S:NP) has (P:VP) a foreign status (O:NP) in Indonesia (A:Prep P). Thus, this is one of the types (SPOA) that has been recognized as the most problematic issue. Another one is the SPOC type with the example 'I imagined your aunt much fatter' which is also the most problematic issues being recognized in the syntactic competence. Its percentage rate is 13.75%. The analysis of this clause type is as follows: I (S:NP) imagined (P:VP) your aunt (O:NP) much fatter (C:Adj P).

The next problematic issue has been classified as the more problematic issue (the fewer percentage of syntactic competence). This issue contains two types: SPOO, and SPO. The problematic issue is the SPOO clause type. The example of this type with its bracketed analysis is 'My uncle (S:NP) will give (P:VP) Mary (O:NP) a dictionary (O:NP)'. This type is called the di-transitive clause one [11]. It has two objects which are distinguishable in terms of a direct object (a dictionary) and an indirect object (Mary). Another classification of the fewer percentage of syntactic competence is SPO (22.50%); e.g. The Indonesian students should understand the English word classes. The functions and categories of this type are as follows: The Indonesian students (S:NP) should understand (P:VP) the English word classes (O:NP).

The last problematic issue has been classified as the problematic issue (the few percentage of syntactic competence). This issue contains three types with their own percentage rates: SPA (30.00%), SPC (31.25%), and SP (36.25%). The example of SPA type in the table above is 'The book is on the table'. The syntactic analysis of this clause type is as follows: The book (S:NP) is (P:VP) on the table (A:Prep P). The next problematic issue in this classification is the SPC type. From the table above, the analysis of the example can be done as follows: They (S:NP) have been (P:VP) full (C:Adj P). And, the last case is the SP clause type. The bracketed analysis of the given example is as follows: A few people (S:NP) cannot sleep (P:VP). The clause has been classified as an intransitive clause one in which the subject does not affect anybody or anything else [11].

Based on the above exposition, it can be concluded that the rating score of the sixth semester students' competence of the seven basic clause types is not so great as it has been expected. The mean score is 2.27 (22.68%). Therefore, it is recommended that the seven basic clause types be discussed in the classroom by giving more examples to the principally problematic issue and applying the syntactic analyses to identify each constituent in any construction, so that the students are able to differentiate among the clause types.

Table 2a and Table 2b below show the quantity and percentage rates of problematic issues concerning the seven basic clause types that the sixth semester students have experienced in their own objective test, respectively.

Table 2a. **The quantity of problematic issues of the seven basic clause types**

No	Types	Key (K)	Correct Options (Q)	Incorrect Options (Q)						
				1	2	3	4	5	6	7
1	SP	1	29	K	31	10	7	1	1	1
2	SPA	2	24	21	K	22	11	0	1	1
3	SPC	3	25	17	25	K	7	4	2	0
4	SPO	4	18	0	12	21	K	4	14	11
5	SPOA	5	3	0	14	6	8	K	5	44
6	SPOO	6	17	1	9	8	23	6	K	16
7	SPOC	7	11	5	8	17	4	23	12	K

Table 2b. **The percentage of problematic issues of the seven basic clause types**

No	Types	Key (K)	Correct Options (%)	Incorrect Options (%)						
				1	2	3	4	5	6	7
1	SP	1	36.25	K	38.75	12.50	8.75	0.13	0.13	0.13
2	SPA	2	30.00	26.25	K	27.50	13.75	0.00	0.13	0.13
3	SPC	3	31.25	21.25	31.25	K	8.75	5.00	2.25	0.00
4	SPO	4	22.50	0.00	15.00	26.25	K	5.00	17.50	13.75
5	SPOA	5	3.75	0.00	16.25	7.50	10.00	K	6.25	55.00
6	SPOO	6	21.25	0.13	11.25	10.00	26.25	7.50	K	18.75
7	SPOC	7	13.75	6.25	10.00	21.25	5.00	28.75	15.00	K

The results indicate that the students get the problems to make a distinction between SP and SPA. The students decide SPA (38.75%) and SPC (12.50%) as the incorrect options for SP. In deciding the SPA type, two incorrect options have greater percentages; they are SP (26.25%) and SPC (27.50%). Besides, the students are also confused of deciding the SPC type because of these two types: SP and SPA. They get a problem to distinguish the SPC type from the SP type (21.25%) and the SPA type (31.25%). This is the main problem of SPO is to distinguish this type from SPC (26.25%), meaning that the difference between object and complement becomes the confusion for the students.

The following problematic issues to be discussed in the above tables are SPOA, SPOO, and SPOC types. The SPOA type is recognized as the SPOC type (55.00%). This recognition means that they still get problem to differentiate between the function of adjunct and that of complement. Then, in recognizing the SPOO type, there are 23 students who choose SPO (26.25%) and 16 students who choose SPOC (18.75%). It means that the students need to know the distinction between the mono-transitive clause type and the di-transitive clause one. Besides, they also need to know the distinction between object and complement. The last problematic issue is that the students fail to gain the greater correct option of SPOC clause type (13.75%) because they choose the SPOA type (28.75%) and the SPC type (21.25%). The two percentages of incorrect options are greater than the correct option of SPOC (28.75% and 21.25% > 13.75%).

The above two tables expose the reasons to make the students fail to achieve great quantity or percentage of syntactic competence in terms of the basic clause types. The choices of incorrect options are greater than the correct ones as the keys. The problem such as this requires the ways to get solutions for the sake of improvement of the students' syntactic competence. Thus, one of the ways to do so is to design a model for the instructional materials of the English syntax.

4. Conclusion

The article analyses the students' syntactic competence for the seven basic clause types and the problematic issues toward this competence. Dealing with syntax at the level of basic clause types, one gets to discover the skills of both language usage and use. We are sure that if the students master these seven clause types, they will be able to identify the phrase categories with their own functions. Anyhow, they still fail to gain greater syntactic competence for the seven basic clause types (22.68%). The principal problem toward this competence is related to the recognition of SPOA (3.75%) and SPOC (13.75%). This case is affected by the problem to distinguish it from the SPOC clause type (55.00%). The other two types which become the fewer competences for them are SPOO (21.25%) and SPO (22.50%). The problematic issues are related to their recognition of making the difference between the mono-transitive clause type and the di-transitive clause one and also related to their recognition of making the difference between the object and complement functions. Thus, the students are suggested that they study the functions of phrase categories with more examples and practice to use the syntactic analyses for identifying each constituent in various constructions.

Acknowledgements

We are indebted to Directoriate General of Higher Education, Ministry of Research and Technology, Indonesia for a grant provision of our research, so that the opportunity to write the article is possible.

References

1. D. Crystal. *A World Language in the Encyclopedia of Language.* Cambridge: Cambridge University Press. (1997)
2. D. O'Neil. *Language Contact.* Wikipedia, the free encyclopedia, the Wikimedia Foundation, Inc. (2010)
3. M. Basri, D., A. T. Ampa, and Junaid. *Syntactic Errors in Descriptive Paragraphs by Native Indonesian-Speaking Students of English.* IJL. 5 (5): 125-137. DOI: 10.5296/ijl.v5i5.4455. (2013)
4. R. Titone and M. Danesi. *Applied Psycholinguistics.* Toronto: University of Toronto Press. (1985)
5. R. T. Bell. *An Introduction to Applied Linguistics: Approaches and Methods in Language Teaching.* London: B.T. Batsford Ltd. (1987)
6. N. Harrison. *Successful Writing.* Peter Francis Publishers, Great Britain. (1987)
7. E. K. Brown. and J. E. Miller. *Syntax: A Linguistic Introduction to Sentence Structure.* London: Hutchinson. (1986)
8. Ravindra. *Morphological Analysis of Verb Phrase in English Teaching.* RJELAL. 4 (3): 558-564. (2016)
9. R. Prasad and M. P. Sebastian. *A Survey of Phrase Structure Learning Methods for Text Classification.* IJNLC. 3 (2): 33-46. DOI: 10.5121/ijnlc.2014.3203. (2014)
10. P. Phopiphat and R. Kongakchandra. *The Recognition System of Sentential Noun Phrase Structure Using Conditional Random Field Model.* IJAIA. 6 (4): 53-62. DOI: 10.5121/ijaia.2015.6405. (2015)
11. H. Jackson. *Analyzing English: An Introduction to Descriptive Linguistics.* Pergamon, Oxford. (1985)
12. G. I. Yusifova. *Syntactic Features of English Idioms.* IJEL. 3 (3): 133-138. DOI:10.5539/ijel.v3n3p133. (2013)
13. M. Basri, D. *An Ecological Perspective on the Uses of the English Verbs in Interactions:* A study at the English department of Faculty of Letters UMI Makassar. A Research Report (June 2005) A research report. Makassar: Faculty of Letters UMI. (2005)

Teachers' classroom assessment and grading practices

Ida Ayu Made Sri Widiastuti[1,*]

[1]English Education Study Program, Mahasaraswati Denpasar University, Denpasar, Bali, Indonesia

Abstract. This study was conducted to investigate the teachers' classroom assessment and grading practices in Junior high schools in Bali. The participants of this study were professional junior high school teachers who were randomly selected as the sample of this study. In depth interviews and classroom observations were intensively conducted to the participants to collect the required data for the purpose of the present study. This study revealed that junior high school teachers made use of formative and summative assessment and employed several ways of grading practices. There were several factors affecting their classroom assessment and grading practices consisting of internal factors and external factors. The most prominent internal factor was the teacher's philosophy of teaching and learning and the main external factors were recognized as mandated statewide learning factors and high stakes tests, government polices, and parents. The findings indicated that junior high school teachers constantly make an effort to establish a reasonable balance between their teaching philosophy and the classroom environment reality. Thus it is recommended that junior high school teachers should review their assessment and grading practices and employ more appropriate assessment and grading practices.
Key words: Teachers, Classroom, Assessment, Grading, Practices

1 Introduction

Assessment is a key component of learning because it provides teachers and students the opportunity to meaningfully reflect how learning is best delivered. When students are able to know how their ability and how they are doing in a class, they will be able to determine whether or not they comprehend the learning material [5], [20, 29]. Teachers, on the other hands, are required to have comprehensive understanding on students' ability in order that they can decide the upcoming lessons.

In determining the students' achievement, teachers should be aware of various factors that may affect their grade. Teachers' strong educational principles and teachers' values and beliefs about teaching and learning also become considerations in doing grading decision-making [13, 24, 28]. Moreover, the reseason why assessment and grading practices are considered to a complex process that should be done by every teacher [11], [14, 28]

Many studies have been conducted related to classroom assessment and grading practices which many researchers found that grading practices are very complex decision-making process influenced by internal and external factors, as studies conducted by [18],

[*] Corresponding author: idaayuwidia@unmas.ac.id

[10, 5]. The phenomenon of assessment discrepancy made most of the teachers found that assessment and grading practices are difficult things to do. Therefore during grading practices, according to [28], [5, 9], the teachers have to employ their critical and logical thinking to accommodate all affecting factors to balance all of those factors.

Understanding the phenomenon that the teachers still encountered problems in assessment and grading practices, and they were still in conflict with their own beliefs and their school policy, therefore, it is very important to study the assessment and grading practices in order to provide valuable contributions to the teachers and other stakeholders for the improvement of assessment and grading practices conducted by all junior high school teachers.

2 Method

2.1 Research Design

Qualitative research design was employed in this study. A qualitative design is a common design whithin social sciences. According to this research method, data is collected from the participant to be coded, analyzed to draw the geralization.

2.2 Participants

The data of this study were collected from 10 English teachers who were randomly selected from junior high school in Denpasar city during the first term of 2017-2018 education year. Random sampling method can be used so that every teacher is equal possibility in the population. The data were collected by the researcher by conducting in depth interviews and classroom observation. In depth interviews and classroom observation give higher reability of data for this study.

2.3 Measures

Triagulation system with comparative analysis was used to measure the realibilty of the data. To ensure all data were agulated correctly, the qualitive data about the teachers' assessment and grading practices were put together and arranged into written form. The transcripts of the data were then coded and condensed into some categories or themes. Each theme is then analyzed argumentatively based on the logical theorical basis and factual phenomena. By employing this measure, the data can be analyzed based on their categories and the result can be easily angulated with other data. Cross data angulation allows the researcher to reveal higher reliability findings.

2.4 Procedure and Data Analysis

In-depth interviews and classroom observation were conducted to get the data regarding to the types of assessment and grading practices which usually conducted by the teachers, and the factors influence the teachers' grading practices. Interview guide were the instruments used to collect the primary data, while informal conversation were conducted to clarify the supplementary information. Observation blank sheets were utilized to note down the required data of the obeservation.The data were then transcribed into written form. The transcripts of the data were coded and condensed into some categories (theme). Using codes, the deconstruction phase began based on categorizing recurring ideas and major themes. The categories data were analyzed descriptively and urgumentatively to obtain the research findings about teachers'assessment and grading practices.

3 Findings

3.1 Types of assessment conducted by the teachers

There are some types of assessments commonly used by the teachers in assessing the students, the following are some excerpts of interviews conducted to the teachers in this study :

> *"I conducted formative assessment to see whether or not my students understand the lesson at the end of every learning unit. I also administer a summative test at the end of every semester to give my students score for one semester"*

> *"I usually give my students formative assessment and then I score my students work straightly. I use the students' exercise book as a test. I score the students and then I return their work as soon as possible "*

> *"I always give summative assessment at the end of semester using multiple choice tests and some completion test. I design my own test by getting most of the items from the internet"*

The data above showed that the teachers carried out two types of assessment to measure the students' learning achievement. Formative assessment is usually conducted at every end of one learning unit and summative assessment is conducted at the end of the semester.

3.2 Grading practices conducted by the teachers

Grading is considered as the most essential parts of assessment process in which a teacher is reguired to provide decision to the student' achivement in learning. The following are some excerpts of the interviews related to grading practices according to the teahers :

> *"I use different way of grading for different activities. If I use the same way of grading I will never be able to see the students' real ability "*

> *"When I am due to grade the students' assignment, I do it straight away because it is not easy for me because it really depends on my mood and my time. Subjectivity often occurred because many things I should consider in"*

According to the excerpt, the teachers in this study carried out several ways of grading which tends to be in different ways to suit their needs.

3.3 Factors influence the teachers' grading practices

There are two factors which influence the teachers' grading practice, those are internal and external factors.

3.3.1 Internal Factors

The following are the excerpts of the data concerning internal factors.

> *"I grade the students based on my own philosophy of teaching and learning. The students' assignments should grade properly because I believe that the assignments are*

very helpful for the students. I grade them based on my philosophy of teaching and learning. I believe that practice makes perfect or by practicing you will get a better result"

"I gave the students assignment to do and then I grade them in order they see their ability and they can make improvement and joining the upcoming learning process easier."

"For me assessment and grading can help the students improve their learning achievement because based on assessment result they know what part of material they have not yet understood".

The excerpts show that the teachers usually carry out grading practices by grading their students assignments to help the students understand their ability and to improve their students' learning.

3.3.2 External Factors

There are many external factors that may influence the teachers' assessment and grading practices.

"I prefer to use my own test which I have been designed according to the standard. It is easier for me to grade because the material are related to what taught.".

"I am free to choose whether I follow the school policy or not. It is up to how I interpret the policy. I feel my personal professional assessment and my own judgment are more appropriate for the students' grade and the students' achievement".

"I am still confused with the school policy and the grading practices. I felt I could not do the grading properly. That is why I just used my own ways in grading".

"I need to show the students' parents considerably comprehensive information on the students' grade to make their parent understand their children's ability"

The result of the interviews showed that the most influential external factors were in the form of statewide mandate testing and high stakes tests; regional grading polices, and parents. The teachers described that the continually tried to establish reasonable balance between their philosophy of teaching and learning and the classroom environment realities and external factors.

4 Results and Discussion

The interviews conducted to the teachers indicated that most teachers carried out formative assessment and a summative assessment. However, it was found that most teachers were lack of comprehension on the primary purposes of formative and summative assessments. Ideally, the result of the formative assessment was mainly utilized for the improvement of teaching and learning processes, that is why feedback is essentially important to be given to the students [5], [1, 19, 22]. Feedback actually leads the teachers for the improvement of teaching [6], [3,15, 25, 26].

The teachers in this study mainly develop their own grading practices depending on their professional ideology, education background, norms and experiences in grading the students' learning performance. Teachers ideally are competent in creating effective grading systems to correctly measure the students' learning achievement [5], [2,4, 22]. Moreover, teachers should have a strong policy understanding to avoid some negative effects from external factors which may influence their grading decision making [2], [13,16].

The grading practices for most teachers are considered to be important to do appropriately because students need to know their real ability in order to be able to make progress in learning [4], [7, 8, 11]. Moreover, the interviews conducted to the teachers showed that teachers have varied responses on the factors that influence their grading practices. The teachers experienced difficulty in avoiding factors which may affect their grading. Those factors varied in their forms including internal and external factors [10], [12, 13, 27].

The most important internal factor that appears to influence teacher decision making concerning classroom assessment and grading practices is the teacher's philosophy of teaching and learning. These factors resulted the various ways of grading undertaken by the teachers and the decision in the same time might be also different. There are five types of teacher beliefs and values were identified internal factors in grading, such as: philosophy of teaching/learning, pulling for students, accommodating individual differences, students' development, and students' engagement and motivation. These factors are considerably influential factors in grading, but they are essential factors to improve and engage the students in learning [18], [22, 28, 23, 29].

The most influential external factors were in the form of statewide mandate testing and high stakes tests, regional grading polices, and parents. The teachers described that the continually tried to establish reasonable balance between their philosophy of teaching and learning and the classroom environment realities and external factors [14], [19, 22, 28]. Many teachers focus on making sure that the students gain a good score in national test. This indicates that the teachers 'assessment and grading decision getting high pressure from external factors especially those related to statewide mandated testing. Teachers more likely to reflect their grading in order to be used to guide the students to be able to do the national test conducted by the government nationally [10], [3, 5, 9]. These factors are considered to be important for the school stakeholders including students, teachers, headmaster and parents [19], [6, 28]

In some cases external factors are still controllable because they indirectly influence the students. Those factors are mainly dealt with how the teachers decide the grading practices undertaken. In a simple explanation, they are outsiders and merely manifested on the teachers' decision rather than the students' attitude or behavior [10]. In this study, three major external factors were identified which may significantly influenced the teachers' assessment and grading practices: (1) externally mandated statewide learning standards, mostly multiple choice, high stakes tests, (2) regional policies, and (3) parents. Here are some excerpts of the interviews related to external factors.

The discussion above showed that teachers' assessment and grading practices are mostly influenced by internal and external factors. In conclusion, the junior high schools teachers tend to employ their own assessment and grading practices based on their belief, knowledge, education background, experiences, school policy, norms and assessment principles. The teachers also preferred conducting their own assessment policy rather than national policy.

References

1. Black, P., & Wiliam, D. Assessment and classroom learning. *Assessment in Education:Principles, Policy & Practice*, *5*(1), 7–74, (1998).

2. Brookhart, S. M. Teachers' grading practices: Meaning and values. *Journal of Educational Measurement, 20,* 123–142, (1993).
3. Brookhart, S. M. Teachers' grading: Practice and theory. *Applied Measurement in Education, 7,* 279–30, (1994).
4. Brookhart, S. M. Classroom assessment: tensions and intersections in theory and practice. *Teachers College Record, 106,* 429–458, (2004).
5. Brookhart, S. M., & Durkin, D. T. Classroom assessment, student motivation, and chievement in high school social studies classes. *Applied Measurement in Education,16,*27–54, (2003).
6. Cheng, L., Rogers, T., & Hu, H. ESL/EFL instructors' classroom assessment practices: Purposes, methods and procedures. *Language Testing, 21,* 360–389, (2004).
7. Cheng, L., Rogers, T., & Wang, X. (in press). Assessment purposes and procedures in ESL/EFL classrooms. *Assessment & Evaluation in Higher Education.*
8. Cheng, L., & Wang, X. Grading, Feedback, and Reporting in Esl/Efl Classrooms. Language Assessment Quarterly, *4*(1), 85–107, (2007).
9. Cumming, A. ESL/EFL instructors' practices for writing assessment: Specific or general purposes? *Language Testing, 18,* 207–224, (2001).
10. Duncan, C.R & Noonan, B. Factors Affecting Teachers' Grading and Assessment Practices. *The Alberta Journal of Educational Research Vol. 53, No. 1, Spring 2007, 1-21,* (2007).
11. Dyrness, R. & Dyrness, A. Making the grade in middle school. *Kappa Delta Pi Record, 44*(3), 114-118, (2008).
12. Guskey, T. Are zeroes your ultimate weapon? *Education Digest: Essential Readings Condensed for Quick Review, 70* (3), 31-35, (2004).
13. Guskey, T. "*It Wasn't Fair!" Educators' recollections of their experiences as students with grading.* Paper presented at the annual meeting of the American Educational Research Association, San Francisco, CA, (2006).
14. Guskey, T. Five obstacles to grading reform. *Educational Leadership, 69(*3), 17-21, (2011).
15. Kulik, J. A., & Kulik, C. C. Timing of feedback and verbal learning. *Review of EducationalResearch, 58,* 79–97, (1988).
16. McMillan, J. H., & Lawson, S. R. Secondary science teachers' classroom assessment and grading practices. *Educational Measurement: Issues and Practice, 20,* 20–32, (2001).
17. McMillan, J. H., Myran, S., & Workman, D. Elementary teachers' classroom assessment and grading practices. *Journal of Educational Research, 95,* 203–213, (2002).
18. McMillan, J. H., & Workman, D. *Classroom assessment and grading practices: A review of literature.* Richmond, VA: Metropolitan Educational Research Consortium, Virginia Commonwealth University. (ERIC Document Reproduction Service No. ED453263), (1998).
19. McMillan, J. *Assessment essentials for standards-based education* (2) Thousand Oaks, CA: Sage Publications, (2008).
20. McMillan, J., & Nash, S. *Teacher classroom assessment and grading practices decision making.* Paper presented at the annual meeting of the National Council on Measurement in Education, New Orleans, LA, (2000).
21. McMillan, J. H. & Schumacher, S. *Research in education: A conceptual introduction.* New York: Longman, (2006).
22. Moeller, A. J., & Reschke, C. A second look at grading and classroom performance:Report of a research study. *The Modern Language Journal, 77,* 163–169, (1993).

23. Patton, M. Q. *Qualitative research and evaluation methods* (3rd Ed.). Thousand Oaks, CA: Sage, (2002).
24. Randall, J., & Engelhard, G. Examining teacher grades using Rasch measurement theory. *Journal of Educational Measurement, 46*(1), 1-18, (2009).
25. Sri Widiastuti, I.A.Md. EFL Teachers' Beliefs and Practices of Formative Assessment to Promote Active Learning. The ASIAN EFL Journal. Volume 3, (2016).
26. Sri Widiastuti, I.A.Md. *Teachers' Understanding of Formative Assessment. Jurnal Bahasa dan Seni.* Vol 45, No 1 Juni, (2017).
27. Stiggins, R. J., Frisbie, D. A., & Griswold, P. A. Inside high school: Building a research agenda. *Educational Measurement: Issues and Practice, 8,* 5–14, (1989).
28. Sun, Y & Cheng, L. Teachers' grading practices: meaning and values assigned. *Assessment in Education: Principles, Policy & Practice, 21:3, 326-343, DOI: 10.1080/0969594X.2013.768207,* (2014).
29. Wormeli, R. Accountability: Teaching through assessment and feedback, not grading. *American Secondary Education, 34*(3), 14-27, (2006).

SHS Web of Conferences **42**, 00053 (2018) https://doi.org/10.1051/shsconf/20184200053
GC-TALE 2017

The Cultural Ideological Contestation in National Examination

Made Kerta Adhi
The Institute of Teacher Education and Training Saraswati

Abstract

The purpose of this research was to find out the forms and the factors leading to the contestation of cultural ideology in National Examination. This research utilized Qualitative Research method and the approach of Cultural Studies. Based on the analysis, it was found that the forms of ideological contestation, specifically, centralistic, prioritizing on the result of cognitive domain, and imaging. The factors which caused consisted of political of education, state policies, and socio-cultural. The implications of this research were (1) There was shifting of educational values into capitalism, (2) The meaning of learning became limited in which it was just purposed at facing National Examination, (3) Educational services have shifted from humanist public services to commodification of education, with the result that honesty was marginalized. Eventhough the government has developed sophisticated system and very tight supervision, if the character of the subjects are not formed and cultured, accordingly, the National Examination stated honest and achievement oriented will be just a slogan. Therefore, the education system needs to be improved. The education paradigm which is more emphasized on the increasing of intellectual intelligence by multiple choice tests and scores needs to be deconstructed by adjusting with Indonesian cultural values based on "Pancasila" as ideological foundation and developing character education to the children from early age.

Key words: contestation,national examination,cultural ideology,character

I. INTRODUCTION

Education as a cultural practice can never be separated from the performance of dominant structure. The role of the country can not be separated in both the process and the educational quality. In relation with this, the government runs national exam as one of the way to enhance educational quality. Tubagus (in Chan, 2010:41) states that the National Examination is highly required and relevant with the fact to find out quality of education. Moreover, Indonesian president, Joko Widodo decided that the National Examination still exists, due as the benchmarking of the students' progress, so the students are not only being local level fighters but also being able to compete on international level (www.pikiran-rakyat.com).

The Goverment Regulation No. 13 Year 2015 and the Regulation of Minister Education and Culture No. 3 Year 2017 are the regulators legitimizing the National Examination still continue until the academic year 2016/2017. Darmaningtyas (2007) states that the authorities certainly do not desire to remove the National Examination because it is a fertile gold mine for material gain, despite giving a huge economic burden on the people and depriving the rights of teachers. The government tends to determine by emphasizing on cognitive domain through multiple choices test with the result in the form of numbers. In addition to computer-based or paper and pencil-based national exams, the students are also required to take School Examination and National Standardized School Examination. This phenomenon is very interesting to be studied in critical education with the approach of Cultural Studies.

This research specifically purposed to find out the forms and causative factors of ideological contestation of cultural dualism and the implications in the practice of National Examination at SMA Duta Pelangi and SMA Pahlawan Bangsa Tabanan. The method used in this research was Qualitative Research method .The problems were eclectically analyzed by utilizing critical theories in the perspective of Cultural Studies Approach, such as (a) Ideological theory, (b) Social Practice theory, and (c) Critical Education theory. The obtained data were analyzed by utilizing Qualitative Data analysis and interpretative, then presented narratively.

II. LITERATURE REVIEW

The ideology behind the cultural dualism is binary opposition as stated by Levi-Strauss (in Ritzer,2009:53) or people of Bali cited by the term "rwa bhineda" (Mantra, 1996: 25; Atmadja, 2010). Cultural dualism are focused on the idea sorting something into two opposite. This sorting can lead to the placement of one thing at the center position, otherwise cause the other to number two or even excluding.

The ideology of cultural dualism which are appear through the texts of laws and regulations dichotomously defined into two categories, specifically success or unsuccessful in National Examination with the slogan " achievement is yes, honesty is a must", as well as tested and untested subject. The implementation is not done by violence, but through the forms of community consent. The dominant class dominate the lower class by using ideology because ideology is an unconscious belief through hegemony. Fakih (2009: 30) states that school and education is not a neutral arena but system and hegemonic structure.

The national examination is deconstructed as important thing. If the exam is not carried out, it will cause bad impact to the students or school, such as students are stated fail or the school can not graduate the students. On the contrary, if the final exam runs with optimal achievement, students, school, and regional area will obtain prestige even an award from Indonesian precident. The achieving schools in the provincial and national level are demanded to be able to maintain the achievement. Bourdieu (in Fashri, 2007: x) states that if the person is on the higher structure, the person will wants to maintain.

When educational institution functionate as hegemonic device, accordingly, it is opened great opportunities to create a form of coercion of ideas, knowledge, and ideology through the mechanism of subtle and invisible violence as symbolic violence (Piliang, 2006: 359; Giddens, 2009: xxii; Bourdieu in Harker, 2009: 120). Behind the hegemony, there is possibility of resistance as a movement to be released from the texts considered discriminatory, inferior to the lower class, and causing injustice.

In the implementation, National Examination constantly sustains dynamics according to the struggle of interests between the dominant structure to the lower class (school community). This condition causes the texts run into adjustment appropriate to the concerns of each structure. The rules which change every year reconstructed based on the wishes and interests of various structures, educational politics, and globalization emancipated and emulated by the quality standard of education. There is interdependence between agents influenced by interests, power, and ideology (Beilharz,2005: 193; Giddens,2009: xviii).

The agents target all students succeed in National Examination practice. This fact occurs continously so it becomes habitus. In achieving the target, there are the performance of capital, ideology, and power. The performance of agents which happen at school involve school principles, teachers, officers, students, school committee whereas the performance of agents outside the school are carried out by local goverment and the other stakeholders, such as family, tutoring agency, and social media. Bourdieu (in Harker, 2009: 9) argues that the texts are stated in the equation of (Habitus x Capital) + Domain = Practice.

In traditional education, it is stated that the implementation of National Examination runs well. However, critical education views that there are hidden missions behind it, specifically representing interests, ideologies, and serving certain power which cause the other items to be marginalized. Henry Giroux (1993) and Paula Allman (1998) believe that there is political content in all educational activities in which the political content does not represent singular and homogenous idea but empowers the oppressed and transforms injustice. National Examination is the representation of power seizing the school right and hegemonizing the schools. The limitation towards success scores and the type and number of subjects examined are the realization of goverment power dominance. Education has been coopted by the power of state hegemony.The power coming from above will remove the culture that tends to uninformism (Nuryatno,2008: 1; Karim,2009: 123; Safa, 2011: 25).

The assessment of National Examination based on academic standardization is harassment on integrity of human dignity. Academic capability is the cornerstone for education, however the character growth is the foundation for person's life. Education is demanded to provide benchmarks to good attitude and behaviour. Education does not just create persons in accordance with forced quantitative specification, so many things of value in the self are lost. Tilaar (2006: 198) argues that the National Examination just prioritizes cognitive achievements on certain subjects, however the aspects of affective and psychomotor are ignored. National Examination has reduced personal wealth to merely production good that can be standardized hence individuals lose the value as unique person. The achievement establishes the equation which becomes means to spread injustice. Freire (2007: xiii) states that education is for freedom, not for assignment. Education must be the process of independence which is not social and cultural domestication. Meanwhile, Fakih (in O'neil 2008: xvi) states that education is humanizing human beings who are dehumanized due to unfair systems and structures.

The result of ideological contestation were represented not just in the written texts of the regulation, but those were able to be viewed concretely in the practice of National Examination, as the agents at the location of research prioritized tested subjects to untested subjects. In achieving achievements, tangible achievement was prioritized more than intangible achievement. The impacts were prioritized more on result to process and cognitive aspect to affective and psychomotor aspects.

The Regulation of National Education System No. 20 Year 2003 defines that education is conscious and planned efforts to realize environment of learning and learning process, so the students develop self- potential actively to hold religion spiritual power, self- restraint, intelligence, noble character, and skills needed for own-self, nation, and state. Based on the regulation, the texts needs to be deconstructed to return to the natural tendency. Similarly, conducting to the assessment, regardless of name and form, the result is demanded to reflect the elements containing in the regulation proportionally and holistically. The deconstruction of the texts utilize the Deconstruction theory of Derrida. Deconstruction refused to the logocentrism and phonocentrism that overall create binary oposition and hierarchical dichotomy ways of thinking (Piliang, 2003:11; Ratna, 2006: 222).

III. RESEARCH METHOD

This research was conducted by utilizing Qualitative Research method and Cultural Studies approach. The Informants who are determined by the sampling techniques of purposive and snowball were students, teachers, leaders and businessmen. The data were collected by using the techniques of participatory observation, deep interview, and document study. The analysis data were conducted with qualitative data analysis which consist of data reduction, data display, conclusion, and verification. The result of this research was presented narratively.

IV. RESULTS AND DISCUSSION

4.1 The Forms of Cultural Ideological Contestation of Cultural Dualism in National Examination Practice at SMA Duta Pelangi dan SMA Pahlawan BangsaTabanan

4.1.1 Centralistic and Prioritizing on the Result of Cognitive Domain

Based on the regulation of National Examination, the National Examination tended to sentralistic which started from tested subjects, tested questions, implementation schedule, forms of examination, and graduation criterias. The tested subjects were reflection of global knowledge examining exact sciences dominantly and prioritizing on cognitive domain. Meanwhile,non-exact sciences were just examined at school. Schools as the lower structure had obligation to follow the established provision, including the subjects examined nationally. The result of the interview with school principle was school in examining subjects tested in national examination refered to the esthablished terms in the Standard Operational Procedure of National Education Standard Agency.

The school did not have authority in determining the subjects tested. The dichotomy between tested and untested subject indicated that there were the separation between public and domestic sector. Untested subjects were as the indication of domestic sector where as tested subjects with "national" label close to public sector valued more dominant and market-oriented. The implication of tested subjects were prioritized and obligated, although the students difficult to answer National Examination (Retnawati, 2017). This resulted untested subjects were marginalized.The comparison of tested and untested subjects in National Examination in Academic Years 2011-2017 can be noticed in table 1.

Tabel 1: The Comparison of Tested and Untested Subjects in National Examination in Academic Years 2011-2017

Untested Subjects in Academic Years 2011 – 2017	Tested Subjects in Academic Years 2011 - 2017	
Natural and Social Science Department	Natural Science Department	Social Science Department
1. Religion	1. Indonesian	1. Indonesian
2. Civics Education	2. Math	2. Math
3. History	3. English	3.English
4. Art and Culture	4. Biology*	4. Geography*
5. Physical Education and sport	5. Chemistry*	5. Sosiology*
6. Balinese	6. Physics*	6. Economics*
7. Character Education		

Source : Standard Operational Procedure of National Education Standard Agency for National Examination in Academic Years 2011-2017

Explanation : *) The subjects must be taken in 2011, however one subject could be chosen according to the department.

The tested subjects as public sectors were given more handling than untested subjects as domestics sectors because the tested subject with national label had more risks on the unsuccessfulness and imaging impact. Moreover, those with hard category, such as Biology, Math, Chemistry, Physics, English, and Indonesian for Natural Science department and Math, Economics, Georaphy, Sosiology, Indonesian, and English for Social Science department (Standard Operational Procedure of National Education Standard Agency for National Examination in Academic Years 2011and 2017). The preferred subjects were indicated by means of National Examination data in 2011 in which the everage scores of school's values for Social Science department were 8.53 and 8.89 for the value of National Examination where as for Natural Science department were 8.79 for School value and 9.17 for the value of National Examination (Kerta Adhi, 2013: 305). The data of National Examination in 2017 indicated that the everage results were 67.68 for School Examination and 88.17 for National Examination for Natural Science department whereas the obtained results for Social Science department were 67.51 for National Examination, 85.17 for School Examination, and 87.83 for National Based School Examination (Archives of SMA Duta Pelangi). Meanwhile, the everage scores of social department at SMA Pahlawan Bangsa reached 43.69 for National Examination, whereas each for School Exam and National Based School Exam were 79.80 and 78.87. The data indicated that the values of National Exam were less than School Exam and National Based School Examination. Those were as the sign of behaviour change in which National Examination did not determained the graduation and the implementation was based on computer.

The tested subjects tended more difficult than untested subjects. Those measured cognitive aspect with the result that the students concentrated more on the tested subjects. National Examination constructed social stratification, specifically the differentiation offered impact to the difference of attitude, behaviour, and students' action to the teachers who teach tested subjects and who teach untested subjects (Damsar, 2011: 141).

The hard subjects were just able to give information in relation to cognitive capability whereas the non-cognitive were not able to be evaluated, moreover the exam was just done temporally and concisely. National examination carried out by the government tended based on formal logical standard than material logical. National Examination was only viewed on the form but it was not accordance with content philosophy and students' need in determining learning result. The purpose of learning at school was constructed to prepare and face National Examination. The educational meaning purposed to develop thinking, soul, and body has been reduced by intellectual western mindset. Karim (2009: 146) and Andyka (in Safa, 2011: 53) state that the National Examination evidently confine the space for potential students intelligence, possibly it opposes human nature. According to this fact, the purpose of learning at school was formed more on preparing and facing National Examination or learning for the test. The spirit of students critical reasoning development became restricted. Future expectations of the students were hanging only on a passing grade or success.

This fact was accordance with Indonesian intellectual people in developing country which are generally grown through schooling system. Consequently, educational system prioritized on exact science and technology and it was not influenced by classical education. Values of education must be perceived as a necessity, not something desirable as an element of personal formation (Drost, 2008: 149).

4.1.2 Imaging

In the reserach location, students or schools carried out maximal efforts to succeed because the success was the interest and cosumtive need. The consumtion was not just to fullfil value for human's utility and basic need but related with symbolic meaning to indicate class, status, prestige, as means to state social's differentiation, and mapping. Irawan (2015) argues that schools understand the mapping for transcribing the extent of National Examination school rankings. Accordingly, the achievement was not merely achievement but also the prestige in stages from students/schools / students' parents and the regions. The prestige of schools and or regions as symbolic meanings to distinguish the capability of students or schools.

4.2 Causative Factors of Ideological Contestation of Cultural Dualism in National Examination Practice In SMA Duta Pelangi and SMA Pahlawan BangsaTabanan

4.2.1 Political Education and State Policy

The occurrence of ideology contestation, particularly achievement and honesty at the research location due to the state policy stating that the National Examination aimed at achieving standarization of national education. National Examination as a public policy is stated in the form of regulation (Masduki, 2008: 8)

The examination was standardized according to logical capitalism. The quality of Indonesian education was established based on capitalistic decision makers, for instance, the Competency Based National Examination. The government always argues that the standard of National Examination by assigned values to execute students' graduation based on the comparison of Indonesian educational quality to the other countries (global capitalism). There was competition of ideology as an implication of education. Ade Hermawan (in Safa, 2011: 25) states that education in globalization era is market-oriented. Globalization leads the shifting function of educational institution to be business, focusing on the issue of National Examination, and viewing educational quality standard just by measuring students'cognition.

The regulation of texts were explained in detail by Educational National Standard Agency in the form of Standard Operational Guidelines of National Examination. The texts contained meaning that there were structurally hegemonic process and cultural dualism ideology. The government, related institutions, communities, and students' parents also participated in successing the examination. Both public and private schools were obligated to follow the National Examination.

4.2.2 Socio-Cultural

Students and other education implementers in the location of research stated of feeling unsuccessful if the obtained values were low, especially the students became inferiority likewise the parents. If the presentage was marginal, the school burdened morally of being claimed to fail to educate children. The failure was often utilized as indicators of school quality. The rank of obtained results were also used as an indicator of educational quality in the region. Accordingly, the agents tried to reach the achievement to enhance the social cultural status, furthermore students can be accepted in the favorite level of education. Zamroni (2007: 239) argues that reaching achievement in the National Examination was backgrounded by unvisible cultural factors. Therefore, the agents conducted guadiance for the students both at school and outside school. In the implementation, students subtly cooperated in answering questions moreover the supervision was loose.

4.3 The Implication of Ideological Contestation of Cultural Dualism in the Practice of the National Exam at SMA Duta Pelangi and SMA Pahlawan BangsaTabanan

The implications of this research were experienced by school principles, the entrepreneurs of the tutoring institution, the agents, and quality of education.

The implication to the school principle was experiencing symbolic violence in implementing the policy, with the result that the principal carried out some efforts, such as prioritizing the tested subjects, making a special budget, and providing guidance to teachers/supervisors and students. The hegemonic teachers had to follow Standard Operational Guideline but gentle resistances were done by tutoring, loosening supervision, and assisting to provide answers. Those were caused by the rights and authority of teachers in accordance with the law need to be noticed, moreover the teachers who had been certified as an educator.

The implications for the entrepreneurs of the tutoring institution were economically benefited. In addition, the students became more prepared normatively. However, beyond that, there were the shift of educational values into capitalism. The involvement of independent monitoring teams in research location indicated that implementation obtained legitimacy.

The implication for the agents continued to the process and result of learning. The meaning of learning became limited and shallow. Educational services had shifted from humanist public services to commodification of education. The orientation of the National Examination has changed from paper and pencil-based test to computer-based test which caused anxiety to the students (Sulistyaningsih, 2016). In addition, National Examination prioritized more on result to process, cognitive aspect to affective and psicomotor aspects, logical and mathematical intelligence, learning for the test to comprehensive material comprehension, uniformity/ standard to diversity, rich individuals to the poor,central government/ sentralistic to school autonomy, and reaching success but ignoring failure.

The implications for the quality of education were biased. In the location of research, the implementation prefered to the tangible achievement in the form of numbers/scores interpreted as achievement than the process in achieving. This condition implicated to the educational quality. A high score was not significant to improve the quality of education when it was obtained by using incorrect ways. Educational quality was not only viewed from the indicators in the form of numbers/ scores, but also from the intangible indicators, specifically honesty, in accordance with the slogan "achievement is yes, honesty is a must".

Globalization by Giddens has brought people to Knowledge Based Society, with the result that knowledge is the standard in all way of life. Cognitive power supported by materialistic attitude without any sense of power and physical power causing human life to dwell only on number by number, the success is numbers, fate is determined by numbers, so the child becomes burdened (Lutfi Lazuardy in Safa, 2011: 85; Chan And Tuti T Sam, 2010: 39).

The texts of National Examination need to be deconstructed because the texts must be constructed from Indonesian cultural values based on "Pancasila" as ideological foundation and not capitalism. The sovereignty of the teachers/schools need to be restored according to the habitus. Beside that, the students have to be valued as subject of uniqueness. Curriculum need to be reviewed and children should not be only stuffed with knowledge in the establishment of intellectual intelligence, but also be equipped with character education from early age.

IV. CONCLUSION

Ideological contestation were represented on written texts of the regulation and viewed concretely in the practice of national examination. In the location of research, the agents prioritized tested subjects to untested subjects because the tested subjects had higher value of indicator level, more risks on the unsuccessfulness, and imaging impact. National Examination as a dominant structure product had power in determining the fate and the future of the agents. National Examination was established by National Education Standard Agency with the slogan "achievement is yes, honesty is a must". However, in the implementation, there were discriminations towards the achievement or the value. The agents prioritized achievement than honesty.The National Examination has become a fetter because the parties involved inside did not really act according to the Standard Operational Guideline but tended to deviate. Therefore, it is necessary that the texts need to be deconstructed by changing the learning system and the judgments not merely seen from the sophistication of technology but the humanistic learning process. In addition, prioritizing character education to the children from early age continuously and consistently need to be carried out.

REFERENCES

[1] Atmadja, Nengah Bawa. "Sekolah (Rintisan) Bertaraf Internasional Sebagai Arena Sosial Melanggengkan Ketidak Adilan Bagi Kaum Miskin (Perspektif Teori Kritis)". *Media Komunikasi FIS.* Edisi Khusus Pendidikan Ekonomi **Vol. 9** No.1 Juni 2010. ISSN 1412-8683. Singaraja: FIS Undiksha (2010)

[2] Beilharz, Peter.*Teori-teori Sosial*. Yogyakarta: Pustaka Pelajar.(2005)

[3] Chan, Sam M dan Tuti T Sam. *Analisis SWOT: Kebijakan Pendidikan Era Otonomi Daerah.* Jakarta: Raja Grafindo Persada. (2010)

[4] Damsar. *Pengantar Sosiologi Pendidikan.*Jakarta: Kencana Prenada Media. (2011)

[5] Darmaningtyas. *Pendidikan Rusak-Rusakan.*Yogyakarta: LKiS. (2007)

[6] Drost, SJ. J.I.G.M. *Sekolah: Mengajar atau Mendidik*. Jogyakarta: Kanisius.(2008)

[7] Fakih, Mansour. *Runtuhnya Teori Pembangunan dan Globalisasi.*Yogyakarta: Insist Press. (2009)

[8] Fashri, Fauzi. *Penyingkapan Kuasa Simbol: Apropriasi Reflektif Pemikiran Pierre Bourdieu.* Yogyakarta: Juxtapose (2007)

[9] Freire, Paulo. *Politik Pendidikan: Kebudayaan Kekuasaan dan Pembebasan*. (Agung Prihantoro dan Fuad Arif Fudiyartanto Penerjemah).Yogyakarta: Pustaka Pelajar (2007)

[10] Giddens, Anthony.*Problematika Utama dalam Teori Sosial: Aksi, Struktur,dan Kontradiksi Dalam Analisis Sosial.* (Dariyatno, Penerjemah). Yogyakarta: Pustaka Pelajar (2009)

[11] Harker, Richard dkk. (Editor). *(Habitus x Modal)+Ranah = Praktik: Pengantar Paling Komprehensif Kepada Pemikiran Pierre Bourdieu*. Yogyakarta:Jalasutra (2009).

[12] http://edukasi.kompas.com."Keberhasilan UN Hanya untuk Kepentingan Politik" Diakses Tanggal 20 Desember (2010)

[13] Irawan, Candra."The National Examination And The Quality Of Education Mapping" Indonesian Journal of Educational Review p-ISSN 2338-2015 | e-ISSN 2335-8407 **Vol.2**, No.1, p 97-105, March (2015)

[14] Karim, Muhamad. *Pendidikan Kritis Transformatif.* Yogyakarta: Ar-Ruzz (2009)

[15] Kerta Adhi, Made. "Pengutamaan Kebijakan UN dan Representasinya pada Sistem Manajemen Pendidikan di SMA Tabanan". *Disertasi.* Denpasar: Program Pascasarjana UNUD (2013)

[16] Mantra, IB. *Landasan Kebudayaan Bali*. Denpasar: Yayasan Dharma Sastra (1996)

[17] Masduki dan Muzayin Nazaruddin (Editor). *Media, Jurnalisme dan Budaya Populer*. Yogyakarta: Prodi Komunikasi Universitas Islam Indonesia dan UII Press (2008)

[18] Nuryatno, Agus. *Mazhab Pendidikan Kritis: Menyingkap Relasi Pengetahuan Politik dan Kekuasaan.* Yogyakarta: Resist Book (2008)

[19] O'neil, William F. *Ideologi Ideologi Pendidikan.* Yogyakarta: Pustaka Pelajar (2008)

[20] Piliang, Yasraf Amir. *Hipersemiotika, Tafsir Cultural Studies Atas Matinya Makna.*Yogyakarta: Jalasutra (2003)

[21] Piliang, Yasraf Amir. *Dunia Yang Dilipat: Tamasya Melampaui Batas-Batas*

Kebudayaan .Yogyakarta: Jalasutra (2006)

[22] Ratna, Nyoman Kutha.*Teori, Metode,dan Teknik Penelitian Sastra.* Yogyakarta: Pustaka Pelajar (2006)

[23] Retnawati, Kartowarigan, Arlinwibowo,Sulistyaningsih. "Why are the Mathematics National Examination Items Difficult and What is teachers Strategy to Over come it?. *International Journal of Instruction.* **Vol.10.** No.3 ISSN 1694-609X pp 257-276.(2017)

[24] Ritzer, George.*Teori Sosial Postmodern.*Yogyakarta: Kreasi Wacana (2009)

[25] Safa, Aziz (Ed.), Tim Kreatif LKM UNJ. *Restorasi Pendidikan Indonesia:Menuju Masyarakat Terdidik Berbasis Budaya*. Jogyakarta: Ar-Ruzz Media (2011)

[26] Sulistyaningsih,Eny. "Student Anxiety Facing Computer Based Test (CBT) System of National Examination". *Proceeding of 3rd International Conference on Research.* ISBN 978-602-74529-0-9 p 113. (2016).

[27] Tilaar, HAR. *Standarisasi Pendidikan Nasional: Suatu Kajian Kritis*. Jakarta: Rineka Cipta (2006)

[28] www.pikiran-rakyat.com. Arie C Meliala. "Presiden Putuskan Ujian Nasional Tetap Ada" dalam Pikiran Rakyat. Diakses Tanggal 10 April (2017)

[29] Zamroni. *Pendidikan dan Demokrasi Dalam Transisi.* Jakarta: PSAP Muhammadiyah (2007)

The Development of Instruments to Measure Student Mathematical Logical Thinking Ability in *Kapita Selekta*

Novaliyosi[1]
1) Pendidikan Matematika, FKIP, Universitas Sultan Ageng Tirtayasa, Serang, Banten, Indonesia

Abstract. The ability of logical thinking mathematically is one of the goals of the learning of mathematics, namely forming ability of reasoning students that reflected his skills in critical thinking, logical, systematic, and objective nature, honest, discipline in solving a problem well in various fields. In addition math can form a mindset into a mindset of mathematical, logical, systematic and critical that can be used to solve problems in daily life. This research aims to develop a test instrument for measuring the ability of logical thinking mathematically students on kapita selekta. The development method used in this research is to include: (1) define the variable; (2) outlines the variables into a more detailed indicators/dimensions; (3) draw up the details; (4) perform validation; (5) conducting trials; (6) analyze of validity and reliability. Based on the results of the test in theoretic instruments that developed included into the category is valid and worthy to be used on next stages as an instrument to measure the ability of logical thinking mathematically.

Keywords: *Instruments, Mathematical Logical Thinking.*

1. INTRODUCTION

Learning mathematics aims at forming ability of students reasoning that reflected his skills in critical thinking, logical, systematic, and objective nature, honest, discipline in solving a problem either in various fields (Widdiharto, 2004). Mathematics can form a mindset into a mindset of mathematical, logical, systematic and critical that can be used to solve problems in daily life (Tiurlina). In line with the opinion, there are two development direction about the nature of mathematics education, development to the needs of the present leads to an understanding of the concepts needed to solve the problem of mathematics and other sciences and the time will come that make learning math can give the ability of reason and logical, systematic, critical thinking, and careful and objective and open (Sumarmo, 2002).

Commitee on the Undergraduate Program in Mathematics (CUPM) (2004) describes the six basic recommendations for majors, programs, and all courses in mathematics. One of them explains about the development of the ability of analysis, critical reasoning, logical thinking, problem solving, communication and mathematical representation can be formed through courses in mathematics. Therefore already for colleges that produce prospective educators (teachers) which is in charge of teaching mathematics is responsible for preparing the student to improve mathematical ability as one of the factors cause the failure of education in building national character are educators.

Students majoring in mathematics education as a prospective educators (teachers) should be aware that they have a strategic role in the formation of the next generation a nation that not only has the intelligence of a brain but also a good character. In this case the brain intelligence indicators which were learned, think logically and critically. Need to be developed so that the ability of logical thinking ability of the students in this case can think logically and systematically.

Think logically give full confidence to say something because everything that is said to be based on a logical reason as well as the support of strong evidence. It makes a person can get the truth, because it has been through the process of formulating and evaluating their own beliefs and opinions are systematically. Logical thinking allows one to evaluate the evidence, logic, and language that underlies the statement of others privately organized. Through math problem solving, logical thinking ability could be developed because math is a tool to develop the ability to think. One of the ability of thinking that can be developed in the process of problem solving student is logical thinking.

Logical thinking ability is closely related to the activities of problem solving. This is in accordance with the opinion of the Tuna, Cagri Biber & Incikapi (2013) stating "activities requiring such critical skills USA, creative problem solving, critical and reflective thinking would also improve teacher candidates ' logical thinking ability". In addition, according to the Yaman (2005) one of the cognitive skills that affect the academic success of students is the ability to think logically. Logical thinking ability based on individual expertise to solve the problem by using its ability to achieve the principles or rules by making generalizations or abstractions. So it needs to be a logical thinking ability students as prospective teachers is enhanced through some activity that demands critical skills include problem solving, creative thinking, critical and reflective. The more frequently involved in the activity of problem solving, logical thinking ability of the students will be getting better.

1 Corresponding author: chi_2409@yahoo.com

Based on that assumption, it will test the ability of the instrument developed logical thinking for students on kapita selekta.

1.1 Research Question

Based on the above description, the reserach question is "How to develop the instrument of mathematical logical thinking ability in Kapita Selekta?" The problems studied are:

a. The developed instrument is the mathematical logical thinking ability.

b. The topic are comparison, probability, combinatorics and **mathematical logic**.

1.2 Aims

Based on the above description, the aims of this study are:

a. To develop instrument mathematical logical thingking ability in Kapita Selekta

b. To develop instrument that can measure students' mathematical abilities on the topic of comparison, probability, combinatorics and mathematical logic of first students at the university.

2. METHODS

This research aims to produce an instrument tests the ability of logical thinking on mathematical subjects was kapita selekta topics comparison, opportunities, combinatorics, logic and mathematics. The instruments developed can be used in the process of the lecture so that the research will be carried out is research development (Development Research) that a heavy emphasis is the development of the instrument. According to Ruseffendi (2005:32), research development (Development Research) is a research to develop and produce products educational material, media, tools and learning strategies, or evaluation, and so on to cope with the problems of education, and not to test the theory.

Research development of procedure consists of two stages, namely to develop instruments and test the quality and effectiveness of the instruments or produced. The first stage is to develop instruments include: (1) define the variable; (2) outlines the variables into a more detailed indicators/dimensions; (3) draw up the details; (4) perform validation; (5) conducting trials; (6) analyze validity and reliability.

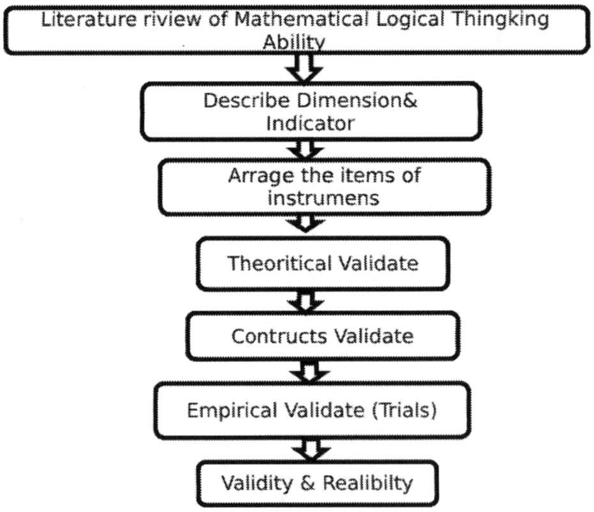

Diagram 2.1 Stage of Develop Instrument

3. RESULTS AND DISCUSSION

This section will describe the results that have been achieved during the study, but the new results are achieved is an instrument that's been validated by experts and panelists theoretically. The following is a breakdown of the results that have been achieved.

Mathematical logical thinking capability of the Instruments

The instruments that have been produced in this research are mathematical logical thinking ability of instruments consisting of 5 packages of instruments the logical mathematical thinking ability. Package instruments this mathematically logical thinking capability will be used to 4-5 times the ability test logical thinking mathematically selekta capita of lectures at the time. As for the definition of logical mathematical thinking ability is the ability of the student in resolving the matter in the form of a description based on several indicators, namely: 1) Think proportional i.e. thinking ability in determining and comparing the ratios. 2) Think the probability of that is the ability to interpret the data obtained in the form of the magnitude of the probability of an event. 3) Think correlational i.e. the ability in determining whether two events/variables are related or not. 4) Think Combinatorial, namely the ability to determine the combination of the likelihood of an occurrence.

These instruments display problems such as logical mathematical thinking ability in accordance with the indicators, which consists of the following topics: 1) Comparison 2) Probability 3) mathematical logic 4) Combinatorics

Here are parts of the display of each topic on the instrument's ability to think mathematically logical.

No	Indikator	Butir Soal	Penilaian					Saran
			1	2	3	4	5	
1	Menyelesaikan permasalahan yang berhubungan dengan perbandingan	1. Umur Tuti 4 tahun lebih muda dari umur Titu. Jumlah umur mereka 16 tahun. Tentukan perbandingan umur Tuti dan umur Titu 2 tahun mendatang!	☐	☐	☐	☐	☐	
		2. Sarah mempunyai 7 apel, 5 mangga, dan 3 jeruk. Berapakah banyaknya tambahan jeruk yang harus Sarah beli?, sehingga perbandingan banyaknya keseluruhan buah dan banyaknya jeruk yang dia miliki menjadi 3 : 1.	☐	☐	☐	☐	☐	
	Menyelesaikan permasalahan yang berhubungan dengan perbandingan	3. Enam tahun yang lalu umur Lisa seperenam umur Dewi. Sepuluh tahun yang akan datang umur Lisa setengan umur Dewi. Berapkah perbadingan umur Lisa dan Dewi sekarang?	☐	☐	☐	☐	☐	
		4. Adi mempunyai permen sebanyak 3 kali banyaknya permen yang dimilik Boni. Boni mempunyai permen yang lebih sedikit dari Coki. Coki mempunyai permen 2 lebih banyak dari Adi. Berapakah perbandingan banyaknya permen yang dimiliki Adi.	☐	☐	☐	☐	☐	

Figure 3.1. Sample of Comparison Topic

	Menyelesaikan peluang dari suatu kejadian	4. Dalam suatu populasi keluarga dengan tiga orang anak, berapakah peluang keluarga tersebut mempunyai paling sedikit dua anak laki – laki?	☐	☐	☐	☐	☐	
		5. Dua buah dadu dilempar bersama – sama. Berapakah peluang munculnya jumlah mata dadu 9 atau 10?	☐	☐	☐	☐	☐	
	Menyelesaikan peluang dari suatu kejadian	6. Sebuah kartu diambil dari seperangkat kartu bridge yang telah dikocok. Kemudian, kartu tersebut dikembalikan dan kartu dikocok kembali. Satu buah kartu diambil lagi dari kocokan kedua. Berapa peluang kartu As pada pengambilan pertama dan kartu King pada pengambilan kedua?	☐	☐	☐	☐	☐	
	Menyelesaikan peluang dari suatu kejadian	7. A, B, C, dan D akan berfoto bersama secara berdampingan. Berapakah peluang A dan B selalu berdampingan?	☐	☐	☐	☐	☐	

Figure 3.2. Sample of Probability Topic

	Menerapkan logika dan penalaran dalam penyimpulan pada pemecahan masalah	3. Toni, Firman, Candra, Bayu, Dandi dan Aji diminta untuk berbaris urut mulai dari tinggi badan yang paling tinggi ke paling rendah. Kondisi mereka sebagai berikut: a. Dandi lebih pendek daripada Candra dan lebih tinggi daripada Aji b. Toni lebih tinggi daripada Aji dan lebih rendah daripada Bayu c. Firman leboh rendah daripada Tony dan lebih tinggi daripada Aji d. Bayu lebih tinggi daripada Candra Maka bagaimanakah urutan peserta dalam barisan dari yang tertinggi sampai terpendek?	☐	☐	☐	☐	☐	

Figure 3.3. Sample of Mathematics Logic Topic

	Menentukan banyaknya jumlah kemungkinan atau cara dari suatu percobaan atau kejadian	7. Sepasang dadu dilempar satu kali. Berapakah peluang jumlah angka yang muncul kelipatan 2 atau 3?	☐	☐	☐	☐	☐	
		8. Berpakah banyaknya bilangan ganjil antara 1000 dan 9999 yang semua angkanya berbeda?	☐	☐	☐	☐	☐	

Figure 3.4. Sample of Combinatoric Topic

The instrument was drawn up on the basis of indicators of the ability to think mathematically logical. From the initial design of logical thinking ability of mathematical instruments that have been produced, then conducted testing or validation of theoritically to the panelists and experts.

Results of the validation of the Panelist and Expert

Table 4.1 Result of Validation Theoretical Instruments

Banyak Rater (N)	**20**
Nilai c	**5**

No. Butir	**Panelis**																				n_1	n_2	n_3	n_4	n_5	**Koef. Validitas**	**Ket**
	1	**2**	**3**	**4**	**5**	**6**	**7**	**8**	**9**	**10**	**11**	**12**	**13**	**14**	**15**	**16**	**17**	**18**	**19**	**20**							
1	5	5	5	5	5	5	3	5	5	4	3	5	4	5	5	5	5	4	4	4	0	0	2	5	13	0,888	Valid
2	5	5	5	4	4	5	4	5	5	5	4	3	4	4	5	5	4	4	4	5	0	0	1	9	10	0,863	Valid
3	5	5	5	4	5	5	4	5	5	4	4	4	4	4	5	5	5	5	4	5	0	0	0	8	12	0,900	Valid
4	5	5	5	4	2	5	4	5	4	5	3	4	4	4	5	4	5	4	5	5	0	1	1	8	10	0,838	Valid
5	5	5	5	3	5	4	3	4	4	4	3	5	5	5	5	5	5	4	5	4	0	0	3	6	11	0,850	Valid
6	5	5	5	4	5	5	4	5	4	4	4	5	3	5	5	5	5	3	5	5	0	0	2	5	13	0,888	Valid
7	5	5	5	3	5	5	4	4	5	5	3	5	5	4	5	4	5	5	5	4	0	0	2	5	13	0,888	Valid
8	5	5	5	4	5	5	4	5	5	4	4	5	4	4	5	4	5	3	5	4	0	0	1	8	11	0,875	Valid
9	5	5	5	4	5	5	5	5	5	5	4	4	4	5	5	4	5	4	5	5	0	0	0	6	14	0,925	Valid
10	5	5	5	5	5	5	4	5	5	4	4	5	4	4	5	4	5	3	5	5	0	0	1	6	13	0,900	Valid

etc ....

Following are the results of the validation test the contents of the instruments carried out by the panelists. Panelists in this test as many as 20 people are comprised of professors and teachers. Based on the calculation of test acquired the above validity coefficients 0.2. This shows the instruments developed as many as 40 grains of matter is valid and deserves to be used in the next step as a test instrument for measuring the ability of logical thinking students.

4. CONCLUSIONS

Based on the results of the expert and expert panelists validation indicates that the instrument is designed easy to read and well understood by the student with some improvements of the sentences and language without reducing the content and meaning. The results of the validation of the content in teoritik indicates that the instrument developed logical thinking ability contains the category is valid and worthy of use for later stages as an instrument to measure student mathematical logical thinking ability in kapita selekta.

Some advice that can be given for improvement the next development is as follows:

1) The material in the development of this instrument is limited to just the courses kapita selekta, should cover other subjects

2) Mathematical logical thinking Ability) who emphasized on 4 topics only, so that in the future should be developed on another topic.

REFERENCES

1. Rachmadi, Widdiharto, *Model-model Pembelajaran Matematika SMP*, (Yogyakarta: Widyaiswara PPPG Matematika, 2004).
2. Tiurlina, Hakikat Matematika dan Pembelajarannya di SD. http://file.upi.edu/Direktori/DUAL-MODES/MODEL_PEMBELAJARAN_MATEMATIKA/HAKIKAT_MATEMATIKA.pdf (diakses 2 Januari 2017).
3. Utari Sumarmo, 2002, "Alternatif Pembelajaran Matematika dalam Menerapkan Kurikulum Berbasis Kompetensi", (Makalah pada Seminar Tingkat Nasional, FPMIPA UPI Bandung, 2002).
4. Tuna, Abdulkadir; Çagri Biber, Abdullah & Incikapi, Lütfi. "An Analysis of Mathematics Teacher Candidates' Logical Thinking Levels: Case of Turkey", *Journal of Educational and Instructional Studies in The World*, February, March, April 2013, Volume: 3 Issue: 1 Article: 10 ISSN: 2146-7463.
5. Suleyman Yaman, *"Effectiveness on Development of Logical Thinking Skills of Problem Based Learning Skills in Science Teaching"*, (The Journal of Turkish Science Education Vol:2 Issue:1;May 2005), hh. 31-33.
6. Ruseffendi, E.T (2005). *Dasar-dasar Penelitian Pendidikan dan Bidang Non Eksakta Lainnya*. Bandung : Tarsito.

Character education based on six value system

Abduloh*

Universitas Singaperbangsa Karawang, Fakultas Keguruan dan Ilmu Pendidikan
Jl.Hs Ronggo Waluyo Teluk Jambe Timur Karawang Jawa Barat Indonesia 41361

ABSTRACT. Humans are social beings who mutually cooperate and interact with each other that have an impact on the changes and dynamics of a complex life, both individually and in groups that may be able to uphold the value and dignity and vice versa, it can be influenced by individual and group characteristics in living his life. The purpose of this study is to describe the application and outcome of character education based on six value systems. This study applied descriptive method with literature review analysis. The results of the analysis show that the character of education based on the six value systems can be applied in various levels of education from basic education to higher education and educational outcomes can have the characteristics of value for use, or benefits for the life of the community worthy both materially and spiritually that can solve the problem, Improvement and change of valuable moral, mental, and spiritual skills.

1 Introduction

Modernization is one of the characteristics of the cultural development which is balanced with the development of science and technology very rapidly, related to the increasingly complex human life behavior so that it can cause share the problem. The problem if not balanced with a valuable education that becomes a character in the learning process that is able to print changes in knowledge, attitude and human skills that are able to solve various problems and challenges of life. To overcome the challenges that are problematic in the learning process is expected learners can have the skills, Benjamin Bloom (1956) suggests knowledge (cognitive skills), attitude skills (affective skills), and Phsicomotoric skills' (Susanti 1995).[1]

Character education should bring learners to cognitive value recognition, appreciation of value affectively, and ultimately to actual value practice. This is the design of moral character education (moral knowing), moral feeling), and moral action (Lickona, 1991: 51)[2] Therefore, all subjects studied by learners in schools should be character-educated characters that can bring human beings who have the character of personality and noble character.

* Corresponding author: abduloh@staff.Unsika.ac.id

Character comes from English: character and in Greek, namely characin meaning "to engrave" (Ryan & Bohlin, 1999).[3] To engrave "means carving, painting, carving, or scratching (Echols & Shadily, 1995).[4] Indonesian dictionary the word "character" means character, psychic traits, morality or character that distinguishes one from another, and character, the person who possesses personality, behavior, nature, character, or character. Thus, the character is a behavior or, character and character of a person who can distinguish a person or individual with others.

Thomas Lickona argues character as " A reliable inner disposition to respond to situations in a morally good way, characteristic of moral knowing, moral feeling, and moral behavior" (Lickona 1991), [5] Good character, including knowledge of goodness (moral knowing), which can generate motivation (commitment) to the good (moral feeling), and finally really do the good (moral behavior).

Learning is a process to achieve a purpose and purpose that can meet the expectations, the character of an individual in fulfilling and maintaining his life, and can provide useful value and benefits for the interests of all living things on earth that is characterized by the inheritance of culture and character that has been owned by society and nation. The process of learning culture and character, actively learners develop their potential, conduct the process of internalization, and the appreciation of the values into their personalities in socializing, develop a more prosperous society, and develop a dignified nation life with various processes of improvement and refinement various resources that support the educational process.

2 Research methods

Based on this the authors are interested and try to explain the theoretical study on the characteristics of education based on six value systems, with descriptive study method and literature review analysis focus on the study of "process of educational character based on six value system as follows: 1) value theologic, which is reflected among other things in the almighty god, pillars iman six pillars of islam five praying, monotheism, charity, forgiveness, prayer, sincere, repentance, ijtihad, focus istiqamah, and jihad fi sabilillah. 2) value ethical-legal, which is reflected among other things in a respectful, kind/humble, loyal, trustworthy, honest, responsible, good faith, fair faithful, peace, patience, forgiveness, helpfulness, tolerance and harmony. 3) aesthetic value, which is reflected among others, in a nice, clean, beautiful, pretty, sweet, charming, harmonious, romantic, and love. 4) value logical-rational, which is reflected in logic/match between facts and conclusions, right, appropriate, clear, tangible, identity/characteristics, process, circumstances conclusion match. 5) physic-physiological value which is reflected clearly its elements, functions, size, strength, changes, location, origin, cause and effect. 6) teleologic value which reflected embodied in a useful, beneficial, according to the function, developing/developed, regular/discipline, integrative, productive, effective, efficient, accountable, innovative. (S.Ahmad 2015). [6]

The goals of education in general at every level, from basic education to higher education, should be designed and implemented programmatically and systematically, so as to achieve goals in accordance with expectations. Formation of the learner character who have knowledge of religion, ethics, morals, and courtesy in interacting with the community, then education should be designed, implemented, and well evaluated (good governance) and education implementers can integrate character education in it to create human- Beings who have noble character. Kementerian Pendidikan Nasional. (2010) [7]

3 Results and discussion

Education is a process of various activities of the body and soul in effort to reach the purpose of educators and learners are expected, through the process of thinking and activity, physically active and creative and productive in solving problems that were able to find something that is useful and valuable (inovative) Abduloh 2017 [8]. Process education caracter based on six value system, Individual characters can be influenced by environmental factors and innate factors of birth. Psychologically behavior is a character of the potential embodiments of Intelligence Quotient (IQ), Emotional Quotient (EQ), Spiritual Quotient (SQ), and Adverse Quotient (AQ), owned by individuals. The context of character totality in psychological and socio-cultural processes can be grouped into 1) spiritual and emotional development, 2) intellectual development, 3) sports and kinesthetic (physical and kinesthetic development), and 4) taste and intention (affective and creativity development). The character of this psycho-social process, holistically and coherently interconnected and complementary in the framework of character formation and the embodiment of noble values in a person. Character formation can be formed through the process of human learning from the environment, family, education, work and nature.

Character is the nature, character possessed by man, good and bad life order of a society can be determined by a character of individual that exist in that group, can be determined by character influencing attitude and behavior of human being. If the character appears positive it may have the benefit but if the existing character can harm others, it becomes a problem for the survival of human and nature. So for that in the learning process needs to be included values of characters that have value for both morally, mentally and spiritually to learners conducted by professional teachers/educators means teachers who are able to perform their duties and functions in accordance with the rights and obligations by applying and developing on six value systems in the education that includes;

1. Value theology, (god's value) in learning can be applied that is to form the character of beliefs (believing skills) on students in doing a good action determination, speech and behavior in life to do that in the learning process that is required though aeart or sense that is to form the skills of spiritual quotient (SQ), and adverse quotient (AQ) owned by individuals.

2. Value ethic-legal, (norms) in the lesson are expressed to establish the character of the disciple's learners, commitment, consistency and responsibility. That is by applying attitude skills (affective skills), among others, respect, good, faithful, trustworthy honest, responsibility, good faith, faithful, peaceful, patient, forgiving, helpful, tolerant, and harmonious. Namely through the application of various examples of educator behavior, and the environment. The ethical value can be applied to the students that is to nurture noble values such as respect for others, good, faithful, trustworthy honest, responsible, good faith, faithful, peaceful, patient, forgiving, helpful, tolerant and harmonious, it is teachers can apply it by giving examples of behaviors in learners which means educators or teachers must be a role model for learners.

3. Value aesthetic, in the implementation of learning teachers can apply attitude skills, behavior (cognitive skills) is by giving examples of values that have artistic value and beauty so that learners can motivate learners formed spiritual quotient intelligence, that is through the application of various models, media methods, materials, and managerial as well as techniques and learning strategies. The aesthetic value, in character teaching, can be applied to the skill-making values of creating something beautiful that can make things beautiful, clean, beautiful, beautiful, sweet, attractive, harmonious, romantic, and loving. These values can be applied in learning by developing ideas, by actively thinking and creatively in developing various models, media, methods, teaching materials, managerial, techniques, and learning strategies that can enhance the aesthetic values of the participants educated, which means teachers should be able to learn, learners so that students are aware of the obligations and rights as learners.

4. Value logic (rational), in learning logical value learners are invited to think from something abstract to concrete and scientific either with the knowledge, belief, or with a characteristic or mark in the form of tangible changes, among others, logic / match between the facts and conclusions, exact, appropriate, clear, tangible, identities / characteristics, processes, states of conclusion match. It is the teacher's educator must be able to master the knowledge and broad knowledge that is by way of continuous learning.

5. Physical-physiological value in the learning process characterized can be applied by conveying something to the learners with something tangible that accompanied by facts and data that can explain elements, functions, size, strength, change, location, origin - cause, cause-effect and so forth. This means that in this case educators or teachers can teach is to convey knowledge (transfer knowledge) to leaners in order to be able to create something in the form of real accordance with the role and function.

6. Teleology value can be applied in the learning activity of the learner is encouraged to do the learning activity to improve the skill in a piece of work, which has useful value and benefit for life according to its role and function, actively, creatively to build and develop by regular, discipline, integrative, productive, effective, efficient, and innovative, meaning that teachers or educators and learners should be sensitive and care about all the development of the environment.

So for that the authors propose the opinion that a teacher who can menghapakkan character education based six system values such as leaders who dare to GILA that is leader 1) Genius means to have ideas, ideas of knowledge and intelligence more than the under it in order to make decisions appropriately and fast. 2) Integrative leadership must dare to come down directly and merge united with subordinates to solve problems regardless of race or class. 3) Loyal leadership must have a devotion to the organization, or institution that is led. 4) Adaptive a leader must be brave and able to adapt to all levels of society in any situation and condition. Abduloh 2016) [9]. Just as in the implementation of learning that is required a teacher who can be a leader figure for learners, who are able to transfer and apply the learning character of six value system. Which can be described in the process of character education based on six systems as follows:

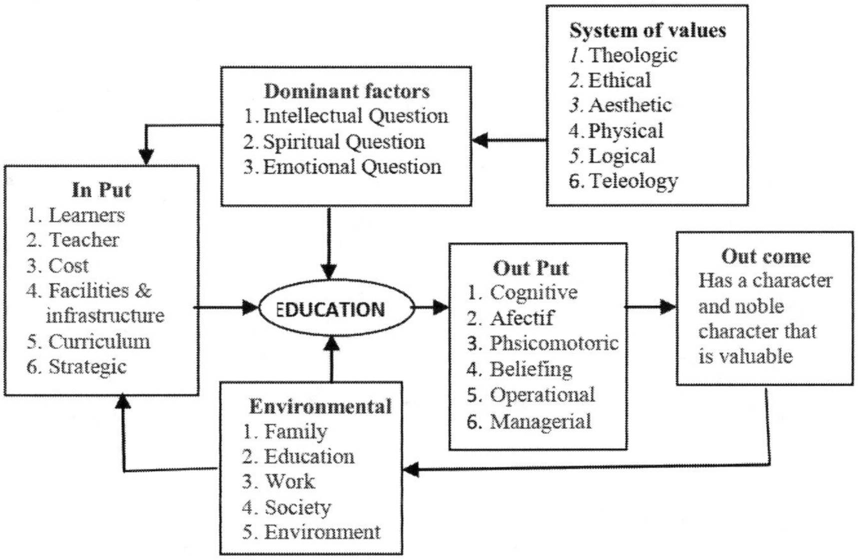

Future. 1 Process character education based on six value systems

4 Conclusion

Individual characters are formed due to a series of skill activities provided to the learner through a systematic and programmed learning process according to established standards including cognitive skills, attitudes skills, and behavioral skills (behaviors skills) , As well as confidence skills, skills of use (operational skills) and managerial skills, especially in the application of the character of the learner can think and act something that refers to the six value systems that include the theological value (divinity) , ethical (norm), aesthetic (beauty), logical (concreet), physiological (tangible), and teleological (used).

5 Acknowledgement

Thanking God for the presence of Allah Subhanahu Wa Ta'ala, regarding the completion of this article, please allow me to thank all those who helped petrify the completion of this article I say, thank you and appreciation submitted wholeheartedly with respect, and proud to:

1. To the Rector of Unsika who has given permission to participate in the Global Conference on Teaching, Assessment, and Learning in Education

2. To Undiksha Rector as organizer of Global Conference on Teaching, Assessment, and Learning in Education 2017

3. To Prof. Dr. Ni Nyoman Padmadewi as Chief Executive of the Global Conference on Teaching, Assessment, and Learning in Education 2017 which has provided an opportunity to participate in conferences activities.

4. The revision team of GCTALE 2017 who has provided feedback for improvement to all conference participants

May all good deeds we always get affection from God Almighty Allah.

References

1. Abduloh. *Leader Of Success In The Modernization Of Education* the 1st International conference on education in Indonesia (ICE). **167** (2016)
2. Abduloh. *Creating Innovation In Education By Using Value Analysis System.* The Fifth south East Asia Development Research International conference (5 Th SEA-DR). **353** (2017)
3. Echols, M. John & H. Shadily, *English Dictionary Indonesia*: An English-Indonesian Dictionary. Jakarta: PT Gramedia. Cet. XXI. **214**, (1995)
4. Kemdiknas, *Character Education Master Design*. Jakarta: Ministry of National Education. **9** (2010)
5. Ryan, Kevin & Bohlin, K. E. *Building character in Schools*: Practical Ways to Bring Moral Instruction to Life. San Francisco: Jossey Bass, **5** (1999)
6. A. Sanusi, *System Value Alternative Wajah-Wajah Pendidikan* Bandung Nuance Cendikia , **35** (2015)
7. Thomas Lickona, *Educating For Character*: How Our School Can Teach Respect. And Responsibility (New York: Bantam Books, **51** (1991),

Financial Learning: Is It The Effective Way to Improve Financial Literacy among Accounting Students?

Nyoman Trisna Herawati*

Accounting Departement, Universitas Pendidikan Ganesha

Abstract. The aims of this study are to determine: the difference of financial literacy level between students who have had experience in financial learning and who have not had experience in financial learning. The data for this study was collected through financial literacy test and questionnaire which was distributed through randomized sampling method. A total of 173 completed and usable questionnaire have been collected. The result shows that the level of financial literacy among accounting students comes under below optimal standard category. Students who have had financial learning experience have a higher level of financial literacy than students who have not. This study provides means to improve financial learning for accounting students in preparation for creating a prosperous future.

Keywords: Financial literacy, financial learning, accounting student

1 Introduction

Global economic development nowadays has brought a change effect onto its society economy system. One of the negative effects caused by it is the consumptive culture in the society, including the teenagers. The consumptive tendency that is actually quite normal for teenagers will turn into a problem if it has developed beyond control. If teenagers manage their money recklessly, it is possible that they carry on that attitude when they grow up. Aside from that, sometimes what the teenager's demand is beyond their parents' capability to provide as their fund source, that can lead to a new economy problem in the family.The matter whether someone is consumptive or not is a result of the minimum financial knowledge (financial literacy). Along with that, some researchers have shown that financial literacy has a significant impact on one's financial behavior [1,2,3]. Someone who has low financial literacy usually has a financial problem in his life.

* Corresponding author: aris_herawati@yahoo.co.id

Learning at university plays an important role in shaping students' financial literacy. Using various teaching methods, media, and appropriate learning sources students are expected to have the proficiency in finance so that they are prepared and capable of handling present as well as a more complex future life [4]. Some studies showed that learning finance at the university has an impact towards financial literacy [4,5,6,7].

However, the research shows that level of financial literacy of college student comes within the low category event students majoring in Business disciplines [6,7,8,9]. Accounting is one of the majors in business disciplines with the most students and the number of enthusiasts is increasing each year. This is a challenge to escalate the quality of the graduates, and one of the means is by providing financial literacy comprehension through financial learning.

Financial literacy learning is not provided independently within the course curriculum. However, a number of finance subjects provided in the Accounting curriculum is relevant to the financial literacy concept. Some of the financial learning subjects are found in Financial Accounting, Financial Management, and Investment and Capital Market. Financial Accounting subject is taught from elementary level namely Introduction to Accounting 1 in the first semester to Advanced Financial Accounting 2 in the final semester, nevertheless, the financial literacy concept is more commonly encountered in Introduction to Accounting. An adequate financial learning process will enhance students' experience in managing their finance.

Based on the problem background above, the formulation of the proposed problem includes: (1) what is the financial literacy level of Accounting students, is it categorized as low, medium, or high? (2) does financial learning experience give an impact towards financial literacy level among Accounting students?

2 Methods

This research design is a quantitative research that reviews financial literacy level among students and the impact of financial learning experience towards student's financial literacy. Research population, in this case, is the entire students of Accounting at Universitas Pendidikan Ganesha. The number of samples of 174 students was taken through *purposive sampling*, by selecting classes with equivalent academic capability. Data was collected through financial literacy test [11] and analyzed using descriptive statistical analysis and ANOVA-OneWay. Descriptive statistic provides financial literacy level illustration whether it is considered as *well literate, sufficient literate,* or *less literate.* Classification of financial literacy is adapted from Chen and Volpe [11], who classified one's literacy level in this range: (1) Low group with the score< 60%, (2) Medium group with the score between 60% and 79%, and (3) High group with the score of≥ 80%.The next analysis is *Anava-One Way* that is used to find out the impact of financial learning experience towards the students' financial literacy level. Respondents are categorized into 3 groups according to their study semester. Group 1 consists of students in their second semester who only learn Introduction to Accounting, in this case, they are treated as a group with no financial learning experience.Group 2 consists of students in the fourth semester and group 3 consists of students in their sixth semester who have had experience in financial learning. Both group have learned accounting subject such as: Introduction to Accounting, Financial Management, Bank and Other Financial Institutions, also Investment and Capital Market. Before conducting hypothesis testing, a data normality and homogeneity test was performed beforehand. The result of normality test showed that all data was distributed normally with a significant rate higher than 0.05. The result of homogeneity test showed that Levene statistic level (0.94) was higher than 0.05 which meant that the intergroup variants were homogenous. Hypothesis test was accepted if the value of *sig*< α 0.05. All statistical methods using SPSS for Windows 20.0

3 Results and discussion

The number of respondents involved in this research is 174 students consists of 55 second semester students as group 1, 58 fourth semester students as group 2 and 60 sixth semester students as group 3. The mean and standard deviation of financial literacy test result in each semester can be found in the following Table 1.

Table 1. Financial Literacy (FL) Test Result

FINANCIAL LITERACY							
Students	**N**	**Mean**	**Min.**	**Max.**	**Std. Dev**	**Std. Error Mean**	**Criteria/ Level**
Group 1	55	48,36	22	82	14,162	1,910	Low
Group 2	58	59,72	26	88	12,214	1,873	Low
Group 3	60	59,24	29	82	11,318	1,461	Low

Data processing outcome in Table 1 showed the mean of financial literacy test on second semester (group 1) students is the lowest compared to students in their fourth (group 2) and sixth semester (group 3). For students in their second semester, the mean of FL test is 48.36 points lower than the mean of FL test on students in their fourth semester, which was 59.72 while the mean of FL test on students in their sixth semester was 59.24. It is visible that the mean of FL test on students in their fourth semester was not that much different from sixth semester students. Based on the classification according to Chen and Volpe (1998), the financial literacy level of Accounting students is considered low yet it is close to medium, which is less than 60%. The financial literacy average score for each component of financial literacy can be found in Table 2.

Table 2. Financial Literacy Average Score for Each Component

Financial Literacy Dimension	Group 1	Group 2	Group 3	Average
1. Financial basic concept	61.01	72.80	71.11	68.31
2. Saving and investment	34.85	43.97	46.39	41.73
3. Credit and debt management	50.30	64.66	58.33	57.76
4. Insurance products	47.27	57.47	61.11	55.29

Based on Table 2, it can be viewed that the savings and investment component reached the lowest score (with the average of 41.49), while financial basic concept reached the highest score (with the average of 67.89). This indicated that students' comprehension in savings and investment concept is still low. Observing these 4 (four) components or financial literacy aspects, only financial basic concept that reached sufficient average of 68.31 (>60). On the other hand, the other financial literacy aspects such as (1) savings and investment aspect with the average of 41.78, (2) credit management comprehension with the average of 57.76, (3) insurance products comprehension with the average of 55.29, which is still considered low. These indicate that students have sufficient comprehension in financial basic concept aspect. Financial basic concept aspect has some indicators, which is comprehension on financial statement concept, time value concept of money, and income concept. Students can differentiate between the concept of "debit" and "credit" regarding good financial management, time value of money and adequate income management. However, viewed from other aspects, students have not comprehended them correctly. For example in investment and savings aspect; students have not comprehended the concept of investment in the capital market as well as its benefits and disadvantages of conventional savings. The case is similar in credit management and wise credit management. Regarding insurance comprehension, average scores were close to the minimum. In this case, it means that students have comprehended the importance of insurance, yet they have not understood the types of insurance and the coverage of each kind of insurance.

For further examination, hypothesis testing with *Anova-One* Way was performed. SPSS output produced calculation F value of 3397.558 for intercept and was significant at 0.05, and so as X variable, in this case, financial learning experience, with F value of 14,676 and was significant at 0.05. It can be concluded that financial learning experience for Accounting students affected financial literacy level.There is a difference in financial literacy level between students who have had adequate financial learning (Group 2 & 3) and those who have not had sufficient financial learning (Group 1). The amount of *adjusted R square* value of 0.137 means that the variability of students' financial literacy that can be rationalized by experiencing financial learning is at 13.7%.

Research shows a significant difference in financial literacy level between students of semester 2 and those in semester 4 and 6. The lowest financial literacy level average was found in students in their second semester with the score of 48.36 and the highest was achieved by students in their fourth semester with the average of 59.72, not so much different from students in their sixth semester, which ranged at the score of 59.24. This proves that students who have gone through financial learning have better financial literacy from those who have not been through financial learning. However, research shows that there is no significant difference between students of the fourth semester and sixth semester. Even financial literacy average score of students of the fourth semester is better than those in the sixth semester (59.72 > 59.24). This indicates those four subjects are highly relevant associated with aspects of financial literacy, while other accounting subjects that are taught in the fifth semester and sixth semester do not provide a significant contribution towards students' financial literacy level.

Financial learning plays a highly important role for students to have the ability to comprehend, judge, and act in their financial interests. Undoubtedly financial knowledge as a result of financial learning, its success theoretically is related to teaching and learning process. Learning process and assessment technique used by lectors also determine students' success in comprehending and implementing material received in their daily life [10]. Learning will achieve a better result when provided with relevant lessons using active learning method. Important skill has to be put into action in order to have it fully embedded within students and become purposeful for their life. This research result comports with the research conducted [6,7], which stated that there is influence between financial learning at university towards financial literacy in the cognitive aspect.

4 Conclusion

According to the exposition above, several items can be concluded as follows.First, financial literacy level of Accounting students is still considered low (< 60). Second, there is a difference in the level of financial literacy among students who have had a financial learning experience and those who have not.Students who have had financial learning experience have higher financial literacy level compared to those who have not.The low financial literacy level among students impacts on the need to have strategic efforts to elevate students' financial literacy. One of them is by upgrading the financial learning quality at the university. A number of financial materials taught does not matter if it is not related to financial literacy aspect since it will not provide a positive impact for students to enhance their financial literacy. Therefore, it is necessary to provide real cases that correspond to personal financial aspect. Besides, this research has also proven that financial learning in Accounting Department, especially in Introduction to Accounting, Financial Management, Bank and Other Financial Institutions and Investment and Capital Market subjects, produce significant contribution towards students' financial literacy.Accordingly, these four subjects are expected to apply financial literacy concept within its learning. This research has limitation in terms of samples used, which only covered one institution. Furthermore, financial literacy level ignores other factors suspected to have a more powerful impact, such as family and social demography status of students.

References

1 Danes, S. M. &Haberman, H. R,Teen Financial Knowledge, Self-Efficacy, and Behavior: A Gendered View. *Financial Counseling and Planning*. 18(2): 48-60 Retrieved from htpp://www.afcpe.org/assets/pdf/7-2866-volume-18-issue-2.pdf, (2007)

2 Vyviyan, Victoria; Blue, Levon; & Brimble, Mark, "Factors that Influence Fianncial Capability and Effectiveness: Exploring Financial Councelors Perspectives" *Australian Accounting, Business and Finance Journal, Vol.8,Iss 4, 2014,3-22*, Retrieved from htpp://ro.ouw.edu.au/aabfj/vol8/iss4/2, (2014)

3 Kozina,Franca Lovsin dan Ponikvar,Nina, "Students' Confidence In Their Financial Manajemen Abilities : The Role Of Socio-Demographic Characteristics And Education". *Journal of Problems of Education, Vol.58.h.108*, Retrieved from htpp:// htpp://www.cgsnet.org/ckfinder/userfiles/files/PracticesofCollege Students_Final.pdf, (2014)

4 Lutfi & Iramani, Financial Literacy Among University Student and Its Impli-cations to the Teaching Method. Paper presented at Annual SEAAIR Conference STIE Perbanas, Surabaya (2008)

5 Jhonson, E & Margaret, S. S, From Financial Literacy to Financial Capability Among The Young. *Journal of Sociology and Social Welfare*, 34:119-146. Retrieved from htpp://www.usc.edu/dept/cheap/HRYANG/publications/110.pdf (2007)

6 Susanti, Pentingnya Literasi Keuangan Bagi Generasi Muda di Surabaya, Paper presented at SNE" Pembangunan Manusia Melalui Pendidikan Dalam Menghadapi ASEAN Economic Community 2015" Surabaya, 3 Mei 2014,h.278, Retrieved from htpp://fe.unesa.ac.id/sne1014, (2014)

7 Widayati,Irin, *Faktor-Faktor yang Mempengaruhi Literasi Keuangan Mahasiswa Fakultas Ekonomi dan Bisnis Universitas Brawijaya.* ASSET :Jurnal Akuntansi dan Pendidikan, Vol.1,No 1, Oktober 2012 (2012)

8 Nidar,S.R.,& Bestari,S, Personal Financial Literacy Among University Students (Case Study at Padjajaran Universitas Students, Bandung, Indonesia). *World Journal of Social Science* (2)4: 162-171 (2012)

9 Layli, N. 2013. Pengaruh Literasi Keuangan terhadap Perilaku Mahasiswa dalam Mengelola Keuangan. *Artikel*. JPA UM Malang. Vol 1. No.4. Hal. 277-285. (2013)

10 Trianto, *Mendesain Model PembelajaranInovatif-Progresif.* Jakarta: Kencana Pre-nada Media Group, (2009)

11 Huston,S.J, Measuring Financial Literacy.Journal of Consumeer Affairs, 44 (2) : 296-316, Retrieved from htpp:// onlinelibrary.wiley.com (2010)

The effects of six beat kick technique training on swimming speed for the 50 meters freestyle

Meiriani Armen[1,*]

[1]FKIP, Universitas Bung Hatta, 25176, Padang, Indonesia

Abstract. Freestyle is one of the main swimming styles taught to swimmers. It has various techniques from stroke and breathing to limbs moving and coordinating. The legs moving technique Six Beat Kick is one of the techniques that could make an athlete swims faster. A quasi eksperiment research with pre-test and post-test program is aimed to recognise the impacts Six Beat Kick technique training on swimming speed for the 50 meters freestyle. This research is carried out in G-Sport Center Swimming Club in Padang city. The sampling method uses purposive sampling with 20 participants. The data are collected from swimming speed test on 50 meters freestyle. Prior the training, a pre-test is conducted. Then, the athletes must follow the training program for 14 times of training. Lastly, a post-test is performed. In addition, an analysis condition testing and a research hypothesis testing use data normality test and t-test with significance level of α 0,05. From the analysis condition testing, it could be assessed that the data follow a normal distribution with $L_{counted}$ 0,108 < L_{table} 0,190. The hypothesis testing has a significant impact on swimming speed for 50 meters freestyle acknowledged from $t_{counted}$ 22,19 > t_{table} 1,73. It can be concluded that Six Beat Kick technique training program could improve the swimming speed of meters freestyle.

1 Introduction

Swimming has been recognized as one of the mosthelpful sports for a long time as the ancient peoples must be able to conquer nature. Since then, swimming has evolved and beenone of the popular water sports [1]. In Indonesia, swimming as a whole has also been gaining its popularity because it has been introduced to the people from kindergarten up to college level.

The objectives of Indonesians to swim are as recreational, educational, therapeutic and competitive events. In addition,the chance for a person to swim also comes in many ways [2]. They could swim by themselves, family and through swimming coaching. "Swimming is a sport done in water through coordinated movement of the limbs, the body, or both"[2]. There are four swimming styles that must be mastered, namely free style, backstroke, butterfly and breaststroke style [3]. Swimming is measured by the time counted in a game which means to measure the time a start system and finish detection system are

* Corresponding author: ria.pjkr12@bunghatta.ac.id

required. The faster an athlete swims the better the time it takes. Defines the speed as the ability to move from one place to another in the shortest time possible [1].

In regional level, the development of swimming in West Sumatra is quite good with the swimming organizations, clubs or groups being fostered and organized by PRSI [4]. The number of swimming clubs in the city of Padang is quite numerous and has lots of talented athletes. But because of some obstacles, there are athletes who cannot practice maximally and cannot perform well. Existing obstacles are 1) Improper training program 2) Monotonous practice methods 3) Inadequate infrastructure facilities. One of the clubs in Padang that continues to actively improve swimming performance is G-sport Center Swimming Club

Swimming exercise should be performed continuously and persistently as it will be seriously affecting the technical changes of the athletes and hopefully the athletes can bring good achievements for West Sumatra. Freestyle is one of the main swim styles taught to swimmers. Logically, the movement of the swimmers' legs is in accordance with the motion of cycling [5]. For freestyle, or front crawl leg kick (a simultaneous up and down kick) the knees bend only slightly to give the legs a relaxed kicking action [6]. But freestyle (crawl) has its own difficulty when taking a breath because the movement of the head should turn to the right side or left. In a simple technique freestyle pool is: 1) Body position, 2) Foot movement, 3) Hand movement, 4) Taking a breather, 5) Coordination [7].

In order to smooth the implementation of the freestyle, then an athlete must a knowledge the appropriate techniques in order to create optimal results in accordance with the style taught in the training program. One of freestyle technique trainings that will be provided is Six Beat Kick.

The Six Beat Kick is nearly always used by short-distance swimmers but is also popular with middle distance swimmers and long distance swimmers but with a somewhat subdued kick. In short, the swimmer should complete six kicks per stroke cycle or three kicks per single arm stroke which allows the movement results a powerful propulsion [8]. Given this, it benefits the swimmers in both the exercises and the race. This form of technique is an inseparable part of freestyle which in this study is observed in 50 meters freestyle (short distance / sprint).

2 Method

This research method uses quantitative research with experimental approach, one group pretest posttest design [9]. Quasi experiments are used because it is difficult to get the control group as a comparison group in this study. Prior the study, a pretest related to Six Beat Kick is conducted and continued with a posttest.

The sampling technique used in this study is Purposive Sampling [9], that consisted of 20 male participants Of the 40 athletes who are sons/daughters of the beginner group category and achievements from G-Sport Center Swimming Club Padang in accordance with certain considerations. The study is conducted in G-Sport Center swimming pool while the treatment is carried out for 16 times with exercise frequency 4 times a week. The exercise is started in the afternoon at 16.00 WIB.

The instrument used in this research is 50 meters freestyle after going through the training process using the technique of Six Beat Kick measured as far as 50 meter which are viewed based on the fastest time result. The hypothesis of this study is there is a significant effect of Six Beat Kick technique exercises on 50 meters freestyle. The testing of requirements analysis and research hypothesis uses data normality test and t-test with significance level α 0,05.

(1)

$$t = \frac{|x1 - x2|}{\sqrt{\frac{\sum D^2 - \frac{(\sum D)^2}{n}}{n(n-1)}}}$$

3 Result and discussion

Deskription data from the results of the pretest, it can be obtained the range 38.55 seconds to 58.10 seconds with a mean time 48.63 seconds and standard deviation 4.98. From this frequency distribution table, it can be seen that the result of the pretest is in Poor category.

Table 1. Frequency distribution in pretest of six beat kick technique

No	Interval	Fi	%	Category
1	38,55 – 42,60	3	15	Very well
2	42,61 – 46,92	4	20	Good
3	46,93 – 50,24	5	25	Medium
4	50,25 – 54,56	5	25	Less
5	54,57 – 58,88	3	15	Less once
		20	100	

As for the posttest, the results obtained range 33.28 seconds to 54.85 seconds with a mean time 43.66 seconds and the standard deviation 5.32. From this frequency distribution table, the posttest result of Six Beat Kick technique is in moderate category.

Table 2. Frequency Distribution in *posttest of Six Beat Kick* Tekhnique

No	Interval	Fi	%	Category
1	33,28 – 37,60	2	10	Very well
2	37,61 – 41,92	6	30	Good
3	41,93 – 46,24	7	35	Medium
4	46,25 – 50,56	2	10	Less
5	50,57 – 54,88	3	15	Less once
		20	100	

From the data of pretest and posttest, it can be observed an improvement, from mean pretest 48.63 to 43.66 at posttest, and the standard deviation at pretest of 4.98 becomes 5.32 at posttest. Research data is tested in the normality test by using the Lillieforstest with the accepted criteria that the data distributed normally if Lo < Ltable for n = 20 with the real level $\alpha = 0.05$.

Table 3. The Summary of Normality Test Results

Training	N	L_o	L_{table}
Six Beat Kick	20	0.108	0.190

From the table above, it can be seen that Lo< Ltable. Then it can be concluded that the research variable is distributed normally in the real level $\alpha = 0.05$.

Discussion of the results of this study is intended as an overview to facilitate draw conclusions research. Freestyle is a way of swimming that resembles an animal so called Crown Style which means crawling and freestyle is a force that performs hand movements over the surface of the water and allows moving in the water faster than with other styles.

Six Beat kick is a technique with six leg beats per arm cycle (i.e. two strokes). This movement technique has a greater frequency of beat, so the freestyle swimmers can learn more quickly and effectively. This technique is very well used or given to freestyle swimming athletes in doing the programmed workout. The increase of the results from before being treated and after being treated shows that the Six Beat Kick technique gives a good effect on the 50 meters freestyle swimmers. The increase can be seen from the time obtained by each athlete that increases 2-4 seconds.

Testing the normality of the Six Beat Kick on 50 meters freestyle uses Lilliefors Lcount < Ltable, and the data is normally distributed and vice versa. From the test, it can be obtained Lcount = 0.108 for Label at significance level $\alpha = 0.05$ is 0.190. From the test results above, it can be concluded that the data is normally distributed. To perceive the available significance data then t-test is used. The test criterion is significant if $t_{counted} > t_{able}$. From the test results, it can also be obtained tcount of 22.19 and ttable of 1.73.

Conclusion

Six-Beat Kick has an effect on the 50 meters freestyle swimmers significantly. It can be seen from the mean time of the pretest (48.63) and posttest (43.66) which there is an increase from pretest to posttest after being treated / exercised.

1. Six-Beat Kick has an effect on the 50 meters freestyle swimmers significantly. It can be seen from the mean time of the pretest (48.63) and posttest (43.66) which there is an increase from pretest to posttest Based on the results of this study, it is advisable to the swimming coaches in Padang to provide Six Beat Kick exercise to improve the technique of the athletes.
2. To the swimming trainers who want to improve the technique and speed of their athletes, it is urged to develop an exercise development program using the Six Beat Kick technique for a freestyle program.
3. It is recommended for the researchers to conduct further studies using much larger samples, longer research time and better exercise program to increase the performance of freestyle swimmers.

Acknowledgment : This study has been tested in front of mentors and examiners in 2008

References

1. Suarta, I Wayan (2013). SWIMMING TRAINING STYLE CHEST 8 TIMES 4 SETS 25 METER POOL SET FURTHER INCREASE SPEED OF 100 YARDS THAN 4 TIMES STYLE CHEST 4 SETS 50 METER POOL SET TO BEGINNERS SON, ISSN : 2302-688X. Sport and Fitness Journal **Volume 1, No. 1 : 45 – 59, Juny**
2. Farokie, Latiful Khobir; Eko Hariyanto, Hariyoko (2016). THE EFFECT OF DRY LAND CIRCUIT EXERCISE MODEL AND SPRINT SPATIAL TRAINING TO CAPABILITIES OF CRAWL 50 METER STYLE SPEED, Jurnal Pendidikan Jasmani, **Vol 26 No 1 April**
3. Kamalia, Ahsanul (2014). THE EFFECT OF MODIFICATION OF TRAINING TECHNIQUES OF BACKSTROKE ENGINEERING ON LUMBA-LUMBA SWIMMING CLUB SURABAYA, Jurnal Kesehatan Olahraga **Volume 02 Nomor 02 Tahun 2014, 106-113**

4. Surahman, Fadli (2016). EFFECT OF REPETITION METHOD AND INTERVAL INTENSIVE METHOD ON THE SPEED FREESTYLE 50 METERS, Jurnal Curricula Kopertis Wil X **Vol 1, No 2 (2016)**

5. Armen, Meiriani (2016). THE EFFECT OF METHOD INTENSIF INTERVAL AND EXTENSIVE INTERVALS TO SPEED FREESTYLE 50 METERS OF PEACE ATLETS TIRTA NGALAU SWIMMING KLUB PAYAKUMBUH, Jurnal Penelitian Dan Kajian Ilmiah MENARA ilmu **Vol. X Jilid 2 No. 64, Februari 2016 ISSN1693-2617**

6. Mielke, Wilhelm. 1990, RENANG, Semarang:effar ofset

7. Sukmawati, Dwi; Setyo hartoto (2015). APPLICATION OF LEARNING FREE STYLES OF FREE STYLE LEARNING RESULT FREESTYLE SWIMMING. **Jurnal Pendidikan Olahraga dan Kesehatan Volume 03 Nomor 02 Tahun 2015, 366 – 370 ISSN : 2338-798X**

8. Kurnia, Dadeng.1987, Pedoman Melatih Renang, Jakarta;Diktat

9. Rubiansyah, Aziz; Agus Rusdiana, R. Boyke Mulyana (20150. THE EFFECT OF EXERCISE PLYOMETRICS ON THE RESULT OF REJECTED START AT SPORTS SWIMMING, **Jurnal Terapan Ilmu Keolahragaan 2015 Vol .02 No.02 Hal 6-11**

10. Strzała Marek, Aleksander Tyka, Piotr Krężałek (2014). **Physical Endurance and Swimming Technique in 400 Metre Front Crawl Race.** Journal of Human Kinetics volume 18 2007, 73 - 86

The Effect of Conceptual Change Model in the Senior High School Students' Understanding and Character in Learning Physics

I Wayan Santyasa[1,*], *I Wayan Sukra* Warpala[2] , and *I Made* Tegeh[1]

[1]Physics Education Department of Universitas Pendidikan Ganesha, Bali, Indonesia
[2]Biology Education Department of Universitas Pendidikan Ganesha, Bali, Indonesia
[3]Educational Technology Department of Universitas Pendidikan Ganesha, Bali, Indonesia

Abstract. Learning physics for senior high school (SMA) students is often coloured by misconceptions that hinder students in achieving deep understanding. So a relevant learning model is needed. This study aims to examine the effect of conceptual change model (CCM) compared with direct instruction model (DIM) on the students' conceptual understanding and character in the subject area of motion and force. This quasi-experimental research using a non-equivalence pre-test post-test control groups design. The population is 20 classes (738 students) of grade X consisted of 8 classes (272 students) of SMA 1 Amlapura, 8 classes (256 students) of SMA 2 Amlapura, and 6 classes (210 students) of SMA 1 Manggis in Karangasem regency in Bali. The random assignment technique is used to assign 6 classes (202 students, or 26.5% of the population). In each school there are set 2 classes each as a CCM group and DIM groups. The data of students' conceptual understanding is collected by tests, while the characters by questionnaires. To analyse the data a one way MANCOVA statistics was used. The result of the analysis showed that there was a significant difference of effect between CCM group and DIM group on the students' conceptual understanding and character. The effect of the CCM group is higher than the DIM group on the students' conceptual understanding and character in learning subject area of motion and force.

Key words: conceptual change model, direct instruction, understanding, character

Introduction

High school students enter the classroom do not bring an empty head (blank mind), but they always bring some ideas. The student ideas commonly in the form of an existing knowledge. The knowledge is derived by previously experience, for example when in junior high school or elementary school, when interacting with the family or society and the media. Existing knowledge quality is often not scientific, but tend to be unscientific or misconception [1, 2, 3, 4, 5]. In the educational praxis, misconceptions caused by improper packaging rote learning that is more about learning [6]. Rote learning in physics is very authoritative to absorb students in building an in-depth understanding of the concepts and

* Corresponding author: santyasa@undiksha.ac.id

principles of physics. On the other hand, rote learning closes opportunities for students to develop good character in learning physics [4].

Physics subject have differ properties if compared with other subjects. He feels the essence of physics as a product, as a process and as an attitude [4]. Physics as a process and attitude requires learning based on the analysis of the difficulties and misunderstandings of students before they learn new material. To build an effective model of learning in physics it should begin by identifying students' difficulties and misconception [7]. Although some misconceptions are difficult to overcome, physics learning have to be aware about the existence of the misconception and keep addressing to overcome students' misconceptions. Thus, physics learning becomes relevant and meaningful as well as providing opportunities for students to remediate their misconceptions and to build scientific concepts and character in learning. In fact, physics learning today still tolerates transmission models, which assume that students are merely the recipients of mechanistic knowledge. Physics learning is conducted by conventional instruction such as direct instruction model (DIM). Is conducted by conventional instruction such as direct instruction model (DIM). This assumption is a barrier to the change of misconception optimally, even though students have moved to higher level of education. Several studies have found that misconception still exist even when the students have attended more physics class, and the same misconception occurs and exists at primary and up to university level [7, 8]. These mean that the DIM which is unable to overcome misconception as a basis of learning have to be left and change to new model of learning which is based on constructivism.

Constructivism-based learning is very coherent with the human education in doing becoming process and growth in leaning. According to the constructivism, learning is the process of integrating new knowledge into the knowledge they have through a process of assimilation and accommodation [9, 10, 11, 12). To speed up the process of integrating the new knowledge to existing knowledge, students need to be assisted with a true model of learning based on the constructivism, as it has been developed before [6], namely conceptual change model (CCM). The learning model serve to accelerate the process of integrating new knowledge into existing knowledge through empower students' prior ability independently. The CCM gives students the opportunity to learn critically, diligently, tenaciously, creatively, and collaboratively. These efforts potentially underlie the formation of a deep conceptual understanding and developing of good character for students in learning physics. This statement is in accordance with the results of previous research [9, 11, 1, 2, 3, 4, 13).

By comparing the description and prescription between the CCM and DIM above, there are shown that models have a different effect on the students' outcome in learning physics theoretically. This research want to analyze the effect of the CCM compared with the DIM in achieving students' conceptual understanding and character in learning physics. Our research question are 1) is there a significant different of student' conceptual understanding and character between the CCM group and the DIM group after the treatment controlled by prior student' understanding statistically? 2) which one of the learning physics models more advance one to another for students in achieving conceptual understanding and character?

Methods

This quasi-experimental research used the nonequivalence pretest-posttest control group design. The pretest scores represent covariate of pre-conceptual understanding that serves as a control statistically. The aim is to eliminate the influence of the nonequivalence groups. The groups constitute an independent variable which are consisted of the CCM as an experiment group and the DIM as a control group. This research investigate two dependent variables (DVs), namely students' conceptual understanding and character in learning physics.

Population and Sample

The research population is 20 classes consisting of 738 students of grade X. The population consists of 8 classes (272 students) from SMA 1 Amlapura, 8 classes (256 students) from SMA 2 Amlapura, and 6 classes (210 students) from SMA 1 Manggis. All of them are located in Karangasem regency in Bali. To determine the sample, a random assignment technique was used to assign 6 classes (202 students, or 26.5% of the population) as a sample. In each school were selected 2 classes each as an experimental group which is learning with CCM and a control group which is learning with DIM. The number of students distribution at each school using CCM compared with DIM as presented in Table 1.

Table 1. Distribution of students' number in each schools who use the conceptual change model (CCM) as compared to direct instruction model (DIM).

Numb	Schools and Number of Classes	Number of Population	Class Samples	Models	Number of Samples
1	8 Classes of SMAN 1 Amlapura Karangasem	272	X-MIA-1	CCM	35
			X-MIA-2	DIM	35
2	8 Classes of SMAN 2 Amlapura Karangasem	256	X-MIA-2	CCM	32
			X-MIA-3	DIM	32
3	6 Classes of SMAN 1 Manggis Karangasem	246	X-MIA-1	CCM	34
			X-MIA-4	DIM	34
TOTAL	3 schools and 20 classes	738	6		202

Data Collection Tools

In this study, students' conceptual understanding in all groups before treatment was measured by test of physics conceptual understanding. The scores of the prior conceptual understanding are used as covariates. The students' conceptual understanding and character after the treatment both in the experimental group and in the control group were measured by test of physics conceptual understanding and questionnaire of students' character respectively.

Physics Conceptual Understanding Instrument

The data of students' conceptual understanding were collected by 35 items of physics conceptual understanding test in the form of five multiple choices with written justification with 0-4 scoring scale. The description of each 0-4 scoring scale is 0 means without choosing answer, 1 means choosing answer but wrong, 2 means choosing right answer but wrong reason, 3 means choosing correct answer true reason but less complete, 4 means choosing right answer and right as well as complete reason. Item-total correlation coefficient of conceptual understanding test are $r = 0.33$ moving up to $r = 0.66$. The reliability coefficient shown by Cronbach Alfa coefficient of $\alpha = 0.85$.

Students' Character Questionnaire

Dimensional of the character refers to the 10-dimensional characters, namely 1) the character of the love of God and all His creations, 2) self-reliance and responsibility, 3) honesty/trust, 4) diplomatic, 5) respect and polite, 6) generous, like mutual help and mutual cooperation/collaboration, 7) confident and hardworking, 8) leadership and justice, 9) kind and humble, and 10) the character of tolerance, peace and unity [19]. The data of the students' character are collected by questionnaire, consists of 55 items list of statements, each equipped with four choices are degradation using 0-4 scale. Description of each scale of degradation 0-4 is 0 means strongly disagree, 1 means disagree, 3 means agree, 4 means strongly agree. The internal consistency of the items moving from r = 0.36 to r = 0.72. The index of Reliability questionnaire are determined by Cronbach Alfa coefficient of $\alpha = 0.80$.

The Procedure of The Implementation

This quasi experiment was conducted for 4 weeks. Prior to treatment, students in all groups were given a pretest on the subject of physics conceptual understanding. In the experimental group with CCM and DIM control group, students studied for (45 minutes x 3) in each week. Every week students work on one type of lesson plan. The subject matter that becomes the object of research is the straight motion.

Implementation of the CCM follows 6 major steps, namely 1) finding and researching conceptual and contextual issues (students discover and examine conceptual and contextual conceptual issues prepared by teachers or other sources) related to material will be learned); 2) confrontation of misconceptions (students put forward ideas that are related to the initial problem that has been discovered and researched); 3) refute of the misconceptions (facing and examining the misconceptions in the form of physics refute text, physics lab activities, demonstration strategies, analogies, and confrontation by counter examples relating to the student's initial ideas or conceptions), 4) proof of concept and principle scientifically (students perform evidence based on scientific concepts and principles for strengthening the misconceptions that students face); 5) enrichment of material and contextual examples (students are invited to enrich material and contextual examples linked to proven concepts and principles); 6) knowledge testing (students conduct self-evaluation concerning their knowledge that has been constructed through problem solving)[4].

Implementation of the DIM follows 6 major Steps, namely 1) the teacher explains the learning objectives; 2) the teacher demonstrates the skills correctly, or presents the information step by step; 3) teachers plan and provide initial training guidance; 4) the teacher checks out the students who have been and who have not been able to perform the task well, and provide feedback; 5) teachers prepare for advanced training opportunities, with particular attention to application to more complex situations and everyday life [4].

After students in all groups get treatment as in accordance with the syntax of each learning model above, they answer a posttest on the subject of conceptual understanding about 90 minute, and then they answer questionnaire concerning their character in learning physics about 30 minute.

Results and Discussion

Results

Test of Normality of Data Distribution

The result of normality analyze of data of the pre conceptual understanding, conceptual understanding, and character of student show that Kolmogorov-Smirnov and Shapiro-Wilk statistic values in each data have significance number move from the sig = 0.220 > 0,05 to sig = 0.841> 0.05. The results of this analysis indicate that the data of pre conceptual understanding, conceptual understanding, and character of each group is normally distributed.

Test of Homogeneity of Variance

Variants tested their homogeneity are variant of the data about students' pre understanding, understanding, and character between CCM group and DIM group. The results show that Levene statistics are either based on mean, based on median, based on median and with adjusted df, or based on trimmed mean having moving significance numbers of sig = 0.397

> 0,05 to sig = 0.975 > 0,05. This indicates that the variant of students' pre understanding, understanding, and character between the two groups is homogeneous. Furthermore, Levene's Test of Equality of Error Variances is done. The results show that for DV of students 'understanding, F = 0.368 with sig = 0.545 > 0,05, and DV of students' character, F = 0.004 with sig = 0.950 > 0,05. This result shows that the error variance of the dependent variable is equal across groups.

Test of Linearity between Covariate and Dependent Variables (DV's)
In this research we tested two linearity events. The first is between pre conceptual understanding and DV of students' conceptual understanding. The test results show that F (linearity) = 51.82 with sig = 0.001 < 0.05. The second is between pre conceptual understanding and the DV of students' character. The test results show that F (linearity) = 0.150 with sig = 0.699 > 0.05. This means that the pre conceptual understanding covariate is only linear with the DV of students' conceptual understanding, but not linear with the DV of students' character.

Test of Collinearity between DV's
As mentioned above, there are two DVs studied, i. e students' conceptual understanding and students' character. The relationship between the two DVs is shown by Pearson Correlation coefficient (r). The results show that r = 0.049 < 0.80. These result indicate that there is no effect of collinearity between the two DVs .

Box's Test of Equality of Covariance Matrices
The result of Box's test of equality of covariance matrices is F = 0.292 with sig = 0.831 > 0.05. These results indicate that the covariant matrices of DV are the same between groups.

Table 2 Multivariate Tests

Effect		Value	F	Hypothesis df	Error df	Sig.
PRE UNDER-STANDING	Pillai's Trace	0.216	27.249	2.000	198.000	0.000
	Wilks' Lambda	0.784	27.249	2.000	198.000	0.000
	Hotelling's Trace	0.275	27.249	2.000	198.000	0.000
	Roy's Largest Root	0.275	27.249	2.000	198.000	0.000
MODEL	Pillai's Trace	0.070	7.422	2.000	198.000	0.001
	Wilks' Lambda	0.930	7.422	2.000	198.000	0.001
	Hotelling's Trace	0.075	7.422	2.000	198.000	0.001
	Roy's Largest Root	0.075	7.422	2.000	198.000	0.001

Table 2 shows two main points. First, that from source of influence of covariate pre conceptual understanding to DV of students' conceptual understanding and character shows statistical value F = 27.249 with significance number 0.001 < 0.05. The meaning that the covariates have a significant effect on students' understanding and character together. Second, that from source of influence of model to students' conceptual understanding and character shows statistical value F = 7.422 with significance number 0.001 < 0.05. Thus Ho who states there is no difference in students' conceptual understanding and character between those who learn with CCM and DIM, rejected. So students' conceptual understanding and character differ significantly between those who studied with CCM and DIM. Different test of each DV is presented in Table 3.

Table 3 Tests of Between-Subjects Effects

Source	Dependent Variable	Type III Sum of Squares	df	Mean Square	F	Sig.
Pre-understanding	Understanding	766.757	1	766.757	54.723	0.000
	Character	6.120	1	6.120	0.117	0.733
Model	Understanding	121.329	1	121.329	8.659	0.004
	Character	269.166	1	269.166	5.150	0.024
Error	Understanding	2788.313	199	14.012		
	Character	10401.068	199	52.267		
R Squared = 0.239 (Adjusted R Squared = 0.231), R Squared = 0.026 (Adjusted R Squared = 0.016)						

Based on the Table 03, the following research findings can be presented. First, from the source of the influence of students' pre understanding on the DV of students' understanding, it was found that the statistical value F = 54.723 with the significance number 0.001 <0.05, but against the DV of students' character found statistical value F = 0.117 with the significance number 0.733> 0 , 05. So students' pre understanding had a significant effect on the DV of students' understanding, but no significant effect on the DV of students' character. Second, from the source of the influence of the model on the DV of students 'understanding found the statistical value F = 8.659 with the significance number 0.004 <0.05, and against the DV of students' character found the statistical value F = 5.150 with the significance number 0.024 <0.05. These findings indicate two things, 1) that Ho stating that there is no difference in DV of students' understanding between those who study with CCM and DIM, is rejected; 2) that Ho stating that there is no difference in DV of students' character between those studied with CCM and DIM, is rejected. So individually both the DV of students 'understanding and the DV of students' character differed significantly between those who studied with CCM and DIM. Students' conceptual understanding who studied with CCM (M = 49.825, SD = 0.527) was significantly higher (Mean Difference = 1,550; sig = 0.004 < 0.05) than those who studied with DIM (M = 48.274; SD = 0.527) . Students' characters who studied with CCM (M = 187.620; SD = 1.018) were significantly higher (Mean Difference = 2,309; sig = 0.024 < 0.05) than those who studied with DIM (M = 185.311; SD = 1.018).

Discussion

Physics misconception phenomena for high school students occur because learning in schools tend to apply direct learning model (DIM). These misconceptions are a major obstacle for them in building a deep conceptual understanding and good character. Physics learning for understanding and developing good character requires a conceptual change model (CCM) to make the learning meaningful. This study aims to analyze the effect of the CCM compared with the DIM on the students' conceptual understanding and character of high school students.

The results showed that the effect of the CCM is higher than the DIM on the students' conceptual understanding and character. Students who learn in the CCM can achieve a higher conceptual understanding and characters than those who learn in the DIM. The results of this study are in accordance with previous research findings [15, 10, 7, 8, 4, 12].

Six CCM steps as mentioned above characterize the model as a derived student centered learning approach. The approach appreciates students' autonomy to flourish on their own [4, 9]. Autonomous student awards provide openness and sharing between them and with teachers, thus motivating the growing awareness and responsibility of learning. Such motivation makes it easier for students to change their views when their misconceptions are denied. The strategy of denial accompanied by proof of scientific concepts and principles can foster students' confidence in choosing a scientific conception rather than defending their misconceptions [2]. The belief in the scientific conception leads to the possibility of constructing a deep understanding. The two CCM steps that support this are the enrichment of materials and self-assessment that leads students to self-reflection based on critical and creative thinking as well as collaborative action on an ongoing basis. These CCM steps accommodate the nature of physics as a product, as a process, and as attitudes [4]. The CCM process that fits the bill of physics as the product provides students the opportunity to develop a deep conceptual understanding [2, 4, 11]. The CCM process that accommodates the nature of physics as a process and as an attitude provides opportunities for students to develop good character [4, 13, 16]. The learning processes provided by the CCM are a factor of excellence in achieving conceptual understanding and character for the students in learning physics.

The superiority of the CCM in physics learning as mentioned above is not provided by DIM. DIMs that are derived from behavioristic approaches tend to be teacher centered, thus accommodating only the nature of physics as a product and are less accommodative of the nature of physics as a process and as an attitude [2, 4]. The learning process provided by DIM appears less humanistic, less involving students in the learning process, less triggers the growth of critical and creative thinking, and does not allow students to engage in collaborative activities. This is a barrier to students not only in building a deep understanding, but also inhibiting them in developing good character in learning physics.

The implications can be drawn from the results and discussion of this study, that physics learning on the material of force and motion should reduce and even leave the DIM, but should prioritize the CCM in order students be able to achieve learning outcome optimally, especially in the level of high conceptual understanding and good character. The implementation of CCM in learning is preceded by the effort of exploration and identification of student misconception before they learn the new material. The results of the misconception identification are used to develop and refine the physics lesson and implementation in school.

Conclusions

Based on the results of research and discussion that has been described above, can be drawn conclusion research as follows. There is a significantly different students' conceptual understanding and character between those learning with CCM and DIM. Students' Conceptual understanding and character both collectively and individually for students learning with CCM is significantly higher than those studied by DIM. Thus, the effect of CCM higher than DIM in achieving students' conceptual understanding and character in learning physics on the material of force and motion.

The result of this research is very important for physics teachers to apply CCM in physics learning. The goal is that students have the opportunity to build conceptual understanding and character optimally. The implementation of the model can be done by first exploring and identifying students' misconception before they learn. It is important for the principal to always provide opportunities for physics teachers for training to explore and identify prior knowledge of students in order to plan and apply the lesson more meaningful and relevant for the students learning.

Acknowledgements

Thanks to the physics teachers and students at SMA 1 Amlapura, SMA 2 Amlapura, and SMA 1 Manggis Karangasem for their participation and collaboration in this research. We also thank to the reviewers of global conference on teaching, assessment, and learning in education for their corrections for the refinement of this article.

References

[1] I W. Santyasa, I K. Suardana, N. K. Tantris, N. N. Suarti, N. P. Paryawati, *JIKA*, 6, 2, (2008).
[2] I W. Santyasa, I N. P. Suwindra, *JRED*, 3, 1, (2009).
[3] I W. Santyasa, AAIN. Marhaeni, I W. Suastra, *JRED*, 5, 1, (2011).
[4] I W. Santyasa, I W. S. Warpala, I M. Tegeh, *IEJ*, 5, 1, (2015).
[5] Y. YYILDIZ, L. TARHAN, *JE*, 42, (2012).
[6] H. Gardner, (Simon & Schuster Inc, New York, 1999).
[7] A. Halim, T. Meerah, L. Halim, *Sains Malaysiana*, 38, (2009).
[8] H. Halim, T. K. Yong, T. S. M. Meerah, *Creative Education*, 5, (2014).
[9] M. F. Ayaz, H. Şekerci, *TOJET*, 14, 4, (2015).
[10] G. Aydin, *IJPES*, 2, 2, (2015).
[11] G. Özdemir, D. B. Clark, *EJMSTE*, 3, 4, (2007).
[12] G. ZHOU, *McGill Journal of Education*, 47, 1, (2012).
[13] T. D. Singh, (Kolkata: The Bhaktivedanta Institute, 2011).
[14] Suyanto, (http://waskitamandirikbk.wordpress.com/2010/06/02/urgensi-pendidikan-karakter/, 2010).
[15] S. Aydin, *JERBSciences*, 1, 1, (2012).
[16] D. Yahya Khan, (Pelangi, Yogyakarta, 2010).

Project-based learning and writing skill in EFL: are they related?

Luh Tiwika Praba[*], *Luh Putu* Artini, and *Dewa Putu* Ramendra

English Education Department, Universitas Pendidikan Ganesha
Jalan Udayana No. 11, Singaraja, Indonesia

Abstract. This study aimed at investigating the effect of Project-Based Learning on students' writing skill of grade 9 at SMPN 3 Kuta Selatan. This study employed embedded mixed method design, in which the quantitative data were considered as the main data. The data were collected through guided interview, observation, and writing test. Paired sample t-test was used for the quantitative data analysis. The result showed that Project-Based Learning has significant effect on students' English writing skill (sig. = .000 < α = .05). Project-Based Learning also revealed potentially promotes students' critical thinking, communication, and creativity through collaborative work. In general, it can be concluded that Project-Based Learning significantly improves junior high school students' writing skill in English as a Foreign Language.

Introduction

One of fundamental learning theories that have been highly considered for the 21st century learning foundation is Constructivism. It is a philosophy of learning that the students' learning occurs when the new knowledge is presented in relation to the previous knowledge within an effective teaching environment [1]. In this viewpoint, the new knowledge is practiced through experiences. The students create their understanding of the world they live in by reflecting on their own personal experiences [2]. They play major role to develop and build their own knowledge through active participation during the process of learning. Regarding this, teachers should take into account what students know. Teachers build on the students' prior knowledge and allow students to put their knowledge into practice [3]. In other words, it is the students themselves who have responsibility for their own learning.

A wide variety of instructional activities claim to be based on constructivist learning theory. One of the applicable teaching and learning model which is based on constructivist learning theory is Project-Based Learning. In its simplest definition, Project-Based Learning is a learning model that focused on activities which attempt to produce learning product through project works. It is considered to be an approach and means to achieve the 21st century skills [4]. As the supports of 21st century education, the concept of Four Cs are considered important in the classroom [5]. The Four Cs include critical thinking and

[*] Corresponding author: tiwika@yahoo.co.id

problem solving, communication, collaboration, and creativity and innovation. Project-based Learning potentially promotes these expected skills due to its learning activities. Project-Based Learning allows the learners work together with hands- on experience in an authentic and meaningful context [6]. When the learner sees the language that can be used in their lives and is applicable to their task or need, they can develop their language skills and communicative competence and gain confidence, co-operation, imagination independence and self-discipline. Moving away from rote learning and memorization, Project-based Learning builds on individual strengths, and allows individuals to explore their interests and ability. It is a process of learning that students are responsible for their own education [7]. Project-Based Learning provides opportunity to create productive and enjoyable classroom atmosphere through the integration of students' knowledge, attitude, and skills.

In its implementation, Project-based Learning is not only expected to create enjoyable classroom, but also give opportunity for students to improve their English skill. One of the skills that should be mastered in English subject is writing. Writing is a cognitive domain-based skill which involves learning, comprehension, application and synthesis of new knowledge [8]. Writing does not only require following the writing conventions, but also involving creative inspiration, problem-solving, reflection and revision that results in a completed manuscript. As a continuous activity, there are two approaches in writing; namely product approach and process approach [9]. Product approach is one approach in writing that expects the students to only analyze texts in terms of what language they used, and how they were constructed. Process writing is a way of looking at what people do when they compose written text. Thus, writing involves a complex process in which the students should become proficient.

Based on the preliminary research, it was found that students learned the concept from the teacher's explanation. In other words, students did not work with the project. The explanation of the material, presenting the text is still dominated by teachers which cause the students is rarely involved in the learning process. This situation confirmed that the learning process happened is not yet as expected. Controlled learning usually teacher-centered in nature, in which with limited resources and facilities to learn, teachers usually become the only source of learning in the classroom, and a textbook is the only learning material [10]. Moreover, in writing activity as a productive skill, the students have to develop their ideas on a paper. It indicates that the teacher role only is not enough. The students are expected to participate actively in the classroom. Regarding the potentials of Project-based Learning and the complexity of writing, a research on this needs to be conducted. This problem leads to the needs for conducting the research in investigating the effect of Project-Based Learning on students' writing skill. Thus, the research question was formulated as: Is there any significant effect of Project-Based Learning on students' writing skill of Grade 9 in SMPN 3 Kuta Selatan?

Research Method

This study employed the Mixed Method design focusing on collecting, analyzing, and mixing both quantitative and qualitative data which was taken on the ninth grade students of SMPN 3 Kuta Selatan. Among the major types of mixed-method, embedded design is considered to be the most appropriate design for the present study. An embedded mixed-method design involves the collection of both quantitative and qualitative data, but one of the data types plays a supplemental role within the overall design [11]. Technically, this design is beneficial when a researcher needs to embed a qualitative component within a quantitative design or vice versa. In the present study, quantitative data was addressed as the core, while the qualitative data provided supplementary one. Quantitative data were obtained from pre-test and post-test. They played major role to provide information about the extent by which project-based learning affects students' writing skill. The quantitative data collection involves several steps: (i) Administration of a pre-test measuring the dependent variable (writing skill), (ii) Application of the experimental treatment to the subjects, (iii) Administration of a post-test measuring the dependent variable (writing skill).

The qualitative data in this study were collected before collecting quantitative data concurrently. It was collected from observation checklist and interview in the preliminary research. Essentially, it was used to support and provide further elaboration related to the results of quantitative data. In this study, the instrument used in collecting the quantitative data is writing test. For the qualitative data, observation checklist and interview are used.

In embedded design, both quantitative data and qualitative data are collected during a single study, the two data sets are analyzed separately, and they address different research questions. This study used bivariate analysis, which indicated the analysis of two variables, dependent and independent variable. Hypothesis testing which was used for bivariate analysis was Paired Sample t-test. Before paired sample t-test was done in this study, normality test need to be conducted. If the data were distributed normally, paired sample t-test can be done. To find the normality of variance of the data, the researcher is conducted the Kolmogorov-Smirnov Normality Test. It is used to know whether the collected data will be normally distributed, the dependent significant value (less than 0,05) will indicate that the distribution of data differed significantly from a normal distribution. In other words, the data can be categorized as having normal distribution if the significance value is the more than 0.05. Moreover, for the qualitative data, the analysis involves three concurrent flows of activities, namely, data reduction, data display and conclusion drawing/verification.

Results and Discussion

Writing pre-test and post-test were conducted to reveal the effect of Project-based Learning on students' writing skill. The test was in form of instruction in which the students were intended to develop a Report Text based on the topic provided. Pre-test was administered to obtain preliminary data at the beginning of the study before the treatment was conducted. Then, post-test was carried out to investigate the effect of Project-based Learning on students' writing skill. The result of the writing pre-test and post-test can be seen in the table 1.

Table 1 Frequency of Writing Test

	Mean Score	Below the Mean Score	Above the Mean Score
Pre-test	68.5	22	10
Post-test	81	19	13

Based on the result of writing test before the implementation of Project Based Learning in the classroom, it was obtained that the mean score of the writing was 68.5. Moreover, from 32 students in class 9.5 there were 22 students (68.75%) who obtained below the mean score. Meanwhile, 10 students achieved above the mean score (31.25%). It means that there were more than half students in the class 9.5 of SMPN 3 Kuta Selatan whose score were below the mean score.

Furthermore, according to the result of writing post-test, it was revealed that the mean score of the writing was higher than in the pre-test, in which 81. Moreover, from 32 students in class 9.5 there were 19 students (59.37%) who obtained the score below the mean score. Meanwhile, the others, 13 students obtained the score above the mean score (40.62%).

Before doing further analysis, a normality test was conducted in order to know whether the obtained data were distributed normally or not. This test was administered through Kolmogorov-Smirnov formula of SPSS 23.0 for windows. The criteria for the test to be used are the data can be categorized as normally distributed if the significant data is more than 0.05. The result of normality test of the writing test can be seen in the table 2.

Table 2. The Normality Test of Writing Test

		pre_test	post_test
N		32	32
Normal Parameters[a,b]	Mean	68.50	81.00
	Std. Deviation	6.720	7.935
Most Extreme Differences	Absolute	.217	.173
	Positive	.217	.173
	Negative	-.103	-.155
Test Statistic		1.228	.980
Asymp. Sig. (2-tailed)		.098[c]	.292[c]

For writing pre-test, the value of Kolmogorov-Smirnov statistic was 1.228 with significance score of .098. For writing post-test data, the value of Kolmogorov-Smirnov statistic was .980 with significance score of .292. Both statistics showed that the significance value of the statistic was higher than 0.05, which can be said that the data were normally distributed.

The next step done after normality test was parametric test. Since the data were normally distributed, Paired-samples t-test was applied. The acceptation criteria of the significant effect was if the significance score less than 0.05. The result of the Paired t-test for students' writing skill can be seen on the Table 3 below.

Table 3. Paired-Samples t-test for Students' Writing Skills

	Paired Differences					t	df	Sig. (2-tailed)
	Mean	Std. Deviation	Std. Error Mean	95% Confidence Interval of the Difference: Lower	Upper			
pre_test - post_test	-12.500	5.902	1.043	-14.628	-10.372	-11.980	31	.000

As mentioned previously, there was an improvement of students' writing test from the mean score 68.5 before the treatment become 81 after the treatment. Moreover, the table above shown that the significance of the effect of Project Based Learning on Students' Writing Skills was .000. If the standard of significance is 0.05, it means that the significance data was less than the standard (sig. = 0.00 < α = 0.05). It can be concluded that there was a significant effect of Project Based Learning on students' writing of the ninth grade students.

In addition, another analysis was conducted to describe the students' writing improvement in terms of content, organization, vocabulary, grammar, and mechanic. Firstly, for the content aspect, it was found that the majority of students showed their development in stating the ideas in form of writing which can be seen from the increased numbers of sentences in the paragraphs. Secondly, in term of organization, it was obtained that the students construct the sentences coherently. For the vocabulary, the students exhibit a satisfying result in which they improve their sentences in paragraph writing. In term of grammar, the students maintained good structure. Lastly, for the mechanic, the students wrote only few errors in spelling and punctuation.

The writing test that the students have in the pre-test was modified in the post-test within the gap of time about 4 weeks for the treatment. In the pre-test, the students were expected to write a paragraph in 30 minutes. Meanwhile, in the post-test, the students were expected to write as many paragraphs as possible within a longer time (45 minutes) to assume that the students had an opportunity to write spontaneously and creatively. However, because the tests were conducted in form of paragraph writing, it can be assumed that the pre-test did not affect the post-test, but the treatment did. In other words, there was not any opportunity for the students to remember the answer of the pre-test when the post-test was conducted.

From the observation during the implementation of Project-based Learning, it was found that the students actively responded to the guiding questions delivered by the teacher. It indicates that critical thinking is also promoted by Project-based Learning. Furthermore, since Project-based Learning activities mostly done in group, the interaction in the classroom was also improved. It makes the communication as one skill which also encouraged in Project-based Learning. In addition, as the students expected to produce learning product, the students get opportunity to explore their creativity.

The previous elaboration indicated that this research found a significant effect of Project Based Learning on students' writing skill. The significant effect can be seen from the improvement of students' writing score after the implementation of Project-Based Learning. This finding was in line with a study conducted by Bas which found out that Project-Based Learning was more effective in the positive development of the students' academic achievement levels [12]. It was also showed a significant difference between the attitude scores of the experiment group and the control group.

Furthermore, the similar result was also found in Foulger and Jimenez-Silva's research which Project-based classrooms integrated technology within an environment where writing development was of concern. The study reported that via project-based learning, writing embedded in natural settings that helped create increased levels of language cognition – attention to topic organization, awareness raising on linguistic features, vocabulary and word choice, genre principles, and sentence formation for a variety of purposes are processed all at the same time [13]. It supported the findings of this research in which Project-based Learning gives opportunity for the students to practice more for their academic achievement.

The findings of this research are also confirmed by Thitivesa. Project work could be viewed as the natural language learning context, in which language learning, writing in particular, is contextualized and presented in the way that learning becomes integrated with the task of communication about some theme [14]. Project-based Learning improves students' writing skill since it place an emphasis on communication, students focus their attention on conveying their message and, at the same time, apply knowledge of conventions to their writing. The conventions are used at various points in the project and practiced in authentic context.

In addition, Project-based Learning potentially promote students' critical thinking, communication, and creativity through collaborative work which benefit on students' writing skill. Project-based Learning creates a pathway for students to explore, inquire, and create [15]. Thus, Project-based Learning provides learning environment that fosters writing skill.

Conclusion

This research serves empirical evidence that Project-based Learning significantly improve junior high school students' writing skill in English as a Foreign Language. It implies that to promote writing skill, which is as part of productive skill, the students should participate actively in the teaching and learning process. It indicates that Project-based Learning is a suggested approach for the teacher to teach writing. As the scope of the research was limited in number of subject and coverage of instruction, it is recommended that further research to duplicate this study with bigger size of subject and more sophisticated method of research so that the result could be generalized.

Acknowledgement

The authors would like to express their gratitude to Post Graduate School of Universitas Pendidikan Ganesha for organizing GC-Tale on which this research is presented.

References

1. H. Karaduman and M. Gültekin, "The effect of constructivist learning principles based learning materials to students' attitudes, success and retention in social studies," *Turkish Online J. Educ. Technol.*, vol. 6, no. 3, pp. 98–112, 2007.
2. L. Fears, "Comparison of Learning Theories: Behaviorism, Cognitivism and Constructivism," pp. 1–9, 2008.
3. N. H. Mvududu and J. Thiel-Burgess, "Constructivism in Practice: The Case for English Language Learners," *Int. J. Educ.*, vol. 4, no. 3, pp. 108–118, 2012.
4. D. H. Y. B. Harun, "Project-based learning handbook: educating the millennial learner," *Malaysia Educ. Technol. Div. Educ.*, 2006.
5. National Education Association, "Preparing 21st Century Students for a Global Society: An Educator ' s Guide to the ' Four Cs ,'" pp. 1–38, 2014.
6. J. W. Thomas, "a Review of Research on Project-Based Learning," 2000.
7. G. Solomon, "Project-Based Learning: A Primer," *Technol. Learn.*, 2003.
8. J. Defazio, J. Jones, F. Tennant, and S. A. Hook, "Academic Literacy: The Importance and Impact of Writing across the Curriculum-A Case Study," *J. Scholarsh. Teach. Learn.*, vol. 10, no. 2, pp. 34–47, 2010.
9. J. Harmer, *How to Teach Writing*. Malaysia: Pearson Education Limited, 2007.
10. L. P. Artini, "Establishing Rich Language Learning Environment at Schools: Preparing Children to Become Autonomous EFL Learners," 2014, pp. 378–386.
11. J. W. Creswell and V. L. Plano Clark, "Choosing a Mixed Method Design," *Des. Conduct. Mix. Methods Res.*, pp. 58–89, 2006.
12. G. Bas, "Investigating the effects of project-based learning on students ' academic achievement and attitudes towards english lesson," *Online J. New Horiz. Educ.*, vol. 1, no. 4, pp. 1–15, 2011.
13. T. S. Foulger and M. Jimenez-Silva, "Enhancing the Writing Development of English Language Learners: Teacher Perceptions of Common Technology in Project-Based Learning," *J. Res. Child. Educ.*, vol. 22, no. 2, pp. 109–124, 2007.
14. D. Thitivesa, "The Academic Achievement of Writing via," *Int. Sch. Sci. Res. Innov.*, vol. 8, no. 9, pp. 2986–2989, 2014.
15. T. Ralston, "Supporting Communication and Collaboration in Project-Based Learning Activities," *P21 Blogazine*, vol. 3, no. 2, 2016.

Indonesian Interference on English Language in VIII Grade Junior High School Students

Km Tri Sutrisna Agustia
trisutrisna.agustia@yahoo.com
Universitas Dhyana Pura, Bali, Indonesia

Abstract. This study aims to determine the type of Indonesian interference found in English language among class VIII SMPK 2 Harapan Untal-Untal students. This research is done by analyzing the result of simple text translation done by the students of class VIII SMPK 2 Harapan Untal-Untal, then from the translation result will be analyzed what kind of interference that appeared. Methods of data collection in this study using observational research methods and literature. The data that have been collected in the form of simple translations from the students will be analyzed descriptively using related theories to find out what kind of interference found in the analysis. Further, in relating to interference, it will find a way out to reduce and even eliminate Indonesian interference itself. The results to be achieved in this research is recognize the level of Indonesian language interference in the use of English among the junior high school students, then find the practical efforts to be done so that interference does not increased. So, the use and English understanding can be done as much as possible. The finding of the paper found that 6 interference such as grammar structure, subject pronoun, adjectives, auxiliaries, preposition and word order.

Keywords: interference, students, language, English, Indonesian

1. Introduction

1.1 Background of the Study

Foreign Language Education is taking a new turn, from seeing mistakes and errors as obstacles to tools that can help to predict and be carefully analyzed in order to understand the construction of a new language system. When a learner is acquiring a foreign language, his native language will have an effect on the production of the language being learned. Interference as errors in the learner's use of the foreign language that can be traced back to the mother tongue. Interference has a negative denotation; that is, the more the native and the target languages are different from each other, the more mistakes and errors will occur. The effect of interference can be on any aspect of language: grammar, vocabulary, accent, spelling and others [5].

In regard to Indonesian, the most salient aspects of interference in the production of English written works can be seen in the excessive use of the definite article the which Indonesian speakers use in most of the cases. Pluralization of adjectives is a characteristic Indonesian feature as in reds cars; the order of adjectives, Indonesian makes use of the noun + adjective syntactic construction while English uses the opposite, adjective + noun. Errors found due to the absence of observable subjects in sentences because Indonesian is a language that has implicit subjects in the conjugation of the verbs in any tense. Sentence word order is another important instance of native language interference. Indonesian can place a subject in any part of the sentence beginning, middle, or end without affecting its meaning, whereas English has a very strict order to construct sentences. Omission of negative auxiliaries (do not, does not, will not, did not) in English is replaced by the use of the sole word not in Indonesian. Indonesian does not need auxiliaries to express tense, tense is implicit in the conjugation specifically in the ending of verbs. Students use phrasal verbs incorrectly in English since this is not definitely a feature of Indonesian.

Students apply false cognates, invent words to avoid the gap of language knowledge, and finally, use inflections inappropriately especially in the present simple tense. All of the aspects described above need to be taken into account and make us; teachers, directors, ministers of education, and even parents, determine new strategies, methods, techniques and teaching materials to change the way foreign language teaching is taught in our schools. This investigation is meant to demonstrate why students repeat the same errors over and over and to recognize why traditional methods may not be productive. We can combine the best traditional methods with new teaching methods and findings. It is very important to be aware and to make the students knowledgeable about the existence of differences between one's native language and a foreign language keeping them from committing the errors and mistakes that hinder their learning of the second language.

1.2 Aim of the study

The aim of the study is to determine the most syntactic and semantic errors in writing in English through the analysis of writing produced by the individuals selected in the sample.

2. Methodology

In order to develop the current work, it was necessary to turn to bibliographic and field investigation. The bibliographic investigation was taken basically from journals, internet and previous findings on the topic. The field investigation was accomplished through the cooperation of a group of 20 junior high school of SMPK Harapan Untal-Untal students who provided the samples for analysis. These students were subjected to written works such as translating Indonesian sentences into English.The following methods were applied: Bibliographic method for gathering the scientific material necessary for the theoretical background. Analytic and Descriptive methods in all other stages in the execution of this research in order to do the descriptive and statistical analysis, and the interpretation of the data gathered through the field research.

With respect to the techniques, they were note-taking to collect bibliographic information and selection of written material. Variables taken into consideration were grammar structure and lexicon. In relation to the variable grammar structure the indicators considered were: subject pronouns, definite and indefinite articles, adjectives, prepositions, auxiliaries, and word order. The considered indicators for the lexicon variable were: use of false cognates, use of phrasal verbs, invented words and inflections. The procedure used for the analysis of the collected data through the field investigation was an approach to conducting Error Analysis [5]. Norrish's approach includes sample collection, identifying errors, describing errors, explaining and evaluating errors. Contrastive analysis was used as part of the procedure to explain the occurrences of errors.

Second language learners develop their capacity and skills to communicate in different environments. Many times, people can recognize a foreigner's native language just by hearing him. These traces will be seen in their physical appearance, pronunciation, word order structures, and even vocabulary. Second language learning is considered successful when a learner masters the structural entities of the new language plus the organization of knowledge into coherent structures that will eventually lead them to effective communication [2]. Second language learning does not happen as easily as the ideal describes it. There appears to be a gap between the two components. Second language learning is influenced by native language and target language [2]. Transfer as the use of patterns of the first language in the production of the second language. When these patterns are identical, the correct language form is produced and positive transfer occurs. On the other hand, if errors happen as a result of such transfer, they are called interference or negative transfer [2].

Lexicons (words) are the most representative feature of a language. Just by knowing the lexicons of a language, a second language learner can understand it, in spite of their sounds or syntactical errors. There is also a natural tendency to assume that there is a word per word equivalence in two different languages, which naturally pushes the learner to do direct translation without considering its contextual meaning. Hence, this tendency produces interference. Furthermore, the second language learner cannot prevent his first language interfering with the production of speech in the target language [3].

Finally, syntactical gaps in the target language and the native language are responsible for the occurrence of interference. When a structure is absent in one of the languages, the learner tries to compensate it by using it in the language that lacks of it. Hence, this indicates interference. This kind of interference also takes place when the learner does not have a good level of mastery of different structures. Interference in the area of syntax could not be only brought about by syntactic gaps, but also by linguistic knowledge of the speakers [1]. The conclusion reached through this study was to place more emphasis on communication itself than on the elimination of errors through a communicative model of teaching, as well as to provide students with varied learning strategies that will enable them to perceive and avoid errors themselves and foster their abilities as independent learners.

3. Result

In the following section the variables grammar structure and lexicon will be analyzed and compared in both languages Indonesian and English in order to point out the existing interference of Indonesian over English writing. This paper demonstrates that in fact, there are similarities and differences between Indonesian and English which affect or interfere the written production of the target language. Both languages share many grammatical components such pronouns, adjectives, articles, prepositions but their differences lie mainly in the way these are used. In addition to what has been mentioned, Indonesian lacks of auxiliary words and phrasal verbs, which are an English language specific feature.

3.1 Grammar Structure

Indonesian has specific syntactic features that will definitely affect a learner's written output such is the case of excessive use of the definite article the, the order and pluralization of adjectives, implicit subjects in verb conjugations and word order. On the other hand, Indonesian lacks the existence of some English syntactic features regarding to the use of the indefinite articles *a* or *an* to provide information about occupations, the mandatory use of personal pronouns or subjects in sentences in English, the use of the pronoun *it* to express impersonal constructions, and complete absence of auxiliary verbs to express negative statements and questions. The absence or existence of the syntactic structures above confuse the native speaker in either the meaning of a word or what the whole idea is meant to express. The native speaker would have to understand the idea based on the context it was used or ask for further explanation. In the following paragraphs the different indicators included in the variable grammar structures will be described, explained and exemplified.

Evidence of native language interference is noted in syntactic structure examples as follow:

a. SL: Saat orang-orang pergi ke negara lain, mereka akan mempelajari bahasa yang lain
TL: When people travel to other countries, have to learn a new language
Revision: When people travel to other countries, they have to learn a new language.

b. SL: Orang tuaku lahir di Ekuador

TL: My parents born in Ecuador
Revision: My parents were born in Ecuador

The mentioned examples and others related to some syntactic and lexicon components will be part of a detailed analysis in this study.

3.2 Subject Pronouns

In Indonesian is optional. This happens because syntactically they are already implicit in the verb endings at the moment of conjugation in any tense. Subject pronouns, in Indonesian, appear in sentences mostly to emphasize as in *yang bertanggung jawab* meaning *I am the responsible one*. The other form of interference found related to the use of the definite article when it appears next to the name of certain places. It can be noticed in the statements such as:

a. SL: Migrasi di Ekuador
TL: Migration in the Ecuador
Revision: Migration in Ecuador

b. SL: Frekuensi yang lebih tinggi di Ekuador
TL: Higher frequency in the Ecuador
Revision: Higher frequency in Ecuador

There is absence of the concept that the definite article the will only appear next to place names which include the words states, kingdom, islands, and republic such as The United States, The United Kingdom, The Galapagos Islands, The Czech Republic.

3.3 Adjectives

The uses of adjectives are another syntactic component present in both languages intended to describe or provide characteristics of people, animals, places, ideas or things. Differences between both languages arise when the concepts of position and pluralization appear as follow:

a. SL: Hal ini menyebabkan berbagai macam penyakit
TL: These cause differents illnesses
Revision: These cause different illnesses

These cause *differents* illnesses very familiar examples of pluralization in Indonesian. Indonesian is a language that expresses adjectives in plural or singular depending on the adjoining noun. So, if the noun is plural then the accompanying adjective will be plural too, similar to the examples posted above. This issue does not happen in English since this language, in general, does not pluralize adjectives.

b. SL: Mereka dapat menyebabkan penyakit yang parah
TL: They can produce illnesses terrible
Revision: They can produce terrible illnesses

Adjective order suffers another change in Indonesian. In the sentences and They can produce illnesses terrible, it is observable that there is Indonesian interference because the order of adjectives in Indonesian is noun first and adjective second, while English acceptance is adjective first and noun second

3.4 Prepositions

English and Indonesian share similar characteristics by using them next to verbs and to express location, time, distance, direction, and origin. The basic concept of the various instances in which prepositions are used is basically the same, but the difference lies when these prepositions are used along with verbs and phrases concerning time and place. In these cases, the English learner will resort to memorizing such phrases and verbs. The examples as follow:

a. SL: Kami mengunjungi sebuah taman
TL: We visited to the park
Revision: We visited the park

b. SL: Pergi ke negara lain untuk menemukan kehidupan yang lebih baik
TL: Go to other countries for find a better life
Revision: Go to other countries to find a better life

The examples are directly translated from Indonesian as well. Once again, the way the phrases are specifically used in English are not well considered. There are not precise rules to help the learner discriminate when to use prepositions in cases similar to the phrases previously mentioned. The sole way to use them correctly is by resorting to memorization.

3.5 Auxiliaries

Auxiliaries are very particular syntactic elements. English relies on auxiliaries to indicate tenses, grammatical persons, negatives and question forms, while Indonesian resorts to verb endings to express tenses and grammatical persons and in negative statements it uses the word tidak which directly translated into English becomes not. The following instances illustrate this last condition as follow:

a. SL: Kami tidak memiliki waktu
TL: We not have time
Revision: We do not have time

In the example the particle not stands alone without an indicator of tense or grammatical person as it is in the English language. This causes confusion because the reader will not be able to comprehend the phrases. He does not know if the writer is trying to express the idea in present, past, or future. Hence, the reader will resort to context. The correct phrases above will be We do not have time. The absence of an auxiliary verb in the sentence such as:

b. SL: Jika kamu memakan kuenya, kamu akan menjadi patung
TL: If you eat the cookie, you become a statue
Revision: If you eat the cookie, you will become a statue

If you eat the cookie, you become a statue shows Indonesian interference. From the context, this sentence was taken the writer really wanted to say, if you eat the cookie, you will become a statue. This construction corresponds to the incorrect formation of the first conditional in English, which explains that when the first part of a sentence is in present simple tense, the second part will need the auxiliary will in order to build it accurately. But if we look at the Indonesian syntactic construction the above sentence would be right.

3.6 Word Order

It is worth mentioning that Indonesian does not constraint its sentences to any special pattern. Such is the case of the place of subjects in sentences. Subjects can be either at the beginning, middle or end of a sentence, whereas English has a rigid pattern of subject + verb + and complement. Also, Indonesian does not limit the way time or place adverbial phrases should be, while English dictates they should go last in a sentence. The following examples illustrate what has been mentioned as follow:

a. SL: Untuk sarapan saya mempunyai telur dan jus jeruk
TL: We visited to the park
Revision: We visited the park

It is noticeable that the adverbial phrase is in the middle instead of being last as determined in English.

4. Conclusion

After observing, analyzing the samples depicting the different syntactic structures, it is possible to notice that the indicators that appear more frequently are those concerning the use of the definite article the adjective order and pluralization, absence of subject pronouns and different word order. The frequency of occurrence of these indicators is highly related to the level of English knowledge and performance. If we look closely at the English level of the students, the students at this level are not aware of the incidence the differences between both languages have.

The data in this document, gathered from the variables grammar structure and lexicon, demonstrate the following conclusions:

a. According to the results, most of the grammar structure and lexicon errors were found in activities performed inside the classroom due to time and teacher's constraints.
b. Most of the errors occurred in indicators related to the variable grammar structure which shows that interference is related to the surface structure of both languages.
c. Subject pronoun omission is the most frequent error students commit when writing in English. This can be traced back to their native language influence where subject pronouns are already implicit in the verb endings.
d. Prepositions related to time and place were used erroneously in every writing activity which indicates this grammar structure differs in English.

Learning a second language, especially after 12 years of age, is a real challenge. Not only for the fact of not being able to acquire the knowledge as quickly as it occurs at younger ages, but also because it is disappointing to know that becoming as fluent as a native speaker may never happen. On the other hand, it is important to recognize that the interference phenomenon is present and the necessity to change techniques, methods, strategies, materials, and institutional policies is mandatory. There is an obligation to encourage language learning awareness from its beginnings; that is, to notice differences and similarities between both languages and to see errors and mistakes as signs of language improvement in order to predict possible future miscommunication.

References

1. Bhela, B. Native language interference in learning a second language: Exploratory case studies of native language inference with target language usage. International Education Journal **1**, 12 (1999).
2. Dulay, H., Burt, M. and Krashen. *Language Two* (New York: Oxford University Press, 1982)
3. Ellis, R. *The study of second language acquisition* (Oxford: Oxford University Press, 1997).
4. Lott, D. Analyzing and counteracting interference errors. ELT Journal **37**, 256- 261 (1983).
5. Norrish, J. *Language learners and their errors* (London: Macmillan Press, 1983).

How consistent is teachers' planning, implementation, and assessment in character education?

Budiarta Luh Gd Rahayu*, *Artini* Luh Putu , *Seken* I Ketut and *Santosa* Made Hery
English Education Department, Universitas Pendidikan Ganesha, Bali, Indonesia

Abstract. This descriptive case study aims at investigating how teachers planned, practiced, assessed students' character development in the classroom. The study involved 29 students of Grade 5, a classroom, and an English teacher. The instruments include classroom observation, checklist, interview guide, field note, and video recording. The obtained data were analyzed by using analysis method by Miles and Huberman (1994). The result of the research shows that there were ten characters values detected in the lesson plans, however, only 55% consistently appeared in the teaching and learning process. Teachers were found to use performance rubrics and diary to keep record on students' character development. The interview revealed that there was reluctance regarding role of teacher in term of assessing students' characters. This fact implies that there should be intensive training and workshop for the teachers to improve the quality of the character education practices in elementary schools in Bali.

Keywords: Curriculum 2013, Character education, Teacher's planning, Practicing, Assessing

Introduction

One of the latest issues in Indonesia's education is the launching of Curriculum 2013 which promotes character building as the highest priority. It can be seen from the core competencies (KI) which are explicitly stated in the document. Character is important manifestation of educational implementation process in every level [1]. It means that character building should be reflected in each learning subject, and not only in certain lessons related to attitude, such as religion, civics, and norms. Character education is very important, not only to help the country's young generations to be 'good', but also help them reach academic achievement [2].

The character education is a deliberate effort aims to cultivate and develop character that are worthy for individual and sociey as a whole [3]. It requires careful and calculated planning for success [4]. Other researcher found character education programs had a positive effect on both character development and academic success [5]. It will increase the nation quality in the future. Creating good a character based education in school is very important to do [6]. Schools, as an education institution, have a very important role in cultivating children character through education [7]. So that, planning the character based education must be done as good as possible by the teachers. Good planning will guide to

* Corresponding author : budiartarahayu@gmail.com

good implementation of the character education [8].At the moment, this curriculum has not been officially regulated yet in all schools in Indonesia. However, some schools have already applied it based on the government decision, or the schools are appointed as piloting schools. The goverment choose one elementary school in Denpasar in Grade 5 as a subject or model in implementing Curriculum 2013. All teachers in the school had joined workshop in implementing the curriculum. Some of them had joined for three times. In addition, the teachers had also given technical books of the curriculum to be learnt. If they find any problems, they can ask the school's supervisor. Here, the supervisor becomes a bridge between the teachers and the government.

However, based on the interview done with the teachers, they have serious problem on the assessment. It might be caused by the teachers' perception of using previous conventional assessment, but in the current curriculum, they are expected to use more authentic assessment. In addition, the assessment system for the character education undergoes frequent changes. Before, the assessment was conducted per theme, and then changed into per sub-theme. Until this research was finished, there is still no fixed assessment for the character education, because it keeps changing. It makes the teachers remain in confusion regarding how to assess the students' characters.

Based on those explanations, the research comes to complete the previous researches by adding some points on research, namely the planning of the character based education, the implementation of the character based education, the way of assessing it, and the encountered problems. According to Curriculum 2013, the subjects are divided into obligatory and local content subjects; therefore this research focused on both of them. The obligatory lesson is in form of thematic, meanwhile the local content one is not. Here, the selected local content subjects was English. It is done in accordance with the researcher's specialty.

Based on the previous explanation, this paper focuses on 1) how the fifth grade teacher plans the character-based education, 2) how the fifth grade teacher practices the character-based education, 3) how the fifth grade teacher assesses the character-based education, and 4) the problems faced by the teachers in implementing the character-based education.

Methods

This research is a descriptive case study whose data were collected as it is. This research was conducted at primary school in Denpasar considering the school as one of piloting schools that implements Curriculum 2013. The subjects of the research were the fifth grade thematic teacher, the fifth grade English teacher, and 29 students of the school under the study; meanwhile the object of the study was the teaching learning process of the character education which involved the way of teacher planning, practicing, and assessing the character education. In order to collect the expected data, three methods of data collection were used, namely observation/recording, interview, and document study. The researcher served as the main instrument who collected the data, observed the phenomena, conducted analysis, and concluded the result of the analysis. To gather the data, the researcher used some devices, such as interview guide, tape-recorder, and video-recorder. The obtained data were analyzed by using analysis method proposed by Miles and Huberman [9].

Results and Discussion

This research was done in one of elementary schools piloting Curriculum 2013 in the municipality of Denpasar. Here, there were two lessons observed, e.g. thematic and English. The finding of the research is presented in Table 1.

Table 1. The summary of character inserted in thematic and English lesson

No	Characters	Thematic and English Lessons					
		LP	Prac-tice	Assess-ment	LP	Prac-tice	Assess-ment
1	Religiosity	√	√	√	√	√	-
2	Tolerance	-	√	-	-	√	-
3	Hardwork	√	√	-	√	√	-
4	Communicativeness	√	√	-	√	√	-
5	Independence	√	√	-	√	√	-
6	Democracy	√	√	-	-	-	-
7	Curiosity	-	-	-	-	-	-
8	Discipline	-	√	-	-	-	-
9	Responsibility	√	√	√	-	√	-
10	Nationality	√	√	-	-	-	-
11	Patriotism	-	-	-	-	-	-
12	Appreciation for Achievement	-	√	-	-	-	-
13	Love for Peace	-	-	-	-	-	-
14	Love for Reading	√	√	-	-	-	-
15	Social Care	-	-	√	-	-	-
16	Creativity	√	√	-	-	√	-
17	Nature care	√	√	-	-	-	-
18	Honesty	-	√	√	-	-	-

Based on Table 1, it is known that among 18 characters established by the government, there were only 10 characters planned by the classroom teachers. In the practice, 14 characters were practiced. It means that the other 4 characters occured simultanously. For the assessment practice, there were only 4 characters which were completed with assessment system. For English teacher, there were 4 characters planned from 18 established characters by the government . However, there were only 7 characters practiced. It means that 3 characters were practiced simultanously. English teacher did not make assessment for the characters.

The finding reveals that the teacher had professionally planned character based education. The experience and knowledge acquired from the workshop are very useful in making the planning. The teacher realized that they must have good decision and creative thinking in order to plan their lesson well [10]. By doing good planning of the character based education, the school can be a place for the students to develop or to cultivate the characters. It allows the school to function as a facilitator of academic and non academic skills to develop [11].

In planning, the teachers included explicit and implicit focuses of character during teaching and learning process. There are three strategies can be used in teaching the character based education, namely a) explicit focus of specified character values/ethics, b) involvement of students in interactive pedagogical activities, and c) provision of characters modelling [12]. The explicit focus is exemplified by giving advice directly to the students regarding what they are allowed and not allowed to do. Meanwhile, the implicit information is done by involving the students in teaching and learning process, i.e. by implementing group discussion technique. Here, the students are taught to be communicative and respectful the others' opinion.

From Table 1, it is known that there are ten characters consistently taught by the teacher, namely religiousity, hardwork, communicative, independence, democracy, responsibility, nationality, love for reading, creativity and nature care. Those characters are consistently appeared in teaching learning process. After knowing how many characters appeared in teaching learning process. Next, it is showed the percentage of each practiced character as Table 2.

Table 2. Frequency of the Character Practiced

No	Character	Frequency in Thematic	Frequency in English (%)
1	Religiosity	22%	22%
2	Hard work	55%	38%
3	Creativity	11%	11%
4	Independence	22%	22%
5	Democratic	55%	38%
6	Curiosity	44%	22%
7	Nationalism	44%	0%
8	Communicativeness	66%	44%
9	Love for reading	44%	0%
10	Nature care	22%	0%
11	Responsibility	44%	0%

Among all, 22 % of the character practiced by both teachers was religious; meanwhile for the classroom teachers, 55% practiced character was hardworking , and the English teacher had the practice by 38%. Relious character was practiced at the beginning and the end of learning; meawhile, the hard working character was practiced through assignment and exercise during the learning process. Next, the communicative character in which the classroom teacher practiced by 66% and the English teacher did by 44%. This character was practiced through the implementation of coooperative learning. In addition, English teacher also applied independent character by the same percentage. By this model, the students were expected to interact and communicate with each other. At the end, they had to be able to do something independently. Next, the classroom teacher practiced creative character by 11%, independence character was inserted by 22% during the planning and the implementation, and the characters of curious, nationalism, love for reading and responsible was integrated during teaching learning process by 44%. The creative character was cultivated by giving the students freedom in accomplishing their assignment based on the instruction; the independent character was practiced by giving individual project or assignment; the curious character was practiced by giving authentic task; nationalism character was practiced by hanging heroes picture and celebrating national day; love for reading character was done by directing the students to visit library; responsible character was practiced by assigning them with group work with responsibility organizing the class in a prepared schedule. The English teacher also did the same thing as classroom teacher. Lastly, democratic character was found by 38%. Democratic was also practiced by carrying out discussion on certain activities, and nature care was done by keeping the environment clean and conducting socialization about the importance of nature.

The application of character education should concern on three interrelated parts, namely moral knowledge, moral feeling, and moral action [13]. Those three interrelated parts of good character can be taught in the classroom during learning activities by involving the students in any projects that can increase students' awareness, teaching the students how to respect and be responsible, analyzing the moral values integrated in the lesson, deciding the moral values aimed to deliver in the learning process, searching for decent learning material, designing effective learning strategies, inviting keynote speaker to discuss particular character values, and doing multicultural education to build students' understanding about different culture [14].

In the implementation of the character education at the piloting school, all parts of the education character have been practiced. The characters have been integrated in the lessons. Besides that, the teachers also applied scientific approach in the class in which the character can be inserted. Here, only "inviting keynote speaker to discuss particular character values" was not done by the school. The knowledge regarding the character values was obtained by the teachers through workshops or seminars.

In term of assessment, the teachers provided lack of assessment for the characters, even the English teacher did not assess the students' characters. Until now, the classroom teacher only used diary and rubric. It is not considered as good, since assessment plays asignificant role in the development of students character [15]. In addition, Character Educational Partnership or CEP stated, in the character education, assessment serves as the source of information about the students' characters [16]. Good assessment leads to appropriate treatment that should be given to the students. If the assessment does not provide holistic information, the teacher may not treat the students' character well [17]. As the consequence, using and developing good assessment instruments can be a way taken to reach the character education's success. There must be two assessment types for effectiveness of the character education, they are pre-assessment of goals and post-assessment of results [18].

Conclusion

Based on the analysis, the character education has been planned by the teachers in the classroom. Based on the result of analysis conducted on four lesson plans, there were ten characters appeared constantly in the learning process. After that, both thematic and English teachers have already practiced all characters planned in the lesson plans.

Until now, the teachers only use rubric and daily journal. The rubric is used in certain subject lesson with clear descriptors for each character. Meanwhile, the daily journal is used to record the students' bad attitudes on daily activities that will be used to score the students in every one semester.

The teacher still have problem in implementing character education, there were four problems encountered by the thematic teacher in implementing character education, namely a) unclear procedure of assessment, b) the different rules applied by the school and parents at home, c) the teachers' lack of IT knowledge, and d) the lack of media, such as books and magazines containing character building. Meanwhile, for the English teacher, there were only two problems encountered regarding the unclear procedure of assessment, and the different rules applied by parents and the school

This research implied that there should be intensive training and workshop for the certainty regarding the character assessment procedure and supervision of the goverment stakeholders given for improving the character education.

References

1. A. Darmansyah,. The development of character education curriculum for elementary student in west sumatera., **2**, 189–198 (2014)
2. L Artini. Integrating universal and local values in character education in indonesian schools. (2016)
3. G.M. Almerico. Building Character through Literacy with Children's Literature (2014).
4. C. Tyra. Bringing books to life: Teaching character education through children's literature. *Rising Tide,* **5**, 1-10 (2012)
5. A. Katilmis., H. Eksi & C. Ozturk, C. (2011). Efficiency of social studies integrated character education program. *Educational Sciences: Theory & Practice* **11**, 36-50 (2011)
6. A. Djailani. Strategy character building of students at excellent schools in the city of banda aceh. **1**, 49–59 (2013).
7. M. Elias. Character Education: Better student better people. (2010).
8. S. Jacques S. Benninga., Marvin W. Berkowitz., Phyllis, Kuehn., Karen, Smith. The Relationship of Character Education Implementation and Academic Achievement In Elementary School. JR. CE. **1**(1). 19–3. (2003)
9. M. A. Miles. In *qualitative data analysis: an expanded sourcebook* , 50–72(1994).
10. Twemlow, S. W., Fonagy, P., Sacco, F., Gies, M. L., Evans, R., & Ewba nk, R. Creating a Peaceful School learning environment: A controlled study of an elementary school intervention to reduce violence. A. J. Psych. **158.** 808-810. (2001)
11. Vaishnavi. R. Kanzal, Subikshalakshmi. G, Lopamudra Goswami. Moral Education: Current Values in Students and Teachers' Effectiveness in Inculcating Moral Values in Students. I. J. I. Psych. **4**(1), No. 81. (2016)
12. M. W. Berkowitz, & Bier, M. C. (2007). What Works in Character Education. *Journal of Research in Character Education*.
13. T. Lickona. What is good character? (2001).
14. F. Clark, P. With Liberty and Justice for All Character Education for America's Future .J. CE. **10**(1). 31–36. (2014) 10
15. Supriyadi, E. *Pengembangan Pendidikan Karakter di Sekolah*. (2012)
16. Ministry of Education and Culture. (2013). Peraturan Menteri Pendidikan dan Kebudayaan Republik Indonesia Nomor 65 Tahun 2013 tentang Standar Proses Pendidikan Dasar dan Menengah.
17. Haryanto. Pendidikan karakter menurut ki hadjar dewantara. **53**, 1689–1699 (2013)
18. K. M. Brooks B. What makes character education programs work? **51**, 19–21 (1993)

Ecological Intelligence Level Of Hight School Students In Cimahi City

Revi Mainaki; *Wanjat* Kastolani ; *Iwan* Setiawan
Master Program of Geography Education, School of Post Graduate Indonesia University of Education, Bandung

Abstract. Indonesia is experiencing serious environmental problems as a negative impact of development. One of the areas that shows the problem of the environment is Cimahi City located in West Java Province. Therefore, efforts are needed to prepare future generations in order to have high ecological intelligence. The intelligence can be prepared and established through the process of education in the School. The purpose of this study is to measure the level of ecological intelligence of high school students in Cimahi City. This research uses survey method based on quantitative approach. The technique of determining the number of samples using stratified ramdom sampling and obtained 35 schools as sample. Data collection techniques used observation, interviews, questionnaires, literature studies and documentation studies. Data then analyzed and presented with descriptive statistic technique. The results showed that the level of ecological intelligence of learners varied between schools, from moderate to high levels. Judging from the components of ecological intelligence, the attitude aspect has the highest score followed by aspects of knowledge, skills and participation. All parties need to encourage schools and learners to improve ecological intelligence through various programs at schools based on environmental conservation efforts.

Key word: Ecological intelligence, environmental degradation, knowledge, skills, participation

1. Introduction

1.1 Backround

When humans are in the century BC where life is highly dependent nature of humans are very appreciative of nature and indirectly become dependent as well. The fertile surface of the earth makes people more likely to think about other things, like science that gives rise to a more advanced civilization. This happened in the time of the Greeks and Romans when the needs of the population were fulfilled so much time to develop the science of the surrounding environment that led to the assumption that the environment as a human residence should be respected, preserved and preserved.

But this becomes a dilemma because it is in line with the progress of science and technology the higher the level of environmental damage, Data Potential kelurahan in Cimahi City (2009) and Puskesmas Cimahi Utara (2016) diseases caused by environmental degradation such as respiratory infections (ARI) About 5000 people, vomiting about 4400 people and dengue about 2,400 people showed the level of environmental degradation that can actually be prevented by growing the ecological intelligence of learners. The volume of vehicles passing through Cimahi City based on data from the Cimahi City Transportation Department (2016) reached 215,488 vehicles of various sizes, indicating high levels of air pollution caused by vehicles. A scientist in the field of psychology that is [2] Goleman (2010) in his book titled *"Ecological Intelligence: The Hidden Impact of What We Buy"* Considers that there is another growing intelligence, known as "Ecological Intelligence". Goleman described many of the human brain and mental maps as part of the intelligence, and how ecological intelligence began to form.

[1]Muhaimin (2015) and [2]Goleman (2010) Explain the concept of human ecological intelligence can be measured through 1) knowledge of the impact of human activities, including know the impact we cause, how to impact a) Geosphere (spatial impact); b) Biosphere (impact on organisms); c) Socio-sphere (social impact); 2) attitude improvements favour ie care for the environment including transmitting his understanding to others (share the new knowledge with others); 3) skills in environmental conservation and 4) involvement in participatory, participatory environmental activities.

If people in this Cimahi City have ecological intelligence, the higher the awareness to use environmentally friendly vehicles or mass vehicles, in order to reduce air pollution in the city of Cimahi. Based on what has been described, showing that the growth of ecological intelligence in the City of Cimahi is very important, and interesting to study in a measurement and research with the title "Ecological Intelligence Level of Hight School Students In City Cimahi". With formulation of the problem in this research is "How is the level of ecological intelligence of high school students in Cimahi City? The puspose of resesrch is for identify and describe the level of ecological intelligence of high school students in Kota Cimahi. Theoretically, it is expected to enrich the study of theories of education and it is practically expected that this research can be input in determining the policy in the field of education and environment.

2. Methodology

The variables in this study are ecological intelligence with indicators that include (1) knowledge; (2) attitude; (3) skills and (4) participation in participating in preserving the environment of the learners themselves. The population in this study is all of the existing high schools in Cimahi City, consisting of 6 State Senior High School (SMAN), 10 Private High School (SMAS), 13 State Junior Secondary School (SMPN) and 24 Private Junior High School (SMPS)) Spread throughout the existing outposts in Cimahi City. Through the Slovin equation described in [3] Kusnendi (2008) using equation (1)..

$$n = \frac{N}{1+N.\alpha^2} \qquad (1)$$

Description: n = Number of samples N = Population size

α = 10% adjustment error rate adjusted by the researcher is the rest of the 90% confidence level

The sample of students of class XII IPS for senior high school and 9th grade proportionally from 6065 was obtained 260 students with equation 2 ([4]Sugiyono, 2010)

$$n = \frac{\lambda^2.N.P.Q}{d^2\,(N-1)+\lambda^2.P.Q} \qquad (2)$$

λ = *Chi Square* dk 1 and the level of error 10% = 2.706
P = Q = 0.5 d = 0.05

Data collection techniques used in this research are 1) observation; 2) questionnaire; 3) interview; 4) literature study; 5) documentation study. The data analysis technique used is by scoring the respondent's answer using measurement scale. Result measurement used Guttman scale [5] for knowllege aspect (0-1 score), Linkert[6-7] scale for attitude aspect (1-4 score) and Ratting scale[7-10] for others aspect (1-5 score), then included in the following average equation (3).

$$\text{Mean} = \frac{\sum Xi}{N} \quad \text{Xi = The value of x to i to n; } \sum \text{ = jumlah; N = amount of data} \tag{3}$$

Results

The level of ecological intelligence of learners in high school country in Cimahi City can be seen as in Figure 1 where students in SMAN 3 are in the first sequence followed by SMAN 2 and beyond SMAN 6 and SMAN 5 Cimahi. Based on the calculation, the level of ecological intelligence of students in SMA Negeri in Kota Cimahi is in the classification of "high" score in order of SMAN 3 (2.74), SMAN 2 (2.71), SMAN 6 (2.69) and SMAN 5 (2.62) Differences in scores occur because there are also differences in scores on each aspect of the formation of the ecological intelligence of learners.

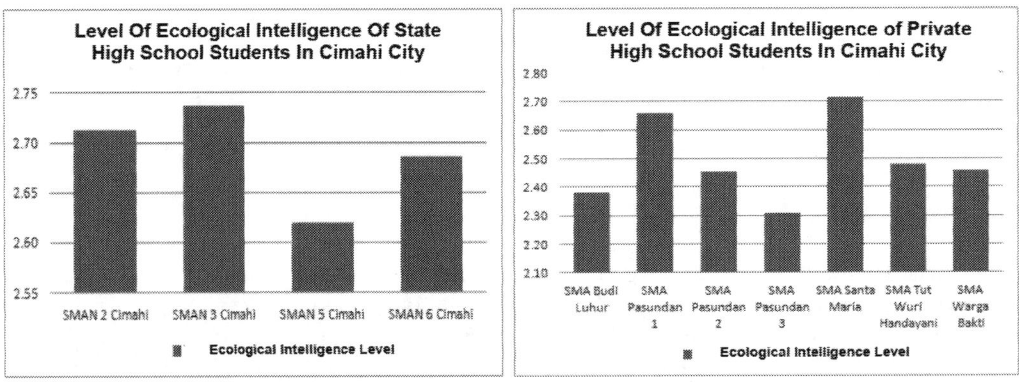

Figure 1 Level of ecological intelligence of state and private high school students in Cimahi City

The score of ecological intelligence level of private high school students in Kota Cimahi is in the "high" classification of SMA Santa Maria (2.71), Pasundan 1 (2.66), Pasundan 2 (2.46), Tut Wuri Handayani (2.48) and Residents Bakti (2.46) and school classification score "medium" SMA Budi Luhur (2.38) and Pasundan 3 (2.31). Overall, all SMP Negeri in Kota Cimahi have students with the classification of "high" score with the order SMPN Terbuka 2 Cimahi (2.72), SMPN 11 (2.66), SMPN 6 (2.62), SMPN 8 (2.55), SMPN 10 (2.55), SMPN 2 (2.55), SMPN 9 (2.52) and SMPN (2.40) score variations but still the "high" score range indicates differences in scores of each aspect.

It is seen that SMP Warga Bakti is first and SMP PGRI 3 Cimahi at the lowest level, the level of ecological intelligence of learners in private junior high school in Cimahi City is in the classification of "high" and "medium" score range. Out of 17 private high schools in the study sample, only 5 schools were classified as SMA Darma Kartini, Kartika Siliwangi, Muslimin Cibeureum, PGRI 3 and PGRI Leuwigajah and the rest were classified as "high" scores.

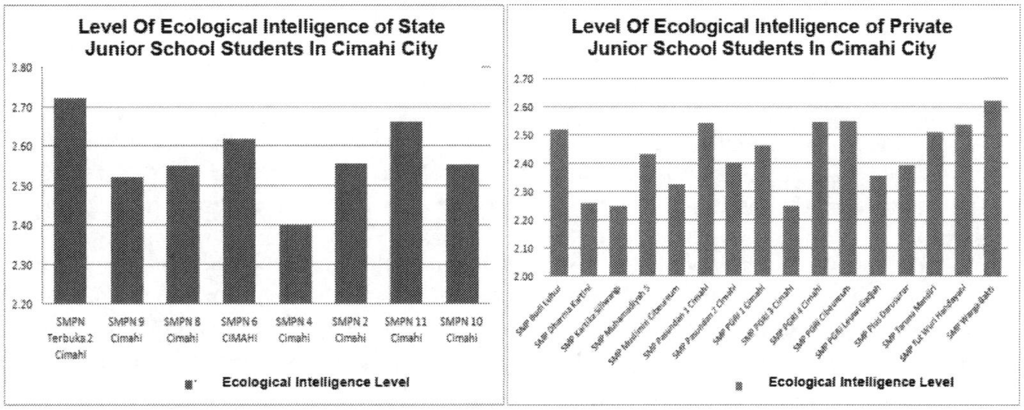

Figure 2 Level of ecological intelligence of students in state and privaet junior high school in cimahi city

Discussion

The cognitive social theory of Bandura and Mischel (in Pervin, 2010)[8] of ecological intelligence is formed depending on the social interaction of learners, it also proves the exemplary aspect in the effort of environmental preservation to shape the ecological intelligence of learners, where exemplary in participating environment for the learners Principals, teachers and educators become the center of social models that shape the ecological intelligence of learners especially in the aspect of environmental preservation skills. According to [11]Loughran (2002, pp. 35) experience alone does not lead to learning more than experience to create ecoloical intelligence. [12]Tsiplakides & Keramida (2010, pp. 22-26) find the correlation of teacher expectations as stated in the learning objectives, on the achievement and behavior of learners, the higher the teacher expectation, the higher the achievement of the learner and also [13]Lynot and Woolfolk (1994, pp. 253-264) found a positive regression of learning towards the formation of learners' intelligence.

Many factors that affect the outcome of the learning process is the facilities or facilities and infrastructure, especially in the learning of geography because the infrastructure will help provide understanding to learners regarding the learning materials especially those related to environmental conservation efforts. [14]Korir and Kikemboi (2014) research results show that intelligence, especially ecological intelligence can be influenced by the school environment in this case the school culture that has the content of the environment, especially aspects of supporting infrastructure. [15]Odeh, et al (2015) also proved the influence of school culture in particular to shape aspects of attitude, skills and train the participation of learners.

School culture does not fully shape the ecological intelligence because the internal factors can also be one of the factors that influence the formation of ecological intelligence. Ecological intelligence is influenced by other factors one of which is an internal factor in accordance with Carl Rogers's phenomenological theory [16] (Pervin, 2010) that intelligence is the perception of the individual and everyone has the ability to perceive the world, so that everyone's views are different as well as intelligence Ecologically everyone is different because of memory and processing factors in different human brains.

Proving Freud's psychoanalytical psychodynamic theory [16](Pervin, 2010) that everyone has a drive to do something, the drive is energy that is the response of environmental conditions, behavior that is influenced by environmental conditions. When receiving a response from the environment everyone will have the expression or discharge of different energy depending on the energy capacity it has, when looking at environmental conditions that are not appropriate environmental conditions then the response of each person will vary whether to leave it alone, whether to participate And damage it or vice versa. Makes the ecological intelligence of each person different depending on the energy in responding to environmental conditions.

According to the cognitive theory of Bandura and Mischel [16](Pervin, 2010) that everyone has different emotional impulses, it is related to the attitude aspect, the pro attitudes towards the environmental sustainability of each person is different depending on his drive to invite others In participating in preserving the environment and to what extent, and directly proportional to the biological theory that intelligence is highly dependent and is also determined by the capacity of the human brain itself.

When viewed carefully the various components of learning include goals, media, strategies and learning resources are all mengaju on making learners to understand the material of learning geography itself so that its contribution is smaller when compared with the school culture on the formation of the ecological intelligence of learners. Although the contribution is small, but geography learning continues to contribute to the formation of ecological intelligence because of the cognitive correlation formed with geographic learning materials that provide spatial understanding and knowledge, but [17]Sylva (1994, pp. 136-171) statement was good output is good input, to create or build ecological intelligence need good input.

Habituation in school culture will shape attitude and skill aspects in environmental conservation efforts, attitudes and skill set will be rationalized by geography learning related to where learners are given understanding about spatial impact of human activities. The success of making students understand is very dependent of the condition of the learners themselves and depends on the content of the environment in various components of the learning of media geography whether it has been able to display the maximum issue, impact, solution and prediction of environmental problems that occur.

Exemplary will indirectly shape the skills aspect and also the learner's knowledge concerning the effort of environmental conservation, participatory become the event in expressing and applying various internal aspect of the learner related to ecological intelligence, participative aspect must be supported with school physical environment so that the participative activity implemented Can be realized properly, re-in this aspect of geography learning to be reinsforment in rationalizing must imitate exemplary principals, teachers and educators in environmental conservation.

Conclusions

Based on what is presented, it can be concluded that the level of ecological intelligence of high school students in Kota Cimahi varies from medium to high. Almost all aspects of knowledge, attitudes and skills have high scores in both public and private schools as well as in public and private secondary schools, while participatory aspects are in the classification of medium and low scores.

The level of ecological intelligence of learners in secondary schools in Cimahi City should be a description of the initial capital in the effort of environmental conservation in Cimahi City the participatory aspect is at the lowest score should be improved through the procurement of various programs regarding environmental conservation efforts. Aspects of knowledge, attitudes and skills with moderate to high scores should be maintained with the support of the school to make rules related to these three aspects.

Acknowledgments are especially for Allah SWT because thanks to his grace this paper can be solved as it should be, also to Rasulullah SAW who has brought us into a period of civilization. This paper is a form of publication of the final assignment in the Master Program of Geography Education, Graduate School, University of Education Indonesia.

Acknowledgments to all parties who assisted the research include lecturers and staff in the Study Program, Cimahi City Education Department, all students, educators and senior high school education personnel in Cimahi City involved in this research and all parties who can not Mentioned one by one.

Reference

1. Muhaimin, Implementasi Model Pembelajaran Berbasis Masalah Lokal dalam Mengembangkan Kompetensi Ekologis Pada Pembelajaran IPS, *Sosio Didaktika: Social Science Education Journal*. **2**, 1, 12-21 (2015).
2. D. Goleman. *Ecological Intelligence The Hidden Impacts of What We Buy*. (New York: Random House, 2010).
3. Kusnendi, *Model-Model Persamaan Struktural Satu dan Multigroup Sampel dengan LISRE,* (Bandung: Alfabeta 2008).
4. Sugiyono, Statistik Untuk Penelitian (Bandung: Alfabeta, 2011)
5. G. Guest. Using Guttman Scaling to Rank Wealth: Integrating Quantitative and Qualitative Data. *Field Methods Journal, SAGE Publishing*, **12**, 4, 346-357, (2000).
6. H. Boone, D. Boone, D. Analyzing Likert Data. *Journal of Extension*. **50**, 2, 1-15, (2012).
7. J.D. Brown. Likert items and scales of measurement?. *Journal ALT Testing & Evaluation SIG Newsletter*. **15**, 1, 10-15, (2011).
8. C.C. Preston, A.M Colman. Optimal Number Of Response Categories In Rating Scales: Reliability, Validity, Discriminating Power, And Respondent Preferences. *Acta Psycoloica Journal Elsevier*. **104**, 1-15. (2000).
9. R.A. Westbrook. *A Rating Scale For Measuring Product/Service Sactifation.* Journal of Marketing. **44,** 4, 68-72. (1986).
10. J. Loughran. Effective Reflective Practice In Search Of Meaning In Learning About Teaching. Monash University. *Journal Of Teacher Education*. **53**, 1, 33-43. (2002).
11. I. Tsiplakides, A. Keramida. The Relationship between Teacher Expectations and Student: Achievement in the Teaching of English as A Foreign Language. *ETL Journal*. **3**, 2, 22-26, (2010)
12. J. Lynot, Woolfolk. *Teachers Implicit Theories of Intelligence and Their Educational Goals*. The Journal Research dan Development Education. **27**, 4, 253-264. (1994).
13. D.K. Korir, F. Kipkemboi. The Impact of School Environment and Peer Influences on Students' Academic Performance in Vihiga County, Kenya. *International Journal Human and Social Science*. **4**, 5, 240, 251. (2014).
14. R. Odeh, O. Oguche, Angelina, E. Ivagher, Influence Of School Environment On Academic Achievement Of Students In Secondary Schools In Zone "A" Senatorial District Of Benue State, Nigeria, *International Journal of Recent Scientific Research Research,* **6,** 7, 4914-4922, (2015).
15. L.A. Pervin, D. Cervon, P.O. John. Psikologi Kepribadian: Teori dan Penelitian Edisi Kesembilan. (Jakarta: Kencana Prenada Media Group, 2010).
16. Sylva, K. School Influences on Children's Development. *J- Child Psychol Journal*, **35,** 1, 136-171. (1994).

Does Educational Disaster Mitigation Need To Be Introduced In School?

Wanjat Kastolani; *Revi* Mainaki
Master Program of Geography Education, Post Graduate School of Universitas Pendidikan Indonesia, Bandung

Abstract. Indonesia is an archipelagic country in subduction of three global plates namely Eurasia, Pacific and Indo-Australia. Make it has many active volcanoes and mountain morphology that are prone to earthquakes, tsunamis and landslides. In astronomical location Indonesia is a country in tropical latitude with high rainfall and irradiation. Makes it vulnerable to atmospheric and hydrosphere disasters such as drought, tropical storms and floods. Also has multicultural population that makes it vulnerable to social conflict. Various factors indicate that Indonesia is a country prone to various disasters. Need disaster mitigation socialization efforts early on. This study describe the need to introduce disaster mitigation education as one of the efforts of disaster risk reduction in schools to students. This study uses literature study method by collecting various written references from books and publications of research results. Use descriftive analysis tecnique for process data. The results show that disaster mitigation education needs to be introduced at the level of schooling in Indonesia. As one of the efforts to prepare the people of Indonesia disaster preparedness. Disaster mitigation education should be included in the curriculum of education, especially on subjects that have a correlation therein.

Key Word: Disaster Mitigation, Educational, Schools

1. Introduction

Indonesia is an archipelagic country that has about 13,500 islands. Is a country that is in tropical latitude climate that makes it vulnerable or at risk of disasters caused by the dynamics of the atmosphere of tropical storms. Being in a subduction zone or encounter between the Eurasian and Pacific Plate making it prone to earthquakes and has many volcanoes due to the formation of two mountainous paths of the world that is the Pacific Circum and Sirkum Mediteranian with morphology of the hills, this makes it at risk of landslides and volcanic eruptions.

In addition, tectonic activity that occurs in the subduction zone located on the seabed make some areas in Indonesia prone to the tsunami. The large number of islands in Indonesia with the geographic conditions of each diverse island also makes the diversity of tribes and languages as well as cultures and religions, it also becomes a potential social catastrophe. [1]Quarantelli (1998) in Lestari (2006) describes disaster as an actual occurrence, more than one potential or more threat is termed as more than an actual threat or termed as the realization of danger.

[2]Suryanti (2010) The disaster occurring in an area will have a direct impact on the surrounding community, the impact is a perceived implication that increases awareness and preparedness in the face of disaster need to be improved to make the community around ready and alert with possible disasters happen. [3] M. Zein (2010) the high level of disaster occurrence in Indonesia, both natural and non-natural disasters in Indonesia that have an immediate impact on the community make people have experience about it and it becomes a capital in socialization efforts related to disaster mitigation

Based on data from ESCAP [4]detailing the Asia-Pacific region of a world-wide economic and social survey body detailing some of the disastrous countries of 1980-2009 Indonesia ranks second in the number of deaths due to natural disasters in the Asia-Pacific region. Statistically the loss of natural disasters has also resulted in a loss of 22.5 billion US Dollars. However, Indonesia is a country with high levels of vulnerability and vulnerability to disasters, so the risk of disaster can not be avoided, but by building disaster preparedness in Indonesian society is expected to reduce the risk of disaster itself.

Each region has different vulnerabilities and types of disasters that need different understanding and handling of disaster preparedness. This is what should be built on the people of Indonesia, aware of the potential and vulnerability of disasters that will occur in the environment around his place of life. So that makes the community more prepared and alert in the face of disaster that will happen, the term is known as disaster mitigation.

The question is how to grow and provide an overview to the community regarding disaster mitigation. One of the areas relevant to it is the field of education. Education based on the law number 20 year 2003 article 1 is an effort to realize the learning atmosphere and the learning process so that active learners can develop the potential in the form of positive values needed for themselves, society, nation and state.

Positive values of awareness of the vulnerability of the surrounding environment are vulnerable and are at risk of disaster. With regard to education, we can not escape from school as an educational institution for the community. Education and growth of public awareness to understand disaster mitigation efforts are closely related to each other. Through schools we can include learning materials regarding disaster mitigation as a form of growing awareness with values inside. But how important is the introduction of disaster mitigation to be done in education? For what is described then the title in this article "Does educational disaster mitigation need to be introduced in school?"

Based on what has been described in the previous background, the problem formulation in this study is "Should disaster mitigation education be introduced in Indonesian schools?" with The purpose of this study is to explain how important disaster mitigation education is introduced in the existing school environment in Indonesia. Value of this research is Theoretically, it can enrich referenasi regarding disaster mitigation education and it is practically expected to be an input for the related institution to determine the policy concerning disaster mitigation education.

2. Methods

This research uses qualitative approach The method used in this study is a literature study that examines a problem that is associated with various relevant theories contained in scientific references in the form of journals, books, print and electronic media then data would analisys with descriftive method.

3. Results

Earth surface hydrosphere layer or commonly known by the term disaster caused by a layer of water on the surface of the earth has its own hazards and vulnerabilities. The dangers of the water itself arise when the volume or amount of water exceeds the capacity, the flow of water so swift that it can attract living creatures into the stream, debris carried by water can injure or damage what passes through, various infrastructure can be damaged by mud Oil and other pollutants carried by the water itself.

Disaster parameters caused by the dynamics of this hydrosphere are the extent in the form of depth, water volume, sediment flow rate, duration of puddle and dasyatnya wave for disaster like tsunami. The disaster caused by hydraulic dynamics is certainly caused by the high rainfall in a place on the surface of the earth, the movement of tectonic plates that cause a tsunami.

This needs to be of particular concern to the community especially those living in vulnerable locations with disasters caused by the dynamics of the hydrosphere itself. Knowledge of this will make people aware of the risk of disasters, the main mitigation that can be done include[5-6] (a) Perform physical construction to with stand the volume and swift flow of water. Preparation of retaining embankments, reforestation of water catchment areas and improvement of drainage system in urban areas; (b) Establish an early warning system such as utilizing the advancement of science and technology that has begun to develop when the capacity of the catchment area exceeds the limits or currents that are around too strong then the system can attract people to immediately move away or perform an action; (c) Increase community participation in disaster mitigation efforts. Conducting development in accordance with the place for example not too close to the shore to reduce the impact in the event of a tsunami.

Disasters due to the dynamics of lithosphere or dynamics of the movement of the earth's shell are included as a result of land movements including earthquakes, landslides and volcanic eruptions. The awesomeness parameters of this disaster for earthquakes of a certain scale (Richter, Mercalli and so on), for landslides are volumes of land undergoing movement and for volcanic eruptions include volumes and types of material removed from the volcano itself. Disaster risk reduction efforts caused by the dynamics of lithosphere such as earthquakes are to build houses with earthquake resistant construction and earthquake warning system, for landslide disaster can be done by vegetative and technical effort, while for disaster volcanic eruption should stay away from center of eruption or caldera and also know sign Will happen volcanic eruptions.

Major disaster mitigation measures that can be undertaken with regard to this type of disaster include[7] (a) evelopment of a qualified infrastructure is certainly earthquake resistant, buildings that resist several types of materials released by volcanoes and avoid areas that are very vulnerable to disasters due to the dynamics of lithosphere; (b) Increasing public awareness recognizes the signs and symptoms of the occurrence of disasters caused by the dynamics of the lithosphere; (c) Community participation that is aware of the hazards and symptoms of this type of disaster.

Disasters caused by atmospheric dynamics include drought, typhoon, hiricane, whirlwind, fohn wind, tornadoes and tropical storms. This is because the air pressure and the suction of wind pressure within a certain time will cause losses, the main cause of this type of disaster is the difference between one place to another. The magnitude level parameters of this type of disaster are the time pages and the higher the scale. This type of disaster can come suddenly so the hall required is the use of technology to provide early warning of the occurrence of this disaster.

Some of the buildings in this area and very vulnerable to the heat should provide a basement for evacuation or emergency of this type of disaster. The main mitigation efforts of these disasters are early warning and physical infrastructure development that is resistant to this type of disaster. Especially for drought humans must realize the importance of the existence of water for their lives.

Technological failures usually begin with explosions that cause death, damage to building infrastructure, releases harmful sanitary substances, accidents cause injury to the victim. The awesomeness parameters of this disaster are the quantity of debt generated, the temperature of the event, the extent of the affected area. Potential to reduce hazards by raising security standards of technology and human consciousness to be more careful in using technology. This type of disaster belongs to a disaster that is unpredictable because it just happens, humans must be fully aware of the risks and dangers of this disaster in order to avoid it.

4. Discussion

[8]Annisa (2011) explains that disaster is anything that happens both due to nature and human impact itself that has the potential to interfere with life, harm human beings, cause natural damage either caused by natural, non-natural or human factors. Another notion of disaster is an event that can threaten the life or human livelihood caused by certain factors.

From this understanding it can be seen that the concept of phenomena is said to be a disaster when there are humans who are harmed, when there is a flood or tsunami in a place that is on the surface of the earth but no harm to humans then it can be ascertained if it is not a disaster. While the term mitigation is an action to reduce the impact of the occurrence of a disaster done either before, during and after the disaster.

Mitigation can also be a human readiness of the occurrence of a disaster event, so that when the disaster occurs man is no longer surprised to lose the things he has, including the loss of his family or loved ones. The following are some of the most vulnerable disasters occurring in Indonesia and the mitigation of disaster mitigation and the causes of such disasters include the following[10] atmosphere, hydrosphere, antroposphere and failed technology.

Based on what has been described previously, disaster mitigation has an urgency to be introduced in education, especially in the school environment in Indonesia, some of the reasons that underlie it are[9-13] (a) each region in Indonesia has certain characteristics and these characteristics also indicate the different vulnerabilities of disaster that occur knowledge of disaster mitigation built through education at least can raise awareness of the disaster mitigation of the population; (b) Indonesia as a country that is very vulnerable to various natural disasters that occur is appropriate to include the value of disaster mitigation in education especially in the environment school; (c) Any disaster mitigation effort required participation and awareness of the population itself means that without the knowledge and awareness of mitigation efforts only a concept; (d) education is an effort to instill positive values, so that through knowledge education and awareness of disaster mitigation is expected to be implanted to the community; (e) Through disaster mitigation education the population at least knows what should be done both before, during and after a disaster; (f) Disaster mitigation education launched early will make people more prepared and alert in the face of disaster, that is ready with all lost property and loss of the nearest person or family.

Those are some of the reasons why the importance of disaster mitigation education is applied in schools with a certain intensity. Disaster mitigation should be included in curriculum units in schools in accordance with potential disasters that can occur, as done in some developed countries and of course also use an approach appropriate to the conditions of child development and growth [14-16]. But to be able to enter school, there must be a centralized policy and we can get some developed countries like disaster mitigation management in Canada, the results of [17]Henstra and Gordonmcbean (2005) research show the measurement of how disaster mitigation policy in the country. The need for measurement before determining the disaster mitigation policy is included in the rules in Indonesia, especially in schools such as the results of [18]Adefisoye (2015) research that conducts an assessment of the disaster mitigation policy that has been done, this will be the first input inputs of how disaster mitigation should be included in government policy and there is a special budget of disaster mitigation result of research of [19]Phaup (2010) et al shows how important budget in disaster mitigation . If clever to see this situation that potential and disaster vulnerability can be a separate business opportunity such as the results of research [20]Dufty et al (2014) that the vulnerability or potential disaster can be a business opportunity, especially in community skills training and procurement various disaster preparedness equipment.

5. Congclusions

Based on what Indonesia has previously described as a country that has a high degree of vulnerability to disasters, whether disaster caused by the dynamics of lithosphere, hydrosphere, atmosphere and technological failure, the need for awareness of the people of Indonesia with regard to disaster mitigation. Growth of this awareness can be done through education and talk about education is not separated from the school as a formal educational institution in Indonesia. The reason why the growth of disaster mitigation efforts is done in the field of education is because Indonesia is a disaster-prone country, education inculcates positive values that can of course be inserted the values of disaster mitigation.

The results of previous exposure show how urgency of disaster mitigation education is important to be introduced in Indonesia. So it should be designed a curriculum in which insert disaster mitigation values in other words the values are included in some or even all subjects in an integrated manner.

Acknowledgments are especially for Allah SWT because thanks to his grace this paper can be solved as it should be, also to Rasulullah SAW who has brought us into a period of civilization. Acknowledgments to all parties who assisted the research include lecturers and staff in the Geography Education Departement, Social Science Education Departement and Geography Education Program, School of Postgraduate, Universitas Pendidikan Indonesia and all parties who can not Mentioned one by one.

Reference

1. F.F. Lestari. *Penerapan Sistem Informasi Geografis Dalam Pemetaan Daerah Rawan Longsor di Kabupaten Bogor*. (Bogor: Institut Pertanian Bogor, 2008).
2. Suryanti et al. *Motivasi dan Partisipasi Masyarakat dalam Upaya Pengurangan Multirisiko Bencana di Kawasan Kepesisiran Parangtritis dalam Penaksiran Multirisiko Bencana di Wilayah Kepesisiran Parangtritis, Yogyakarta*. (Yogyakarta: PSBA Universitas Gadjah Mada, 2010)
3. M. Zein. *Community Based Approach to Flood Hazard and Vulnerability Assessment in Flood Prone Area: A Case Study in Kelurahan Sewu, Surakarta City, Indonesia*. (Thesis, ITC, The Netherland, 2010)
4. M. C. Ulum. *Governance dan Capacity Building Dalam Manajemen Bencana Banjir Di Indonesia*. Jurnal Penanggulangan Bencana, **4**, 2, 5-12, (2013).

5. W. Adiyoso, H. Kanegae. The Effect Of Different Disaster Education Programs On Tsunami Preparedness Among Schoolchildren In Aceh, Indonesia. *Disaster Mitigation of Cultural Heritage and Historic Cities Journal*. **6,** 165-172 (July 2012).
6. J. O. B. Rotimi, S. Wilkinson. Improving environmental management legislation to facilitate postdisaster reconstruction. *International Journal of Disaster Resilience in the Built Environment*, **5**, 1, (2014).
7. M. Moore, H.R. Trujillo, B. K. Stearns, R. B. Davila, D. K. Evans. Learning from Exemplary Practices in International Disaster Management: A Fresh Avenue to Inform U.S. Policy?. *Journal of Homeland Security and Emergency Management,* **6**, 1, (2009).
8. Annisa. *Zonasi risiko bencana banjir akibat sea level rise.* (Surabaya: ITS, 2011)
9. A.W Conburn, RJS Spent, A. Pomonis. *Mitigasi Bencana, Program Pelatihan Manajemen Bencana*. Second Edition. (Cambrige Architectural Research Limited, 1994)
10. M. Tanudidjaja. *Bumi dan Antariksa*. (Jakarta : Depdikbud, 1995)
11. P. Putra. *Penataan Ruang Berbasis Mitigasi Bencana Di Kabupaten Kepulauan Mentawai*. Jurnal Penanggulangan Bencana. **2**, 1, 11-20 (2011)
12. A.W Conburn, RJS Spent, A. Pomonis. *Mitigasi Bencana, Program Pelatihan Manajemen Bencana*. Second Edition. (Cambrige Architectural Research Limited, 1994)
13. L. Gibbs, M.D. Pietro, A. Harris, G. Ireton, S. Mordech, M. Roberts, J. Sinclair, R. Wraith. Core Principles For A Community-Based Approach To Supporting Child Disaster Recovery. *Australian Journal of Emergency Management I.* **29**, 1, 17-24, (2014).
14. B. R. Guerdan. Disaster Preparedness and Disaster Management: The Development and Piloting of a Self-Assessment Survey to Judge the Adequacy of Community-Based Physician Knowledge *American Journal of 32 Clinical Medicine.* **6**, 3, (2009).
15. H. Boon, L. Brown, P. Pagliano. Emergency planning for students with disabilities: a survey of Australian schools. *Australian Journal of Emergency Management I.* **29**, 1, 45-49, (2014).
16. T.Nadu, I. Asharose, I.Saizen, P. Kumar, C. Sasi. Awareness Workshop as an Effective Tool and Approach for Education in Disaster Risk Reduction: A Case Study from Sustainability. *Sustainability Journal.* **7**, 8965-8984, (2015).
17. D. Henstra, Gordonmcbean. Canadian Disaster Management Policy: Moving toward a Paradigm Shift?. *Canadian Public Policy/ / Analyse de Politiques Journal.* **31,** 3, 303-318, (2005).
18. T. Adefisoye. An Assessment Of Nigeria's Institutional Capacity In Disaster Management. *Scientific Research Journal (SCIRJ),* **3**, 1, (2015).
19. M. Phaup and C.Kirschner Focusing on the Good Times.. *OECD Journal on Budgeting.* **1,** 1-22, (2010)
20. R.A. Quero, D.L. Salle. Framework for Building Skills in Disaster Risk Management for Businesses *Journal of Global Business,* **1**, 1, (2012).

Task-based Language Teaching and Text Types in Teaching Writing Using Communicative Approach

Ni Nyoman Riyana Sari*, *Dewa* Komang Tantra , and *Made* Hery Santosa
Universitas Pendidikan Ganesha Bali Indonesia

Abstract. One of the most important language competencies in teaching learning process is writing. The present study focused on investigating the effect of communicative approach with task-based language teaching and communicative approach on the students' writing competency at SMP N 2 Kediri viewed from text types(i.e. descriptive, recount, and narrative). To analyze the data, the design of the experimental study was posttest-only comparison groups by involving 60 students that were selected as the sample of the study through cluster random design. The sample's post tests were assessed by using analytical scoring rubric. The data were then analyzed by using One-way ANOVA and the post hoc test was done by computing Multiple Comparison using Tukey HSD Test. The result showed that there was significant difference of the effect of communicative approach with task-based language teaching and communicative approach on the students' writing competency. These findings are expected to give contribution in teaching English, particularly writing.

Keywords: Tasks-based language teaching, text types, writing competency, and communicative approach

Introduction

According to Coffin, Curry, Goodman, Hewings, Lillis and Swan (2003), the main point of teaching and learning process in higher educational level is writing. Moreover, writing is also the centre of a range of purposes fulfillment based on the various contexts where writing should be done. The purposes of writing as a higher educational context changer can be divided into three, assessment, learning and entering particular disciplinary communities.

The previous statement is also supported by Hyland (2003, p. xv). He states that writing's position is very fundamental among the other skills such as listening, speaking and reading. The development of writing skill must be further increased because the skill of teaching writing will show how skillful is the English teacher. Teaching writing is not a piece of cake. The teacher must be able to elaborate the teaching style with the teaching

Corresponding author. nyomanriyana@yahoo.co.id

material. When the teacher has a good capability in writing, it can be assumed that the teacher should be able to become a good model for the language learners.

Besides the above theories, the important of writing also can be seen from the researcher that did a study about writing. Marashi and Dadari (2012) say that nowadays, writing has shown its importance especially in daily lives. It can be seen from a side, specifically where written form has been becoming a dominant language among the world's population and speakers. Moreover, Tilfarliogu and Basaran, (2007) advised that writing is categorized as the trend centre of research topic especially in the relation of second language learning. In addition, they also said that writing has a role in language learning as a tool in the teaching learning process. Therefore, it can be synthesized that writing is one of the most important competency that should be mastered by the students as every single thing in teaching is related to writing.

From the above explanation, we can see how important writing is for the students. Those make the researcher eager to conduct a study about writing and implement a writing strategy for the students under the study. In order to know how the implementation of teaching writing in SMP N 2 Kediri, the researcher did a preliminary observation towards the teaching learning process of English subject, especially writing. In addition, an interview with the English teacher and the students was also done by the researcher to get further information about the teaching learning process.

After doing observation and interview, the researcher found that most of the eighth grade students of SMP N 2 Kediri are still unable to write a good text in English. According to the English teacher, most of the students could not pass the standard minimum score criteria in writing which is 74.00. The students are still not able to develop their idea properly in writing text. As a result, they are not able to write a good text. They were given a chance to write a text in English, but they did not know what to write because the teacher only asked them to write a text that has similar type with the one teacher had. There is no further instruction what should be there in the text. At this stage, we can see that the task given by the teacher is not clear so the students couldn't know how is the correct process of writing. Where they should be able to produce a good writing product especially writing there were three kinds of texts to be mastered by the eighth grade students, they are descriptive, recount and narrative text.

At this stage, something must be done to solve the students' problem in learning writing. As what have been mentioned previously that the tasks given by the teacher is not clear, the researcher believes that Task-based Language Teaching might help the students as it has already been proven by the other researcher that conduct related studies and already mentioned by some experts about Task-based Language Teaching.

The present study was intended to compare two different learning approaches and they are to be implemented in two different classes to see the significant difference. Both of the classes must be still treated by using Communicative Approach as the mandatory approach in 2006 curriculum. At this stage, one of the classes would be treated by using Communicative Approach with Task-based Language Teaching and the other one class would be treated by using Communicative approach only at this stage. Moreover, as the syllabus of the eighth grade students advises that there are three text genres to be learnt, they are descriptive, recount and narrative, the researcher chose them as the moderator variable.

TBLT is leaded by a theory of learning rather than a theory of language (Richards & Rodgers, 2001). They also state that the central role of meaning in language use is the one that is being emphasized in TBLT. TBLT allows the language learners to learn something focusing on both, process-oriented and product-oriented. The language learners are also given a chance to discuss their work later after the tasks done. This will be becoming very interesting for the students because they can also work with their peers to discuss their tasks at the end of the discussion.

Cao (2012) mentions on her study about Task-based Approach that was applied in EFL writing classroom. The result showed that Task-based Approach was considered to be really feasible for college EFL learners to improve their writing competence. She was emphasizing on shifting the learning activities from teacher-centered writing instructions to students-centered writing instruction. This gave a good impact towards EFL learners' interest to a large extent to participate in a series of communicative activities. Moreover, TBLT also allowed the students to work in peers spontaneously and they are able to interact to others talking about the readers' needs and expectations.

The second study that proved that TBLT is appropriate to be implemented to teach writing comes from Zhaochun (2015). He was concerning on applying TBLT towards English majors in Chinese settings. The focus on TBLT was product –oriented and process-oriented. The result of his study showed that TBLT is more effective than the other approaches to improve learners' writing competence. It was not only showing the improvement on writing competence seen from the test scores, but also from the length of the essays and the awareness of collocations and word accuracy. Moreover, it brought to the students' study motivation and interest. The stereotyped of teacher-centered is changed to students-centered approach.

The present experimental research used text types as the moderator variable to control the effect of Communicative Approach with TBLT and Communicative Approach only on the students writing competency. Additionally, Putra, Tantra, and Padmadewi (2016) on their study mentioned the use of text types as the moderator variable to control the treatments given to the students. Therefore, the researcher was confident to use text types as the controller in other to make sure the study was not going too far away from the objective of the study.

In relations to the previous studies about the use of TBLT in teaching writing, the researcher would like to get something new on her research. Therefore, the researcher was eager to conduct something new by applying Communicative Approach with TBLT in this present experimental research. Moreover, the existing use of text types as the moderator variable to control the effect of Communicative Approach with TBLT was also considered as a new thing in the present experimental research.

Based on the above explanation about the students' problem in writing and related theory and studies about TBLT, the writer would like to see the effect of TBLT on the students' writing competency but it would be applied with Communicative Approach. The writer believes that TBLT would definitely help the students in solving their problem related to writing three kinds of texts; they are descriptive, recount and narrative. At this stage, one group would be taught by using Communicative Approach only and the other one would be taught by using Communicative Approach with TBLT.

Methods

The present study was an experimental study by focusing on two groups comparison post-test only design. The population of the present experimental study was 232 of the eighth grades students of SMP N 2 Kediri in academic year 2016/2017 which in 60 of them were selected as the sample of the study through random sampling lottery technique.

Teaching scenario, posttest, and writing rubric were the instruments of the study. In addition, the posttest was administered after the treatments done in two meetings for each text types. After posttest was done, it was then scored by using scoring rubric. There were seven criteria of the rubric adapted from Davis and MacKay (1996). The criteria are title;

topic sentence; development; arrangement; diction; grammar; and mechanic. The data were then analyzed by using one-way ANOVA.

Finding and Discussion

The result showed there was significant difference between the effect of Communicative Approach with TBLT and Communicative Approach on the students' writing competency as shown in Table 1 and 2. In accordance to the result of descriptive analysis, it can be seen that the effect of Communicative Approach with TBLT was more significant than the effect of Communicative Approach. Moreover, based on the analysis of one-way ANOVA, the significant value was a figure of 0.000 which was lower than 0.05. this indicated that the directional hypothesis which was formulated previously was accepted.

Table 1. The Summary of the Descriptive Analysis of Students' writing Competency

Group	Text Types	Descriptive Statistic		
		Mean	Standard Deviation	Variance
Communicative Approach + TBLT	Descriptive	82.73	5.40	29.16
	Recount	82.46	4.63	21.44
	Narrative	82.23	4.10	16.81
	Grand Mean	82.47		
Communicative Approach	Descriptive	79.90	3.37	11.36
	Recount	77.23	3.00	9.00
	Narrative	73.40	5.62	31.58
	Grand Mean	76.84		

Source: Data analysis on SPSS 22 Program for Windows, April 2017

Table 2. The One-way ANOVA Analysis of Communicative Approach with TBLT and Communicative Approach

	Sum of Squares	df	Mean Square	F	Sig.
Between Group	466.154	1	466.154	49.818	0.000
Within Group	542.714	58	9.357		
Total	881.695	59			

Source: Data analysis on SPSS 22 Program for Windows, April 2017

Communicative Approach with TBLT allowed students to learn based on the tasks given by the teacher and discussion at the end of the task section. They would need to complete each task by showing their best in writing text types. Moreover, the approach gave opportunities to the students to analyze the indictors of writing and do the tasks individually before discussing it classically. Meanwhile, the students who were thought by using Communicative Approach were only focusing on what students have known about writing the texts. Therefore, it could be said that Communicative Approach with TBLT gave significant effect upon students' writing competency compared with Communicative Approach only.

The result of the data analysis was also based on the theory from Richards & Rodgers (2001). They stated that TBLT can be defined as the unit of planning and instruction in language learning. Moreover, the implementation of applying the approach was also based on their theory. of As in the present study was also focused on giving tasks to the students, Communicative Approach with TBLT allowed the students to work on the tasks given by teacher. The students would need to complete each task to show their understanding. At this point, the teacher gave pre-tasks section to the students. They were given such a brainstorming to build their concept about what was being taught. The second step was giving the tasks and the instruction what should they do to complete the task. If any issues, students were allowed to ask as many questions as possible until they catch the point. The students mentioned that by using this technique, they were able to more understand the material given. It was because the tasks were clear and the instruction was also clear for them. Moreover, at the end of the tasks, they were able to do classical correction to make sure their writings were correct or not according to writing indicators. This was the last steps of TBLT which was post-tasks section.

Moreover, the result that showed Communicative Approach with TBLT gave more significant difference was supported by the theory of Communicative Approach. According to reasons Jarvis and Atsilarate (2014), the English language teaching has been dominated by communicative approach which has been established firmly by itself on a worldwide basis and it has some good historical. The steps of applying Communicative Approach with TBLT was a combination betewen the steps of Communicative Approach and TBLT. The steps of Communicative Approach were exploration, elaboration and confirmation. It was based on 2006 curriculum. Therefore, in this study, the result showed that group which was treated by using Communicative Approach with TBLT gave better achievement in terms of mean score.

The application of Communicative Approach with TBT was also supported theoretically by an empirical study by Littlewood (2004). He mentions that rather than accept common communicative and tasks types, better to show the continuity between TBLT and Communicative Approach and their development. If we refer to interface outside the classroom, communications would reflect the students' needs and interest. However, if connect it to inside-classroom area, it would provide the language components such as grammar, vocabulary and so on that the students must learn to build the communication with the other. This study was a library study so it could support the present study theoretically. This showed that, the combination of Communicative Approach with TBLT was making sense to be implemented.

To conclude, Communicative Approach with TBLT was categorized to be an effective approach to improve students' writing competency. The experimental study also proved that the effect of Communicative Approach with TBLT was more significant than the effect of Communicative Approach upon students' writing competency especially for the eighth grade students.

Closing

Communicative Approch with TBLT provided a lot of benefits for the students to practice their writing competency. They could completely finish the task by considering the criteria of a good task. Moreover, they also gave chance to do much more discussion at the end of the task session. This finding was also supported by Littlewood (2004) that confirmed on his library research that Communicative Approach with TBLT is appropriate to be combined.

The writer wishes her great-sincere and heartfelt gratitude to the Almighty of God, *Ida Sang Hyang Widhi Wasa* for His everlasting blessing so that this thesis could be accomplished.Greatest gratitude goes to all lecturers of English Education Department of Postgraduate Program Undiksha, the writer's family especially her parents and sisters, and all friends for the everlasting support and motivation during the study process.

References

1. L. Cao. A feasibility study of task-based teaching of college english writing in chinese EFL context. *English language teaching , 5 (10)*. 80-91. (2012)
2. C. Coffin, M. J. Curry, S. Goodman, T. M. Lillis, & J. Swann. *Teaching academic writing: a toolkit for higher education*. New York: Routledge. (2005).
3. L. Davis & S. McKay. *Structures and strategies: An intrduction to academic writing*. South Melbourne: Acmillan Education Australia PTY LTD. (1996).
4. K. Hyland. *Second language writing*. New York: Cambridge University Press. (2003).
5. H. Jarvis & S. Atsilarate. Shifting paradigms: from a communicative to a context-based approach. *Asian EFL Journals , 1 (1)*. 1-23. (2014).
6. W. Littlewood. The task-based approach: some questions and suggestions. *ELT Journal , 5 (4)*. 319-326. (2014).
7. H. Marashi & L. Dadari. The impact of using task-based writing on EFL learners' writing performance and creativity. *Theory and Practice in Language Studies , 2 (1)*. (2013).
8. P. M. Putra, D. K. Tantra & N. N. Padmadewi. The effect of simulated communicative approach on students' writing competency of the seventh grade student of smp widhya bhuana munggu based on text type. *Jurnal Penelitian Pasca Sarjana Undiksha , 5 (1)*. (2016).
9. J. C. Richards & T. S. Rodgers. *Approaches and methods in second language teaching*. Cambridge: Cambridge University Press. (2001).
10. F. Tilfarlioglu & S. Basaran. Enhancing reading comprehension through Task-Based writing activities: an experimental study. *The Rreading Matrix , 7 (3)*. 7-15. (2013).
11. S. Zhaochun. A tentative study on the task-based teaching of writing to english majors in chinese settings. *English Language Teaching , 8 (3)*. 71-79. (2015).

Ethnopedagogy And Local Genius: An Ethnographic Study

I Ketut Ngurah Ardiawan*[1]
STAHN Mpu Kuturan Singaraja, Primary Teacher Education Department, 81112, Bali, Indonesia

Abstract. Preserving local genius is one of the ways to keep values existed in a society. In relation to that this present study aims to identify the procedures of traditional games called *megoak-goakan* , and reveal the ethno-pedagogy values contained within *megoak-goakan*. This study employs a qualitative view and utilizes ethnographic study. The setting of the study is in Buleleng regency, Bali. In order to collect the data, the researcher utilizes interview, observation sheet, and field notes. The interviewees were decided through purposive s sampling. Further, there are three main components of the data analysis comprising data reduction, data modelling, and conclusion. Based on the investigation, it is revealed that the procedures of traditional game of *megoak-goakan* are as follows: form a group, decide the snake group and the crow groups, line up and holding each other, determine the tail of the snake, the crow starts hunting the snake tail, while the head of the snake prevents it, the crow and the snake move freely as the agreement stated, the crow shouts as it catch the tail, and winning and losing are decided by whether or not the crow is able to catch the tail. In addition, this game is also expected to bring philosophical, anthropological, sociological, and psychological values.

Keywords: Ethno-pedagogy, Local Genius, Traditional Game

Introduction

Lately, as the rapid changes in this globalization era, local genius faces great challenges to maintain itself. This is due to modern societies are interconnected globally and a culture is intervened whereby processes of globalization and identity clash each other [9, 21, 14]. In addition, [20] the exposure of globalization scrapes local culture due to an acculturation process between native culture and foreign culture. Further, [10] the effects of globalization can be destined to institutional/organisational, process, system or value driven, ideological or a combination of some of the aforementioned aspects

[14] Globalization has a direct effect to local culture. In addition, he noted that the Nigerian cultures are slowly but surely degrading. However, the effect of globalization also makes cultural identity play essential roles for the peaceful cooperation of civilization.

*stahnmpukuturansingaraja@gmail.com

Cultural identity needs nurturing so that people in a certain society believed in where they belong and become more appreciative to their own culture while learning other cultures.[11]

In Balinese context, its multicultural tradition undergoes a number of cultural issues in its historical journey. Local Balinese ethnic who dominantly lived there has already made contact in a very long migration process with migrants ethnic, like Javanese, Buginese, Sasakese, Floresian, Timor, Minang". In reality, the current citizens have been applying educative tradition including character education. Tradition is an ancestral cultural heritage that is implemented and preserved. In this respect, one of the traditions that become focus in this study is *Megoak-goakan*.

A brief about *megoak-goakan*, it is a traditional culture and also being categorized as a local genius. This is called so due to its role to maintain moral values. Historically, this was done firstly by Panji villagers of Buleleng District, Buleleng Regency. This tradition has been played among kids and adults in Panji village. It is different from other regions; *Megoak-goakan* in Panji village is done in *Ngembak Geni* day; a series ceremonial day of Nyepi. To local villagers, *Megoak-goakan* is believed to have a purpose in maintaining the harmonic relationship among the people who play *Megoak-goakan*. However, as the time flies the implementation of this game is degrading bit by bit.

As more issues on ethnic problems are rising, ethno-pedagogy emerged as one of the approaches to deal with the aforementioned issues. In the beginning its emergence, it is considered as a branch of pedagogy. It is underpinning instructional process as non-separated aspect with culture education [15, 8]. Ethno pedagogy itself is defined as the model of cross cultural learning [1]. The teachers can teach by designing local culture as the media of learning that may be different from one place to another one. It is none but one way to preserve local culture in a society. Education through ethnopedagogic approach sees that indigenous knowledge as the source of innovation and expertise that can be empowered [16].

There are number of studies that supporting the use of ethnopedagogy to sustain culture by inserting it in educational process. Orshaak and Orshaak [4] focus their study on analysing Tuvanese ethnic culture in education as well as its compliance toward board knowledge. They found out that ethnopedagogy allows students to identify the degree of importance of aspects of their own culture and value orientations with the help of techniques. Focusing on Indonesian context, [1] analyse the function of Kalosara as ethnopedagogic media in nation character-building in South Sulawesi. They revealed that inserting ethnic or local culture in education let students to be more appreciative of their own culture and strengthen the relationship among people.

Similar studies are also conducted to the population in Siberian [18,19,13]. All of those experts focus on revealing the national peculiarities of physical cultures, original physical education, functions and meaning of traditional games and sports. The authors listed above express a great concern about the lack of attention to the problem of studying, preservation and implementation of traditional forms of motion activities of indigenous populations of west Siberian into education process.

As efforts should be done in preserving local genius and culture as well as how it is implemented in real situation, the researcher narrows this study to investigate the ethnopedagogical values in *Megoak-goakan* tradition in Buleleng regency. Likewise, the researcher focuses on two major things as follow; 1) how *megoak-goakan* tradition is conducted in order, and 2) what ethnopedagogical values that this tradition holds. It is expected that the result of the study will give contributions in enhancing the insight about Indonesian cultural diversity and suggestions for the government about the importance of local genius-based education.

Method

This study employs a qualitative approach in investigating the ethnopedagogical aspects of *Megoak-goakan* tradition as local genius values. The study was conducted in two months starting from February to April 2017. Some stages are done covering 1) identification of problem; 2) review of data sources; 3) preparation of instrument and proposal. The second was research implementation stage, it includes: 1) data collection; 2) tabulation and data analysis; and 3) data interpretation. The third stage was construction of research report.

The subject and the scope of this study are limited to the analysis of ethnopedagogical values in *Megoak-goakan* tradition. The data was gathered by obtaining the information about *Megoak-goakan* from religious figures, society figures, and the local people that were relevant with this study. Then, the data derived from the interview was used to analyse the procedures of playing *Megoak-goakan*. Other than interview, this study also used observation to obtain the data. The observation method was used to see directly the procedures of playing *Megoak-goakan* regarding Balinese tradition ritual. This allowed the researcher to analyse the ethnopedagogical values in the procedures of playing *Megoak-goakan*. The researcher also documented the activities in the form of pictures. In observation process, the researchers also used field notes to record and note all of the activities during *Megoak-goakan* process.

Regarding the setting of the study, this study was conducted in Panji village, Sukasada district, Buleleng regency. This place was chosen because it was where *Megoak-goakan* historical tradition firstly occurred. Panji villagers were consistent in maintaining and preserving the local genius values in *Megoak-goakan* as the heritage from their ancestors. Further, the data collected then was analysed by using inductive analysis technique that as proposed by Miles and Huberman. There were four stages in the analysing process; data collection, data reduction, data presentation, and conclusion.

Results and Discussion

Based on the study that has been conducted, some result of the data will be explained as follows. *Megoak-goakan* was done in *Ngembak Geni* started from 3 p.m. until finish. The purpose of *Megoak-goakan* is to build good character among the society who got involve in the game. It was also an attempt to preserve the culture inherited from their ancestor.

Based on the observation and interview process, it was found that the society would firstly gather in Panji village field. After that, the people who got involve in the preparation process would drain water around the field of the game to avoid some bad injuries. There was no people who specifically organized and managed the game, therefore the teenagers would start the game by moving and running around the area of the game while inviting their friends to join them playing *Megoak-goakan*.

There were no specific requirements for the people who wanted to join the game. The main point was, there must be two opposite groups; one group would be seeker and another group would be the rival. There were various version of *Megoak-goakan*; it could be played individually or classically. When *Megoakan-goakan* was played individually, there would be one person acted as *goak*, while the opposite group made a line like a snake consisted of more than two people. On the other hand, the classical game was joined by two groups consisted of five people or more for each. The more people joined in the groups, the more fun the game would be.

This game consisted of male or female group, or it can be mixed between them as one group. There was no limitation of age; all people could join the game. Yet, it was suggested that each group should have contained of people in the same level of age and gender and the opposite group must have the same character with the rival group. The biggest and strongest person would stand in the frontline to lead his members. His duty was to keep his members from *Goak's* attack in catching the members. One person would be *Goak*. His duty was to catch the rival's tail as fast as he could. The time limitation was 5 minutes, if *Goak* could not catch the tail in 5 minutes, he would lose. *Goak's* rival group made a line behind the leader. They also required to hold the hip of a person who stood in front of them tightly. During the game, it was not allowed to take his friends' hip off.

There were some procedures in playing *Megoak-goakan*:

1) Forming a group
   The group consisted of 10-20 of Panji villagers. Each group can be mixed by the gender and age.
2) Deciding the snake group and *Goak* group
   After making two groups consist the same number of people, next step was to decide which group would be the snake and crow or *Goak* by using a lottery or pointed directly to the groups.
3) Making a line and holding on to each other
4) Defining the tail of the snake
5) *Goak* must seek and touch the tail of the snake, while the head (the leader) of the snake must block *Goak's* efforts.
6) *Goak* and snake could freely move wherever they wanted as long as it was still in the area of the game.
7) *Goak* players would scream *goak...goak...goak....*
8) The winner of the game was defined by the capability of *Goak* in touching or catching the person who became the tail of the snake.

After conducting the research and gathering the data systematically, it can be derived from the data analysis that *Megoak-goakan* tradition had ethno pedagogical values. It included philosophical value, anthological value, sociological value, and psychological value.

Megoak-goakan tradition had philosophical value for the society in Panji village in the context of their daily life. *Magoak-goakan* tradition was associated with Hindu philosophy that was Tri Hita Karana. Tri Hita Karana (the three causes of humans' blissfulness in life) included *parahyangan, pawongan, palemahan*. The philosophical values in Tri Hita Karana suggested the society to respect the whole aspect in life, it included respecting the nature, the Creator, and the humans. This also meant respecting the diversity values as one of the aspect of local genuine [3, 7].

In the implementation, Tri Hita Karana philosophy was showed by the attempts of the society to get closer with God and Goddess, and their ancestors. This was begun with the praying in Pajenengan Panji temple before playing Megoak-goakan.

In Manawa Dharmasastra III book, 76 it was stated that:
agnau prastahutih samsyag
adityam upatistate,
adityajjayate vrstir
vrsterannamtatah prajah. [5]

Translation:

The offerings put into the fire will pursue the sun, from the sun, it is followed by rain
From the rain, foods arise on which all the creatures get their life from

It was suggested that what we got from the nature should be returned to the nature. *Megoak-goakan* aimed at building the stability, harmony, and conformity in ourselves. It also became the way to get closer to our Creator. *Megoak-goakan* as a heritage was believed having meaning to Panji society in the terms of its togetherness and harmony.

Anthropology is the study of mankind and his way of living [6]. This statement told that humans' way of life could not be separated from their basic character. Thus, anthropology is the study of various aspects of humans by learning about their various physical, society, and culture in a social-cultural interaction.

Megoak-goakan became one proof of cultural and traditional richness in Bali. Its sustainability has been maintained until now on. It did not only have a high historical value, but this tradition also provided togetherness and hardworking aspects for the players. The most recognized character in *Megoak-goakan* was competitiveness aspect. This allowed the society to preserve the culture and do the sport all at once. To enhance the spirit, enthusiasm, and fun atmosphere, *Megoak-goakan* was accompanied by *bleganjur*, mask usage, and traditional cloth. It implied that *Megoak-goakan* contained of art and cultural values in it.

The main focus of sociology study was the humans in the context of society, it included the pattern of common life or the pattern of social interaction [12]. In *megoak-goakan*, the players were required to work hard in attacking and defeating the rivals. This was showed by the agile movement done by the players; they ran, danced, squatted, spun, or jumped around. All of the activities that were done during the game built relationship among the players on which the local genuine values were also created. This in line with [2] who stated that sociology studies the society as a whole, it included the relationship between human and another human, human and a group, group and group, formally and materially, statically or dynamically.

Megoak-magoakan became a tool to build the sense of unity, kinship, and responsibility. These aspects were the main factors in a development and a just society on which it created a responsible sense and loyalty among the villagers. From sociological point of view, *Megoak-goakan* emphasizes on the socializing process. This discussion was aimed at explaining the nature of *Megoak-goakan*, the procedures in conducting *Megoak-goakan,* the values in *Megoak-goakan*, and the relation of the socialization process among the kids in the society

The Panji village society believed that there was a psychological kinship relationship built in *Megoak-goakan*. This psychological kinship relationship was built because of there was a consanguinity derived from the inhabitation and daily intercommunication. Therefore, [17] there were some grouping of psychological kinship that were made; psychological relationship based on consignation, psychological relationship based on the kinship, and psychological relationship based on the same inhabitation.

Consignation relationship meant the players or the society who got involved in this tradition were a family derived from patrilineal line of their family or they were the local people born in that village. The kinship relationship was built from the marriage between the man's family and woman's family who lived in Panji village. Psychological relationship based on the same inhabitation occurred because the people were born in Panji village.

Conclusion

Based on the result of the study, *megoak-goakan* traditional game has some procedures, it includes forming the groups, deciding the snake group and Goak group, making a line and holding each other, *Goak* must seek and touch the tail of the snake while the player who became the head of the snake must block snake's effort to catch the snake's tail, *Goak* and the snake freely moved around the area of the game, *Goak* players screamed 'goak' word, and the winner of the game was defined by the capability of *Goak* in touching or catching the person who became the tail of the snake

Megoak-goakan had local the genius-based ethnopedagogical values which covered the philosophical values, anthropological values, sociological values, and psychological value. Philosophically, *Megoak-goakan* tradition had philosophical value for the society in Panji village in the context of their daily life since the living society is in direct touch with Hinduism. Anthropologically, *Megoak-goakan* becomes one proof of cultural and traditional richness in Bali. These results become interesting to be elaborated more to give new insight in educational aspect.

The success and final outcome of this research requires a lot of guidance and assistance from many people and the researcher fortunate to have got this all along the completion of the research within the time given. I respect and thank the committees for giving the researcher to revise and complete the research based on the revision given.

References

1. A.Hafid, Amiruddin, K.Suardika,Sartono,Peribadi, Inter. Res. Jour. of Emer. Tre. in Multi. **1**, 9 (2015)
2. E. Tukijan, *Bahan Ajar Cetak Sosiologi Pendidikan*, (Jakarta: Dirjen Dikti 2009)
3. E.E.Etta, A.E. Asira,O.O. Asukwo, Amer. Jour. of Soc. Scien. **1**, 2 (2011)
4. H.D.N.Oorshak, S.Y. Oorshak, Jour. of Sib. Fed. Uni, **3**, 7 (2014)
5. I.N.S. Buda, E-Journ. Fils.Pub, (2013)
6. I.W. Rasna, Jurn. Kaj. Bali, **6**, 1(2016)
7. J.E. Ukpokolo,. *Philosophy Interrogates Culture*, (Ibadan, 2004)
8. K. Klara, O. Baktiyar, K. Sandygul, U. Saikhan, J.Gulzhiyan, Med, Jour. of Soc. Scien. **6**, 1 (2015)
9. K. Otasevic, Rec. Res. in Appl. Eco. and Man. **2**,1 (2011)
10. L.E.A Aimiuwu, Jour.of Nig. Inst. of Man, **39**, 1 (2004)
11. M. Jones, *The development of Cultural Identity in Early Childhood Settings*, (De Montfort University, 2009)
12. M.C.N.R. Sri, Jurn. Kom, **7**, 1 (2010)
13. N. Sinavskiy, *Ethno-pedagogic of physical training of the peoples of the North (Khanty and Mansy)* (Moscow, 2005)
14. O.C. Ugbam, B. Chukwu, A.Ogbo. IOSR Jour. of Buss. and Man. **16**,1 (2014)
15. P. Dunbar-Hall, Act. Crit. & Theo. for Mus. Edu. Jour. **8**, 2 (2009)
16. P. Surya, *Leadership of Ethnopedagogy in School* (Dynamic Scientific Magazine, Yogyakarta State University, 2011)
17. S. Ndiung, S, *The 1st International Conference on Language, Literature and Teaching (2017)*
18. V. Krasilnikov, *Folklore and national traditions in physical education of Siberian children* (Chelyabinsk, Institute culture, 1998)
19. V. Krasilnikov, *Games and competitions in traditional physical training of Khanti*. (Yekateringurg, 2002).
20. W. I. Lasmawan, *Perspektif Global dan Problematika Pendidikan Dasar*. (Singaraja,Mediakom, Indonesia. 2016)
21. Y. Wang, Inter. Comm. Stu. 16, 1 (2007)

SHS Web of Conferences **42**, 00066 (2018) https://doi.org/10.1051/shsconf/20184200066
GC-TALE 2017

Politeness strategy preference of male and female teachers in classroom interaction during English classes

Lestari, I Gusti Ayu Agung Yuni[1], *Seken*, Ketut and *Putra*, I Nyoman Adi Jaya

Universitas Pendidikan Ganesha

Abstract. Politeness is the use of right word or phrase in the proper context, which is determined by the rules that are prevalent in society. This study aimed to identify the politeness strategies used by male and female teachers; and analyze the differences in terms of strategies used by teachers. This study was designed in the form of qualitative research. The data were taken from observed interactions between teachers and students during English classes and interviewed the teachers of SMPN 2 Abiansemal. The obtained data were in form of utterances produced in teaching learning process during English classes. The utterances were identified and analyzed descriptively by using [9]'s theory in order to know the politeness strategy used by male and female teachers. The result of this study showed that male and female teacher implied four types of politeness strategies, namely; bald on record, positive politeness, negative politeness, off-record; male and female teachers were different on strategy preference depended on context and situation.

Key Words: Politeness Strategies, Gender, Classroom Interactions

1 Introduction

Politeness is a common social phenomenon, and regarded as a moral code in human communication and social activities. It is one of the basics of human interaction. Many researchers such as [4, 9, 10] focused on politeness in their studies and proposed different theories about politeness. Just like any other interaction, teacher-student interaction is positively affected by politeness. It is important to know about the extent in how teachers apply politeness strategies to their language use in EFL classroom. EFL context is a special place where nonnative language is used more frequently. The higher students' overall level of English proficiency, the better students can understand teachers' talk, and the higher awareness students may have in teachers' politeness strategy. Considering by the politeness strategy used in classroom and difference gender explained, it reflects that discussing politeness as strategy used by the speaker is also important in classroom interaction. Politeness strategy can be used as politeness behavior to the students by the teacher or vice versa as the function of the politeness strategy is to maintain a good relationship of to save hearer's face. Usually, the teacher wants to save the students' face in order to make a meaningful teaching learning process and keep the students pay their attention. In addition, SMP Negeri 2 Abiansemal is one of junior high schools in Abiansemal sub district that taught English as a foreign language. This school is located between city and village that affects the way teachers and students behave, think and communicate. The location of this school is good because it is in transition position which is not in a city and inland neither. In addition, it is also important to know the differences of politeness strategy preference by male teacher and female teacher. The teachers come from different background and they have different experience as well, and those influence the politeness strategies preference by the teachers. Men and women differ psychologically in the way they act, from the style in which they communicate [1].This study aims to identify politeness strategies used by male and female teachers and to analyze the differences in terms of politeness strategies used by male and female teachers of SMP Negeri 2 Abiansemal. This present study used [9]'s politeness theory as the main theory to analyze types of politeness strategy used by male teacher and female teacher. This theory was one of popular theories in politeness strategies. They state that politeness is the speaker's strategy to minimize the effect of an FTA (Face Threatening Act) in communication. [9] Sum up human politeness behaviour in four main types of politeness strategies: positive politeness, negative politeness, bald on record, and off-record-indirect strategy. According to [9], positive politeness consist of 15 sub strategies: notice, exaggerate, intensify interest to hearer, use in-group identity markers, seek agreement, avoid disagreement, presuppose common background, joke, assert speaker's knowledge of and concern for hearer's wants, off or promise, be optimistic, include both speaker and hearer, give or ask for a reason, assume or assert reciprocity and give gift to hearer. According to [9] there are ten strategies include negative politeness: be conventionally indirect, question/ hedge, be pessimistic, minimize the imposition, give deference, apologies, impersonalize S and H, state the FTA as a general rule, nominalize, and go on record as inquiring a debt or as not indebting H. These strategies presume that the speaker will be imposing on the listener and there is a higher potential for awkwardness or embarrassment than in bald on record strategies and positive politeness strategies. Negative face is the desire to remain autonomous so the speaker is more active to include an out for the listener, through distancing styles like apologies. Bald on-record strategies usually do not attempt to minimize the threat to the hearer's face, although there are ways that bald on-record politeness can be used in trying to minimize face-threatening acts implicitly. Often using such a strategy will shock or embarrass the addressee, and so this strategy is most often utilized in situations where the speaker has a close relationship with the audience, such as family or close friends. Moreover, [9] defined the criteria of bald on-record as; the one who speak directly in emergence situation, with maximum efficiency, and usually be used to talk with someone you know well. The fourth strategy is off record. Off record utterance is usually something general (containing less information in the sense that it rules out fewer possible states of affairs) of actually different from what it is really meant. There are15 sub-strategies conveying off record: give hints, give association clues, presuppose, understate, overstate, use tautologies, use contradiction, be ironic, use metaphor, use rhetorical question, be ambiguous, be vague, over generalize, displace H, and be incomplete.

2 Research Method

The study was designed in qualitative research. This type of research was chosen because it aimed at describing a situation or phenomena. The term qualitative research referred to research that was based on description data that do not use of statistical procedures. Observation and interview were used for data collection. The subjects of the study were a male English teacher and a female English teacher of SMP Negeri 2 Abiansemal. The interaction between teachers and students were observed and video-recorded for two hours of class. After the observation, the researcher transferred the data from video into transcriptions in order to make the analysis easier and clear. The data analysis was done after making the transcriptions by categorizing the teachers' utterances based on politeness strategy by [9]'s theory. After the data were taken from the result of transcription, observation and interview, then the data were analyzed by using the three steps of data analysis by [8]. Those three steps were: First was data reduction, in this step the researcher scrutinized, selected, and simplified politeness strategies used by male and female teacher during English classes in junior high school from super strategies to sub strategies. Next step, the teachers' utterances were evaluated by using [9]'s politeness theory. Second, data display which in this step the researcher putted the data into table, matrix and pie chart then the data interpreted based on the display to make the readers understand the result of the study easier. Third, conclusion drawing or verification was not a one-blow process and did not necessarily take place just at the end on an episode of the data analysis. It was drawn base on the interpretation of the data, interview and observation notes. This present study concluded the important data to be used to answer the research questions.

3 Results

This present study found that there were some politeness strategies found to be employed at English classroom interaction in SMP N 2 Abiansemal. The data were collected by conducting observation in four classes which were handled by two teachers. After collecting the data, the researcher classified the data based on the [8]'s theory. The data showed that 90 (36.14%) expressions were classified as Bald-on record, 93 (37.35%) were classified as positive politeness, 56 (22.49%) were classified as negative politeness, and 10 (4.02%) were classified as bald-off record. The findings of this study showed some differences of strategy used by the male and female teacher in teaching learning process in class interaction. Those differences came out because of some reasons such as the distance of their relationship and teacher's confidence. This result was strengthen by another study from [7] showed that evidently the adoption of politeness strategies shortens the teacher-students social distance, makes the class interesting, and in turn facilitates English teaching and learning. In addition, as [5] in his research stated that there are three parameters which influence the chosen of politeness strategy, those are: social distance, power, and imposition. In addition, context of situation influenced the differences politeness strategies used by teachers. As [2] produced a model for analysis of speech event: setting and scene, participants, ends, act sequence, key, instrumentalities, norms, genres. Besides, the differences of politeness strategy used by male and female teacher happened because there was indication of linguistics' strategy stated that men and women's communication style different. Psychologically, men and women are different in the way they act such as from the style in which they communicate to the way in which they attempt to influence others. On the other hand, [3] showed that while men view conversations as a way of establishing and maintaining status and dominance in relationship. Male teacher used bald on record as the dominant politeness strategy especially to give command or instruction.

e.g (Selected example from data transcription 1)

Excerpt 68

Students : *Sebuah kejadian peristiwa ataupun kenangan dimasa lampau.* "A past event or it happened long time ago" (89)

Teacher : *Apa?**ulangi!***"What? Repeat!" (90)

Students :*Sebuah kejadian peristiwa ataupun kenangan dimasa lampau.* ""a past event or it happened long time ago" (91)

The excerpt 68 above showed that male teacher used his superiority in the interaction. The command was direct and clear to get all students did what the teacher wanted. Men are goal oriented [6]. Male teacher defined his sense of self through his ability to achieve result. Besides, female teacher used positive politeness strategy as the dominant strategy during classroom interaction because she wanted to show her solidarity in the interaction. In addition, female teacher sees the purpose of conversation to create and foster an intimate bond with the other party by talking about topical problems and issues they are communally facing [3].

e.g (Selected example from data transcription 3)

Excerpt 35

Teacher : *Merek aapa?* "They what?" (571)

Students :*Mereka suka berenang!* (lebih keras) "They like swimming"(louder)(572)

Teacher :*Iya.* ***Ada yang suka berenang disini?*** "Yes. Is there anyone like swimming here?" (573)

Student : *Nggak..* "No"(574)

The expression on excerpt 35 intensified interest to H by asking about their likes especially about swimming. In this case, teacher's interest to students' interest which related with the lesson at that time about like and dislike. Asking students' interest gave positive impression to teacher that could lift up teacher-students relationship. Women are relationship oriented [6]. Female teacher defined her sense of self by her feelings and by the quality of their relationships.

4 Conclusion

This research was investigating and analyzing kinds of politeness strategy used by male and female English teachers during English classes at SMP Negeri 2 Abiansemal in academic year 2016/2017. The researcher examined and analyzed the politeness strategies used by the teachers in relation to their effort to manage the relationship between the teachers and the students. The result of this study showed that female teacher use positive politeness as dominant strategy, and followed by bald on record, negative politeness, off record. Besides for male teacher use bald on record as dominant strategy and followed by positive politeness, negative politeness, off record. The differences in politeness strategies preferences were performed by some motives from the teachers when it was seen by the similar context. Similar context or situation of the teachers could employ different strategies used of male and female teacher. The context or situation referenced to some motives. Those motives were found based on observation and interview with the teachers. Besides, the differences of politeness strategy used by male and female teacher because there was indication because of linguistics' strategy stated that men and women's communication style different. Men and women differ psychologically in the way they act such as from the style in which they communicate to the way in which they attempt to influence others.

References

1. A. Sulu, Teacher's Politeness in EFL Class, **2**, 216-221 (2015)
2. D. Hymes,*Foundation in Sociolinguistics: An Ethnographic Approach* (Philadelphia. University of Pennsylvania Press, 1974)
3. D. Tennen, *You Just Don't Understand: Women and Men in Conversation* (Ballantine Books, New York, 1990)
4. G. Leech, *Principle of Pragmatics* (London: Longman, 1995)
5. I K. Seken, *Kesantunan Linguistik dan Pembelajaran Bahasa Kedua Naskah Orasi Pengenalan Jabatan Guru Besar Tetap Dalam Bidang Pendidikan Bahasa Inggris*. (Singaraja: Ganesha University of Education, 2007)
6. J. Grey,*Men are from Mars, Women are from Venus:a Practical Guide for Improving Communication and Getting What You Want in a Relationship* (HarperCollins, New York, 1992)
7. L. Peng, F. Xie, Cai. A Case Study of College Teacher's Politeness Strategy in EFL Classroom. Theory and Practice in Language Studies, **4** (2014)
8. M. B. Miles, A. Huberman, *Qualitative Data Analysis: an Expanded Sourcebook (2nd. Ed.)* (London: Sage, 1994)
9. P. Brown, S. C. Levinson, *Politeness: Some Universals in Language Usage.*(Cambridge: Cambridge University Press, 1987)
10. R. Lakoff, *The Logic of Politeness: or Minding your p's and q's.* (Chicago: Linguistics Society, 1975)

The Implications Of Politeness Strategies Among Teachers And Students In The Classroom

Widiadnya, I Gusti Ngurah Bagus Yoga [1], *Seken,*Ketut and *Santosa,* Made Hery
Universitas Pendidikan Ganesha, Singaraja, Bali-Indonesia

Abstract. Politeness was needed to be implemented since rudeness creates conflict between teacher and students. Politeness also used in order to teach students the way of being polite, and redress the conflict in conversation. This study aimed at analyzing the implication of using politeness strategies on teaching and learning process. The subjects of this study were the tenth grade teacher and students of SMK Nusa Dua Bali. The data were in the form of conversations among the subjects in their interactions during the teaching and learning process in the classroom. The data were collected through observations and interview. The result of this study showed that, there were some implications of the politeness strategy employed by the teacher and students at SMK Nusa Dua, such as politeness created efficient teaching and learning process, respected communications between teacher and students. Besides that cooperation interaction between teacher and students were found improving, and less imposition and indirectness in teaching and learning process. Those implications motivate students and develop a meaningful teaching and learning process.

Key Words: Politeness Strategy, Teaching Learning Process

1 Introduction

Politeness in communication, courtesy, or etiquette is a procedure, custom, or custom prevailing in the society. The people live together in a society and do available conventions or norms in the social society where the politeness as a strategy to avoid conflict that might be happened and also in order to develop good relationship or togetherness in social interaction [8,9]. When talking about politeness in the communication, language will be as a behavior of human in the social interaction [4]. Means that politeness is rules of behavior which is agreed together by the society where it is applied. In the real life that is occurred in the society politeness that can be seen from the various facets of daily interaction. First, politeness contains the value of demonstrating good manners in daily interactions. When someone says something in a polite way to others, the hearer will consider that person is polite or not. When it has good values, the hearer will continue those manners, and vice versa. Second, politeness is very contextual; it is valid in a society, place, or a particular situation, but not necessarily applicable to other communities, place, or situations. Third, politeness is always bipolar, such as between children and parents, between young people and older people, between the hosts and guests, among men and women, between students and teachers, and so on. Fourth, politeness is reflected in the dress code (dress), how to do (acting), and how to speak (language) [9]. Based on those facets, the researcher put this attention to the implication of using politeness in the communication among students and teacher. As stated by [6], the use of appropriate politeness strategy will motivate the students. In other word, students have good spirit when they learn English in classroom

Politeness is relevant and important in education, such as in classroom teaching and learning practice. During the teaching and learning process the teacher uses many utterances to communicate with the students which show politeness strategy. Considering the politeness, it is consisted of strategy to make the politeness appropriate in a situation of communication among students and teachers. In a definite community like in classroom, politeness is needed to be implemented since rudeness creates conflict between teacher and students. It means that politeness is used in classroom interaction in order to build good relationship among teachers and students. By considering the politeness strategy in communication among students and teacher, it is obvious that discussing of implication of politeness strategy used by the students and teachers is important in the classroom interaction. The politeness strategy can be chosen as politeness behavior of the students by teacher or by the students to their teacher.

This research study aimed at analyzing the implication of politeness strategies that applied by students and the teacher in their interaction in the classroom. Furthermore, the researcher wants to analyze the implication of how does politeness strategies are use in teaching and learning process. Learning process is the main process in the school, when the teacher starts to use the polite utterance with the student in their interaction, it could give the student knowledge of how the politeness must be used in all the context of situation. It would change the student attitude when they make an interaction with other. SMK Nusa Dua Bali is the school that concerns about hospitality industry, so it means that the teacher and the students should have a good manner and attitude in communication. According to [5], politeness is a strategy used by the speaker to avoid conflict with the hearer. Besides, [2] groups the linguists' theories that have been introduced in the world. There are four categories of politeness theory. The first is politeness theory that believes with social norms. The second one is politeness theory that believes relationship or togetherness as a principle in a communication. The third type is politeness theory that believes about saving face. Then the fourth is politeness strategy that believes with conversational contract. This theory believes that all of participants have prepared the contract which is obtained the obligations. On the contrary, [7] claim that particular kind of act can threaten face, that is, the act that contras to what the face needs from the speaker. This is call face threatening act. In accordance to FTA, [10] gave the definition to face. According to Yule, face is the public self-image of a person. It refers to the emotional and social sense of self that everyone has and expects everyone else to recognize. Certain communicative acts might represent a threat on the face needs of interlocutors. FTA is seen as utterance or action that threatens person face. Some acts can threaten the hearers' negative face. Negative face is the desire of every competent adult member to have freedom of action, freedom of imposition, and not to be impeded by others.

Furthermore, some others can threaten positive face of the hearer. Positive acts is the need of every member that his wants to be appreciated and accepted, to be treated as the member of the same group, and to know that his or her wants are shared by others. Basically, negative face as the basic claim to territories, personal preserves, right to non-distraction, such as: freedom from imposition, therefore, negative face emphasize other's rights and freedom. In the other hand, positive face as the positive consistent self-image or personally aimed at establishing strong ties between individuals. These to distinction of FTA may lead to do the FTA or not to do the FTA. The strategies of doing FTA depend on the circumstance whether the speaker wants to reserve the hearer face to any degree. In order to minimize the FTA, [7] presented an analysis strategy which lead to the theory of linguistic politeness. In line with the explanation above, it can be presented the politeness strategies to face the FTA's : Bald-on Record strategy usually do not attempt to minimize the threat to the hearer's face, although there are ways that bald on-record politeness can be use in trying to minimize face-threatening acts implicitly. Positive Politeness is tried to minimize the threat to the audience's positive face. Negative politeness involves expressing oneself appropriately in term of social distance and respecting status differences. Off-record, according to [4] do an FTA by off record means that the speaker conveys his FTA, by using indirect and ambiguous utterance. The researcher follows the theory from [7] as the main theory.

2 Research Method

The design of this research was descriptive qualitative research where all of aspect which related to the interaction between teacher and students involved and the strategy that related to the politeness were investigated in depth and explained in details. Descriptive-qualitative research was a research which investigated the quality of relationships, activities, situations, or materials [1]. The subject in this study was the tenth grade EFL teachers and the students of SMK Nusa Dua consisted of 5 classes and 2 teachers that taught in the grade tenth. The class divided into two course, hotel accommodation class and cookery class. For hotel accommodation class there were three classes namely class 10 AP1, 10 AP2, 10AP3 which consisted of 30 students in each class. Besides for cookery class there were two classes namely 10 TB1 and 10 TB2 which consisted of 35 students in each class. This study used non-participatory observation. It was based on the consideration that the researcher did not belong to the subject of the data. The researcher only took the role as the observer in the attempt of collecting the data. The data that collected in the form of conversation of the subject in their interaction during the teaching and learning process in the classroom collected through non-participants observation, where the researcher collected the data by recording the teaching and learning process as a video file, which have recorded by video camera or handy-cam and created the transcription of all the subjects' activities and utterances employed in the classroom interaction of tenth grade EFL students and teacher of SMK Nusa Dua. In addition, the researcher created a transcription from the video of teaching and learning process in order to analyze the utterances that containing the politeness strategy easier. The interview conducted directly to the tenth grade EFL teacher and student of SMK Nusa Dua. In the interview process was documented in the form of writing a notes and audio visual, this was done to support the valuable of the data obtain through observation of the video.

The researcher was the key point in conducting this qualitative research study. The researcher who was the informant that supported and developed another instrument of data collection in conducting this research study. The focus of the research defined clearly, and it might be possible for the researcher to develop other instruments which were expected to support and complement the data collection. So, the researcher had done the analysis of politeness strategy by [7] in classroom interaction that employed by the teacher and students. Thus, in conducting the observation in this research study, the researcher used video recorder to record the interaction that happened in teaching and learning process. The observation was focused on the theory from [7] where the criteria of each strategy of politeness have been stated, so that the utterances produced by teacher and students identified precisely. Besides, the implication here viewed from [3] theory. The observation conducted through the video recorder that helped the researcher to gain the occurrence of politeness strategies produced by teacher and students. Video record was a recording of both visual and audible components. Besides used video record, the researcher also used note to support the result of video record. Note was a brief record of something written down to assist the memory.

The analysis was conducted interactively and continuously in order to create the corpus data of the utterances in classroom interaction among teacher and students by following some steps. The first step was data analysis start as soon as the transcription was completed, and data analysis was administered in terms determining the politeness strategies and the implication. Second step was choosing the data which was needed to answer the research problems and throw out data which was not needed, identify the politeness implications. The last step, the data analysis was presented based on the research questions staging in order to get a good of the explanation of the implications of politeness strategies in this research study. The research questions were answered based on the information took basically from the linguistic behavior, especially the pragmatic aspect that focused on the implication of politeness strategies. Interpretation was based on the total context of the utterance under analysis. Based on the result the conclusion drawing was as the final activities. The conclusion was started to draw from the description of the data that obtained through analysis

3 Results

This present study found that there were some implications of using politeness strategies in teaching and learning process. The implications here were viewed from theory of [3] cooperative principle. First implication is efficient communication interaction in the process of teaching and learning. The teacher and students fulfill the efficient communication in teaching and learning in the term of interaction in communication by using bald on record as the strategy for creating the efficient communication in the classroom interaction for example,

(Selected example from data transcription 2)

Note:	
T	: Teacher
S	: Student
TD	: Transcription Data
#	: Number of Utterance
Teacher	:*Kemudian tadi tentang apa kesalahannya? Jalan - jalan. Jadi kalau kalian jalan-jalan bukan walking - walking* ***tapi have a****?* 'Next, what is the mistake? 'jalan-jalan'. So, if you have a walk do not say walking walking, but have a? (T1; TD2; #76)
Student	: walk. (S2; TD2; #77)

Teacher : have a walk. *Atau* 'or' take a walk, *jadi kamu sesuaikan kalau jalan-jalannya sudah berlalu 'have'nya jadi had, bentuk kedua have itu had.'Take'nya jadi?**Verb duanya take apa?*** 'So you should adjust if you have a walk in the past, the word 'have' become 'had', the second form of have is had. The take become? Verb two of take is? (T1;TD2;#78)
Student : Took. (S2; TD2; #79)
Teacher : *coba semuanya ucapkan 'take'* 'Please all of you say take' (T1; TD2; #80)
Students : Take. (S2; TD2; #81)
Teacher : *Lebih keras* 'speak loudly' (T1; TD2; #82)
Students : Take (S2; TD2; #83)
Teacher : *Bentuk duanya?* 'The second form is?' (T1; TD2; #84)
Students : Took. (S2; TD2; #85)

In that example, it could be seen that teacher gave the students a clue. By giving a clue the teacher guided the students to answer the desired answered. In this case the teacher tried to create the communication during teaching and learning process more efficient which can be proved by the condition where the students was able to get the teachers' wanted by guessing the clue. So politeness strategies made the teaching and learning process efficient.

The second implication was respect communication between teachers and students. Bald on record used by the teacher and students fulfill the respect behavior in their classroom interaction, for example:
(Selected example from data transcription 4)
Students : Good Afternoon. (S4; TD4; #1)
Teacher : Good Afternoon, (T2; TD4; #2)

From that example, the students and teacher used greeting, welcoming, and farewell. They expressed to respect behavior in order to create good relationship between teacher and students in classroom. Good relationship made good learning atmosphere in the process of teaching and learning. In this case, both of the teacher and the students should respect each other.

Third implication was cooperate interaction between teacher and students. In this case, the cooperation could be found in teaching and learning process, for example:
(Selected example from data transcription 5)
Teacher : Look at page 127, in your book. (T2; TD5; #30)
Students : Okay. (Finding) (S5; TD5; #31)

That example showed that teacher gave an instruction to the students then students looked responding the teachers' instruction. Teacher could design their instruction by choosing politeness strategies at the time; however, students also responded the teachers' instruction by choosing the strategy of politeness. In this case, politeness strategies created cooperation between teacher and students.

The last implication was less imposition in teaching and learning process. Teacher should be able to decrease imposition to students and also students should decrease imposition to the teacher. Less imposition was performed by using negative politeness. For example,
(Selected example from data transcription 1
Teacher : *Silahkan dibaca dengan keras* 'please, read loudly' (T1; TD1; #26)
Teacher : *Ayo, baca sendiri* ***aja.*** 'Come on, you just read it by yourself. (T1; TD1; #27
Students : *Baca sendiriya?* 'read by myself?' (S1; TD1; #28)

Teacher : *Baca sendiri* ***aja*** 'just read by yourself' (T1; TD1; #29)

From that example teacher used word 'aja' in order to minimize the imposition or intrinsic of the seriousness of the students. So, the students more relaxed in doing something or not to be imposed in doing something.

4 Conclusion

Based on the result of the data analysis and discussion, it could be drawn several conclusions that there were some implications of the politeness strategy that employed by the English as a foreign language teacher and students performed in four aspects where those aspects were very influential to the process teaching and learning, the four aspects such as efficient teaching and learning process, respect communication between teacher and students, cooperation interaction between teacher and students, less imposition and indirectness in teaching and learning process and togetherness between teacher and students. By implementing them in the teaching and learning process, it became meaningful and create a good atmosphere which all of the implication previously is much needed in the teaching and learning process of English as a foreign language classroom interaction.

References

1. Fraenkel, J. R. and Wallen (1993). *How to Design and Evaluate Research in Education*. New York: McGraw-Hill.
2. Fraser, B. (1990). Perspective on Politeness. *Journal of Pragmatics, 3*.
3. H. Grice, *Logic and Conversation. In Leech, Principle of Pragmatics*. (New York: Longman, 1975)
4. I K. Seken, *Kesantunan Linguistik dan Pembelajaran Bahasa Kedua Naskah Orasi Pengenalan Jabatan Guru Besar Tetap Dalam Bidang Pendidikan Bahasa Inggris*. (Singaraja: Ganesha University of Education, 2007)
5. Lakoff, R. (1973). *The Logic of Politeness: or Minding your p's and q's*. Chicago: Linguistics Society.
6. N. P. Carniasih, *The Politeness at English Classroom Interaction and Its Implication Toward English Teaching-Learning in Wearnes Education Centre Bali* (Denpasar: Udayana University, 2011)
7. P. Brown, S. C. Levinson, *Politeness: Some Universals in Language Usages*. (Cambridge: Cambridge University Press, 1987)
8. Rash, F. (2004). Linguistics Politeness and Greeting Rituals in German Speaking Switzerland. *Journal of Curriculum and Supervision, 7*.
9. R. Watts, *Key Topics on Sociolinguistics Politeness*. (Cambridge: Cambridge University Press, 2003)
10. 10.Yule, G. (1996). *Pragmatics*. New York: Oxford University Press.

DESIGNING AFFECTIVE INSTRUMENT BASED ON SCIENTIFIC APPROACH FOR ENGLISH LANGUAGE LEARNING

Ira Maisarah[1*], *M.Zaim*[2*], *Hermawati* Syarif[2*], and *Alwen* Bentri[3*]

[1] Dosen Program Studi Pendidikan Bahasa Inggris, STKIP YPM Bangko, Jambi, Indonesia, dan mahasiswa Program Doktor, Program Pascasarjana Universitas Negeri Padang, Sumatera Barat, Indonesia
(e-mail : ira_bundadzaki@yahoo.co.id)
[2] Guru Besar dan dosen Fakultas Bahasa dan Seni, dan Program Pascasarjana Universitas Negeri Padang, Sumatera Barat, Indonesia.
(e-mail : mzaim_unp@yahoo.com, hermawati_sy@yahoo.com)
[3] Dosen Fakultas Ilmu Pendidikan dan Program Pascasarjana Universitas Negeri Padang, Sumatera Barat , Indonesia
(e-mail : alw3n_b3ntri@yahoo.com)

Abstract. This research was describing the designing of instrument for affective assessment in English language teaching. The focus of the designing was only for observation sheet that will be used by English teachers during the teaching and learning process. The instrument was designed based on scientific approach that has five stages namely observing, questioning, experimenting, associating, and communicating. In the designing process, ADDIE Model was used as the method of research. The designing of instrument was considering the gap between the reality and the teachers' need. The result showed that the designing was also notice to the affective taxonomy such as receiving, responding, valuing, organization, and characterization. Then, three key words were used as the indicator to show the five levels of affective taxonomy such as seriously, volunteer, and without asked by teacher. Furthermore, eighteen types of affective such as religious, honesty, responsible, discipline, hard work, self confidence, logical thinking, critical thinking, creative, innovative, independent, curiosity, love knowledge, respect, polite, democracy, emotional intelligence, and pluralist were put on each stage of scientific approach. So, it is hoped that can be implemented in all of context of English language teaching at schools and can assess the students' affective comprehensively.

Key Words: Designing, Affective, Instrument, Scientific Approach

Introduction

Assessment is an ongoing process that encompasses a much wider domain [1]. According to Nitko, an assessment is a process obtaining information that is used for making decision about students, curricula and programs, and education policy [2]. In education, a teacher has an important role in assessing the progress of students' competences in learning such as cognitive, affective, and psychomotor competence. Bloom, et.al. describe that cognitive competence deals with the recall or recognition of knowledge and the development of intellectual abilities and skills. Affective competence includes objectives which describe changes in interest, attitudes, and values, and the development of appreciations and adequate adjustment. Then, psychomotor competences attempts at the manipulative or motor-skill area [3]. These three competences are related each other.

How are those three domain related each other?

A positive and stable emotion can motivate learners to be more active in learning. They will not easy to get depression when they fail to learn something. LeDoux in Keblowska states that minds without emotions are not really minds at all [4]. Piasecka also found that positive emotions and feelings can help students in developing their communication skills [5]. So, it can help the students to improve their competence in mastering English.

Furthermore, affective is not only affects the cognitive, but also affects the psychomotor domain. Harrow states that desire to move are commanded by attitude [6]. Thus, if the learner has no desire to move, then he will get nothing. So, in order that to force the students want to move, it needs a motivation. Moreover, the motivation itself is part of affective. In other words, if the students have no motivation, they won't make a movement at all. It is also can be said the successful of students in learning is not only caused by their intelligence, but also the goodness of their affective.

For that reason, Indonesian government asks teachers, included English teachers, to assess these three domains in teaching and learning process [7]. The purpose of assessment is to view the students' learning progress, the students' learning outcome, and to detect the problems of learning in order to improve the quality of learning in the future. That's why, the English teachers need an appropriate instrument to assess the goodness and badness of the students' affective.

In fact, based on the preliminary study that was conducted at Junior Highs Schools in Merangin District of Jambi Province, it was found that the English Teacher hadn't an appropriate instrument to assess the students' affective during the teaching and learning process. Moreover, the English teachers said that the progress of students' affective is very important to be known in order to rank the best students in a classroom. It means, the first rank will be given to the student who has higher intelligence and best attitude only. In other words, even thought there is a student who is good in cognitive, but has a bad attitude, the first rank can't be given to him/her.

To overcome this problem, Brown explains that a teacher can use journal, interview, observation, self-assessment, and peer-assessment to assess the students' affective [1]. But, Indonesian government only provides a general guideline in assessing students' affective. Its' purpose is the teachers are hoped can develop the affective instrument based on the subject which is taught by the teacher. In fact, the English teachers were still having the difficulties in designing an affective instrument. It was caused by they must relate the assessment based on the approach of teaching that is scientific approach. Hosnan describes that there are five stages of scientific approach, i.e. observing, questioning, experimenting, associating, and communicating [8]. So, the English teachers must assess the students' affective in these five stages.

In line on that reasons, the designing of instrument for affective assessment was based on scientific approach. In this case, ADDIE Model was used as the process of designing this instrument. While, the focus of the designing was only for observation sheet that will be used by English teachers during the teaching and learning process. The designing of the affective assessment was also notice to the affective taxonomy which is suggested by Krathwohl, et.al. It consists of five levels that are receiving, responding, valuing, organization, and characterization [9]. So, three key words are used as the indicator to show the five levels of taxonomy such as seriously, volunteer, and without asked by teacher. Then, 18 (eighteen) types of affective such as religious, honesty, responsible, discipline, hard work, self confidence, logical thinking, critical thinking, creative, innovative, independent, curiosity, love knowledge, respect, polite, democracy, emotional intelligence, and pluralist were put on each stage of scientific approach. .

Finally, the product of the designing is hoped can be used to assess the progress of students' affective time by time. It is also hoped can give impact toward the improvement of students' affective, because they will always feel under the teachers' control during the teaching and learning process. Moreover, it is hoped can assess the students' competences comprehensively in all domains.

Method

In designing the instrument of affective assessment based on scientific approach for English language classroom, the researcher used ADDIE Model. This model is suitable because it has five stages such as Analyze, Design, Develop, Implementation, and Evaluation [10]. So, the researcher analyzed the English teachers' needs toward the affective assessment that was developed. The researcher took six English teachers from three Junior High Schools in Merangin District of

Jambi Province as the informants for this research. The consideration of choosing these six English teachers was because they taught English at school by using scientific approach.

Furthermore, in gathering the data, the researcher used open-ended questionnaires. The six English teachers were asked to fill the questionnaire in order to know their needs. Then, the researcher was sorting out the English teachers' needs in accordance with the objectives or the results to be achieved from this research. After that, the researcher designed the instrument of affective assessment based on scientific approach for English learning that is connecting the theories and needs in the field. This is referred to as the first draft.

Results and Discussion

First step in designing the instrument for affective assessment was doing need analysis. So, before designing the instrument for affective assessment, the researcher analyzed the teachers' need in order to find out the instrument for affective assessment which is appropriate for English language teaching. It is important to figuring out the current condition to the desired condition in order to define the problems and to determine the specific product [11].

To find out the teachers' needs, the researcher used open-ended questions. Raths, Harmin, and Simon in Bloom, et.al. view that the open-ended question has been effective in getting at attitudes, beliefs, activities, and values [12]. In open-ended question, the teachers' concern and points of misunderstanding can be seen. So, the researcher used *what*, *how*, and *why* concepts to reach valuable insight from the English teachers.

The concept of *what* is aimed to know the understanding of teachers related to affective judgments. The teacher's understanding commonly will affect the implementation of using the instrument for affective assessment. Furthermore, the concept of *how* is aimed to explore the real conditions that occur in the process of learning English in the classroom. By knowing the realities or current conditions that occur in the English classroom during the teaching and learning process, it can be the main source in designing the instrument for affective assessment based on a scientific approach. Meanwhile, the concept of *why* is aimed to explore the teacher's reasons or the reflection of the teacher's own expectations related to affective judgments.

Based on the answers given by the informants, it is known that there are 18 (eighteen) types of affective that are hoped can be assessed by the English teachers during the teaching and learning process in the classroom i.e. religious, honesty, responsible, discipline, hard work, self confidence, logical thinking, critical thinking, creative, innovative, independent, curiosity, love knowledge, respect, polite, democracy, emotional intelligence, and pluralist. These types of affective will become an input to the learning outcomes which is called as the affective entry characteristics as described by Bloom [13]. Specifically, these types of affective are divided into:

1. First stage, observing, the affective orientation are religious, respect to others, curiosity, hard work, logical thinking;
2. Second stage, questioning, the affective orientation are curiosity, love science, logical thinking, critical thinking, emotional intelligence, responsible;
3. Third stage, experimenting, the affective orientation are polite, honest, hard work, creative thinking, innovative thinking, independent, self-confidence;
4. Third stage, associating, the affective orientation are honest, hard work, creative thinking, innovative thinking, logical thinking, critical thinking, self-confidence, discipline, responsible; and
5. Third stage, communicating, the affective orientation are honest, democracy, logical thinking, critical thinking, polite, self-confidence, respect to others, responsible, pluralist.

Designing Instrument for Assessing Affective in English Language Learning

Based on need analysis, it can be concluded that the English teachers need a simple instrument for assessing students' affective. They are also need an easy instrument to be used during the teaching and learning process in the English classroom. It is expected that the instrument of affective assessment based on a scientific approach which has designed will be more comprehensive, easier or simple, effective, and useful in assessing students' affective during the learning process.

Then, the eighteen types of affective can be associated into attitude, self efficacy, values, self concept, and interest [14]. These affective cannot be directly observed. It other words, assessing affective domain is not easy, because the affective of someone can be changed time to time. Besides, the affective is something abstract and inner that can't be seen by eyes. But, it doesn't mean that the affective can't be assessed. We can only infer that a person has good or bad affective by her /his words and actions [15]. Moreover, nonverbal behaviours have also been measured affective [16]. So, in assessing affective of students in the classroom, an English teacher must see what the students do or what are they respond in some over way toward something.

Furthermore, based on the objective of this research, the designing of instrument for affective assessment is only for observation sheet that is used by English teachers during the teaching and learning process. The observation which is done by teachers is argued will be more effective than other instruments such as journal, interview, self-assessment, and peer-assessment. According to McCoach et.al direct observation is a very popular method of data collection in education and the social sciences, and it is the preferred method for collecting certain types of data, included affective assessment [17]. That's why, it is hoped that the English teachers can assess the students' affective objectively.

Simply, then, the basic design of instrument for affective assessment based on scientific approach can be seen in table 1 as follow:

Table 1. The Basic of Designing Instrument for Affective Assessment

Aspect	**Current Affective Assessment at School**	**Desired Affective Assessment**
Affective Assessment Technique	It is used a blank paper to assess the student's affective	It is formed into observation sheet with the detail assessment
Affective taxonomy	It is not based on affective taxonomy	It is divided into five levels i.e. receiving, responding, valuing, organization, and characterization
Scientific Approach Stages	It is not based on scientific approach stages	It is divided based on scientific approach i.e. observing, questioning, experimenting, associating, and communicating
English Learning Process	Assessment is only once for one basic competence	Assessment for each material in one basic competence
Types of Affective	It focuses on one affective for one meeting such as politeness only, or others	There are 18 (eighteen) types of affective were spread out in every stages of scientific approach in one meeting

Based on table 1, there are five aspects to be the basic in designing the instrument for affective assessment such as types of affective assessment technique, affective taxonomy, scientific approach, English learning process, and types of affective. In the current condition, it can be seen that the English teachers were only using a blank paper to assess the students' affective. Besides, they can't divide the assessment into levels in affective taxonomy and in every stage at scientific approach. Even, they were only assessing once for one basic competence and focus on one type of affective. So, the English teachers hope that there is an observation sheet with detail assessment such as consist of five levels of affective taxonomy, divided into five stages of scientific approach, assess the affective for each material of teaching English, and comprehend for all types of affective.

In accordance with the teachers' hope, the researcher took model that is suggested by Joyce, et. al. [18]. This model consists of five elements such as syntax, the social systems, principle of reactions, support system, and instructional and nurturant effect. In detail, the components of theses five elements can be seen as follows:

1. Syntax consists of two aspects. First, the guidance of using and filling the instrument of affective assessment based on scientific approach. Second, the indicators and scoring rubric for instrument of affective assessment based on scientific approach. In this case, the English teachers can give score in rank 1 (one) until 4 (four). Score 1 (one) is the lower score, and score 4 (four) is the higher score.
2. Social system consists of two aspects. First, the user of instrument for affective assessment based on scientific approach is the English teachers at Junior High School. Second, the target of application the instrument for affective assessment is the students who learn English in the classroom at Junior High School.
3. Principle of reactions consists of three aspects. First, the affective assessment will be done in every stage of scientific approach. Second, the affective assessment will be aimed into five level of affective taxonomy. Third, the affective assessment will be focused on into some types of affective.
4. Support system is instrument of affective assessment based on scientific approach
5. Instructional and nurturant effect hope that the students' affective can be controlled during the teaching and learning process.

In addition, the researcher made detail assessments for each stage of the scientific approach. It had designed separately in different space for observing, questioning, experimenting, associating, and communicating. Furthermore, for taxonomy of affective, the researcher designed some statements as the keywords that reflect the level of affective taxonomy such as 'seriously', 'volunteer', and 'without being asked by the teacher'. Meanwhile, for English language skills, it consists of speaking, listening, reading and writing. Finally, the material of teaching is considered into the syllabus of English language teaching for Junior High School.

This effort is suitable with the researched done by Markle dan O'Banion [19] who found that it is needed a teacher to assess the students' affective in order to create positive attitudes and improve their learning outcomes. Moreover, Olalekan [20] argues that students who have good affective will have more power in facing the problem in 21st century. That is why the affective assessment is needed to be done.

Conclusion

Assessing affective has its own unique problem. Assessing affective is not a simple one, because a teacher infers the relationship between behaviours and what the students' say. The affective itself is the result of many complex factors such as previous experience, expectations of other, and consequences of a particular act. Unfortunately, sometimes, a teacher found inconsistently of affective that is showed by students every day. But, it doesn't mean that the affective can't be assessed. Absolutely, in teaching and learning process, everything can be assessed. It needs the responsible of the teachers, being patient, and creative in carrying out the roles as the teachers.

References:

1. **Brown, H.D.**, *Language assessment: Curriculum and classroom practices*, **4**, **251-280** (2003)
2. **Nitko, A.J.**, *Educational assessment of students (3rd)*, **4** (2001)
3. **Bloom, B.S.** [ed.], *Taxonomy of educational objectives*, **7** (1956)
4. **Keblowska, M.**, *The place of affect in second language acquisition*, **157-167** (2012)
5. **Piasecka, L.**, *What does it feel like to use english? Empirical evidence from EFL students*, **219-237** (2013)
6. **Harrow, A.J.**, *A taxonomy of the psychomotor domain*, **14** (1971)

7. **Peraturan Menteri Pendidikan Dan Kebudayaan Nomor 104 Tahun. 2014,** *Penilaian Hasil Belajar oleh Pendidik pada Pendidikan Dasar dan Pendidikan Menengah.*
8. **Hosnan, M.**, *Pendekatan saintifik dan kontekstual dalam pembelajaran abad 21.*, **31-90** (2014)
9. **Krathwohl, D.R, Bloom, B.S, and Masia, B.B.**, *Taxonomy of educational objectives: The classification of educational goals*, **95-174** (1964)
10. **Branch, Robert Maribe.** *Instructional design: The ADDIE approach,* **1-5** (2009)
11. **Gupta, K.**, *A practical guide to need assessment*, **15** (2007)
12. **Bloom, B.S, Madaus, G.F and Hastings, T.J.**, *Evaluation to improve learning*, **315-316** (1981).
13. **Bloom, B.S.**, *Human characteristics and school learning*, **11** (1982)
14. **Gable, R.K and Wolf, M.B.**, *Instrument development in affective domain (2nd)*, **5-24** (1993)
15. **Henerson, M.E, Morris, L.L., and Fitz-Gibbon, C.T.**, *How to measure attitudes*, **12** (1978)
16. **Fishbean, M.**, *Belief, attitude, intention, and behafior: an introduction to theory and research*, **3** (1975)
17. **McCoach, D.B., Gable, R.K., and Madura, J.P.** *Instrument development in the affective domain (3rd).* **33** (2013)
18. **Joyce, B., Weil, M., and Showers, B.**, *Models of taching (4th)*, **14-16** (1992)
19. **Markle, R. and O'Banion, T.**, *Assessing affective factors to improve retention and completion.* Vol. 17, **1-16** (2014)
20. **Olalekan, M.**, *Teaching and assessing of affective characteristics:A critical missing link in online education*, Vol. 4, **96-107** (2013)

Improving The Creativity Of Learning Food Production Through Portfolio Assessment In The Project Learning Model

Ni Desak Made Sri Adnyawati[1,*]

[1]PKK, Engineering and Vocational Faculty, 81116 Undiksha Singaraja, Bali

Abstract. The implementation of the portfolio assessment in the project learning model is intended to identify the improvement made by the students in the food production learning, and the response given by the students to the implementation of the portfolio assessment in the project learning model. This classroom action research involved 23 students from the study program of food production of Ganesha University of Education during 2014/2015 academic year. The study was conducted in two cycles through the stage in which it was designed, the stage in which it was implemented, the stage in which observation was made, and the stage in which reflection was evaluated. The data were collected using the observation technique in which the learning process and product were observed. The observation sheet of process included preparing, processing, and serving dishes. The observation sheet of products included appearance, texture, aroma and taste aspects through the organoleptic test. The data were descriptively analyzed. The result of the study showed that the learning creativity of food production in cycle 1 averaged 79.15 and could be categorized as good, in cycle 2 the learning creativity averaged 87.08 and could be categorized as very good. This means that the improvement of the learning creativity of food production made by the students through the implementation of the portfolio assessment in the learning project was 7.93. The response given by the students to the implementation of the portfolio assessment in the project learning model could be categorized as good and averaged 81.27. **Keywords:** portfolio assessment, food production creativity, project learning model.

1 Introduction

Vocational education is education which provides skills through trainings as part of the learning process. The food production learning, as one of the vocational educations, is established to concentrate on the food production education. It emphasizes the stage in which foods are processed and the stage in which foods are ready to be served. There are three stages in the food production learning activity; they are the process of planning, processing, and serving foods. In the planning stage the materials and tools needed were planned, the processing process included the tools and technique used to process foods, and

* Corresponding author: adnyawatisri@yahoo.com

the serving stage included the tools and technique needed to serve foods, and the quality of what was served.

So far at the Engineering and Vocational Faculty of Ganesha University of Education, assignments have been collaboratively divided between the students and the teaching staff in the food production learning. The orientation of the food production learning is that there will be final learning products in the form of foods. From what was observed it can be stated that the learning process and products could not be clearly controlled; as a consequence, the learning outcome had not been optimal yet. The laboratory activity took a long time, the materials and tools needed were improperly prepared, causing the food products not to fulfill the criteria. Therefore, the learning activity should be packaged through accurate assessments in order to maximize the learning process and learning outcome of food production.

The learning process and learning outcome of food production were observed through portfolio. According to Fajar (2004), the word *portfolio* is derived from the English word 'portfolio', meaning documents. It can also be defined as a collection of valuable papers of a particular job. Portfolio is often collocated with learning; as a result, the portfolio-based learning has appeared. However, if it is collocated with the assessment concept the term portfolio-based assessment has appeared (Budimansyah, 2002). The steps in the portfolio-based learning method, aaccording to Center For Civic Education (in Depdiknas, 2004) include: 1) identifying the problems in society; 2) sorting the problems to analyze in the classroom; 3) collecting information on the problem to analyze in the classroom; 4) making portfolio; 5) presenting portfolio (showing case).

According to Cole, Ryan and Kick as quoted by Wahyudi (2010), there are two types of portfolio, namely the process portfolio and the product portfolio. The process portfolio shows the learning stages and presents the notes of the development of learners from time to time. The product portfolio is a form of the portfolio assessment emphasizing the mastery of the material required by the assignment in the standardized competence, basic competence, and indicator of the achievement of the learning outcome shown through documents.

Portfolio is a collection of artifacts showing the development and what is achieved by learners in a learning program (Marhaeni, 2007). It means that the real work in a project can be controlled using portfolio. According to Surapranata and Hatta (2004), portfolio is defined as a collection of systematically organized works or documents of learners taken during the learning process. It is used by the teaching staff and learners to assess and evaluate the development of the learners' knowledge, skill and attitude in a particular learning process.

Utami Munandar (1992: 21) stated that creativity is the ability to create something new, give new concepts which can be implemented in problem solving. In the learning process, learners are demanded to learn creatively in order to create something new. Creativity refers to the ability of someone to produce composition, product or concept, whatever they are; the important thing is that they are new and who create them has not been identified before. Creativity can be in the forms of an art product, a literary work, a scientific product, or it can be in the form of a procedure or methodology. Suarni (1996) stated that "creativity is the ability to create something. What is created is not necessarily entirely new; what is new can be its combination whose elements are already available before". Creativity is not always in the form a new creation. Creative products often appear as a result of the combination of what has been already available or can adopt the superiority of the elements of what has been created before. The observation of the design is finally in the form of a real work made using portfolio. Portfolio is a collection of artifacts showing the development and what is achieved by the learners in a learning program (Marhaeni, 2007). That means that the real work of a project can be controlled using portfolio. Creative learning is the ability reflecting fluency, flexibility, and originality in the way of thinking and the ability to collaborate a concept (Utami Munandar, 1992: 42). According to Triffenger, as quoted by Conny Setiawan (1990: 37-38), there are three reasons why creating learning is important; they are 1) being able to help children learn more successfully in the learning process, 2) being able to create the possibilities of solving the problem which we cannot predict, 3) being able to lead to great satisfaction and enjoyment.

The creativity as an indicator in the food production learning is shown in the learning process and outcome. Suarni (1996) stated that creativity is the ability to create something. What is created is not necessarily entirely new; what is new can be its combination, and the elements which are combined have already been previously available. Creativity is the ability to develop new ideas and new ways of solving problems and thinking about new things. The application of creativity in the framework of solving problems and discovering new things requires ability referred to as innovation, meaning that there is newness. The creative and innovative abilities are really reflected in the ability and desire to start up, the ability to do something new (creative), the ability to search out opportunities, being brave to bear risk and the ability to develop ideas. Such abilities and desires are required to: (1) apply new techniques; (2) produce new products or services; (3) produce new added value; (4) clear a path for a new business which is oriented towards the market; (5) develop a new organization (Rizqidiaz, 2011).

Such new and different things can be in the forms of a product and service, and a process such as an idea, method, and manner. The new and different things created through the process of creative thinking and creative action are the added value and valuable superiorities which are aimed at by the food production learning. The food production learning can be made to be a more interesting and meaningful learning experience for adults, or a traditional training in order to support the work skill (Gaer, 1998). In the food production learning, learners become motivated to learn more actively; the teacher is only the facilitator; the teacher evaluates the product as the result of the learners' performance including the outcome which can be displayed from the result of the project done. This learning emphasizes the environment in which students learn actively and collaboratively, and the authentic technique assessment is applied. The project learning is a comprehensive learning in which learners are continuously involved and cooperatively observed. The project learning focuses on the core concept and principle, facilitates learners to investigate, solve problems, and do other assignments and produce real products. It is student-centered learning (Santyasa, 2007). Then Nanang Hanafiah and Cucu Suhana (2009) state that the project learning model is the learning approach which allows leaners to work independently and construct their own learning which is culminated in a real product. This type of learning emphasizes the environment in which students learn actively and work collaboratively, and the authentic assessment technique is applied. The portfolio is one of the authentic assessments. It is not wrong if it is used in the project learning model in order to improve the creativity of learning food production.

The superiorities of the application of the portfolio assessment in the project learning are : a) it can be used to assess students based on all the assignments and the products which have been done and related to the performance assessed, b) students participate in assessing the progress they have made in completing different assignments, and the development which has taken place during the learning process, c) it can be used to assess every student based on what has been respectively achieved by taking individual differences into consideration, d) it can be used to realize the collaborative assessment process, e) students can assess themselves as what is aimed at, f) the assessment focuses on the progress achieved, the attempt made, and what has been achieved, g) it cannot be separated from the assessment, teaching and learning processes.

Based on what is described above, this present study is intended to identify 1) the improvement of the creativity in the food production learning through the portfolio assessment in the project learning, and 2) the response given by students to the application of the portfolio assessment in the project learning model.

2 Research Model

This present study is a classroom action research involving 23 students of the study program of food production education during the 2014/2015 academic year. It was conducted in two learning cycles; each was made up of four stages. They are (1) the action design, (2) the action implementation, (3) the observation and evaluation, and (4) the reflection. Each cycle was conducted in four learning times.

The data which were collected are 1) the data on the students' learning creativity including preparing, processing, serving collected through observation sheet; 2) the data on the response given by the students collected through questionnaire. The data were descriptively analyzed with the criterion that the students' learning outcome improved from cycle I to cycle II. The data on the response given by the students were descriptively analyzed with the criterion that the improvement made by the students in the response they gave was under the positive category.

3 Result and Discussion

This classroom action research involved the students who had programmed the course of food creativity '*Cipta Karya Boga*', as one of the courses under the food production learning. The students totaled 23; they were grouped into five groups, two groups were made up of four students, and three groups were made up of five students. The learning process was packaged into two cycles; each was implemented in four sessions with five steps of the project learning. Each session was held once a week and took place for four lecturing hours.

In the beginning of the learning process, the teaching staff, in this case, the lecturer informed that the learning activity in the class room for the course food creativity '*Cipta Karya Boga*' was delivered using the project learning model. The lecturer informed that the assessment model applied during the learning process included design, processing, and serving using the portfolio assessment. Then the lecturer gave a general picture of the portfolio assessment and the types of what was expected from them which were then used as the students' portfolio. What was expected would be used as the portfolio including the design of the modern menu structure, the process of realizing the design, the product which was expected based on the criterion, and the report containing the food creativity.

The learning creativity was adjusted to the steps taken in the project learning. Step (1) started with a big question; this was done by offering a problem with the theme the use of the local material in the design of the modern menu structure. In step (2) a plan for the project was designed; in this step the students designed the modern menu structure using the local material; preparing, processing and serving activities were innovated. The modern menu structure was made up of the appetizer, entrée, main course, and dessert. In step (3) a schedule was created; in this step the schedule containing when the material was prepared, the time needed for that, the tools needed, the process, and the time when the project would be completed was made. In step (4) the students and the progress made by the students were monitored; the design and the real design made in the production laboratory were discussed in and outside the classroom. In step (5) the outcome was assessed; the completion of the menu was observed so that the product produced would meet the criterion already determined. In the step when the experience was evaluated, the aspects of appearance, aroma, texture, and taste of the menu as the final product were tested using the organoleptic test.

The students' creativity obtained from the result of observation and evaluation in cycle I showed that the portfolio averaged 79.15. Based on the criterion of classification already determined, the criterion of the students' criterion in cycle I was under the category *good*. In this cycle the students' weakness in the menu design to label the menu was weak. The application of the principle of variation in how to arrange the menu was still poor, giving the impression that the menu design was monotonous. This could be seen from the material and the technique which finally affected the texture, taste and serving variation. It seemed that the time needed for completing the menu product was not maximally controlled and this contributed to the quality of the result.

The result of evaluation in cycle I was deeply reflected both in the preparation, process and completion. Improvement was made by emphasizing the learning process with the portfolio assessment through the steps in which the project learning model was applied. The innovation made for the menu in the second menu structure was observed. It was obtained that in cycle II the students' creativity averaged 87.08, which was under the category *very good*. This showed that the improvement made by the students in the learning process was significant. The menu designed through the project in which the local material was used in the modern menu structure had been maximally produced. The original menu as the project innovation, the material used, the technique used in the steps in which the product was realized showed the maximum attempts which had been made by the students. How the appetizer, entrée, main course and dessert were arranged varied depending on the material used, the processing and serving techniques; therefore, the quality of performance, aroma, texture, and taste were under the category *good*.

The data on the response in the food production learning using the portfolio assessment in the project learning model collected through the response given by the students in the end of cycle II averaged 81.27 and under the category *good*. Based on the result obtained from cycle I and cycle II, it can be stated that the action taken could improve the food production learning activity, resulting from the application of the portfolio assessment in the learning project. This gives wide opportunities to the students to learn food production creatively. The students have opportunities to develop their attitude and skill in the learning process. The optimal mastery of the process can help them to be creative. The balance between the process and product are two sides which support each other in the food production learning. The portfolio assessment in the project learning model can also give wide opportunities to the students to be innovative in designing, processing and serving foods as the periodic observation can motivate them to work better. Being happy to learn food production through the application of the portfolio assessment in the project learning model is important to support the learning success.

The result of the present study is also supported by the study previously conducted by Lucy Sri Musmini (2012) that the portfolio assessment could also contribute to the students' positive way of thinking. In addition, the results of the studies conducted by I A Made Wedasuwari (2013) and Abdorreza Tahriri (2014) showed that the implementation of the portfolio assessment could improve the students' knowledge and skill in writing.

From what is described above and the reflection made, the superiorities of the portfolio assessment in the project learning model are as follows; 1) the students become the learning center; 2) the students can evaluate themselves; 3) the lecturer can use it for controlling the students' learning activity; 4) it gives both the students and lecturer responsibility; 5) the learning process becomes more systematic and guided causing the learning process to be effective.

4 Conclusion

Based on the problems of the study and the result of analysis, it can be concluded that the implementation of the portfolio assessment in the project learning model could improve the activity of learning food production. In cycle 1 the learning creativity averaged 79.15 and could be categorized as *good* and in cycle 2 the learning creativity averaged 87.08 and could be categorized as *very good*, meaning that the creativity of leaning food production through the implementation of the portfolio assessment in the project learning improved by 7.93. The response given by the students to the implementation of the portfolio assessment in the project learning model could be categorized as good and averaged 81.27.

References

1. Abdorreza Tahriri. The Effect of Portfolio Assesment on Learning Idioms in Writing. *International Journal of Education & Literacy Studies*. **Vol.2**. (2014)
2. Budimansyah, Dasim. *Model Pembelajaran dan Penelitian Portofolio*. Bandung: PT. Genesindo. (2002).
3. Dedi Supriadi. *Kreativitas Kebudayaan dan Perkembangan IPTEK*. Bandung: ALFABETA. (1994)
4. Fajar, Arnie. *Portofolio dalam Pembelajaran IPS*. Bandung : PT. Remaja Rosdakarya. (2004)
5. Gaer, S. What is Project based learning? (Online), (http://members.aol.com/Culebra Mom/ pblprt.html, diakses 26 Juli 2017). (1998)
6. Irma Sumarsih. Implementasi Pembelajaran Kooperatif Tipe STAD untuk Meningkatkan Kreativitas Belajar Matematika Siswa SMP. (2008).
7. Julius Chandra. *Kreativitas*. Yogyakarta: Kanisius. (1994).
8. Marhaeni & Budi Adnyana. Asesmen Berbasis Kelas. *Materi PLPG*. Undiksha. (2007).
9. Muhammadiyah 3 Depok. *Skripsi tidak diterbitkan*. FMIPA-UNY.
10. Musmini, Lucy Sri. Pengaruh Asesmen Portofolio dan Gaya Kognitif Terhadap Kemampuan Berpikir Produktif Mahasiswa. *Jurnal Pendidikan dan Pengajaran*, **Jilid 45, Nomor 2**. (2012).
11. Nanang Hanafiah & Cucu Suhana. *Konsep strategi pembelajaran*. Bandung: PT. Refika Aditama. (2009).
12. Rizqidiaz. *Kreativitas dalam Wirausaha*. Diakses hari Selasa, tanggal 25 Juli 2017 dari: http://rizqidiaz.blogspot.com/2011/12/kreativitas-dalam-wirausaha. html. (2011).
13. Santyasa, I W. Model-model pembelajaran inovatif. *Pelatihan Sertifikasi Guru di provinsi Bali*. Undiksha. (2007).
14. Slameto. *Belajar dan Faktor – Faktor yang Mempengaruhinya*. Jakarta: Rineka Cipta. (2003).
15. Suarni, K. *Perkembangan dan Belajar Anak*. Singaraja: STKIP. (1996).
16. Sudewi. Pendekatan Penilaian Berbasis Produk pada Praktikum Cipta Boga Sebagai Upaya Meningkatkan Kualitas Pembelajaran. *Jurnal Universitas Pendidikan Indonesia*. **Vol 5**. 13. (2008).
17. Surapranata, S., & Hatta, M. *Penilaian Portofolio*. Bandung: PT Remaja Rosdakarya. (2004).
18. Sutikno, M. Sobry. *Menuju Pendidikan yang Bermutu*. Mataram : NTP. (2004).

19. Utami Munandar. *Mengembangkan Bakat dan Kreativitas Anak Sekolah*. Jakarta: Grasindo. (1992).
20. Utami. *Pengembangan Kreativitas Anak Berbakat*. Jakarta: Rineka Cipta. (2009).
21. Wahyudi. Asesmen Pembelajaran Berbasis Portofolio di Sekolah. *Jurnal Visi Ilmu Pendidikan*, **Vol. 2**. 1. (2010).
22. Wedasuwari, IA Made & Sukanadi, Ni Luh. Implementasi Asesmen Portofolio Untuk Meningkatkan Kemampuan Menulis Siswa. *E-Journal Santiaji Pendidikan*. **Volume 3**. 1. (2013).

Some Effects of Explicit Grammar Instruction and Syntactic Priming on Students' Written Language Production

Muhammad Asfah Rahman[1], and *Indrawaty* Asfah[2]

[1]Faculty of Languages and Literature, Universitas Negeri Makassar, Kampus UNM Jl. A. P. Pettarani, Makassar, 90222, Indonesia
[2]Faculty of Languages and Literature, Universitas Negeri Makassar, Kampus UNM Jl. A. P. Pettarani, Makassar, 90222, Indonesia

Abstract. Structural or syntactic priming is a phenomenon in which prior exposure to specific language structures either facilitates or interferes with a learner's subsequent language production [1]. Exposure to English structures through explicit instruction is reported to have inconclusive results. [2] reported that explicit and implicit grammar instruction ends up with automatization. This study reexamines the effect of syntactic priming and explicit grammar instruction on students' writing. Specific grammatical features frequently appeared on TOEFL (Written Expression Section) test were intensively practiced and then the students took a test whose items were specifically collected from TOEFL practice tests. Finally, the students were assigned to write a short essay. Sentences with similar structures which the students had been exposed to were extracted from the students' essays. Out of 40 test items, only 59.86% in average could be answered correctly, and all of the grammatical features to which the students were previously exposed were contained in their essays. However, in average only eight out of 18 sentences were grammatically constructed. It can be concluded that although priming method with explicit instruction leads the students to use similar syntactic features in their writing, it seems to have little impact on students' grammatical knowledge for immediate use in written language production.

Keywords: *syntactic priming, grammatical competence, EFL writing*

1 Introduction

English as a required subject at junior and senior secondary schools in Indonesia seems to be an uninteresting subject for the majority of the students. Generally, English is formally introduced at the first year of junior secondary school. During this period, the students build up their interest and attitude towards the English subject. Positive attitude towards and high interest in the subject are to some extent shaped by their impression on the beginning English instruction. Their learning achievement may greatly be influenced by their attitude and interest. A survey conducted by [3] reveals that Indonesian junior secondary students had very poor mastery of their English materials, that is only around 30% for each of the four language skills. In 2016, [4] reported that after one year of English instruction in Junior secondary school, the students were still very poor in reading (only 39.54% letters correctly spelled, about 35.77% words correctly read/pronounced, only about 4.29% correct answers of reading comprehension questions, and only about 46.56% of dictated words correctly written). The English mastery that the students achieve at junior secondary becomes the basis for their further study at senior secondary schools and colleges.

The poor skills as foundation carried over from junior to senior secondary school level to a large extent influence the achievement of the students. [5] reported that senior secondary students had great problems when assigned to produce even simple sentences. They were unable to select appropriate words and put them into grammatical sentences. Most of the time, their sentences did not have any predicates, contained incorrect word order and subject-verb agreement.

The poor mastery of English of the secondary school graduates might be attributable to such factors as limited time of exposure to the language and lack of explicit instruction of particular language items, especially vocabulary and grammar. English is heard, spoken, or used only in the classroom. In six years, the students have four classroom hours every week or about 138 hours for one school year. Meanwhile, during the English lessons the teachers and the students do not use English all the time—they use both English and Indonesian. The limited time for English as dictated by the curriculum does not allow the teachers to practice explicitly such language items as forms and vocabulary which are the basis for developing the students receptive and productive language skills. It is realized that the adoption of communicative-based approaches emphasizes attention to communication with little or no attention to grammatical form.

Senior secondary school graduates who are admitted to university might have poor level of English skills. To determine the English language proficiency of the senior and vocational high school graduates who entered colleges, especially those who enrolled at English Department at the State University of Makassar, an assessment of the newly admitted students' English entry level behavior was carried out. The test was developed based on 1994 English curriculum for senior high school [6]. The study revealed that (1) the English mastery level of the freshman majoring in English (which once regarded as the entry level knowledge) only 45.31 per cent; (2) generally, the basic elements of grammar were not yet mastered by the students; the most problematic grammar items were, in the order of difficulty: gerund, verb followed by causative (e.g. "have"), passive sentence construction, use of tenses, sentence modality (conditional sentences), construction of compound sentences with conjunctions and sub-ordinator, and the relative and noun clauses; (3) the level of language skills of new students majoring in English in general was still weak. The mastery level of the material for each of the language skills was respectively speaking, listening, writing, and reading, while their level of vocabulary mastery was relatively the same, i.e. only 40.15 percent. Referring to [7], this mastery level belongs to frustrating level.

At the university, especially at the English Education Department the students are exposed to the language more intensively. They have particular courses on English structure and basic writing. Explicit instruction of English structure and writing is offered during their first four semester course of study. Therefore, they are expected to have sufficient grammatical knowledge and skills when they already complete the four semester courses. However, a recent study [8] reported that the students majoring in English did not apply correctly the grammatical items that they had learned in four consecutive courses when they were assigned to write. Among the most frequently grammatical errors commited by the students in their writing are verb tenses, sentences structure, subject-verb agreement, word order, pepositions, and determiners.

Similar studies of linguistic errors are also widely practiced in other countries where English is taught as a foreign language. For example, [9] indicated that the first year students of English at Udon Thani Rajabhat University committed errors especially in the use of verb, noun, article, adjective, adverb, and sentence structure. Similar studies about students with different native language background documented relatively similar types of errors; for example, students of Saudi Arabia [10], students of of Iraq [11], students of Turkey [12], and students of Malaysia [13].

Investigation of the effects explicit grammar instruction on oral or written language production has revealed that there is no high quality evidence that teaching of traditional grammar or syntax (or the direct teaching of formal or generative/transformational grammars) is effective with regard to writing development [14]. This review article deals with students of ages 5 to 16 where English is their first language. Some recent studies investigating the role and impacts of grammar instruction in the context of English as a foreign language report some positive impacts of grammar teaching on writing. [15] reported that explicit teaching of collocation outperfomed implicit grammar teaching in paragraph writing of Iranian students. Another study by [16] and [17] found that explicit grammar instruction and grammar consciousness-raising tasks significantly improved students' grammatical knowledge, accuracy and fluency of writing. In Korean context, [18] reported that there was a positive impact of explicit grammar instruction on students' perception of their writing and editing abilities.

The discussion above may question the effects of explicit grammar instruction and the concept of priming method which suggests that language users' prior experience with language shapes their subsequent language use [19]. Despite long exposure to the language structure through explicit teaching and syntactic priming, the undergraduate students are still unable to produce the previously practiced syntactic structure in their written language production. The present study aimed to investigate the effect of explicit instruction and syntactic priming on graduate students' writing. The graduate students were required to have a minimum score of 500 on TOEFL test as a prerequisite for taking graduate courses. Therefore, they took TOEFL preparation course for three months. While attending the preparation, the researcher classified the syntactic structures most frequently appeared in the practice tests. For the purpose of this study, ten structural items were selected. They were (1) active-passive construction, (2) relative clauses, (3) suber-verb agreement, (4) tenses, (5) gerund and infinitive, (6) prepositions, and (7) word order, (8) conjunction, (9) verb form, and (10) word forms.

2 Method

Eighteen graduate students participated in the study. For three months, they were exposed to the ten selected syntactic structures through a variety of activities such as doing exercises from the TOEFL practice books, identifying the items from the reading passages provided by the researcher, intentionally learned and practiced the specified structural items from the assigned book (*Structure and Meaning in English: A Guide for Teachers by Graeme Kennedy*). At the end of the course, a structure test was administered to the group. (The test was a collection of 40 test items related to the ten selected syntactic/structural features the students had been exposed to from various TOEFL practice tests). This test was intended to assess whether the structural items were correctly identified as an indication that they already had knowledge and skills about the items. The following week, the students were assigned to write a short narrative essay of about 250 words about one of the most memorable experiences or events in their lives. The essays were examined whether they contained the ten tested syntactic structures that the students had been exposed to.

3 Results and Discussion

This section consists of two parts. First, description of the result of the structure test. The second segment focuses on the narrative essays written by the participants.

3.1 Structure test

The result of the structure test is quite disappointing. None of the students answered all the test items correctly. The highest percentage of correct answers was 72.5% and the lowest was 45% items were correctly answered. The average correct answers was about 59.86%. Fig. 1 shows the percentage of correct answers for each participants. The figure indicates that the students' performance on the test is far from satisfactory.

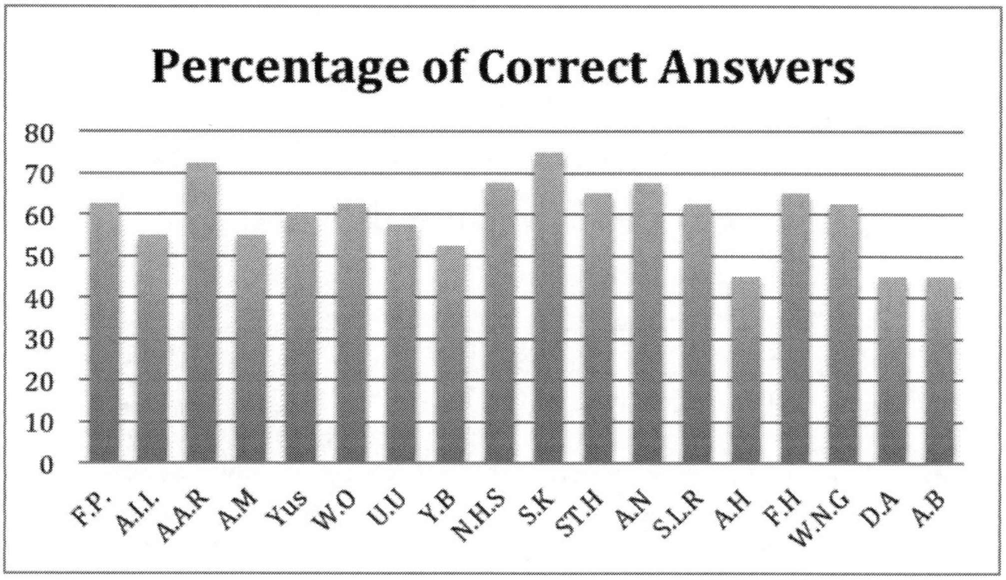

Fig. 1 Percentage of Correct Answers by Participants

Table 1. Number of Students' Correct Answers by Structural Features

Structural Items	Q1	Q2	Q3	Q4	Q5	Q6
Active-passive	14	18	15	8		
Conjunction	7	13	9	8		
Gerund/Infinitive	14	13	8	8	15	
Relative clause	8	10	9	14	13	7
S-V Agreement	10	6	11			
Tenses	11	8	16	6	10	
Word forms	16	15	4			
Verb forms	10	13	7	15	15	
Preposition	3	11	15			
Pronoun	16	6				

Table 1 shows the number of students that answered correctly the test items on each structural features. The table indicates that the students' performance is not consistent yet. For example, all students answered correctly question 2 (Q2) but some of them answered incorrectly the other three questions for the same active-passive construction. The figure also indicates that there might only be eight out of 18 students who consistently answer correctly the active-passive items. Similar phenomenon can be observed in all types of structural items.

3.2 Short Narrative Essay

The students' narrative essays contain all the structural aspects to which they have been exposed. The number of sentences and the number of correct sentences that each participant produced can be seen in Fig. 2.

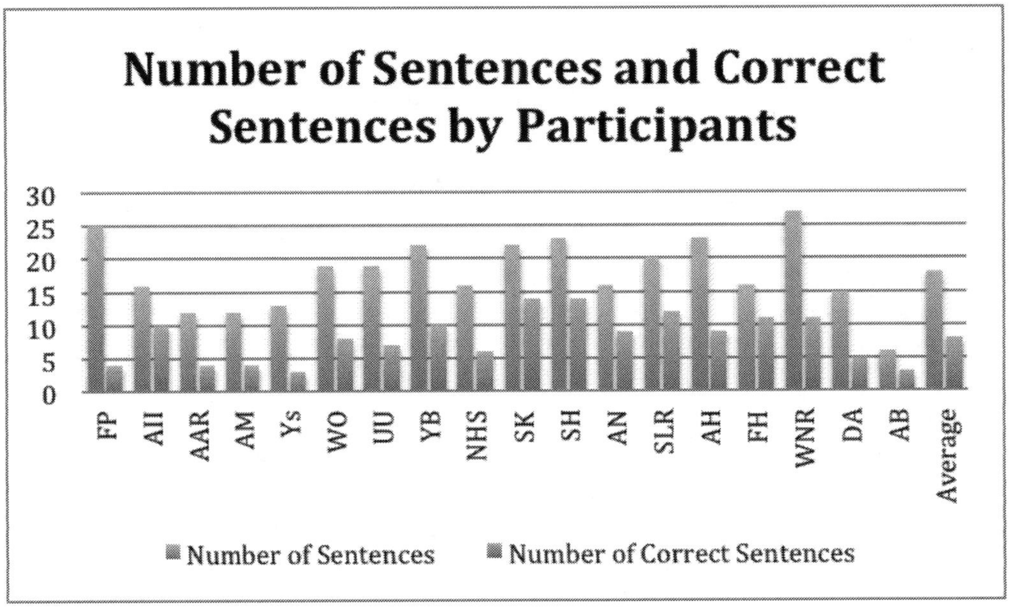

Fig 2. Number of Sentences and Correct Sentences by Participants

Fig. 2 shows that in average number of sentences that the students produced is about17.89, of which only 8 (44.72%) sentences in average are grammatically constructed. As expected the students used almost all of the syntactic features that they had been exposed to through various ways. The students tend to produce the recently practiced syntactic structures in their subsequent written expressions as found in their essays. However, many of their sentences are not well-formed—do not fully attend to the rules of the intended syntactic structure.

Following are some sample sentences extracted from the students' essays.

A student participant who correctly identified the answer of a passive form on the structure test (Question: *Aspirin used to lower fever, relieve pain, reduce inflammation, and thin the blood;* Answer : *Aspirin is used to lower fever, relieve pain, reduce inflammation, and thin the blood.*) writes the following sentence in his essay :
**This program called "BRIDGE" Building Relationship Intercultural Development and Growing Engagement.*
It should be : *This program is called "BRIDGE" Building Relationship Intercultural*

Development and Growing Engagement.
*My school SMPN 7 Pattallassang in Gowa region was choose as one of the school in Indonesia to have a sister school in Australia. (*was chosen*)

Examples of other syntactic structure are shown below.

Subject-verb agreement:
*Development of any nations is determined by leader who are in charge of managing the economic, political and social aspect of the country. *(leaders who are) or (a leader who is)*
An education equips a person with the necessary attributes with enables them to be compatible with the society. (... the necessary attributes which enable ...)*

* Education gives people skill and knowledge that helps them in solving the problems which affect them. (*skill and knowledge* that *help*)

Word form :
*It was an embarrased moment. (*embarrassing*)
*I felt very pain on my leg. (*painful*)

Relative clause:
*A country will choose its leader whom has good understanding about government and the regulation. (*who*)
*It was an international school where was located on Jl. Botolempangan No. 45 Makassar. (*that* or *which*)

Conjunction (complex sentences)
* It was very pleasure, because I enjoyed learning and practicing.
(*It was so pleasing that I enjoyed learning and practicing.*) or
(*Since it was very pleasing, I enjoyed learning and practicing.)*
* I have no worry about the finance, Australian government funding this program fully.

(I have no worry about the finance because (or since) Australian government funds this program fully.)

This study reveals that explicit grammar instruction together with syntactic priminng encourages the students to produce sentences which contain the syntactic structure they are exposed to. Consequently, their essays contain a variety of sentence types. Yet, the kinds of exposures including practices, identifying the structures out of the reading passages, and self-study on particular grammar books seem to have not created sufficient grammar knowledge for practical application on writing tasks. The use of a wide range of syntactic structures does not follow each structure item used in the essays. Thus explicit instruction and priming method in English as foreign language produce relatively the same results as those studies reviewed by [20]. Grammar teaching does not provide significant evidence that it improves quality and accuracy in students' writing in English as a first or a foreign language.

4 Conclusion and Implication

The result of this study confirms that explicit grammar instruction plus syntactic priming encourages the students to use the structures that they are exposed to. All of the syntactic structure items to which the students are exposed to are found in their essyas. This indicates that priming and explicit instruction have positive effect on students' writing in terms of language use or sentence varieties. However, the accuracy of expressions in the written forms is not yet guaranted. For accuracy, there should be other activities which emphasize the development of rule accuracy, especially for complex sentence construction. The activities should raise the students' awareness of the grammatical rules, their functions and meanings.

This study involves too many syntactic structure items in the priming and therefore it does not meet the principle of mastering one item before moving to another item. Immediate application of the syntactic structure in productive language tasks should always follow to ensure accurate use of the structure. Tasks should be designed which require the students to apply the newly primed syntactic structure which will in turn become part of the students' tacit knolwedge.

REFERENCES

1. P. Trofimovich, K. McDonough. Priming Research. In C.A. Chapelle (Ed.) *The Encyclopedia of Applied Linguistics.* Blackwell Publishing Ltd. (2013).
2. B. Tutun. Int. J. Hum. Sos.Sc. **2,** 5 (2012).
3. E. Sadtono, et al. *Survei Diagnostik Bahasa Inggris dan Rekomendasi untuk Program Pelatihan In-Service untuk Dosen/guru SLTP. Technical Report No. 8b.* Central Program Coordination Unit (CPCU), Junior Secondary Education Project, Ministry of Education and Culture, Indonesia (1997).
4. M. A. Rahman. Reading in English as a Foreign Language: A Case of First Year Students of Junior Secondary Schools in South Sulawesi, Indonesia. *Asia Pacific Journal of Contemporary Education and Communication Technology Year 2/Volume 2/Issue 03 (Special Issue),* pp. 15-21 (ISSN: 2205-6181. (2016).
5. M.A. Rahman & Mustamir. Word-Attacks in Action : Improving the Reading Comprehension Skills of the Students of SMU Negeri II Makassar. *Laporan Penelitian, Lembaga Penelitian UNM.* (2003).
6. M.A. Rahman, M.A. Rasyid. Bekal ajar awal mahasiswa angkatan 2001/2002 Jurusan Bahasa Inggris Fakultas Bahasa dan Seni Universitas Negeri Makassar. 2001. *Laporan Penelitian*: Lembaga Penelitian UNM (2001).
7. W.H. Miller *Reading Diagnosis Kit.* USA: The Center for Applied Linguistic in Education, Inc. (1978).
8. M.A. Rahman, M.N. Sainu, I. Asfah. Revisiting the Linguistic Errors in the Writing of the Students Majoring in English. *Paper presented at the 2nd ICEST.* (2016)
9. K. Nonkukhetkhong. Grammatical Error Analysis of the First Year English Major Students, Udon Thani Rajabhat University. *The Asian Conference on Language Learning 2013: Official Conference Proceeding*: pp. 117-126. (2013).
10. M. H.M. Sawalmeh. Error Analysis of Written English Essays: The case of Students of the Preparatory Year Program in Saudi Arabia. *English for Specific Purposes World, ISSN 1682-3257, http://www.esp-world.info, Issue 40, vol. 14* (2013).
11. M. Q. Mohammed. Error Analysis: A Study on Grammatical Errors in the Writings of Iraqi EFL Learners. *European Academic Research. Vo. III, Issue 11/February 2016*: 11648-11668. (2016)

12. I. Abushihab. An Analysis of Grammatical Errors in Writing Made by Turkish Learners of English as a Foreign Language. *International Journal of Linguistics. Vol. VI, No. 4.* pp.213-223. (2014).

13. A. Y.H. Tse. A Case Study of Grammatical Errors Made by Malaysian Students. *International Journal of Science Commerce and Humanities: Volume No 2 No 5,* July 2014: pp. 154-160 (2014).

14. Andrews R, Torgerson C, Beverton S, Locke T, Low G, Robinson A, Zhu D. The effect of grammar teaching (syntax) in English on 5 to16 year olds' accuracy and quality in written composition: Review summary. University of York, UK. (2004).

15. M. Karami. Exploring Effects of Explicit vs. Implicit Teaching of Collocations on The Writing Performance of Iranian EFL Learners. *International Journal of Language Learning and Applied Linguistics World (IJLLALW) Vol. 4(4)* (2013).

16. P. Rajabil, E. Dezhkam. The Effect of Explicit Grammar Instruction on Improving Writing Accuracy of Iranian EFL *Learners. Journal of ELT and Applied Linguistics (JELTAL) Volume 2—Issue 1*, (2014).

17. M. Nosratinia, S. Roustayi. The Effect of Grammar Consciousness-Raising Tasks on EFL Learners' Reading Comprehension and Writing Ability. *IOSR Journal Of Humanities And Social Science (IOSR-JHSS) Vol 19(3). (2014)*

18. G.H. Wang, S-D. Wang. Explicit Grammar Instruction for EFL Writing and Editing: An Exploratory Study at a Korean University. *Linguistics and Literature Studies 2(2): 65-73.* (2014).

19. P. Trofimovich, K. McDonough. Using priming methods to study L2 learning and teaching. In P. Trofimovich, K. McDonough (Eds). *Applying Priming Methods to L2 Learning, Teaching, and Research.* John Benjamin Publishing Co. (2011).

20. Andrews R, Torgerson C, Beverton S, Locke T, Low G, Robinson A, Zhu D. The effect of grammar teaching (syntax) in English on 5 to16 year olds' accuracy and quality in written composition: Review summary. University of York, UK. (2004).

An Evaluation on Library Services Using *Servqual* Method

Made HeryWihardikaGriadhi[1]
herygriadhi1234@gmail.com

[1]Library Sciences, Library Department, Undiksha

Abstract.after The aim of this study was to identify the quality and effectiveness of the services provided by the library using Servqual Method, which covered several variables, namely: tangible, reliability, responsiveness, assurance and empathy. The reached population were UNDIKSHA students who were already members of the library, this study was designed as quantitative evaluative research which analyzed the quality and effectiveness of the services provided by the library by analyzing the role of each factor based on Servqual Method. Data were collected using questionnaire with Likert Scale. The data obtained were analyzed in a descriptive quantitative way using Glikman's Quadrant Analysis. The impact of this research for Undiksha Library is able to provide information about the quality and effectiveness of existing service programs and can be used as input to develop strategies for developing the Undiksha library services. The results showed the quality of each variable is as follows: overall the result of quality measurement service in Undiksha library based on servqual variable is in good quality, but still need improvement in tangible variable especially in terms of physical facility and library collection, reliability variable still need improvement on indicator accuracy of information. Responsiveness Variable still need improvement especially in terms of hospitality librarian, Assurance Variables still need improvement in indicator of knowledge and skill of librarian. Level of effectiveness of library service of University of Education Ganesha, Based on the direction of score T of indicator in each variable; tangible variable show effective category, variable of reliability less effective, responsiveness variable show effective, anssurance vriabel are effective and emphaty variables show less effective.

1. Introduction

Servqual method is a deskrtiptive method developed to know the quality of service presented to the customer. The method was developed in 1985 by Parasuraman, Valarie A. Zeithaml, and Leonard L. Berry, then the method was revised back to them in 1988 published in the Journal of Retailing which concluded that servqual is a measure of how well a service meets expectations customer. The very poluler concept of service quality measurement is the Serv Qual concept developed by Parasuraman. There are five dimensions that influence to the quality of services consists of: (1) tangible (direct evidence) includes: physical facilities, equipment, employees, and means of communication; (2) reliability, includes: the ability to provide promptly, accurately, and satisfactorily promised services; (3) responsiveness, includes: the willingness of staff to assist customers and provide responsive services; (4) assurance, including: knowledge, competence, courtesy, and credibility of staff, free from harm, risk, or doubt; (5) empathy, including: ease of relationship, good communication, personal attention, and an understanding of the individual needs of the customer.

From the results of preliminary observations found statistical data utilization of Undiksha library by students are as follows; the average daily student visit to the Undiksha Library is in the range of 150 students. this can be seen from the data of the quantity of visits in 2014 as much as 36,457 people and in 2015 as many as 37,625 people (statistics visit library Undiksha 2014-2015). For the membership of the library, in 2014 as many as 7681 and in 2015 as many as 7862 people (library member statistics 2014-2015) of the total quantity of students in 2014 as much as 13472 people and in 2015 as many as 13878 people, it means that the library members are approximately half of the total quantity of students. From this data can be explained that the level of visits to the library by students is still relatively low. Based on such conditions it is necessary to assess the level of service quality and effectiveness of library services with servqual method. Problems that arise related to the application of servqual method to measure the quality and effectiveness of the library service in accordance with the variables contained in the servqual method is; How is the quality and effectiveness of tangible variables, reliability, responsiveness, assurance, and empathy of services organized by Undiksha library?

The purpose of this research is to know the quality and effectiveness of servqual variable, including tangible variable, reliability, responsiveness, assurance, and emphaty of service organized by Undiksha library. The population in this study is all students of Undiksha as the target and the affordable population is the Undiksha students who have become members of the library with the assumption that the students who have become members of the library have visited and utilize the library facilities Then the selection of students who visit the library in six month of each month is more than 4 (four) times. From 7862 the quantity of the students who are members of the library will be sampled by 350 students. This is done with the opinions of Krecjek and Morgan (in Fernandez, 1988)

This research uses ex post facto approach with servQual model. Methodologically, the research carried out includes quantitative evaluative research, which shows the procedures and processes of program implementation. In this study analyze the quality and effectiveness of library services by analyzing the role of each factor in accordance with servqual model (tangible, reliability, responsivenes, assurance, and emphaty). Quantitatively, the evaluation process is done by emphasizing the objectivity, reliability, and validity of measurement, which is focused on data in the form of numbers. For that data collection using a questionnaire-shaped instrument with Likert scale.

2. Results

2.1 Data Description of Tangible Variablea.

a. Facilities Physical Library

Table 2.1 Conversion The physical facility of the library Score

Score	Criteria	F	P (%)
41 – 50	Very high	166	47.43
31 – 40	High	155	44.29
21 – 30	Medium	29	8.29
10-20	Low	0	0
0 – 10	Very Low	0	0

Based on the conversion table, 47.43% of samples stated that the quality of library physical facilities in the category is very high, 44.29% of samples stated the physical facilities of the library in the category of high, 8.29% of sample stated the quality of library physical facilities in moderate category, and 0% there are the sample states the quality of the physical facilities of libraries in low and very low categories. The average score of the quality of library physical facilities is 39.52 is in high criteria.

b. Library CollectionsTable

Tabel 2.2 Conversion of library collection scores

Score	Criteria	F	P (%)
41 - 50	Very high	128	36.57
31 - 40	High	190	54.29
21 - 30	Medium	32	9.14
10-20	Low	0	0
0 – 10	Very Low	0	0

Based on the conversion table, 36.57% of samples stated the quality of library collection in very high category, 54.29% of samples stated the quality of library collections in high category, 9.14% of samples stated the quality of library collection in moderate category, and 0% (none) the sample states the quality of library collections in the category of low and very low. The average score of library quality is 36,27 in high criteria.

c. Supporting Facilities Library Services

Table 2.3 Conversion indicator of service support facilities score

Score	Criteria	F	P (%)
41 - 50	Very high	167	47.71
31 - 40	High	154	44.00
21 - 30	Medium	29	8.29
.10-20	Low	0	0
0 – 10	Very Low	0	0

Based on the conversion table, 47.71% of samples stated that the quality of library service support facilities in the category is very high, 44.00% of samples stated the quality of library service support facilities in the category of high, 8.29% of sample stated the quality of library service support facilities in moderate category, 0% the sample states the quality of library service support facilities in the low category, and 0% sample states the quality of library service support facilities in the category is very low. The average score of quality of library service support facilities that is 41.24 is in very high criteria.

2.2. Data Description Of Reliability Variable

a. Library Service Procedures

Table 2.4 Conversion of service procedure indicator scores

Score	Criteria	F	P (%)
41 – 50	Very high	199	56.86
31 – 40	High	123	35.14
21 - 30	Medium	28	8.00
.10-20	Low	0	0
0 – 10	Very Low	0	0

Based on the conversion table, 56.86% of samples stated the quality of service procedures in the category is very high, 35.14% of samples stated the quality of service procedures in high category, 8.00% sample stated the quality of service procedures in moderate category, and 0% service in low category and very low. The average score of service procedure quality of 41.59 is in very high criteria.

b. Accuracy of Information

Table 2.5 Conversion indicator accuracy of information score

Score	Criteria	F	P (%)
41 – 50	Very high	180	51.43
31 – 40	High	101	28.86
21 – 30	Medium	69	19.71
.10-20	Low	0	0
0 – 10	Very Low	0	0

Based on the conversion table, 51.43% of samples stated the accuracy of the information in the category is very high, 28.86% of the sample stated the accuracy of the information in the high category, 19.71% of the sample stated the accuracy of information in moderate category, and 0% there are the samples state the accuracy of the information in the low and very low categories. The average score of information accuracy of 34.93 is in high criteria.

2.3. Data Description of Responsiveness Variable

a. Librarian's Readiness

Table 2.6 Conversion of librarian readiness indicators score

Skor	Criteria	F	P (%)
41 – 50	Very high	192	54.86
31 – 40	High	126	36
21 – 30	Medium	32	9.14
.10-20	Low	0	0
0 – 10	Very Low	0	0

Based on the conversion table, 54.86% of samples stated the librarian readiness quality in very high category, 36,00% sample stated librarian readiness quality in high category, 9.14% sample stated librarian readiness quality in moderate category, and 0% Librarian readiness quality is low and very low. The average score of librarian readiness quality of 39.96 is in high criteria.

b.Librarian's Sensitivity

Table 2.7 Conversion of sensitivity Librarian scores

Skor	Criteria	F	P (%)
41 – 50	Very high	183	52.29
31 – 40	High	121	34.57
21 – 30	Medium	46	13.14
.10-20	Low	0	0
0 – 10	Very Low	0	0

Based on the conversion table 52.29% of samples stated the sensitivity of librarians in very high category, 34.57% of samples stated the quality of library service implementation in high category, 13.14% of samples stated the sensitivity of the librarian in moderate category, and 0% (none) the sample states the sensitivity of librarian libraries in low and very low categories. The average score of sensitivity of librarian quality is 37,43 is in high criterion.

c.. Hospitality of Librarians

Table 2.8 Conversion indicator of hospitality librarian score

Skor	Criteria	F	P (%)
41 – 50	Very high	91	26.00
31 – 40	High	136	38.86
21 – 30	Medium	123	35.14
.10-20	Low	0	0
0 – 10	Very Low	0	0

Based on the conversion table, 26.00% of samples indicate the quality of hospitality of librarians in very high categories, 38.86% of samples expressed hospitality librarians in high category, 35.14% of samples stated the quality of librarian hospitality in moderate category, and 0% the sample states the hospitality of librarians in the low and very low categories. The average score of hospitality of librarians of 32.21 is in high criteria.

2.4. Data Description of Assurance Variable

a. Librarian's Knowledge

Table 2.9 Conversion of indicators of knowledge of librarians score

Score	Criteria	F	P (%)
41 - 50	Very high	154	44.00
31 - 40	High	123	35.14
21 - 30	Medium	73	20.86
.10-20	Low	0	0
0 – 10	Very Low	0	0

Based on the conversion table, 44.00% of samples stated that the quality of librarian knowledge in the category is very high, 35.14% of samples stated the quality of librarian knowledge in high category, 20.86%

of samples stated the knowledge of librarian knowledge in moderate category, and 0% (not there is) the sample states the quality of knowledge of librarians in low and very low categories. The average score of librarian knowledge quality of 32.57 is in high criteria.

b. Librarian's Skills

Table 2.10 Conversion of Librarian skill indicator scores

Score	Criteria	F	P (%)
41 - 50	Very high	152	43.43
31 - 40	High	124	35.43
21 - 30	Medium	74	21.14
.10-20	Low	0	0
0 – 10	Very Low	0	0

Based on the conversion table, 43.43% of samples indicate the quality of librarian skills in the category is very high, 35.43% of samples express the quality of librarian skills in the high category, 21.14% of the sample states the quality of librarian skills in moderate category, and 0% Librarian skills are low and very low. The average score of skill quality of librarian is 38,29 is in high criterion.

c. Courtesy of librarian

Table 2.11 Conversion of courtesy indicator score

Score	Criteria	F	P (%)
41 – 50	Very high	130	37.14
31 – 40	High	165	47.14
21 – 30	Medium	55	15.71
.10-20	Low	0	0
0 – 10	Very Low	0	0

Based on the conversion table, 37.14% of samples stated the quality of courtesy of librarians in very high category, 47.14% of samples stated the quality of civility librarians in high category, 15.71% of samples stated the quality of librarian politeness in moderate category, and 0% the sample declares the librarian's modesty in the low and very low category. The average score of courtesy quality of librarians is 38,29 are in high criterion.

d. Insights of Librarian

Table 2.12 Conversion Score Indicator Insights Librarian

Score	Criteria	F	P (%)
41 – 50	Very high	144	41.14
31 – 40	High	154	44.00
21 – 30	Medium	52	14.86
.10-20	Low	0	0
0 – 10	Very Low	0	0

Based on the conversion table, 41.14% of the sample stated that the insight of librarians in the category is very high, 44,00% of samples stated the quality of librarian insight in high category, 14,86% sample stated librarian insight quality in moderate category, and 0% the sample states the insight of librarians in low and very low categories. The average score of librarian insight quality of 42.33 is in very high criteria.

3.5. Data Description of Emphaty Variable

a. The sincerity of the Librarian

Table 2.13 Conversion of sincerity indicator scores

Score	Criteria	F	P (%)
41 – 50	Very high	231	51.33
31 – 40	High	167	37.11
21 – 30	Medium	52	11.56
.10-20	Low	0	0
0 – 10	Very Low	0	0

Based on the conversion table, 51.33% of samples stated that the quality of the librarian's seriousness in the category is very high, 37.11% of the sample stated librarian's seriousness in high category, 11.56% of samples stated the quality of the librarian's seriousness in moderate category, and 0% the sample states the sincerity of the librarian in the low and very low category. The average score of the quality of the librarian's sincerity of 38.29 is in high criteria.

b. Librarian's Concern

Table 2.14. Conversion Concern indicator score

Score	Criteria	F	P (%)
41 – 50	Very high	196	49.25
31 – 40	High	161	40.45
21 – 30	Medium	41	10.30
.10-20	Low	0	0
0 – 10	Very Low	0	0

Based on the conversion table, 49.25% of samples stated that the quality of Concern of librarians in the category is very high, 40.45%% of samples express the quality of the librarian's care in the high category, 10.30% of samples stated the quality of the librarian's care in moderate category, and 0% the sample states the concern of librarians in low and very low categories. The average score of the quality of care of librarians is 42.33 is in very high criteria.

2. Discussion

The tangible variable for the physical facility quality of the library based on the average score of 39.52 is in the high criterion. this means that students as library users mostly feel comfortable in the library and utilize all the facilities of library services that already meet the requirements of service standards and yet still need improvement especially in terms of the provision of reading room and collection space are still many who feel less widespread Then to the quality of the library collection shows the average score of the quality of the library collection that is equal to 36.27 are in high criteria. It means that the quality of the collection has been able to meet the needs of students but still needs to be improved especially in terms of diversity of collections including books collection of entertainment that needs to be improved. As for the supporting facilities Library services based on the average score of the quality of library service support facilities that is equal to 41.24 are in very high criteria. It can be said that the facilities and infrastructure owned by the undiksha library already meet the adequate standards in support of providing excellent service,

The Data description of reliability variable for library service procedure, shows the average score of service procedure quality that is equal to 41,59 is in very high criteria. it means the service procedure provided is very good. Excellent procedures that include the mechanism of library services has been done quickly precise and cheap, done professionally, referring to the existing library rules, implemented with orientation on the interests of users effectively and efficiently, referring to the standard library services. For the indicator of the accuracy of information The average score of 34.93 is in high criteria. this indicates the accuracy of information provided by the library services is good but needs to be improved in many ways, the accuracy of information can be improved in terms of provision of information on OPAC (online public acssess cataloque) which ensures the students know with certainty the condition of library materials, while much information is presented in OPAC is not in accordance with the fact both in terms of the number of books that still exist and the existence of the required collection of students, accuracy in terms of recording lending transactions and return of books in some cases still encountered mismatch between the data base with print out data brought by students. Therefore it is necessary to improve the quality of service in the field of information accuracy

Responsiveness variable for librarian readiness indicator shows the average score of librarian readiness quality that is equal to 39,96 is in high criterion. This means that librarians are ready in organizing library services. Readiness is meant among librarians in implementing library services have been done quite professional, but still need to be enhanced through the training or technical guidance of the field of library services, For sensitivity indicator Librarian, shows the average score of sensitivity of librarian quality that is equal to 37.43 are in the criteria high. which means librarians or librarians have shown sensitivity to the meaning in various matters concerning the smoothness of service implementation, librarians are always sensitive to the needs of students will the information required is usually done by providing direction and instructions and explanation of the use of the library. However, it still needs to be improved. Medium for the indicator of hospitality Librarians show the average score of hospitality librarian that is equal to 32.21 is in high criteria, it illustrates that in terms of hospitality librarians or librarians already meet the expectations of users but still need to be improved, need to civilize the librarian to greet visitors and treats visitors with hospitality so that the visitors feel comfortable and cared for, the habit of smiling and giving polite reply to the visitors need to be cultivated

The Data description of Assurance Variable for Librarian Knowledge indicator shows the average score of quality of 32,57 is in high criterion. The meaning of librarian knowledge in various fields concerning library services is good enough but still needs to be improved. For Librarian Skills indicator shows an average score of 38.29 are in high criteria. this illustrates the skills of librarians is sufficient enough in organizing library services. However, there are still opportunities to be improved, increasing the skill of information technology midwife is very relevant. As for indicators of Courtesy Librarian, based on the average score of courtesy quality librarian that is equal to 38.29 are in high criteria. in the sense that the courtesy of the librarian in organizing the library services received appreciation from the students as the user is good enough, but still need improvement. For insight indicators Librarians based on the average score of insight quality librarian that is equal to 42.33 are in very high criteria. this illustrates that in carrying out librarian libraries libraries already have a very good insight into various things about libraries with different dimensions of service

Empathy variable for the indicator of sincerity Librarian based on the average score of the quality of the librarian's sincerity of 38.29 is in high criteria. this illustrates that in carrying out library services, librarians have shown seriousness in carrying out their duties, but still need to be improved. For the Librarian care indicator based on the average score of the quality of care of librarians that is equal to 42.33 are in very high criteria. This means that the concern of librarians in providing services to students as library users has been able to meet the expectations of users, every librarian in managing library services already has concern for the problems faced by students in utilizing the library.

To know the effectiveness level of library service of Undiksha, beside conducted with quantitative univariate descriptive analysis, also using T score analysis. T score analysis is done to tangible, reliability, responsiveness, assurance and empaty variable according to the following table:

Table 3.1 The results of data analysis of variable tangible, reliability, responsivesess, assurance, foury Undiksha library service in 2016

.Variable	Indicator	f +	F -	Score T Directions	Epl
Tangiible	Library Physical Facilities	185	165	Positive (+)	+ - + Effective anough
	Library Colections	158	192	Negative(-)	
	Library Support Services	215	135	Positive (+)	
Reliabelity	Library Service Procedures	184	166	Positive (+)	+- Less effective
	Accuracy of information	123	227	Negative(-)	
Responsivenes	Librarian's Readiness	187	163	Positive (+)	+++ Effective
	The Librarian Sensitivity	194	156	Positive (+)	
	Librarian's Hospitality	177	173	Positive (+)	
Assurance	Librarian's Knowledge	188	162	Positiive(+)	++++ Effective
	Librarian's Skill .	176	174	Positive(+)	
	Librarian's Courtesy	191	159	Positive(+)	
	Librarian's Insight	181	169	Positive (+)	
Empathy	The Sincerity of the Libraian	180	170	Positive(+)	+ - Less Effective
	Librarian's Concern	166	184	Negative (-)	

Based on the direction of the T score of the indicators in each variable, the tangible variables show effective categories, less effective reliability variables, responsiveness variables, effective. Anssurance variables are effective and emphaty variables show less effective.

3. Conclusion

From result of analysis and discussion hence can be drawn conclusion as follow:

1) The quality of tangible variables from library services with indicators of library physical facilities, library collections and service support facilities are high to very high with effectiveness levels being effective.
2) The quality of variable reliability of the library service with the indicator of service procedures and the accuracy of information is in the category very high with the level of effectiveness are in less effective category,
3) The quality of variable rensponsiveness with indicator readiness, sensitivity and hospitality of librarians are in the category of high with the effectiveness level in effective category,
4) The quality of assurance variables with indicators of knowledge, skills and modesty of librarians are in high category while insight librarians are categorized very high in effectiveness of variable assurance is in effective category.
5) Foury variable quality with indicator of sincerity and care of librarians are in high and very high category, while for the effectiveness level is in effective category.

Bibliography

Asogwa, Brendan E; Asadu, Boniface U; Ezema, Jonas U; Ugwu, Cyprian I, PhD; Ugwuanyi, Ferdinand C. Library Philosophy and Practice; Use of ServQUAL in the Evaluation of Service Quality of Academic Libraries in Developing Countries Lincoln (Aug 2014): 0_1,1-25. Proquest (https://e-resources.perpusnas.go.id:2171/docview/1738007975?pq-origsite=summon)

Arshad, Alia; Ameen, Kanwal, 2011. Reality versus Expectations: A Survey of University of the Punjab's Libraries Using SERVQUAL .Library Philosophy and Practice; Lincoln (Sep 2011): 1-13. Proquest (https://eresources.perpusnas.go.id:2171/docview/900093231/fulltext/6E157656F244F1CPQ/1?accountid=25704

Dantes, Nyoman. 2012 , Metode Penelitian, Yogyakarta ; Andi.

Margono. 2005. Metodologi Penelitian Pendidikan. Jakarta: Rineka Cipta

S.M. Zabed Ahmed; Md. Zahid Hossain Shoeb. Measuring service quality of a public university library in Bangladesh using SERVQUAL, Performance Measurement and Metrics; Bradford 10.1 (2009): 17-32. Proquest (https://e-resources.perpusnas.go.id:2171/docview/223663749?pq-origsite=summon)

Tripalupi,LE. &kadek Rai Suwena,2014. Statistika Dilengkapi dengan Pengenalan Statistk dalam Analisis SPSS, Yogyakarta: Graha Ilmu,

Tabasi, M. Ali 2015 dkk, Evaluation of Quality Services in Zahedan Public Library, European Online Journal of Natural and Social sciences 2015 Vol 4 No 1.

Integrated Learning in History Using Babad Banyumas to Improve the Empathy of Fellow Students

Darmawan Edi Winoto[1*] , Sariyatun[2] , and Warto[3]
[1]Master of History Education, Universitas Sebelas Maret, Surakarta 57126, Indonesia
[2] Universitas Sebelas Maret, FKIP, Surakarta 57126, Indonesia.
[3] Universitas Sebelas Maret, Faculty of Cultural Science, Surakarta 57126, Indonesia.

Abstract. The article focused on the analysis of *Babad Banyumas* which is integrated in historical learning contents. The purpose of this integration was an effort to improve the sympathy of fellow students. Practicing this empathy was for the sake of realizing a harmonious friendship. This research was conducted in explorative way due to data collection techniques using observation, interview, document analysis, and questionnaire. The selection of the object was based on its position as a standard text. This paper was based on text analysis with qualitative research. Checklist was used as assessment instrument. Validity was done through discussion and triangulation technique. Data analysis was done through interactive analysis technique based on *independent samples t-test using* SPSS 15. The results of the study showed that the integrated history of *Babad Banyumas* could improve the empathy of fellow students.

1 Introduction

In this globalization era, technology has developed to bring a huge impact on human being and society nowadays. The most important thing is fast, easy, and unlimited way of information society. The information society is a result of the production system globalization, of distribution and use of information. This means that only based on globalization processes and in their wake the information society gets established on a global scale, turns into a world-wide human community [1]. Information technologies in the educational process are of a great importance. However, this practice has not yet been mastered in full in the sphere of culture [2]. Due to educational field, globalization brought a great influence either. Globalization has a significant positive potential and it could help to overcome poverty, reduce inequalities between people, and strengthen cross-cultural, economic, educational and other contacts [3]. Besides, interfaith, intercultural, and cross country interactions have been grown so fast. So, intercultural interaction skill is basically needed to survive in anywhere. Cross-cultural interaction with representatives of various

* Corresponding author: ediwinotodarmawan@yahoo.co.id

cultures forms in the personality such qualities as tolerance, empathy, ability to respect and accept other point of view [3].

In the other hand, there is negative impact due to globalization which is the increasing of individualistic. Individualistic caused less solidarity, tolerance, sympathy, and empathy between one and another. Based on several researches, it was showed that tolerance and understanding skill in society have been decreased day by day. Anti social behaviors such as bullying, indifferent, egoistic, and apathetic toward others then appear.

To be able to decrease and prevent those unwanted behaviors, education to enhance the importance of empathy. Education can be used as proper way to overcome modern lifestyle in society nowadays.Under modern conditions educational activity becomes increasingly important for the society, since knowledge is developing faster and tends to become obsolete faster than before [4]. Besides, proper education can reduce any anti social behaviors and enhance pro-social behavior. Recently studied show that empathy training as an essential element in bullying prevention and intervention programs [5]. Education mentioned is paradigm based of education which helps to improve affective and psychomotor skill rather than cognitive aspect. Basically, Eisenberg stated that individual process to show empathy is including affective and cognitive skill.

Empathy (in cognitive or effective aspect) is basically changing in time due to teaching and learning process. A small number of study have examined how empathy levels might change through life. An early study indicated that empathy increased with age for the normal controls during adolescent period, whereas the cases did not. Recently, Dadds measured both cognitive and affective empathy of a group of Australian children and adolescents aged 4-16. The results suggested that cognitive empathy, as judged by parents increased with age, but affective empathy did not [6].

Education with good representative of affective aspect is mostly found in historical learning. Yet, historical teaching and learning at school is brought to national or common learning subjects so that affective or psychomotor aspect could not be taken by student in specific local wisdom. This phenomenon can trigger some differences of perspective and behavior of students in certain areas due to national historical learning. Thus, learning history at school is only aimed to cognitive aspect. It cannot lead to affective aspect of each student properly.

Educational and cultural problems which are connected with globalization can't be solved only by pedagogical means, integration of education and culture, interrelation of education and all subjects of a sociocultural environment are important here [3]. So, learning history integrated on local wisdom is truly needed. Babad Banyumas, as one of local wisdom, consists of values that can encourage student's empathy. Meanwhile, Babad Banyumas can be used as an effective way to behave at school. So, it is effective to be integrated in learning history in order to improve student's empathy.

2 Literature Review

Teaching and learning history is an effort in growing values and norms which are included in historical events so that students could learn to be wise in a proper way. Teaching and learning history is a process through values internalisation based on past events, namely origin of theory, genealogy, collective experience, and exemplary of characters in history. This learning process is held to preserve human being to be wise and kind hearted since historical learning tends to produce good quality outpout with comprehension in nation history, examplary in wisdom, and clever character of characters in history [7].

Babad, which is well known as traditional historiography, has brought the soul of past time and the thought of human being from past events. Babad is one of meaningful historical's opuses since each event in the past contains a lot of wisdom characters. Babad Banyumas as one of traditional historiography can reflect to five genealogy circle in structure, which consists of Pangiwa history, Majapahit family tree, Banyuamas dynasty family tree, Mertadirejan genealogy, and new family genealogy [8]. Babad Banyumas tells the origin of Banyumas itself.

Empathy is typically conceptualized as a multidimensional construct with both cognitive and affective components. Its cognitive component refers to a person's ability to comprehend another person's emotions (e.g., Hogan 1969); its affective component refers to capacity to experience another person's emotions (e.g., Mehrabian and Epstein 1972). Davis (1983) and Cohen and Strayer (1996) have defined empathy as including the ability to understand and experience how another person feels [5]. Empathy can conduct togetherness. Empathy can also prevent conflict as if we can listen, understand, and feel the related person [9]. In the other hand, empathy is truly related to social awareness. Silvera stated that social awareness describes the ability to recognize culture and values related aspects of different social groups and how these aspects influence individual's actions and behaviour [10].

Empathy, according to its correlation to education, is very important in student learning success. Therefore, empathy contributes to student's achievement and performance in classsroom. Recently research showed that young children with higher levels of empathy (the ability to empathize with others) tend to be less aggressive, more socially adapted and more successful in learning than children with low levels of empathy [11]. Goleman argue that there are 3 indicators of sympathy, (1) listening to other people talk about. It means that the individual can be a good listener. They listen to other people who are talking, (2) Accept others' point of view. It means that more look the problems from the other student's perspective than personal's, (3) Be sensitive to other people's feelings. It means that individual feels what the other people feel about [12].

3 Method

The study conducted is using qualitative research method, derived from field research. Sample of research consists of 60 students as experiment class and 60 students of control class. The methods used in collecting data are observation based on behavior, questionnaire, interview, and documentation during teaching and learning process in classroom held, starting from the beginning until the end. Yet, assessment instrument used is checklist with several indicators of empathy attitude which is developed from empathy aspect of Willams, Berard & Barrchard (2005). To be able to get data validity, several techniques are used based on discussion with teachers of history subject in Senior High School in Banyumas and deeper talks with experts. Triangulation technique of data source is also used to enhance the validity by checking the data from several different sources due to the same problem. Data analysis is held by analysis interactive technique. Then, data is analyzed using independent samples t-test through SPSS.

4 Result and Discussion

Treatments used for this experiment class consist of initial action, repetition, and recall. Treatments use six values of empathy attitude in Babad Banyumas. These values are arranged by indicators of empathy attitude and the influence will be seen after posttest in the end of experiment. The details of indicators can be seen in table 1.

Sympathy Indicators	Babad Banyumas values
Listening to other people talk	Politeness
Accepting other point of view	Respect, egalitarian
Being sensitive to other people's feelings	Compassion, mutual help, forgiving.

Table 1. Sympathy indicators

Normality test through Kolmogorov Smirnov analysis technique is used in prerequisites test. Besides, homogeneity test is using Levene's Test for Equality of Variances as analysis technique through SPSS 15. Each result of normality test and homogeneity test show $\rho >$ 0,05, which means the data used is normal (population representation) and homogeny (have the same variant).

Then, hypothesis test is conducted using independent samples t-test to compare mean of group between experiment class and control class. Based on hypothesis test result, it can be concluded that: (1) when pretest is held, experiment class hasn't gotten any treatment yet. The result shows that there is no difference in empathy attitude between experiment class and control class. (2) when posttest is held, experiment class has already experienced the treatment. The result shows that there are differences in empathy attitude between experiment class and control class. The details of hypothesis can be seen in table 2.

Test	Class	M	ρ
Pretest	Experiment	5.96	0.419
	Control	5.56	
Posttest	Experiment	8.55	0.003
	Control	6.80	

Table 2. The Result of Hypothesis

Based on hypothesis test in table 1, it can be showed that mean score of empathy attitude of experiment class is increasing from 5,96 to 8,55. From this result, there are several discoveries. These discoveries have been identified as an improvement factor of student's empathy attitude.

First, when the treatment is conducted, several students give responses and say that the story in Babad Banyumas is similar to their daily life. It is caused by their background which is appropriate to Babad Banyumas. The character in Babad Banyumas is described as the student's characteristics. In example, they learn about Ki Dipati who likes to help others. Besides, several students point to some characters that are similar to them visually as well and the characters are devoted to elderly. A lot of students who give good feedbacks by asking about the character of the story for further information. This means visual and graphic concept which is similar to real life can attract students to reflect on specific character in the story. In the other hand, the same cultural background help students to understand the plot story itself. For the researcher, similar cultural setting between the story and subject helps to gain the goal of this study.

Second, when the experiment is conducted, the learning subject used is a character of Raden Baribin. Treatment is given by telling about Raden Baribin who was driven out from his kingdom and no one who was willing to help. Almost of all students express their sincerity for Raden Baribin with sadness cue. Several students give feedbacks such as a question about Raden Baribin's condition that was already starving day by day. After the treatment was delivered, students show an improvement of sympathy and empathy by doing a lot of research's indicators. They help their friend who doesn't bring any pen and ask their friend to join in a game (especially who doesn't have any friends before). Moreover, a student comes to erase the whiteboard without any convenience so that teacher can use it again.

Third, when the treatment is conducted, students seem to enjoy the story since the teacher includes some videos, music, and pictures. During the treatment, students are active in giving feedbacks like questions and statements toward Babad Banyumas story. Students are not engaging in boredom during ten minutes treatment. They are enthusiast when listening to the story. When the bell of switched lesson is ringing, students are still craving for more stories.

Fourth, students look like that they are having more comprehension to act or react toward real situation which is similar to the story. Repetition and recall give students an extra time to understand the story better. When repetition and recall are held, students are active in giving a lot of questions. This means recall lets students to build retention of knowledge toward empathy attitude which can be implemented in some sort of real phenomenon.

Based on observation of behavior which is supported during experiment conducted, mean score of experiment class is increasing from 5,96 in pretest to 8,55 in posttest. As many as 48 students or 80% students show that there is an improvement of empathy attitude after treatment is held, while 12 students or 20% students haven't showed any improvement.

Based on conducted observation, most students showed anti social behaviors such as apathetic toward others, avoiding friends with different background, and ignorant whether they see dirty blackboard or not. There was also bullying toward students who are considered as the weakest, the ugliest, and else. After treatment was applied, empathy attitude appears in students. These attitudes like caring in others and initiative idea to help erasing whiteboard have been developed so far. Moreover, bullying has been decreased significantly. Some of empathy attitudes are categorized as appearance and some others are enrichment.

By using Babad Banyumas in classroom, it can lead to experience and knowledge that will be used by students to enhance their empathy attitude. It is strengthened by discovery of researcher. Students who join in treatment less than six times (once or twice) will not express any improvement of empathy. Presence is one of the essential factor which influences toward information and knowledge in students. So, student's presence is identified as one of big impacts that can help students to grow their empathy, while it has been already one of important thing in education.

Students who attend less treatment show an ignorant attitude toward others who already express the actual knowledge from Babad Banyumas. In example, one of Babad Banyumas stories is brought by helping other people. After action, repetition, and recall phase, there are several tendencies in classroom. Students who interpret the story of Babad Banyumas will help their friends happily, while others are ignorant about this.

After that, there are several things that are considered by researcher. One of picture visualizations must be interpreted as well. The discovery of researcher tells that students are focusing in visual learning. This fact influences on information analysis in students based on story teller. Several facial cues which are not properly recognized by students will give a little confusion. Thus, several students will ask and criticize the story.

In the other hand, Babad Banyumas that has been told in the beginning of teaching and learning process can be considered as one of subject in history class as well. It is because teaching and learning process of students is arranged in one fine theme and cannot be separated. Empathy attitude in Babad Banyumas, which is being new knowledge in the beginning of classroom activity, will be suitable subject in the next meeting, such as daily conversation or evaluation.

5 Conclusion

This research shows that the using of Babad Banyumas can influence student's empathy attitude in Senior High School in Banyumas. The influence is proved by improvement of mean score in experiment class, increasing from 5,96 to 8,55 with 0,003 of significance value in posttest of experiment and control class. The result shows that there are differences in student's empathy attitude between the ones who receive treatment and the others who don't get any treatment. From this result, it can be concluded that Babad Banyumas can improve student's empathy attitude.

Babad Banyumas can be used for teacher in teaching history to improve student's empathy attitude at school. Then, teacher can give maintenance program to defend the growth of empathy attitude based on habituation from Babad Banyumas. This program will be delivered effectively by getting any support from school in line with teaching and learning innovation held by the teacher in order to encourage students toward optimum development.

Acknowledgments

The author would like to thank all the participants for their support, especially Prof. Dr. Warto, M.Hum and Prof Dr. Sariyatun, M.Pd., M.Hum who give a lot of time for the author to guide and correct this research due to several advices and criticism.

References

1. Trubina, Iriana I. & Braines, Anna A. Imp Ed. *SHS Web of Conferences* **29**, 1 (2016)
2. Novikova, Nataliya. Comp Edu. *SHS Web of Conference* **29**, 2 (2016)
3. Gukalenko, Olga V & Borisenkov, Nadimir P. Edu Glob. *SHS Web of Conferences* **29**, 1 (2016)
4. Karpov, Alexander., Kharin, Alexander & Kharina, Olga. Edu Envy. *SHS Web of Conferences* **29**, 1 (2016)
5. Noorden, et al. Emp Involv Bul Child. *J Youth Adolescence*, **44**, 637 (2015)
6. Geng, Yaoguo., Xia, Dan & Qin, Beibei. Emp Sca. *Child Psychiatry Hum Dev* **43**, 501 (2012)
7. Hermanu Joebagio dalam Garvey, Brian & Krug, Mary, *MPSSM* (Ombak, Yogyakarta, 2015).
8. Sugeng Priyadi, *Sej Lok* (Pustaka Pelajar, Yogyakarta, 2011)
9. Hutagalung, Inge. Func Int. *SHS Web of Conferences* **33**, 3-5 (2017)
10. Sudraba, V., Martinsone, K. & Amis, V. Chang Ind. *SHS Web of Conferences* **10**, 2 (2014)
11. Naydenova, Irina S. & Naydenova, Natalia N. Teac Port. *SHS Web of Conferences* **29**, 3 (2016)
12. Goleman, Daniel, *Social Intelligence: New Scie* (Bantam Dell, New York, 2006)

The Effect of the Instructional Media Based on Lecture Video and Slide Synchronization System on Statistics Learning Achievement

I Gede Partha Sindu[1,*], and A.A. Gede Yudhi Paramartha[2]

[1]Informatics Engineering Education, Universitas Pendidikan Ganesha, Singaraja, Bali 81116, Indonesia
[2] Informatics Management, Universitas Pendidikan Ganesha, Singaraja, Bali 81116, Indonesia

Abstract. The purpose of this study was to determine the effect of the use of the instructional media based on lecture video and slide synchronization system on Statistics learning achievement of the students of PTI department . The benefit of this research is to help lecturers in the instructional process i to improve student's learning achievements that lead to better students' learning outcomes. Students can use instructional media which is created from the lecture video and slide synchronization system to support more interactive self-learning activities. Students can conduct learning activities more efficiently and conductively because synchronized lecture video and slide can assist students in the learning process. The population of this research was all students of semester VI (six) majoring in Informatics Engineering Education. The sample of the research was the students of class VI B and VI D of the academic year 2016/2017. The type of research used in this study was quasi-experiment. The research design used was post test only with non equivalent control group design. The result of this research concluded that there was a significant influence in the application of learning media based on lectures video and slide synchronization system on statistics learning result on PTI department.

Keywords: video and slide synchronization system, lecture slide and video, statistic learning achievement.

1. Introduction

Success in a lecture activity is among other things determined by the instructional media used which is capable of arousing interest in the students to learn so that they can learn effectively. A lecturer has to be able to plan a lecture, to process relevant information into learning material, to translate the program tht has been designed into learning objectives , to present the material through varioius methods and instructional media, and to evaluate learning achievement. So far lecture process generally uses slide presentation, which generaly contains the summary of the material which can take the form of brief

* Corresponding author: partha.sindu@undiksha.ac.id

defintios of a topic, illustrations, pictures and other learning contents. The students so far in learning through a lecture process have not yet optimally understood the material taught. Based on this problem there is a need for a technological innovation in helping the learning process, one of the innovations is the use of video media. Conventional teaching in the classroom can be recorded in a video format video which then can be viewed again by the students. Lecture video can helpp students in replaying the lecture that has taken place before. The result of a study by Paramartha, Sindu, and Aryanto about the development of lecture video and slide synchronization as a media to support electronics teaching was aimed to help students' learning process in understanding a lecture [1]. The result of this study was that a lecture video and slide presentation in which the system can be used to produce a new electronic instructional media in the form of a lecture video and a slide presentation which are presented side by side on a screen and are sychronized based om transition time of each sup-topic. This instructional media is designed so that it can be accessed through computer, tablet, and smart phone through web browser. In addition, the instructional media which is capable of being produced by this system is made in such a way that it is portable and can be transferred from one gadget to another without internet connection.

Research in video and slide synchronization has only reached the stage of usability testing, i.e., to test an interaction between the user and the system based on effectiveness, efficiency and customer satisfaction in using the system. Thus, it is very important to continue the study by doing an evaluation , i.e., by trying out the system with the students to find out to what extent this system affects the students' learning achievement. The previous study which is related to this study is the study done by Primavera and Suarna, which state that an instruction which uses audio-visual media (video) has the supporting capacity toward learning process of a good category. In addition, audio-visual media (video) has proven to be more effective in enhancing the ability to understand (C2), to apply (C3), and to analyze (C4) [2]. Other works that are related to this study are Iwantara, Sadia and Suma, who state that the use of video youtube media in Science instruction can increase students' learning motivation and concept understanding [3]. The result of the study done by Lestari and Maspiyah showed that there was the effect on the students' learning achievement of video media in the teaching of sub-competence in decorating a stage appearance at SMKN 2 Boyolangu Tulungagung [4].

Based on what has been elaborated above, it is felt that there is a need to investigate further whether the use of an instructional media based on lecture video and slide synchronized system has an effect on the students' learning achievement in a study entitled " The Effect of theeee Instructional Media Based on Lecture Video and Slide Synchronization System on Statistics Learning Achievement of the Students of PTI Department." This study aimed to find out the effect of the instructional media based on lecture video and slide synchronization on Statistics learning achievement of the students of PTI Department.

Media is an intermediary or something that sends a message from the sender to the reciever of the message [5]. Instructional media in general is an aid in the teaching and learning process. Instructional media has a very important function, that is, as the carrier of information and something that prevents obstacles in the learning process, so that information or message from the communicator can reach the communicant effectively and efficiently. The effectiveness of the use of video media as the teaching aid cannot be separated from the demand of the development of technology. The use of video media in teaching can help in giving meaningful experiences to the students in terms of cognitive, affective and psychomotor domains and can overcome the time limitation in the classroom and the teaching process becomes more interesting. The lecture video and slide synchronization system is used to produce an electronic instructionjal media in video and slide format that are synchronized based on the transisiton time of every sub-topic. This system can be used by the lecturers to make an interactive electronic instructional media which then can be used by the students to support electronic instruction. The lecture video and slide system which can be used to produce a new electronic instructional media in lecture video and slide presentation format which is presented side by side on a screeen and is syncrhonized. Based on the transition time of each sub-topic [1]. The instructional media is designed to be accessible through computer, tablet, and smart phone through web browser. In addition, the instructionalmedia that can be produced by this system can be made for portable use, in which the instructional media can be transferred from a gadget to another without the aid of internet connection.

Statistics is a study that studies how to plan, collect, analyze, and present data. In brief, statistics is a study which is related to data. The term statistics is different from statistic. Statistics os a study that is related to data, while statistic is data, information, or a result of the application of statistical algorithm to a data. From a collection of data statistics can be used to conclude or describe data; this is called descriptive statistics. Most of the basic concepts of statistics assume the theory of probability. Some statical terms include population, sample, sample unit, and probability.

Learning is a process done by an individual to gain a new change in behavior which is entirely the result of his or her experience in interaction with the environment. Learning achievement shows a change in the individual knowlege structure as the result of learning situation [6]. Learning achievement in declarative knowledge form consists of conceptual knowledge and logical knowledge. In addition, learning achievement can also take the form of procedural skill. The desired learning achievement has to be parallel with the learning objectives. If the learning objectives are in cognitive, affective, and psychomotor domains, then the learning achievement lies in those domains. According to behaviorist perspective, learning is a behavior that makes connection between a stimulus (S) and a response (R), and then the connection is reinforced. The behaviorists believe that learning achievement will be more fully mastered if better if the material is memorized over and over again. According to the cognitive perspective, learning occurs in mind so that a behavior can only be called learning if it reaches understanding. Bloom divided learning achievement into three domains, namely cognitive, affective and psychomotor [7]. The cognitive domain is related to learning objectives that are related to the ability to think. The affactive domain is related to the learing objectives that are related to feeling, emotion,value, and attitude that show aceptance or refusal towards something. He psychomotor domain is related to motor skilll and manipulation of materials and objects

2 Methods

The type of research method used in this study was experiment research method. Meanwhile, the experiment design used was quasi-experiment. This method is used in an activity of research which proceed in a condition in which the controling of variables is felt to be difficult if it is done in 24 hours so that a pure experiment is difficult enough to conduct. The research design used was Post Test Only With Non Equivalent Control Group Design. In this design sampling was done by selecting classes which are used as sample at ramdom. This design was selected since during the course of the experiment there was no possibility to change the exisring classess. In terms of procedure, this study followed a pattern as shown in Table 1.

Table 1. Post Test Only With Non Equivalent Control Group Research Design [8]

Experiment	E	X	O_1
Control	C	-	O_2

Note: E: Experiment Group; C :Control Group ; X : (treatment) Instructional media based on video and slide synchronization system; - : no treatment of the experiment instruction; O_1 : Post Test Only/observation of experiment group; O_2 : Post Test Only/ observation of control group.

The variables used in this study were as follows.

1. Independent Variable
   In this study the independent variable was instructional media based on lecture video and slide synchronization system
2. Dependent Variable
   In this study the dependent variable was Statistics learning achievement of the students of PTI Department.

This study was conducted at Jurusan Pendidikan Teknik Informatika, Universitas Pendidikan Ganesha. The population consisted of all of he sixth semester students of Jurusan Pendidikan Teknik Informatika. Since the number of individuals / classes was small, saturated sampling was used , thus all of the population was used as sample. Then based on the result of equality testing it was assumed that the conditions and abilities of the classes that were taken as samples were equal.

In this study the form of cognitive test used was a subjective test of essay items. Before the instrument was used, it was tested to know its level of validity, reliability, index of difficulty of the test items, and index of degree of discrimination of the test itrms. The validity of an essay test can be computed by using Product Moment formula. In this study, the testing of the reliability of the essay test was done by using statistical formula, that is, Alpha Cronbach formula. For a standard test, the items tolerated have the range of reliabilies between 0.31 and 0.70. The next step was determining index of discrimination to know to what extent a test item can discriminate between high ability testees and low ability testees.

Before doing the hypothesis testing, the data that had been collected were subjected to the prerequisite testing first. This prerequisite testing was done to prove that the data had met the prerequisite of parametric statistical analysis. The prerequisite testing done consisted of normality testing and variance homogeneity testing.

The normality testing was done to determine the step in testing by using parametric or non parametric statistics.

The hypotheses to be tested are :

H_0 = Data have a normal distribution.

H_a = Data do not have a normal distribution.

To test the homogeniety of variance for both groups F-test was used, The hypotheses to be tested are

H_0 : There is no difference of variance between control group and experiment group.

H_a : There is a difference of variance between control group and experiment group.

After testing the homogeneity of variance and the normality of data, then it was continued with hypothesis testing. The hypothesis testing was specified into Null Hypothesis (HO) versus alternative hypothesis (HA). Statistically, the hypotheses can be written as follows.

H_0 : that is, there is no signifcant effect in the application of the instructional media based on lecture video and slide syncrhonization system on Statistical learning achievement of the students of Jurusan PTI

H_a : that is, there is a significant effect in the application of the instructional media based on lecture video and slide synchronization system on Statistical learning achievement of the students of Jurusan PTI

The study used analysis of mean difference between two groups of scores to test the hypotheses.. If it turns out that the data collected have a normal distribution and are heterogeneous, then t-test at sig. 5% would be used. The t-tes used was pooled variance. The hypothesis was tested by using t-test with the formula used by Sugiyono as follows [8].

$$t = \frac{\overline{X}_1 - \overline{X}_2}{\sqrt{\frac{(n_1 - 1)s_1^{\ 2} + (n_2 - 1)s_2^{\ 2}}{n_1 + n_2 - 2}\left(\frac{1}{n_1} + \frac{1}{n_2}\right)}} \quad (1)$$

3 Results and Discussion

Before collecting data in the field, the instrument was judged first by two experts who were experts in their field, that is, experts in statistics and instructional media. After the expert judgement, then the instrument was revised and the essay posttest was tried out. Based on the result of the try-out, items whose validity, reliability , index of difficulty, and index of discrimination had been found. There were 5 items which had passed the test and were ready to be distributed to the students after the study ended as posttest items about the students' learning achievement. Based on the whole data of the posttest about the students' learning achievement which had been processed, the average score of learning achievement of the experiment class was found. That is, 67. 5 and that of control class was 49.5 which shows the experiment class group had a higher mean than the mean of learning achievement of the control group. The computation of data to test normality, homogeneity, and hypotesis testing beside using Microsoft Office Excel, the computation also used SPSS Statistics 16, and both produced the results of data processing which were not much different, that is, that the two sample classess were normally distributed and homogeneous.

Table 1 Result of Hypothesis testing by SPSS

Independent Samples Test										
		Levene's Test for Equality of Variances		t-test for Equality of Means						
									95% Confidence Interval of the Difference	
		F	Sig.	t	df	Sig. (2-tailed)	Mean Difference	Std. Error Difference	Lower	Upper
HB	Equal variances assumed	0.356	0.554	4.880	38	0.000	18.000	3.689	10.533	25.467
	Equal variances not assumed			4.880	37.675	0.000	18.000	3.689	10.531	25.469

The data of learning achievement of the two classes which had been tested were then used to test the hypothesis using t-test aided by Microsoft Office Excel. Based on the comparison of $t_{obs.}$ and $t._{c.v.}$, the criterion used was if $t_{obs.} > t_{c.v.}$ then Ho was rejected and Ha was accepted. From the computation it was found that $t_{obs.} = 4.880$ and $t_{c.v.} = 1.69$, which means that $t_{obs.} > t_{c.v.}$ (4.880 > 1.69). While the hypothesis testing with the aid of SPSS Statistics 16 showed that $t_{obs}. = 4.880$ and $t_{c.v.} = 1.69$, thus, $t_{obs.} > t_{c.v.}$ (4.880 > 1.69). Thus, H_a was accepted and H_0 was rejected. The result of study showed that there was a significant effect on learning achievement, thus, Ho was rejected and H_a was accepted and it can be concluded that there is a significant effect of the implementation of the instructional media based on the lecture video and slide synchronization on Statistcs learning achievement of the students of Jurusan PTI

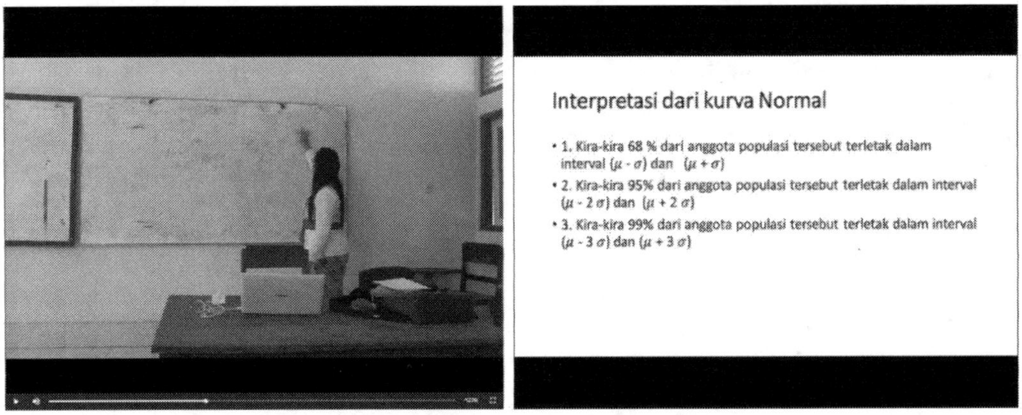

Figure 1. The Display of Video and Slide Syncrhonization System

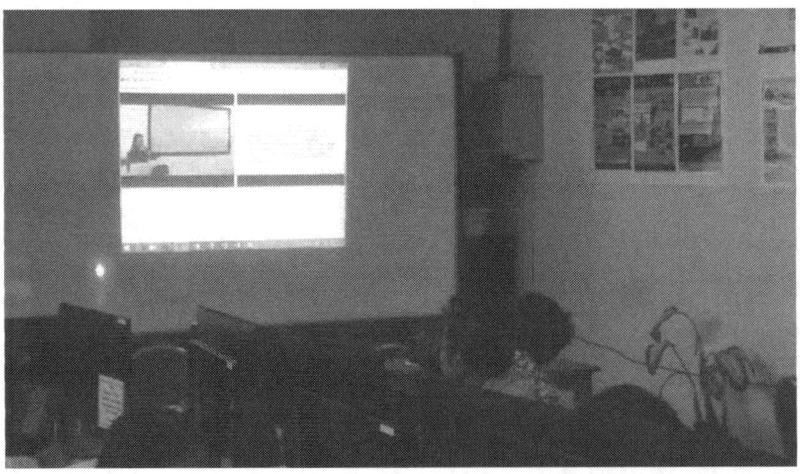

Figrue 2. The Process of Video and Slide Implementation in the Classroom

The instructional media based on video and slide synchronization system gives an added value to the students' progress in learning achievement and increase their motivation to learn. The use of instruction based on video and slide synchronization in the classroom helps very much in the experiment class in understanding the instruction. This result agrees with the result of the study done by Primavera and Suwarna, who state that the instruction that uses audio-visual (video) media has a supporting capacity for the learning process and is categorized into a good category [2]. The result of this study is in line with the result of the study entitled "Effects of different video lecture types on sustained attention, emotion, cognitive load, and learning performance," where the effect of video teaching media results in a significant increase in attentionm emotion, and learning achievement of the students [9].The result is also consistent with the result of the study entitled "Learning process and learning outcomes of video podcasts including the instructor and PPT slides, The result revealed that the participants using the video podcast with both the instructor and PPT slides gained the best learning outcomes [10]. Another researh that is related to the instructor's image size in lecture video effect on learning outcomes showed that the students felt the same social presence and cognitive load as in the lecture video with different instructor's size [11]. Based on the theoretical and operational considerations, the implication of this study viewed from constructivist learning theory is the belief in student-centered instruction, in which the student will construct knowledge obtained and modify it to develop new knowledge when he or she is learning, which has an impact on the development of his or her skill and thinking pattern. Thus, the source of the knowledge is an instructional media based on the system of video and lecture slide presentation synchronization.

4. Conclusions

On the whole, the finding of the study from the descriptive analysis and hypothesis testing can be concluded that from the descriptive analysis and hypothesis testing the class in which the lecture video and slide synchronization was used gave a better effect compared to the class which was not treated with this media. Theoretically, the use of the instructional media based on video and slide synchronization system facitates the students at the time of learning, that is, at the time of material discussion and it makes the use of instructional time more effective. The constraint in this study lies in the the fact that the lecture video and slide syncrhonization system has not been used optimally in e-learning for the students to access it well. The solution which can be offered to overcomoe this constraint is to optimize the use of e-learning at Universitas Pendidikan Ganesha

Acknowledgments

The research team would like to extend their gratitude to the Institute of Research and Community of Universitas Pendidikan Ganesha who has supported and approved this research. Our gratitude also goes to the Faculty of Enginnering and Vocations and the Department of Informatics Engineering Education which always support lecturers in doing researchers.

References

1. A. A. G. Y. Paramartha, I G. P. Sindu, K. Y. E. Aryanto, ICIC. (2016)
2. I. R. C. Primavera, I. P. Suwarna, *Prosiding Seminar Nasional Pendidikan IPA FITK UIN Syarif Hidayatullah Jakarta*, 122 (2014)
3. I W. Iwantara, I W. Sadia, I K. Suma, *E-Journal Program Pascasarjana Universitas Pendidikan Ganesha*. 4. (2014)
4. R. C. Lestari, Maspiyah, e-Journal UNESA, 2, 3 (2013)
5. A. S. Sadiman, R. Rahardjo, A. Haryono, Rahardjito. *Media Pendidikan; Pengertian, Pengembangan dan Pemanfaatannya*, (2002)
6. I W. Santyasa, *Buku Ajar Jurusan Pendidikan Fisika. Universitas Pendidikan Ganesha*. (2005)

7. W. N. Nasution, Jurnal Penelitian IAIN, 5 (2006)
8. Sugiyono. *Metode Penelitian Kuantitatif Kualitatif dan R&D*, (2013)
9. C. M. Chen, C. H. Wu. Computers & Education, **80,** 108 (2015)
10. Z. Pi, J. Hong. IETI, **53**, 135 (2016)
11. Z. Pi, J. Hong, J. Yang. JCAL, 1 (2017)

Promoting autonomous learning in English through the implementation of Content and Language Integrated Learning (CLIL) in science and maths subjects

Putu Fika Andriani [*], *Ni Nyoman* Padmadewi, *I Gede* Budasi
Universitas Pendidikan Ganesha, Singaraja, Bali, Indonesia

Abstract. Autonomous learning is a concept in which the learner has the ability to take charge of their own learning. It becomes a notable aspect that should be perceived by students. The aim of this research is for finding out the strategies used by grade two teachers in Bali Kiddy Primary School to promote autonomous learning in English through the implementation of Content and Language Integrated Learning in science and maths subjects. This study was designed in the form of descriptive qualitative study. The data were collected through observation, interview, and document study. The result of the study shows that there are some strategies of promoting autonomous learning in English through the implementation of CLIL in Science and Maths subjects. Those strategies are table of content training, questioning & presenting, journal writing, choosing activities, and using online activity. Those strategies can be adopted or even adapted as the way to promote autonomous learning in English subject.

Key words. autonomous learning; CLIL

1 Introduction

Learning languages other than our mother tongue is the proof that human being has tried to cross one of the boundary in communication called language. English is one of the language which need to be mastered in order to communicate in this globalization era. Related with that fact, many strategies have been found in order to teach English as a second or foreign language. However, sometimes it is not only about how the teacher use those strategies to teach the language but also about the students' effort to learn that language. Being successful in learning is not merely about score. It is about making the effort to learn something. The same thing goes for language learning. No matter how much students have learnt at school, in the future they will need to learn other things by

[*] Corresponding author: andrianiputufika@gmail.com

developing and relating new things with their background knowledge. The point is, sooner or later they will have to study on their own, and they need to learn how to learn. "Critically aware learners" are those who will do well in a fast-changing world [1]. Thus, in this fast changing world, there is a need to shape our students to be autonomous learners as they will need the skill not only for language learning, but also for variety of work and environments in their future lives. Students need to become capable in adapting and developing strategies to do individual learning, so they have critical awareness about what and how they should learn. The aim of this research is for finding out the strategies used by grade two teachers in Bali Kiddy Primary School to promote autonomous learning in English through the implementation of Content and Language Integrated Learning in science and maths subjects.

Autonomous learning is defined as "the ability to take charge of one's own learning" [2]. Other definition emphasizes that autonomous learning itself is an act of learning that consciously motivates learners to make best decisions about their own learning [3]. Learning can be more focused and purposeful for students when they are involved in decision making processes regarding their own language competence [4]. There are some reasons for helping learners to take on more responsibility for their own learning. The first reason is because of the learning effectiveness. It is improved when learners take charge of their own learning. It happens because the learners learn what they want to learn. The second reason is the learners who take responsibility for their own learning can continue learning outside the classroom, and the third reason is that learners who know how to learn can transfer their learning strategies to other subjects [5]. On the other hand, Content-and-Language-Integrated-Learning (CLIL) refers to the use of a language other than the students' mother tongue as the medium of instruction in an educational setting [6].

There are some researches which focused in strategies to promote autonomous learning. Padmadewi mentioned some techniques to promote autonomous learning in the classroom for young learners. Those techniques, namely maximizing the use of reading log; using reading rocket system, and using reading response journal system [7]. Hafner & Miller described the potential of the collaborative digital video project in a technological learning environment to provide students with opportunities to exercises their capacities as autonomous learners [8]. In their view an autonomous learner is not only the one who has intrinsic motivation and learns outside the classroom with no need for support from the teacher, but also develop in the structured learning environment of the classroom and become part of the pedagogical objectives of a language course. Ramamuruthy and Rao found that the use of smartphone could boost learners' critical thinking, creative thinking, communication, and collaboration skills [9]. The result showed that learners have moved toward autonomous learning, however, they were still reliant on the teachers to achieve their learning goals. The used journal writing [10] and weekly report [11] also can stimulate students' autonomy in writing as the students are given an opportunity to reflect on their new knowledge, ask questions about unclear ideas, and explore the value of question asking itself.

Those researches show that autonomous learning can provide the students with a better chance for learning foreign language, in this case, English. Students' English proficiency was significantly and positively related to their learner autonomy [12]. However, there is a condition where the autonomous learning in English can be promoted while learning other subjects which use English as the language instruction. CLIL provides an intensive chance to use English not only for the language but also for the subject content, such as science and maths. CLIL become the platform for the students to use English, and the need to master the content in the subjects, in this case, maths and science, can be used as the tools to promote autonomous learning in English, as English is the language instruction of those subjects. By being autonomous learners, the students can take responsibility for their own learning. Moreover, when the students are armed with the knowledge of how to learn, they can transfer their learning strategies to other subjects.

2 Research method

This research was designed in the form of a descriptive qualitative study. The data were collected through classroom observations, and document study, and then interviews with the teachers were conducted to obtained clarification regarding the result of observation. Finally the data were analysed descriptively. The observation was conducted three times in each subject. It was meant to find out the strategies to promote autonomous learning in Maths and science class. The interviews were conducted to confirm the result of observation to the teachers. Document study was done to support the result of the interview and observation. The subjects of this study were 20 students of grade 2, and 2 teachers who teach Science and Maths in Bali Kiddy Primary School.

3 Findings and discussion

The findings in this study revealed that there are five strategies used for promoting autonomous learning in science and maths class. Four strategies were used by the science teacher to promote autonomous learning in English. Those teaching strategies namely (a) Table of Content Training, (b) Questioning & Presenting, (c) Journal Writing, and (d) Choosing Activities. Meanwhile, in maths subject the teacher use educational web for online learning as the way to promote autonomous learning in English. Those strategies are commonly used by any teachers in the classroom. However, the science and maths teachers in Bali Kiddy Primary School designed those activities to train the students so they are able to do self-directed study.

Being able to read and use table of content is a simple but meaningful skill. By training the young learners to read and find the information through table of content, the teacher believed that it would provide guidance for them in their learning. This skill is useful when the students do independent study at home. The teacher wanted the students to realize that all books have parts and each part contains information for the reader. They can find the information by themselves in the book only if they know how to find it. In other words, by having this skill the students can learn how to learn. It is not only about bring and open the book based on teacher's instruction. The teacher wanted the young learners to have the awareness that their books are actually the assistant for their learning activity, and they do not need to wait for teachers or other people to help them to study. This skill is useful when the students want to do independent learning.

Questioning is the tool which is used by grade two teacher in Bali Kiddy Primary School to trigger the students' curiosity. The questions are expected to stimulate the students to use English and communicate their thought. Presentation activities gives the students a chance to perform in front of peers and teachers. The students usually just simply present the result of their work or demonstrate something to teach their friends. When the students are presenting something in front of their friends, they take the responsibility as the one who know-it-all, and are ready for the audience's questions and responses. Besides, they are stimulated to show more effort in order to perform better.

The next strategy used by the science teacher to promote autonomous learning in English is by writing science journal. The teacher asked the students to write down what they had learnt in their science class regularly. This journal was made as a project. The students were challenged to make the journal as informative and interesting as possible. The science teacher provided them with opportunities to modify and adapt classroom tasks.

They were given choices about how they wanted to make the journal. They could make illustration, table, diagram, or mind map. Figure 1 shows the example of student's journal. It shows that the student made the journal in the form of simple diagram along with its illustration.

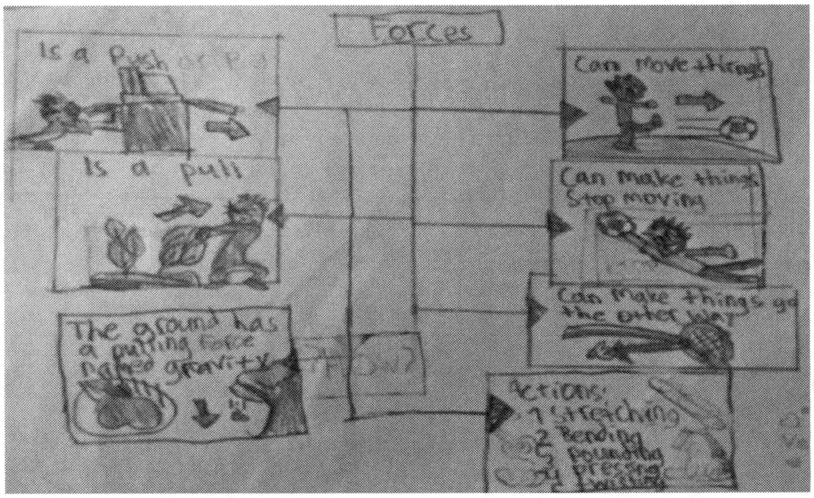

Fig. 1. Example of student's journal.

By making the journal writing as a project, the students could show their conscious effort to learn and find out more information and vocabulary to be written on the journal. They were free to add information about the science topic from various sources. It could be from other books or internet sources. Students can set their own goal about how informative and interesting their journal will be. This journal project can also train their skill in taking note. Teaching learners to take notes helps them to be independent learner [13]. Making their own note Figure 2 below shows the steps of training the students to write the science journal in order to promote autonomous learning in English.

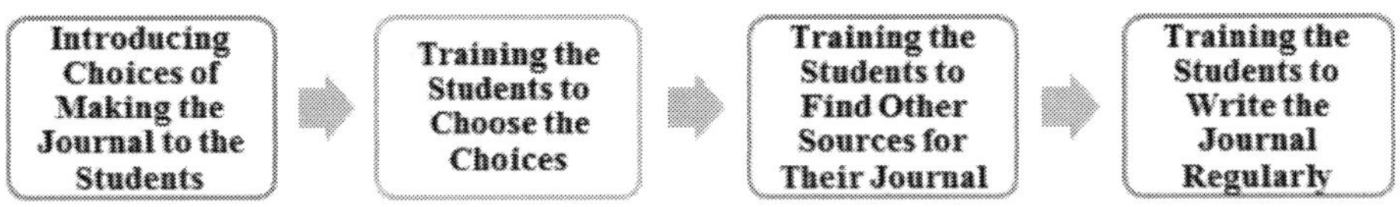

Fig. 2. Step of training the students to make the journal.

By letting the students to make choice, the teacher trained the students to develop their autonomy in learning. They made the choice and be responsible toward the choice they made. The feeling of success can bring confidence to the students; meanwhile, the feeling of failure can lead to an evaluation in making choices. The reward for the students are given in the form of titles, such as "The best journal illustrator", or "The most informative journal".

The teacher also let the students to choose their activity based on their preference. In review time, the students could choose to study alone or with a group of friends. They could study by sitting on the floor or combining their tables together. Instead of asking the student to follow a similar learning procedure, the teacher let the students choose their own way in learning. The important aspect that should be emphasized here is the reason behind the choice. The students should know in what way they learn best. The choice made by the students should be the conscious act to study in the way which they know they can learn best. By doing so, the teacher not only promoting autonomous learning but also using differentiated instruction. That was done as the attempts to combine teaching and learning to a learner's unique strength and needs [14]. Letting the students to choose also means that the teacher involves the students in their own learning which gives the sense of ownership of their achievement.

The freedom of making choice and the feeling of being involved in the learning process can motivate the learner to their study. At the initial step in training the students to make choices, the teacher still need to supervise the students' choices, and communicate the reason or the benefit behind the choices which were introduced by the teacher. This was done in order to make sure that what the students want is what they need.

Different from the science teacher, the Maths teacher used online learning as the way to promote autonomous learning in English. There is a list of topics that can be chosen by the students in their online account which use English as the language instruction. The students can make choice regarding what topic/ what activity they want to do in their online account, and the teacher is still able to track the students' progress. Stars chart was used to give reward for the students only at the beginning of the online learning introduction. The chart can give stimulation to initiate the online learning and do self-directed study.

The online account can be accessed by the students at home or at school in the computer laboratory. It provides the students with easy access to various information and knowledge regarding the topic they want to learn. In other words, the online account becomes the platform for the students to do self-access in learning. The students were trained to be critical in analysing what they need to learn and doing the online learning independently, but the teachers still has the role in helping the students to master the maths concept in the classroom, and check their understanding toward the concept in maths subject.

In conclusion, the steps in promoting autonomous learning in both subjects and the analysis of both teaching techniques which are used in science and maths and are formulated in table 1.

Table 1. Steps in Promoting Autonomous Learning in English through CLIL in Science and Maths Subjects.

No.	Steps	Description
1	Giving choices to the students	The initial step of promoting autonomous learning is by introducing students with choices. Teachers give a chance for the students to see all possibilities for their learning so they have the basic knowledge to do self-direct in learning. The choices can be about what to learn and how to learn it.
2	Training the students to make a choice	The second step of promoting autonomous learning in Science and Maths subjects is stimulating the students to make a choice. The freedom to make a choice and the feeling of being involved in every decision related with their study can motivate the learner to their study.
3	Training the students to find learning sources	The aim of training the students to find learning sources is for stimulating the students' awareness that teacher is not the only source of learning. The students are expected to be able to find sources and learn from it. In that way the students will learn how to learn
4	Habit formation in reading and writing	The fourth step is making the students getting used to with all the activities that can shape them to be autonomous learners.
5	Establishing habit for	Establishing habit for being autonomous learner is done

	being autonomous learner	by giving reward in order autonomous learning activity to be repeated.

4 Closure

Based on the result of the research, it can be seen that some common learning strategies can be designed to promote autonomous learning in English. Table of content training, questioning & presenting, journal writing, choosing activities, and using online activity emphasize the students to make choice and take control of their own learning in science and maths. Control and responsibility are the combination through which the students can have the sense of belonging in learning. Through science and maths they consciously learn English as the language instruction of both subjects. As the need to be autonomous learner is inevitable in the global competition, the teacher should consider the strategies to promote autonomous learning not merely as strategies for teaching but as the mean to get the real meaning of learning.

References

1. J. Godwin, LLT, **15**, 4-11 (2011)
2. H. Holec, *Autonomy and foreign language learning*, (Oxford, Pergamon, 1981)
3. J. Xu, ELT, *A survey study of autonomous learning by Chinese non-English major post-graduates*, **2**, 25-32 (2009)
4. A. Littlejohn, ELT Journal, **39**, 253-261 (1985)
5. G. Ellis, B. Sinclair, *Learning to learn English,* (Cambridge University Press, United Kingdom, 1989)
6. Dalton-Puffer, *Discourse in content and language integrated learning (CLIL) classrooms*, (2017)
7. N. N. Padmadewi, JESOC, **3**, 45-52 (2016)
8. C. Hafner, L. Miller, LLT, **15**, 68-86 (2011)
9. V, Ramamuruthy, S. Rao, MOJET, **3** (2015)
10. L. Kamberi, Proceeding, 1st Annual International Interdisciplinary Conference. AIIC, *Promoting learner autonomy in foreign language learning by using student journals*, (2013)
11. Wardah, Nurwahida, Proceeding, The 62nd TEFLIN International Conference 2015, *Stimulating students' autonomy in writing through weekly report*, (2015)
12. D. Davei, Asian EFL Journal, (2007)
13. F. Mollaei, M. Riasati, JSLTE, **1**, 105-120 (2013)
14. D. Cruickshank, D. Jenkins, K. Metcalf, *The act of teaching (4th ed)*, (McGraw Hill, New York, 2006)

The effect of conceptual change text structure on concept understanding and misconception reduction of dynamic electricity

Gde Parie Perdana[1*] , *Ketut* Suma[1] , and *Ni Made* Pujani[1]

[1]Science Education Department, Post-Graduate Program, Universitas Pendidikan Ganesha, Singaraja, Indonesia

Abstract. This study investigated the effect of conceptual change text on student understanding and misconception reduction of dynamic electricity concept. A quasi-experimental research with pre-test/post-test non-equivalent control group design was used. The subjects for this study consisted of 90 tenth-grade students. The three-tier test, the Dynamic Electricity Concept Test (DECT), was developed as pre-test and post-test to access the student conceptions. While the experimental group (n=45) received a conceptual change text, the control group (n=45) received an expository text. MANCOVA analysis was used to know the effect of conceptual change text structure on both dependent variables. The results of the study indicated that the student in the experimental group showed significantly higher understanding and higher misconceptions reduction compared to the students in the control group. Both groups had increased concept understanding and misconceptions reduction, however, the experimental group's result was better than the control group.
Keywords: Conceptual change text, concept understanding, misconceptions, dynamic electricity

1 Introduction

Students have conceptions, but not all of them are scientific concepts, there are also misconceptions. The misunderstanding that occurs will disturb the process of forming new concepts. There are various researches that have investigated misconceptions in the students in understanding science concepts. Especially in physics, the result of research found that many students have misconceptions rather than understand the concepts [1, 2]. This shows that concept understanding is very low and the misconception level of the students is high category.

In literature, the student's personal experience, knowledge and language of science used by teachers in the classroom, and the textbooks are listed as sources of students' conceptions or reasoning [3]. Physics textbooks are generally designed in conventional text format known as expository text (ET) which contains knowledge encyclopedia. However, many studies found that expository texts cannot give cognitive conflicts to the students [4, 5, 6] so that it is difficult to reconstruct scientific concepts in the students.

* Corresponding author : edgparie@gmail.com

To overcome the weaknesses of the conventional text, a text or instruction can be developed using conceptual change approach that causes cognitive conflicts [7]. Therefore In the present study, it is aimed to investigate the effect of conceptual change text on student understanding and misconception reduction.

2 Literature Review

2.1 Conceptual Change

The conceptual change was proposed for the first time by Posner [7]. It is defined as a mental process in which the students can change their concept by adding new concepts to their knowledge or by changing their prior concepts into appropriate scientific concepts. Conceptual change is aimed at helping the students to understand concepts that they have not understood and to correct their misconceptions [7, 8].

Conceptual change consists of two phases, namely assimilation and accommodation [7]. The assimilation phase is the phase where the concept that the students have acquired is not sufficient to understand some new phenomena completely. The accommodation phase is the phase where the students have to change or restructure their major concepts. In addition, conceptual change also needs four stages of the process of understanding before the students can accept it, namely, (1) dissatisfaction, (2) intelligible, (3) plausible, and (4) fruitful [7].

At the understanding level, the student's haves be able to organize and interpret what they have learned, it means that they can visualize and express it in a different way using their own words each piece of information given to them [9]. Although there is a difference in the way of expressing the information in the process of understanding, the content of the information has to follow its scientific rule.

2.2 Prior Knowledge

The perspective of constructivists recognizes the existence of prior knowledge possessed by the student before learning in the classroom. The student is seen as a subject of learning who has possessed prior knowledge before learning [10]. The student's prior knowledge strongly supports the formation of scientific concepts in the student to achieve learning achievement [11]. It does not only support the student's learning, but it can also become an obstacle in learning since some prior knowledge is probably contradictory with the information that he/she will learn [10]. Hence, to know the student's prior knowledge becomes an important step in the teaching and learning process.

2.3 Conceptual Change Text

Conceptual change text (CCT) is a teaching material that is based on conceptual change approach and is designed to correct the student's misconceptions [12]. CCT can cause conceptual change and make the conceptual change that occurs in the student become permanent [3]. CCT can also be used in a remedial program [9]. A CCT is more effective than a conventional text (expository text) in restructuring the student's schemata that have undergone misconceptions [4, 9]. The structure of CCT consists of 5 parts, namely (1) problem presentation, (2) common misconceptions, (3) explanation by using a scientific concept, (4) students' opinions, and (5) evaluation [13].

3 Method

This study was a quasi-experiment. The subjects were grouped based on their classes and were selected by random selection. The study used pretest-posttest non-equivalent control group design [14]. Every group was given a pretest and a posttest to measure the dependent variable before and after the treatment.

3.1 Participants

The subjects consisted of 90 tenth grade senior high school students. Two physics teachers volunteered to teach in this study. One teacher and the class that was taught were selected at random to become control class, and the other teacher and the class to become experiment class. The physics lesson lasted for 5 periods each week (45 minutes for each period). The researcher and the teachers held a discussion in the beginning of research about how to use CCT in the class.

3.2 Text

The texts were written by the researcher and reviewed by 2 lecturers (a professor and a doctor). The structure of the CCT text followed the structure developed by Ozkan and Selcuk [13]. There are five parts in the text. The first part is aimed to identify the possibility of students' misconceptions. The second part presents misconceptions and common answers that are scientifically false. The third is the scientific truth about the subject being explained. The explanation has to be very clear and can be understood. In the fourth part, when the students feel there is a difference between the misconception and the true scientific explanation, they are asked to express their own opinions. At the last part, evaluation, the aim is to know whether the students have understood the text well or not.

3.3 Instrumentation

The Dynamic Electricity Concept Test (DECT) was developed and used to categorize students' conceptions as understand the concept or misconception. The DECT is three-tier test includes 20 multiple-choice, reason, and student certainty of their understanding. In developing the test, 15 questions of the multiple-choice type were adopted from literature [1, 15, 16, 17, 18, 19] and the rest were developed by the researcher. DECT consists of 6 scopes of concepts, namely, (1) current, (2) resistance, (3) series & parallel circuits, (4) electromotive force & voltage difference, (5) energy & electric power, and (6) electric measuring instrument.

Four lecturers (2 professors and 2 doctors) who teach Physics at Physics Education Department had reviewed the instrument to test its content validity. The four lecturers agreed that the test items were suitable to be used to achieve the aim and scope of the study. The testing of reliability had been done with the students who once learned the concept of dynamic electricity. The reliability of the test was 0.71 obtained by using Sperman-Brown equation.

The answer of the multiple-choice test was categorized as correct and incorrect, the students' reasons into a scientific and unscientific explanation, while the students' certainty of their answers into sure (if >3) and not sure (if < 3). All student's answer categorized into 4 conceptions categories from 8 possible response types, see Table 1. The student conception of SK category is the students' concept understanding. The difference score of students conception categorized M from pretest-posttest is misconception reduction.

Table 1. Conception Categories of Response Type [20]

Conception Categories	Response Types
Scientific knowledge (SK)	correct response + scientific explanation + sure
Lack of knowledge (LK)	correct response + scientific explanation + not sure incorrect response + scientific explanation + not sure correct response + unscientific explanation + not sure incorrect response + unscientific explanation + not sure
Error (E)	incorrect response + scientific explanation + sure
Misconception (M)	correct response + unscientific explanation + sure incorrect response + unscientific explanation + sure

3.4 Data Analysis

The data analyzed by IBM SPSS 24 software. Statistical analysis of the study was completed in four phase, (1) frequencies and mean value, (2) Kolmogorov-Smirnova and Shapiro-Wilk normality distribution of data analysis, (3) Levene test of homogeneity of variance, and (4) MANCOVA analysis by controlling the students' prior knowledge.

4 Results and Discussion

Table 2. Percentage of student's conceptions

Concept	Conception (%)			
	SK	M	E	LK
current	36.4	43.2	5.0	15.4
resistance	34.2	43.9	2.0	19.9
parallel and series circuit	22.3	46.6	5.1	26.0
electromotive force & voltage difference	19.1	35.3	0.0	45.6
energy and electric power	16.7	50.0	0.2	33.1
electric measuring instrument.	13.6	44.9	7.4	34.2
Average	23.7	44.0	3.3	29.0

The student's conceptions in dynamic electrical concepts are presented in Table 2. This shows that most of the students' concepts were misconceptions.

4.1 Effect of Text Structure on Understanding and Misconception Reduction

The MANCOVA analysis for text structure variable yielded the level of significance is below 0.05 for Pillai's Trace, Wilks' Lambda, Hotelling's Trace and Roy's Largest Root. Thus, it can be concluded that there was a difference in concept understanding and misconception reduction between the group of students who learned with CCT and the group who learned with ET.

Descriptively, the group who learned with CCT ($\bar{x}$ =9.8) had a higher average of concept understanding than ET structure group ($\bar{x}$ =8.7). Students' misconception reduction of CCT group ($\bar{x}$ =6.3) also had a higher average score than ET structure group ($\bar{x}$ =3.7). From the result, it is indicated that CCT structure is superior in improving concept understanding and misconception reduction. CCT is more effective than conventional text in restructuring the students' schemata that undergo misconceptions [5, 9, 13]. The effectiveness of CCT because of its advantages that gives the chance to the students to construct concepts through cognitive conflicts which occur in learning.

Table 3. Tests of Between-Subjects Effects of MANCOVA

Source	Dependent Variable	Sum of Squares	df	Mean Square	F	p
Corrected	CU	44.654	2	22.327	7.333	.001
Model	MR	156.898	2	78.449	9.387	.000
Intercept	CU	888.100	1	888.100	291.673	.000
	MR	383.346	1	383.346	45.868	.000
Prior	CU	16.876	1	16.876	5.542	.021
Knowledge	MR	2.187	1	2.187	.262	.610
Treatment	CU	29.900	1	29.900	9.820	.002
	MR	152.509	1	152.509	18.248	.000
Error	CU	264.902	87	3.045		
	MR	727.102	87	8.357		
Total	CU	7964.000	90			
	MR	3134.000	90			
Corrected	CU	309.556	89			
Total	MR	884.000	89			

*CU = Concept Understanding; MR =Misconception Reduction

Table 4. Mean difference score for concept understanding and misconception reduction

Dependent Variable	(I) Text Structure	(J) Text Structure	Mean Difference (I-J)	Std. Error	p
Concept	CCT	ET	1.154	.368	.002
Understanding	ET	CCT	-1.154	.368	.002
Misconception	CCT	ET	2.607	.610	.000
Reduction	ET	CCT	-2.607	.610	.000

4.2 Student Concepts Understanding

Based on the result of the testing of the hypothesis in table 4 it was apparent that there was a difference in concept understanding between the group of students who learned with the CCT structure and the group of students who learned with ET structure. The result of the hypothesis testing showed the value of $F_{obs.} = 7.333$ ($p < 0.05$). Thus, it can be concluded that there was a difference in concept understanding of dynamic electricity between the students who learned with CCT structure and those who learned with ET structure.

A further testing is done by LSD method. The statistical score for the boundary of LSD rejection of the students' concept understanding is obtained 0.839. The difference of the mean scores was $\Delta\mu_{(CCT\text{-}ET)} = 1{,}154$ ($p<0{,}05$). It shows that $\Delta\mu >$ LSD and the significance level is below 0.05. Thus, the mean score of the group of students who learned with CCT differs significantly from that of the group of students who learned with the ET structure.

4.3 Students' Misconception Reduction

Based on Table 3, the result of the study showed that there was a difference in students' misconception reduction between the group of students who learned with CCT structure and the group of students who learned with ET structure. The result is shown with the value of $F_{obs.}$ = 9.387 ($p<0.05$). From a further test of the data of students' misconception reduction with $t_{c.v.}$ = 2.281 and $MS\varepsilon$ = 8.357 for the dependent variable of students' misconception reduction, the boundary of LSD rejection was obtained 1.390.

Based on the data in Table 4, it is apparent that the difference in the mean scores of the group who learned with CCT structure and the group who learned with ET structure was $\Delta\mu_{(CCT-ET)}$= 2.607 ($p<0.05$). The result shows $\Delta\mu$ > LSD and the level of significance is below 0.05. Thus, the mean score of students' misconception reduction of the CCT group of students differs significantly from that of the ET group of students.

5 Conclusion

Based on the results obtained in this study and the discussion, some conclusions can be drawn as follows (1) There was a difference in concept understanding and misconception reduction in the topic of Dynamic Electricity between the group of students who learned with conceptual change text (CCT) structure and the group of students who learned with conventional text structure (ET). (2) There is a difference in the students' concept understanding between the group of students who learned with CCT text structure and those who learned with ET text structure. CCT structure can improve the students' concept understanding better than ET text structure after controlling the students' prior knowledge. (3) There was a difference in students' misconception reduction between the group of students who learned with CCT structure and those who learned with ET structure. CCT structure can increase the students' misconception reduction better than ET structure by controlling students' prior knowledge.

In the light of the findings of the study, there are several suggestions. First, the teacher has to be able to probe the students' conceptions. Second, the school has to provide books with conceptual change text structure. Thirdly, the teacher has to be able to arouse the students' interest to read so that the texts can be used optimally.

References

1. Ismail, A. Samsudin, E. Suhendi, I. Kaniawati. *Prosiding Simposium Nasional Inovasi dan Pembelajaran Sains*, 381 (2015).
2. K. Suma. *Prosiding Seminar Nasional MIPA* (2015).
3. D. K. Gürel, A. Eryilmaz. *Hacettepe Üniverst. Eğitim Fakült. Dergisi*, **2**, 235 (2013).
4. B. J. Guzzetti. *Reading & Writing Quarterly*, **16**, 89 (2000).
5. M. Akpinar, M. Tan. *Western Anatolia J. Edu. Sci.*, 139 (2011).
6. C. D. Tippett. *Inter. J. Sci. Math. E.*, **8**, 951 (2010).
7. G. J. Posner, K. A. Strike, P. W. Hewson, W. A. Gertzog. *Sci. E.*, **66**, 211 (1982).
8. N. W. Adnyani, I W. Sadia, I N. Natajaya. e-Journal Program Pascasarjana Universitas Pendidikan Ganesha, 3 (2013)
9. S. Aydin. *J. Edu. Res. Behavioral Sci.*, **1**, 1 (2012).
10. J. R. Cordova, G. M. Sinarta, S.H. Jones, G. Taasoobshirazi, D. Lombardi. *Contemporary Edu. Psych.*, **39**, 164 (2012).
11. F. Calisir, M. Eryazici, M. R. Lehto. *Compt. in Human Behavior*, 24, 439 (2008).
12. N. Ultay, U. G. Durukan, E. Ultay. *Chem. E. Rsrch. and Pract.*, **16**, 22 (2014).
13. G. Ozkan, G. S. Selcuk. *J.of E. Res.*, **3**, 981 (2015).
14. C. Robson, K. McCartan. *Real world research*, (2016).
15. S. K. Chamber, T. Andre. *J. Res. Sci. Teach.*, **34**, 107 (1995).
16. S. Sencar, A. Eryilmaz. *J. Res. Sci. Teach.*, **41**, 603 (2004).

17. P. V Engelhardt, R. J. Beichner. *Am. J. Phys.*, **72**, 98 (2004).
18. H. Peşman, A. Eryılmaz. *J. Edu. Res.,* **103**, 208 (2010).
19. E. Taslidere. *Creative Edu.*, **4**, 273 (2013).
20. D. Kaltakçi, N. Didiş. *AIP Conference Proceedings*, **899,** 499 (2007).

The Use of Mobile Phone in English Language Lesson: A Shift from Teacher-Centred to Student-Centred through Mobile Learning

Mozes Kurniawan[1,*]
[1] Universitas Kristen Satya Wacana, Faculty of Teacher Training and Education, Indonesia

Abstract. Most area of life requires changes. One of the things changes from time to time is learning method or strategy. In educational sector, especially in Teaching English to Speaker of Other Language (TESOL), teachers are demanded to have creative and innovative ways to teach language other than their own mother tongue in order to maintain pupils' interest, creativity and learning outcomes. One alternative way to teach English is by integrating technology particularly mobile phone. However, there still questions on what is going on if teachers allow student to access mobile phone during class and how it can be done. This study aims to enrich teachers' knowledge and broaden the paradigm of using mobile learning method serving the digital native generation in TESOL. A reflective analysis was done in collaboration with literature support to provide a clear image of mobile learning done in English class at a particular study program of Universitas Kristen Satya Wacana, Indonesia. The result showed that students' need a transformation of teaching learning process, students can be engaged more actively on learning by using mobile phone and some possible activities are proposed for supporting student-centred instruction.

1 Introduction

Many people considered as modern citizen in this global era are agree with statement mentioning that English is an international language and is important to be mastered [1]. One of countries who believes in that statement is Indonesia. Recently, Indonesia have been a part of ASEAN Economic Community meaning that its people should have an additional competence called English language proficiency for English is a global language used by many countries in the world and one of requirements for people to compete [2].

The importance of English in this modern era seems to encounter problems. Some said that the problems come from the learning materials and subject knowledge (curriculum) while some others said it was because of teacher's lack of instructional approach, strategy and technique (pedagogy) [3]. Perceiving those issues, practitioners as well as common people might have various perspective and reactions. However, one of perspectives which

* Corresponding author: mailbox.mozeskurniawan@gmail.com

is better to describe is by using analogy of Walt Disney's story in 'Alice through The Looking Glass'. The main character, Alice, was demanded to borrow a tool which could be used to travel back time and save her friend named Hatter, a hat maker. Due to her kindness and love to Hatter, Alice went to a castle owned by Time, the personification of time itself and begged to borrow the tool. In a brief, she took the tool without permission and was commanded to put it back to its place since she cannot change the past. Time who taught her strictly without any gentleness and innovation did not attract her attention. Alice travelled back time and changed nothing but she learned something valuable from the trip. Finally, she realised that what Time said was right and she got valuable things to learn by experiencing rather than listening to Time's prohibition.

This story represents educational problems happening in recent years, i.e. pupils tent to ignore classroom lesson, teacher's lecture and instruction by arguing that they get bored by the materials given and the way teacher conduct the lesson. Pupils cannot develop their English proficiency since the curriculum and pedagogy match nothing to their needs and interest even it makes them frustrated. The old-fashioned English teaching where teacher standing in front of class, lecturing and giving instruction resists their motivation to learn [4]. Furthermore, improper material design in teaching English contributes to the problems. If there is no way out for the problems, there will be a stagnancy of Indonesian people's English proficiency and Indonesia will suffer a big lost in this global era.

On regard to the problems, there appears an idea of using mobile phone in English language learning, a.k.a. mobile learning. This method has caught researchers' attention for years and become a fresh alternative to conquer problems particularly in English language education although practitioners still have questions on how the implementation will be. Therefore, this study aims to enrich teachers' knowledge and broaden the paradigm by presenting perspectives, real practice and reflective analysis on the use of mobile learning method serving the digital native generation in Teaching English to Speaker of Other Language (TESOL).

2 Literature Review

2.1 Perspectives on English Language in Indonesia

English as mentioned previously is perceived as a widespread language used by people all over the world. Due to its importance, many people among countries urge to learn and master English in order to get positive chances. Since English can be a good medium to communicate, teach subject knowledge in higher education and enhance social status, many countries have tried to develop their curriculum by adding English as one required subject or just a compulsory one.

Indonesia did similar action responding to the popularity of English in this modern era. English was considered as the most important and significant foreign language in Indonesia around 1980s and, starting from that time, English grew quickly from 1990s ahead. It was also determined as one of subjects taught in formal schools based on the national curriculum of Indonesia [5].

For Indonesia, English is a foreign language which enables its speakers to shape their identity to be a global citizen or join international community. In a simpler way, someone is considered as a part of Indonesia when the one is able to communicate in Indonesian language and internalize it as a part of his/her daily life. The same concept applies for English language. Indonesian people are recognised as global citizen, for example in ASEAN Economic Community, when they are able to communicate and interact to people in the same community. It will be possible if Indonesian people have proficiency in using

English. Thus, language, especially English, is not only functioned as medium of communication but also an identity shaper to have broaden connection and chances [6].

By considering those ideas that English, nowadays, becomes crucial for Indonesian people in order to compete and take benefits from global opportunities, people are looking for effective English education since no language is sufficient to aid communication and open the opportunities among countries [7]. Even though English had been introduced in educational setting for decades ago and believed as important language to master, its implementation in Indonesia is still out of expectation. Researchers found that English in Indonesia is still in a trouble such as teachers' way in delivering suitable English material, activities and assessment for Indonesian context need to be improved. Teacher are required to develop their basic English knowledge for being effective English as a Foreign Language (EFL) teachers. In other words, teachers should develop and/or innovate effective and even creative method so that pupils are motivated to learn English full-heartedly [8].

Besides, the matter of large number of pupils in class that need to be managed and put attention on, conventional way of teaching where teacher is standing in front of class lecturing and less availability of instructional media also become point to stress [9]. Those problems lead to an unconscious teaching learning process called by Teacher-Centred Learning where teacher is the main focus while students are less dominant and tent to be passive rather than Student-Centred Learning where students are the focus of learning which are permitted to design, innovate and develop an attractive leaning atmosphere.

2.2 Teacher-Centred vs. Student-Centred Learning

Teacher-Centred Learning (TCL) is an instructional approach which puts teacher at the centre of learning process such as: teacher leads the learning process, prepare for activities, becomes source of knowledge and make the determination in classroom activities [10]. TCL is believed to make pupils more passive for just listening and writing teacher's speech and instruction. TCL has some characteristics as follows: 1) Teacher is the sole of direction and control, 2) Teacher makes the rule to be followed, 3) Tasks and responsibility owned by teacher [11], 4) Teacher passes the knowledge and information to students, and 5) Teaching and assessment are two separate entities done by scored test system [12]. This approach is perceived to be problematic particularly in EFL since pupils' innovation and interest are hindered.

The other strategy which have been caught as a flexible and fresh to meet pupils' needs and wants is Student-Centred Learning (SCL). SCL is clearly defined as instructional approach focused more on shifting students' passive role to active role in discussion, curiosity and participation. SCL places students in the centre of teaching learning process while developing personal learning skill, knowledge, critical thinking and reflective thinking [13]. SCL characteristics are including 1) The responsibility of learning is delegated to students, 2) Topics are constructed relevantly for learners, 3) Learners are encouraged to engage to real learning experience, 4) The collaboration between learners is enhanced, and 5) There is a shift from teacher's function as learning resource [14].

Recently, there are a great change from TCL to SCL since the characteristics are best suit to English language lesson from preparing appropriate materials, conducting active learning and assess students' development through their performance. Nonetheless, teachers still get confused and trapped on how SCL can be conducted in innovative ways. Researchers proposed an alternative adopting SCL approach through familiar technology, mobile phone. Later on, mobile phone-based learning process is called by mobile learning.

2.3 Mobile Learning in Education

Mobile learning (M-Learning) defines as a process to get information and knowledge through interactive learning environment with multiple contexts using a portable technology called by mobile phone. M-Learning serves learners by using personal gadgets namely smartphone, personal digital assistant (PDA), tablet and similar technologies [15].

M-Learning is characterized by having accessibility where students are free and easy to access learning material; immediacy where students do not need to wait for long just to have/generate learning materials and tasks; interactivity that encourage students to engage in discussion based on the lesson [16], flexibility that gives students wider option of technologies to be used in accessing learning material; and mobility where students are permitted to learn anytime and anywhere.

English materials that previously perceived as something difficult to learn especially those whose first language is not English might be reduced and even solved by integrating technology through M-Learning. Various type of technology was believed changing teacher's way of teaching. It also reported as a successful way to help students develop their English proficiency from vocabulary mastery, grammar acquisition and some subject-specific knowledge comprehension [17].

However, teachers still question and demand about detail exploration and explanation on how mobile phone or similar technologies are used in English language lesson and what kind of activities are possible to do with such technology. The following discussion reveals transformation in English curriculum and pedagogy, mobile learning for engaging students' participation and some possible mobile-based activities.

3 Methods

This was a qualitative study aiming to enrich teachers' knowledge and broaden the paradigm of using mobile learning method serving the digital native generation in TESOL. This qualitative study used reflective analysis in collaboration with literature review so that the substance of this research was presented empirically based on real practices. The main data were 6 international journals discussing mobile learning in English education from 2012 up to 2017. The data complementary was gained from reflection of real experience on English language education course conducted in Faculty of Teacher Training and Education of Universitas Kristen Satya Wacana, Indonesia.

4 Discussion

4.1 Transforming English Curriculum and Pedagogy through M-Learning

Based on 6 international journals analysing about the use of mobile learning in English language education, there were changes of former instructional material and method. Researchers found that English required its speaker to learn large amount of vocabulary, pupils did not only learn new words but also the relation among words, and there were various learning style in learning English that should be accommodated wisely. Those challenges brought pupils to difficulty and anxiety [18]. The action showed in journal 2017 (Indonesia) was changing students' vocabulary learning through mobile application and content while pupils build their learning style that fits to their needs.

Another point of view from journal 2015 (Ethiopia) revealed that English education might encounter classroom limitation in conducting learning activities and improper context while demanding students to demonstrate language features. Those challenge let to the shift of conventional teaching learning process to mobile-assisted English learning. The result was that utilizing mobile devices could promote interaction during learning process and situate learning based on intended context and purpose. It also brought advantages such as: portability and connectivity where students can access the learning material inside and outside classroom.

In addition, journal 2016 (Saudi Arabia) and 2012 (USA) found English language learning problem such as: English taught in traditional-based classroom (face-to-face) lessen students' vocabulary retaintion and engagement. Students also tent to avoid physical labour namely visiting library and borrowing books [19]. By integrating iPad in classroom task and submission of works, students were attracted and committed to engage and collaborate in a boundless access. Moreover, the use of M-Learning improved students' English writing skill, new words and spelling knowledge and helped them translating words as well as saving learning resource immediately.

Furthermore, journal 2013 (USA) and 2014 (Hong Kong) mentioned that there were activities done by using small mobile device such as: communicating to family, shopping, mapping, finding a good living but less of them used it to improve learning in their education [20]. There were also teachers who resist to change their teaching and learning process with new technology since it was not perceived as part of learning culture. To have a better change, the researchers implemented mobile technology in English education and found some eye-catching changes. There were a large number of positive views toward the use of M-Learning particularly that it increased access to resources, improved communication with teachers and classmates, and was convenient in using time effectively.

4.2 Some Possible Mobile-Based Activities in English Lesson

At the beginning, there was question asking on how to use mobile devices or similar technology in English language lesson. Based on the analysis of 6 international journals, there were some suggestion in applying mobile learning in English lesson.

4.2.1 Instant Messaging Application for Vocabulary and Grammar Training

Vocabulary and grammar were developed by using instant messaging such as: BlackBerry Messenger, WhatsApp, LINE and others. The mechanism was just preparing topic and/or instruction that students need to do then chose one media as the medium. Students did discussion, works and/or tasks through the chosen media and built their habit of using English by such instant messaging. The result showed that students learn more new English vocabulary and simple pattern. The point is habituating pupils in using English as frequently as possible. Pupils have gradually been engaged in English lesson by using instant messaginng application more frequently.

4.2.2 Mobile Device for Virtual Sharing and Engagement

As previously stated about connectivity, interactivity and immediate access, teachers might use mobile learning through its virtual access namely: E-mail, Dropbox, Google Drive, e-Portfolio and so on to enlarge students' access to resources and teacher-students' distant interaction. The researcher used this way to provide English instructional module prior to course schedule in order to be learnt. In class, where students have accessed given module, discussion seemed to be more alive without pause and repetition on topic discussed. The flow of discussion and brainstorming went smoothly. Students admitted that file sharing and teacher-students' contact prior to course schedule helped them a lot. It built engagement between teacher and students as well as the material prior to class meeting.

4.2.3 Video Application for Practicing Speaking & Communication

The most interesting part of mobile device was multimedia feature. These included YouTube, Recorder, audio video application such as: HOOQ, Netflix and other similar tools. Naturally, students frequently accessed those tools for their own sake and got something to learn from what they did in pleasure. In researcher's English course, students were demanded to watch instructional videos days before the lesson. They watched how to teach English to children, how to pronounce a string of words in a certain topic and how to deal with interlocutors. Those videos might be a model for them to imitate and inspire a new insight. Most of them learnt to speak well by following the ideal flow of speaking, pronounce difficult words correctly and practice communication with classmates.

5 Conclusion

To sum up the result of this reflective analysis, there are three things highlighted. First, researcher and English language practitioners should be aware that technology in this case mobile devices are very close to students' life and can be utilize to enhance English language learning effectively. Then, to provide students with an interesting, innovative and engaging English language learning, teachers need to consider transformation of their learning method by integrating mobile technology in English lesson since it benefits students as well as teachers to have a meaningful teaching learning process. The last but not least, there are some possible English language tasks and activities which utilize mobile devices to have a fresh and creative learning atmosphere resulting to the improvement of English proficiency namely vocabulary, grammar, speaking and development of instructional management such as file sharing, resources accessibility and connectivity.

References

1. M. Lak, H. Soleimani, F. Parvaneh. J. Adv. in ELT. **5**, 1 (2017)
2. F. Copland, S. Garton, A. Burns. TESOL Q. **48**, 738 (2013)
3. P. Goss, J. Sonnemann. G. I. R. **01**, 6 (2017)
4. M. Kurniawan. Prodi PBI FBS UKSW. 24 (2013)
5. S.S. Mappiasse, A.J.B. Sihes. ELT. **7**, 113 (2014)
6. S.S. Sukandi. *ICLLCE*. 24 (2015)
7. B.P. Sinha. TC: Int'l. J. in En. **2**, 1 (2011)
8. H. Wati. IJI. **4**, 79 (2011)
9. Abid, P. Mercieca, T. Dobinson. TACLL. (2015)
10. O.A.S. Idris. AJER. **4**, 1317 (2016)
11. T. Garret. JCI. **43**, 34 (2008)
12. S. Rafi. M.Phill. Edu. (2015)
13. B. McCombs, J.S. Whistler. *The Learner-Centered Classroom and School: Strategies for Increasing Student Motivation and Achievement*. J-B Publish. CA. (1997)
14. S. Hodge. Occa. Paper on Learn. & Teach UniSA. **3**, 1 (2010)
15. M. M. Bachore. J. Edu. Pratc. **6**, 50 (2015)
16. J.P. Rossing, W.M. Miller, A.K. Cecil, S.E. Stamper. JSTL. **12**, 1 (2012)
17. H. Cahyani, B.Y. Cahyono. TEFLIN J. **23**, 130 (2012)

18. M. Kurniawan. ASIAN EFL J. **8**, 155 (2017)
19. G.M. Nalliveettil, T.H.K Alenazi. JLTR, **7**, 264 (2016)
20. D. Kin, D. Rueckert, D.J. Kim, D. Seo. J. Lang. Learn. Tech., **17**, 52 (2013)

Development of e-learning for English class

Noni Agustina[1,*], and *Nugroho Budhisantosa*[2]
[1]Primary Teacher Education Department, Education and Teacher Training Faculty, Esa Unggul University, 11510, Arjuna Utara Street, Jakarta, Indonesia
[2] Information Systems Department, Computer Science Faculty, Esa Unggul University, 11510, Arjuna Utara Street, Jakarta, Indonesia

Abstract. In the twenty first century, teachers are required to have a digital literacy skill. They must be able to integrate technology in learning process. It was already conducted by a teacher in one of public junior high schools in Jakarta. She searched the materials from the internet but she had a problem to adjust the learning materials to her students' needs and characteristics. Therefore, this study was undertaken to explore deeply how to develop e-learning in English class based on her students' needs and characteristics. This study employed research and development methodology. The participants were an English teacher and the first graders of junior high school. The result showed that this e-learning used PHP, Framework Bookstrap and MySQL. The e-learning also used Adobe Flash and Adobe Illustrator to make some animated videos. It consisted of listening, reading, and grammar which the topic focused on descriptive texts comprising some themes referring to 2013 curriculum. Those themes were people, animal, thing, place, and occupation. The development of e-learning deserved to be implemented after it was validated by the expert, evaluated by the teacher and tested to the students. It is expected to contribute as supplemental learning media in English learning process.

Keywords: English skill, e-learning, research and development

1 Introduction

The proliferation of information communication and technology grows rapidly nowadays. That phenomenon can be seen from the high use of cell phone and the internet around the world. 3.15 billion people use the cell phone and 3 billion people use the internet [1, 2]. Internet is highly demand in our life [3]. The use of the internet in Indonesia increases 100 percent each year [4]. It offers all information needed around the world [5]. It affects the education world [3]. It influences the teaching and learning process in the class [6, 7, 8]. When the previous observation was conducted, one of teachers at public junior high school in West Jakarta, Indonesia used the laptop and projector as well the speaker

* Corresponding author: noni@esaunggul.ac.id

when teaching English in the class. She taught her students using video downloaded from the internet. Sometimes, she used the online learning such as Edmodo to teach her students. She was a portray of a teacher in the twenty first century that she had the digital literacy. The teacher needs the ability to use the information and communication technology in their classroom such as the internet use [9]. The Ministry of Education and Culture of Republic of Indonesia also determine the national literacy movements. One of them is the digital literacy. It means that Indonesian government has been supporting schools to implement the digital literacy. Nowadays, they applied computer-assisted test for the national test in some schools for some areas around Indonesia.

Based on the previous interview, even though the teacher integrated the technology in teaching English in the classroom, she had problem to adjust the material she found from the internet to meet her students' needs and characteristics. She also had difficulty to find resources such as listening. Although, information and technology centre of the Ministry of Education and Culture provides the e-learning in website (http://edukasi.kemdikbud.go.id/), it cannot be accessed because it always comes up *"file or directory not found'* command. The e-learning also has not been much more developed in Indonesia [10, 11]. To address those problems, this study is important to undertaken by developing English e-learning. There are some various terminologies used addressing the e-learning: web-based learning, online learning, computer-based learning or computer-based instruction. Morrison and Aparicio et al. call it online learning [12, 13]. Even though it has different names, it is used in teaching and learning process involving the technology such as computer network (internet/ intranet) or multimedia [13, 14, 15]. It has some functions: supplementary and substitution functions. Supplementary function means that it can be combined with the conventional learning which emphasizes on face-to-face learning in the classroom. It is used to enrich the learning process. On the other hand, substitution function means using e-learning fully in learning process [11]. A study showed that the e-learning and traditional education are still used together. They complete each other in order to reach the educational result [16].

Some studies found that the e-learning was more effective than traditional instruction [17]. It has some advantages [14]. It can support the learning process [3, 18, 19, 20]. It gains the academic achievement, motivation, performance and attitudes [14, 15 ,17]. It facilitates the teachers to provide learning materials and they ask their students to accomplish and submit the assignments or exercises. The learning process is also fun, interactive and effective by using it [19]. It provides the flexibility of time, place, learning speed, instructional standardization and effective instructional. The slow students can learn the material or tutorial provided in the e-learning. It can help them [4]. The students also can learn autonomously [14].

The previous studies were undertaken by some researchers in developing the e-learning. Yaniawati and Muis et al. conducted the e-learning for mathematic class. Munir dan Prabowo developed the e-learning using SMS Joomla for vocational high school students in Bantul, Indonesia which its content was about computer sciences. Turino et al. developed e-learning for teaching listening skill for senior high school (SMA 4) in Semarang, Indonesia. He designed it using Adobe Flash Professional CS3 in *swf* format. Setyowati focused on reading skill for senior high school students, thus she designed the e-learning for reading. Sutisna developed the e-learning for teaching grammar class at English Class Department, Pakuan University, Indonesia. The result showed that students' understanding on grammar and their motivation increased when they used e-learning and they could access it easily. It's a useful tool for the lecturer too when teaching grammar [21]. Siswono and Liona also developed e-learning which emphasized on phonology, syntax, semantic, and pragmatic. They provided stories with proper grammar and the result showed that the teacher was easier to teach grammar to the students and it helped students to understand the grammar well. Solak & Cakir developed the course of e-learning consisted of basic English skills which contain animation and interactive presentation for higher education level. The students also were able to discuss with their teachers in a synchronized way. Rymanova et al. developed e-course using Moodle as the Learning Management System which its contents were lexical, grammar, and speaking skill. From the studies, it could be concluded that the e-learning for English class concerning on some English skills. It is developed for senior high school level and mostly for higher education level [22]. In USA, almost 90 percent of universities use it [4] and in Korea 70 percent [23]. Thus, this study develops listening, reading, and grammar for English learning at junior high school level especially for first grade. The study aims to explore and describe deeply the development of the English the e-learning by applying a need analysis. It is required to design the instructional material for the e-learning [24].

2 Method

This study employed research and development methodology. Its aim was to yield a product [25]. It had many kinds of model used. Those models had different steps and processes [6]. This study applied model developed by Gall, Gall and Borg but did not fully complete all processes due to the time constraint. The process consisted of conducting need assessment, planning (identifying skills that will be learnt, generating the objective and learning activities, and having expert's validation), revising the products referring to the feedback given by the experts, conducting the field testing of the product to three students and teacher, and revising the product based on students and teachers' feedback.

This study was conducted from January to June 2017. The participants of this study were first graders and an English teacher of 89 Junior High School in West Jakarta, Indonesia. Interviewing the teacher was undertaken to obtain the data what she and her students needed. The document analysis (the syllabi) was also required to write the objective so it could be used a supporting data. The instruments used to validate media and content of e-learning were given to the experts. The questionnaire for teacher and students are also provided to get their feedback.

3 Result and Discussion

Based on the need analysis, students required to learn listening and reading skills which focused on descriptive texts. Those descriptive texts had some themes such as people, animal, place, thing, and occupation. The materials referred to the 2013 curriculum. Not only listening and reading skills, they were also necessary to learn grammar such as simple present tense, adjective, adverb, and preposition. After analysing the needs, the next steps were making the e-learning content, determining the name of e-learning website, making the logo, making the login for the students and teacher (administrator), developing English Gate by using PHP, Framework Bootstrap, and MySQL, making Entity Relationship Diagram (ERD), inputting the content into database system, hosting all database systems created in mini Raspberry Pi 3, sending English Gate as English e-learning to be validated, and testing English Gate to three students.

The content of e-learning consisted of listening, reading, and grammar. The listening contained the audio by recording an American native speaker's voice. It provided the audio with the animation, vocabulary (English-Indonesia), and multiple-choice questions. When students clicked the words in vocabulary, they had sound. The reading provided some texts with the sound, pictures, vocabulary, and short-answer questions. The students had to type the answer in the box provided. The grammar had the animated video which aimed to help the students to learn adjective, adverb, preposition and simple present tense. After they watched the animated video, they could play the game by accomplishing some exercises related to the grammar they learnt before. The exercises were in fill-in-the blank format. The students also could see their result and the answer key after they finished answering the exercises. The animations were created by using some software such as Adobe Flash and Adobe Illustrator.

English Gate was chosen as the name of the English e-learning. The logo was created using Adobe Illustrator. It provided login for the students and teacher (administrator). Students' login could be used to learn the material, do the exercises and see their scores. Administrator's login could be used to see the students' score, assess her/his students' score, register students' data, edit and delete the questions and answers. It used PHP as a back-end, Framework Bootstrap as front-end and MySQL as a database system. Entity Relationship Diagram (ERD) was used to describe the logical structure of database.

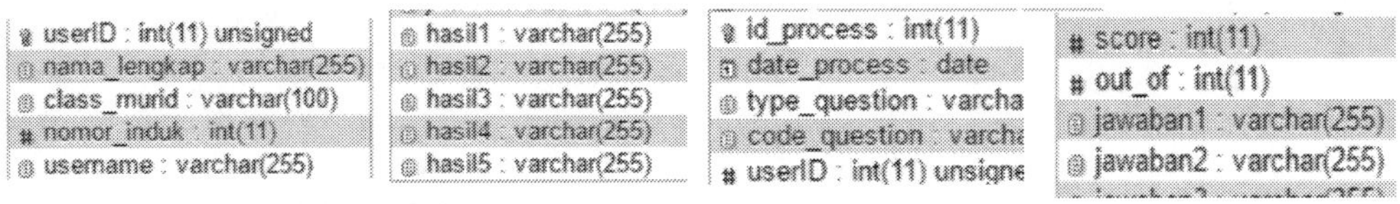

Fig. 1. Example of ERD

The contents were inputted into the database system. They were the homepage of English Gate consisting of home, login, about us, listening, reading, and grammar; username and password of administrator for login in the administrator dashboard; the questions by clicking 'manage question' in the administrator dashboard. All those data base systems created were hosted in mini Raspberry Pi 3 that functioned as web server. Thus, the IP address of the web server was 192.168.1.200.

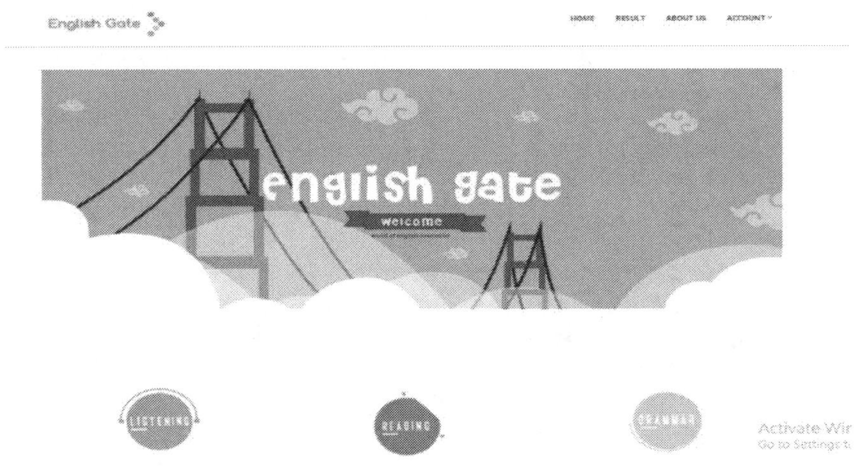

Fig. 2. English Gate homepage

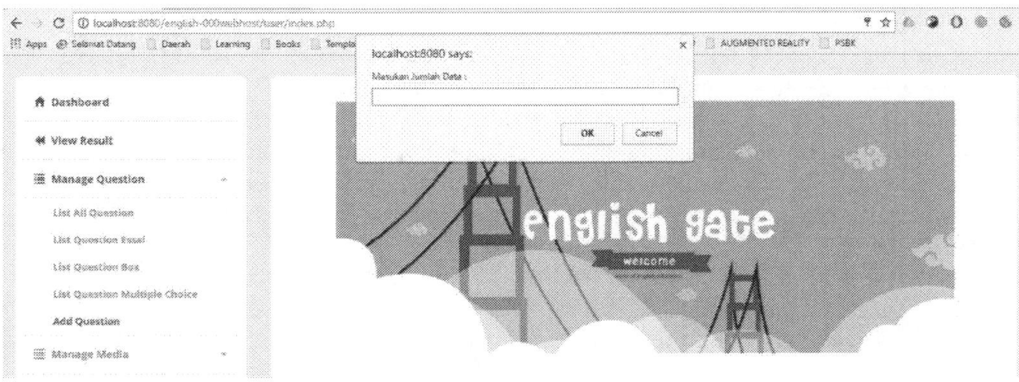

Fig. 3. Administrator dashboard

English Gate was sent to two experts and teacher to validate its media and content. The result showed that the content was good and the content expert and the teacher did not give any suggestion about the content. On the other hand, the media experts gave some feedback. The feedback was the homepage display should be added and provided the manual how to use English Gate. Referring to those suggestion, it was revised and sent back to be validated again. The second validation showed the e-learning deserved to be applied and tested to some students.

These results of this study were not fully similar to the previous study. The development of the e-learning (English gate) consisted of login, homepage, interactive learning material (listening, reading and grammar) consisting of animated video, and quizzes with the feedback. The previous study found that the components of the e-learning contents were varied. This result study showed that Morrison developed it consisting of register, login, personal profile, material consisting of pre-assessments, texts, pictures, simulations, interactive exercises, downloaded materials, quizzes, bookmarks, post-assessments, and feedback[19]. It's a little bit different from Morrison, Munir and Prabowo developed their e-learning content comprising curriculum, syllabus, materials and tests. Turino, et al. also provided login, homepage, syllabi, evaluation, and online consultation and chatting forum. Login was for the administrator, students, and teachers. The homepage consisted of main menu, register, and learning menu which offered downloaded material for students and teachers. The materials were in audio, Microsoft Word, Microsoft PowerPoint, PDF, and animation format. The listening materials developed used Adobe Flash Professional CS3. The syllabi were also provided, if the teachers wanted to delete or add them, they could do it. The evaluation comprised the questions and answers with various format such as multiple choice and true-false. The online consultation and chatting forum were designed to facilitate the students' interaction and participation. Almost similar to the previous researchers, Siswono dan Liona developed their e-learning consisting of login, learning plan, lesson, story, exercise, and dictionary. The learning plan provided a year learning planning. The lessons were the materials given to the students. The story feature consisted of animations with different themes. The dictionary was useful for the students to find the meaning of words [27]. Rymanova et al. developed the e-learning, namely e-course, which consisted of the general data (teachers' personal data), unit feature (instructional unit, materials such as vocabulary and grammar, glossary, activities and supplementary resources from internet, and assessment feature. The materials were video, animation, text or images. They also provided Forum, Chat, and Wiki which facilitated students to interact with their instructors or teachers. The tests of units had some variations such as multiple choice, matching, short answer and other forms [14].

English Gate was tested to three students. They were observed by recording them when they used it. The result showed that they could use it well. Then they were given a questionnaire sheet to know their response in using it. All of them urged that they were happy to use it and more understand to learn English. The vocabulary also helped them to understand the meaning of words. Some of them gave suggestion to improve the assessment of reading because they thought their answers were correct but they had unsatisfying scores. Therefore, the teacher had to rechecked their reading answers and scores. Some studies showed that the content of the e-learning was not absolutely same one another. The important thing was it gave a significant impact for the students. It could be accessed autonomously by the students, flexible use, effective and efficient. Siragusa (2000) revealed that the main components in online learning were content, interaction, feedback, and interface design. The content had to align to the purpose and students' ability. The interaction referred to the communication between the students and teachers or instructors. The feedback referred to the teachers' comment or suggestions given to the students both individually or in group. The interface referred to easy navigation that could help and respond the students. The use of image, colour, sound, animation could entice the students [17]

4 Conclusion

English Gate is a web-based learning that can be used to learn English especially listening, reading, and grammar which it is developed for first graders of junior high school. Its contents are login, homepage, interactive learning material (listening, reading and grammar) consisting of animated video, and quizzes with the feedback. It is also equipped with the vocabulary (English-Indonesia) as aid tool for the students. It is expected to give a contribution as a learning source for students to enhance their English ability. This study needs some further improvement in reading assessment because students' possibility answers are varied. Further research is also required to investigate the effectiveness of English Gate as e-learning if it is applied in some classes.

This study is supported by grant of Ministry of Research, Technology and Higher Education of the Republic of Indonesia.

References

1. A. Koller, G. Foster, M.Wright, *ACM International Conference Proceeding Series*, 131–138 (2008)
2. M. Kende, Internet Society, 146 (2014). http://www.internetsociety.org
3. E. Solak & R.Cakir, E-Learning Digital Media, **12,** 107–120 (2015)
4. E. Effendi and H. Zhuang, *E-learning Konsep dan Aplikasi*, 4 (2005)
5. D.H. Eldardiry and Z.A. Elmoghazy, Int J Technol Des Educ, 1-16 (2016)
6. H. Cai, *Procedia IERI*, 841-846 (2012)
7. J. Loizzo and P.A. Ertmer, Education Tech Research Dev, 1-20 (2016)
8. N.Tuncay & Uzunboylu, *Procedia - Soc. Behav. Sci,* **47,** 131–137 (2012)
9. H. L. Chou, & C. H. Chen, Comput. Educ, **103,** 124–133 (2016)
10. Setyowati, Lib UI, 1-8 (2009). http://lib.ui.ac.id/
11. Turino et al., Jurnal Teknologi Informasi, **5**, 725-728 (2009)

12. D. Morrison, *E-learning strategies. How to get implementation and delivery right first time*, 10-11(2003)
13. M. Aparicio *et al.*, Journal of Educational Technology & Society, **19,** 292–307 (2016)
14. I. Rymanova, N. Baryshnikov & A.Grishaeva, *Procedia - Soc. Behav. Sci,* **206,** 236–240 (2015)
15. N. Mohammadi, V. Ghorbani & F. Hamidi, *Procedia Comput. Sci.,* **3,** 464–468 (2011)
16. C. J. Kim & R. Santiago, *Educ. Technol. Res. Dev,* **53,** 108–115 (2005)
17. R. Khodabandelou, & S. A. A Samah, *Procedia - Soc. Behav. Sci.,* **67,** 545–552 (2012)
18. M.B. Triyono, *Procedia Social and Behavioral Science*, 55-61 (2014)
19. Muis et al., Pelaksanaan Pembelajaran Berbasis E-Learning Bidang Studi Matematika Pada SMP Negeri 1 Magelang, 1-17(2012). http://eprints.ums.ac.id/
20. M. Munir and K. Prabowo, Pengembang E-Learning Berbasis Multimedia Sebagai Multimedia Center Untuk Meningkatkan Kualitas Pembelajaran TKJ di SMK, 1-9 (2011). http://staff.uny.ac.id/
21. E. Sutisna, Model Pembelajaran Grammar Berbasis Website, 10, (2013). http://digilib.upi.edu/
22. L. Xiangqian, & G. Fuqing, *IERI Procedia* **2,** 854–858 (2012)
23. D.Josic *et al., Educ. Technol. Res. Dev,* **59,** 445–464 (2011)
24. H. Hussin, F. Bunyarit, R.Hussein, *Campus-Wide Inf. Syst.* **26,** 4–19 (2009)
25. Sugiyono, *Metode penelitian kuantitatif kualitatif dan R & D*, 297 (2012)
26. N. Putra, *Reseach and Development. Penelitian dan Pengembangan : Suatu Pengantar,* 119-121(2015)
27. Siswono and Y. Liona. *Jurnal ComTech,* **1**, 300-308 (2010)

Teacher's behaviour towards students' motivation practice

Sondang Pondan Perlindungan Leoanak[1,*], and *Bonik* Kurniati Amalo[1]
[1]Politeknik Pertanian Negeri Kupang –Nusa Tenggara Timur, Indonesia

Abstract. Education goal is not to produce higher exam scores, but to educate learners to become responsible people with well-developed minds and good character. Fact says, most of the students who graduated from Islamic elementary school in Kera island (a small island in the Gulf of Kupang city, East Nusa Tenggara Province, Indonesia) prefer to work as fishermen instead of continuing their study to higher education levels. This practice has been continued for several years. In relation to such fact, the present study aims to reveal the teacher-student interaction in connection to teacher's behaviour in students' motivation practice. For this purpose, a qualitative research design was applied to gain deeper understanding on such practice naturally. The data collection process was part of authors' Community Service Project (IbM RISTEKDIKTI for Bajo tribe) on the island. Data were collected using classroom observations and semi-structured interviews. This study reveals that the teacher performed positive roles in motivating the students during classroom interaction. Hopefully, the present study may help teachers and educators in Indonesia concerning teacher behaviours in elementary classroom environment.

KEYWORDS: *Teacher's behaviour, students' motivation.*

Introduction

It is truly believed that when students' needs are met, they will have a good engagement in classroom activities with their teachers and peers [1]. When teachers treat students with care and affection, students will have feeling of belonging to school. As a result, a connection between students and teachers would be well established and feeling safe at school would also be increased [2]. When such things occur, the success of teaching and learning process would be successfully achieved and the goal of education would be easily obtained.

In contrast, when poor relationships occur in classroom, not only hard on students but also as source of stress for teachers [3]. Students could have difficulties in learning lesson materials and teacher could have problem in delivering lesson materials. In addition,

* Corresponding author: sonleo2002@yahoo.com

when students face problems and difficulties, they would have neither the personal nor social resources to bounce back and are prone to utilize maladaptive coping strategies and to give up [4]. As a matter of fact, in general terms, these patterns are related to lower academic achievement, dropout, and other disadvantages for students [5]. Thus, it can be argued that, motivational and disciplinary problems with students are among the most upsetting issues teachers face [6].

Teacher and student positive relationship have effects on both side. This means that, when teachers have positive relationship with their students, this will raise better satisfaction with their job and with preventing of burnout [7]. On the other side, when students experience teachers' positive interpersonal behavior, this will lead to their positive motivation and performance in all subjects [8]. Thus, a positive connection is necessary needed between teacher-student relationship and learning activities in classroom. In addition, it is agreed that a healthy interpersonal relationship would create better results for students to engage in learning activities [9].

Many researchers believe that good and strong relationships between teachers and students are essential components to make healthy academic development of students at school environment [11, 12]. In addition, many studies have shown that the nature and quality of teachers' interactions with children have significant effects on their learning [13, 14]. Hence, teachers' engagement in classroom interaction with students is necessary needed to make learning process occur actively. It also provides a unique entry point for teachers to improve the social and learning environments of schools and classrooms [11].

Brackett *et al.* [15] used a multi-method, multi-level approach to examine the link between classroom emotional climate and student conduct, including as a mediator the role of teacher affiliation, i.e., students' perceptions of their relationships with their teachers. They found that there was a direct positive relationship between classroom emotional climate and conduct mediated by teacher. They also highlighted that teacher has the role of emotionally supportive classroom environments in promoting teacher affiliation and better conduct among students.

Downey [16] conducted a study about synthesizing educational research on factors that affect academic success. For him, students need teachers to build strong interpersonal relationships with them, focusing on strengths of the students while maintaining high and realistic expectations for success. Downey [16] added that the interactive relationships should be based on respect, trust, caring, and cohesiveness. Classroom interaction is the communication between teachers and learners in the classroom [17]. By means of this interaction, both teachers and learners receive input and produce output. Teachers have opportunity to know their students performance during the interaction and produce teaching practices that fit their needs. For students, interaction increases their knowledge and competence and produces better skill as the output.

Teachers' motivation in classroom interaction is in one important factor that educators can target in order to improve learning. Fact says that teachers play important role in students' motivation. Many scholars suggest that when teachers treat students with respect and seek out, listen to, and value their opinions, students are more willing to commit themselves to the hard work entailed in learning [18, 19]. If teachers can see classroom interaction as valuable resources, this will increase students' motivation and encourage them in experiencing the value of learning in the classroom. Hence, learning objectives are achieved and academic goals are obtained successfully.

When students are motivated, they pay attention, begin working on tasks immediately, ask questions and volunteer answers, and appear to be happy and eager [20]. Students need motivation as it is an essential element needed for thier education quality improvement [21]. Williams and Williams [21] propose five key ingredients impacting students' motivation, they are: student, teacher, content (should be relevant and useful), method/process (an approach in which content is presented in classroom instruction), and environment (must be available and accessible, safe, open and free to learn from mistakes).

Williams and Williams [21] say that, as one of the key ingredients impacting students' motivation, a teacher must possess some characteristics. The teacher must be well trained (knowledge, qualifications and skills), must focus and monitor the educational process (test administration, scientific management and human relations), be dedicated and responsive to his or her students (be conscious, reach out to students, and build relationship and strength) and be inspirational and enthusiasm. In addition, William and William [21] say that motivation is optimized when students are exposed to a large number of these motivating experiences and variables on a regular basis. That is, students ideally should have many sources of motivation in their learning experience in each class.

Efforts to promote students' motivation need to be done by teachers at classroom level and school administrators at school or district level. Brewster and Fager [22] suggest some strategies for both teachers and administrators seeking to increase students' motivation in classroom as follows: giving rewards, performing clear and consistent classroom interactions, knowing each student, giving positive responds, promoting mastery learning, breaking tasks into smaller goals and evaluating students' work and providing feedback.

To measure the perceptions of interpersonal teacher behavior, Wubbels *et al.* [23] developed an instrument called questionnaire on teacher interaction (QTI). The QTI was constructed to describe teacher interpersonal behavior along two dimensions: influence (the degree of control of the teacher over the communication process) and proximity (the degree of cooperation or opposition between the teacher and the students). These dimensions are then subdivided in terms of the eight scales or sectors as: leadership, helpful/friendly, understanding, student freedom, uncertain, dissatisfied, admonishing and strictness.

Hence, for this purpose, students need structured interactions, where standards and clear expectations are set for them, as to achieve positive behaviors and performance and consistently follow through on teachers' demands [3]. By having such optimal structured interaction, students are able to figure out how to interact and behave to reach better performances at school. Such interaction includes teachers' confidence in helping students to reach high levels of understanding and performance [10]. At school environment, teachers can explain to students regarding ways to improve their performances by giving feedback and motivation. These will increase students' perceptions of themselves as academically competent, including their beliefs about what it takes to do well in school and their own capacities to enact those strategies [3]. Based on these concepts, this paper aims to find out how teacher-students interaction occurred in an elementary school classroom situation. In a narrow scope, this paper investigates the practice on teacher's behaviour in motivating students during teaching and learning process in Rijalul Ghad Islamic elementary school in Kera island, Kupang regency, East Nusa Tenggara Province, Indonesia. The results of the present study may help teachers and educators in Indonesia concerning teacher behaviours in elementary classroom towards students' motivation practice.

Research method

The present study used a qualitative research design. In addition to find out how teacher-student interaction in connection to teacher's behavior on student motivation practice, some methods of data obtain were also applied. The data was obtained by classroom observation, questionnaire and semi-structured interview. The concepts of William and William [21] key ingredients impacting students' motivation, Brewster and Fager [22] strategies to increase students' motivation in classroom and Wubbels *et al.* [23] QTI were also applied as research framework for the present study.

In classroom observation stage, the writers observed how the interactions between teacher and students occurred in the classroom. The teacher's behaviors and strategies to motivate the students during the interaction were noted down and analyzed. After that, the writers used the notes to carry out semi-structure interview to get a clear picture about such interaction happened.

During semi-structure interview, a set of question was used as guidelines. These questions were administrated based on Williams and Williams [21], Brewster and Fager [22] and Wubbels *et al.* [23] classroom motivation strategies. The question was a self-reporting questionnaire, design to assess teacher's behavior and interaction (perceptions and responses) in the classroom. The questionnaire was divided into two main parts. Part one was about respondents' personal information and part two was about the application of teacher's motivation practice during the interaction. The question in part two consisted of 20 items of a five-point Likert scale from "Never "to "Always". In responding to this questionnaire, the respondents were also asked to give additional information with example regarding the implication of the practice. The participants in this study were 15 students and a teacher in Rijalul Ghad Islamic elementary school.

Results and discussion

Generally, this paper talked about teacher's behavior in students' motivation practice during classroom interaction. Specifically, this paper was focused on teacher's leadership practice in teaching technique and administrating questions. Teacher's feedback were measured and analyzed using classroom observation and a questionnaire. In addition about the school, it was found that, this school only has one teacher and he teaches in all grades.

During the classroom observation, it was found that the teacher demonstrated positive practice teaching behavior in motivating the students during classroom interaction. However, it was also found that, the teacher talked too much and less interaction among the students. In general it was seen that the teaching technique applied in the classroom was teacher-centered model, and most of the classroom activities were done by the teacher. It was teacher's dominant classroom typed. In some instances, there were some shared activities between the teacher and the students such as when giving demonstration and examples. In contrast, when explaining lesson materials, most of the activities were done by the teacher.

In relation to questioning procedure, most of the classroom interaction was devoted to ask questions. The teacher asked questions about lesson content as meant to seek students' answer to promote classroom interaction. It was also found that, when students could not answer teacher's question, it was the teacher who gave the right answer. The teacher gave the correct answer and the students wrote down the answer. Examples of question applied were: requesting factual information, yes/no questions, short answer (one or two words) questions.

In relation to the use of teaching aids in the classroom, it was found that, the teacher well applied teaching aids during his classroom interaction. For example, he used his family pictures when explaining about family relationships topic. The use of teaching aids may increase interaction in the classroom, especially when the teaching aids are related to students' real life. Making the content relevant to real life can increase a student's motivation [21]. In addition, Palmer [20] suggests that, teachers should emphasize the links between real life and school subject that is by designing assignments and experiments which have connection with students' everyday life. In addition, Frey and Fisher [24] add that tasks which have meaningful relationship with students' real life can motivate them.

The practice of giving reward to students was carried out very well by the teacher. When the students accomplished their assignment and gave correct response to his questions, the teacher used positive verbal statements to praise students as encouragement. Positive statements can strongly influence student motivation and increase students' self-confidence [21].

Conclusion

When students' needs are met, they will have a good engagement in classroom. It is also believed that when students are treated with affection, it will increase their feeling of belonging to school. However, when poor relationships happen in the classroom, students could have difficulties in learning the lesson materials and teacher could have problem in delivering and motivating the students. Hence, connections between teacher and students need to be established in the form of structured interactions. The present study revealed that teacher's behaviour contributing students' positive motivation practice in the classroom. The teacher's behaviours applied are: questioning the students, giving examples, using teaching aids and giving rewards by using positive verbal.

Acknowledgement

We would like to express our gratitude to the Ministry of Research, Technology and Higher Education (RISTEKDIKTI) for the support via Community Service Project (IbM) 2017, Center of Research and Society service (P3M) of POLITANI Kupang and to all the students, teacher, and the community in Kera island for their support and contribution. Thank you for all the valuable supports.

References

1. E.A. Skinner, C.J. Furrer, G.C. Marchand, T.A. Kindermann. J. Educational Psychology, **100**, 765–781 (2008)
2. A.J. Martin, M. Dowson. Rev. of Ed. Research, **79**, 327–365 (2009)
3. J.A. Furrer, E.A Skinner, J.R. Pitzer. National Society for the Study of Education, **113, 1,** 101-123 (2014)
4. E.A Skinner, J.R. Pitzer. Developmental dynamics of engagement, coping, and everyday resilience. In S. Christenson, A. Reschly, & C. Wylie (Eds.), *Handbook of research on student engagement,* 21–44 (2012)
5. M. Janosz, I. Archambault, J. Morizot, L.S. Pagani. J. Social Issues, **64**, 21–40 (2008)
6. M.L. Chang. Ed. Psych Rev. Educational, **21**, 193–218 (2009)
7. D. Ben-Chaim, U. Zoller. Research in Sci. Ed, **31**, 437–454 (2001)
8. P. Den Brok, M. Brekelmans, T. Wubbels. School Effectiveness and School Improvement, **15**, 407–442 (2004).
9. M. Brekelmans, P. Sleegers, B.J. Fraser. Teaching for active learning. In P.R.J. Simons, J.L. van der Linden, T. Duffy (Eds.), *New Learning.* 227–242 (2000)
10. D. Stipek. *Motivation to Learn: Integrating Theory and Practice*. Boston, Massachusetts: Allyn and Bacon. (2002)
11. B.K. Hamre, R.C. Pianta. Child development, **72** (2), 625-638 (2001)
12. J.S. Eccles, A. Wigfield. Annual review psychology, **53**, 109-32. (2002)
13. D. Dickinson, J. Brady. Toward effective support for language and literacy through professional development. In M. Zaslow & L. Martinex-Beck (Eds.), *Critical issues in early childhood professional development*, 141-170 (2006)
14. R.C Pianta, W.S. Barnett, M. Burchinal, K.R Thornburg. Psychological science in the public interest, **10**(2), 49-88 (2009)

15. M.A Brackett, M.R. Reyes, S.E. Rivers, N.A. Elbertson, P. Salovey. J. of Classroom Interaction, **46**(1), 27-36 (2011)
16. J.A. Downey. Preventing school failure, **53**, 56-63 (2008)
17. H. Brown. *Teaching by principle an interactive approach to language pedagogy*, White Plains, NY: Pearson Education. (2001)
18. J. Reeve, H. Jang. J. of edu. psych, **98**, 209–218 (2006)
19. C.R. Stefanou, K.C Perencevich, M. Di Cintio, J.C. Turner. Edu psych, **39**, 97–110 (2004)
20. D. Palmer. The J. of the Australian sc. teachers association, **53**(1), 38-42 (2007)
21. K.C. Williams, C.C. Williams. Research in Higher Education Journal, **1**, 1-23 (2011)
22. C. Brewster, J. Fager. *Increasing student engagement and motivation: From time-on-task to homework.* Northwest Regional Educational Laboratory, (2000)
23. T. Wubbels, M. Brekelmans, P. den Brok, J. Tartwijk. An interpersonal perspective on classroom management in secondary classrooms in the Netherlands. In C. Evertson & C. Weinstein (Eds.), *Handbook of classroom management: research, practice, and contemporary issues*, 1161-1191 (2006)
24. N. Frey, D. Fisher. English Journal. **100**(1), 30-36 (2010)

The Effect of Teams Games Tournament on Mathematics Self-Efficacy in Junior High Schools

Pradipta Annurwanda
Pamane Talino Teaching College
West Kalimantan, Indonesia

Abstract. Teams Games Tournament is one cooperative learning method which actively involves students to solve their problems through an interesting game. The game consists of questions that have content relevant to the main topic and to boost up students' self-confidence in their ability to exert their self-control over motivation, behavior and social environment. This research aims at investigating the effect of Teams Games Tournament toward students' self-efficacy on mathematics. The study was conducted toward 64 seventh graders in Landak Regency on social arithmetic material, selected using cluster random sampling. The experimental design used the one group pretest posttest experimental design that was analyzed by quantitative method. Data collection employed "Mathematics Self Efficacy Questionnaire" and was analyzed by statistical method using SPSS-20. The results show that Teams Games Tournament has a significant effect toward students' self-efficacy on mathematics. The result is drawn from t-value = -12.369 and sig.(2-tailed) = 0.00. Therefore, it can be concluded that Teams Games Tournament has positive effect toward students' self-efficacy on mathematics. The study implies that teachers should consider the implementation of Teams Games Tournament in classroom teaching.

Keywords: Teams Games Tournament, Self-Efficacy, Mathematics.

1. Introduction

Currently, mathematics is widely used in many areas of life. Mathematics is also taught in every level of education starting from primary, middle and higher levels. Mathematics as a way of thinking, a language of symbols, terms, definitions and principles, does not develop naturally like a mother tongue, but must be learned, and demands cognitive and meta-cognitive effort [1].

However, the decline in mathematics learning achievement is of particular concern to students, teachers and parents. One reason for the decline in mathematics achievement of students in schools is the students' assumptions of mathematics which is a difficult and tedious lesson. This occurrence should be addressed immediately by the teacher. Teachers should pay attention to the needs of each student. In order to achieve success in learning mathematics, students must be given the opportunity to communicate mathematically, reasoning mathematically, develop self-confidence to solve mathematics problems. One of the ways this is through cooperative learning.

Further, cooperative learning helps students to foster their more complex thinking and problem-solving skills which are demonstrated by groups of students taught using cooperative learning [2, 3]. Cooperative learning is achieved by dividing the class into small groups that work together to achieve the best group results by means of mutual assistance among the group members to create powerful learning opportunities [4, 5]. In a cooperative learning context, there are many interactive tasks that would naturally stimulate the students' cognitive, linguistic, and social abilities.

There are some advantages of cooperative learning approaches, especially for mathematics learners, including: 1) promotion toward deep learning of materials, 2) achievement of better grades as compared to competitive or individual learning, 3) better learning of social skills and civic values, 4) acquisition of higher-order and critical thinking skills, 5) promotion of personal growth and 6) development of positive attitudes toward autonomous learning [5, 6, 7, 8].

One of cooperative learning techniques widely used by teachers is Teams-Games-Tournaments (TGT). The purpose of TGT is to create an effective classroom environment in which students are actively involved in teaching process and are consistently receiving encouragement for successful performance [10]. TGT has three basic elements: 1) teams, students are assigned to equal teams categorized by equivalent academic levels, 2) games, skill exercises relating to content material are played during weekly tournaments, 3) tournament, students represent their teams and compete individually against students from other teams. The winnings are brought back to their teams. Total winnings are counted and team champions are announced.

TGT technique evidently provides several positive influences toward student achievement [11, 12, 13]. Among them are to bring more practice, to create greater and earlier awareness of learning outcomes, to give extra motivation for students and to build a more enjoyable learning experience. Besides, the use of TGT can enhance students' motivation, learning outcomes and especially self-efficacy [12].

Self-efficacy is a person's self-confidence about their ability to accomplish a task [14]. Self-efficacy is strongly related to learners' expectation about what they can actually accomplish within their learning. The outcomes people expect are largely dependent on their judgments of what they can accomplish [15]. Researchers reveal that students' self-efficacy in their academic capabilities is closely related to academic motivation and performance outcomes in domains that include science, mathematics, and language arts [16, 17].

There are several previous studies which have been carried out on the effectiveness of cooperative learning toward junior highs school students' self-efficacy, such as technique jigsaw [18], Think Pair Share [19] and Group Investigation [20]. Teams Game Tournament itself has been widely used in some studies. However, most of the studies are not specifically addressing the issue of TGT effect toward self-efficacy [11, 12, 13]. Therefore, this present study is conducted to investigate the effect of TGT toward mathematics self-efficacy of junior high school students.

This research aims at investigating the effect of the use of Team Games Tournament toward students' self-efficacy on mathematics. The specific objectives of the study are: 1) How are mathematics self-efficacy scores of 7th grade junior high school students before and after applying TGT technique?; 2) Is there any significant difference between the pretest and posttest mathematics self-efficacy scores of 7th grade junior high school students to whom the TGT has been applied?.

2. Methodology

This research was a quantitative research with experimental design of one group pretest posttest experimental design. The population in this study was 7th graders in a public junior high school in Landak Regency, Indonesia in the academic year 2016/2017 with total number of 94 students (47 men and 47 women). The sample in this research was the students of grade 7A and 7C in that school as many as 64 students (31 men and 33 women). The study employed cluster random sampling. The study was conducted to the students who were currently receiving social arithmetic material. The materials were taught using the TGT technique. The learning activities were conducted in 5 classroom meetings.

To measure students' mathematics self-efficacy scores, a "Mathematics Self Efficacy Questionnaire" (MSEQ) developed by May [21] was used. This 14 item questionnaire used Likert scale with 5 options as follows: (1) Never, (2) Seldom, (3) Sometimes, (4) Often, (5) Usually. Test reliability for MSEQ using Alpha Cronbach and Alpha Cronbach coefficient of 0.93 was obtained. Therefore, MSEQ was reliable to be used to measure students' mathematics self-efficacy scores. The maximum value that can be obtained by students in filling the MSEQ was 70 and the minimum value was 14. In this study, the interval of Likert scale was calculated to be (5-1)/5 = 0.80.

In the early stages of data collection, seventh grade students of SMP were given the MSEQ to measure the students' mathematics self-efficacy score before to treatment. Then, it was followed by teaching social arithmetic material using TGT technique. After completion of the teaching, the students were asked to complete the MSEQ again to measure the post-test value of students' mathematics self-efficacy score as the final stage of data collection.

Descriptive statistics methods and paired samples t-test were applied for the analysis of data. For all of the statistical decoding, .05 significant level was taken as the basis. The data were analyzed using the SPSS 20.0 software.

3. Result and Discussion

Table 1 exhibits students' mathematics self-efficacy scores of 7th grade junior high school students before and after the implementation of TGT technique. The table is used to understand the mean score of each student's self-efficacy.

Table 1. Students' mathematics self-efficacy scores before and after the treatment

No.	Items	Mean	
		Before	After
1.	I feel confident enough to ask in my mathematics class	2,625	3.344
2.	I believe I can do well on a mathematics test	3.141	3.516
3.	I believe I can complete all of the assignments in a math course	3.266	3.578
4.	I believe I am the kind of person who is good at mathematics	3.234	3.656
5.	I believe I will be able to use math in my future career when needed	3.203	3.625
6.	I believe I can understand the content in a mathematics course	2.953	3.625
7.	I believe I can get an "A" when I am in a mathematics course	3.250	3.719
8.	I believe I can learn well in a mathematics course	3.203	3.609
9.	I feel confident when taking a mathematics test	2.813	3.531
10.	I believe I am the type of person who can do mathematics	3.406	3.656
11.	I feel that I will be able to do well in future mathematics courses	2.906	3.531
12.	I believe I can do the mathematics in a mathematics course	3.125	3.594
13.	I believe I can think like a mathematician	2.938	3.500
14.	I feel confident when using mathematics outside of school	2.922	3.641

Based on the above table, before the implementation of TGT technique, it is noted that item 10 (I believe I am the type of person who can do mathematics) has the highest mean score. And item 1 (I feel confident enough to ask in my mathematics class) has the lowest mean score. The finding shows that students generally believe that they are the learner type who can learn mathematics, yet they are not confident to ask question in mathematics class.

However, after the implementation of TGT, it is seen that item 7 (I believe I can get an "A" when I am in a mathematics course) has the highest mean score and item 1 (I feel confident enough to ask in my mathematics class) has the lowest mean score. This finding shows that the students believe that they can get a good score in mathematics but they are still not confident to ask question in mathematics class

Table 2. Frequencies and percentages values related to average scores of students.

Average Points Before and After Study										
Interval of Likert Scale	Between 1.00-1.80 points		Between 1.81-2.60 points		Between 2.61-3.40 points		Between 3.41-4.20 points		Between 4.21-5.00 points	
Frequencies and Percentages	f	%	F	%	f	%	f	%	f	%
Before	0	0	15	23.44	31	48.44	16	25.00	2	3.13
After	0	0	1	1.56	22	34.38	36	56.25	5	7.81

Based on Table 2, before TGT is applied, no student has responded "never". Students' self-efficacy on mathematics response before applying TGT shows that there are 15 students (23.44%) respond "seldom", 31 students (48.44%) respond "sometimes", 16 students (25%) respond "often" and 2 students (3.13%) respond "usually". The result shows that the majority of students choose "sometimes" as their response to MSEQ. Students still have anxiety in finishing their mathematics assignment in class.

After TGT is applied, no student has responded "never" in filling MSEQ. Students' mathematics self-efficacy scores response after applying TGT shows that 1 student (1.56%) respond "seldom", 22 students (34.38%) respond "sometimes", 36 students (56.25%) respond "often" and 5 students (7.81%) respond "usually". The result shows that majority of students choose "often." The students are able to decrease their anxiety and they are more confident in finishing their mathematics assignments after receiving TGT.

The responses of students vary in four items, namely "seldom", "sometimes", "often" and "usually". However, based on the table, it can be concluded that the students' self-efficacy after having TGT is improving. The students have better confidence in finishing their assignments of mathematics in class and the students also receive better score. The analysis of pretest and posttest scores shows the positive change toward students' self-efficacy. The finding is in accordance with several previous studies on the effect of cooperative learning toward students' self-efficacy which reveal that cooperative learning has positive effect toward students learning achievement, attitude and self-efficacy in mathematics [3, 8, 19, 20, 22].

Related to the significant difference between the pretest and posttest mathematics self-efficacy scores of 7th grade junior high school students after TGT is applied, the is presented in table 3.

Table 3. Results of the paired samples t-test of students' mathematics self-efficacy scores.

	N	Mean	Std	df	Correlation	t-value	Sig. (2-tailed)
Pretest	64	42.984	8.08	63	0.823	-12.369	0.000
Posttest	64	50.125	6.15				

Based on table 3, it can be concluded that there is a significant difference between the value of pretest and posttest of self-efficacy mathematics of 7th grade junior high school students after the TGT technique is applied. The result is drawn from t-value = -12.369 and p-value < 0.05. In addition, there is a strong and positive linear relationship (0.823) between the pretest and posttest valued students' mathematics self-efficacy scores. The average students' mathematics self-efficacy score before the TGT technique is applied is 42,984. The average students' mathematics self-efficacy score after TGT technique is applied is 50.125.

The result shows that TGT has significant effect toward self-efficacy. The finding is in accordance with several studies which show that TGT has several positive features in improving students self-efficacy, among them are 1) fostering the role of students as a group to improve their team performance, 2) creating a fun and attractive learning environment [11], 3) improving students' motivation to work harder in winning the tournament [23, 12, 24]. Such features will generally lead them to be more successful, more deserving of reward, and happier about their team outcome and learning process.

4. Conclusion

In conclusion, there is an increasing number of students' mathematic self-efficacy score after applying TGT. The results indicate that TGT gives positive effect toward students' self-efficacy in studying mathematics. The performance of groups who win the games makes the students to be more confident. Besides, the students from the winning team also show better individual performance in finishing their individual work. The improvement is because TGT has several positive features such as cooperative game playing and interpersonal competitive game playing.

As a suggestion, teachers should consider using TGT technique in their classrooms to improve their students' self-efficacy in mathematics which has positive correlation with their learning achievement. The present study only investigates the effect of TGT toward students' self-efficacy in mathematics without further analyzing what factors which causes such effect. Future research is suggested to start considering and investigating such factors to gain more precise understanding of TGT effect toward students' self-efficacy.

5. Acknowledgement

I would like to send my gratitude to those who have contributed important roles in the research process. I specifically address my infinite gratitude to Landak Bersatu Foundation which has provided research funding. I also offer my sincere happiness for my colleagues in Mathematics Education Program of the Pamane Talino Teaching College for always supporting me along the process.

6. References

1. Katz, Sara. Enhancing Self-efficacy of Elementary School Students to Learn Mathematics. Journal of Curriculum and Teaching. **4**, 42-55. (2015).
2. Kirik, Ozgecan Tastan & Markic, Silvija. The Self-efficacy of pre-service Elementary Teachers Using Cooperative Learning in Science Teaching. Procedia Social and Behavioral Sciences, **46**, 5005-5009. (2012).
3. Gillies, Robyn M. The Effects of Cooperative Learning on Junior High School Students' Behaviours, Discourse and Learning During a Science-Based Learning Activity. School Psychology International. **29**. 328-347. (2008).
4. Johnson, D.W., & Johnson, R.T. *Learning Together And Alone*. London: Allyn and Bacon. (1994).
5. Azizinezhad, Masoud., Hashemi, Masoud., & Darvishi, Sohrab. Application Of Cooperative Learning In EFL Classes To Enhance The Student's Language Learning. Procedia Social and Behavioral Sciences. **93**, 138 – 141. (2013).
6. Shimazoe, J. & H. Aldrich. Group Work Can Be Gratifying: Understanding And Overcoming Resistance To Cooperative Learning. Journal College Teaching., **58**, 52-**57**. (2010)
7. Aziz, Zahara & Hossain, Md. Anowar. A Comparison of Cooperative Learning Anf Conventional Teaching on Students' Achievement In Secondary Mathematics. Procedia Social and Behavioral Sciences. **9**, 53-62. (2010).

8. Hossain, Anowar., & Tarmizi, Rohani Ahmad. Effects Of Cooperative Learning On Students Achievement And Attitudes In Secondary Mathematics. Procedia Social and Behavioral Sciences. **93**, 473 – 477. (2013).
9. Zakaria, Effandi., Chin, Lu Chung., & Daud, Md Yusoff. The Effects of Cooperative Learning on Students' Mathematics Achievement and Attitude towards Mathematics. Journal of Social Sciences. **6**, 272-275. (2010).
10. Veloo, Arsaythamby., Md-Ali, Ruzlan., & Chairany, Sitie. Using Cooperative TGT in 11 Religious School to Improve Mathematics Understanding and Communication. Malaysian Journal of Learning and Instruction. **13,** 97-123. (2016).
11. Gonzales, Arturo., Jennings, Davis., & Manriquez, Loreto. Multi-Faceted Impact Of A Team Game Tournament On The Ability Of The Learners To Engage And Develop Their Own Critical Skill Set. International Journal of Engineering Education. **30**(5), 1213-1224. (2014).
12. Frianto., Soetjipto, Budi Eko., & Amirudin, Achmad. The Implementation of Cooperative Learning Model Team Game Tournament and Fan N Pick To Enhance Motivation and Social Studies Learning Outcomes. IOSR Journal of Humanities and Social Science. **21**, 74-81. (2016).
13. Salam, Abdus., Hossain, Anwar., & Rahan, Shahidur. Effects Of Using TGTs (TGT) Cooperative Technique For Learning Mathematics In Secondary Schools of Bangladesh. Malaysian Online Journal Of Educational Technology. **3**, (2015).
14. Bandura, A. *Self Efficacy: The Exercise of Control*. NY: Freeman & Company. (1997).
15. Usher, Ellen L, & Pajares, Frank. Sources of Self-efficacy in Mathematics: A Validation Study. Contemporary Educational Psychology, **34.** 89-101. (2009).
16. Lavasani, Masoud Gholamali., & Khandan, Farah. The Effect Of Cooperative Learning On Mathematics Anxiety And Help Seeking Behavior. Procedia Social and Behavioral Sciences. **15**, 271-276. (2011).
17. Hossain, Anowar. and Tarmizi, Rohani Ahmad. Effects of Cooperative Learning On Students' Achievement and Attitudes In Secondary Mathematics. Procedia Social and Behavioral Sciences. **93**, 473 – 477. (2013).
18. Sengul, Sare., & Katranci, Yasemin. Effect Of Jigsaw Technique On Mathematics Self-Efficacy Perceptions Of Seventh Grade Primary School Students. Procedia Social And Behavioral Sciences. **116**, 333 – 338. (2014).
19. Nuyami, N. M. S., Suastra, I. W., & Sadia, I. W. Pengaruh Model Pembelajaran Kooperatif Tipe Think-Pair-Share Terhadap Self-Efficacy Siswa SMP Ditinjau Dari Gender. E-Journal Program Pascasarjana Universitas Pendidikan Ganesha. **4**. (2014).
20. Anita, N. M. Y., Karyasa, I. W., & Tika, I. N. Pengaruh Model Pembelajaran Kooperatif Tipe Group Investigation (GI) Terhadap Self-Efficacy Siswa. E-Journal Program Pascasarjana Universitas Pendidikan Ganesha. **3**. (2013).
21. May, Diana. K. *Mathematics Self Efficacy and Anxiety Questionnaire*. A Dissertation Submitted to the Graduate Faculty of The University of Georgia in Partial Fulfillment of the Requirements for the Degree. Doctor of Philosophy. Georgia. The University of Georgia. (2009).
22. Tran, Van Dat. The Effects of Cooperative Learning on the Academic Achievement and Knowledge Retention. International Journal of Higher Education. **3.** 131-140. (2014).
23. Pangestuti, Ardian Anjar., Mistianah., Corebima A D., & Zubaidah, Siti. Using Reading Concept Map-TGT (Remap-TGT) to Improve Reading interest of Tenth Grade Student of Laboratory Senior High School State University of Malang. American Journal of Educational Research. **3.** 250-254. (2015).
24. Zimmerman, Barry J. Self-Efficacy: An Essential Motive to Learn. Contemporary Educational Psychology, **25.** 82-91. (2000).

Incorporating Local Culture in English Teaching Material for Undergraduate Students

I Gusti Ngurah Agung Wijaya Mahardika*
Balinese Language Education Dept Institut Hindu Dharma Negeri Denpasar. Jalan Ratna 51, Denpasar 80115, Bali, Indonesia.

Abstract. This paper discusses the incorporation of local cultural material in a teaching material developed for the students of the Hinduism Education Department of IHDN Denpasar. Teaching material plays an important part in teaching learning process, yet inappropriate teaching materials may become more harmful than useful. The unique nature of the HED students warranted the need for a tailor-made teaching material. The study found that the use of culturally familiar materials is beneficial for the students learning process. The result of the study also highlighted students' needs and prior knowledge as the main factors to be considered when developing teaching material.

1 Introduction

Teaching material is an integral part a teaching and learning process. It plays a vital role in improving students' achievement and motivation in learning. With the shift of teaching learning focus from teacher centered to student centered, teaching materials plays an increasingly focal part in teaching [1]. A good material may become the tipping point which decide the result of a learning. In order to better facilitate students' learning teacher often have to develop their own material that will suit the students' needs [2], [3], [4], [5]. A good teaching material will be able not only to facilitate students' learning process, but also provide enjoyment for the students [6]. Furthermore, a good teaching material, which provide enough room for the students' development, will prompt and sustain students independent learning [5].

To be able to produce a good material, the teacher should always start from the students [2], [3]. The material should be able to capture students' need and, both present and future needs [4], [5]. The teaching material should also take into account students culture and background, for language is inseparable part of culture. Therefore, when teaching English in the context of English as a Foreign Language, then both the English language culture and the students' own culture must be taken into account [7], [8]. Local culture can be a great help in teaching English for non-native English speaker students [9]. [9] integrated local Filipino literature (*cordilleran*) into English material for the first graders in The Philippines. Her result showed that the local literature helps the students to learn English, as well as keeping the students rooted to their culture.

Institut Hindu Dharma Negeri Denpasar or IHDN (Denpasar State Hindu Dharma Institute) is a state university under the Ministry of Religious Affairs of The Republic of

* *I Gusti Ngurah Agung Wijaya* Mahardika: wijayamahardika@ihdn.ac.id

Indonesia. One of the main objective of the university is to train and educate future Hinduism teachers. The future Hinduism teachers are trained and educated in the Hinduism Education Department, hereinafter HED. English is one of the important competences to be mastered by the students of the HED, both for their present and future needs. English as the language of academics serves as the means to understand the ideas in foreign journals and literature for their study [10], [111, p.279].

English teaching in the HED poses several unique conditions for the lecturer. The first is the specialized nature of the students. The students need teaching materials which provide them not only with English learning material but also Hinduism materials. The second condition is that the existing material for English course, is considered to be too heavy and relied heavily on reading, with no activities for listening and very little activities for speaking. The third condition is the inherent mindset of the students, that they do not need nor want to study English, basing on the assumption that they do not nor will not need English for their study. These conditions thus prompted the researcher to develop a totally new material which will help and encourage the students to learn English [12], [5].

2 Discussion

The study was designed and conducted as Research Development study using a simplified version of the ADDIE development model by Dick and Carrey [13]. The object of the study was to develop English teaching material for the students of the Hinduism Education Department of IHDN Denpasar. The subject of the study were the 82 students of the first semester from HED IA and HED IB at the Denpasar Campus. In developing the material, there were four phases of development. The first is the Blueprint Phase, the second is the Draft Writing Phase, the third is the Draft Testing Phase, and the fourth is the Final Production Phase.

The Blueprint Phase aimed at observing the teaching learning process and the existing material, to identify the weaknesses of the existing material. The weaknesses of the existing material, as well as the findings of the observation, was then used as the basis of writing the blueprint for the new material. The new material was written by observing the blueprint, and then referred to the experts on English teaching and religious education to be reviewed. The criticism and suggestion from the experts were then used to improve the material. after the material was deemed acceptable, it was used in teaching learning activities in the Draft Testing phase. The students were also asked to give criticism and opinion regarding the material. the result of the criticism and students input were then used to revise the material during the Final Production Phase. The result of the final Production Phase was a ready to use, tailor-made English teaching material for the students of Hinduism Education Department.

During the Blueprint phase, it was found that the existing material had three main weaknesses. The first weakness was the topics of the materials which was considered to be too foreign by the students. There were five reading texts, and the content of the texts were considered to be too advanced for thy students' level. The second weakness, was that the level of difficulty. 88% of the students considered the existing material to be very difficult. It was found that the reading texts were too long, used advanced level of diction, and used numerous vocabulary from foreign language notably Sanskrit. The third weakness was caused by the layout and presentation of the material. The students considered the material as boring, since the material relied heavily on structural exercises, and little else. Furthermore, the material was printed in black and white with very little pictures.

To overcome the weaknesses of the existing material, the researcher developed the material based on the needs and the criticism from the students. To do this, the researcher decided to use materials which are culturally familiar for the students. Learning a language also means learning a culture. Culture is so intertwined with language that in order to be able to use the language correctly and accurately, one must also understand the culture of said language. Scholars [14], [15], [16] agreed that culture of the target language should be taught to promote a situation where the students are aware of other cultures and can accept the other cultures.

Many scholars also put emphasis on the local culture or the culture of the learners. These scholars believe that the local culture is just as important as the target language culture. The students' local culture will act as an identity for the learners, this identity will enable the students to construct their own meanings and to reflect on their own culture as well as the culture of the target language [7], [8]. On the other hand, [17] stated that the local culture should be integrated into the teaching of a target language, in this case, English. [17] argues that by integrating local culture in teaching English as a foreign language, will reduce the "foreign" sense of the language. Due to the local culture integration to the teaching materials, students will learn English as if they are learning something familiar since the materials deal with things which are familiar to the students [18], [19], [20].

The improvement started with the reading text. In order to provide students with culturally familiar reading texts, which will reduce the sense of "foreignness", the researcher decided to use simple reading text taken from various sources. The sources of the reading texts were newspaper articles, official briefs from the Indonesian Hindu Dharma Parisadh (PHDI), internet articles provided by travel agents, as well as material from Hinduism textbooks.

The texts were selected based on their cultural familiarity with the students. There were eight texts selected for the eight chapters of the course material. The texts were, 1) The Gods of Balinese Hinduism, 2) Balinese Offerings, 3) Besakih Temple, 4) Traditional Tooth Filing Ceremony in Bali, 5) The Sacred Mount Agung, 6) Balinese Marriage, 7) Galungan and Kuningan, and 8) Nyepi. All these texts discuss the aspects of Hinduism and the life of the Hindus in Bali, something which the students are familiar with.

To ensure that reading texts in the new material were appropriate to students' level of English mastery, the reading text selected were modified in terms of difficulty. The modification was not done simply by shortening sentences or omitting difficult words, for such actions does not always reduce the difficulty level of the text [21]. Instead, the texts used were selected based on students' interest, needs, and prior [22], [5].

The difficulty level of the texts was further reduced by using redundancy of ideas by providing extra-textual support. The extra-textual support was in form of pictures which provide information in line with the content of the reading. The pictures support the explanation in the text by providing more concrete representation of the ideas in the text [23]. the pictures were used as facilitating tool to help understanding the text. The presence of the pictures is very important, because lowering the difficulty of a reading text should be achieved by providing facilitation tools instead of simply using simpler words [24], [21]. Another step taken was deleting unnecessary information from the reading text. Such unnecessary information within reading text were usually found in articles taken from newspaper. These steps ensured that the reading texts were appropriate to the students' level [21], [22], [25].

Appealing colors and pictures in a material is important, as stated by [26]. A teaching material should be able to achieve impact on the students. The term impact here means that the material is able to arouse students' interest and curiosity. This is achieved by using appealing pictures and color scheme. The pictures and colors will reduce students' boredom when studying the book. The pictures and color scheme can also act as a sort of landmark, which helps the students navigating the material. students can use a certain picture to mentally bookmark certain part of the material, such as the last part that they read or worked on.

As stated before, one of the weaknesses of the existing materials was the monotonous repetition of very limited types of activities in the material. To overcome this weakness, the new material was developed with numerous activities. Besides reading text, the new material incorporated relevant pictures and colorful presentation, structural points, speaking activities, listening activities, writing tasks, and language games.

Other learning activities in the new material were added to provide students with activities which can be used by the students when learning independently at home. Each of the chapters in the new material contains a section called structural point which consists of grammatical explanation and examples, which followed by grammatical exercises. The listening sections were added and designed with ample activities, to help students learn independently at home. For the writing activities the students were provided with a number of writing exercises, ranging from retelling a story, to writing a complete story based on a series of pictures. All these activities were included to enable students to practice all the aspect of English. Because mastery of English does not stop at the mastery of reading skill alone, but also other skills of English, such as speaking, listening, writing and the structure of English [27], [28].

After a lengthy process of revision and rewriting, the new material was then put into field testing. The students were asked to review the new course book while studying with them. The students were also asked to fill a questionnaire about the material. The data from the questionnaire were the triangulated through observation and interview. The result of the interview is presented in the following table.

Table 1. The questionnaire results (in percent).

NO	QUESTION	Strongly Agree	Agree	Neutral	Disagree	Strongly Disagree
1	This material more interesting than the existing material	77	17	5	1	0
2	This material is easier to understand than the existing material	68	28	4	0	0
3	This material is more fun to learn than the existing material	72	23	4	1	0
4	This material helps me learn English better than the existing material	85	15	0	0	0
5	I prefer to study English using this material than the existing material	91	9	0	0	0

The majority of the students, prefer the new material rather than the existing material. the students favored the existing material in terms of appeal, level of difficulty, enjoyment during studying, material content, and overall preferences. The interview with the students showed that the students love the fresh and colorful look of the new material, and they also love the pictures included in the new course book. The students claimed that the pictures, colors, language games, and language activities in the material help them reduce their anxiety and stress level.

The above results highlighted two things. The first is that the students' needs must be the most vital factor in developing material. Because material that directly correspondent with said needs will automatically be useful and meaningful for the students. This finding is similar to the results of [3], and in line with the conclusion of [2]. With appropriate material the students will be prompted to learn well because they realize that they are learning not for the sake of exams or grades, but more for the mastery of skills that they need in the first place. This realization is further supported by the material ample activities which enable students to learn more outside the classroom own their own pace which allows students' path to autonomous learning [5].

The second is that the students preferred the new material over that existing material due the culturally familiar materials. The familiar material is in line with the students' prior knowledge, thus helping the students in understanding the new material. This result is similar to [9] results. In both researches, local culture plays important role in improving students' involvement in English learning. Therefore, local culture should always be considered when developing English material. Furthermore, as noted in the observation the familiarity of the material enables the students to learn more and achieve more. This achievement boosted students' motivation and involvement in learning. This result confirms various research results from around the world [18], [19], [20].

3 Conclusion

Based on the above explanation, it can be concluded that the incorporation of local culture material helps the students in learning English, the use of cultural material has decreased the foreign nuance of the material, it reduces the stress level of the students due to the familiarity of the material. The result of this study confirms the notion that a culturally familiar material will help students in learning a foreign language. It is recommended for other researcher to also inquire the effectiveness of culturally familiar material compared to less culturally familiar material in teaching English in different level or learning.

Acknowledgement

I would like to express my most sincere gratitude to the Rector of IHDN Denpasar, Prof. Dr. Drs. I Nengah Duija, M.Si, and the Head of the Hinduism Education Department, I Made Arsa Wiguna, S.St. Par, M.Pd.H., for their blessing and support during my research in HED. My warmest gratitude goes to my esteemed colleague, IGA Purnamika Utami, S.Pd. M.Pd., without whose inspiration and untiring criticism this work would have been impossible. Finally, I would also like to extend my warmest appreciation and gratitude to all the students of HED IA and HED IB, for all their efforts and hard work.

References

1. K. Kitao, S. K. Kitao. Selecting and Developing Teaching/Learning Materials. The Internet TESL Journal, **Vol. IV**, No. 4. (1997).
2. F. Rochman, Yuliati. The Development of The Indonesian Teaching Material Based on Multicultural Context by Using Sociolinguistic Approach at Junior High School. Procedia- Soc. and Behav. Sci. **Vol. 9.** pp. 1481–1488 (2010).
3. Y. M. Harsono. Developing Learning Materials for Specific Purposes. TEFLIN Jour. **Vol. 18**, Number 2. pp. 169-179 (2007)
4. Sismiati, M. A. Latief. Developing Instructional Materials on English Oral Communication for Nursing Schools. TEFLIN Jour. **Vol. 23**, Number 1. pp. 44-59 (2012).
5. P. Whacob. Methods and Materials for Motivation and Learner Autonomy. Ref. on Eng. Lang. Teach. **Vol. 5**, No. 1, pp. 93-122 (2006).
6. E. Illes. What Makes a Coursebook Series Stand the Test of Time? ELT Jour. **Vol. 63**/2. pp. 145-153. (2009)

7. C. Kramsch, A. Whiteside. *Language and Culture*. Oxford, U.K.: Oxford University Press. (1998).
8. H. Straub. Designing a Cross-cultural Course. English Forum, 37(3). (1999).
9. L. O. Florentino. Integrating Local Literature in Teaching English to First Graders under K-12 Curriculum. *Theory and Practice in Language Studies*, **Vol. 4**, No. 7. pp. 1344-1351 (2014).
10. D. P. Mahu. Why Is Learning English So Beneficial Nowadays? Inter. Jour. Of. Comm. Res. **Vol. 2**. issue 4. pp. 374-376 (2012).
11. S. R. Ahmad. Importance of English communication skills. Inter. Jour. of App. Res. **Vol. 2**(3). pp. 478-480 (2016).
12. R. T. Jiménez, B. C. Rose. Knowing How to Know: Building Meaningful Relationships Through Instruction That Meets the Needs of Students Learning English. Jour. of Teach. Edu. **Vol. 61** (5). pp. 403–412 (2012).
13. W. Dick, L. Carey. *The Systematic Design of Instruction* (4th Ed.). New York: Haper Collins College Publishers. (1996).
14. M. Canale, and M. Swain. Theoretical Bases of Communicative Approaches to Second Language Teaching and Testing. Applied Linguistics, **Vol. 1**, no. 1, pp. 1-39. (1979).
15. D. Pecks. Teaching Culture: Beyond Language. Retrieved January 18, 2017 http://teachersinstitute.yale.edu/curriculum/units/1984/3/84.03.06.x.html. (2016).
16. M. Byram. Cultural awareness in vocabulary learning. Lang. Lear. Jour. **Vol. 16**. pp. 51-57. (1997).
17. M. Regmi. The Role of Local Culture and Context in English Language Teaching. Retrieved from https://neltachoutari.wordpress.com/2011/04/01/the-role-of-local-culture-and-context-in-english-language-teaching on 7 January 2017. (2011).
18. C. S. Barfield, and J. Uzarski, Integrating Indigenous Cultures into English Language. English Teaching Forum, 2009, Number 1. Retrieved from https://americanenglish.state.gov/files/ae/resource_files/09-47-1-b.pdf on 4 February, 2017. (2009).
19. K. M. Yin. The Role of Prior Knowledge in Reading Comprehension. Reading in a Foreign Language. **Vol. 11**. pp. 375-380 (1985).
20. A. S. Sase. and N. M. Abdelaal. Relationship Between Prior Knowledge and Reading Comprehension. Adv. in Lang. and Lit. Stu. **Vol. 5,** No. 6. pp. 125-131. (2014).
21. A. Hetherington. Assessing the Suitability of Reading Materials for ESL Students. TESL Canada Journal/Revue Tesl Du Canada. **Vol. 3**, No.1. pp. 37-52. (1985).
22. I. J. Arias. Selecting Reading Materials Wisely. *LETRAS*. **Vol. 41**. pp. 131-151 (2007).
23. C. P. Yi, and C.P. Yi. The Effects of Pictures on The Reading Comprehension of Low-Proficiency Taiwanese English Foreign Language College Students: An Action Research Study. VNU Jour. of Sci. Foreign Languages. **25.** pp. 186-198. (2009).
24. F. Yu. An Analysis of Pictures for Improving Reading Comprehension: A Case Study of the New Hanyu Shuiping Kaoshi. The Nebraska Educator: A Student-Led Journal. **Vol. 27.** (2015).
25. M. R. Ghahroudi, and E. Sheikhzadeh, Selecting Reading Texts for University Iranian EFL Students. Jour. of Adv. in Eng. Lang. Teach. **Vol 5,** No 3. pp. 25-30. (2017).
26. B. Tomlinson. *Materials Development in Language Teaching*. U.K: Cambridge University Press. (1998).
27. S. M. Ahmadi. The Importance of Listening Comprehension in Language Learning. Inter. Jour. of Res. in Eng. Edu. **Vol. 1**, No.1. pp 7-10, (2016).
28. F. Bano. Towards Understanding Listening Comprehension in EFL Classroom: The Case of the Saudi Learners. English Language Teaching. **Vol. 10**, No. 6. pp. 21-27 (2017).

The effectiveness of using portfolio assessment in lecture by controlling mathematical logical intelligence

Mahayukti, G.A,[*1]*,Dantes, N*[2]*, Candiasa, I.M*[3] *and Marhaeni, A.A.I.N*[4]
[1] Pendidikan Matematika FMIPA Universitas Pendidikan Ganesha, Indonesia
[2] Program Studi Pendidkan Dasar Pascasarjana Universitas Pendidikan Ganesha Elementary
[3] Program Studi Ilmu Pendidikan Program Pascasarjana Universitas Pendidikan Ganesha Educational
[4] Program Studi PEP Pascasarjana Universitas Pendidikan Ganesha

Abstract

The purpose of this research aims to examine the effectiveness of using portfolio assessment in the integral calculus lecture after logical mathematical intelligence is controlled. The population of this research are all students in the department of mathematics education who program integral calculus courses on the 2015/2016 academic year. Samples in this research are 88 students, who were selected randomly. The data were analyzed by one-way covariance analysis. The result of this research showed that student's learning outcomes of Integral Calculus who were taught Integral Calculus with portfolio assessment, was higher than those of the students who were tought with essay assessment. The findings of this study show that using portfolio assessment is effective to improve student's learning outcomes of Integral Calculus.

Keywords: learning outcomes of integral calculus, logical mathematical intelligence, portfolio assessment.

1. Introduction

Assessment of the portfolio is an assessment which measure the extent of students in constructing and reflect on a task by collecting information materials relevant to the purpose and course material and the results of its construction can be assessed by lecturers in a certain period of time [1]. Kaino [2] revealed that the assessment of the portfolio provides an opportunity for the lecturer to think about what the student will do, why it should be done and assess the actions that have been done, current and expected results.. The characteristics of the portfolio assessment are consistent with De Lange's assessment [3].

Integral Calculus courses should be liked and well studied by the students and the learning process is also supposed to be going well, interesting and meaningful. Mutakin [4] provide that some of the causes difficulty studying Calculus is a lack of mastery of the concept of Differential Calculus, because in the process of learning and assessment applied still conventional. Mahayukti [5,6] reveals that the low learning outcomes of Student Calculus are caused by 1) student learning independence is still low, 2) student learning pattern is still like in high school, 3) weakness of mastery of concepts underlying Calculus,

* gustiayumahayukti@undiksha.ac.id

4) the difficulty of students understanding Integral Calculus material because it is not supported by varied learning resources, and lecturers rarely use MFIs or mathematical software.

The assessment system used Middle Semester Exam and Final Exam after students experience the learning process, have not measured the student's capabilities comprehensively. Problems arising from inappropriate judgment can decrease the individual's motivation to learn which results in poor learning outcomes [7]. This reality suggests, there is gap between expectation and reality in Integral Calculus lectures. Learning outcomes are influenced by external factors and internal factors. According to Kaino [2], one factor that comes from outside the student is the form of assessment. Gunay [8] and Abidin [9] conducted a study using a portfolio assessment as a formative assessment and found that students whose learning using a portfolio assessment achieved higher learning outcomes than student learning outcomes using conventional assessment. However, the previous study compare between authentic assessment and paper test.

The students whose learning using the portfolio assessment achieved better learning outcomes, because they are accustomed to discipline and able to assess their own work in proving a theorem or in solving a problem. Slavin [10] states that, lecturers can not only provide knowledge to students, but students must build knowledge in their own minds. Lecturers can facilitate this process by teaching in ways that make the information meaningful and relevant for students, and by providing opportunities for students to find or apply your own ideas.

Other factors that contribute to student learning outcomes, either directly or indirectly, are students' mathematical logical intelligence. Gardner [11] states that logical mathematical intelligence is the ability associated with the use of numbers and logic effectively. Logical mathematical intelligence is required in mathematics learning that prioritizes numeracy and logic.

According to Campbell [12], logical mathematical intelligence involves components: Mathematical calculation, logical thinking, problem-solving, deductive and inductive considerations, and sharpness of patterns and relationships. While Iskandar [13] states that mathematical logical intelligence includes a person's ability to think inductively and deductively, the ability to think according to the rules of logic, understand and analyze patterns of numbers and solve problems by using the ability to think. Thus the characteristics of intelligent people logically mathematically include abilities in reasoning, sorting, think in patterns of cause and effect, creating a hypothesis, sought order conceptual or numerical patterns, and a rational outlook on life in general.

Suparta [14] said that in studying, understanding and solving math problems especially Integral Calculus which examines the theorems of integrity, the fundamental theorem of Calculus, the application of integral integral does not necessarily require logical mathematical intelligence because one of the objects in mathematics education is the theorem. While Safranj [15] said that Learners whose learning refers to mathematical logical intelligence get better results than the conventional group.

Based on the background of the problems mentioned above, this study use Posttest Only Control Group Design in Mathematics Education Department FMIPA to examine and analyze the effectiveness of portfolio assessment in improving Integral Calculus learning outcomes, after logical mathematical intelligence is controlled.

2. Method

The research design used was Posttest Only Control Group Design which was held on semester Even Academic Year 2015/2016 in Mathematics Education Department FMIPA Undiksha. The population of this study are all students of the Department of Mathematics Education who were attending lectures Integral Calculus. The population consists of 153 students who are distributed into 4 (four) classes with relatively homogeneous characteristics. Through random sampling obtained class B and C as experimental group and class A and D as control group. Members of the sample from each group were taken randomly with the same number as many as 44 people, So there are as many as 88 samples.

The instrument used is the Integral Calculus achievement test in the form of essays, and mathematical logical intelligence test. By using the Lawshe [16] test involving 8 experts, obtained indeks content validity from the Integral Calculus achievement test of 0,95 and the logical mathematical intelligence test of 0,88.

Results of empirical test, the 10 items are valid with reliability coefficient of 0.87. Meanwhile, the logical mathematical intelligence tests, multiple choice with 45 items. The empirical test results show 40 items of valid question and 5 fall problems, with a reliability coefficient of 0.85.Data were analyzed using one path one-way covariance analysis (Ancova Test).

3. Results and Discussion

3.1. Results

3.1.1. Testing Requirements Analysis

All requirements analysis includes a data distribution normality test, homogeneity of variance test, linearity test, and test the significance of the regression line alignment fulfilled. Therefore, continued to test the hypothesis

3.1.2. Hypothesis Test

The following results are presented as in Table 1.

Table 1. Tests of Between-Subjects Effects

Source	Type III Sum of Squares	df	Mean Square	F	Sig.
Corrected Model	1108,721[a]	2	554,361	23,433	0,000
Intercept	2600,080	1	2600,080	109,904	0,000
covariat	907,710	1	907,710	38,368	0,00
faktor1	141,894	1	141,894	5,998	0,016
Error	2010,904	85	23,658		
Total	499771,000	88			
Corrected Total	3119,625	87			

The result of analysis on line 4 (factor 1) shows F arithmetic = 5,998 with sig.<0,05. Which means there is a difference in learning outcomes Integral Calculus between students who use portfolios with students who use essay assessment, If seen from corrected average it turns out that average A1 (76,55) is bigger than average A2 (73,70). So it can be concluded that the learning outcomes Integral calculus of students who follow the learning with a portfolio assessment is higher than students who use the essay assessment, after logical mathematical intelligence controlled.

3.2. Discussion

Suparta in [10] states that logical mathematical intelligence in this case must be controlled its influence, because logical mathematical intelligence is the ability that is associated with the use of numbers and logic effectively and this intelligence is very important because it helps develop people's thinking and logic skills and deals with Mathematics especially Calculus. Supported also by Safranj in [15] who found that students whose learning refers to mathematical logical intelligence achieve better results than conventional groups.

The application of portfolio assessment in learning also helps students in solving more difficult and more complex mathematical problems [3], because mathematics could not be separated from the theorems interrelated and abstract.

Therefore, in order Integral Calculus concepts can be understood correctly, it must be through a continuous process. This means that to master high-level math, then the underlying concepts must be mastered by students. The hierarchical mathematical sciences fit the characteristics of a sustainable portfolio assessment, and students have the opportunity to assess their work through self-evaluation. Students also have the opportunity at least twice in the tasks given, So if there is a mistake the concept can be improved immediately and the student's retention of the concept is getting stronger, As revealed by Rose and Nicholl [17] that the more a person can see, hear, write or do something, the easier something is learned, so the results are also optimal.

The findings of this study are supported by the theory that, in the learning process there are two important things that must be considered by students. First, what materials are taught and how the material is delivered, it is none other than the competence given to students should be conducted with respect to the characteristics of competency to be controlled by the student and student characteristics. Given the Integral Calculus is part of a highly abstract mathematical material, The material is interconnected like a spiral and is a discipline that not only contains concepts, formulas, or principles But also contains the process by which such concepts, formulas or principles are obtained, So it takes extra effort for students to have a basic mastery of the concept well to understand and master the material in accordance with the expected competencies.

Second, how to assess or access student's learning progress. In this context, the assessment should be in accordance with the characteristics of learning achievement competency indicators. Therefore, appropriate forms of assessment are needed, and when appropriate time to access student learning progress. The most important is how the results obtained through the assessment are followed up for improved learning.

Compared to the learning process along with an essay assessment that tends to be limited and the same, then the essay assessment could not assess student competence comprehensively. Weldeana & Sbhatu in [18] said that effective integration portfolio of evidence in learning college geometry in the instructional process helped students develop reflective thinking and other metacognitive skills and solve real-life problems that demand higher order thingking.

So far in the course Integral Calculus at the end of the lecture students tend to be given an essay-like assessment that tends to be the same, And judging from limited material tests. Students in completing the assessment, follow the rules already given. Assessment of the given essay tends to be well structured so that it is clear the concept will be used to answer the assessment.

This causes the students do not have the opportunity to conduct self-evaluation of the work they make, they tend to use the principles or concepts that have been taught only as a solution to solving the problem so that mistakes are made they do not know, consequently the error is repeated. The application of these two assessment forms will provide different conditions and environments. This will also have an impact on understanding the concept. Learning conditions with the right assessment will make the learning process become more optimal where students can use the ability to reason optimally in solving problems and make decisions. This is in accordance with the opinion of Aiken [19] which states that the advantages of using a portfolio assessment can measure students' ability to organize, connecting his mind well, which is not easily obtained from other assessments. It is confirmed by Cakan et al [20], in that the use of portfolio assessment makes learning meaningful and can help students learn better.

The findings of this study also support by Cakan, et al in [20] groups of students who use the portfolio gained higher learning outcomes than the group that did not use the portfolio, they are also positive attitudes towards science. Meanwhile, Damirel & Duman in [21] study revealed that the portfolio assessment has a significant effect on writing mastery. Gunes, et al [22] also revealed that the use of the portfolio assessment had a significant effect on the success of students as prospective teachers and reduced their anxiety in the learning and evaluation process. And [23] the experimental group received the treatment 1.e portfolio assessment while the control group underwent the traditional approach of writing assessment. Thefindings suggest that portfolio assessment technique improves writing ability of the students. Therefore, it is acceptable that the use of portfolio assessment in learning can improve student learning outcomes after logical intelligence is controlled mathematically.

4. Conclusions

Based on the results of hypothesis testing, the findings in this study is the learning result Integral calculus of students who follow the process of learning accompanied by portfolio assessment, higher than students who use the essay assessment. Thus it can be concluded that an effective portfolio assessment improves students' Integral learning calculus results, after logical mathematical intelligence is controlled. Referring to the results of research that has been stated above, the use of portfolio assessment can be developed in other subjects.

5. References

[1] Zainul, A. 2001. *Alternative Assessment*. Jakarta: UT.
[2] L. M. Kaino. Journal of Social Sciences. **5 (15)**. (2014)
[3] J. De Lange. 1997. *Mathematics Insight and Meaning*. Uttrech: OW & OC.
[4] T.Z. Mutakin. Jurnal Formatif. **III (1)**. 49-60 (2013)
[5] G. A. Mahayukti. *Pengembangan Perangkat Pembelajaran Berbasis Masalah Berbantuan Pertanyaan-Pertanyaan Pembimbing untuk Meningkatkan Kualitas Perkuliahan Kalkulus*. This article was presented in Senari (2014)
[6] G. A. Mahayukti. *Pengembangan Perangkat Pembelajaran Berbasis Masalah Berbantuan Software Matematika dengan Asesmen Autentik untuk Meningkatkan*

Efektifitas Perkuliahan Kalkulus Integral dan Kemandirian Belajar Mahasiswa. This article was presented in Senari.(2016)
[7] Javanmard. Y. & Faharani, H. "Investigating Using Portfolio Assessment and Learning English Language in Qom Secondary Schools." *Global Journal of Human Social Science* 12(12): 52–60.(2012)
[8] A. Gunay and F. O. Bekiroglu. EJMSTE. **10 (6)**. 667-680. (2014)
[9] Abidin, Z & Walida, S. "The Model of Mathematics E-Portfolio Assessment for Senior High School." *ARPN Journal of Science and Technology 10*: 4-11.(2014)
[10] R. E. Slavin. *Psikologi Pendidikan : Teori dan Praktek.* Jilid 2. Jakarta: Indeks. (2009)
[11] Howard Gardner, *Multiple Intelligences, The Theory in Practise* (New York: Basic Book, 1993).
[12] Campbell et al. *Metode Praktis Pembelajaran Berbasis Multiple Intelligences*. Depok: Intuisi Press. (2006)
[13] Iskandar. *Psikologi Pendidikan : Sebuah Orientasi Baru.* Jakarta: Referensi. (2012)
[14] I. N. Suparta. *Kontribusi Pendidian Matematika Dalam Pengembangan Nilai Kemanusiaan.* An Oration on Profesor I. N. Suparta for the honorary degree of Profesor of Mathematics Department, Faculty of Mathematics and Natural Sciences, Ganesha University of Education in 2014.
[15] J. Safranj. *Procedia*-Social and Behavioral Sciences, **232**. (2016)
[16] Lawshe, C.H. 1975. "A Quantitative Approach to Content Validity." Personal Psychology. 28: 563-575
[17] Rose, C. & Nicholl, Malcolm J. *Accelerated Learning for The 21ST Century* (New York: Delacorte Press, 1997), h. 142.
[18] Weldeana, H. N & Sbhatu, D.B. Portfolio of Evidence: An assessment Tool in Promoting Geometry Achievement Among Teacher Education College Students, 13, issue 6: 1981-2004. (2017).
[19] Aiken, L. R. 1988. *Psychological Testing and Assessment.* Boston: Allyn and Bacon, Inc.
[20] M. Cakan, G. Mihladiz, and Taskin. Journal of Educational Science. **2(2).** 362-377. (2010)
[21] M. Damirel and H. Duman. Procedia- Social and Behavioral Sciences. **191**. (2015)
[22] Gunes, M. H, Demir, S., Balaban, M. 2015. "The Effect of Portfolio Assessment Application On Academic Achievement and Test Anxiety in Teaching Animal Tissue." *Necatibey Faculty of Education Electronic Journal of Science and Mathematics Education*. 9. Issue 1: 1-22.
[23] Tabatabaei, O. "The effect of portfolio Assessment Techinique on Writing Performance of EFL Learnes." *English Language Teaching, 5(5):* (2012).

The integration of character education in local tradition: an ethnographic study

I Made Ariasa Giri[1], and *I Ketut Ngurah* Ardiawan[2]
[1]STAHN Mpu Kuturan Singaraja, Primary Teacher Education Department, 81112, Bali, Indonesia
[2]STAHN Mpu Kuturan Singaraja, Primary Teacher Education Department, 81112, Bali, Indonesia

Abstract. The significance of character education has becoming an appealing issue to be discussed. Its significance has been brought to school through a number of practices. In fact, character education is not merely the result of schooling but this is a widespread from integration of character education in local culture. This present study attempts to explore character values contained within a local genius called *subak pancoran*. Further, this study is a qualitative study by utilizing ethnographic design. It was conducted in Buleleng regency, Bali. There are three ways that the researcher uses to collect the data, namely interview, observation sheet, and field notes. The subjects are selected purposefully. Accordingly, the collected data are analysed qualitatively. This study reveals that *subak pancoran* tradition emphasis on the following characters; tolerance, religious, democratic, communicative, hard work, responsible, mutual help, and environmental awareness. The children character development cannot be separated from the culture where children live and grow. The contribution of this study to the board of knowledge will also be added.

Keywords: Integration, Character Education, Local Genius

1. Introduction

Character education has emerged as an appealing issue for years. This term was initially proposed by Lickona [11] that sees the importance of character education. He later sees character education is not merely teaching what is right and what is wrong to children, but more than that character education inculcate habituation of the good that students understand, able to feel, and want to do good. In addition, Dantes [8] closely looks at the components of good characters as the following; moral knowing, moral feeling, and moral action.

The emergence of character education to the field as it is now is also much influenced by the fact that family nowadays often provide less supervision to their children [5,4].

[1] tiyagiri@yahoo.com
[2] stahnmpukuturansingaraja@gmail.com

Because of technology and the advancement of education, families are leaving the core of development of their children up to the schools. Hence, there must be a way to insert character education in any way possible. One of the ways is that by integrating it in local culture.

In Indonesian context, preservation on local cultures in Indonesia has been degrading and there is a tendency that people neglect culture as a local genius, which in fact, is the main characteristic that holds Indonesian diversity. This has concerned the Indonesian government which results in developing Indonesian present curriculum (Curriculum 2013). It emphasizes on education process should be originated from present and future national culture [9]. The government seems to be quite aware of the importance of culture which functions as living well each other [6].

Specifically in Bali, Balinese society has seen local genius as an inseparable part to ways of life. A difference in religion is no longer a matter and obstacle to insert moral code and local genius as well as culture as a part of building a good character in daily application. Rasna [3] find out that character education in Bali emphasizes on individual ethic dimension and checking on how right and wrong are developed. One of the well-known local geniuses in Bali is *subak pancoran*.

Subak pancoran is a pluralistic area where it contains uniting factor to create a social integration within it. The members of *subak pancoran* consist of Hindus members and Muslim members. It is located in the northern part of Bali, precisely in Panji Anom village, Sukasada district, Buleleng regency. A general description of Panji Anom village is presented as follows. This village is one of the villages which are located in Sukasada district, Buleleng regency. In order to reach this village, a visitor should take a route from the town to this village for 12 km long. This village is divided into four sub areas, namely *Dusun Batu Pulu, Dusun Leabah Siung, Dusun Pancoran,* and *Dusun Abasan*. From these four areas, *Subak Pancoran* is located at the areas of *Dusun Pancoran*. The villagers in this area are heterogeneous in terms of the religion and the occupation. Most of them work as farmer. *Subak Pancoran* has four adjacent, they are *Subak Cempaka* on the southern part, *Pangkung Seka Lalang* on the eastern part, *Subak Lebah Siung* on the northern part, and *Tukad Sari* on the western part.

In this present study focuses on identifying the character values contained in the *subak pancoran* tradition from ethno pedagogy perspective, investigating how the character values are integrated in primary school instructions. Further, the findings of this study are expected to enrich the knowledge regarding the character values and local genius as well as giving valuable suggestions to practitioners on how to design localization based instructions.

A number of studies have been conducted in relation to ethnopedagogy and local genius. Orshaak and Orshaak [2] focus their study on analysing Tuvanese ethnic culture in education as well as its compliance toward board knowledge. They found out that ethnopedagogy allows students to identify the degree of importance of aspects of their own culture and value orientations with the help of techniques. Focusing on Indonesian context, [1] analyse the function of Kalosara as ethnopedagogic media in nation character-building in South Sulawesi. They revealed that inserting ethnic or local culture in education let students to be more appreciative of their own culture and strengthen the relationship among people.

Similar studies are also conducted to the population in Siberian [7,12,13]. All of those experts focus on revealing the national peculiarities of physical cultures, original physical education, functions and meaning of traditional games and sports. The authors listed above express a great concern about the lack of attention to the problem of studying, preservation and implementation of traditional forms of motion activities of indigenous populations of west Siberian into education process.

2. Method

The researcher identifies the character values contained within *Subak Pancoran* tradition from ethno pedagogy perspective as well as investigating the integration of character values in *Subak Pancoran* tradition into learning in schools by using ethnographic in qualitative approach. The scope of this present study is limited to *Pancoran* by analyzing the character values from the general description and activities exist in *Subak Pancoran* tradition. Regarding the activities, the researcher searches for information from public figures, community leaders, local residents who are relevant to the field of study.

This research was conducted in Buleleng Regency. The setting of this research was in Pancoran Village, Panji Anom Village, Sukasada District, Buleleng Regency. This research was conducted on October to December 2016. This study covered preparation phase including: 1) stating problems, 2) review of data sources, and 3) instrument and proposal construction. The second phase is the implementation of the study including: 1) data collection, 2) tabulation and data analysis, and 3) data interpretation. Thereunto, the third phase is report construction of the research findings. The data were collected through several techniques such as interview, observation, and field notes. In order to analyze the data of the study, the researcher needed a valid instrument. Thus, the researcher constructed an interview guide and observation sheet as well as the field notes form. The collected data were further analyzed inductively as adapted by Miles and Huberman. The analysis comprises four stages, namely data collection, data reduction, data presentation, and conclusion remarks.

3. Findings And Discussion

From the result of observation as well as interview to the leading figure in the community, the researcher found out that *Subak Pancoran* was formed in 1820 in the same location as it is now in Panji Anom Village, Sukasada District, Buleleng Regency which is 15 Km from the sub-district city, about 12 Km from the capital city, and about 90 km from the center of the province. There are four boundaries of this area, starting from the (1) southern part which is *Subak Cempaka*, (2) *Pangkung Seka Lalang* in the eastern part, (3)*Subak Lebah Siung* in the northern part, (4) *Tukad Sari* in the western part.

The land in *Subak Pancoran* is in the form of gray regusal soil, with frequent terraced rice field, with a height of 250 m to 450 m. In addition, the large of *subak* area is 61.00 Ha, divided into two smaller areas, namely: *Tekad Sari tempekan* and *tempekan Seka Lalang*.

Each religion encompasses a comprehensive essence of peaceful life, including a tolerant life in a plural society. Religion contains the noble order and rule, each of which avoids disputes and prefer the road of peace. This is because every religion has fundamental values underlying it. The Balinese people believe that the fundamental and universal values are contained in the *Tri Hita Karana* which covers three causes of harmony in life that underlies every traditional organization in Bali such as *pakraman* village, traditional *banjar*, *sekehe* including *Subak*. The same thing was expressed by Sudarta (2004) who says that the philosophy of *Tri Hita Karana* is not only applied in *subak*, but has popularized, entrenched and animated almost all the living and livelihood of Hindu people in Bali.

A harmonious life requires a person full good of character. From an anthropological point of view, character education is a noble character that everyone nurtures. This statement contains a message that human way of life cannot be separated from its basic character. In the context of education, if teachers can control the character, such as anger, the education process is expected to run better. Conversely, if the basic character of anger is put forward, it is possible that there will be disagreement as a social phenomenon [3]. The visualization of character values in the real life based on the philosophy of *Tri Hita Karana* can be seen as follows.

Table 1. The Elaboration of Character Values from Subak Pancoran Tradition

A Harmonious Relationship between Human and the Almighty God (*Parahyangan*)		
Aspect	Implementation	Character Values
Temple and Mosque as the symbol of *Subak Pancoran*	The Mosque and temple are built side by side	✓ Religious ✓ Tolerance ✓ Environmental awareness ✓ Hard work ✓ Responsibility
	The *Subak* temple consists of Ulun Suwi Temple and Bedugul temple	
	Subak temple was built in 1960 and finalized in 1970	
	The mosque was built in 1992	
Worship	Every Hindu's and Muslim farmer prays in his/her house of worship, even it was built side by side and hold the prayer at the same time, the worship is done solemnly	
Pest restriction, rice plantation, harvesting	Each religion holds its own ways of ceremony	

Table 2. *The Elaboration of Character Values from Subak Pancoran* Tradition (*Pawongan*)

A Harmonious Relation between Human and Human (*Pawongan*)		
Aspect	Implementation	Character Values
Informal leadership	There was once Mr. Husaini who was a Muslim elected as the chief of *Subak Pancoran*	✓ Responsible ✓ Discipline ✓ Fair ✓ Friendly ✓ Honest ✓ Democratic
Membership	The membership of *Subak Pancoran* is divided into to	
Rules in *Subak Pancoran*	Balinese language is obligatory in daily communication	
Mutual watering	This activity is about lending water flows to other farmers in turn	
Money lending	This activity is about saving and lending money amongst farmers	

Table 3. *The Elaboration of Character Values from Subak Pancoran* Tradition (*Palemahan*)

A Harmonious Relationship between Human and Environment		
Aspects	Implementation	Character Values
Pest restriction, rice plantation, harvesting	Member of *subak* both from Hindu's and Muslim help each other to catch pests using crab carcass	✓ Mutual Help ✓ Responsible ✓ Fair ✓ Hard Work
Water distribution	Water distribution is administered through three systems, they are *Ngulu, Maongin,* and *Ngasep*	

In relation to the aforementioned matter, *Subak Pancoran* tradition is actively contributing to enrich the life of the nation. This is reflected by the implementation of local genius which are religious, tolerance, environmental awareness, hard work, responsible, discipline, fair, friendly, honest, democratic, mutual help within *Subak Pancoran*. In this respect, people discover many benefits within culture to spread awareness on how important character values are. *Subak Pancoran* tradition also contributes to philosophical values, anthropology, sociology, psychology as the source of study of ethno pedagogy. In addition, ethno pedagogy contributes to character education is a basis and pillar toward better education in 21st century. Moral values are lived in a form of good action, whether in the process of education or social living. Basically, those activities are the manifestation of character values contained in *Tri Hita Karana*. It is also clear that its implementation has gone deep within society ways of life and has become a tradition preserved by the people.

Taking into account to the integration of character values, a harmonious live should occur in both Hindu's and Muslim's adherents. Socio-cultural living in Subak can be viewed from historical, structure, and the functional, and participation of Subak members. This can be seen from the relationship pattern or interaction that occurs among Subak members. In relation to the interaction, character values are implemented in form of fundamental activities which aims at uniting all members.

Primary school is an institution which aims to deliver basic needs toward learners such as knowledge, skills, proper attitude education, and prepare them to continue to the next step of their education (Suharjo:1). Schools do not only function as establish learners to have cognitive aspect, but also form students' attitude. Therefore, character values or character education is one that can be developed in school. It can be inserted through instructional process. Teacher may utilize local genius to be inserted thematically within any subject. By integrating it, students are expected to understand their own local genius, which by far will affect them to love their own culture. Nonetheless, teacher should adjust the character values insertion to children development, materials, strategy, method as well as the teaching media. For instance, teacher may come up with a theme, "the place where I live". This theme can be integrated to all subjects, i.e. 1) mathematics; students are asked to count the members participated in *Subak*, 2) Natural Science; students are asked to participate in rice plantation, harvesting, watering, etc., 3) social science; students are asked to explain the activities conducted by farmers in *Subak Pancoran*, 4) civil education: students are asked to explain how to live in harmoniously and with tolerance, and 5)Indonesian language; students are asked to catch the meaning of the vernacular language used by farmers and explain it in Indonesia. Those examples are some ways to integrate character values of *Subak Pancoran* into instructional process in primary schools. By this integration, students' are expected to understand and be able to build the character of local genius to be their own.

4. Conclusion

Subak Pancoran which is located in Panji Anom village, Sukasada district, Buleleng regency plays a role in educating the society. This is shown by deep application of local genius through religion, tolerance, environmental awareness, hard work, responsibility, discipline, fair, friendly, honesty, democratic, and mutual help. *Subak Pancoran* tradition positively contributes to philosophical, anthropological, socialogical, psychological aspects of ethno pedagogy. This further emphasizes that character education are the basis and pillar toward education in 21st era.

In relation to the aforementioned explanation, *Subak Pancoran* plays a key role in integrating local genius in instructional process in primary schools by adjusting it to children development, strategy, method, and teaching media used. This integration process can be done for all subject matter in schools.

5. Acknowledgement

The success and final outcome of this research requires a lot of guidance and assistance from many people and the researcher fortunate to have got this all along the completion of the research within the time given. I respect and thank the committees for giving the researcher to revise and complete the research based on the revision given.

References

1. A.Hafid, Amiruddin, K.Suardika,Sartono,Peribadi, Inter. Res. Jour. of Emer. Tre. in Multi. **1**, 9 (2015)
2. H.D.N.Oorshak, S.Y. Oorshak, Jour. of Sib. Fed. Uni, **3**, 7 (2014)
3. I.W.Rasna, Jurn. Kaj. Bali. **6**,1(2016).
4. K, Larson. *Understanding the Importance of Character Education*. (Stout, 2009)
5. M, Benigni. Amer. Sch. Boa. Journ. **193**, 3 (2006)
6. M,A, Reinfeld. Journ. of Pub. and Inter. Aff. **14** (2003)
7. N. Sinavskiy, *Ethno-pedagogic of physical training of the peoples of the North (Khanty and Mansy)* (Moscow, 2005)
8. N.Dantes, *Techno-humanistic-based Character Education (A. Series of Perspektives and Education Policies for Meeting Global Challeng*es, (2012)
9. S. Ndiung, *The 1st International Conference on Language, Literature and Teaching (2017)*
10. Supraptiningrum, Agustini. *Jurn. Pend. Kar.* **2**, 2015
11. T, Lickona, *Educating for Character: How Our School Can Teach Respect and Responsibility*. (Sydney, 1991)
12. V. Krasilnikov, *Folklore and national traditions in physical education of Siberian children* (Chelyabinsk, Institute culture, 1998)
13. V. Krasilnikov, *Games and competitions in traditional physical training of Khanti*. (Yekateringurg, 2002).
14. W.Sudarta, *Beragam Nilai Tradisional Subak (Konsepsi yang Relevan dengan Inovasi) Dalam Revitalisasi Subak Dalam Memasuki Era Globalisasi* (Yogyakarta, 2004).

A need analysis of development of implementation model of regional school operational assistance (Bosda)

Zulkifli[1], *Suheni*[1], *and Jaenam*[1]
[1]STKIP PGRI Sumatera Barat, Padang, Sumatera Barat, Indonesia

ABSTRACT

It is found widespread problems about the implementation of Bosda of Padang City, like the weakness of policy legality and delays in disbursement of funds. This research was carried out to reveal the needs analysis of real conditions in the field for the development of implementation model of Bosda in Padang City. This research was a preliminary study of research and development to get the model of development of Bosda implementation consisting of three stages; library study, field survey and modeling. However, this report was only up to the field survey stage in the form of needs analysis on implementation of Bosda. The findings indicated that the planning has not been referring to the material and non-material planning aspects, the implementation of Bosda is not supported by adequate socialization, cooperation of managerial, and commitment of Bosda recipient, and the evaluation has not fully covered the work program aspect and the implementation process of Bosda. The paper concludes that Bosda of Padang City is not implemented well. It was recommended that the managers should plan, implement, and evaluate the implementation of Bosda thoroughly and sustainably, so that the Bosda program can be realized well.

Key Words: Need Analysis, Bosda of Padang City

INTRODUCTION

The government has realized the invaluable contribution of teachers in improving the quality of education, because they are the components related directly to the process of education. In this regard, Verspoor as cited by V.T. Fadeyi, et.al. [1] found that teachers are the initiators, facilitators of teaching and learning activities. They act as agent of changes in any school system because of these roles they perform; they can be regarded as the heart of quality improvement strategy. It means, as one of the educational staff, teachers have a decisive role in the success of educational goals. In addition, teachers are given additional tasks in the structural positions of schools such as principals, vice principals, and homeroom teachers. The additional tasks have an impact on the decrease of teachers' task in implementing face-to-face teaching in the classroom. To address it, some of the schools recruit non-permanent teachers. Therefore, teachers need to carry out their duties and responsibilities well which are not only limited to the comparison of actual achievements that can be achieved with the work performance set but shows the various quality of the work done.

To be able to carry out their duties and responsibilities properly, the government has to give more attention to the teachers such as the improvement of welfare in the form of giving the right reward. A. D. Timpe [2] stated that to change and improve someone's performance there must be rewards, because people will change if they see any rewards, such as increases of salary, increases of responsibilities, praise, and special assignments.

Adequate income will affect someone's performance then there is satisfaction in working. As E. A. Locke and G. P. Latham [3] theory opined that a satisfied worker will produce more and use this as a basis for relating motivation, satisfaction and job performance. The need to motivate workers is a fundamental issue in all organizations that are aimed at achieving set goals. Teachers like workers in other organizations, need the renewed energy brought by incentive from time to time if they are to perform effectively.

To improve teachers' welfare as an effort to improve the quality of education is required a sustainable education development strategy from the government. In this regard, V. T. Fadeyi, et.al. [1] found that there is sifnificant relationship between teacher salary, promotion and students academic performance. If necessary attention is given to teachers' welfare, this will bring about positive development in the educational system. It is hereby recommended that teachers' welfare scheme should be of paramount focus to the educational policy makers and administrators. They also recommended that teachers should be placed on meaningful salary scale and given opportunities to further receive training in their various disciplines for better performance to be achieved. In line with V.T. Fadeyi, et.al., F. Luthans [4] posited that welfare schemes are the strategies employed in motivating teachers for better job performance. These are in form of salary increment, gratuity, regular promotion, ensuring job security, and establishing cordial relationship among teachers.

In a related study, N. Frances, et.al. [5] examined how school welfare provision influences teacher performance. The findings are that school administrators provide only those welfare programmes that have a direct bearing on task accomplishment. Second, teachers' performance is high mainly on examination management, punctuality, and co-curricular activities. Third, school welfare provision is however too insignificant to cause a remarkable teacher performance. They concluded that school welfare provision will positively influence teacher performance if teachers are reciprocally committed to work and administrators meet teachers varied needs.

The concept of welfare is concerned with the total wellbeing of employees both at school and at home. H. Dale-Olsen [6] specifically defines school welfare as referring to teachers' health status and happines, as well as their safety, although he does not provide what constitutes teachers' happiness. In his research, H. Dale-Olsen [6] observed a positive correlation between wages and fringe benefits. Higher wages and more fringe benefits reduce the worker turnover rate. According to B. J. Keitany [7], the provision of well-being to teachers is a source of earning and satisfaction which is likely to increase their productivity because they are motivated and happy.

As a form of awareness of the government of Padang city in improving the quality of education through the improvement of teachers' welfare, it has been established the Regional Medium Term Development Plan (RMTDP) of Padang City by 2014-2019 which is primarily aimed at improving the quality and equity of education. With the RMTDP, the government of Padang City has done regulation in the field of education, by issuing Regulation of Mayor of Padang Number IA by 2015 on the amendment to the Regulation of Mayor of Padang Number 17 by 2014 on the Regional School Operational Assistance (*Bantuan Operasional Sekolah Daerah*, Bosda), and Decision of Mayor of Padang Number 82 by 2015 about the amount of Bosda. Bosda is the funds provided by the local government that is not accommodated by the School Operational Assistance (*Bantuan Operasional Sekolah*, BOS) from the central government that is especially prioritized for improving the welfare of principals, vice principals, homeroom teachers, non-permanent teachers and non-permanent employees with the aim of improving performance and quality of education services. The implementation of Bosda is more emphasized on improvement activities of education quality in the form of additional income for principals, vice principals, homeroom teachers, non-permanent teachers and non-permanent employees.

Based on the opinions above, the purpose of this study is to find the data about the implementation of Bosda in Padang City.

Theoretical Study

Educational policies are initiatives mostly by governments that determine the direction of an educational system and increase the quality of life of a people. A policy or regulation does not have to be rigid, as there should be room for adjustment if necessary after its formulation. Hoy and Miskel as cited by N. S. Okoroma [8] believe that "policies are not only formulated but also programmed, communicated, monitored and evaluated". The non-rigid nature of policies is confirmed by Lindblom as quoted by N. S. Okoroma [8] when he describes policymaking as a "process of successive approximation to some desired objectives in which what is desired itself continues to change under reconsideration". In fact, a good policy is one that can be reviewed as the need may arise. Lindblom believes that a wise policy maker cannot expect all their policies to achieve a one-hundred percent success. Regardless of how good a policy may be, its implementation may introduce some element of imperfection.

The realization of a regulation is mostly influenced by the level of implementation in the field. The implementation here is how to manage the funds of Bosda in order to be implemented well. In managing a program, there are various concepts presented by experts as in the following table:

Table 1. Aspects of Management

Fayol	Terry	Koontz	L. Gullick
Planning Organizing Leading Controlling	Planning Organizing Actuating Controlling	Planning Organizing Staffing Directing Controlling	Planning Organizing Staffing Directing Coordinating Reporting Budgeting

Source: S. Danim [9]

For this study, the management discussed is based on the management applied by the Bosda managers, that is adopting the concept stated above, with aspects of planning, implementation and evaluation of Bosda implementation. Planning is a process of thinking, selecting, and linking knowledge, facts, imaginations and assumptions for preparing a set of desirable decisions in the future. According to Henri Fayol as cited by C. W. L. Hill and S. L. McShane [10], "planning is a formal process whereby managers choose goals, identify actions to attain those goals, allocate responsibility for implementing actions to specific individuals or units, measure the success of actions by comparing actual results against the goals, and revise plans accordingly". And according to Beishline as quoted by M. Manullang [11] Planning determines what should be achieved (timing qualitatively) and when it should be achieved, where it should be achieved, how it should be achieved, who is responsible, and why it should be achieved. Planning activities focus on maintaining goals.

From the above opinion, it can be concluded that the planning is useful to provide clear direction for each activity, so that every activity can be cultivated and implemented as efficiently and effectively as possible. To be able to provide a clear direction, the planning should refer to the aspects of determining needs, formulating goals, setting goals, and revising planning.

The aspect of planning is more related to the abstract aspects of the management process, while the implementation emphasizes the activities directly related to the people in the program. Implementation of the bosda is an effort to make the planning become a reality, through the socialization, cooperation and commitment so that every activity can be implemented optimally in accordance with the roles, duties and responsibilities of program implementers.

Practically, the implementation is an effort to create a climate of cooperation among the staff of the program implementing so that the organization's goals can be achieved effectively and efficiently. The last aspect of Bosda implementation is evaluation activity. Evaluation is a systematic and planned activity to measure, assess and classify the implementation and success of the program. In general, the term of evaluation can be equated with the appraisal, scoring, and grading; the words that indicate attempting to analyze the policy outcomes in terms of unit of value. Specifically, evaluation is concerned with the production of information on the value or benefits of policy outcomes (W. N. Dunn, [12]). While, S. Arikunto [13] stated that evaluation is an activity to collect information about the operation of something, then the information is used to determine the right alternative in making decisions.

The main function of evaluation in this case is to provide useful informations to determine the policy to be taken based on the evaluation that has been done. Evaluation activities are the fairest way to determine the rewards for employees. Evaluation of work aims to ensure the achievement of goals and objectives. Besides, the purpose of evaluation is to know the position of activities both associated with the program that has been established and the ongoing process.

Evaluation can provide good and correct information. Giving criticize on clarification of a value of a goal and target, then create a policy method to achieve performance so that evaluated programs and activities contribute to reformulating the policy of an activity in the implementation of Bosda. The most important thing in evaluating the program and the process is how to know the deviations that occurred as early as possible, both in the planning and the implementation stages so that it can be done anticipation, correction and adjustments as soon as possible in accordance with the situation, condition and progress of the program.

METHODS

This research was a preliminary study of a research and development proposed to obtain the needs analysis in finding hypothetical model of Bosda implementation of Padang City. Sukmadinata [14] said that preliminary study in a research and development consisted of three stages, namely library study, field survey and modeling. However, this report was only up to the field survey stage. Library studies were conducted to examine the concepts and theories related to the model of Bosda implementation, while the field survey was conducted to collect the data. Furthermore, from the data analysis obtained from the field survey and refer to the concepts and theories that have been studied, it can be found problems or needs that will be used to find hypothetical model of Bosda implementation.

To obtain quantitative data through questionnaires, the sample was representative of headmasters, vice principal, homeroom teachers, non-permanent teachers and non-permanent employees receiving Bosda assistance, determined by purposive random sampling. While, for the qualitative data through interview, the informant was the manager of the Bosda. Data analysis performed on each component was different, in accordance with the objectives of the research to be achieved systematically. In this regard, Bogdan as cited by L. J. Moleong [15] stated that data analysis is a systematic process of searching and compiling transcripts of interviews, field notes, and other materials that have been collected to improve understanding and allow someone to present what he has found to others.

RESULTS AND DISCUSSIONS

The empirical data found in this research indicated that the Bosda policy had been implemented optimally, but there were still many technical obstacles faced. The obstacles came from the implementation process, both in the aspects of the planning, implementing, and at the evaluation stages. These conditions had an impact on the achievement of Bosda goals.

Description of Planning Data

In planning aspect, it is known that the indicator of determination of needs was not good, the indicator of formulating the objective was good enough, the indicator of setting target was not good, and the indicator of revising planning was good enough. It means that the overall indicator of the planning has not been implemented well. Problems in the planning aspect are the factors that influence the implementation of the Bosda. According to Beishline as quoted by M. Manullang [11], planning determines what should be achieved and when it should be achieved, where it should be achieved, how it should be achieved, who is responsible, and why it must be achieved.

From the above opinion, it appears that planning is essentially an activity of thinking of a better future, which describes the change, both quantitatively and qualitatively compared with the previous situation or current conditions. The Bosda planning process should be based on the real issues that occur in the field. The problems in planning are directly related to the problems of achieveing the objectives that have been formulated. It is very clear that the planning process is the starting point for the implementation of Bosda program in improving the quality of education.

Description of Implementation Data

The findings of the research on the implementation aspect showed that from the three indicators found the answer that the socialization activities were not implemented optimally, the cooperation was not done well, and the commitment was not good category. Socialization activities in the implementation of Bosda aims to equate perceptions in the implementation of activities so that the set targets can be achieved in a timely, effective and efficient manner. Socialization refers to all factors and processes that make each program to be aligned in implementation. It is important because if socialization is not done optimally, the goals of the program will be difficult to be achieved.

Socialization can realize solid cooperation. However, because the socialization activities were not implemented optimally, the synergic cooperation can not be achieved. Cooperation in the implementation of the Bosda is defined as an activity conducted jointly to achieve common goals. It means a job or action undertaken by each party in order to achieve the goals or targets that have been previously planned and also mutually agreed. Cooperation will arise if a person or group is aware if they have the same interests and at the same time. People who work together will have the knowledge and awareness to fulfill all responsibilities. It means that cooperation is a combination of various different areas in implementing a program that has the same goal. This cooperation will usually take time because the different areas have different workflow, but because it has the same purpose so that cooperation can be implemented.

Commitment was one of the factors that influence the level of achievement of Bosda implementation. It can be seen from the findings that the respondent's answer was in not good category. Based on the above conditions, the commitment should be a special attention because the commitment is a demand to each individual in accordance with his duties and functions in implementing Bosda.

Description of Evaluation Data

In determining a decision on what has been and will be implemented in the implementation of Bosda required program and process evaluations. The findings of the study indicated that the evaluation of the programs undertaken in the implementation of Bosda had not been done well and the evaluation of the process had not been implemented well, so every year the bosda recipient receives the same complaints.

The systematic evaluation need to be done in order the program can be implemented well. H-T. Chen [16] argued that program evaluation is the application of models, techniques and knowledges of evaluation, to assess and refine planning, implementation and effectiveness of a program systematically. In line with that, S. Arikunto [13] defined that the evaluation program is a process to determine whether the objectives of the program has been realized. Evaluation program is the process of determining the value, purpose, effectiveness or compatibility of something in accordance with the criteria and goals set previously. Further, the evaluation of the program aims to collect information regarding the implementation of the program used to conduct follow-up or decision-making activities. Thus, the process of program evaluation in the implementation of Bosda should be based on careful comparison to the data being observed by using certain standards that have been standardized.

Process evaluation involves targeted, planned, and systematic efforts to examine the process of implementing Bosda. Based on the findings indicated that the evaluation process was not performing well. A well coordinated program is mostly influenced by the extent to which the activity process can be evaluated thoroughly. Process evaluation should be done as well as possible. This is important because the results of process evaluation will be used as a consideration in future decision making. Process evaluation should be directed to control the achievement of the objectives and level of task control by the bosda implementers. By evaluating the process, it can be known the level of efficiency of the implementation activities, both in terms of strengths and weaknesses, so as to provide feedback in order to improve the implementation of Bosda.

CONCLUSION

Based on the findings and discussions that have been described above, it can be concluded that the needs analysis of the field conditions in the implementation of Bosda, found that in the aspect of planning with indicators of determining needs was in not good category, the indicator of formulating the objectives was in good enough category, the indicator of setting target was not good, and the indicator of revising planning was good enough.

The aspect of Bosda implementation on the indicators of socialization were not implemented optimally, the cooperation was not done well, and the commitment was not good category. Meanwhile, on the evaluation aspect, both program evaluation and process evaluation have not been implemented well.

It means that the overall indicators of planning, implementation, and evaluation of the implementation of the Bosda need to be addressed thoroughly and sustainably, so that the Bosda program of Padang City can be realized well. The implication is with the findings of this field condition, will be a reference in developing hypothetical model of Bosda implementation of Padang City.

ACKNOWLEDGEMENTS

I wish to acknowledge the management of STKIP PGRI Sumatera Barat and education department of Padang City for providing a chance for me to undertake this study. I also thank to the participants in this study for accepting to provide me with useful information that has greatly determined the success of this study.

REFERENCES

1. V. T. Fadeyi, A. O. Sofoluwe, and R. A. Gbadeyan, APJEAS, *Influence of Teachers' Welfare Scheme on Job Performance in Selected Kwara State Secondary School*, **2**, 4, 89-93 (2015)
2. A. D. Timpe, , *Manajemen Sumber Daya Manusia.* Jakarta: Media Komputindo (1994)
3. E. A. Locke, and G. P. Latham, Academy of Management Review, *What Should We do about Motivation Theory? Six Recommendations for the Twenty-First Century*, **29**, 3, 388-403 (2004)
4. F. Luthans, , *Organization Behaviour* (10th Edition), Boston: Mc Graw Hill/Irwin (2005)
5. N. Frances, S. Denis, and O. Vincent, IJEPRR, *The Interplay of School Welfare Provision and Teacher Performance: The Case of Ugandan Secondary Schools*, **3**, 1, 6-13 (2016)
6. H. Dale-Olsen, Labour Economics, *Wages, Fringe Benefits and Worker Turnover*, **8**, 1, 87-105 (2006)
7. B. J. Keitany, *Perceived relationship between Employee Welfare Programmes and Employee Performance at Kenya Pipeline Company.Unpublished Masters of Business Administration Thesis, Nairobi University Press* (2014)
8. N.S. Okoroma, Australian Journal of Adult Learning, *Educational Policies and Problems of Implementation in Nigeria*, **46**, 2, 243-263 (2006)
9. S. Danim, *Profesi Kependidikan*, Bandung: Afabeta (2010)
10. C. W. L. Hill and S. L. McShane, *Principles of Management*, New York: McGraw-Hill/Irwin (2008)
11. M. Manullang, *Dasar-dasar Management*, Yogyakarta: Gajah Mada University Press (2008)
12. W. N. Dunn, *Pengantar Analisis Kebijakan Publik*, Jakarta: UGM Press (2003)
13. S. Arikunto, *Dasar-dasar Evaluasi Pendidikan*. Jakarta: Bumi Aksara (2008)
14. Sukmadinata, *Landasan Psikologi Proses Pendidikan*. Bandung: Rosdakarya (2005)
15. L. J. Moleong, *Metodologi Penelitian Kualitatif*: Bandung: Rosdakarya (2007)
16. H-T. Chen, *Practical Program Evaluation: Assessing and Improving Planning, Implementation and Effectiveness*, London: Sage Publications (2005)

Do Learners Like Project in Their English Classes?: Impact of PBL on Students' Attitudes

Ni Luh Putu Ning Septyarini Putri Astawa[1], *Luh Putu* Artini[1] , and *Putu Kerti* Nitiasih[1]

[1]English Education Department, Ganesha University of Education, Bali, Indonesia

Abstract. This article presents the results of a study on the effect of Project-Based Learning (PBL) on students' English language learning (ELL) attitudes and how the activities influence teaching-learning process in a junior high school in Bali. This research applied an embedded mixed-method design in which the quantitative data were collected using close-ended questionnaires test, and the qualitative data were collected using interview, observation checklist, open-ended questionnaire, and field note. Paired-samples t-test was used to analyse whether or not there is a significant effect on students' attitude after being taught using PBL, while the qualitative data were analysed descriptively. The analysis shows a significant effect on students' attitude. PBL enhances students' learning quality in term of enthusiasm, confidence, and creativity learning ability while it also promotes teacher's teaching satisfaction. This study supports the implementation of PBL which enhances students' English Language Learning attitude and teaching-learning process.

1 Introduction

In 21st century learning, collaboration, critical thinking, creativity, and communication ability are the keys that students should have in order to be able to compete in the future [1, 2]. Students are demanded to be able to effectively analyse, evaluate facts, and communicate their thoughts or ideas using oral and written communication skills. Those abilities are being activated through collaborative learning which acquired the students to demonstrate their skills to work effectively in diverse society. The involvement of students' creativity is expected in order to achieve the aims of the study to help students to create new ideas or innovation.

Students' attitude is also considered as one of the important factors that can influence students in achieving the knowledge [3, 4, 5]. Learners' attitudes reflect their emotional involvement such as values and believes about certain educational situation [6]. Attitude is also classified into three different components, such as behavioural, cognitive, and affective attitude [7]. In language learning context, the behavioural component of students' attitude is defined as the tendency of the students to follow or adopt some particular learning behaviours. The cognitive competent refers to the students' beliefs, thoughts, and point of view about the learning while affective competent is stated as the students' feelings and emotion towards the language learning process.

PBL is one of the strategies that are potentially used in 21st century learning activities. In PBL, students learn the knowledge by working on the project, while they are demanded to plan, implement, reflect, and evaluate their own learning by working on task that reflects real-life issues [8, 9]. Students enquire some specific objects to answer challenging problems that are designed to give opportunities to the learners to activate their critical-thinking. In PBL, students' innovation is promoted through the activity of obtaining data by using multiple sources of information, such as technology tools [10].

PBL is one of the keys of strategies for creating independent thinkers while also promotes students to collaborate with others [11]. Students take responsibility by pursuing, reviewing, and responding to the feedback of their own learning [12]. Their communication skill is activated by some activities such as discussion and reflection. In discussion, each of the students shares and negotiates his/her idea [13]. While figuring on how well they work collaboratively, the students also reflect on how well they express their contribution and listen feedback to reconstruct their project [11].

The theory of 21st century learning acquires a new way of learning especially in Indonesia. Instead of focusing on teacher-centred, the students-centred learning should be carried out in teaching and learning process [14]. It is very unfortunate that in fact there are some teachers in Indonesia who do not implement any kind of strategies that reflect the 21st century learning such as PBL. Instead of using an effective learning strategy, the teachers choose to implement a conventional strategy which is based on teacher-centred learning. The facts were found by many researches that were administered in Indonesia. Some previous researches [15, 16] stated that the lack of teachers' knowledge and implementation on alternative teaching strategies caused students' low attitude in learning. The students were found unmotivated and had less creativity in joining the learning process.

In the preliminary research, teacher-centred was carried out throughout the learning process. The learning was conducted monotonously using single source (i.e. students' worksheet). The teacher mostly used direct instruction method in guiding students to answer some questions which were written in the book. From the open-ended questionnaire, the researcher found that the students considered the class activities as boring. The interview with the teacher confirmed the teaching strategies that mostly used were repetitive and lack of varieties.

Those problems inspired the study to implement PBL as a reflection of 21st century learning. Previous studies found the relation between PBL and students' motivation and attitude, for example a research which was done by Konrad [17] who found a significant effect on students' achievement and motivation towards algebra learning. The data of her research showed substantial increase in students' attitude on learning math. Another study which was done by Tseng, et al [18] showed that students' attitude towards engineering learning was increasing significantly after being taught by using PBL.

Those studies prove the benefits of using PBL on students' attitudes in learning. The studies that have been conducted previously were not specified in teaching English class and investigated PBL instructions effect on teaching and learning process. This study focuses on answering two research questions: (1) Is there any significant effect PBL on students attitude towards English language learning ?, (2) How PBL instruction influence teaching and learning process?

2 Method

This research administered embedded mixed-method design. Both of quantitative and qualitative data were collected separately approximately at a single phase of study. Twenty-eight students were the participants of this study. The effect of PBL on students' attitudes towards English Language Learning (ELL) was analysed by using quantitative test. The quantitative data play superior roles which are collected through close-ended questionnaire before and after intervention, while interview, observation checklist, open-ended questionnaire, and field note were used to collect the qualitative data. The close and open-ended questionnaire was constructed based on the 3 components of attitude towards ELL namely behavioural, cognitive, and affective attitude. The influences of PBL on teaching and learning process are presented by the qualitative result of this study. In analysing the data, paired-samples t-test was used on the quantitative result while the qualitative data was analysed by using Miles and Huberman's flow model.

3 Findings and Discussion

In embedded mixed-method design, quantitative and qualitative data were collected in single phase of study. The first question was answered by the result of quantitative data. The close-ended questionnaire as quantitative data were analysed in order to find out whether or not there is a significant effect of students' attitude on ELL after being taught by using PBL. The questionnaire consisted of 35 points of questions which were developed based on Wenden's [7] theory namely behavioural, cognitive, and affective attitude. The findings are presented below.

3.1 The Effect of PBL on Students' English Language Learning

Paired-samples t-test was administered to analyze the first question of the study. The close-ended questionnaire was distributed twice, before and after the PBL implementation in order to find out the significant difference of students attitude on ELL. The result of the paired-samples t-test are presented below.

Table 1. Result of Attitude Questionnaire Paired-samples t-test

Questionnaire Test	Paired Differences					t	df	Sig. (2-tailed)
	Mean	Std. Deviation	Std. Error Mean	95% Confidence Interval of the Difference				
				Lower	Upper			
Before Treatment – After Treatment	-28.679	19.226	3.633	-36.134	-21.224	-7.893	27	.000

Paired-samples t-test was used to analyzed the data from close-ended questionnaire in order to answer the first question of the study. As displayed in Table I, t= -7.893 and the significant value of 2-tailed is .000. If the standard of significance is 0.05, it indicates that there is a significant effect on students attitude towards ELL after being taught using PBL.

The open-ended questionnaire was also distributed in order to find out students' attitudes towards ELL after being taught using PBL. The set of open-ended questionnaire covers 3 components of attitude on ELL namely behavioral, cognitive, and affective attitude. There are 3 questions which represent every component of attitudes itself. It was found that PBL positively influences students' attitudes towards ELL in term of behavioral, cognitive, and affective attitude which are presented in the following quotation: *"I want to learn more about English because it becomes more interesting for me. In the future, I will try my best to give a better presentation about my group work." (S08/M)*

The previous quotation indicates that there is a change of students' behavior on ELL. Before the PBL was carried out in the classroom, the students demonstrated the unwillingness to learn by showing their passiveness in the classroom. It was also reflected through the open-ended questionnaire answers which defined that more than one quarter of the total students had willing to join and do the activities (i.e. listening to the teacher's lecture and answering the questions on the book). After being taught using PBL, the learners tended to adopt positive particular learning behavior as evidence on the previous quotation.

It was also found out that the students percieved English learning as a hard thing to learn before the intervention of PBL. The strategies which were used previously by the teacher seemed to give students some disadvantages. The evidence of the open-ended questionnaire result which represents by the following quotation proved it: *"Before, I thought I would never have any interest to learn English because it was boring and definitely not attracting for me. However, now I do think that I will have different perspective on English class, because it becomes more fun and interesting and there are so much things that I am able to do beside just answering questions like usual. I am also very glad because it seems like it is fine when I accidentally make mistakes." (S17/F).* From the previous quotation, it was also discovered that the students' perspective became more positive towards English learning after being taught by using PBL as presented on quotation above which indicates that the students see the learning activities as a fun process. It also changes students' perspective, instead of just acquired to answer the task correctly, the students also perceive the project as a whole process of learning and try their best effort in doing project without really focusing on some mistakes that appear along the process of learning itself.

PBL was also proven to affect students' feeling on ELL. Before PBL was implemented in the classroom, the students' open-ended questionnaire result shows that they tended to feel weary towards the learning. The students was used to learn by focusing on the text-book in which they were asked to listen to their teacher's lecture and presentation about a subject, and finally answered the questions related to the subject. Those conventional method clearly did not work to promote students' attitude on ELL. After being taught using PBL, students feel differently by demostrating their positive feedback on the affective attitude. They experience PBL as a fun and exciting activities to do in learning English. The finding is represented by the following quotation: *"I feel so excited when I made public figure's biography because I had chance to do some interesting activities such as cutting, gluing, decorating the biography, and sharing my idea."*(S07/M)

Those findings are in line with Wanchid and Wattanasin's [19] study. Their research showed that the students achieved high to moderately positive attitude after being taught using PBL in their English class. The students gained their own autonomy in learning. It means that the students were able to have more power to direct their own learning without being neglected or disregarded.

The data presented above show the significant value of 2-tailed was less than 0.05 ($0.000 < 0.05$). The findings are also supported by the students' open-ended answers which indicate the students' positive feedback on their attitude towards ELL. Therefore, it can be conlcuded that PBL has a significant effect of students' attitude toward ELL.

3.2 The Influence of PBL on Teaching-Learning Process

In order to answer the second research question on how PBL influences teaching and learning process, the observation checklist was used to check if the teaching and learning procedures had been in conformity with the priciples of PBL implementation. These data were also supported by the field notes.

From the observation checklist, it was found that the students showed positive attitudes toward the teaching and learning process which was conducted based on PBL instruction. They showed active participation on the discussion by giving their attention, ideas, and constribution to complete the project. They were also able to appropriately place themselves in a group by doing and finishing each of their tasks as a member of the group.

The findings showed that PBL could promote students' enthusiasm which was noted on observation checklist. The observation checklist result shows that students were actively participated in the learning process. The significant evidence was also shows students' response to the open-ended questionnaire which is represented by following quotation: *"The activities are so challenging and also exciting for me. I wish we have more project to do like this in the future"(S19/M)*

The following picture demonstrates the students' enthusiasm in joining the clasroom activities. The students actively contributed in doing group work by completing their own responsibility as a team member in order to produce the final result of the project. They were all actively engaged and involved in the process of creating the final product of the project.

Fig. 1. Students' enthusiasm in joining in the classroom activities.

Rochmahwati [20] also found out that there was a positive effect of PBL implementation on students' enthusiasm. She analyzed students' enthusiasm during teaching and learning process through the interview on the students and the lecture of speaking II subject at STAIN Ponorogo. The students significantly improved their attitude and skill along the process of learning.

Also, the result shows that PBL could promote students' confidence in their learning process. The learners were placed in the condition where they needed to work as a group. They were given chance to contribute in the group work by stating their ideas, comments, and addition in order to complete the project. The finding is represented by the following quotation: *"I feel more brave and confident in stating my idea to my teammates, the teacher, and every one in the class."* (S27/M)

Fig. 2. A student bravely presented her work in front of the classroom.

The image illustrates students' presentation in front of the classroom. All of the students were able to present their work without hesitation. They did not seem to worry about making mistakes in the process of learning. The active participation of all students during the discussion and presentation supports the previous data which is proposed to analyze the influence of PBL on students' teaching and learning process in term of their confidence. Students' level of confidence is one of the important aspects that support their communication (oral and written) skills during the process of learning itself [21, 22].

Based on Pinzon Castaneda [23] who explored that PBL was able to enhance students' self-confidence. It showed that students felt more confident when they were exposed to the speaking activity which acquired them to communicate on different topics during several presentations about the project. Self-confidence is also stated very beneficial for the students because it normalize the mistakes that students would posibbly make. Even though students were struggling and making some mistakes in the process of learning, they would accept that as normal and appropriate along the process of mastering the knowledge.

Students' creativity is also influenced by PBL instruction. The students were given chance to develop their higher thinking skills through the projects which are implemented in the classroom. The students were able to express their idea which full of creativeness and carried it on the project that they worked on. The finding is represented by the following quotation: *"I am so happy to able to make the project as creative as I could by using every information that I have about my family and relative identity. I am also able to write many things about them in my album. The decoration is precisely like what I want it to be." (S03/F)*

Fig. 3. Students' final products of a project activities (Family Album).

The students' creativity is represented by the image of the result of a project above. The students were given chance to explore their higher thinking skills by being allowed to express their idea about how the project should be done. The products which were created from the activities reflected the students' effort to produce an excellent product based on their perspective and interest despite the mistakes they could possibly make.

The students were also given chance to use multiple resources in order to finish their work. As found on students' product which is presented in Figure 4, it represents the students' biography works. They were required to use internet, magazines, newspaper, etc in order to finish the product. They were allowed to see examples of some innovative biographies as wall displays on the Internet in order to make the draft and layout of their own biography.

Fig. 4. Students' final products of a project activities (Biography)

The simillar finding was also found by Mihardi, Harahap, and Sani [24] on their research entitled "The Effect of Project Based Learning Model with KWL Worksheet on Student Creative Thinking Process in Physics Problems". It was found that the students were creatively working on the project when they were taught using PBL. The findings showed by the result of analysis on the students' average value who were in the experimental class was higher than those who were in the control class. PBL was stated as a successful method to be carried out to activate students' creative thinking in finishing the project.

The qualitative result also shows that PBL influences teacher's teaching satisfaction on the students' learning experience result. The teacher was also satisfied by other factor such as the teaching material (projects), activities, and students' feedback on the activities itself. The following quotation represents the influence of PBL on teacher's teaching satisfaction, *"I feel so glad by the students positive response on the activities. They are actively engaged and willing to join the classroom. It fulfils my excpectation on how the learning process should be done in my English class." (Teacher)*

Hixson, Ravitz, and Whisman [25] also found consistent findings on their research. The teacher experienced a satisfying process while developing their professionalsm to become familiar with the implementation of PBL. It gave them a feeling of satisfaction towards their profession as a teacher.

4 Conclusions

Quantitatively, this research found that PBL significantly affects on students' attitude towards ELL. Qualitatively, it also shows the influence of PBL instruction on the teaching and learning process. The statistical findings show that there is a significant effect of PBL on students' attitude towards ELL. In addition, PBL also promotes students learning process by enhancing their enthusiasm, confidence, and creative learning ability. From the teacher's point of view, it enables him to enhance his satisfaction in teaching English.

Practically, the result of this study presents practical reference for effective teaching. Firstly, considering on the significant effect of PBL on students' attitude towards English language learning, PBL is recommended to be carried out by the teachers in the English classes. PBL also successfully enhanced the teaching and learning process by promoting the students' enthusiasm, confidence, and creativity learning ability while it also promotes teacher's teaching satisfaction. The Indonesian government is also expected to facilitate teachers' with more knowledge about valuable methodologies that can be implemented in the classroom which promotes the 21st century learning skills such as PBL. Finally, for the future study, it is expected that there will be more research on PBL in different level of English as Foreign Languge setting.

References

1. C. D. Jerald, *Defining a 21st century education*. (USA, Center for Public education, 2009)
2. V. Greenhill, *21st Century Knowledge and Skills in Educator Preparation*. (USA, AACTE, 2010)
3. W. J. Popham, Education and Urban Society, **26(4)**, 404-416 (1994)
4. N. Reid, *Getting started in pedagogical research in the physical sciences* (UK, LTSN Physical Sciences Centre University of Hull, 2003)
5. D. Fakeye, Journal of Social Sciences, **22(3),** 205-211 (2010)
6. R. C. Gardner, Language learning, **30(2),** 255-270, (1980)
7. A. Wenden, *Learner strategies for learner autonomy* (New York, Prentice-Hall, 1991)
8. F. Stoller, Forum, **35(4)**, 4-24 (1997)
9. S. Boss. Extracted on 12 june 2017 in http://www.edutopia.org/project-based-learning-history (2011)
10. J. Simpson, *Integrating project-based learning in an English language tourism classroom in a Thai university (*Doctoral dissertation, Australia, Australian Catholic University, 2011)
11. S. Bell, The Clearing House, **83**, 39-43 (2010)
12. J. Ravitz, N. Hixson, M. English, J. Mergendoller, Proceedings of American Educational Research Association Conference, **16** (2012)
13. Y. C. Gai Mali, Indonesian Journal of English Education, **3(1),** 89-105 (2016)
14. D. Kember, Higher Education, **58(1),** 1-13 (2009)
15. E. B. Ferti, Adiyalmon, Ranti, Jurnal STKIP PGRI Sumatera Barat Padang (2014)
16. L. A. Hastuti, Doctoral dissertation of Universitas Muhammadiyah Surakarta (2016)
17. T. F. Konrad, *The effects of project-based learning on student achievement and motivation in remedial high school algebra* (Thesis, USA, Montana State University, 2014)
18. K. H. Tseng, C. C. Chang, S. J. Lou, W. P. Chen, International Journal of Technology and Design Education, **23(1),** 87-102 (2013)
19. R. Wanchid, K. Wattanasin, Mediterranean Journal of Social Sciences, **6(5),** 581 (2015)
20. P. Rochmahwati, Kodifikasia, **9(1),** 199-222 (2016)

21. A. Heyde, *The relationship between self-esteem and the oral production of a second language* (Unpublished doctoral dissertation, USA, University of Michigan, 1979).
22. N. L. P. N. S. P. Astawa, L. P. Artini, P. K. Nitiasih, Journal of Language Teaching and Research. (to be published)
23. R. J. Pinzon Castaneda, Cuadernos de lingüística hispánica, (**23**), 151-170 (2014)
24. S. Mihardi, M. B. Harahap, R. A. Sani, Journal of Education and Practice, **4(25),** 188-200 (2013)
25. N. Hixson, J. Ravitz, A. Whisman, *Extended Professional Development in Project-Based Learning: Impacts on 21st Century Skills Teaching and Student Achievement.* (USA, West Virginia Department of Education, 2012)

The Implementation of Character Education at Senior High School

*Julia** and *Tedi* Supriyadi

Program Studi PGSD, Universitas Pendidikan Indonesia Kampus Sumedang, Jl. Mayor Abdurachman No. 211 Sumedang, Jawa Barat, Indonesia

Abstract. This paper is aimed at analyzing the implementation of character education at Senior High School in Sumedang Regency, West Java, Indonesia. A content analysis method was employed to collect the data with interviews with six teachers from six different senior high schools, which represent the rural, transitional, and urban areas. The findings revealed that: (1) not all teachers understand the concept of character education; (2) the character education has not been done systematically or has not had the specific design/model for the teaching and learning process. Most teachers embedded the character values during the teaching and learning process as a form of character education. For example, through Qur'an recitation, learning tasks, group discussion, lecture, socio-drama, observation and admonition, and even through teachers' model. Meanwhile, outside the class, character education was implemented through competition and extracurricular activities; (3) the evaluation of character education was relatively varied, such as an observation followed by admonition, group guidance, and also focusing on behavior and affective assessment in the classroom learning process. This research implied that it is needed to make a policy on a program development through the Bureau of Education to improve Senior High School teachers' knowledge and skills in implementing the character education. **Keywords:** character education, senior high school, implementation, evaluation.

1 Introduction

Character education is now one of the main topics in the implementation of education in Indonesia. Since its inception in January 2010, which was reinforced by the president's speech in the commemoration of the National Education Day on May 2, 2010 [1], character education continues to be developed through various education programs both within the Ministry of Education and Culture, such as research grants from Center for Education Policy Research (*Puslitjakdikbud*), as well as within the Ministry of Research, Technology and Higher Education (*Kemenristekdikti*).

On the local government level, such as Sumedang Regency in West Java in 2013 released a Regent Regulation that provides guidelines to the Sundanese people to behave within the norms of the *kasundaan* (Sundanese) culture and character [2, 3]. The Sundanese character consists of 10 main characters that include *taqwa* (pious/religious), *someah*

* Corresponding author: juli@upi.edu

(friendly), *surti* (empathetic), *jembar* (broadminded), *brukbrak* (transparent), *guyub* (getting along with others), *motekar* (creative), *tarapti-taliti-ati-ati* (meticulous, careful), *junun-jucung* (work well, consistent), and *punjul-luhung* (high competence) [2]. The ten characters, in addition to having distinctive differences, also have similarities to the 10 pillars of character education set by Lickona, namely trustworthiness, respect, responsibility, fairness, caring, honesty, courage, diligence, integrity, and citizenship [4, 5]. The creation of various programs or regulations on character education, one of them is triggered by the behavior of the Indonesian people, particularly school students, which increasingly show behaviors of anticulture and anticharacter of the nation which have been reported by various mass media [6-9]. However, a person's characters are essentially influenced by the intercommunication within the surrounding community. Psychologists show that as part of socialization, people internalize the norms and values of their society [10-12].

Since the national culture and character education have been implemented for approximately seven years, it is necessary to see the application in the field, specifically in schools. In general, based on the results of preliminary studies, teachers in schools do not have a design in applying character education, especially high school teachers in Sumedang Regency. According to Suryadi [13], character education is a value education program that is packed not in the form of subjects, but in the embodiment of the school as a character institution. Therefore, this study attempts to analyze three aspects, namely: (1) how is the teacher's understanding of character education?; and (2) what is the form of character education implementation?

2 Methods

This study employed a content analysis method. The data were taken by interviews with six teachers from six high schools in Sumedang Regency, West Java, Indonesia, in the 2015/2016 school year. The selected schools represent two schools from the urban area, two schools from the transition area, and two schools from the rural area. The first respondent (R1) was academically qualified as a civics education teacher, the second respondent (R2) was an Islamic religion teacher, the third respondent (R3) was a Mathematics teacher, the fourth respondent (R4) was a Physics teacher, the fifth (R5) was an Indonesian language teacher, and the sixth respondent (R6) was a Guidance and Counseling teacher. The data analysis was conducted based on Miles and Huberman [14] flow model, as translated by Rohidi [15], which consists of data collection, data reduction, data display, and conclusion.

3 Results and Discussions

3.1 Understanding of Character Education

The presented findings are translated results of the interviews with the six respondents.

Interviewer	: How is your understanding of character education?
R1 responded	: "(I) have read and tried to understand the culture and character education program."
R2 responded	: "(I) have understood the cultural education and character program in KTSP (school-based curriculum)."
R3 responded	: "(I) have only understood the cultural education and character program a little."
R4 responded	: "(I) have only understood a part of the cultural education and character program."
R5 responded	: "(I) have understood the cultural education and character program, because at school where I teach, there was a teacher that had been

trained in culture and character education and it had been socialized."

R6 responded : "(I) have understood the cultural education and character program."

Based on the aforementioned data, it can be identified that the understanding of character education among teachers at the senior high school level was varied. From the example, three teachers had understood culture and character education, two teachers had not fully understood the culture and character education program, and one teacher had just tried to understand the culture and character education program. Of the six respondents, only one respondent mentioned the training on culture and character education, thus the information about culture and character education from the training was obtained which was then disseminated to the teachers. In order to provide better learning, the teachers should improve their understanding and skill as it has done by the teachers in Russia. Hauer [16] stated that Russian teachers worked with a trainer to improve their understanding and skills in character and moral education. Dengan demikian, guru bisa memberikan perlakuan yang berbeda terhadap peserta didik, bahkan guru bisa memiliki karakter sebagai a caring teacher as a navigator, gardener, leader, or servant [17].

3.2 Implementation of Character Education

The following are translated results of interviews with the six respondents.

Interviewer : What is the learning model used in applying character education?

R1 responded : "As a Civics teacher, culture and character education is already integrated in the learning of civics education, and it is taught to learners starting from their arrival at school, in the classroom, outside the classroom, and outside the school. In the model of culture and character education that I apply, there must be mutual respect for people who are speaking, asking a question, and listening, and there is even a lesson about discussion procedure, how the procedure of good discussion, I think... at least as a Civics teacher, I have standards about the procedure of deliberation in general and in the Sundanese way."

R2 responded : "As a teacher of religion (Islam), then religious education is automatically very relevant to culture and character education, because the goal of religious education is to foster the children to have noble characters. The implementation model of culture and character education (that I use), in addition to being fed with various sciences, may be with examples of the performance of a teacher as an educator and how a teacher must behave in front of the children so that the child can see and, 'Oh, this one must be followed.'"

R3 responded : "If any students committed violations, they would be immediately reminded or called to the Guidance and Counseling room or to the teacher's office to receive counseling as a form of character education. Sometimes, in the classroom, things that are related to it (character education) were included. For example, if there was a child breaking a rule, we would talk in front of the classroom for a minute or two for everyone."

R4 responded : "During classroom learning, some aspects of culture and character education have been implemented, such as instilling honesty, discipline, whether in submitting tasks or in doing exercises. There are no specific models, but the attention to character education has existed... sometimes there are students who go forward, bringing their friend's book when they did not do their own task... because I observe and eventually know which students are being honest or dishonest"

R5 responded : "The results of the teacher training on culture and character education are socialized and implemented in an integrated manner in every subject, including Indonesian language; for example, every student is required to memorize 10 verses of the Qur'an on the first hour of the day. In learning Indonesian language, there is a lesson to criticize sentences... I think it is already there... the philosophy of Sundanese language in every attitude and behavior of students either during the learning or outside the learning."

R6 responded : "Cultural education and character are included in Guidance and Counseling material, such as the values of life, which connect the daily life of the students with society life... as well as about ethics and norms, either written or unwritten which are usually discussed in the classroom or in group guidance. For the learning model, once there were examples (presented in) a socio-drama about life values... such as saving ten magical people from a sinking ship."

There are various opinions regarding the implementation of character education [18-34]. In other words, in applying character education, there are various references that can be implemented, such as referring to Kohlberg [34] who divides students' character development in three stages: pre-conventional reasoning, conventional reasoning, and post-conventional reasoning, or referring to Lockheed and Verspoor [35] who formulate the concept that character education is performed through four stages; the initial stage, the formal stage, the transitional stage, and the stage of meaning. Otherwise, referring to Hendarman, Saryono [36] who state how to strengthen character education on the basis of classroom, school culture, and society.

The above data shows that the implementation of character education programmed in national level have not had a special form or design in the learning process yet. As from the case examples, four teachers believed that culture and character education was relevant or integrated with the subjects they taught, one teacher implemented it through learning tasks, and one teacher implemented it by giving reprimands or direct behavioral guidance. The implementation of Sundanese character education had been programmed by the local government by the obligation to use Sundanese language and traditional outfit. However, since it had not been integrated into the curriculum, the program did not run well, especially in the use of Sundanese language by learners. The other forms of implementation were quite diverse, such as teaching *panta-panta* (levels of politeness) in Sundanese language, organizing competitions related to Sundanese culture, integration with school vision and mission, and integration with 3S programs. All of the teachers also expressed that they had not had specific models or designs for Sundanese character education, and they implemented character education as supplementary in the learning process, such as making the students to get used to read Al-Quran before the first lesson begins, and to get used to kiss the teachers' hand palm. The forms of the implementation of character education in the process of learning in the classroom were embodied in several forms, such as group discussion with the emphasis on the particular character, teacher exemplary, short lectures on the norms, observations and warnings about the learners' characters, and socio-drama about life values. Without employing the character education model, at least, the teachers should be the model of the character education as Abebe and Davis [37] point out that teachers are expected to be the moral model in school. Therefore, Waterson [38] notes that teachers teach students not only methodology and content, but also values.

4 Conclusions

The conclusion drawn from the results of this study is that high school teachers in Sumedang Regency did not show their understanding of the concepts of character education that have already existed globally. Therefore, both the form of implementation and evaluation tend to be in a form of guidance or daily activities that are usually delivered to the learners in the learning process at school, both routine and incidental.

The authors would like to thank the *Bappeda* of Sumedang Regency and *Puslitjakdikbud* who had facilitated this study, and to the Board of Education of Sumedang Regency who assisted in the study process.

References

1. Marzuki. Revitalisasi Pendidikan Agama di Sekolah dalam Pembangunan Karakter Bangsa di Masa Depan. Jurnal Pendidikan Karakter. **1**:64-76 (2013).
2. Bappeda. *Peraturan Bupati Sumedang Nomor 113 Tahun 2009 tentang Sumedang Puseur Budaya Sunda*. Sumedang BK. (2009).
3. Fahdini R, Julia, Suhandani D. *Implementasi Pendidikan Karakter Kasundaan Sebagai Upaya Membangun Budaya dan Karakter Bangsa (Survei Pada Anak SMA di Kabupaten Sumedang)*. Sumedang: Bappeda Kab. Sumedang. Tidak Dipublikasikan. (2015).
4. Lickona T. *Educating for character: How our schools can teach respect and responsibility*. New York: Bantam Books (2009).
5. Wahyu. *Masalah dan Usaha Membangun Karakter Bangsa*. In: Budimansyah D, Komalasari K, editors. Pendidikan Karakter: Nilai Inti Bagi Upaya Pembinaan Kepribadian Bangsa. Bandung: Widya Aksara Press (2011).
6. Kompasiana.com. Anak SD Melakukan Pembunuhan Berencana Terhadap Temannya. (2012). Accesed on: 20 October 2016. Available from: http://www.kompasiana.com/iraannisa/anak-sd-melakukan-pembunuhan-berencana-terhadap-temannya_552bb5b36ea8344f6f8b457b.
7. Sindonews.com. Masih di Bawah Umur, Terdakwa Kasus Penganiayaan Dibebaskan. (2016). Accesed on: 10 December 2016. Available from: https://metro.sindonews.com/read/1104013/170/masih-di-bawah-umur-terdakwa-kasus-penganiayaan-dibebaskan-1461596823.
8. Liputan6.com. Kedapatan Mencuri HP, Tiga Anak SD Diringkus. (2010). Accesed on: 13 Mei 2016. Available from: http://news.liputan6.com/read/292114/kedapatan-mencuri-hp-tiga-anak-sd-diringkus.
9. Pikiran-rakyat.com. Masih Ada Pelajar SMP Pesta Miras Oplosan di Sumedang. (2014). Accesed on: 18 April 2015. Available from: http://www.pikiran-rakyat.com/jawa-barat/2014/12/19/309139/masih-ada-pelajar-smp-pesta-miras-oplosan-di-sumedang.
10. Campbell EQ. The internalization of moral norms. Sociometry.391-412 (1964).
11. Henrich J, Boyd R, Bowles S, Camerer C, Fehr E, Gintis H, et al. In search of homo economicus: behavioral experiments in 15 small-scale societies. The American Economic Review. **91**:73-8 (2001).
12. Mazar N, Amir O, Ariely D. The dishonesty of honest people: A theory of self-concept maintenance. Journal of marketing research. **45**:633-44 (2008).
13. Suryadi A. *Pendidikan Karakter Bangsa: Pendekatan Jitu menuju Sukses Pembangunan Pendidikan Nasional*. In: Budimansyah D, Kamalasari K, editors.

Pendidikan Karakter: Nilai Inti Bagi Upaya Pembinaan Kepribadian Bangsa. Bandung: Widya Aksara Press (2011).
14. Miles MB, Huberman AM. *Qualitative data analysis: An expanded sourcebook*: sage (1994).
15. Rohidi TR. *Analisis Data Kualitatif*. Jakarta: Universitas Indonesia (2007).
16. Hauer J. The Impact of Character Education on Russian Teachers: An Informal Survey, Spotlight on Josephine Hauer. Journal of College and Character. **1**:null-null (2000).
17. Waterson R. The Art of Teaching: Ernest Shackleton as Navigator/Teacher. Journal of College and Character. **12**:null-null (2011).
18. Allers R. *Character education in adolescence*. New York: Wagner (1940). 188 p. p.
19. Lickona T. Eleven principles of effective character education. Journal of moral Education. **25**:93-100 (1996).
20. Bronfenbrenner U. Soviet methods of character education: Some implications for research. Religious Education. **57**:45-61 (1962).
21. Johnston CH. National Institution For Moral Instruction. The Journal of Education. **77**:243- (1913).
22. Dreikurs R. *Character education and spiritual values in an anxious age*. Boston: Beacon Press (1952). 23 p. p.
23. Kuntz LF. Religion and Character Education. Religious Education. **49**:255-63 (1954).
24. Yulish SM. *The search for a civic religion : a history of the character education movement in America, 1890-1935*. Washington, D.C.: University Press of America (1980). xii, 306 p. p.
25. Neall BS. *The concept of character in the Apocalypse, with implications for character education*. Washington, D.C.: University Press of America (1983). xii, 224 p. p.
26. Chinawa PO. *The teacher's role in character education of the youth following the mind of Rudolf Allers*. Rome: s.n. (1984). 135 p. p.
27. Gill (Aberystwyth) C. Plato and the Education of Character. Archiv fur Geschichte der Philosophie. **67**:1-26 (1985).
28. Beswick R. Character Education. OSSC Bulletin. **35** (1992).
29. Kelsey IB. *Universal character education*. Edinburgh: Pentland Press (1993). xvi, 100 p. p.
30. Brooks BD, Goble FG. *The case for character education : the role of the school in teaching values and virtue*. Northridge, CA: Studio 4 Productions (1997). 161 p. p.
31. Lockwood AT. *Character education: controversy and consensus*. Thousand Oaks, California: Corwin Press (1997). viii, 71 p. p.
32. Mulkey YJ. The history of character education. Journal of Physical Education, Recreation & Dance. **68**:35-7 (1997).
33. Lickona T. The return of character education. Educational leadership. **51**:6-11 (1993).
34. Kohlberg L. Revisions in the theory and practice of moral development. New directions for child and adolescent development. **1978**:83-7 (1978).
35. Lockheed ME, Verspoor AM. *Improving primary education in developing countries*. Washington, D.C: Published for the World Bank, Oxford University Press (1991).
36. Hendarman, Saryono D, Supriyono, Kamdi W, Kartadinata S, Latipun, et al. *Konsep dan Pedoman Penguatan Pendidikan Karakter*. Jakarta: Pusat Analisis dan Sinkronisasi Kebijakan (2017).
37. Abebe S, Davis W. Transcendence in the Public Schools: The Teacher as Moral Model. Journal of College and Character. **5**:null-null (2004).
38. Waterson RA. What is the Role of the Teacher? To Guide the Studies of Others. Journal of College and Character. **10**:null-null (2009).

Instructional materials development through 4D model"

*Ade Gorbi Irawan**[1], *Ni nyoman Padmadewi**[2], and *Luh Putu Artini**[3].
Postgraduate program English Education Department, Universitas Pendidikan Ganesha.

This research aimed at developing Instructional materials for the second semester students in economy faculty of Panji Sakti University. This research was a Research and Development (R & D) study. The type of the research was 4D model The steps of this study were: (1)Define,(2) Design, (3)Develop, and (3)Disseminate. In the "Define" step, There were two questionnaires employed in the research: (1) needs analysis questionnaire (2) expert judgment questionnaire. The needs analysis questionnaire was distributed to the students to identify the target needs and the learning needs of the students. The expert judgment questionnaire was used to evaluate the material. The data from the needs analysis questionnaire were analyzed using frequency and percentage and the results were then used as the basis to design and develop the materials, while the data from the expert judgment questionnaire were analyzed using descriptive statistics. Based on the findings from the materials evaluation, the content, the language, the presentation, and the layout of the materials are appropriate. The mean score of all aspects of the developed materials was 4.18. This indicates that the developed material is categorized as excellence material.

Keywords: *Research and Development , 4D model*

Introduction

English stands as a foreign language. Indonesia has also decided English as a formal subject, that has been taught as a formal subject from primary school to university. In the University, English course that introduced as the general English course and English for Specific purpose. In the English teaching and learning process in the university, the lecturer has to prepare many things such as the material, classroom management, and many other aspects.

Materials take an important part in teaching and learning process [2]. Good and appropriate materials will give positive influence to the students' learning process. From that reason, the researcher conclude that, developing appropriate English learning materials is necessary. It can help the students to learn English more easily. It is because they are learning useful materials that are related to their future job. Nowadays, some universities give opportunity to their students by having special policies, one of those policy was tailored program in which a group of students be able to request the convenient schedule, because they mainly work during the day and continue the study after their finish their work. This situation made the material must mastered in the limited time. The lecturer has to design the learning media which give the learner possibility to study independently. The lecturer does not only have to design the media but also has to find the way to improve the learner motivation, make the learner more active and improve the learner ability.

In the process of developing the appropriate intructional materials, the researcher has to determine the material needed to be developed by using the needs analysis[3][4], there were some steps in developing the instructional materials such as: determining the materials needed, determining the method and evaluate the developed materials to find it's quality . In this study, the main purposes of this research were:

1) To investigatee the materials needs for General English course II for Second semester student of Economic Faculty in Panji Sakti University.

Adegorbi@gmail.com

2) To develop Instructional materials for the second semester students of Economic Faculty in University of Panji Sakti Singaraja
3) To describe the quality of the designed materials.

In the process of developing the appropriate the instructional materials, the researcher must understand the basic principles materials developments are clearly proposed by [5] the principle explained are as follows:

1. the researcher has to consult the syllabus to identify the theme, sub-theme, the objectives that are prescribed in it and the researcher must consider some sources, such as books, magazines, newspaper, and other authentic materials that related to the topic
2. Materials should be workable for the students and teachable for the lecturer.
3. Material should relevant to the students' level and their real world.
4. Material should suit the students' needs and interests.
5. Material should be feasible to be developed for language skills or language components.
6. Materials should stimulate interaction.
7. Materials should encourage learner to apply their developing language skills to the world beyond the classroom.

Based on the principles of development materials described above, it seems that developing materials should contains 4 language skills namely: listening, speaking, reading and writing which at the same time gives space for language elements, such as structure, vocabulary, pronunciation and spelling in the teaching process in classroom setting. One important thing should be considered in developing materials is the using authentic materials, in which are originally used in real situation and are not designed for use in language teaching, such materials are used in the classroom to expose the learners to language in real use. The using authentic materials are significant because it increase students' motivation for learning, makes the learners be exposed to the real language. The main advantaged of using authentic materials have pointed by [6] are (1) they have positive effect on learners' motivation, (2) they provide authentic cultural information, (3) they provide exposure to real language, (4) they relate to some closely to learners' needs, and (5) they support a more creative approach to teaching.

Method and Instrument

Here, the researcher using 4 D models [1]. The Four- Door (4D) is a simple instructional design model that helps the researcher to design the product that will help the learner to improve their ability in learning process. 4 D models divide into 4 steps such as: Define, Design, Develop and Disseminate. The participants of the study are 30 second semester students in Economy faculty of Panji Sakti Singaraja.

There were 4 methods of data collection in this study such as observation, interview, questionnaire, and document study. The first question was "what the materials are needed to be developed for the second semester students of economy faculty of Panji Sakti Singaraja? To answer this question, there were some instruments used in this step. Observation sheet, interview, the questionnaire and document study were conducted in this step. To find out The data of needs analysis, the researcher using the observations sheet, questionaire and the interview to find out the material needed by the students.

In this study, the data will be analyzed descriptively. The data from observation, interview, and document study conducted in preliminary observation will be described qualitatively in order to know the problem faced in real situation. While data from the questionnaire will be described quantitatively. In measuring the quality of the developed prototype, the data gathered through some instruments which will be analyzed quantitatively. The validity of the prototype was judged by 2 experts.

RESULT AND DISCUSSION.

4D models was a simple method in the process of developing a product. The model of research divided into some simple steps namely: "Define", "Design", "Develop" and "Desseminate". The method started by defining the students and material needs. Based the interview and the questionaire result, the Instructional Materials developed based on the student's interest, student's English ability and student needs for supporting the student's future job. The strenght of 4D model was the researcher has to decide the appropriate media and format selection in the step of design. The materials obtained the illustration, and pictures to make the students understand the concept of the material. The design of the materials gave the effect for materials delivery process. [3][7]Materials evaluation was conducted by giving an expert judgment questionnaire to a materials expert in English language teaching. There were five aspects evaluated by the experts such as:

Adegorbi@gmail.com

The Objectives, Material aspect, methods, learner's aspects, character building, and Instructions of the Material. The result of the questionnaire was explained below: The judge evaluated the material by using instrument in form of evaluation rubric provided by the research.

Table 1: *The result of expert judges on the product*

	Judge I		Judge II	
The objectives of materials	4	High	4	High
Materials	4.15	Highest	4.07	Highest
Methods	4	High	4.75	Highest
Learner	4	High	4	High
Character Building	4	High 4	5	Highest
Instructions	4	High	4	High
Total Score	4.12		4.15	

The table above presented the result obtained from the expert judges toward the quality of the product. Based on the table, it could be calculated that the mean score from two experts' judge was 4.18, totally there were 24 items examined by the experts' judges. The maximum score could possibly give to each item was 5 while the minimum score was 1. The materials evaluated by six learning aspect.

Table 3 : The interval score of the category of product quality

Criteria	**Interval Score**	**Category**
$\overline{X} \geq Mi+1,5Sdi$	$\overline{X} \geq 4.05$	**Excellent Material**
$\overline{X} \geq Mi+0,5Sdi \leq \overline{X} <$ Mi + 1,5 *Sdi*	$3.35 \leq \overline{X} < 4.05$	**Good Material**
$\overline{X} \geq Mi$ - $0,5Sdi \leq \overline{X} <$ Mi + 0,5 *Sdi*	$2.65 \leq \overline{X} < 3.35$	**Average Material**
$\overline{X} \geq Mi$ - $0,5Sdi \leq \overline{X} <$ Mi - 0,5 *Sdi*	$1.95 \leq \overline{X} < 2.65$	**Below Average Material**
$\overline{X} < Mi$-$1,5Sdi$	$\overline{X} < 1.95$	**Poor Material**

Based on the data above, it could be seen that the mean score of the developed material was 4.18 It indicates that all two judges mostly gave high score to each of item in the instrument. Furthermore, if it was converted into the criteria and category of the product quality, the main score was in the first criteria in which the interval score was $\overline{X} \geq 4.05$. Since the main score obtained from the experts' judgment was 4.18, the developed was categorized as excellent material, or in other words, it was a high quality materials. The last step was "desseminate". Here, the materials formed in the form of module and converted into "Pdf". It format makes the materials be able to opened in the student's smartphone. This strategy gave the students oppurtunity to focus of the process of learning.

Conclusion

This article has emphasized the significance of instructional materials development in the university. The instructional materials that are appropriate for a particular class need to have an underlying instructional method and

Adegorbi@gmail.com

technique which suit the students needs[8]. Based on the explanation about the problems faced by the lecturer in teaching English for economy students, it could be concluded that developing instructional materials was very important. it was because the students will use the target language in their future job. The result of developing instructional materials based on the need analysis [3] and the principle of material development [5] will create the appropriate materials which cover the objectives of the English course and student's weaknesses

Acknowledgement

The result of this research requires a lot of feedback from some supervisor. For their suggestion, supports, guidance, criticism, and motivation during the research. . the researcher could not thank them enough for their dedication, consistency, and commitment.

REFERENCES

1. Thagarajan, Dorothy S. Semmel. *Instructional Development For Training Teachers of A Extional Children A Sourcebook.* Bloomington Indiana,Indiana University. (1974)
2. Tomlinson, B. ,Masuhara, H. *Developing language course material.* RELC **,1-2**(2004).
3. Hutchingson, Tom and alan Waters. *Engslish for Specific Purpose: A Learning Centered Approach*,Cambridge university press (1987)
4. Richards, Jack C . *Curriculum Development in Language Teaching*. USA:
   a. Cambridge University Press. (2001).
5. A. Kosasang, .*The effectiveness of instructional module for teaching*
   a. *culture related in holiday of speakers of English for Mattayom three students*,l Choochon Tessaban **pp. 143-145.**(2003,)
6. Nunan, D. *Principles for Designing Language Teaching Materials.* Guidelines **10, 1-24** (1988).
7. Tennyson, R, D *et.al. Instructional Design*; Instructional Perspective, Theory, Research and Model. London: LEA Publishers. (1997).
8. Alan, Maley. Review: *Materials evaluation and design for language*
   a. *teaching. ELT Journal,* **58**(4).394-396. (2004).
9. Sombat Rahmat *Using English Instructional Module by B-SLIM Model to*
   a. *Promote English Reading Comprehension of High School Students.* International Journal of Information and Education Technology (2016)
10. Pakkan, G. *Language Teaching Materials Evaluation and Selection,*
   a. *Adaptation and Development.* Istanbul: Sürat ELT (1997).

Adegorbi@gmail.com

Evaluation on the use of animated narrative video in teaching narrative text

Rahmat Soe'oed*, *Noor* Rachmawaty, and *Mohamad* Huzzin As'ari.

English Department, Mulawarman University, Jalan Kuaro, Gn. Kelua, Samarinda Ulu, Kota Samarinda, Kalimantan Timur 75119, Indonesia

Abstract. In the 21st century, our life is strongly affected by the information technology. Educational technology has been rapidly improved by the development of audiovisual tools. Teachers may choose a number of different types of resources for teaching purposes, including videos and movies. Therefore, this study is aimed at evaluating animated narrative videos from YouTube for the teaching narrative text and identifying potential factors which influence the quality of educational videos. The videos were examined by using assessment rubric to see the quality and suitability of animated narrative videos which might be used in the teaching narrative text. The rubric was adapted from Prince Edward Island (PEI) Department of Education: Evaluation and Selection of Learning Resources. It consists of four criteria, content, structure, instructional design, and technical design In addition, the study presents critical awareness of how these aspects can be interpreted to measure animated narrative videos and at the same time the engagement of the teachers in exploring animated narrative videos used in classroom.

Keywords: Animated Narrative Video; Video Evaluation; Teaching Narrative Text.

1. Introduction

Information and communications technologies have significant roles in today's educational case [1]. Nowadays technology cannot be separated from educational activities. Information technology strongly influences the life of the people and plays a significant part in today's human society development. It is necessary to take advantage of the modern technological facilities in improving the development of English language teaching [2]. Studies have shown various advantages to the students related to the general use of technology for EFL, such as the improvement of motivation, self-concept and mastery of the basic skills, the increase of student centered learning and engagement in the learning process [3]. In this case, teachers may use technologies and learning resources for teaching purposes.

Many teachers use various learning resources such as educational YouTube videos.. Founded in 2005, the YouTube has quickly become a leader in online media. It is an internet application through which people can upload, share, and watch. There are millions

*Email Address: mr.soeoed@yahoo.com

of messages being presented in this media every day [5]. Educational video is not a specific genre but various genres of video, integrating a variety of production styles, techniques and conventions for the purpose of education [6]. One of the main advantages of using video in the classroom is that the students can both hear and see the language [7].

This study was related to the evaluation of educational YouTube videos. They were very limited in number and most of them were made for non language educational purposes such as medical, patients' health, veterinary and animal sciences [8-9]. The evaluation of educational videos uploaded by Indonesian on YouTube has not been widely exposed. However, the benefit of using YouTube videos in EFL classes has been documented in several studies [10-12].

The aims of this study are to evaluate the educational YouTube videos and identify potential factors which influence the quality of educational videos.

2. Method

This study adopts a mixed method design. Quantitative method was applied in evaluating the quality of the animated narrative videos. Video selection criteria were formulated based on several points. Theoretically, the videos fulfill the classification of educational videos and their duration lasted between 1 to 6 minutes. The videos in this study were produced by Indonesian people. The last criterion was the genre of the narrative text. The videos were selected approximately within a month. The evaluation of the videos was conducted by three experts in EFL and Informational Technology. 13 out of 25 collected videos were classified as educational ones and evaluated by the raters. The instrument used in this study was a video evaluation rubric adapted from PEI of Department of Education: Evaluation and Selection of Learning Resources [4]. They consisted of four aspects, 5 items for content, 4 items for structure, 16 items for instructional design, and 9 items for technical design. The validity of the instrument was administered through content validation. The collected data of the evaluation rubric were analyzed by descriptive statistics using SPSS. This study was also backed up by qualitative data. Since only few studies on video evaluation were conducted particularly in Indonesia, thus qualitative method was applied [13].

3. Research Findings and Discussion

The research findings and discussion are presented in accordance with the formulated research questions. The first part of this section addresses the evaluation of the animated narrative videos and discusses the aspects of content, structure, instructional design and technical design. The second part discusses the potential factors which influence the quality of the animated narrative videos.

3.1 Content

Content is the first aspect and it consists of five items. They were features reflecting the accuracy and appropriateness of the content towards curriculum and students' needs. 2 out of 5 items have averaged rating of 3.0 which indicates that the content of the videos goes with the curriculum and difficulty level of the content. The average mean of the rating was 2.72 which indicate that the content quality of the videos was categorized into "Disagree."

It indicates that the content of the videos still needs to be improved particularly for item number 5 in which real world activities are integrated in the content.

Figure 1. Raters' rating of animated narrative videos for content aspect

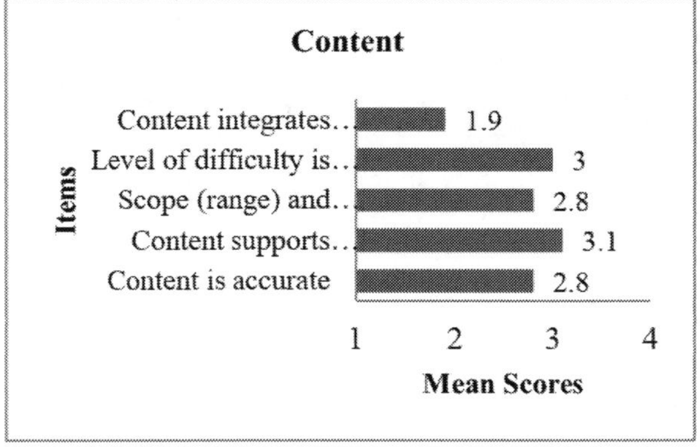

3.2 Structure

Structure was the second aspect addressed in the video evaluation. The structure aspect concerns with arrangement of the videos as well as length and its alignment with the planned activities in the videos. The evaluation was made by the raters using 4 items. Only items 3 and 4 have average rating of 3 and the others have average mean 2.72 which indicates that most items of the structure of the videos do not reach the well-structured videos.. This indicates that the structure of the videos was categorized into 'Disagree'.

Figure 2. Raters' rating of animated narrative videos for structure aspect

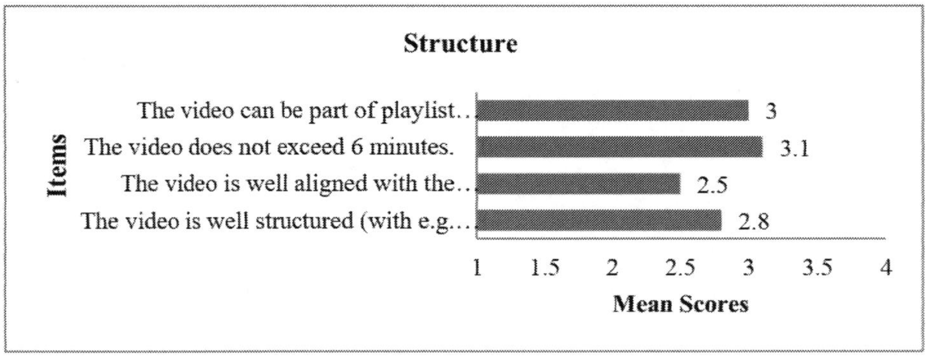

3.3. Instructional Design

The evaluation of the instructional design of the videos covers the examination of its goals, objectives, teaching strategies, and assessment provisions. 16-items were used in rating the instructional design aspect.

As shown in figure 3, the highest average mean was 3 for item number 2. Meanwhile, the lowest average mean was 2.2 for item number 1. The lowest rating score indicated that most of the videos did not clearly state the learning goal and neither adequate assessment. The average mean rating for instructional design was 2.61 categorized into 'Disagree'.

Figure3. Raters' rating of animated narrative videos for instructional design aspect

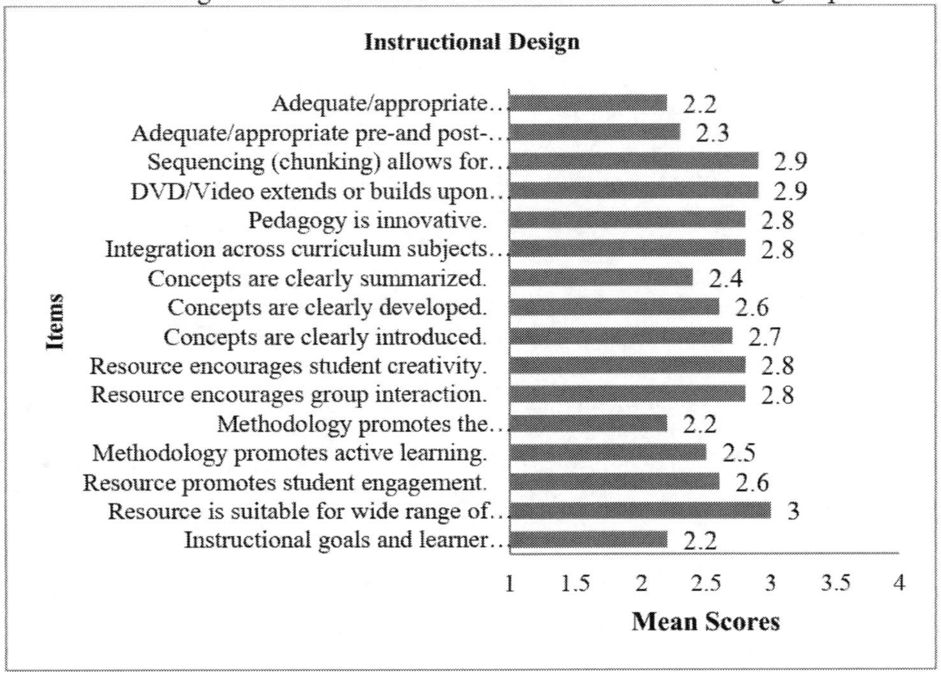

3.4 Technical Design

The last aspect in the evaluation process was technical design which addresses the design of visual and auditory information. Its role in enhancing the learning and developing efficient mental processing is very important. The technical features of this aspect are used as the indicators of good quality videos. As shown in table 4, 3 out of 9 items have averaged mean '3' which indicates that the sound's quality, music and visual effects were appropriate for the topic presented in the videos. The remaining items have average mean less than '3'. The lowest score was 2.6 for item 2. This indicates that improvement on the tone and clarity of the voice should be made. The average mean for technical design was 2.88 which was categorized into Disagree.

Figure 4. Raters' rating of animated narrative videos for technical design aspect

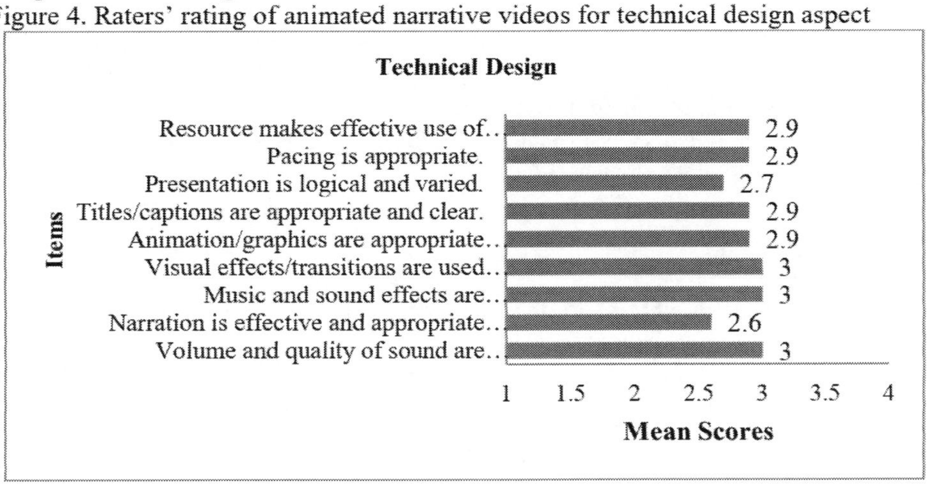

Based on the evaluation made by the raters, various considerations should be made regarding the use of the 13 animated narrative videos in English classroom. The results revealed that none of the aspects under the evaluation reached the average rating 3 (Agree) nor 4 (Strongly Agree). The rating score for content, structure, instructional design and technical design were all under '3'. Each aspect of the evaluation contributes to the overall quality of an educational video. The content of the video must be useful. The video should stimulate, motivate and inform the learner to act on the information. Ideally, learners should consider and incorporate the ideas presented [15].

Another point is the instructional design and structure as they are used to control the design process. Looking at the average rating results of the videos particularly for structure and instructional design aspects which were still below 'Agree' category, it can be inferred that thorough and careful planning was still insufficiently performed. Educational videos should in line with teaching and learning purposes as Clark claims that learning is only influenced by the instructional method, irrespective of which medium is used [16].

With regards to the technical design, the visual and sound quality of the video is an important point as video with poor quality may distract learners' engagement and interest [18]. The results in this study revealed that the narration in the videos was still considered ineffective and inappropriate as it did not correspond to the visuals. In order to overcome this issue, authors of animated narrative video should equip themselves with adequate knowledge on audio and video software and how to use them in order to produce a good educational video as by using appropriate teaching media to present relevant information may increase the efficiency of the self-learning process [19].

Based on the results of the ratings for each video as well as their individual rating descriptions, it was found that the main factors which possibly influence the quality of the animated narrative videos were; lack of preparation, unclear learning goals and objectives, limited knowledge and skills in creating effective and appropriate educational videos. Referring to the identified factors mentioned previously, therefore it is necessary for teachers to carefully select which videos that is countable to be used as learning resource in their teaching. To help them selecting the effective and appropriate videos, Catlin (2013) provides list of points to consider in evaluating the quality of videos. The points are the video has a clear objective, the content is accurate, the creator of the video is an expert or enthusiast on the topic, a balance between educational and entertainment content, the media used in the video should enhance the material presented and the video presents a brief summary in the end.

Teachers could minimize the issues related to the low quality videos by producing their own videos. The following are some considerations to be taken into account while producing videos such as follow three steps production procedure, they are planning, producing and publishing, understand the elements of video design and implementation and master the integrated knowledge or the integrated knowledge is known as Technological Pedagogical Content Knowledge (TPACK).

4. Conclusion

Available online videos on YouTube or other platforms have high possibilities to be used as teaching aid in the classroom. However, an evaluation regarding its quality is necessary to performed in order to maximize its potential benefits as learning tool for students. In addition, results from video evaluation will improve the quality of the next educational videos which will be produced by experts, teachers or people who are enthusiast on particular fields.

5. Acknowledgment

This work was partially funded by the Postgraduate Program on English Education, Mulawarman University, East Kalimantan. Result of this work was presented at The Global Conference on Teaching, Assessment, and Learning in Education (GC TALE) which was conducted in Bali on 28-29 August 2017.

References

1. Z. Kazanci, Evaluating English Language Teaching Software For Kids: Education or Entertainment or Both?, **8** (2009)
2. B. Nomass. The Impact of Using Technology in Teaching English as a Second Language, **3** (2013)
3. J. Stepp-Greany, Technological Environment: Student Perceptions On Language Learning In A Implications For The New Millennium, **6** (2002)
4. *Evaluation and Selection of Learning Resources: A Guide*. Charlottetown, Prince Edward Island, 8 (2008)
5. YouTube. Upload New Video (2007)
6. G. Winslett. What counts as educational video?:Working toward best practice alignment between video production approaches and outcomes. **30** (2014)
7. J. Harmer, *The Practice of Language Teaching*, 282 (2001)
8. S. A. Azer, et al. Evaluation of the Educational Value of YouTube Videos About Physical Examination of the Cardiovascular and Respiratory Systems. **15** (2013).
9. C. Klupiec. Development and evaluation of online video teaching resources to enhance student knowledge of livestock handling, **92** (2014)
10. N. Rulia, M. Al-Hafidz. Teaching Writing A Recount Text by Using Mr. Bean Video to Junior High School Students. Unpublished Thesis (2014)
11. N. Indrasari. Improving Students' Writing Skill of Narrative Texts by Using Short Videos. Unpublished Thesis (2010)
12. S. Nur. A. Optimizing The Use of Youtube Video to Improve Students' Competence in Writing Procedure Text. Unpublished Thesis (2012)
13. Strauss, A., & Corbin, J. The Basics of Qualitative Research: Techniques and Procedures for Developing Grounded Theory. (2nded.). Thousand Oaks, CA: Sage. (1988)
14. B. P. Beaudin, D. Quick. Instructional Video Evaluation Instrument. Vol. 34 N0. 3 (1996). https://joe.org/joe/1996june/a1.php
15. F. G. Krink,., & Gustafson K. L. Instructional technology: A systematic approach to education. New York: Holt, Rinehart, & Winston (1986)
16. Clark, R.E. Reconsidering research on learning from media. Review of Educational Research, 53(4), pp. 445-460 (1983)
17. Melissa N.L.Y.A., Fong, S.F. & Ong, S.L. An Evaluation of Instructional Videos in EDUWEBTV: Technical Qualities, Pedagogical Aspects, Engagement and Perceived Impact on Learning. Malaysian Journal of Educational Technology, 10(2), pp. 101-113. (2010)
18. J. Harmer. The practice of English language teaching (3rd ed.) Edinburgh Gate: Longman. (2001)
19. Ruiji, L. The development on multimedia teaching resources based on information processing theory. International Journal of Advancements in Computing Technology, 4(2), 58-64. (2012)

Penalty kick skill through knee tuck jump exercise and barrier hops exercise

Lingling Usli Wargadinata[1*], and *Muchamad* Ishak

[1]STKIP Pasundan Cimahi, Pendidikan Olahraga, Jln. Permana 32 B Cimahi, West Java

Abstract. Football always attracts society's attention. Unfortunately in practice, at school for example, students have difficulties in mastering penalty kick. This research aimed to know the influence of knee tuck jump exercise towards penalty kick result in football and the influence of barrier hops exercise towards penalty kick result in football. The design of this research is pretest–posttest design. The population taken was the entire students from class XI SMK 3 LPPM-RI Batujajar which consisted of 126 students. Purposive sampling technique was used to determine the sample. Intentionally, the writer chose as many as 30 students who joined football extracurricular as sample then divided them into two groups namely group A (knee tuck jump exercise) and group B (barrier hops exercise). Based on the result, there was significant difference in mean score between pretest and posttest in group A (29.00) than group B (26.00) towards the result of penalty kick in football. The result of compared t from the difference of two results is 4.92 bigger than t table 1.70. Therefore, knee tuck jump exercise gives more significant result than barrier hops exercise towards penalty kick result in football to the students of football extracurricular in SMK 3 LPPM-RI Batujajar.

1 Introduction

Football is one of famous sports that is played in team. It attracts society's attention start from children to adult. Many clubs that is associated with football for recreation or achievement purposes. Football is a team game that is played by two teams whereas each of team consists of eleven players and one of them is goalkeeper. [1]

A good team consists of players who have teamwork skill, master the basic technique and be able to control the game. There are some basic techniques that must be mastered in football. Basic techniques in playing football such as kicking, shooting, dribbling, stopping, controlling the ball and goal keeping techniques that are adjusted to its function and situation in the field. [2]

Foot is a dominant part of body in doing basic techniques in football. Passing the ball means transfer the ball from one's foot to another player's foot by kicking. [3] Besides, kicking can be used as a technique to score a goal. A player must master kicking technique and develop shooting technique. The player is expected to do shooting and score a goal from various positions in field. [4]

Kicking technique can be executed when the ball come to a standstill or move. Learning kicking technique is not easy. Kicking technique is usually done using arch in order to control the ball. Another factor that should be noticed is leg power. Leg power is a power produced by the contraction of leg's muscles in short time. Power is muscles' ability in making maximal strength in fast time. [5] Power as neuromuscular system ability in producing strength as great as possible in the shortest time. [6] It can be said as neuromuscular system ability to cope with resistance by contraction speed as high as possible.

Leg is the foundation of human's body to support human's activities and it has crucial role in in every sport field. The power of leg's muscles is very determining since its function is to swing the leg from back to forward which is done fast in short time. In obtaining fast and tough kick needs power. Power has big influence towards kick's strength and velocity. [7]

Many students especially in SMK 3 LPPM-RI Batujajar lack of mastering penalty kick technique because they have less intensity in learning kicking technique whereas kicking technique is the most basic technique that must be developed. Based on the researcher's observation to the eleventh grade students in SMK 3 LPMM-RI Batujajar, the students still could not do penalty kick well. They tended to kick the ball weakly and slowly. They did not have enough power to do penalty kick.

A big power of leg is needed to produce good kicking technique. To exercise leg's power, plyometric method must be implemented. Barrier Hops (jump using barrier) and knee tuck jump (jump without barrier) can be implemented to increase the power of leg's muscles.

Knee tuck jump is one form of jump exercise which is started with standing position and folding knee's joint. Then, the arms are placed beside the body. After that, the arms are swung up and jump vertically. When jumping vertically, both of thighs almost touch the chest and back to ready position. This is done continually.

Barrier hops is the exercise that is done on goal post or obstacle which has length between 30-90 cm. The barrier is placed in some lines with certain spacing depend on athlete's ability. The barrier will fall if the athlete makes mistake. [8]

To increase the result of penalty kick, the researcher tried to give some exercises that improve students' ability in doing penalty kick such as knee tuck jump and barrier hops. It is hoped that this implementation will increase the result of penalty kick to be stronger, faster and tougher especially to the students of SMK 3 LPPM-RI Batujajar.

2 Research method

The method used in this research is experimental method. This method is experimentation to see the result of a learning process. In the other word, experimental research is research which is done by manipulating research object and there is a control. In this research, the researcher reveals the illustration of the influence of knee tuck jump exercise and barrier hops exercise towards penalty kick in football. The design of research is pre test – post test design. The design can be seen in following figure 1.

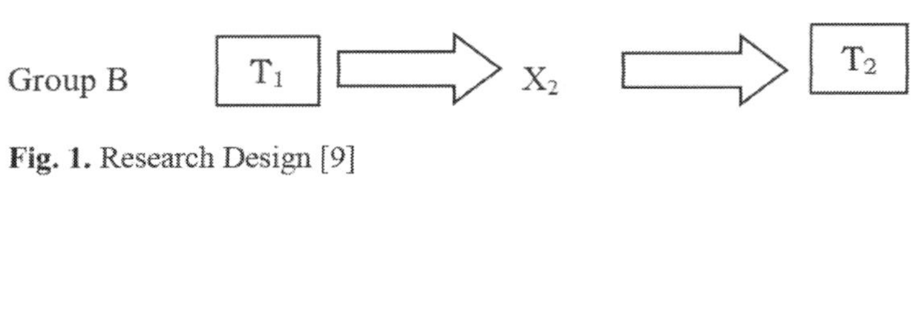

Fig. 1. Research Design [9]

Note:
T1 : pretest
X1 : knee tuck jump exercise
X2 : barrier hops exercise
T2 : posttest

There are two groups of sample. Group A is group which is given knee tuck jump exercise and group B is group which is given barrier hops exercise. The population was the eleventh grade students of SMK 3 LPPM-RI Batujajar that consisted of 126 students. Then, from the population it was chosen as many as 30 students as the sample. The sample was divided into two groups.

The learning process is done three times in a week on Tuesday, Thursday and Saturday for two months. This is based on Sumosardjono (2009, p. 12) who said the frequency of exercise must be done at least three times in a week because of one's endurance will decrease after 40 hours if he doesn't do exercise.

3 Result and discussion

3.1 Mean score and standard deviation

Table 1. The result of mean score and standard deviation calculation

Group	Test Period	Mean score	Standard Deviation
Group A	Pretest	22.80	2.88
	Posttest	51.80	2.40
Group B	Pretest	22.67	2.90
	Posttest	48.67	3.18

Based on the table, it can be seen that in group A, the mean score of pretest is 22.80 and the mean score of posttest is 51.80. Besides, the standard deviation of pretest is 2.88 and the standard deviation of posttest is 2.40. In group B the mean score of pretest is 22.67 and the mean score of posttest is 48.67. Besides, the standard deviation of pretest is 2.90 and the standard deviation of posttest is 3.18.

3.2 Normality test

Table 2. The result of normality test

Group	Test Period	Lo	L table	Result
			0.05 : 15	
Group A	Pretest	0.167	0.220	Normal
	Posttest	0.145	0.220	Normal
Group B	Pretest	0.169	0.220	Normal
	Posttest	0.137	0.220	Normal

According to the table, it can be concluded that in group A, it was obtained Lo pretest as many as 0.167 and Lo posttest 0.145. They are smaller than L table 0.220. So, the data of pretest and posttest in group A is normal. Besides, in group B, it was obtained Lo pretest as many as 0.169 and Lo posttest 0.137. They are smaller than L table 0.220. So, the data of pretest and posttest in group B is normal.

3.3 Homogeneity test

Table 3. The result of homogeneity test

Group	F	F table	Result
Group A	1.45	2.48	Homogenous
Group B	1.20	2.48	Homogenous

From the table, it can be seen that the value of derived F in group A is 1.45 and the value of derived F in group B is 1.20. They are smaller than F table 2.48 which means that the hypothesis is accepted. In other words, the data from pretest and posttest from both of groups are homogenous.

3.4 Hypotheses test

Table 4. The calculation of significance increase the study result of group A and group B

Group	Test Period	Mean Score	Standard Deviation	t test	T table 0.95 (14)	Result
A	Pretest	22.80	3.64	30.81	1.76	Significant
	Posttest	51.80				
Difference		29.00				
B	Pretest	22.67	4.44	22.68	1.76	Significant
	Posttest	48.67				
Difference		26.00				

Based on the table above, the value of derived t in group A is 30.81 bigger than t table 1.76 which means the implementation of knee tuck jump exercise gives significant influence. Then, the value of derived t in group B is 22.68 bigger than t table 1.76 which means the implementation of barrier hops exercise gives significant influence.

Also, it is found that there is differences in studying result between group A (30.81) and group B (22.68). The value of derived t from group A is bigger than group B (30.81 > 22.68). It means there is more significant increase in study result using knee tuck jump exercise rather than barrier hops exercise in SMK 3 LPPM-RI Batujajar.

Table 5. The calculation of Significance difference the study result of group A and group B

Group	Mean score	Standard deviation	T test	T table	Result
A	29.00	3.64	4.92	1.70	Significant
B	26.00	4.44			

Based on the table, there is difference in mean score between pretest and posttest that appear significantly in group A which is implemented knee tuck jump exercise (29.00) rather than group B which is implemented barrier hops exercise towards the result of penalty kick in SMK 3 LPPM-RI Batujajar. The calculation of t is 4.92 bigger than t table 1.70. Therefore, the hypothesis is accepted stated knee tuck jump exercise gives more significant result than barrier hops exercise towards penalty kick in football in SMK 3 LPMM-RI Batujajar.

According to data analysis above, it is known that the data from pretest and posttest are normal and homogeneous. Besides, the result from the calculation of pretest and posttest from both of group shows that the implementation of knee tuck jump exercise and barrier hops exercise give significant result. From the calculation of significance difference, it is known that knee tuck jump exercise has more significant influence rather than barrier hops exercise with the value of derived t 4.92 bigger than t table 1.70.

4 Conclusion

Based on the result of research, the conclusions are follows:

4.1 In group A, after being given treatment knee tuck jump exercise as long as two months, there is significant increase towards penalty kick in football to the students of SMK 3 LPMM-RI Batujajar.

4.2 In group B, after being given treatment barrier hops exercise as long as two months, there is significant increase towards penalty kick in football to the students of SMK 3 LPMM-RI Batujajar.

4.3 Group A which was given treatment knee tuck jump exercise gives more significant influence rather than group B which was given treatment barrier hops exercise towards the result of penalty kick in football to the students of SMK 3 LPPM-RI Batujajar with the calculation of t test 4.92 bigger than t table 1.70.

References

1. Sucipto, *Sepakbola Latihan dan Strategi,* 7 (2005)
2. U. Lingling, *Pembelajaran Sepakbola*, 36 (2013)
3. Koger, *Latihan Dasar Andal Sepakbola Remaja*, 19 (2005)
4. Mielke, *Dasar-dasar Sepakbola*, 67 (2008)
5. H. Awan, *Metode Melatih Fisik Pencak Silat,* 200 (2006)
6. Persunay, *Latihan Kecepatan dan Kekuatan*, 2 (2005)
7. Scheunemann, *Dasar Sepak Bola Modern*, 147 (2012)
8. C. Donald A., *Jumping into Plyometric*, 40 (2007)
9. Sugiyono, *Manajemen Penelitian*, 10 (2011)

Integrating *tri hita karana values* in teaching reading: students' and teachers' opinions

I Gede Sukarma[1*], *Putu* Kerti Nitiasih[1], and *I Gede* Budasi[1]

English Education Department, Postgraduate Program, Ganesha University of Education, Jl. Udayana, No. 11, Bali 81116, Indonesia

Abstract. This paper aims at explaining: students' and teachers' opinion on reading materials integrating *Tri Hita Karana* values and students' reading comprehension before and after the implementation of that reading materials. The data of the study were collected through questionnaire, scoring rubric, pre-test, and post-test. The data were analyzed qualitatively and quantitatively. The result of the study shows that the students and teachers have positive opinion toward the reading materials that integrate *Tri Hita Karana* values. It can be seen from the result of the questionnaire that shows both the students and the teachers gave more positive comments than negative comments. The result of the scoring rubric that aims at measuring the quality of the reading materials shows that both the students and the English teachers gave a high score that indicates the reading materials have a high quality. Furthermore, the implementation of reading materials that integrate *Tri Hita Karana* values also improved students' reading comprehension. It is proven by the result of pre-test and post-test that were delivered before and after the implementation of the reading materials that integrate *Tri Hita Karana* values.During the pre-test, the students'mean score was 65,18 and it improved into 74.29 on the post test.

Keywords: *tri hita karana*, reading materials, character values

1 Introduction

Juvenile delinquency is still a big issue in Indonesia. The common juvenile delinquency in Indonesia includes students brawl, unwed pregnancy, drug abuse, motorcycle gang, and theft [1]. In 2013, there were 1.200 cases of unwed pregnancy, while for students brawl and motorcycle gang, according is increasing in 2016 [2]. Realizing those problems, Indonesian government considers it is very important to build students' character through character education. Therefore, the new curriculum that is applied in Indonesia, which is known as Curriculum 2013, gives a special attention to character education. The new curriculum requires all teachers in Indonesia to insert character values in their teaching and learning process. The main purpose of this curriculum is to shape the individuals who are faithful in God, good in characters, confident, successful in learning, responsible citizens and positive contributors to the civilization [3].That condition makes it important to have other materials as supplementary materials to support the Indonesian government's program. This study aims at developing character-based reading materials for teaching English, which is limited to develop reading materials that insert one of the Balinese local wisdom concepts, which is known as *Tri Hita Karana (THK)*. This study was conducted at *Sekolah Menengah Kejuruan Negeri 3 Singaraja*. Reading materials are chosen to be developed in this study because reading is an important skill in mastering English as a foreign language. Rreading can improve students' vocabulary mastery [4] and reading also could improve students' speaking ability [5]. Furthermore, reading brings positive effect on foreign language mastery [6]. Besides, in terms of character building point of view, reading also can be a way to deliver character education [7].Therefore, it can be said that reading a text that contain character values is effective to build students' character.

While the values of *Tri Hita Karana* concepts are taken to be inserted in the materials because those concepts are common concepts which are already understood by the students. Integrating something that is already known by the students, e.i. students' prior knoweledge, in the instruction will help the students to understand the lesson, especially reading comprehesnion. Students' prior knowledge has positive correlation toward students' reading comprehension [8]. Prior knowledge helps the learners to understand the reading text better [9]. If the reading materials contain characer values and the students easy to comprehen the content of those reading materials, it means that they will also undertsand the character values in that text easily. Sekolah Menengah Kejuruan Negeri (*SMKN*) 3 Singaraja was selected as the setting of the study because since 2014 has been implementing *Tri Hita Karana* concepts. By developing *Tri Hita Karana*-based reading materials, it is expected that this study could support the Indonesian goverment to conduct character-based education in order to build students' character and at the same time helping the students to improve their reading comprehension. Moreover, the result of the study also can be used to support the school program to implement *Tri Hita Karana* concepts at *SMKN* 3 Singaraja.

2 Methods

This study was conducted at *SMKN* 3 Singaraja in the academic year of 2016-2017. The subjects of this study were the teacher and the students of the tenth grade of *SMKN* 3 Singaraja. Numbers of students involved in this study were 28 students. It consisted of 7 females and 21 males. The object of the study was *Tri Hita Karana*-Based Reading materials. This study followed the steps in conducting Research and Development model [10]. This model is known as ADDIE model that consists of five steps, namely: Analyze, Design, Develop, Implement, and Evaluate.

In order to collect the data the researcher used some instruments, namely: Observation Sheet, interview guide, expert judgement sheet, questionnaire, and reading comprehension test. The results of the observation and interview were analysed qualitatively through interactive data analysis model. This interactive data analysis model consists of three steps, namely: data reduction, data display, and conclusion drawing/verification [11]. While the result of the test and questionnaire were analyzed using descriptive statistic.

3 Results and Discussion

3.1 Teachers' Opinion on Values of *Tri Hita Karana* Concepts which are Appropriate to be Inserted in the Reading Materials

Based on the result of the study, the *Tri Hita Karana* values that should be inserted in the materials, namely: (a) from Parahyangan point of view, i.e. good relationship between human and God: be cautious of God, be faithful of God, be grateful of God, and praying diligently; (b) from *Pawongan* point of view, i.e. good relationship between human: like to

igedesukarma@yahoo.co.id

work together, Help others who need an assistance, tolerant to others, treat others equally, polite and courteous to others, appreciating others' opinion; (c) from *Palemahan* point of view, i.e. good relationship between human and nature: preserve the forest and its animal, maintaining the cleanliness of the environment, manage garbage properly and wisely, , and maintain water channel. *Tri Hita Karana* is the core values that the Balinese people should maintain [12]. Therefore, since this study aimed at developing English materials which are going to be used in Bali the moral values of *Tri Hita Karana* is really appropriate to be inserted.

If it is seen from the character education point of view as stated in the Guideline of Character Education, those values are similar to the values that should be taught for the senior high school students. Those character values are religious, friendly, communicative, social concern, democratic, independent, hard work, and discipline [3]. *Tri Hita Karana* is one of local cultures of Balinese people. The local culture that was inserted in the English language teaching is only for building students nationalism, so that they will not focus on western culture only during the teaching and learning process of English [13]. Integrating universal and local values is important in conducting character-based education [14]. Threfore, knowing the local culture and the wetern culture are both important in studying English. By knowing both cultures, it will raise the cross cultural understanding and it is very important to aviod any conflicts when a foraign language is used to communicate.

3.2 Teachers' and Students' Opinions toward Tri Hita Karana-based Reading Materials

Teachers' opinions toward the Tri Hita Karana-based reading materials were collected through a questionnaire. The questionnaire was developed based on the theory of good learning material by Tomlinson [15]. Four aspects were taken into account in the questionnaire, namely : feature, content, language used, and assessment. That questionnaire was distributed to the six English teachers and twenty eight students at *SMK N* 3 Singaraja.

The result of the questionnaire show that, from the feature aspect, all of the teachers and students (100%) agree that THK-based reading materials: are presented using interesting images and illustrations on each text ; and are presented using a clear and interesting typeface to read. Based on those two indicators, it can be said that the English teachers and students at *SMKN* 3 Singaraja have a positive opinion toward the new materials from the feature aspect point of view. From the content aspect, all of the teachers and students (100%) agree that the materials: make students feel good about learning; are able to stimulate the students' imagination to make it easier to understand the content of the reading; are presented in relation to local wisdom in the student area; stimulate students to be more confident in learning; encourage students to think critically; encourage students to be creative; presented in accordance with the competency standards, basic competencies, and achievement indicators; contain values that govern human relationships with God, values that govern human relationships with humans, values that govern human relationships with the environment; are presented according to the student's daily life; provide benefits for everyday life to students; contain the character values; and are presented using original/authentic sources contained in the student's real life. Therefore, considering the indicators of the content aspect it can be concluded that the teachers and students have positive opinions toward the materials from content aspect.

Seeing from the language used aspect, the all of the teachers and students (100%) agree that the sentences presented in THK-based reading materials: use correct and correct grammar; correspond to the students' level of understanding; and are simple and easy to understand. In other words, it can be said that the English teachers and students have positive opinions on the new materials from the language used aspect. Considering the

igedesukarma@yahoo.co.id

assessment aspect, the all teachers and students (100%) agree that the assessments: stimulate students to learn independently and/or in groups; are made from simple to more complex; are presented to facilitate student learning strategies; encourages students to develop learning strategies; and provide an opportunity for students to disclose their opinions or assessments of the material provided. Thus, it can be said that the teachers and students have positive opinion toward the assement aspect.

From the findings above, it can be conluded that the English teachers and students at SMKN 3 Singaraja have positive opinions toward the THK-based materials. In order to confirm the result of the questionnaire, the researcher conducted pre-test and post-test. The result of those tests were used to see whether or not the materials are practically good. From the result of the pre-test and post-test before and after the implementation of THK-based reading materials it was found that the implementation of THK-based reading materials could improve students' reading comprehension.

If we see from the result of descriptive statistic of the pre-test, it can be seen that the lowest score that was obtained by the students was 45, while the highest score was 85. The mean score was 65.18 with standard deviation was 12.58. From the result of the analysis, it can be said that from the mean score, in general, most of the students still got low score and only few students got high score. That was supported by the standard deviation that was 12.58, which means the data or the students score were vary, however in those variations low scores were dominating.

Table 1.The descriptive statistic of result of pre-test.

	N	Minimum	Maximum	Mean	Std. Deviation
pre_test	28	45.00	85.00	65.1786	12.58174
Valid N (listwise)	28				

Based on the result of post-test, it was found that the lowest score was 60 and the highest score was 90. While the mean score was 74.29 and the standard deviation was 9.09. Therefore, it can be interpreted that students' reading comprehension after the implementation of THK-based reading materials is better than before the implementation of THK-based reading materials. The mean score shows that the students got higher score than the mean score during the pre-test. It means that, in general, the students' reading comprehension was getting better.

Table 2. The descriptive statistic of the post-test.

	N	Minimum	Maximum	Mean	Std. Deviation
post_test	28	60.00	90.00	74.2857	9.09968
Valid N (listwise)	28				

If the results of pre-test and post-test were compared, it shows that most of the students' score were improved during the post-test. Only three students did not get higher score after the implementation of THK-based reading materials. From that data it can be said that most of the students could improve their reading comprehension through THK-based reading materials.

From the results of pre-test and post-test above, it can be said that the result of the test support the opinions that the teachers and students have stated. There are some reasons why the implementation of THK-based reading materials could improve students' reading comprehension. The first reason is that the materials provide prior knowledge to the students before they read the text. It makes the students become aware of what they are going to read and enable them to see the context and link the context to the information in

igedesukarma@yahoo.co.id

the text. Building students' prior knowledge has significant positive correlation toward students' reading comprehension [8]. Prior-knowledge helps the learners to understand the reading text better [16] and can improve students' reading comprehension [9]. From the study that they were conducted they found that prior knowledge help the students to retell the information that they got in reading English text.

The second reason why THK-based reading materials could improve students' reading comprehension because they provide exercises that build students' vocabulary mastery, espeacilly for new vocabulary that is available in the text, so that they are ready to read the text. Some studies have proven that vocabulary has important role in reading comprehesnion. Reading comprehension and vocabulary mastery are interconected [17]. When students have a good vocabulary mastery they will have a good reasing comprehension and vise versa. In addition, the vocabulary mastery of learners will influence their understanding on a text [18].

The third reason is that THK-based reading materials are designed to be implemented through scientific approach, e.i. a type of student centered approach. By involving students more during the teaching and learning process the students become active learners who are responsible for their own learning. The implementation of student centered approach could improve students' reading comprehension [19]. Students centered approach improves learners autonomy in learning and this learning autonomy makes students' reading comprehension better [20].

4 CONCLUSION

From the results of the study, some important points can be concluded. First, the English teachers at *SMK N* 3 Singaraja are aware that it is important to promote character values, especially the values that come from the local wisdom of the students. Second, both the English teachers and the students at *SMK N* 3 Singaraja have positive opinions toward the implementation of THK-based reading materials. Third, the implementation of THK-based reading materials support the students' and teachers' positive opinions toward THK-based reading materials, since the implementation of THK-based reading materials could improve students' reading comprehension.

The writer would like to express his gratitude to the Almighty God for His blessing so that the writer could complete this thesis. In addition, the writer also would like to thank several parties who already supported the writer in completing this thesis. Those parties are:

1. Prof. Dr. Putu Kerti Nitiasih, M.A.,and Dr. I Gede Budasi, M.Ed., as the first and the second supervisor, who patiently guided me and gave some improving comments and suggestions during the study and writing processes of this article.
2. The English teachers and Hindu Religion teachers at SMK Negeri 3 Singaraja who were willing to spend their time to give some information that the writer needed to complete this article.
3. The students of SMK Negeri 3 Singaraja who supported the writer in doing the study as the subjects of the study.

References

[1] N. Unayah, M. Sabarisman, Sosio Informa **1**(2) 121-140 (2015)

[2] Yanuar, "Remaja, Klithih, dan Hal-Hal yang Belum Selesai," 16 Maret 2017. [Online]. Available: http://regional.liputan6.com/read/2888555/remaja-klithih-dan-hal-hal-yang-belum-selesai.

igedesukarma@yahoo.co.id

[3] Badan Penelitian dan Pengembangan, *Panduan Pelaksanaan Pendidikan Karakter* (Kementerian Pendidikan Nasional, Jakarta, 2010)

[4] F. Teng, English as a Global Language Education Journal **1**(2) 63-90 (2015)

[5] A. D. Mermelstein, Reading in a Foreign Language **27**(2) 182-198 (2015)

[6] M. A. Alzu'bi, English Language Teaching **7**(1) 28-35 (2014)

[7] G. M. Almerico, Research in Higher Education Journal **26**(1) 1-13 (2014)

[8] I. M. Alfaki, A. G. Siddiek, World Journal of English Language **3**(4) 45-66 (2013)

[9] Y.-H. Lin, C.-L. Chern, Contemporary Educational Research Quarterly **22**(4) 149-186 (2014)

[10] R. M. Branch, *Instructional Design: The ADDIE Approach* (Springer, New York, 2009)

[11] M. B. Miles and A. M. Huberman, *Qualitative Data Analysis* (Sage Publications, Inc., *California*, 1994)

[12] P. Sudira, *The 8th APNME Anual Conference*, Yogyakarta (2013)

[13] S. Sudartini, Jurnal Pendidikan Karakter, **2**(1) 45-54 (2012)

[14] L. P. Artini, 3rd International Conference on English Across Culture, Singaraja (2016)

[15] B. Tomlinson, *Material Development in Language Teaching* (Cambridge University Press, Cambridge, 1998)

[16] W. Pei-shi, Sino-US English Teaching **9**(9), 1516-1523 (2012)

[17] S. Kara, International Journal on New Trends in Education and Their Implication **4** (1) 39-53 (2013)

[18] H. Sidek, H. A. Rahim, Procedia - Social and Behavioral Sciences **197**, 50-56 (2015)

[19] M. Lak, H. Soleimani, F. Parvaneh, Journal of Advances in English Language Teaching, **5** (1), 1-10 (2017)

[20] P. Channuan and P. Wasanasomsithi, *"Promoting Learner Autonomy through an Extensive Reading Program among Second Year Udergraduate Students,"* pp. 1-23, (2012)

igedesukarma@yahoo.co.id

Development of science practicum performance assessment in junior high school

Ni Luh Rin Riantini[a*] , *I Wayan* Suastra[a] , and *Putu Budi* Adnyana[a]

[1]Science Education, Post-Graduate Program, Ganesha University of Education, Singaraja, Indonesia
[*]rinriantini93@gmail.com

Abstract. The purpose of this research was produce a performance science practicum assessment instruments based science process skill which is valid, reliable, practical, and effective for measuring the skill of science process in junior high school. The type of this research is development research designed with the assessment development model from Doran et al., 1998. The average result of the instrument validation test is 3.50 and can be categorized very valid and can be applied in the learning after going through several revisions. Reliability test results show that reliable instrument with reliability coefficient of 0.78. The average result of the instrument practicality test is 3.38 and it is practically categorized and can be applied in the learning after going through several revisions. From the average student score can be stated all students get the value above the minimal mastery criteria, with students the average score is 83.98 and from the t-test results obtained t count 30.376. Based on these findings, it can be concluded that the instrument assessment of science practicum performance based on science process skills can be used to measure students' science process skills, easy to use in learning, and effectively used in science learning activities.

Keywords: assessment, performance, science process skills

1 Introduction

The curriculum in force in Indonesia for the last three years is the Curriculum 2013. The learning process in Curriculum 2013 is based on a scientific approach. The scientific approach directs students to the learning process in accordance with the scientific method. Changes in the learning process applied to the Curriculum 2013 led to changes in learning tools including the assessment process. Assessment is done on the competence of attitudes, skills, and knowledge. Assessment is the process of gathering qualitative and quantitative information about the achievements and qualities of a science program [3]. According to [4], assessment is the process of gathering information about learners, regarding what students know and what students do. In the information gathering teachers can observe learners, test the skills, knowledge, and products produced by learners. Assessment activities are not limited to the characteristics of learners only, but also include the characteristics of teaching methods, curriculum, facilities and school administration. Assessment instruments may be formal or informal methods or procedures, to generate information about learners: written tests, oral tests, observation sheets, interview guides, homework assignments, etc. [12].

The assessment emphasized in the implementation of the 2013 curriculum is authentic assessment. An assessment is authenticated when the assessment involves learners on useful, important, and meaningful tasks [4]. Authentic assessment will materialize when student learning activities involve an activity that demonstrates high thinking skills as well as coordination of broad knowledge as per contextual learning objectives. Assessment becomes authentic when the learning measured by the assessment has value beyond the classroom and is meaningful to learners. Authentic assessment mandates the skills and abilities needed to perform the actual task. In this connection authentic assessment leads to each goal such as: (1) Demanding learners develop a response rather than simply having a pre-determined option; (2) Raising high-level thinking in addition to basic skills; (3) Directly evaluate a holistic project (task); (4) Integrating with teaching; (5) Using examples of students' work (portfolio); (6) Derived from clear criteria for learners; (7) Allows the emergence of many opinions; (8) Closely connect with classroom learning; (9) Teaches learners to evaluate their own work. (10) is comprehensive and holistic [12]. The judgment used to assess the outcome of the learning process is authentic assessment. Authentic assessment has relevance to the scientific approach in learning according to the demands of the Curriculum 2013 which is able to describe the improvement of learning outcomes of learners. Authentic assessment aims to measure skills in various contexts that reflect real-world situations [6]. One type of authentic assessment that can be used to measure skill competency is performance assessment. In school learning, one of the many subjects using performance assessment in the learning process is science.

Science is the knowledge of the natural surroundings, in this case relates to how to find out about nature systematically, so that science in the school called the Natural Science is not only the mastery of a collection of knowledge in the form of facts, concepts, or principles only , But also a process of discovery [2]. Therefore, science learning in schools should be applied with a scientific approach and students are expected to have skills tailored to the learning process of science. Performance assessment is an assessment of the performance or performance or performance of learners from assignments given teachers where learners perform or display or do something to show achievement in learning [7]. Performance assessment requires students to complete their performance tasks using their knowledge and skills embodied in the form of actions, actions or performance. The Performance Test asks students to embody actual tasks that represent the overall performance to be assessed, such as preparing tools, using tools or assembling tools, writing data, analyzing data, summarizing, reporting and so on. In particular, performance appraisals explain students' abilities, conceptual understanding, ability to apply knowledge and skills, perform performance and ability to perform a process. Science as a product involves a collection of knowledge consisting of facts, concepts, and principles of science. Science as a process includes skills and attitudes possessed by scientists. Aspects of science process skills include eight aspects: observing, interpreting, using tools and materials, predicting, applying concepts, planning activities, communicating, and asking questions [8]. Process skills are skills involving cognitive or intellectual, manual and social skills. Cognitive skills are involved because by doing the process skills the students use their minds [11].

Many teachers who have not implemented a good performance assessment in the learning process can be seen from the results of several studies. In the performance appraisal is also shown in the research conducted by [1], This research researcher about execution of science practicum of class VIII at Junior High School in Lubuk Begalung, Padang City. The results of this study indicate at the stage of practicum planning, where one of the observed aspects of formulating the form and type of explanation clearly and completely get less criteria with a value of 59%. In addition, research conducted by [9], regarding teacher difficulties in the implementation of the assessment in the 2013 curriculum in Yogyakarta City, one of which is the assessment of skills. The study shows that teachers do not understand the concept of skills assessment, it is difficult to make a skill assessment rubric, and it is difficult for teachers to determine indicators to assess student skills. This is due to the lack of insight and preparation of teachers in assessing the process of practicum. Teachers need to be given an intensive training related to the designing and implementation of authentic assessment both by schools and other institutions [10]. Therefore, research is conducted to produce a valid, reliable, practical, and effective performance assessment instrument.

2 Method

2.1 Research design

The research development design used in this research is the assessment development model designed by [3]. The steps of assessment development model that state the purpose, select appropriate task format, write or modify the task, clarify administrative procedures, develop the scoring rubric, trial the task, analyze result, revise task. The product trial is done with one shoot case study, Validation test obtained from the average assessment results of two experts, use reliability test inter rater agreement, The practicality test obtained from the average assessment of three science teachers on the instrument, and the effectiveness test using one sample t test.

2.2 Participants

Research subjects in this study are two experts, two appraisal teachers, three junior high school teachers, 34 students of grade VIIA SMPN 2 Kuta. Two experts aimed to test the validity of the product, two teachers as assessors aimed at testing product reliability, three junior high school teachers aimed at product practicality test, and students aimed at product effectiveness test. The object of this research includes the assessment instrument of science practicum performance based on science process skills.

2.3 Instrumentation

To test validity, each expert is given a validation sheet, which on the validation sheet contains several aspects including: presentation method, and physical form of the instrument. In the validation sheet, the validator's opinion is categorized into four scoring scales, excellent (score 4), good (score 3), not good (score 2), and very bad (score 1). The reliability of performance assessment science practicum based on science process skill can be known from the results of the assessment conducted by two assessors. The results of the assessors will be calculated the number of similarity scores so that it can be known level of reliability. The response questionnaire is a data collection instrument in the form of a number of statements filled by the teacher. The list of statements used in the instrument test phase aims to collect teacher opinion data related to the practicality of the instrument. Questionnaires used in this study using four scales. The results of this questionnaire will be used to test the practicality of the instrument. Scientific performance appraisal sheets are used to measure students' science process skills in practicum work. The performance assessment sheet contains indicators of the science process skills with a scale of 1-4. The practice performance appraisal rubric contains an explanation of the criteria for each score. Student performance appraisal instrument consists of two types of assessment, namely: assessment of student activities and student practice report reports. Rubrics with a scale of 4 to 5 are effective for defining every aspect that will be used to assess student performance and direction from teachers enabling to improve student discipline as well as maximizing student knowledge development [5]. With the use of analytic rubrics the teacher can be more effective in assessing each student's performance and the results of the assessment more objectively.

3 Result and discussion

The results and discussion of the validation test results, reliability, practicality, and effectiveness of the performance assessment tools science-based science process skills are as follows.

3.1 Validity of performance assessment instrument

The data of validation result of performance appraisal apparatus of science as follows. Instrument tested validity consist of three instruments that is, first observation sheet of student activity with result of validation 3.47 and get valid predicate, both sheets of report of practice report with result of validation 3.55 and get predicate very valid, third worksheet of student with validation result 3.47 and get a valid predicate. From the result of validity test of each instrument, it can be obtained the average of validation result of the assessment of science practicum performance based on science skill in 3.50 with very valid predicate and can be used to measure students' science process skill so that it can be applied in learning with do some previous revision.

3.2 Reliability of performance assessment instrument

In this research, reliability test is done using 34 students from class VIIA. From result of analysis of reliability test obtained result of reliability instrument observation activity of student equal to 0.737 and can be said that instrument have reliable. The result of reliability analysis of appraisal report instrument is 0.82 and it can be said that the instrument has been reliable. The average reliability test results obtained by 0.78 indicating that the developed instrument has been reliable because it has met the minimum reliability of 0.7. In other words, developed instruments are consistent and are viewed the same when used by more than one teacher or assessor.

3.3 Practicality of performance assessment instrument

From the results of practical tests conducted by the three practitioners obtained the average of practicality test results for student activity observation sheets and sheets of practice report report of 3.47 with a practical predicate. Practical test results for student worksheets obtained an average value of 3.28 with a practical predicate. It can be concluded for all the assessment instruments of science science-based science-based practice performance achievement of 3.38 practicality test results with practical predicate. It can thus be concluded that the science-based science performance assessment appraisal instrument is easy to use without finding any significant difficulties in its use.

3.4 Effectiveness of performance assessment instrument

Table 1. One sample t test result

	Paired Differences					T	df	Sig. (2-tailed)
	Mean	Std. Dev.	Std. Error Mean	95% Confidence Interval of the Difference				
				Lower	*Upper*			
(1)	(2)	(3)	(4)	(5)	(6)	(7)	(8)	(9)
Performance score of practicum of students - value of minimal mastery criteria	83.98	1.72	0.295	8.37	9.58	30.376	33	0.00

At the first meeting conducted practicum of water pollution. In the water pollution practice the students have not fully understood the student worksheets that have been distributed to each group so that the teacher should provide much guidance to the students. The second meeting is the practice of soil pollution and the third meeting of air pollution practice runs more smoothly than before because students have clearly understood what to do. The value used in the effectiveness test comes from the value of practicum and practicum reports. From the results of effectiveness test at each practice can be obtained results that all students get the value above the minimum mastery criteria with the highest value of 89.55 and the lowest score of 76.25. Judging from the result of one sample t test obtained t count equal to 30.376 and t table equal to 2.042. Thus, t count> t table H0 rejected and Ha accepted, and can be expressed student value bigger or equal with minimum mastery criteria. It can be concluded that the assessment instrument of science practicum performance based on skills in the science process is effective in measuring students 'science process skills and can improve students' science process skills.

4 Conclusion

Based on the exposure of the results and discussion of the research, then obtained the following conclusions. (1) Scientific performance appraisal of science practicum based on science process is very valid and can be used in science learning process because instrument can be used to measure students' science process skill. (2) The assessment instrument of science practicum performance based on the skills of the science process can be said reliably with a reliability coefficient of 0.78, so that the statement on the assessment is consistent. (3) Scientific performance appraisal of practical science-based practical and easy-to-use science-based practice assessment instruments. The instrument has been completed with an appraisal rubric so that it can objectively assess. (4) Scientific performance appraisal tools for science-based science process sciences are effective in measuring students 'science process skills and can improve students' science process skills that are seen from the student's grades in each practicum already exceeding the minimum completion criteria and from one sample t test result where t count> t table.

Reference

1. W. N. Atnur, Lufri, S. Ramadhan, *Analisis Pelaksanaan Praktikum IPA Biologi Kelas VIII Semester I di SMP Negeri Se-Kecamatan Lubuk Begalung Tahun Pelajaran 2014/2015* (2014).
2. Depdiknas. *Laporan IPA* (2004).
3. R. Doran, F. Chan, P. Tamir, *Science Educator's Guide to Assessment* (1998).
4. D. Hart, *Authentic Assesment: A Handbook for Educators* (1994).
5. Johnson, C. Carla, D. Zang, J. B. Kahle, Research in Middle Level Education **35** (2012).
6. D. A. Lindayani, *Penerapan Penilaian Autentik Dalam Kurikulum 2013* (2014).
7. A.A.I.N Marhaeni, L.P. Artini, Jurnal Pendidikan Indonesia **4** (2015).
8. I. W. Suastra, *Pembelajaran Sains Terkini Mendekatkan Siswa dengan Lingkungan Alamiah dan Sosial Budaya* (2013).
9. H. Retnawati, *Analisis Kuantitatif Instrumen Penelitian (Panduan Peneliti, Mahasiswa, dan Psikometrian,* 2016).
10. I.W. Suastra, N. P. Ristiati, IRJEIS **3** (2017).
11. N.Y. Rustaman, *Strategi Belajar Mengajar Biologi* (2003).
12. S. Widodo, E. Irmayanti, Sulistyono. *Asesmen Autentik* (2011).

Learning history integrated local wisdom values "babad Banyumas" to build a student's national identity

Ngismatul Khoeriyah[1,*], Warto[2], and Sariyatun[3]

[1]Post Graduate in History Education, faculty of Teacher Training and Education, Universitas Sebelas Maret, Surakarta, Indonesia
[2]Professor of History Science, Universitas Sebelas Maret, Surakarta, Indonesia
[3]Professor of History Education, Universitas Sebelas Maret, Surakarta, Indonesia

Abstract: This study aims to build a national identity and collective memory of students through learning history that is integrated with the values of local wisdom Babad Banyumas, which is considered ideal teaches values in learning history. The study employed a qualitative descriptive research method through literature review. The results show that in Babad Banyumas have values that match the character's learning, such as: 1) Social values, it is illustrated in the Babad Banyumas that is Ki Dipati Kaleng's attitude in entertaining his guest; 2) the value of morality, this is demonstrated by Raden Baribin's attitude of confronting a power conflict with his brother; 3) The value of religiosity include spiritual attitude shown by Dipawijaya the form of asceticism beg a boon to the Supreme Court and tradition Sang Hyang every Friday night visit to the palace followed by reciting Qur'an in Surau (mosques king).

Keywords: learning history; local wisdom, babad Banyumas, national identity.

1. Introduction

The phenomenon of low character education in school inevitably we have to think about it. Because the education of values that concern the formation of the character of learners has not been applied maximally. Considering how the current state of education in Indonesia is illustrated, cases that are inconsistent with ethics, morality, courtesy or behaviour that indicate low morality have become so prevalent in society. More worrying, the behaviour is not shown by few people who are educated. This proves that education is less successful in shaping the character is good. Under such conditions, it would be relevant to recount the old paradigm of education, namely education as a legacy of values.

In relation to that according to Thomas Lickona there are 10 signs of human behavior that indicate the direction of destruction of a nation that is: 1) increasing violence and vandalism among adolescents; 2) dishonest dishonesty; 3) increasing height of disrespect to

* Corresponding author: ngismatulkhoeriyah3@gmail.com

parents, teachers and leaders; 4) the influence of peer group against acts of violence; 5) increased suspicion and hatred; 6) the use of deteriorating language; 7) decline of work ethic; 8) decreasing the sense of responsibility of individuals and citizens; 9) the heightened self-destructive behavior, and 10) the escape of moral guidance [1].

Considering how the current state of education in Indonesia is illustrated, cases that are inconsistent with ethics, morality, courtesy or behaviour that indicate low morality have become so prevalent in society. More worrying, the behaviour is not shown by few people who are educated. This proves that education is less successful in shaping the character is good. Under such conditions, it would be relevant to recount the old paradigm of education, namely education as a legacy of values. The legacy of cultural values of the past is not small that contains the values of character and moral education. This is honestly caused on January 10, 2010, kemendiknas organized a workshop so the emergence of the idea of the declaration on "Culture and National Character Education" as a national movement [2].

Ideally, the application of character education in educational institutions is integrated with subjects that have the content of local wisdom as part of the nation's character formation. Character education that must be developed in schools is the religious values and cultural values of the nation (local wisdom). The material content taught in character education is inseparable from the local wisdom that is held by the people. The local wisdom that exists in society is one of the material content that must be contained in character education. History is a subject that teaches about value education is expected to have a significant role in the strengthening of education in Indonesia.

The history lesson aims to create historical insights or historical perspectives. Historical insights further highlight the continuity of everything. Being is the result of the process of becoming, and being itself is in the process of becoming a point. Meanwhile, the sociocultural nature of our environment is a product of history, including the territory of RI, national state, national culture. National History multidimensional functions include prevention of determinism, expand intellectual horizons, to prevent the occurrence of synchronism, which ignores the determinism [3].

Babad Banyumas as an ancestral cultural heritage, in this research, is integrated with learning history to build national identity and collective memory of students. Babad Banyumas is considered ideal in teaching values in the model of teaching history because it contains many values of character that is learning about manners, tolerance, fairness, honest, responsible, integrity, discipline, courage, mercy (*welas asih*), the generosity shown by the figures in the babad as Raden Baribin, Raden Kaduhu, Adipati Warga Utama, Adipati Mrapat, Adipati Janah, dan Tumenggung Yudanegara, where overall indicators are exemplary by learners. Through the overall values, students are expected to learn and understand the noble culture that comes from the region and become the inspiration in reflecting himself, the surrounding environment and nation.

2. Methodology

The study employed a descriptive-qualitative research method derived from literature reviews by searching for written sources, whether in the form of books, archives, articles, and journals, or documents relevant to the subject matter studied.

3. Result and Discussion

The results of this study found that learning history based on Babad Banyumas can build national identity and collective memory of students. In the Javanese literary tradition, tembang books generally contain teachings about noble values or guidance. This is different from the gendhing-gendhing dolanan whose content is more entertainment. Babad Banyumas as an ancestral cultural heritage, in this research, is integrated with the learning of history to build the identity and collective memory of students either in the realm of cognitive, affective, psychomotor or connotation.

The results of the research findings in the field of SMK case studies as Banyumas (with a sample of three schools: SMK N 1, SMKN 2, SMKN 3 Banyumas) with the update of curriculum 2013 which emphasizes on values and character education. With the existence of KI (Core Competence), scientific approach and with models that can build students' activeness and criticality in learning as well as the existence of historical lesson material integrated with local wisdom of Babad Banyumas, so it makes it easier for students to understand the local culture and heritage. When compared with previous historical learning (not yet integrated with local wisdom), history learning is only limited to studying events within the national scope. It is reinforced by Nadlir, that this educational model, learning materials must have meaning and high relevance to their empowerment of life in real, based on the reality they face. The curriculum must be prepared is a curriculum that is appropriate to the environmental conditions, interests, and psychological condition of learners. It should also take into account the sociological and cultural constraints they face. Local wisdom-based education is education that teaches learners to always stick with the concrete situation they face. Because faced with the problems and concrete situations encountered, learners will be more challenged to respond critically [4].

History-based learning design Babad Banyumas is based on the needs and problems of low identity and awareness of student character values. According to researchers, Babad Banyumas must be actualized in the contemporary context. The actualized Babad Banyumas is part of the planning process which includes a learning design process that produces products in the form of draft models or syntax, learning tools and learning modules. The product is then implemented in classroom teaching through a scientific approach with supporting modules and learning media. After going through the implementation phase then the learning model in the evaluation whether it has been in accordance with the goals and indicators. The resulting product is then a historical learning model based Babad Banyumas that contains aspects of cognitive, affective, and psychomotor.

One of the regeneration processes for the delivery of local wisdom as a reinforcement of the Banyumas identity is through literary works, such as Banyumas diverse Babad version. Knowledge of past community in Indonesia is very close to the oral tradition and writing tradition, from here also perhaps the intellectual roots of the Java community can be tracked. In a society that has not known the writing of oral tradition has been used as one way to pass down the collective memory or knowledge of the past gained from previous generations.

Wolf Kansteiner defines collective memory as representations of the past that are collectively adopted [5]. Collective memory is a general category that includes several main kinds of memories. First is the popular memory, defined by David Midelton and Derek Edwards as representations of the past held by the society's members, best manifested directly by public opinion surveys [6]. This memory significantly influences the psychological dynamics mentioned above and the behavioral reactions of the people holding it, and therefore, it is accorded great importance [7]. The second kind is the *official* memory, according to Jeffrey Olick: the representations of the past adopted by the institutions of the state. This memory is manifested, for instance, by publications of state ministries and the army, exhibitions in national museums, and textbooks approved for use in the educational system [8]. Third is the *autobiographical* memory, that of the people who directly experienced the events, manifested typically in memoirs and oral histories. This is a primary source for the past (in addition to documents) and is therefore usually accorded importance [9]. Fourth is the *historical* memory, the way the research community mostly academics as well as independent scholars views the event in its studies of the past [10]. Fifth and final is the *cultural* memory defined by Ian Assmann as the way the society views its past via newspaper articles, memorials, monuments, films, and buildings [11].

Local wisdom education also has a very important role in the development of national identity. As Vikers says, nation building involves more than the simple promotion of a language. It also aims at socializing youngsters in a particular national culture and historical narrative [12]. Babad as one source of history until now still not used by many historians. Perhaps because theoretically and methodologically the babad has many shortcomings, especially when it is associated with temporal, factual or spatial issues. In addition, because it is a literary work, the Babad using literary language elusive hole common people. Babad Tanah Jawi, for example, is still not fully understood about the origin, intent, material and its components. De Graaf mentions that Babad Tanah Jawi is a strange writing. There are allegations that the babad was written by several people aimed at strengthening the legitimacy of the ruling king [13]. Last but not least the fill factor that is sometimes unacceptable to common sense, the more distanced historian's attention to this literary work. Apart from all these weaknesses, babad actually also contains some historical facts. In this case, Taufik Abdullah states that through literary works we can understand the procession of past events and recapture the time structure of reality. Taufik Abdullah further states that literary work is a collective experience of the author and reflects the mood of time when the work was created [14].

Babad Banyumas is a living tradition of the text that gave rise to new and varied texts. The creativity of the authors and copyists shows high productivity [15]. Babad Banyumas versions are in fact very necessary in enriching the repertoire of local wisdom Banyumas. Babad Banyumas is considered ideal in teaching values in history learning. Because in the Babad Banyumas contains many good values that is learning about manners, tolerance, fairness, honest, responsible, integrity, discipline, courage, compassion, generosity shown by the figures in the Babad such as Raden Baribin, Raden Kaduhu , Adipati Warga Utama, Adipati Mrapat, Adipati Janah, and Tumenggung Yudanegara, all of which indicate that the indicators are worthy of being followed by the students. In the babad Banyumas it basically contains a lot of life values grouped into core values such as the value of morality, social value, and the value of religiosity.

3.1. Value of Morality

Research on moral education has drawn on ethical philosophy, developmental, psychology, and constructivist pedagogy to show that individuals can develop the capacity for moral reflection and judgment, which becomes increasingly inclusive, principled, and independent of the dictates of established authorities [16]. In this tradition, critical judgment consists of an active process in which participants a) recognize multiple moral dilemmas and contested issues that cannot be resolved relying simply on personal preferences, formed habits, and social traditions, and b) engage with these dilemmas and controversies through reasoned dialogue, seeking to recognize different view points and coordinate them in judgments and choices that are comprehensive, fair, and responsive to rights and needs of different parties in conflict. Moral judgment becomes critical in so far as it takes reflective distance from one's egocentric and socio-centric prespective, and is self-directed, yet sensitive to and inclusive of others [17].

Babad Banyumas teaches moral guidance as a part of character education that can be classified as personal ethics because it is a more self-directed ethical guide. As the teachings in Javanese culture emphasize patience. This is demonstrated by the journeys of Raden Baribin in his struggle for power. Where he had to fight for his life expelled from his own home due to a dispute with his ambitious brother in power. This can be known from the contents of the babad

// warnanen ingkang lumaris / wus medal sangking nagara / lesu forgot sarirane / tillar spirits opponent nendra / tiyang saurut marga / tanana ngaturi suguh / ajrih undhange san nata //
Translation:
This is all due to the threat of the prabu who has strictly prohibited to accept or entertain them. Until finally, Raden Baribin stumbled due to tired and hungry he suffered. Then came in the pagagan area, where he found the fruit the cucumber.

However, the babad banyumas also alludes to the importance of the development of reason, thought, rationality or intellect. About the ethos of learning (science), that science can be possessed with the practice that begins with a strong will.

// Raden Kedhuhu kang wewasi / wus diwasa pamit mring kang Rama / tuwin mring ibu sorine / arsa anganyut tuwuh / kesah sangking praja nireki / sumeja anglelana / ngetan kang dinunung / prapteng tanah lurung tengah / pan ing wirasaba ingkang den enggoni / jumeneng Adipati //

The translation:
Raden Kedhuhu son Raden Baribin as an adult, he permits to rama's mother to wander looking for science to flee to the eastern country and come in wirasaba, he became there later with the name of Adipati Wirautama.

3.2. Social Value

Babad Banyumas teaches some moral guidance as part of a character education that can be classified as a social value because it is an ethical guideline more expressed to others and to its social environment. In it is taught to act decently in the meeting. Because in Babad Banyumas illustrated that the good person is usually sociable to get along with various circles. This is explained in the stanza:

// ri sampunya sinuguh / ing wau kyai Dipati / parentahi dadya mira / kinen akaryaha sami / pasareyan ing pandhapa / bade gene Kyai Tolih //

The translation:
After ending the conversation then Ki Dipati Kaleng entertained her guests with food and joy, then prepared the bed for Kyai Tolih special in pendhapa.

3.3. Religiosity Value

Babad Banyumas teaches some moral guidance as part of character education that can be classified as the value of religiosity. That religion is a precious grip of life. Javanese religious views which often use the terms of Islam. It is depicted in a stanza that reads:

// amung antepana sira kaki / suwita ing katong / aya anak putu nira tembe / antuk kanugrahanning Hyang Widi / kawruhaning ngaji / witana sireku //
// sawusnya ntuk kawan dasa hari / gya luwar sang anom / pan kacathet ing driya wisike / langkung sukur donga ing Hyang Widi / samono wus lami / genya suwiteng riku //
// saben malem jumuah ing wengi / sowan ing karaton / sareng lawan perdikan lebete / darus aneng suranatan masjid / duk sawiji wengi / kangjeng kang sinuwun //

The translation:

Doing tapa pendem for forty days, then he got a *wangsit* at the will of Sang Hyang Widhi that Dipawijaya will not get the position he wants, name to continue to serve the sang katong. Then after forty days, he ended his khalwat. Since the incident, every Friday night Dipawijayapun visit to the palace then followed by reciting al-qur'an in Surau (king's mosque).

4. Conclusion

Learning history integrated local wisdom values "Babad Banyumas" as a treasure of local wisdom of Banyumas community can certainly build the identity and collective memory of students. Because in it contains many values that can be the handle of the learner's life as a form of character education and Banyumas community as the heir and guardian of culture, which values are summarized in moral values, social values, and the value of religiosity shown by the figures in the babad such as Raden Baribin, Raden Kaduhu, Duke Residents Main, Duke Mrapat, Duke Janah, and Tumenggung Yudanegara, which overall core values are exemplary by Banyumas community and the Indonesian people, in general, to build a national identity through such wisdom. By taking these values, students are expected to learn and understand the noble culture that comes from the region and become an inspiration in reflecting himself and his environment in order to have an identity. Thus many ways that can be done through the exploration of local wisdom values for awakening identity and collective memory of learners. One of them through Babad Banyumas.

References

1. T. Lickona, *Ed for Char,* (Bandung, Penerbit Nusa Media, 2014).
2. Z. Darmiyati, *Pendidikan Karakater*, (Yokyakarta, UNY Press, 2010)
3. S. Kartodirdjo, *Pendekatan Ilmu Sosial*, (Jakarta, Gramedia Pustaka Utama, 1993)
4. Nadlir. Urgensi Pembelajaran. *Jurnal Pendidikan Agama Islam*, **02**, 306 (2014)
5. W. Kansteiner, Finding Meaning in Memory, *History and Theory*, **41**, 179– 197 (2002)
6. D. Midelton, D. Edwards, *Collective Remember*. (London, Sage Publications, 1997).
7. E. Halperin, D. Bar-Tal, R. Nets-Zehngut, E. Drori, Emotions in Conflict, *Peace and Conflict*, **14,** 233–258 (2008); I. Shelter, Historical Memory as a Foundation for Peace, *Journal of Peace Research*, **47,** 649–650, (2010)
8. Jeffrey Olick, What does it Mean to Normalize the Past, *Social Science History*, **22,** 547–571 (1998); W. Zheng, National Humiliation, History Education, and the Politics of Historical Memory, *International Studies Quarterly*, **52,** 783–806 (2008)
9. K. Jennings, N. Zhang, Generations, Political Status, and Collective Memories, *The Journal of Politics*, **67,** 1164–1189 (2005); H. Schumann, H. Akiyama, B. Knauper, Collective Memory of Germans and Japanese, *Memory*, **6**, 427–454 (1998)
10. J. Winter, E. Sivan, *Setting the Framework, in War&Remembrance in the Twentieth Century*, J. Winter, E. Sivan, (Cambridge: Cambridge University Press, 1999), 6–39.
11. J. Assmann, Collective Memory&Cultural, *New German Critique*, **65**, 125–133 (1995)
12. Vickers, E. Introduction: History, Politics and IEA, *International Journal of Educational Research* **37**, 537–44 (2002)
13. H.J. De Graaf, *De Regering van Panembahan Senapati*, (Jakarta, Grafitti Pers, 1985)
14. B, Purwanto, *Gagalnya Historiografi Indonesiasentris*, (Yogyakarta, Ombak, 2006)
15. S. Priyadi, *"Babad Banyumas,* (Purwokerto: FKIP UMP, 2013)
16. L. Kohlberg, *The psychology of moral development,* (San Francisco, harper&row 1984)
17. A. Bermudez, Four tools for critical, *Revista de estudios sociales*, **52**, 102-118, (2014)

The Effect of Value Clarification Technique (VCT) using Contextual Problem Content on Social Attitude and Social Science Learning Achievement of the Elementary School Students

Desak Putu Parmiti[1,*]

[1]Educational Technology Department, Universitas Pendidikan Ganesha, Singaraja, Indonesia

Abstract. This study analyzed and described changes in social attitude and social science learning achievement of the students who learn through value clarification technique (VCT) with contextual problem content and the students who learn through conventional teaching. This experiment or research was conducted at SD Gugus in Buleleng Regency with the population of 503 students. The number of sample for the experiment group was 40 students, and that for the control group was 40 too. The study used social attitude questionnaires and social study learning test to collect the data, while descriptive analysis and inferential analysis were used for analyzing the data. In this study, it was found (1) there was a significant difference in social attitude between the students who learned through VCT with contextual problem content and those who learned through conventional teaching, (Fobs.=6.370), (2) there was a significant different in learning achievement between the students who learned through VCT with contextual problem content and those who learned through conventional teaching, (Fobs.=4.040), and (3) there was a significant difference between social attitude and social science learning achievement simultaneously between the students who learned through VCT with contextual problem content and those who learned through conventional teaching. It means that the effect of teaching with VCT with contextual problem content in social science teaching produce a higher social attitude and social science learning achievement than the conventional teaching.
Keywords: Value Clarification Technique (VCT), contextual problem, Social attitude, and learning achievement

1. Introduction

Social science teaching is a strategic means in developing the students to become sociable citizens (Hasan, 2005). Thus, social science teaching will help the students to have an independent personality, physically and mentally healthy, and to become useful members of society. An independent person is someone who is able to grow holistically and harmoniously from all aspects of

* Corresponding author: dskpt_parmiti@yahoo.co.id

humanity and is able to respect humanity of everyone. There are also people who say that social science knowledge is the field of social science that is used by the students to solve problems that he/she faces. Social science teaching has the function and role which are very strategic in the effort of developing good and reliable citizens with the goal of building the nation (Lasmawan, 2010).

As an educator, a teacher is one of determiners in the success of every educational effort. That is why when there is an education innovation, especially in curriculum and human resource improvement resulted from educational efforts always lead to teacher factor. This shows how strategic the teachers' role is in the world of education. Thus, there is a need for teachers who are able to design teaching process, so the teaching can run well, joyfully, effectively, and efficiently. The statement above is supported by Sanjaya (2010) who states that teacher is a very important component in the success of the implementation of educational process. To create a joyful learning process, a teacher is not allowed to transfer whatever is explain in the text book to the students, it is a wrong action. The current teaching process focuses on how students find themselves through direct and real experience, can be proven based on theories or facts. Social science teaching is also one of the subjects that are not only based on the theory but also on facts.

Some research findings also state the condition of social science teaching at elementary schools today, it turns out that many students find difficulty in learning in the classroom because the technique of teaching selected on used by the teacher is felt to be less appropriate (Sumantri, 1999), thus the teaching and learning process will run awkwardly and is not supported by the development of knowledge, attitude, moral, and skill students development (Hasan, 1997). The problem is also felt in the students learning achievement in social science which has not reached the maximum learning mastery level that is it only reaches 69%, while the target of mastery is 75%. The low quality in the process and product of social science teaching was seen from the methodic and DIADECTIC perspective is assumingly caused by some factors, both internal and external. The problems faced in the field are: (1) the teaching method used by the teacher is still conventional, largely using lecturing method, (2) the number of students is too big for one class, (3) the equipments and media for learning are still very minimal, (4) the supplementary books are still lacking, (5) advancement in science and technology has not been used optimally, (6) infrastructure and facilities are not yet representative, (7) the evaluation done by the teachers focuses more on cognitive concept mastery without considering attitude and skill, (8) the classroom atmosphere is less conducive, (9) the environment and school stick holders do not care much about the condition.

With the implementation of school based curriculum in 2006, teachers are required to be able to develop the curriculum according to the potentiality, development of the students, their needs, school environment, and the development in science and technology (BSNP, 2006). The teaching expected in school based curriculum is (1) stresses more on students centered learning, (2) competency based, (3) and the newest one is that the teacher has to I2M3. Thus, teachers have to be able to develop the model of teaching that can change the teaching and learning process: from teacher-centered to student-centered, from content based to competency based. In this context Sommers (1993) states one of the important method in moral education is value clarification method (value clarification). By using this method, the teacher does not directly present to the students about the correctness or the incorrectness of something but on the contrary the students have to be given the opportunity to states their values in their own way. Furthermore, it is suggested that (1) schools have to have rules for behavior that stress the important of politeness, virtuous, self discipline, and honesty; (2) teacher should not indoctrinize the students if they want to ask something firmly on the basis of politeness, honesty, and openness; (3) the students have to be told stories that focus on good principals and the students should be fond of READING, learning and discussing or issues.

In Values Clarification Technique (VCT) development in this context of study, the focus of problems will be directed to the VCT testing of the students' social attitude and learning achievement.

Based on the empirical and conceptual studies above, it seems that VCT is one of techniques in social science teaching that is visible to be investigated in depth and scientifically especially in relation to social science teaching at elementary school level. This discussion enables us to combine VCT teaching and contextual teaching in social science teaching. Such kind of teaching will produce new and innovative characteristic, so that teaching will be more productive, joyful and inspiring. Teaching with contextual problems is the teaching that occurs in close relation the real experience. Contextual approach is concept that combines the findings of many researches recently in cognitive science. Viewed from the students' aspect, teaching with contextual problems has the probability for fostering knowledge strongly in the students, developing the student cognitive ability, and full with the development of social aspects in the students.

Based on the above explanation, the teaching with contextual problem content and conventional teaching have different characteristic. The characteristic different will cost consequences in the way of the result of teaching that the students get, and this is assumed to influence social attitude and learning achievement. However to what extends thus the teaching with VCT with contextual problem content in social science teaching has an effect on the students social; skill and learning achievement, especially the student of 5th grade at SD Gugus VI in Buleleng regency. Therefore, the aim of this study was to analyze and describe (1) difference in social attitude between the students who learn through VCT with contextual problem and those who learn through conventional teaching, (2) difference in the students social learning achievement of those who learn through VCT with contextual problem content and those who learn through conventional teaching, (3) simultaneously, the difference in social attitude and social science learning achievement of the students who learn through VCT with contextual problem content and those who learn through conventional teaching.

2. Researches Method

The dependent variables which were used as the object in this study were social as attitude and students social achievement. The independent variables to be tested in this study were with the teaching using VCT with contextual problem content and conventional teaching model. The study used an experiment design, that is, posttest only controls group design.

The population in this study was all students of the 5th grade of SD Gugus VI, Buleleng regency with the total of 12 classes and the number of students were 503 students.

The sampling tech use in this study was classroom sampling technique. The classroom sampling tech was selected in this study because the object or the targets of the sampling in this study were classes in Gugus VI Buelelng regency, rather than individuals (students in group).

Based on the characteristic of the population, the sample in this study was drawn by random sampling technique. The selection and determining experiment and control groups was done by using random sampling technique of the equivalent class pairs selected as sample.

The data collected in this study consisted of (1) social skill, (2) learning achievement. This social attitude questionnaire was written and developed by researcher herself with the approval from some judges. In the same way the learning achievement test improvement was made and developed by researcher herself with the approval from some judges. Social learning achievement in this study focuses on cognitive domain that is to meter social science learning achievement.

The methods of analyzing the data used in this study were descriptive analysis method and inferential analysis method to test the hypothesis was MANOVA analysis. Edited by SPSS-16.0 for-windows and the hypothesis testing is done by Hotelling's Trace (Hair, et all, 1998: 334) since it involved two dependant variable (multivariate) by comparing two different groups, that is, experimental and control group. The testing criterion: if the value of Hotelling Trace yielded a significant level below 0.05, then the null hypothesis was rejected and otherwise the null hypothesis was accepted (Santosa, 2002:219).

3. Results And Discussion

The description of the results in mean modus and median and the data spread can be seen in table 1.

Table 1. Recap of the result of computation of the student's social studies learning achievement score and the student social attitude.

Statistic	Learning Achievement		Social Attitude	
	VCT	Conventional	VCT	Conventional
N	40	40	40	40
Mean	27.15	24.90	109.95	100.05

Median	27	25.50	112.00	97.00
Variance	25.669	24.451	292.356	323.126
Mood	25	23	120	110
Satandard deviation	5.066	4.945	17.098	17.976
Minimum score	18	13	78	72
Maximum score	35	34	140	140
Range	17	21	62	68
Class number	6	6	6	6
Class length	3	4	11	12

Before the hypothesis testing was conducted by using the statistical method with the one way ANAVA and MANOVA prerequisite test was conducted first. Based on the statistical test with the aid from SPSS, the variable of social attitude and learning achievement in the teaching using VCT with contextual problem content and conventional teaching were normally distributed and had homogenous variants or the data came from populations with the same variant. The result of computation showed between groups probability value for students social attitude was 0.14 lower than 0.05. It means there was a significant difference in social attitude between the students who learn through the teaching using VCT with contextual problem content and that of student who learn through conventional teaching model. The students who learn through teaching using VCT with contextual problem content had scored in social attitude with the means of 105.95; while the group of students who learn through conventional teaching had the means of 100.05 for social attitude. This means that the students' social attitude for those who learn through the teaching using VCT with contextual problem content differs from those who learn through conventional teaching model. Furthermore, the calculation of one tailed T test yielded Tobs.-5.18 and tc.v with df = 94 at the 0.05 level of significant was 1.658. this means Tobs. > tc.v, thus it can be concluded that social attitude is the groups of students who learn through teaching using VCT with contextual problem content was higher than the group of students who learn through conventional teaching model at the 0.05 level of significant.

In the testing of the second hypothesis, one way ANOVA was used. The result showed that between groups probability or social science learning achievement was 0.048 smaller than 0.05. This means that there was significant different between the social science students who learn through the teaching using VCT technique with contextual problem content and that of the students who learn through conventional teaching model. The result of analysis also showed that the group students who learn through the teaching using VCT technique with contextual problem content had a means of 27.15; while the group of students who learn through conventional teaching had the means for 24.90. furthermore, the result of one tailed T test competition showed that Tobs. = -5.18 and tc.v at df = 94 at the 0.05 level of significant = 1.658, which means that tobs > tc.v, thus it can be concluded that social science achievement of the students who learn through the teaching using VCT technique with contextual problem was better that that students through conventional teaching model at 0.05 level of significant.

The simultaneous difference of the students' social science learning achievement and the students' social science attitude of those who learn through teaching following VCT with contextual problem content and those using conventional teaching model was done using ANAVA.

Table 2. Recap of the Result of Computation of Social Learning Achievement and Students' Social Attitude by One Way MANOVA Edited by SPSS 16.0 for-windows.

Effect	**Statistic**	**Value**	**F**	**Hyphothesis df**	**Error df**	**Sig.**
Intercept	*Pillai's Trace*	0.983	2224.210	2.000	77.000	0.000
	Wilks' Lambda	0.017	2224.210	2.000	77.000	0.000

	Hotelling's Trace	57.772	2224.210	2.000	77.000	0.000
	Roy's Largest Root	57.772	2224.210	2.000	77.000	0.000
A	*Pillai's Trace*	0.107	4.614	2.000	77.000	0.013
	Wilks' Lambda	0.893	4.614	2.000	77.000	0.013
	Hotelling's Trace	0.120	4.614	2.000	77.000	0.013
	Roy's Largest Root	0.120	4.614	2.000	77.000	0.013

Based on the result of analysis in table 2, the F values of Pillai's Trace, Wilks' Lambda, Hotelling's Trace, and Roy's Largest Root had the level of significant smaller than 0.05. It means the F values for Pillai's Trace, Wilks' Lambda, Hotelling's Trace, and Roy's Largest Root were significant. This means that the students' social attitude and social science learning achievement of student learning through teaching using VCT technique with contextual problem content was better than that use conventional teaching model.

4. Discussion

This study was supported by the research done by Winata (2010) entitled the effect of implementing VCT and achievement motivation on the civic learning achievement of the students of class V SD Gugus 1 Mendoyo and Jandut Gregorius (2011) in Education Journal sixth edition No. 4 2011 entitled Implementation of Modified of VCT Model as An Effort To Develop An Awareness of Students in Civic Lesson in SDN Sumur Welut.

The result of analysis showed that there was difference in social learning achievement and social attitude simultaneously between the students using VCT technique with contextuall problem contenta and those who learn through conventional teaching model. These student social science attitudes were affected by various factors, which include the tech of teaching use by the teacher. Since the teacher has to select teaching technique that is suitable for the certain topic in order to achieve the teaching objective that will be reached. The teacher in teaching is expected not only to develop the cognitive domain, but he or she has to develop effectiveve psychomotor domain too. This is in line of the nature of social science as the product and social science as process. Based on the result of this study, teachers need to be aware that not all topics can be taught by the same teaching technique especially in relation of the development of student social attitude and improvement of the students' social science learning achievement. The teaching by using VCT technique with contextual problem content stresses the important of the students improvement holistically to define the material to be learn and to relate it to the real life situation that motivate students to be able to apply it in real life. From that concept there are 3 things that can be observed. First, the teaching using VCT with contextual problem content stresses the important of direct experience process. The learning process in teaching using VCT with contextual problem content does not expect the students to only receive lesson but also put an emphasize on the process of seeking and finding the material by the students themselves.

Second, the teaching through VCT with contextual problem content encourages student to be able to finds the relation of material being learn and the real life situation, that means students are required to be able to catch the relation between the students learning problem at school and real life, the material will not only be meaningful functionally but it will be learn in a close relation deeply in students memory, it cannot be easily forgotten.

Thirda, the teaching through VCT with contextual problem content encourages the students to be able to apply what they learn in real life, it means the teaching through VCT with contextual problem content does not to only expect students to be able to understand the material learned but also how the material to be colored with their behavior in real life. The material in the teaching through VCT with contextual problem content is not aimed at collecting the material in the brain and are forgotten but as their asset in living in the real life.

5. Conclusion

Based on the result, it can be concluded that (1) the teaching approach using VCT with contextual problem content in social science teaching produces higher social skill than conventional teaching approach, and (2) the effect of the approach with VCT with contextual problem content in social science teaching produce a better learning achievement than the result of conventional teaching model, (3) the effect of teaching through VCT with contextual problem content on social science teaching produce better social attitude and social science learning achievement compared to conventional teaching model. Based on the findings on this study, then some suggestion can be made, namely (1) the teaching through VCT with contextual problem content has a comparative advantage in improving the students social attitude and students' social learning achievement, thus it is recommended to the elementary school social science teacher to use the teaching through VCT with contextual problem content to improve students' social attitude and learning achievement. (2) To the decision makers in education, especially in social science education, it is suggested that they consider the teacher tech applied in this study as an innovation in teaching, so that the social science teaching objectives can be reached optiomally.

References

[1] BSNP. 2006. Panduan Penyusunan Kurikulum Tingkat Satuan Pendidikan Jenjang Pendidikan Dasar dan Menengah. Jakarta: Pusat Kurikulum.

[2] Hasan. 1997. Pendidikan Ilmu-ilmu Sosial (Buku I). Bandung: Jurusan Sejrah fakultas PPKn IKIP Bandung.

[3] Hasan, S.H. 2005. Pembaharuan PKn di Era Otonomi Pendidikan. Makalah, Disajikan dalam seminar sehari Fakultas PPKn IKIP Negeri Singaraja.

[4] Lasmawan, Wayan. 2010. Menelisik Pendidikan IPS Dalam Perspektif Kontekstual-Empiris. Singaraja : Mediakom Indonesia Press Bali.

[5] Sanjaya, Wina. 2010. Strategi Pembelajaran Berorientasi Standar Proses Pendidikan. Jakarta: Prenada Media Group.

[6] Santoso, Singgih. 2002. SPSS Versi 11.5 Cetakan Kedua. Jakarta: Gramedia.

[7] Sommers, Kathleen. 1997. Core Competency-Based Strategy. London: International Thomson Business Press.

[8] Sumantri, E. 1999. Desentraisasi Pendidikan dan Wacana Demokrasi dalam Konteks Pendidikan Nasional. Makalah disajikan dalam Seminar Lembaga Penelitian UPI Bandung.

[9] Winata. 2010. Strategi Belajar Mengajar. Jakarta: Universitas Terbuka.

EVALUATION OF CAREER GUIDANCE PROGRAM IN VOCATIONAL HIGH SCHOOL

Sri Tutur Martaningsih[1].
Faculty of Teacher Training and Education, Ahmad Dahlan University, Yogyakarta
Email: smart.ningsih@gmail.com

Abstract

Career success is the achievement and self-actualization of the individual in his chosen career path. Understanding self-potentials and self-weaknesses, choice suitability, readiness and decisions, and understanding career development are expected to support individual career success. Facilitating the preparation of individual career development needs to be pursued for individual careers optimal development to benefit themselves and many others. Vocational high school students armed with relevant job competences, prepared for working after graduation. On the other hand, vocational high school graduates are still in their adolescent age with the development stage of search for identity, not yet thinking about career, so they needed more intensive career insight briefing through career guidance. Career guidance in vocational high school should be able to build students' career readiness, and the maturity in their adolescent age which is relatively unstable in their socio-emotional.

This review of career guidance program evaluation is conducted qualitatively through surveys, interviews and leiterature studies to provide an overview of evaluation program and its relevance to the necessity. Understanding the quality, weaknesses, obstacles to service implementation, and potential utilization are expected to improve career guidance effectiveness services in vocational high school. An evaluation on the overall career guidance program, will provide feedback for ongoing improvement. Various evaluation models are available, it needs to be selected about the relevance to the career counseling program characteristics, so that evaluation feedback is more optimal.

Keywords: program evaluation, career guidance, and vocational high schools.

1. Introduction

The number of unemployed workers who come from vocational high school graduates with a large percentage, proving that there is an unconformity between the purpose of vocational high school and the reality in the field. In addition to the factor of job competencies pursued through learning in schools and industrial work practices, it also needs the participation of various parties, especially guidance and counseling in schools to develope the career readiness and maturity of vocational high school students.

Vocational high school graduates are still in the adolescent age range, with development characteristics of socio- emotional that is still relatively unstable. The psycho-sociological development of vocational high school graduates needs to be conditioned in order to be able to face various challenges at the beginning of entering the workforce. Not only the preparation of cognitive skills and job skills, preparation of the softskill readiness is needed for career success.

Conditioning for vocational high school students to have a good career readiness is not enough just by mastering competencies that school has developed in cooperation with the business and industry, but it needs guidance and counseling role to help socio-emosional development and softskill ability. Programs developed by the Directorate of Vocational High School Education and Schools are often more dominant in preparing job readiness competencies in terms of job skills, giving less attention to softskill development for career preparation, and not paying much attention to the role of counseling guidance.

Guidance and counseling play a role in helping to optimize the development of individual learners in various development aspects including in career development. Guidance and counseling can play a role through career guidance services with various programs designed, and of course there must be service continuity from basic education level to higher education level. Implementation of guidance and counseling that has been done, often not in accordance to the field of duty, but rather to administrative routines activities, handling cases of discipline in a narrow view. Some guidance and counseling teachers are not educated in Guidance and Counseling courses, so their understanding of guidance and counseling needs to be developed from time to time.

To observe the services effectiveness that have been done so far, it needs to be done through evaluation, in order to observe the direction suitability and achievement level of goals. Evaluation feedback becomes an important input for the ongoing improvement efforts of career guidance programs. There are various obstacles to the absence of optimal guidance and counseling program evaluation. The results of Triyono's research, Afrizal Sano, and Fitria Kasih (2013: tp) stated that 31.1% of teachers in Padang experience internal constraints (knowledge and skills, perceptions, and sense of responsibility); 28.4% of external factors (criteria, training and upgrading, time, and cost). In addition, Myrick in Badrujaman (2011: 22) stated that there is no obligation to carry out evaluation (program) and teachers of Guidance and Counseling do not feel guilty if not do it.

There are many obstacles that resulted in the implementation of guidance and counseling program evaluation not maximal.

2. Problem Formulation

2.1 What is the urgency of career guidance program evaluation in vocational high school

2.2 What are the considerations for selecting the guidance and counseling program evaluation model

2.3 How to optimize the implementation and evaluation role in improving the success of career guidance in schools

3. Review of Literature

3.1. Career development of high school students.

Career development is essentially every stage of life development demands mastery of various emphases that include self-characteristics awareness and life choice (Suherman, 2009: 278). Career guidance should be tailored to the individual development level, especially in the career field. Career development of high school students. In terms of high school graduates age is at the age of 19 or 20 years. At this age of 16 to 20 years according to stated that socially independence from parents increases and cognitive reasoning and moral reasoning begin to develop. Understanding career field needed by adolescents includes self-understanding, diverse life values understanding, environment understanding, and planning for the future (Sukardi 1987: 341-342).

Career development level in high school age requires conditioning from educators, both parents and school educators. Conditioning on understanding can be done by various strategies inside and outside school. Especially with the information development and communication systems, the variety of information available and accessible from the environment (very broad and diverse) can be utilized as a learning resource through innovative service strategies development. Learning program and general policy in vocational high school relatively more dominant in the development of cognitive abilities and vocational skills, less support the career development required by students. Career guidance as part of counseling guidance service in school can play a more optimal role.

3.2 Career guidance in Vocational High School

Career counseling is an effort to help the individual to know and understand himself, knowing the working world, and developing a future which is in accordance to the expected form of life. Furthermore, by career guidance, individual capable of making decisions appropriately and responsibly or decisions taken so as to realize themselves meaningfully (Nurihsan, 2006: 16-17). Career guidance field provides direction, provision, and ability to students to have career readiness, able to plan, making career decisions and achieving career maturity.

Career guidance emphasizes on the planning of a person's life by taking into account the circumstances of himself and his environment in order to gain a broader view on the positive role influence which is feasible in society (Hatari, 1983). Super (1957), as quoted by Sukardi, 1987: 24) stated that career guidance is a process to help individual to develop acceptance of self-unity as well as its role in the working world. On the other hand, mastering the competence of job skills and its relevance to the needs in the field is done through learning by teachers in the normative, productive, and adaptive subjects supported by industrial work practices through school cooperation with the business world and industry.

Career guidance purposes above shows that the career guidance work field is very supportive toward education in vocational high school in general, preparing graduates ready to enter the working world. Career guidance works on strengthening softskill field, for career readiness, and exploration capability development, career planning to career decision making and career maturity of vocational students. To carry out these tasks a careers training program is developed which is designed, implemented and reflected on an ongoing basis.

Program is a unity of activities which is a policy realization or implementation. Program is a series of carefully designed activities and in its execution takes place in a continuous process and takes place within an organization that involves many people (Widoyoko, 2010: 8). Similarly, career guidance service program is designed as a continuous activity in helping to support the career of vocational high school individual students development in particular. Implementation on an ongoing basis should always be improved from time to time, in order to be able to achieve the goal effectively.

3.2. The importance of program evaluation

The important role of career guidance in supporting the education success in vocational high school, especially in helping career readiness, career planning, until career decision making requires seriousness in its activities. Activities undertaken by counseling guidance in schools are often still perceived wrongly by some service providers as well as by external guidance and counseling.

To prove career guidance services performance, there needs to be an accurate observation and measurement, through program evaluation activities. Evaluation results are expected to provide input for program improvement effectively.

Program evaluation is an effort to assess the efficiency and effectiveness of any attempt to achieve a goal by using a certain standard or criteria as a benchmark (Winkel and Hastuti, 2013: 821). Evaluation determines the degree of activity progress quality (Sukardi, 1990: 47). Program evaluation can be done by reviewing or analyzing the programs that have been or are being implemented to develop and improve the program.

Evaluation is defined as the process of collecting data systematically, drawing conclusions based on data obtained objectively, conducting interpretations, planning steps of improvement and development (Salahudin: 2010: 217). With the evaluation, it is expected to know the implemented program effectiveness, followed by the decision-making process. Another definition of this evaluation is an effort to obtain various information on regular, continuous and comprehensive way about the process and results of the program activities that have been implemented. The results of career guidance programs can be the development of attitudes and behavior, or students development tasks completion.

Program complexity from the component side, as well as the role of each related party, requires a relevant evaluation system and quality, so that the program accountability can be obtained through a reliable process and evaluation procedures. Fitzpatrick, Sandra, and Worthern (2011: 491) stated that with evaluation, it will become an increasingly useful force in the following ways: improving programs, improving policy making, improving organizational learning and decision making,

improving societies through improving the social conditions program address, improving even itself.
The evaluation of career guidance in vocational high school will be very beneficial for schools and related policy makers to prepare better graduates so that they will be success entering and developing their career.

3.3. Variety of program evaluation

Program evaluation is a process to provide reliable information to be considered in determining the program success rate. The program evaluation results are expected to help program implementers to prove the program's accountability. Many program evaluation and evaluation approaches can be used and selected to evaluating the program. Understanding on the program characteristics, the desired evaluation approach, and the appropriate evaluation model become abilities that can not be ignored in evaluating the program.

There are many program evaluation variations with different steps and components of evaluation. The career counseling program as a form of a program with special characteristics, requires an appropriate form of evaluation. Guidance and counseling teachers' need to understand the process of planning, implementation, evaluation, analysis, interpretation, follow-up and formulation of recommendations based on evaluation results.

Stufflebeam (1969) is evaluation that prioritizes evaluation objective on four aspects: context, input, process, and product CIPP). The UCLA model (Alkin, 1969) defined evaluation as the process of preparing information for proper analysis and decision making. Brinkerhoff model et al (1983), is grouping the evaluation into three groups which are fixed vs emergent evaluation design, formative vs summative evaluation; experimental and quasi experimental design vs natural/unobtrusive inquiry. In short, the grouping can be explained as follows. Fixed design is an evaluation that from the beginning there has been systematic planning; while the emergent design is an evaluation with a design that can be develop during the process, adjusting to the situation and conditions.

Kitzpatrick, Sandra, and Worthern (2011: 127-251) described several alternative approaches as follows: expertise and consumer oriented approaches; program-oriented evaluation approach or the objectives oriented evaluation approaches; decision-oriented evaluation approaches; participant-oriented evaluation approaches; other current considerations: cultural competence and capacity building; and a comparative analysis of approaches.

The selection of models and approaches to be used in the program evaluation considering the evaluation objectives, the program characteristics, and compliance with the feasibility of the implementation of the evaluation process and procedures within.

4. Discussion

There are constraints faced in the implementation of student career preparation. False perceptions of internal and external parties toward roles, duties and obligations of guidance and counseling in schools, resulting in the of career counseling services unoptimal implementation, not yet corresponding to the expectations. Program evaluation has not been properly implemented, so feedback for continuous improvement is not obtained. Understanding on the program evaluation system still needs to be improved, the provision of models and evaluation tools may help teachers, principals and various parties to be able to evaluate the program.

Urgency of career guidance program evaluation in vocational high school. The evaluation of a career guidance program is needed to provide a comprehensive, complete, and detailed overview of the program; knowing the level of goal achievement; examine the program's constraints, weaknesses and shortcomings; provide input or feedback to properly improve the program. Furthermore, the evaluation results forms the basis of recommendation to related parties in order to improve the program based on the finding of weaknesses in the evaluation.

Consideration on the selection of the career guidance and counseling program evaluation model in schools has characteristics as a service program that has a variety of components, procedures, and specific roles of each party. The object of career counseling is individual students with many characteristics variety, resulting in variations in the needs of career guidance services as well. Services characteristics that have diversity in these aspects require complex program evaluations. Although limitation can be made in the scope of evaluation, but it required a complete picture and program detail.

Program recommendations and follow-up evaluation. Recommendations are suggestions for improvements based on the finding on the evaluation results. Accurate and accountable evaluation implementation is expected to provide detailed improvements that must be done. Whereas quality evaluations are sustained, also accurate recommendations serve as the basis for program improvement; then continuous quality improvement of the program can be done, that give more benefit to all parties widely.

Optimize the implementation and evaluation role in improving the career guidance success in schools (a) Improving guidance and counseling teachers' and various stakeholders' understanding on quality evaluation processes and procedures. (b) Improving various parties skills (evaluators) on the evaluation process and procedures; from planning to analysis, interpretation, and the recommendations forming. There needs to be a well-tested evaluation model complete with evaluation tools that can practically guide the evaluator in the evaluation process completly.

5. Conclusion and Suggestion

5.1. Conclusion

Vocational high school graduates are expected to enter the workforce in adolescent age, therefore need assistance in order to have career readiness and maturity which formed simultaneously with the process of mastering the work skills competence. Efforts to facilitate individual career success can be done by counseling and guidance in schools through career guidance services. Constraints faced in the implementation of career guidance need to be find the right solution to overcome it, so that the service program is effective in achieving the goal. Detection of constraints, weaknesses and objectives achievement levels can be obtained through the implementation of quality program evaluations, so that appropriate improvement recommendations can be obtained.

Complete program evaluation model practically equipped with the necessary tools, up to an overview on how to use, till analysis, results and recommendations interpretation; urgently needed to guide the program evaluation process. This optimization step is needed considering the understanding on evaluation process implementation in field is still not as expected.

5.2. Suggestion

For schools. Schools need to make arrangements to evaluate career guidance programs in schools, to implement and obtain periodic improvement feedbacks from recommendations made based on evaluation results. *Guidance and counseling teachers, principals, and implementing parties.* Insights regarding program evaluation need to be improved, from planning to results analysis based on data obtained in the field. Including make recommendations which are complete, accurate, and detailed feedback based on evaluation results.
For other researchers. A quality career guidance evaluation model is required, complete with instrument tools and analytical tools as well as guidelines for interpreting the results and formulating appropriate recommendations.

References

Badrujaman, Aip. (2011). *Teori dan aplikasi evaluasi program bimbingan dan konseling.* Jakarta: PT. Indeks.

Brinkerhoff, Robert O, Brethover, D.M,, Hluchyj. T., & Nowakowski, J.R., (1983). *Program Evaluation. Practitioner"s Guide for Trainers and Educators.* Boston: Source book & Casebook.

Fitzpatrick, Jody L., Sandra, James R., Worthern, Blaine R. (2011). *Program Evaluation Alternative approaches and practical guidelines.* United States: Pearson Education Inc.

Hattari.P.M. (1983). Suatu strategi bimbingan karir dalam pembangunan nasional. *makalah* Konvensi Nasional V IPBI. 24 – 26 Pedruari 1083 di Yogyakarta.

Nurihsan, Achmad Juntika. (2006). *Bimbingan dan Konseling dalam Berbagai Latar Kehidupan*. Bandung: refika Aditama.

Salahudin, Anas. (2010). *Bimbingan dan Konseling*. Bandung: Pustaka Setia.

Stufflebeam,D.L., (1969). *Evaluation and Enlightment for Decision Making*. Columbus, OH: Ohio State University, evaluation Center.

Suherman, Uman AS. (2009). *Konseling Karir Sepanjang Rentang Kehidupan*. Bandung: Universitas Pendidikan Indonesia.

Sukardi, Dewa Ketut (1987). *Bimbingan Karir di sekolah-sekolah.* Jakarta: Ghalia Indonesia

______________. (1990). *Pengantar Pelaksanaan Program Bimbingan Sekolah*. Jakarta: Rineka Cipta.

Triyono, Afrizal, dan Fitri Kasih 2013. Faktor penghambat pelaksanaan evaluasi Program BK di SMA N Kota Padang. *E-journal*: PPS Undhiksa Prodi Administrasi Pendidikan .Vol. 2 number 1 Year 2013

Widoyoko, Eko Putro. (2010). *Evaluasi Program Pembelajaran.* Yogyakarta: Pustaka Pelajar.

Winkel, WS. dan MM. Sri hastuti. (2013). *Bimbingan dan konseling di Institusi Pendidikan.* Yogyakarta: Media Abadi.

Learning of social studies in elementary school as a medium to strengthen multicultural education in the curriculum era 2013

Yulfia Nora[1,*]

[1]FKIP, Universitas Bung Hatta, 25171 Padang, Indonesia

Abstract. The writing of this article aims to describe the learning of social studies in elementary schools as a medium to strengthen multicultural education in the era of the curriculum 2013. Writing methods used are qualitative descriptions and literature studies. Social studiesis one of the subjects found in the basic education curriculum.Social studies essentially examines human relationships with the environment. The environment in question here is one of the ethnic, religious, racial, ethnic and linguistic diversity in Indonesian. Become, the context of social studies learning is multicultural education.Multicultural education is a process of developing human potential that values heterogeneity as a consequence of diversity based on the principle of equality, mutual respect, and mutual acceptance and understanding of the moral commitment to bring about social justice. But in fact, social studies learning has not been applied as a medium of multicultural education.Therefore, through the implementation of curriculum 2013 Social studies subjects as a medium to strengthen multicultural education.This is because the implementation of the 2013 curriculum is oriented towards the formation and development of learners' character in diversity.

1 Introduction

Indonesian is a country of diverse people, composed of various religious, cultural, linguistic, ethnic, racial and ethnic groups. In other words, Indonesian is a multicultural country. It cannot be denied that such diversity can divide unity and unity, even the worst can dispel the sense of nationalism. This is because the community is more concerned with the group than the interests of the nation. The diminishing sense of unity and the decay of nationalism can be seen from the life of Indonesian society today, especially among the young who are strongly influenced by outside cultures, resulting in many behaviors that are inconsistent with the national identity and the number of narrow fanaticism groups that can lead to disharmony in society. In response to these problems, the government and society must overcome and anticipate early one of them through multicultural education.

Multicultural education is not a subject in the curriculum, but multicultural education is a strategy and concept of education based on the utilization of diversity that exists in

* Corresponding author: yulfianora@yahoo.com

society through learning materials, especially those that exist in learners such as religious diversity, culture, language, social status, gender, ethnicity and race. Multicultural education is education that always upholds values, beliefs, heterogeneity, plurality and diversity, whatever its aspects in society. Thus, multicultural education does not make all human beings homogeneous. Multicultural education is not business-oriented education. Yet as is known that the actual education is not an education that produces mere skills, but education that must accommodate all types of intelligence known as multiple intelligence. In other words, multicultural education is integrated in the subject matter of every subject at school, especially in elementary and secondary schools.

Government efforts to overcome and anticipate problems that occur, because as a multicultural country that makes education as one of the strategic means in efforts to build, develop and maintain the identity of the nation in the current era of globalization. Application of the 2013 curriculum is one of the main means to form and build the character of learners in accordance with the national identity of Pancasila and the 1945 Constitution. This is because the internalization of Pancasila values and the 1945 Constitution is further expanded not only in the subjects of Civics alone but rather integrated in all subjects in school. Thus, that each subject has an important role to shape the character of the learner. Some integrated subjects can also bear a character. One of the compulsory subjects in the elementary school curriculum is the subject of Social Studies.

Social study is a subject that integrates various branches of social science such as economics, sociology, geography and history. Social studies essentially examines events, facts, concepts and generalizations on issues, phenomena and social phenomena. In other words, Social studies examines human relationships with the environment, related to what causes and effects that arise after a social process conducted by humans both individually or in groups, be it cultural, religious, ethnic and other groups. This means that social studies material nuanced or directly related to multicultural education. Thus, social studies can serve as a medium for shaping and developing multicultural education.

2 Writing method

The method of writing this article using descriptive qualitative and literature study. In essence descriptive qualitative research is a method in researching the status of a group of people, an object with the aim of making descriptive, picture or painting systematically, factually and accurately about the facts investigated. This descriptive qualitative research aims to describe what is happening today. This means that this study describes, notes, analyzes and interpret the current conditions occur. In other words descriptive qualitative research is aimed to obtain information of existing circumstances. While the literature method is a method of data collection is done by taking the necessary data from the relevant literature.

3 Discussion

3.1 2013 curriculum

The curriculum of 2013 is a change of curriculum structure of KTSP. This change is made of many problems that occur in education and is one effort to improve the curriculum that is less precise to the needs of public education today. Curriculum 2013 or abbreviated as "kutilas" has the main character in which every competence that must be achieved by students are character-filled.

Competence of SD/MI graduates that are contained in the curriculum of 2013 that is, 1) Attitudes, through curriculum 2013, students are expected to have behaviors that reflect the attitude of believers, noble character, confidence and responsible in effectively interaction with social and natural environment around the house, school and playground. 2) Knowledge, through the 2013 curriculum, students are expected to have the ability of factual and conceptual knowledge in science and technology, art and culture. 3) Skills, through the curriculum 2013, students are expected to have the ability of thought and action effective and creative in the realm of abstract and concrete in accordance with the assigned to learners.

The implementation of the 2013 curriculum, is expected to support better education in the future that considering the many challenges found in the current globalization. This means that 2013 curriculum is expected to filter out the outside culture that goes into the country, so that the identity of the nation can still be maintained. As mentioned earlier that 2013 curriculum has a special characteristic in which each learning material is character-filled. Characters that are formed and developed in the participants include the mutual acceptance and respect for differences in social life.

In general, the 2013 curriculum has the following characteristics:

a. Develop a balance between the development of spiritual and social attitudes, curiosity, creativity, cooperation with intellectual and psychomotor abilities
b. Schools are part of a community that provides a planned learning experience in which learners apply what is learned in school to community and utilize the community as a learning resource
c. Develop attitudes, knowledge and skills and apply them in various situations in schools and communities
d. Give enough time to develop various attitudes, knowledge and skills
e. Competence is expressed in terms of core competencies of the class detailed in the subject's basic competencies
f. The core competencies of the class are elements of basic competence organizing, in which all basic competencies are developed to achieve core competencies
g. Basic competencies are developed based on accumulative principles, mutual reinforcement between subjects and levels of education.

Based on the above characteristics, the objective of the 2013 curriculum is to prepare Indonesian people to have the ability to live as individuals and citizens who are faithful, productive, creative, innovative and affective and able to contribute to the life of society, nation, state and civilization.

Social studies is one of the subjects contained in the curriculum of 2013, which has an important function to achieve curriculum objectives, especially in shaping and developing the character of learners in accordance with national identity (Pancasila and 1945). Thus the subject becomes an appropriate means for multicultural education.

3.2 Social studies learning in elementary school

According to [1] Social studies is a science that examines various disciplines of social sciences and humanities and basic human activities are packed scientifically in order to provide insight and deep understanding to learners, especially at the level Basic and intermediate.

Social studies is a subject that is expected to nurture its members really developed as a rational and responsible social human, so that therefore created values. Social studies learning in elementary school is subjects related to human life, educate, provide stock and train attitude, values, morals, and skills for learners, so that learners can become democratic and responsible Indonesian citizens and citizens of a peace-loving world [1]. In line with the opinion of [2] social studies education is an educational program that builds learners to become good citizens in an atmosphere of peace. Indicator, after studying social studies, learners will have a number of competencies that is communicate, adapt, synergize well and always think positive to others.

Social studies learning is a learning that concerns all aspects of relationships in human life. The purpose of IPS learning is the change of behavior and positive behavior of learners with the culture, values, customs and traditions that exist in society. Thus, the context of IPS learning is multicultural education. Multicultural education is a process of developing human potential that values heterogeneity as a consequence of diversity based on the principle of equality, mutual respect, mutual acceptance and understanding of moral commitment to bring about social justice.

Social studies learning in elementary schools should consider the needs and characteristics of learners aged between 6-12 years. Children in the 7-11 years age group according to Piaget are in the development of cognitive abilities at concrete operational level. Concrete operations are the cognitive level by which the child will understand what is learned when looking at concrete objects. This means that even though the subject matter of Social studies in elementary school is the concept material, the teacher must relate the material with everything that can be seen and felt or experienced by the children. Thus, the message of multicultural education can be easily conveyed through social studies learning.

3.3 Multicultural education

The context of multicultural life is a multi-dimensional dimensioned understanding that is presented to broaden the discourse of human thought which has been maintaining cultural, religious, and group "egoism". Maintaining a plurality of cultural diversity is a social and political interaction between people of different ways of life and thinking in a society. Maintaining a plurality of cultural diversity is a social and political interaction between people of different ways of life and thinking in a society. Thus, the attitude of multiculturalism should be owned by everyone. The attitude of accepting and appreciating the difference is not automatically developed by itself. Especially because in a person there is a tendency to expect others to be like him [3].

[2] argues that in essence multicultural education is meaningful for preparing all students to work actively toward structural similarities in school organizations and institutions. Multicultural education is not a policy that leads to the institutionalization of inclusive education and teaching through curricula that contribute to individual cultural competition. In addition, multicultural-based learning seeks to empower students to develop respect for people of different cultures, ethnic or race directly. Multicultural education is one of the strategies presented in education to understand the reality condition of Indonesian society that is rich in diversity in various dimensions of life.

The mutual acceptance and attitude for different cultures, religions, races and ethnic groups will develop if trained and taught to learners through multicultural education. Multicultural education is not a subject taught in schools, but integrated multicultural education in every subject. Through the education system, then multicultural education is planned well that is contained in the curriculum and elaborated in the form of competence for each subject matter that contains character according to the values of national identity. In other words, multicultural education is an effort to teach the diversity that exist in Indonesia such as the existence of Sundanese, Javanese, Minang, Batak, Ambon and others. And because of its cultural contents, among others arts, traditional clothing and custom homes as which are taught in the subjects of social studies, multicultural it is teach various instruments, clothing and custom homes every region in Indonesian [4].

Multicultural education is a proper and proper education for all learners. This is because the main purpose of multicultural education are; 1) eliminating educational discrimination means giving equal opportunity for each learner to develop his potential, 2) make learners can achieve academic achievement according to their potential, 3) make learners aware of social and active as citizens of local, national and global community. Obviously this goal must be achieved given the enormous external cultural influences on the development of behavior of the younger generation (students). To achieve the goals of multicultural education, the role and skills of teachers is needed in teach any learning materials that contain aspects of multicultural education. Hanum and Sisca [5] argue that multicultural education at least concerns three things: 1) awareness of the important value of cultural diversity, awareness of diversity contributes to the personal development of students, 2) educational reform movement, meaning that multicultural education is not merely an educational program, but covers all aspects of education.

Among the skills of teachers that can be applied of the teach multicultural is the ability of teachers to find and convey multicultural messages through learning materials in each subject, especially on social studies materials. Social studies is a subject where the material is directly related to multicultural, especially in primary school. In order for teachers to convey multicultural messages properly, teachers need to use varied and innovative learning models. Sagala (2010, p.176) suggests that the learning model is a conceptual framework that contains systematic procedures and organizes certain learning experiences that serve as teacher guides in the teaching and learning process. In addition to using the learning model, teachers can also use the module as a tool to convey the message of multicultural education.

The research of [5] on the implementation of multicultural learning models in elementary schools using a module, concluded that most teachers already have readiness to convey multicultural messages to students. And even most teachers argue that multicultural education can be applied in everyday life as well as developed modules integrated with social studies subjects can be used as a tool to cultivate multicultural education in elementary schools. Further research results [6] also concluded that "multicultural learning model for elementary school is integrated multicultural learning using modules can be accepted and established by teachers as a model of multicultural learning applied in schools integrated with social studies material and supported by material modules Teach as a relevant material supplement ". Further research results revealed that the project-based multicultural education learning module can also improve learning outcomes of students [7].

It can be concluded that the subjects of social studies is a medium to strengthen multicultural education considering the social studies learning materials in elementary schools is directly related to multicultural education. As an example of concrete social studies material learning such as the role material and position of a person in the family, community and country. The material of cultural diversity teaches learners to recognize and appreciate other cultures. The material about heroes can be used as role models for students to live their lives.

4 Conclusion and recommendation

4.1 Conclusion

Based on the discussion, it can be concluded that (1) the main characteristics of the 2013 curriculum are characterized. One of the characters in question is the attitude of accepting and respecting religious, cultural, racial and ethnic differences. (2) multicultural education is a solution for the people of Indonesia to maintain and uphold the national identity of the Pancalisa and the 1945 Constitution in the current global era. (3) Social studies learning is a learning that is directly related to multicultural education, especially in elementary schools. (4) A teacher must be able to find and convey the message of multicultural education contained in every learning material especially on social studies learning materials. (5) Teachers can use various models of learning, approach or module as a tool to achieve learning objectives.

4.2 Recommendation

Based on the conclusion, the authors suggest to teachers in elementary schools to relate learning materials social studies with concrete situations so that learners can easily understand what the teacher submitted. Teachers are also skilled at discovering and conveying the message of multicultural education through social studies learning materials. In addition, teachers are also advised to use various learning models, approaches or modules as a tool to achieve learning objectives.

References

1. A. Susanto, *Proses Belajar Mengajar* (Kencana Prenadamedia Group, Jakarta, 2014).
2. Sudrajat, Jurnal JIPSINDO **1** (2014).
3. R. Ibrahim, Jurnal Pendidikan Islam El-Carbawi **1**(2008).
4. Tatang M. Amirin, Jurnal Pembengunan Pendidikan: Fondasi dan Aplikasi **1** (2012).
5. F. Hanum, S. Rahmadonna, Jurnal Penelitian Ilmu Pendidikan **1** (3) (2010).
6. F.Hanum, S.Raharja, Jurnal Penelitian Ilmu Pendidikan **2** (6) (2013).
7. L. Prastyawati, F. Hanum, Harmoni Sosial: Jurnal Pendidikan IPS **1** (2) (2015).

An Estimation of Variances and Factors in Mathematics Attainment in Upper-Primary Schools

Laurens Kaluge[1,*]

[1]Graduate Program, Universitas Kanjuruhan Malang, East Java, Indonesia

Abstract. This study aimed at identifying variances related to school and pupil levels, and the performance indicators measuring the learning progress in mathematics. A longitudinal design was carried out and the stratified random sampling was used to obtain a sample of 5118 pupils from 60 primary schools. In each school three upper grades were selected, including their 180 class teachers and 60 headteachers. The data collection was carried out through questionnaires and school archives. Using multilevel analysis, the results showed that the proportion of total variance in pupil attainment for mathematics at the school level ranged from 20 to 29 per cent. When intake characteristics were taken into account as the baseline, the variance proportions were less than the prior attainments. Occupation and education of parents, besides reading facilities at home were significant explanatory variables for children progress in mathematics achievement. The outcomes have important consequences for school effectiveness research in determining the baseline and performance indicators.

Keywords: Variance estimation, achievement factors, mathematics attainment, primary schools

Introduction

Schools exerted a more powerful influence on achievement than pupils' family background factors. It was the conclusion of most studies in developing countries up to mid-1980. This led to the inference that developing countries, schools were more effective than schools in well developed countries, both in raising achievement and in producing a route for social mobility [1, 2]. This conclusion had often been justified by the claim that in many developing countries, operating in communities [3] where written literacy and numeracy was a recent phenomena, and that as a result, schools with a few material inputs and only a moderate quality nonetheless might have a significance influence on academic attainment.

It was known from the survey that family background and school factors on pupils' achievement in developing and developed countries [1] as follows. The variation in school factors explained only small portions of variance in achievement in the developed countries. However, in developing countries the block of school factors explained significant portions of the variance in achievement. They also found a significant correlation between a nation's wealth and the amount of variance explained by the school factors and concluded that the lower the income of the country, the weaker the influence of pupils' social status on achievement. On this basis the conclusion was that a different paradigm of school effectiveness existed in the developing countries where schools had a greater impact on achievement than home background factors.

The Indonesian Primary Education Quality Improvement Project (PEQIP) published internationally [4, 5, 6] might be recognized as the first Indonesian research using multilevel approach. This study examined the PEQIP project in two provinces (from six provinces), Aceh and North Sulawesi. Each province included 72 to 78 schools with 9 clusters (each one nuclear school and six to nine satellite schools).The project was focused in teacher professional development, educational management (including community participation), books and learning materials, and evaluation and monitoring. Using two-level analysis they found that 46 per cent variance between schools for mathematics, 39 per cent for Indonesian Language, and 40 per cent for science in primary schools. In North Sulawesi where the PEQIP had a statistically significant influence, the pupil, classroom, and school variables were found for the three subjects as the following. By and large, PEQIP had positive effects on mathematics attainment and science where satellite schools made more progress than nuclear schools. The

* Corresponding author: laurens@unikama.ac.id

 https://doi.org/10.1051/shsconf/20184200095

teacher professional development comprised innovative teaching and homework frequency were positively significant. The influence of evaluation and monitoring were not significant.

This current study search further by using multilevel techniques for primary schools in East Java, particularly in the City of Malang - Indonesia. In reality, the typical social data have multilevel characteristics [7]. Due to only a handful study in developing countries using the multilevel techniques analysis, this was an effort to figure out the right reality by splitting the student and class levels then compared different models in explaining the educational attainment. The main research problem was that how much the school and pupil variances and what part taken into account among class/schools?

Methods

This study was designed as a longitudinal survey. The collection of pre-test data was carried out in the beginning of the school year. Then post-test at the end of the school year. While the background data on family and on teacher were obtained before the post-test. All the data related to children were nested into classroom and school for the sake of multilevel modeling. Data on teachers were parts of class level variables.

The sample in this study, was based on 5 subdistricts in the City of Malang (Blimbing, Kedungkandang, Klojen, Lowokwaru, and Sukun), the Province of East Java – Indonesia. The sample consisted of 5118 pupils, and 180 teachers from 60 primary schools. Three upper grades (4-6) were selected in each school. As a longitudinal survey, data were collected through questionnaires and archives in schools. Taking pupil prior year attainment as baseline, multilevel statistical techniques were used on assessment at the end of the school year to examine factors related in pupil, classroom and school levels which were related to pupil attainment a year later, described later in Table 1.

The response variable was the mathematics scores obtained by the pupils at the end of the school year. During this school year children had moved one grade higher i.e. grade 4. Tests at the end of grade 3, before moving to grade 4, were used as the baseline tests.

Table 1. Description of Pupil and School Variables

	Grade 4		Grade 5		Grade 6	
Variable	Mean	SD	Mean	SD	Mean	SD
Pre-test	64.9	14.9	58.4	13.27	61.9	14.11
Post-test	64.2	16.25	58.3	12.99	54.4	14.0
Pupil age (in months)	116.7	8.38	129.95	9.49	141.01	7.58
Textbooks owned at home	2.31	0.87	2.42	0.93	2.63	1.0
Teacher age (in years)	37.78	8.22	37.27	5.73	40.73	6.97

	Grade 4	Grade 5	Grade 6
Father occupation			
clerical & professional	40.3%	38.6%	39.5%
unskilled manual,	34.5%	36.8%	34.9%
skilled manual	25.2%	24.6%	25.6%
Father education			
Primary level	45%	45.2%	45.2%
Secondary level	40.4%	40.5%	40.8%
Tertiary level	14.6%	14.3%	14%
Mother education (in years)			
Primary level	56.6%	56.5%	57%
Secondary level	34.2%	35.6%	34.2%
Tertiary level	9.2%	7.9%	8.8%
Home language (Indonesia language)	35.9%	33.3%	32.4%
Reading newspaper	23%	26.7%	41.7%
Teacher gender (%male)	25%	18%	27%
Teacher's maths training	41.7%	30%	45%

The frequency distributions of the raw outcome measures were examined and tested the skewness and kurtosis were conducted as well. In order to facilitate comparison of the different model results it was considered desirable to transform the outcome measures to a standard format. After the response variables were transformed to normal scores and centering the baselines around the average to improve model fit [7, 8], the strategy consisted of 4 possible consecutive models were planned. *Model 0*, known as the null model, disaggregated the variance components of the two levels. In this model, the mathematics scores were regressed on the constant term (coded 1 for every student). In addition, the constant term was set at random at both the student and school levels. The aim of this analysis was to estimate the overall mean achievement at both intake and at the end of school year and also to see whether there were any school differences in mean achievement.

Then the next was exploring the following models at the fixed part. A model reduction procedure of omitting predictors that did not contribute to the variation in achievement would refine the model into a 'minimal' model. Levels of significance under the 95 per cent confidence interval were obtained if the estimate was equal to or greater than double standard error [9]. *Model 1* was a simple value-added analysis of pupil progress over one school year after controlling for prior attainment only. Language and numeracy were closely related each other as shown in many studies,. Therefore in this study, both pupils' scores in Indonesian Language and in Mathematics were included as baseline measures of prior attainment. The quadratic terms of each baseline were also tested in order to allow for any non-linearity in the relationships with the response variable.

The contextualized model, *Model 2*, included measures of child background factors such as individual characteristics and family/home background without prior attainment. Model 3 was a more complex longitudinal value-added analysis which includes both significant background factors and the prior attainment measures.

Results

Pupil samples at Grade 4 for every school ranged in size between 14-41, at Grade 5 between 17-45, and at Grade 6 between 16-44 pupils. These conditions of range satisfied the requirements of multilevel modeling [9]. After performing the null model, the modeling stages dealing with explanatory variables were consecutively entered: initial scores as baseline, child background and characteristics, class and school variables. Then the variances attributable to school and pupil were checked.

Table 2. Proportions of Unexplained Variance at Pupil and School Level

	Model 0	Model 1	Model 2	Model 3
Grade 4				
School/class level	20.4%	26.4%	15.8%	24.1%
Pupil level	79.6%	73.6%	84.2%	75.9%
Variance accounted for	-	31.3%	12.8%	35.0%
Grade 5				
School/class level	21.4%	27.6%	21.1%	27.4%
Pupil level	78.6%	72.4%	78.9%	72.6%
Variance accounted for	-	33.8%	7.6%	35.3%
Grade 6				
School/class level	29.2%	38.6%	32.3%	39.1%
Pupil level	70.8%	61.4%	68.7%	60.9%
Variance accounted for	-	26.2%	7.0%	32.9%

The null (empty) models of the two-level analysis, as in Table 2, demonstrated that a significant proportion of the variance of each outcome was related to between class/school differences (as expected in all cases the majority of the variance was found to lie between pupils). From the total in the empty models, the proportion of class/school variances ranged between 20.4 to 29.2 per cent. The proportion of pupil variances ranged between 70.8 to 79.6 per cent. An interesting difference interesting result was that the higher the grade the larger the variance attributed to the class/school level.

Before entering the prior attainment into model 1, the correlations between pre- and post-test scores were checked. The correlations at grade 4, 5, and 6 were 0.54, 0.55, and 0.54 significantly. Model 1 revealed that the prior scores were non-linear predictors as shown by the significance of the squared variables. The results still existed up to the last model as in Table 3.

Results from stepwise procedures, discovered in Model 2 that pupil gender, father's occupation, mother's education, and textbooks at home were significant predictors of Mathematics attainment. Girls outperformed boys, children whose mothers were in clerical and professional work attained higher scores than those whose mothers were in skilled-manual or unskilled manual or non-earning categories. The higher the educational level of mother, the better the child achieved in Mathematics achievement. The greater the provision of textbooks in pupils' homes, the better the progress made in Mathematics. The prior scores were better predictors as well. In total variance for *Model 1* was only 12.8 per cent the reduction. Of the remaining unexplained variance 84.2 per cent was attributed to the pupil and 15.8 per cent to the class/school level.

Table 3. Regression estimates and standard errors of explanatory variables per grade (the significant ones from model 3)

Variable (in *Fixed Part*)	Estimate (Standard Error)		
	Grade 4	Grade 5	Grade 6
*Intercept	-.53 (.14)	-.092 (.092)	-.22 (.15)
*Mathematics	.029 (.0014)	.03 (.0017)	.03 (.0015)
*Sq-Mathematics	.00021 (.000075)	.00027 (.00009)	.00032 (.00007)
*Language	.013 (.0021)	.023 (.002)	.02 (.002)
*Sq-Language	.00053 (.00011)	.00037 (.000094)	
*Pupil gender	-.16 (.036)		
Age1		.052 (.045)	-.061 (.045)
Age2		-.091 (.049)	-.12 (.045)
*Age3		-.13 (.05)	-.31 (.049)
*Father's occup.1	-.14 (.052)	-.15 (.048)	
Father's occup.2	-.049 (.051)	-.12 (.047)	
*Mother's education	.02 (.0078)		
*Home language		.058 (.02)	
*Books at home	.059 (.022)		.038 (.019)
*Maths training			.024 (.18)
*Teacher gender (male)	.15 (.044)	.32 (.12)	-.08 (.034)

Note: * $p < 0.05$; Standard Error in brackets;
Pupil gender (0=female pupil, 1=male pupil);
Father occupation (0=clerical & professional, 1=unskilled manual, 2=skilled manual)

After controlling for measures of the personal and family background of pupils and prior attainment, in the model 2, the range of proportion attributed to school and class factors were between 24.1 to 39.1 per cent.

A more complex value-added analysis appeared in Model 3 by including the initial scores and background factors provided a better fit. In all 35 per cent of the total variance was accounted for. The unexplained variance 75.9 per cent was attributed to the pupil level and 24.1 per cent to the school level. Apart from pupil gender, all the other findings differed in comparison with Indonesian Language attainment for the same cohort. Similar to the previous analysis, the coefficients of exploratory variables in Model 3 became smaller after prior attainment was controlled for. We might say that the prior attainment measures showed a stronger relationship with later Mathematics results.

Discussion

Data on pupil attainments over a one year period (using a pre- and post-test framework) were analyzed using multilevel statistical techniques as discussed early. These provide the most appropriate basis for the analysis of school effects on mathematics, a cognitive outcome in basic skills. Although the use of basic skills as a measure of school performance had been criticized, they can be defended for two reasons [10, 11]. Firstly, the historical view of effectiveness concerns the improvement of the basic skills of disadvantaged children. Secondly, basic skills provide the foundation for further learning and therefore can be seen as lying at the core of subsequent educational outcomes and progress.

From the empty model, the proportion of class/school variance was 20.4% for mathematics. The findings were in line with those found by other researchers working in primary schools in developed countries in terms of the effect size of the subject. Researchers using multilevel analysis in developing countries have found that the school variances tend to be considerably higher although many of the studies suffered from limitations in designing [12]. For this Indonesian study the smaller variance results were likely to reflect the particular nature of the sample in which only urban state schools were included. In interpreting the results this particular context must be remembered and it was likely that if schools in other areas of Indonesia were included in the study the level 2 (described as a combination of class/school effects which could not be separated as only one class per grade were sampled) variance estimates would be expected to be greater.

After controlling for measures of the personal and family background of pupils and prior attainment, shown in Model 4, the proportion attributed to school and class factors was 24.1%. The proportion appeared to be different from those in developed countries i.e. 12-18% [13]. Few exceptions were, for examples, a study England [14] where in the lower grades of British primary schools, 40% was attributed to the school level variation and a study in Scotland for Mathematics controlling for prior attainment 32% in the primary study [15].

It should be noted in interpreting this study's findings that the percentages had a role in setting the boundaries for the potential impact of factors in different levels only [3, 16]. To discover more about the explanatory factors in each level, further analyses were needed.

The relationship between outcome measures and pupils' prior attainment was assessed in order to examine the question of linearity. Non-linear relationships were found between grade 4 mathematics' and Indonesian language's prior with post scores. By controlling for these non-linear relationships, more appropriate results were achieved in terms of model fit [7, 9]. After discovering the typical relationships, the prior scores were treated as baselines for examining the impact of other significant predictors.

The child background measures i.e. gender, parental occupation and education, and books at home were found to be statistically significant predictors for mathematics. Girls were performing better than boys. Children whose fathers had skilled manual and unskilled manual jobs had poorer achievement than those with fathers in clerical and professional works. Mother's education was the stronger predictor for child's attainment. The availability of textbooks at home affected the mathematics achievement as well. These factors were confirmed in the previous studies [17-19].

Conclusions

The variances of children attainment at upper grades in primary schools were much higher at the individual level than the class level. Even low, the classes variances expressed how important the process of the teaching. The findings had important consequences for school effectiveness research in determining the baseline and performance indicators. Of course, the implication of this study was on the policy for educational quality improvement. The educational quality could be misleading if someone ignoring the possible variances in terms of controlling the predictors and the hierarchical aspects of schools. The present study was designed to assess a set of performance indicators for schools by distinguishing the characteristics of levels inside the schools that might be of use to subsequent educational policy development.

Acknowledgment

The author would like express the sincere appreciation to four colleagues from the State University of Malang, Head of Sub-district Offices of Education in Malang, and school principals in the City of Malang for helping the data collection. The author received no financial support for the research and authorship of this article.

References

1. S.P. Heyneman, W.A. Loxley, Amer. J. Sociol. **88**, 1162-94 (1983).
2. B. Fuller, Rev. of Ed. Res. **57**, 255-92 (1987).
3. A.G. Wilhelm, C. Munter, K. Jackson, The Elem. Sch. J. 117, 345-70 (2017).
4. B. Creemers, G.v.d. Werf, Sch. Eff. Sch. Imp. **11**, 361-84 (2000).
5. G.v.d. Werf, B. Creemers, H. Guldemond, Sch. Eff. Sch. Imp. **12**, 447-66 (2001).
6. G.v.d. Werf, B. Creemers, R.d. Jong, E. Klaver, Comp. Ed. Rev. **44**, 329-56 (2000).
7. H. Goldstein, *Multilevel Statistical Models.* (John Wiley & Sons, 2010).
8. J.J. Hox, *Multilevel Analysis: Techniques and Applications.* (Routledge, 2010).
9. T.A.B. Snijders, R.J. Bosker, *Multilevel Analysis: an introduction to basic and advanced multilevel modeling*. (Sage Publications, 2012).
10. B.P.M. Creemers, J.A.A.P. Scheerens, Int. J. Ed. Res. **21**, 125-40 (1994).
11. B.P.M. Creemers, G.J. Reezigt, Sch. Eff. Sch. Imp. **8**, 396-429 (1997).
12. H.N. Kiwanuka, J.V. Damme, W.V.D. Noortgate, D.N. Anumendem, G. Vanlaar, C. Reynolds, S. Namusisi, Sch. Eff. Sch. Imp. **28**, 1-21 (2017).
13. A.A.M. Houtveen , W.J.M.V.D. Grift, B.P.M. Creemers, Sch. Eff. Sch. Imp. **15**, 337-76 (2004).
14. P. Tymms, C. Merrell, B. Henderson, Ed. Res. Ev. **3**, 101-18 (1997).
15. P. Sammons, R. Smees, S. Thomas, P. Robertson, J. McCall, P. Mortimore, *The impact of background factors on pupil attainment, progress and attitudes in Scottish schools*. (Paper presented at the British Educational Research Association Annual Conference. University of York, 11-14 September 1997).
16. R.E. Slavin. Sch. Eff. Sch. Imp. **24,** 383-91 (2013).
17. C. Özdemir, Eur. Ed. Res. J. **15**, 193-217 (2016).
18. K. Schenke, T. Rutherford, A.C. Lam, D.H. Bailey, AERA Open, **2**, 1-16 (2016).
19. S. Close, G. Shiel, Eur. Ed. Res. J. **8**, 23-33 (2009).

Students' acceptance of song lyrics containing national characters

Prana Dwija Iswara[*1], *Sandie* Gunara[2] , and *Julia*[3]

[1,3]Elementary School Teacher Education, Universitas Pendidikan Indonesia Kampus Sumedang, Jl. Mayor Abdurahman 211 Sumedang
[2]Music Education, Faculty of Arts and Design Education, Universitas Pendidikan Indonesia, Jl. Dr. Setiabudi 229 Bandung

Abstract. This paper aims at analyzing the acceptance of elementary school students of song lyrics containing national characters. The study was employed using research and development method, and this paper is the result of the interview related to the learning of nine song lyrics containing national characters. The results showed that all students had no difficulty in learning the vocabulary in the lyrics. However, the students were generally passive when asked about the meaning of the lyrics, hence the ability of students in singing the songs does not represent their understanding of the content of the lyrics. The implication of this research is that students always love songs and can sing songs, but it does not guarantee that students really understand the contents of the song.
Keywords: national character, song lyrics, elementary school.

1 Introduction

This research was conducted to create songs that contain national character for elementary school children. As the phenomenon that occurs in Indonesia today, primary school-age children tend to consume songs for adults, which lyrics are mainly about adult love stories. We cannot directly blame the children, because the parents and Indonesian people in general do not filter song contents that are suitable for children. Meanwhile, the electronic media does not always consider this issue in broadcasting their content because their goal is the commercialization of songs or the entertainment industry. Therefore, they pay less attention to who watch their contents. This song commercialization is inseparable from how the lyrics are made. Lyrics can be used to predict personalities [1]. Hence, one's personality can be used to identify a genre and song classification, and vice versa.

Song writing is basically following the current social situation [1]. From time to time, they have changed relatively according to the social situation and the economic condition of the society. Therefore, the political situation will also affect the song lyrics theme development and will be influenced by the growing social situation in society.

Nowadays, writers of children songs are increasingly disappearing [2]. Moreover, nationwide writers of children song are currently quite difficult to find. Therefore, children's songs that have new nuance and contextual to world of children, and that are

* Corresponding author: iswara@upi.edu

popular in the community are hard to find. In fact, to some extent, songs that are part of musical activities have a role in building children's character.

On a study from Lee [3], it was found that, when musical activities are combined with character development, children are better equipped to proactively address problems in their daily lives. Likewise, de Vries [4] suggests the use of music education as a medium to teach value to children. Therefore, we can be sure that the inception of good behavior to children can be done using songs as media.

In Indonesia, in the last few years, character education is emphasized to build 18 characters, consisting of religious, honest, tolerant, disciplined, hard working, creative, independent, democratic, curios, nationalistic, loving his/her country, appreciative for achievements, friendly/communicative, peace loving, avid reader, caring for the environment, have social awareness, and responsible [5]. Thus, to inculcate these characters to the children through songs, it is necessary to write songs that contain or relate to the characters, and to teach them to elementary school children. This is the main purpose of this study, which is to create songs containing national characters for elementary school children. The songs were tested on elementary school children to see their ability to sing the melody of the songs and their understanding of the lyrics. This paper is limited to the limited trial stages, and what is described here is focused on the aspect of the song lyrics, especially on students' acceptance of the song content with a total of nine song lyrics.

2 Methods

This paper is the survey part from the research and development method [6]. The participants of the study subjects were nineteen chosen students of Wargaluyu Elementary School. The researchers collected data by survey complemented by observation and question and answer sessions with the students. The analysis of the data that has been obtained through survey, observation and question and answer session was employed by analyzing the attitude and understanding of the song lyrics. The data that has been analyzed were then described, categorized with relevant theories, and concluded.

3 Results

The results of this study through observations and discussions of the nine song lyrics taught to 19 students of Wargaluyu Elementary School are described as follows.

Table 1. The Understanding, attitude and acceptability of the song lyrics

No	Theme	Lyrics	Understanding	Students' Attitude
1.	Religion	*Mengenal semesta* *Alam dengan seluruh ciptaNya* *Merenung pencipta* *Diri kecil di tengah semesta* *Ku ingin mengucapkan tasbihku padaNya* *Bertakbir di bawah ke agungan namaNya* *Syukur hamba pada diriNya* *Sujud dengan tangis padaNya* *Dosa terasa besar* *Berat hamba memikul bebanNya* *Tanpa ampunanNya* (Knowing the universe Nature with all His creations Contemplating the Creator Small self in the middle of the universe I want to say my *tasbih* to Him	No student question the vocabulary in the lyrics	All students are passive in the Q&A, but all students accept the lyrics well

		Say takbir to His name Thankful to Him Head down with tears for Him The sins feel heavy This servants can't carry the burden Without His forgiveness)		
2.	Honesty	*Sungguh beruntung berkata jujur* *Menjadi benar dan terpercaya* *Tempat menitip segala urusan* *Dan mencari amanat yang disimpan* *Betapa indah menjaga lidah* *Menghindar dari bohong dan fitnah* *Memelihara dalam hati yang bersih* *Dan bertekad menjadi teladan* (How lucky to be honest Be true and trustworthy A place of all sorts of affairs And to search for messages How wonderful to keep the tongue From lying and slander Keep a clean heart And determined to be an example)	No student question the vocabulary in the lyrics	All students are passive in the Q&A, but all students accept the lyrics well
3.	Tolerance	*Bangun toleransi dalam beragama* *Ayo kita lakukan* *Jangan perbedaan menjadi alasan* *Untuk pertengkaran* *Paham dan aliran rahmat perbedaan* *Hindari pemaksaan* *Lepaskanlah fanatisme* *Damai hidup bangsaku* (Build tolerance in religion Let's do it Do not make differences into a reason For a fight Ideologies and factions are the merits of differences Avoid coercion Let go of fanaticism Peace live my nation)	No student question the vocabulary in the lyrics	All students are passive in the Q&A, but all students accept the lyrics well
4.	Discipline	*Mari disiplin ayo kerjakan* *Bersama teman hargai waktu* *Patuh aturan dan ketentuan* *Tertib jangan ada yang terlewat* *Di sisiku ada waktu 'tuk dijaga* *Tutup waktu kosong dengan pekerjaan* *Tetap gembira saat beban di pundak* *Berdisiplin selalu* (Let discipline let's do it Together, friends, appreciate time Obey rules and regulations	No student question the vocabulary in the lyrics	All students are passive in the Q&A, but all students accept the lyrics well

		Orderly, do not miss a thing Beside me, there is time to keep Fill idle times time with work Stay happy even when there is burden on the shoulders Always be disciplined)		
5.	Hard work	*Kerja keras melakukan kewajibanku* *Itulah tugasku* *Belajar dan menaati orang tuaku* *Harus kulakukan* *Kerja keras merapikan pekerjaanku* *Demi masa depan* *Terus bekerja terus belajar* *Jangan sampai lalai* (Work hard to do my duty That's my job Learning and obeying my parents I have to do it Hard work to tidy up my work For the future Keep working continuously Do not get negligent)	No student question the vocabulary in the lyrics	All students are passive in the Q&A, but all students accept the lyrics well
6.	Creativity	*Kerja kreatif* *Kerja daya akal pencipta* *Semesta dalam alam rumah dan sekolah* *Bangun bangsa raharja* *Renungkan bahan jadi karya* (Creative work Work of mind power Universe at home and school Build the prosper nation Reflect on materials into a product)	No student question the vocabulary in the lyrics	All students are passive in the Q&A, but all students accept the lyrics well
7.	Independence	*Dalam buaian kasih sayang ayah ibu* *Kita semakin besar beranjak remaja* *Belajar tanggung jawab mandiri* *Tidak lagi pantas menyusahkan* *Mencoba mengerjakan tugasnya* *Berharap sukses saat nanti dewasa* (In the cradle of mother's and father's affection We are becoming teenagers Learn self-responsibility No longer troubling Trying to do his job Expecting success in adulthood)	No student question the vocabulary in the lyrics	All students are passive in the Q&A, but all students accept the lyrics well
8.	Democracy	*Demokrasi di kelasmu* *Ayo pilih KM mu yang terbaik* *Kita jaga perbedaan* *Kalau beda itu biasa* (Democracy in your class Let's choose your best chairman We maintain differences Difference is normal)	No student question the vocabulary in the lyrics	All students are passive in the Q&A, but all students accept the lyrics well
9.	Curiosity	*Luas alam getarkan akal* *Panca indra ingin merasakan*	No student	All students are

		Rasa hati ingin ku belajar *Dengan guru di dalam bimbingnya* (The vast universe shakes the mind The senses want to feel The heart wants me to learn With a teacher guiding)	question the vocabulary in the lyrics	passive in the Q&A, but all students accept the lyrics well

4 Discussions

The results showed that all students were able to sing both the melody and the words in the lyrics quite well. However, during the discussion of the lyrics, students tended to be passive, so it is basically their real understanding of the lyrics were questionable. This behavior is one of the common phenomena in humans in general, that is although they did not understand the lyrics, they were happy and enjoyed the song [7]. In other words, a person can sing, even sing a song in a foreign language without really understanding its meaning. However, even if one does not understand the lyrics of the song, one may be happy to sing the song. Meanwhile, the songs could have contained lust, sin, evil, mistakes, or meanings that lack good value. People just love the song without understanding the meaning of the lyrics [8].

The passivity of the students in the discussions does not represent the students' lack of understanding. Various questions regarding the song lyrics vocabulary were responded with signs that represent the absence of difficulty. However, the form of their understanding of the contents of the lyrics could not be expressed in words. This can also be caused by the students' lack of ability to express their thoughts (opinions) or feelings. Another possibility includes the unwillingness of the student to share their difficulties in understanding the song lyrics to the teacher.

Students preferred to sing and memorize the lyrics of the song. It is indeed quite positive to develop their ability to memorize the whole song, both melodies and lyrics. Even other research find out that people find it easier to remember songs (melodies) and titles better than the lyrics. However, lyrics can help people to remember songs and titles. In contrast, titles and songs cannot help people remember lyrics.

The melodies of songs evoke emotions better compared to lyrics [9]. Therefore, lyrics are actually neutral or being paid less attention by the students in term of their meaning and message. The students favor songs in a way that they only accept the lyrics based on how they like the melody. The students did not pay attention to the lyrics because they pay more attention to the songs. If the students like the songs, they will like the lyrics. Songs can indeed impact people's behavior [10]. On the other hand, studies of one's understanding related to their experience were conducted [11] and [12].

5 Conclusions

Based on the limited trial and the findings, students tended to be passive in discussing the contents of the song, they preferred to sing and memorize the lyrics. Therefore, their ability to sing does not necessarily represent their understanding of the vocabulary of the lyrics or the meaning of the song. However, all students accepted the presence of new song lyrics well and they did not show any rejection or confusion when studying it.

In this study, we would like to extend our gratitude to a number of parties who specifically contributed without underrating others who also gave major or minor contributions. Our first gratitude is sent to the Directorate General of Higher Education who provided funds for this study. The second goes to the Institute for Research and Community Service (LPPM) of Universitas Pendidikan Indonesia that helped facilitating this study.

References

1. Pettijohn TF, Sacco Jr DF, Journal of Language and Social Psychology. **28** :297-311 (2009).
2. Hendrayana D, *Anak-anak Kita Butuh Lagu* (Pikiran Rakyat, 26, 9 Juni 2016).
3. Lee A, International Journal of Music Education **34**:340-51 (2016).
4. de Vries P, Journal of the Australian Council of Orff Schulwerk **12**:14 (2007).
5. Hasan SH, *Pengembangan Pendidikan Budaya dan Karakter Bangsa. Materi disajikan sebagai bahan pelatihan Penguatan Metodologi Pembelajaran berdasarkan Nilai-Nilai budaya untuk Membentuk Daya Saing dan Karakter Bangsa* (Badan Penelitian dan Pengembangan Kemendiknas, Jakarta, 2010).
6. Sugiyono. *Metode Penelitian Pendidikan (Pendekatan Kuantitatif, Kualitatif, dan R&D* (Alfabeta, Bandung, 2006).
7. Peynircioğlu ZF, Rabinovitz BE, Thompson JL, Psychology of Music **36**:47-61 (2008).
8. Madanikia Y, Bartholomew K. Themes of lust and love in popular music lyrics from 1971 to 2011. SAGE Open. **4**:2158244014547179 (2014).
9. Ali SO, Peynircioğlu ZF, Psychology of Music **34**:511-34 (2006).
10. P. Fischer, T. Greitemeyer, PSPB **32** (9): 1165–1176 (2006).
11. J. P. Maxwell, Violence **16** (7): 640–661 (2001).
12. N. Gueguen, C. Jacob, L. Lamy, Psychol. Music **38** (3): 303–307 (2010).

Interactive teaching materials based on scientific approach: triangles and quadrilaterals

Heni Pujiastuti *

Mathematics Education Department, Universitas Sultan Ageng Tirtayasa, Jl. Raya Jakarta KM 4 Serang, Indonesia

Abstract. One of the Indonesian government's efforts to improve the quality of education is by changing the curriculum. Currently the Curriculum 2013 is being implemented in schools. Implementation of Curriculum 2013 is require teaching materials in accordance with the characteristics of the students, utilizing computer technology facilities, and contains the components of the scientific approach. Therefore, it is necessary to develop teaching materials in accordance with the Curriculum 2013. In this research we developed the Interactive Teaching Materials based on Scientific Approach (ITMSA). The research method is research and development (R&D) which consist of ten steps. The product design validation performed by multimedia, mathematics, and mathematics education expert involving lecturers and mathematics teacher. Utility testing of product conducted on junior high school students in Serang City, Banten Province, Indonesia. Based on research known that the ITMSA obtained a total score 85,30% from mathematics expert, 87,80% from mathematics education expert, 83,60% from multimedia expert, and 89,40% from students. In addition, students mathematical concept understanding who learn by using ITMSA better than student who learn without ITMSA. From these results concluded that the ITMSA is considered feasible and can be used in mathematics teaching in schools.
Keywords: Interactive teaching materials, Scientific approach, Triangles, Quadrilaterals

1 Introduction

Curriculum is a set of plans related to the content, ways, objectives and teaching materials used for guidance in conducting teaching and learning activities in order to achieve the goals of national education. The changing of curriculum by the Government of Indonesia is one form of efforts to improve the quality of education in Indonesia. Since 2013, the government began to enact the Curriculum 2013 which replaces the previous curriculum of the Curriculum 2006. Through the Curriculum 2013 is expected to build a generation of people who are able to compete globally in the future. The enactment of the Curriculum 2013 certainly gives a new nuance in the world of education, especially the nuances of learning held in the classroom by the teacher. The curriculum 2013 more

* Corresponding author: henipujiastuti@untirta.ac.id

focuses on the use of scientific approaches and technology in classroom learning activities. Therefore, teachers are expected to create innovations that can bring new things, both in terms of methods, media, and learning resources are more adequate and meaningful.

However, the undeniable changes in the curriculum itself has many consequences, one of which is a lack of understanding of educators practitioners in implementing the new curriculum. In addition to training, To help the implementation of the Curriculum 2013 is needed the availability of learning resources such as the books or other teaching materials that can facilitate teachers and students to learn. To support the implementation of the Curriculum 2013, the government has published companion textbooks for teachers and students. The book contains the material presented based on a scientific approach that includes five steps: mengamati (observing), menanya (questioning), menalar (reasoning), mencoba (trying), and membuat jejaring (make a networking)[1].

Through scientific approach the expected learning becomes more meaningful for students and they become more active in the class. In the implementation, of course, required textbooks or other types of teaching materials that can guide teachers and students in learning, so that learning is more directed. However the presence of the companion text books that exist nowdays it is still not enough. In addition, government-published book is still in text book type, so it has not been able to facilitate students to learn through the use of technology. In the structure of Curriculum 2013, information and communication technology is no longer a subject taught, but as a learning tool contained in all subjects. This means that the skills of the use of information technology and communication facilities is absolutely necessary for students to make the learning process work well and learning objectives are achieved. Therefore, the aim of this study is to develop the Interactive Teaching Materials based on Scientific Approach (ITMSA). The developed of material in this research is the Triangle and Quadrilateral concept for seven grade of junior high school students.

2 Methods

This research used research and development method. In this method including 10 steps, namely: Preliminary Study, Planning Research, Development Design, Preliminary Field Test, Revised Results of Field Test, Main Field Test, Revised Result of Field Test, Feasibility Test, Revised Final Results of Feasibility, and Dissemination and Implementation Final Products[2].

- Preliminary Study: Perform needs analysis, literature studies, and small-scale research to determine the potential for product development and other information related to the planned product development.
- Planning Research: Formulate research objectives, estimate the budget, personnel and time, determine the qualifications of the researcher and the forms of participation in the research.
- Development Design: Determining the product design (hypothetical design), determine the research facilities and infrastructure needed during the research and development process, determine the stages of the implementation the trials design, and determine the description tasks of the parties involved in the research.
- Preliminary Field Test: Testing the product on a limited basis.
- Revised Results of Field Test: Revision of product design based on limited test.
- Main Field Test:
- Revised Result of Field Test: Testing the product on a wide scale.
- Feasibility Test: Conducting product experiments on large scale
- Revised Final Results of Feasibility: Refining the final product

- Dissemination and Implementation Final Products: Publishing the results of the research through scientific forums. Product will distribution after the quality control stage.

The instruments used in the data collection include of validation sheet, student questionnaire, interview guidelines, and observation sheet.

3 Results and Discussion

The aim of the study is to develop the ITMSA. The development of the interactive teaching materials is conducted by researchers and assisted by four others as a team. The research team in this study is tasked to make the design of interactive teaching materials using a number of hardware and software required. The hardware used are Personal Computer with Intel Pentium 4 or AMD Athlon 64 processor specification, 2 Gb of RAM memory with 64 Bit VGA Video Card, and Android Smartphone version 2.2 Froyo. Some of the software used in developing this application is Adobe Flash Professional CS6, Adobe Illustrator Professional CS6, Notepad ++, and video needs software using ProShow Producer.

The material are developed in this interactive teaching material are the Triangle and Quadrilateral concept for seventh grade junior high school students. Until now, the ITMSA that are being developed are still in the design development stage. Figure 1 and Figure 2 are examples of the ITMSA view.

Fig. 1. The cover page view of ITMSA

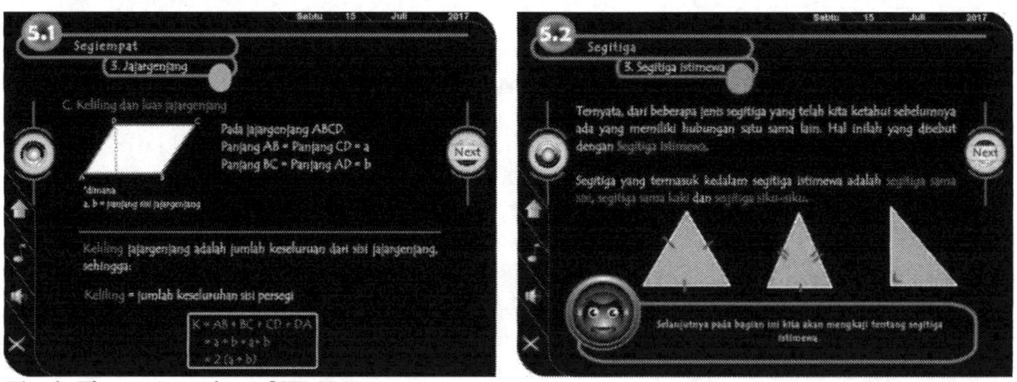

Fig. 2. The content view of ITMSA

Interactive teaching materials that have been developed further validated by experts and tested in limited scale to a number of seventh grade junior high school students in Serang City. Based on the validation results, it is known that the ITMSA obtained a percentage of total score 85,30% from mathematics experts, 87,80% from educational experts, and 83,60% from multimedia experts. The results of limited trials to 30 junior high school students are obtained a percentage of total score 89,40%. The results of validity test and limited test can be seen more clearly in Figure 3 below.

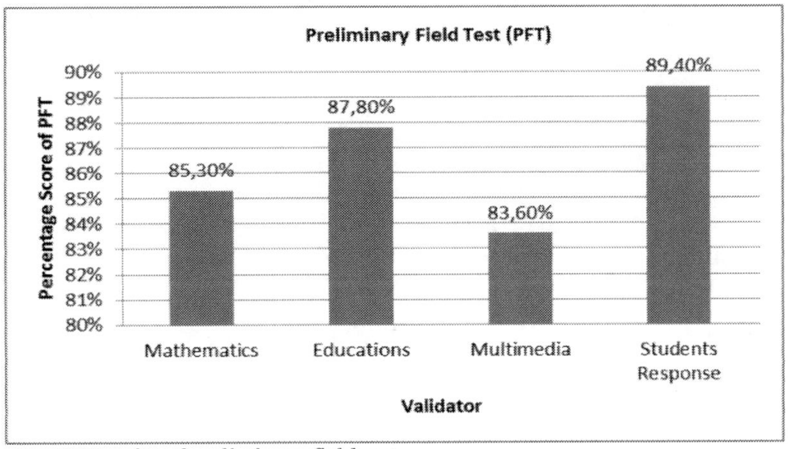

Fig. 3. Results of preliminary field test

After going through a process of expert validation and limited testing, subsequent trials were conducted on a wide scale to students in three different schools. Two classes were take from each schools, one class learning with the ITMSA and another class is learning without ITMSA. From the overall results of large-scale experiments, it is known that the scores of students mathematical understanding concept abilty who learn with ITMSA 85,38%, while students who do not use ITMSA only 77,54%. This is can be interpreted that the students mathematical understanding concept abilty who learn with ITMSA better than those who do not use the ITMSA. This result is in accordance with previous research that students mathematical understanding concept abilty in the Number concept who learn with ITMSA better than those who do not use ITMSA[3]. The recapitulation of large-scale test can be seen in the Figure 4 below.

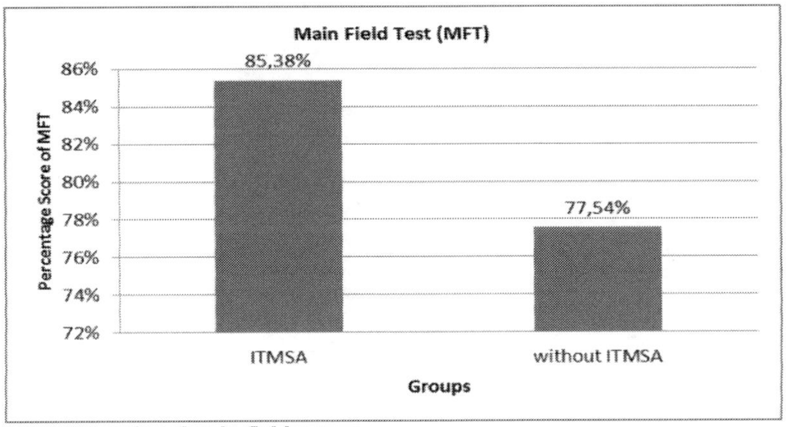

Fig. 4. Results of main field test

From all of results test known that ITMSA obtained a percentage total score over than 70%. It means that the ITMSA is considered feasible and can be used in teaching mathematics in school[4].

4 Conclusions

In this study it is known that the percentage of validation and test results is more than 70%. In addition, the experimental results show that students mathematical understanding concept ability who learn with ITMSA is better than students who learn without ITMSA. These results indicate that ITMSA deserves to be a resource for students and teachers in the classroom. From all of results test it can be concluded that ITMSA is considered feasible and can be used in teaching mathematics in school.

The suggestions that can be given to further researchers are: 1) Developing interactive teaching materials for other mathematics concepts; 2) Testing interactive teaching materials in larger groups so as to know the effectiveness of teaching materials in improving students mathematics skills; 3) to be more focus on the high order mathematical thinking skill so that students mathematics skills are more improved.

We would like to thank to the Ministry of Research, Technology and Higher Education for the research grant (Penelitian Produk Terapan). Also similarly to the Institute of Research and Community Service, University of Sultan Ageng Tirtayasa who helped the process of research implementation.

References

1. Kemdikbud, *Matematika SMP/MTS Kelas VII* (Kementrian Pendidikan dan Kebudayaan, Jakarta, 2013)
2. W. R. Borg, M. D. Gall, *Educational Research: An Introduction* (Longman Inc, London, 1987).
3. H. Pujiastuti, *Proceedings International Seminar on Mathematics, Science, and Computer Science Education* (Faculty of Mathematics and Science Education, UPI, Bandung, 2015).
4. Puslitjaknov, *Metode Penelitian Pengembangan* (Pusat Penelitian Kebijakan dan Inovasi Pendidikan Badan Penelitian dan Pengembangan Departemen Pendidikan Nasional, Jakarta, 2008).

The Dynamics of Social Capital in Global Era of Rural Communities (Case Study Poor People in Nagari Rambatan, Tanah Datar District)

Marleni (marleni@stkip-pgri-sumbar.go.id)[1]
Adiyalmon[2], *Elvawati* [3]

[1,2,3]Sociology Education Department of STKIP PGRI Sumatera Barat, Indonesia

Abstract. This article is based on the change of social capital in society triggered by globalization. According to Soetomo (2012) social resources (social capital) is a medium that enables development to take place successfully because of cultural elements have traditional legitimacy that symbolically the most valuable communication model, and have a number of functions then make it a means for change. Based on that, the article will explore about the dynamics of social capital with a case study in Rambatan Sub-District, Tanah Datar District, where is Nagari Rambatan as sample. Data collection using mixed methods, there are quantitative (questionnaire) and qualitative (interview). The results of research indicate the shifting of social capital in the society in Nagari Rambatan seen from the level of participation in social organization, the leaders, nagari institutions, relatives and neighbors as well as social relationships with relatives or neighbors, and also the level of dependence in economy.

Keywords: *Social Capital, Rural Communities*

1. Preliminary

Development influenced by globalization, which is not only in the urban areas but also in rural areas that affect the change of society, especially rural communities. The villagers are a strong society with values and norms. The entry of globalization of the countryside resulted in changes in values, social structure, institutions, functions, norms and all aspects related to rural society. The flow of globalization also leads to more modern villages that puncture life and society think rationally.

Globalization with its development slogan has eliminated the purpose of community development itself. The first natural resources that play an important role in community life that should be utilized for the benefit of society and community welfare but only utilized by a handful of people. Every natural resource development should take into account its conservative aspects but it is not realized. Natural resources are only depleted for development but conservatives are not implemented by those who use them. Natural resources are only made as money and capital for economic growth Both human resources, which are not seen from economic but also non-economic role in development. Rill of human resources at the present moment in the development always thinking about the economic and production aspects so as to ignore the socio-cultural aspects so that all of natural resources are depleted to produce commodities, in addition to human resources awareness of quality is still low, especially in rural areas that have an impact on poverty. These three social resources, or so-called social capital. This resource which has been forgotten by the society when according to Soetomo in his book of community development, social resources is a media that enables development can be place successfully because the cultural element has traditional legitimation, cultural element symbolically is the most valuable communication model, the number of functions that make it a means for change. [1]

Tanah Datar District, specially in Rambatan Sub-District is also affected by globalization. This was seen from the changes that occur in the community where is the results of observation and initial interviews in the past, the community in carrying out work always work together but at the present where it supported by technology as a result of modernization then rural communities began to work independently. The impact is social relations in society will be more tenuous because inter-community interaction is not intensive anymore.

In addition, the impact of modernization for the poor on the positive side , they can compete globally, but in the other side in the global era of free trade is enforced so that the poor become poorer because they are not able to compete. For example the agricultural field of cheap rice prices due to the proliferation of imported rice. Home industry will die because people prefer to import goods because of good quality with low price and followed by consumptive society. Pursuant to petmasalahn above purpose of this research is to describe dynamics of social capital of global era in rural society with case study of poor society in Kecamatan Rambatan Kabupaten Tanah Datar.

2. Literature review

Social capital itself refers to the values and norms that are believed and carried out by most members of society in daily life that directly or indirectly affect the quality of life of individuals and the sustainability of community communities [2]. According to Bardhan in Putnam 1993, social capital is also understood as a series of norms, networks and organizations in which people have access to power and resources and where decision-making and policy are made [3].

The research on social capital is also done by Areekul et al, 2015 with the title of his research Development For Strengthening Capital Capital Social Capital For Being A Sustainable Lifelong Learning Society, where he sees the components of social capital comprising humans, institutions, knowledge of cultural wisdom, public funds and resources natural. The social capital strengthening component consists of five stages: credible participation in learning problems, credible participation in decision making, credible participation in implementation, credible participation in benefits, and credible participation in evaluation [4].

Social Capital also defines resources that can be cultivated as an investment to acquire new resources. Social capital places greater emphasis on the potential of groups and patterns of interpersonal relationships within a group and between groups with an attention space on social networks, norms, values and beliefs among peers born from group members and to being a group. [5] According to the World Bank (1999) defines social capital as referring to institutional dimensions, established relationships and norms that shape the quality and quantity of social relations in society. In Hasbullah (2006: 6), Cohen and Prusak (2001) provide the notion that social capital is the stock of an active relationship between communities. Each relationship pattern that occurs is bound by trust, mutual understanding, shared values that bind members of the group to make possible joint actions can be done efficiently and effectively. Social capital is also defined as a set of processes of interpersonal relationships supported by networks, norms, and social beliefs that enable efficient and effective coordination and cooperation for mutual benefit and virtue. According to Putnam social capital consists of five principal characteristics, namely:

1. Community Network voluntary number and density, country and private network.
2. Civic engagement: the participation and use of civil networks.
3. The identity of local citizenship: a sense of belonging, of solidarity and equality with other members of the Community.
4. Reciprocal and cooperative norms: a sense of duty to help others, along with confidence that the aid will be restored.
5. Trust in the community.

3. Method

This research uses qualitative and quantitative approaches, with data collection techniques (writings), observation or observation, in-depth interviews and questionnaires. Data analysis is done by quantitative approach (descriptive and inferential numerical analysis) and qualitative (descriptive and analytical texts or images systematically) or between these two approaches. The initial strategy is to collect certain stages or themes of participation (qualitative stage) and then use the stages as specific items and themes as scales to create quantitative survey instruments. In the third stage validate the instrument with a representative sample of the population.

4. Finding and Discussion

Social Capital Dynamics in Global Era of Rural Communities

Development is influenced by globalization, not only in the urban areas but also in rural areas that affect the change of society, especially rural communities. The is a strong society with values and norms. The entry of globalization in the rural resulted some changes in values, social structures, institutions, functions, norms and all aspects related to rural society. The flow of globalization also leads to more modern villages that puncture life and society think rationally.

People's thinking rationally affects social capital, where there is dynamics or changes in social capital in the poor in Rambatan Sub-District. Social change is changes in the social institutions that affect their social systems, including social values systems, attitudes and patterns of behaviour between groups of people [6]. Non-social determinants of social change such as climate, availability of resources and biological forces. While external factors refer to the influence that comes from the presence of other communities in intercultural environment. Internal factors refer to the linkage of values, social structures and classes as instituted in a given society [7]. The change of social capital is seen from the level of participation in social organization, and also to leaders, nagari institutions, families and neighbours and social relationships with family, neighbours, and economic dependency.

4.1. Change of Participation in Organization

Participation is the participation of the community in various activities that aim to overcome problems and produce something useful. While the organization is a container used by the community to achieve these goals. The globalization affecting the village community, especially Rambatan's nagging also affects the involvement and participation of the community within the organization. This can be seen from the low level of participation of the poor compared to five years ago in social organizations such as farmer groups, religious and cooperative.

On the other hand this is triggered by the development of nagari and the level of community needs is increasing, so the poor focus more on the job and the fulfillment of daily needs. Shifting values and norms due to modernization and globalization also make people more focused on personal affairs than the organization. So far, the organization used as a place of friendship by the community, but now the organization used as a place to gain profit so that people always think about profit or loss when following it. This factor makes the participation of the poor increasingly decreasing in following the organization's activities in this warehouse.

It is also disclosed by RM, the cause of the participation or inactivity of the poor in social organizations that exist in this kenagarian caused by the way of thinking society has advanced, where they think rationally and take profit or loss. The advantages they think are more to material gain, this is because to meet the needs of their lives are increasingly increasing while the conditions of land and their income is not fixed

4.2. Trust Against Leaders, Nagari Institutions, Family and Neighbours/Friends

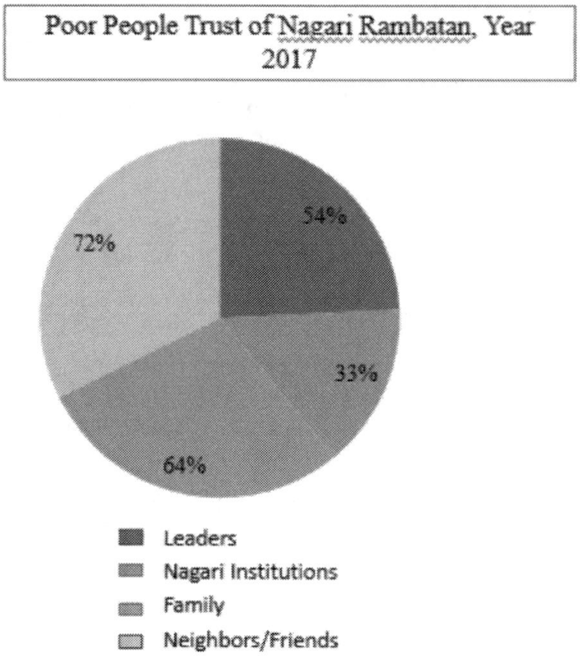

Graph 1. The Level of Trust for the Poor Country of the Year 2017

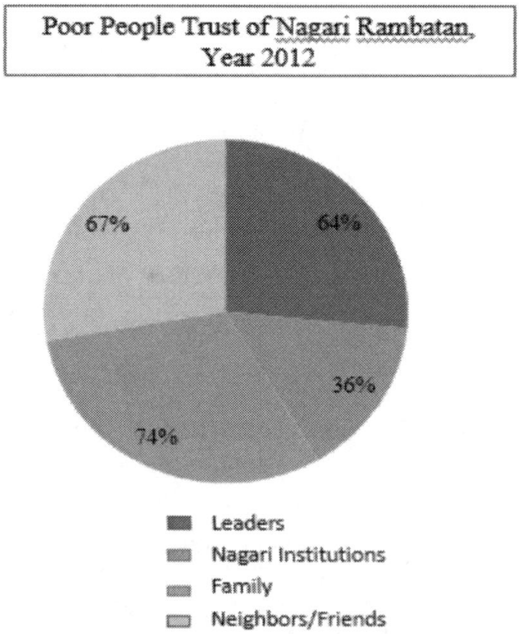

Graph 2. Level of Trust of the Poor People of the Year 2012

Based on the data above, it can be compared to changes or shifts in the level of trust of the poor to the leaders, nagari institutions, relatives or friends or neighbors. The level of confidence in the leadership experienced a 10% shift. This is caused by the meaning of the poor community leaders who only think about their own interests such as corruption without thinking about the lives of their people. Thought arises from the many corruption cases that occurred in Indonesia which involves various elements either exposed region or center. The same thing is expressed by JH, that his level of trust in the leadership is little or can be said no, because many current leaders are dishonest and understand the nature and needs of society. The position given by the community is only used as a place to find his own wealth, while the people are still many who live in poverty.

This means that ignorance of the leadership of the poor makes the level of public confidence to be reduced, the increased information and knowledge of the public will be more rational corruption cases of public thinking that affect the level of confidence in the leadership. On the other hand, globalization strengthens the level of trust of the poor towards relatives or friends / neighbors. Increased level of community trust in *nagari* Rambatan to friends or neighbors by 5%, this is because the neighbors always provide assistance and help in economic and non economic.

In addition to the economic side, the great confidence given by the poor to friends and neighbors is the help of morale given in the form of assistance when there is a thanks giving activity or when death. For example neighbors or friends come to help in taking care of all the needs voluntarily without expecting any rewards. Therefore relationships with friends and neighbors are getting stronger even though modernization and globalization enter the lives of rural communities.

4.3. Social Relations

According to Durkheim, social change occurs as a result of ecological and demographic factors, which change people's lives from the traditional conditions that mechanistic solidarity binds, to the conditions of modern society bound by organizational solidarity. Increased population and increased education and community knowledge result in changes in society, especially in social relations. Based on the results of research changes in social relationships that occur in *Nagari* Rambatan experienced a very significant shift due to the entry of globalization that has an impact on the decline of cooperation that has formed the network within the community. This can be seen in figure 3 below.

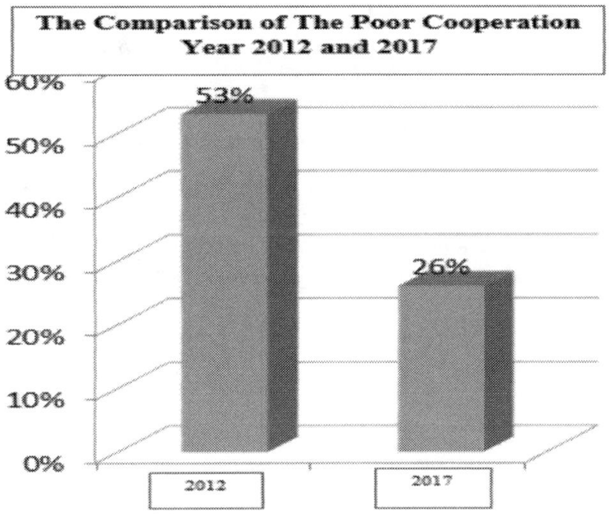

Graph 3. Comparison of Poor Cooperation in 2012 with 2017

Based on graph 3, it can be seen decreased cooperation of the poor by 27% for 5 years. This cooperation could be collective land management and other social activities. As for their reasons because the level of economic needs is increasing so that they put personal interests ahead of the public interest. In other cases, the rational mindset of society also affects the decline of cooperation among these communities. As expressed by AW; His absence in social activities such as street clean-up will not entail the performance and the completion of the road, the urgent economic interest leaves him with such activities. This means that rational thinking makes a decision more preceding economic interests because it involves family life.

Relationships with relatives also decreased by using aid and loan indicators from siblings, where the grants and liabilities imposed by respondents on relatives for 5 years decreased by 18% (see figure 4). As for their reasons that you already have their own responsibilities and are busy with activities to meet the needs of their core family. Supported also by life at the present time everything needs money so time is precious and hard to share with others. On the other hand, in terms of material economic life is almost the same while the level of demand is higher and the price purchased is also more expensive.

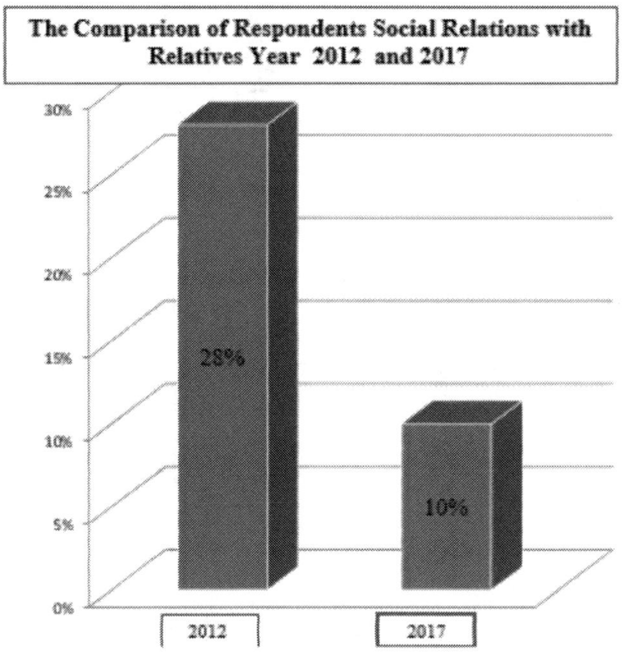

Graph 4. Comparison of Respondent's Social Relations with family of 2012 & 2017

Viewed from the relationships with neighbors especially in asking for help and the loans also decreased about 15% from 2012 to 2017. In picture 5 below;

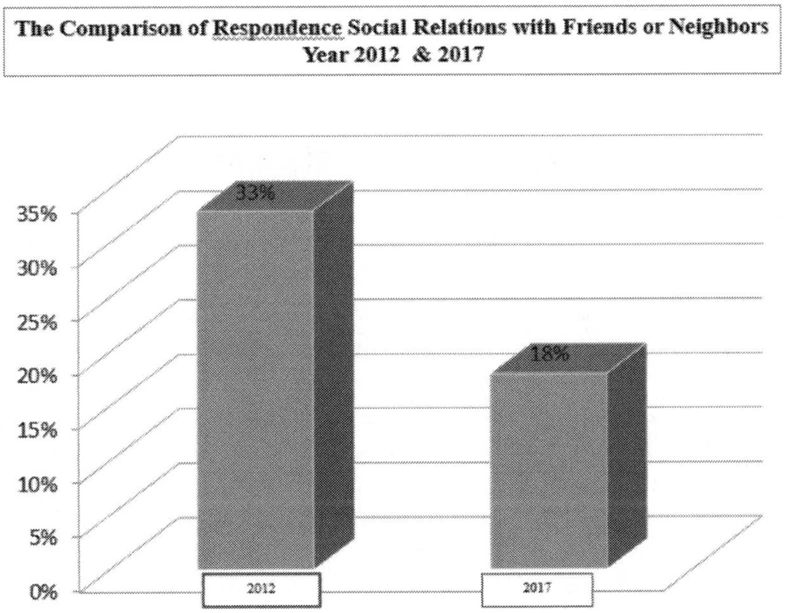

Graph 5. Comparison of Respondent's Social Relations With Friends or Neighbor Year 2012 & 2017

4.4. Economic Dependency

The fourth form of social capital dynamics is the dependency between individuals or groups, especially in the economic field. Dependence is the life condition of a person depends on others. Based on field data of globalization and modernization that enter the countryside under the positive impact where the dependence of society on the economic side has decreased in the last 5 years that is 11%. In Figure 6 below can be seen,

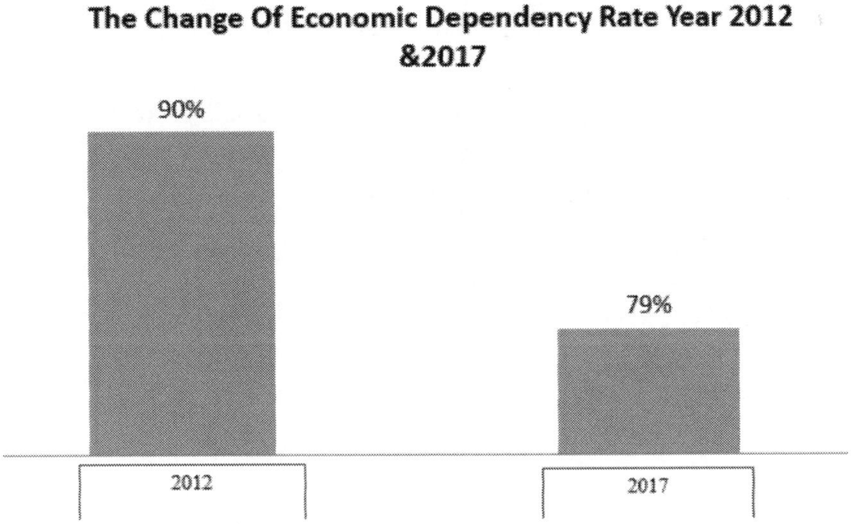

The decline in the economic dependence of the poor towards neighbours or neighbours, is a positive outcome of modernization and globalization. This position makes the

community more persistent and rational in improving economic life. Dependence has been caused by the poverty that impose them, the poverty makes the dependence on the poor economy high on the help of others, especially family, friends and neighbours.

5. Conclusion

Conflict in the Raya market Padang City post-earthquake in 2009 is one form of conflict asymmetrical. The warring parties basically structurally and power flaw. The government as parties who have power or power fight with traders who in structurally have no power to attempt conflict resolution of the role city Padang council is important in effort to resolution. The council has relation by the community as its constituents or the council is the representation of the community. And that city council has representatives function of council of the society sued to do their job as an institution control and escort the government. Here the supervisory function of DPRD to municipal the policy in line legislation and regulations. With the implementation of this parliament interests market traders of Raya market victims 2009 well accommodated.

Reference

1. Soetomo. Pemberdayaan Masyarakat. Yogyakarta: Pustaka Pelajar, (2011)
2. Dwipayana, dkk. Membangun Good Governance di Desa. Yogyakarta : IRE Pres, (2003)
3. Putnam, R. Making democracy work: civic traditions in modern Italy. Princeton, NJ, (1993)
4. Areekul, Chananporn dkk. Modal Development For Strengthening Social Capital For Being A Sustainable Lifelong Learning Society. Journal Procedia Social and Behavioral Sciences 191, (2015) 1613 – 1617.
5. Hasbullah, Jousairi. Social Kapital (Menuju Keunggulan Budaya manusia Indonesia). Jakarta: MR-United Press.
6. Soemardjan, Selo. Perubahan Sosial di Yogyakarta. Yogyakarta: UGM Press, (1981)
7. Haferkamp, Hans & Neil J. Smelser. Social Change and Modernity. Los Angeles Oxford: University Of California Press, (1992)

Teaching Practice in Distance Education Context

Rhini Fatmasari and Suripto

Faculty Education of Indonesian Open University (UT)

Abstract. Teaching practice is an activity designed to enable students in mastering teaching skills. This activity is very important to improve the ability of prospective teachers, because theoretical knowledge has not been able to equip the teaching practice. The practice of teaching at schools provides a real experience for students in the real class, because theoretical knowledge through the process of learning in the classroom can not equip the teaching practice.
Faculty of Teacher Training and Education of Indonesia Open University (UT) with distance learning system has designed Teaching Practice through Praktek Kemampuan Mengajar (PKM) course with 3 credit score. PKM is implemented in the schools where they teach and are guided by two supervisors. The PKM exam is conducted at the designated centre schools at each UT Regional Office .
This study is aimed to evaluate the activities of PKM students of Economic Education Studies Program FKIP UT in UT Regional Office. The study was conducted at 12 UT Regional Office samples involving respondents consisting of UT students and UT Regional Office staff.
The results showed the level of independence of students doing PKM was quite high (72%). This level of independence needs to be supported by a 43% improvement in preparation and administrative services at UT Regional Office. In the implementation of teaching practice in schools, in average 53.14% Supervisor 1 and 63.86% Supervisor 2 always provide direction in accordance with UT standards. While 62.5% of Headmaster Sentra has role in accordance with its function.

Key word: teaching practice, UT, UT Regional Office

1. Background

The teaching capability is main requirement teacher must have in their professionalism. Since teaching is not only routine and mechanical activities, but needs a series of skills that teacher must possess. In teaching activities consist of capabilities to analyse students' needs, decision making what should be done, planning the effective and efficient learning, activating students through both extrinsic and intrinsic motivation, evaluating the learning result, also revising the next learning to be more effective and to increase students learning achievement. Therefore teaching is a managerial activity that must be implemented professionally.
Theoretically, Professionalism capabilities can be obtained by prospective teacher through learning process in a lecture. But these capabilities were not enough to equip for doing teaching learning process in the class. So, a briefing of gradually designed teaching practice, is needed by students to master teaching skills.

Universitas Terbuka as long distance university owned 279.873 (70,50%) students of FKIP that it's graduations would have professions as a teacher. One of the subjects to be taken by students in finishing their study at FKIP UT is Praktek Kegiatan Mengajar (PKM) subject. PKM is a program to train students as a teacher in managing learning to be effective and efficient. PKM essentially trains and equips students with organizing capabilities (preparation, planning, implementing, evaluating and improving) learning at middle schools (Junior and Senior High school or others in the same level) based on learning principles that can be accounted for by the truth. PKM implementation is an activity of teaching practice done by students of FKIP UT at Centre determined by a local UT Regional Office . Through strengthening teaching capabilities (PKM), a teacher is expected to be able to increase educational qualification and professional capabilities of the teacher.

Based on the importance analysis of PKM for the students prospective teachers and research results in the field, UT needs to make an evaluation on the effectiveness of PKM implementation, especially for Student non basic education. This evaluation is addressed to UPBJJ as administrator of learning in general, Students Study Group as PKM Organizers, and school where practice activities take place. The evaluation is done to see (1) PKM's Plan and Administration at UPBJJ, and (2) recruitment system of PKM potential participants, system of RP development guidance and PKM practice and also teaching training system and teaching practical exams at Study Group (Kelompok Belajar).

2. LITERATURE REVIEW

2.1 Teaching Practice

Planned, directed and sustained educational reforms can be implemented and running well if teachers are able to increase and develop their qualities and qualifications. Those improved qualities and qualifications are expected to be able to increase the teacher professional capabilities in order to produce a better generation in the future days. Teaching is an activity that aims to create a possible condition that students can study. According to Oemar Hamalik (2001), teaching had several important definition: a) deliver knowledge to the students in schools; b) passed a culture down to young generation through school educational institution; c) the effort to organize environment to create learning condition for students; d) gave learning guidance to students; e) prepared students to be a good citizen accordance to society demands; f) a process to help students in facing daily life of society. While Sardiman (2004) explained that teaching was an activity that organized or managed the environment well and connected to a child, so learning process was happened. It could be said that teaching was an effort to create conducive situation to the ongoing learning activities for students. This condition was created in a way to support children development optimally both physically and spiritually, also both physically and mentally. Dadang Suhardan (2006) state that teaching was an academic activities consisting of communication interaction between educator and students.

2.1 Practice of Teaching Ability

Students of S1 FKIP UT are teachers who implement learning activities, were considered important to stabilize their teaching skills (PKM). Through PKM teachers were trained and equipped with habits to prepare, plan, implement, evaluate and improve learning at schools based on teacher's knowledge and education that can be truly responsibled.

Stabilization of teacher abilities (PKM) is a program to train student as a teacher in managing effective and efficient learning. At the end of stabilization of teaching abilities

(PKM), students are required to make PKM report. PKM reports is made as one of requirement to take exams of Stabilization of teaching abilities subject (PKM) PDGK 4209 Program S1 PGSD FKIP Universitas Terbuka.

2.2 Implementation of Teaching Practice at Universitas Terbuka

a. Exercises/practices of teaching, Students who implement exercises/practices of teaching, must create : (1) 10 sets of Learning Implementation Plan (RPP) and it's teaching practice in school where they work; (2) 8 RPP + their practices were assessed by supervisor 2 with using observation sheet.
b. Assessment of exercises include 1 RPP + its practice which was assessed by supervisor 2 and available teacher's colleague at school and 1 RPP – its practice was assessed by supervisor 1 and their college students in the tutorial class.
c. PKM report component includes Introduction, Findings of learning, conclusion, also its attachment which consist of 8 RPP, 8 observation sheets, 8 forms of reflection sheet.
d. The exams of teaching practice, (a) consist of 2 RPP and 2 teaching practice; (b) the exams of teaching practice is conducted at school determined by UPBJJ as PKM center; (c) The examiner recaps the scores by using Form PKM PGSM 2; (d) exams score contributes 40% towards final scores of PKM course/subject.
e. Pattern of PKM tutorial; (a) activities of PKM tutorial includes guidance and independence activities; (b) guidance by Supervisor 1 is implemented as much as 8 meeting times; (c) one class of PKM tutorial consist of 15 students.

3. METHODOLOGY

3.1 Design, Population and Research Sample

This research is a monitoring and evaluation that used design of evaluative description through a survey, observation and interviews. Population in this research were students who took PKM course/subject and UPBJJ that conducted PKM. UPBJJ sample was taken by using purposive sampling with considering the total students implementing the practice at the UPBJJ. The instrument developed in this research were questionnaires form and interviews guidance. Data of this questionnaires would be analysed descriptively by using the middle value trend and percentage. At the same time, data got from interviews would be analysed by content analysis.

4. ANALYSIS AND DISCUSSION

4.1. Respondent

Respondent in this research were students of S1 Economics Education, Head of UT Regional Office and Academic Coordinator in UT Regional Office. Total respondents and distribution of students per UPBJJ can be seen in Figure 1 as follows :

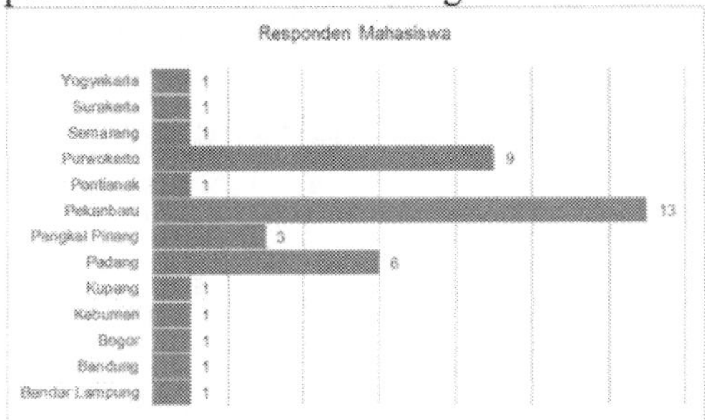

Figure 1 Total and distribution of students per UT Regional Office

4.2 Discussion of Research Result

Questions asked to students were related to (a) students independency in taking PKM course/subject, (b) tutorial by supervisor 1; and (c) Co tutorial by Supervisor 2. The indicators of PKM Practice and result of questionnaire processing on the whole can be seen in Table 1 and Figure 3 as follows:

Table 1. Indicators of PKM Practice

NO	Category
Students' independence in taking PKM /course/subject	
1	Students registered by him/herself to take PKM course/subject
2	Students could decide what the best time they must take PKM course/subjects.
3	Students made 10 sets of Plan of Learning Implementation 9RPP) and teaching practice at school where students taught.
Tutorial/gudance of Supervisor 1	
4	Students got tutorial by Supervisor 1 at PKM Centre determined by UT Regional Office
5	Tutorial by Supervisor 1 implemented in 8 times meeting, located in tutorial class determined by local UPBJJ.
6	Supervisor 1 appointed had the same background of knowledge with the students
7	Supervisor 1 appointed could guide students well
8	Supervisor 1 guide, discuss and give inputs of RPP and PKM report made by students
9	Supervisor 1 discussed and gave inputs to learning implementation done by students
10	Supervisor 1 guided students in fill their own reflection sheet.
11	Supervisor 1 guided and evaluated PKM reports which were written by students based on evaluation guidance which was avaliable in PKM Guidance Modul
12	Recommendation of exams to students provided by Supevisor 1
13	Recommendation of exams to students provided by Supevisor 1
Tutorial Supervisor 2	
14	Students chose Supervisor 2 and teacher who were in the same school
15	Students chose Supervisor 2 from teacher who were in the same school
16	Supervisor 2 guided, discussed and gave inputs to students in making RPP
17	Supervisor 2 evaluated RPP using observation sheet
18	Supervisor 2 evaluated RPP using observation sheet
19	Based on observation result, Supervisor 2 gave inputs to students about RPP and teaching practice that have been done
20	Based on observation result, Supervisor 2 gave inputs to students about RPP and teaching practice that had been done
21	Supervisor 2 filled journal of guidance appropriate with journal of PKM activities that had been done under guidance Supervisor 2, known by Supervisor 1
22	Tutorial by Supervisor 2 was done in 8 times
23	Supervisor 2 guided students in teaching practice in their own class by using observation sheet, then discussed the result
24	Supervisor 2 filled journal of guidance which was signed by Supervisor 2 and students and submitted to Supervisor 1 and UPBJJ
25	Supervisor 2 evaluated students in Teaching Practice 1 time by using APKG PKM 1 and 1
26	Supervisor 2 appointed having the same field of science with the students

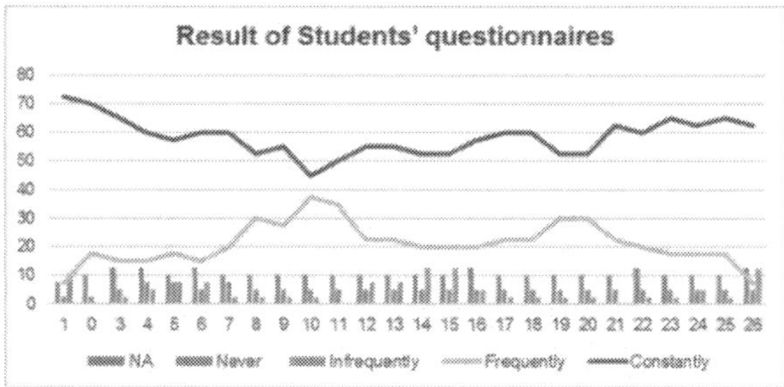

Figure 2. Result of Student Questioners

Figure 2 showed that UT students were very independent in taking PKM Course /Subject. Students could determine accurately when they must take PKM course/subject and made their own Learning Plan. Beside that, on the side of Supervisor 1, students got well enough guidance. Practice that must get attention from UPBJJ and UT was an existences of PKM Centre and the background of Supervisor 1 that was not similar with the guided students. This happened if UPBJJ did not have Supervisor 1 who had similar educational background appropriated with the determined requirements. This condition was also happened if at

UPBJJ whose the total students took PKM course/subject did not reach 1 PKM tutorial class (10 persons)

Questionnaire results also showed that UPBJJ must increase roles of Supervisor 1 in guiding students, especially in filling self reflection sheet and made PKP report. This was happened since students took PKM at every UT Regional Office that spread throughout the region. Supervisor 1 and Coordinator of UT Regional Office explained that it was very difficult to gather students regularly every weeks.

Problems in managing PKM in long-distance educational context especially for FKIP regular students were how to control the involvement of Supervisor 2 actively. Result of the research at Figure 4 showed that 47,5% Supervisor were not always giving inputs to students about RPP and teaching practice that had been done. While 42,5% Supervisor 2 were not always guiding and giving inputs to students in making Learning Improvement Plan. Managing PKM at UPBJJ showed that there were some activities that needed improvement and special attention. Result of questionnaire process about managing PKM at UPBJJ can be seen at Figure 3 and Tabel 2 as follows :

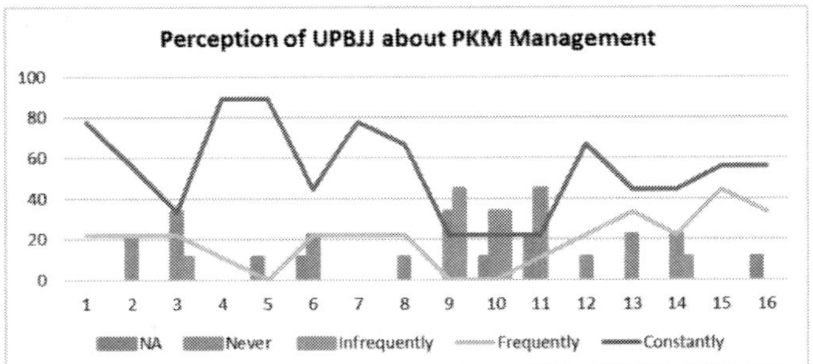

Figure 3. Perseption of UPBJJ

Table 2. Indicators of PKM Management in UPBJJ

No	Category
	Plan and PKM administration at UPBJJ
1	UT Regional Office prepared students data of prospective PKM participants based on Registation Data
2	UT Regional Office decided location of PKM Guidance
3	Name of Supervisor 1, Supervisor 2< Examiner and PKM Evaluator were prepared by UT Regional Office
4	Selection of Supervisor 1 refered to Procedure of Selection and Evaluation of academic staff Non- UT
5	UPBJJ adjusted the competency of Supervisor 1, Supervisor 2, Examiner and PKM Evaluator
6	UPBJJ implemented orientation of PKM tutorial implementation, by inviting prospective supervisor 1 and Study Group (Pokjar) Organizer
7	UPBJJ/UT prepared Practice Material, Report Forms and Recapitulation format of PKM practice assessment.
8	UPBJJ submitted technical guidance of PKM tutorial and file completeness of PKM guidance (included APKG, PKM 1, APKG PKM 2, Tutorial Journal, APL, PKM PGSM to supervisor 1 and supervisor 2 via students.
9	UPBJJ did monitoring of PKM guidance implementation PGSM peridically.
10	Monitoring was done in 3 times
11	UPBJJ prepared PKM centre, location of teacing practice examination
12	UPBJJ received PKM final score/assessment and submitted to UT exam center.
13	Location of PKM practice exam decided by UT Regional Office accordance to Prodi requirement.
14	UPBJJ evaluated Supervisor 1 through questionnaire
15	UT Regional Office ensured the guidance took place accordance to schedule and conditions.
16	UPBJJ identified if some problem deviation happened in implementing PKM

Figure 3 showed PKM organizeer/management at UT Regional Office that needed attention such as : (1) Preparation name of Supervisor 1, Supervisor 2, Examiner and PKM evaluator by UT Regional Office . Implementation in the field up to this day, Supervisor 1, Supervisor 2, examiner and PKM evaluator were prepared by study group. While

Supervisor 2 searched by students was their own senior teacher at their school. The problems happened were the educational background Supervisor 2 that was not similar with the students who took PKM subject; (2) UPBJJ conducted orientation of PKM tutorial implementation by inviting prospective supervisor 1 and study group (pokjar) management. This activity was rather difficult implemented by UPBJJ since the spread of students and prospective supervisor was uneven. At some UPBJJ that had students domiciled in the island region where this activity was difficult to be implemented.

Only 22,2% UPBJJ sample that did monitoring the PKM implementation, 33,3% did not do this activity regularly. Interview results with UPBJJ staff explained that monitoring activities did not implement accordingly, since UT policy determined that only 30% tutorial location which were monitored every semester. Therefore not all PKM practice activities in every location could be monitored by UPBJJ staffs. Supervision of PKM implementation by students was monitored by Supervisor 2, supervisor 1 and Study Group Management (4) Monitoring was done in 3 times. By the similar reason with monitoring activities, so UPBJJ could not do PKM monitoring in 3 times. 33,3% UPBJJ never did 3 times monitoring directly (5) UPBJJ prepared PKM Centre where the location of teaching practice exam was.

Preparation of PKM Center and location of teaching practice exam were indirectly handled by UPBJJ. This activity was handled by Group Study (pokjar) organizer. 44,4% UPBJJ did not handle the organizer which related with practice location and PKM exams. Organizer of study group usually establish cooperation with schools located around the tutorial location to serve as a partners in conducting the teaching practice exams. Meanwhile activity of teaching practice during 8 weeks done by students at school where they work. (6) Location of PKM practice determined by UT Regional Office was appropriated with determined Prodi. Determination of teaching practice exams was implementation by 44,4% UPBJJ, 22,2% UPBJJ could not determine that appropriate with Prodi requirement. This determintation was very dependent on condition in every region. In some regions even it was very difficult to be done, since there was no school that met the determined criteria, UPBJJ provided policy to implement PKM practice exam at school available in those region; and (7) UPBJJ evaluated Supervisor 1 through questionnaire. Evaluation activity of Supervisor 1 through questionnaire only implemented at 44,4% UPBJJ, while 22,2% other UPBJJ did not ever do evaluation through questionnaire. Interview Result with staff UPBJJ explained that they did evaluation of Supervisor 1 through study group organizer. This evaluation was felt more optimally since directly done in the field

Activities of other PKM organizer/management by UPBJJ were considered good enough and have met the requirement determined by ISO UT related to PKM management of non Basic Education. Although there were not 100% activities done by UT Regional Office .

5. CONCLUSION

Based on result of research about implementation of Teaching abilities Practice of students S1 UT economic Education could be concluded that : (1) UT needs to increase monitoring toward implementation of Teaching Practice in the field, especially related to management by UPBJJ. There were still 5 activities which its implementation level in the field were very low; (2) the involvement Study Program in overseeing the teaching practice activities are needed, therefore one mechanism need to be made in order Study Program could directly involve in management and PKM practice.

 https://doi.org/10.1051/shsconf/20184200099

References

1. A.M, Sardiman, *Interaksi Dan Motivasi Belajar Mengajar*, Jakarta: Rajawali Pers (2004)
2. Dadang Suhardan, *Pengawasan Profesional*, Bandung: Dewa Rhuci (2006).
3. Hamalik, Oemar, *Proses Belajar Mengajar*, Jakarta : Bumi Aksara (2001).
4. Irna Fitriana, Yudha Prasetyawan, ST, M.Eng *Analisis Dampak Rencana Pembangunan Busway terhadap Kemacetan Lalu Lintas pada Jalur Utara – Selatan Dengan Pendekatan Sistem Dinamik.* , digilib.its.ac.id/.../ITS-Undergraduate-11032-Paper.pd.
5. *Pemantapan Kemampuan Mengajar (PKM) PDGK 4209 program S1 PGSD FKIP* Universitas terbuka.
6. Wahyuni Kadarko, Pelaksanaan Praktek Mengajar Bagi Guru Peserta Pendidikan Jarak Jauh Yang Tinggal Di Wilayah Terpencil, Jurnal Pendidikan, Volume 11, Nomor **2**, September 2010, 78-87.
7. *Standar JKOP BB07 tentang Prosedur Penyelenggaraan Bimbingan Praktik PKM Dan PKP Program Pendas dan Nonpendas FKIP*
8. *Undang-Undang No. 20 Tahun 2003 Tentang Sistem Pendidikan Nasional*

SHS Web of Conferences **42**, 00100 (2018) https://doi.org/10.1051/shsconf/20184200100
GC-TALE 2017

The power of mindful learning in professional development course

Kadek Sonia Piscayanti

Ganesha University of Education

Abstract. Mindful Learning has been widely used in education nowadays, due to its significant and valuable impacts towards learning. Mindful learning is an effective tool to enhance students' awareness of learning, students' engagement with the context of learning and students' flexibility towards new ideas in learning. The three main characteristics of mindful learning above raise students' motivation and ownership of learning. It is shaping the students new perspective and leading them to powerful impact of learning. This is a descriptive qualitative research on the use of mindful learning in Professional Development Course in English Education Department Ganesha University of Education. The subjects are 73 students of sixth semester in academic year 2016/2017. Mindful Learning is used as a powerful strategy to enhance the students' achievement in learning. It is implemented through 16 meetings consisted of lectures, discussions, presentations and final projects. They are assessed by reflective journal, presentation notes, and final projects. At the end, the result showed that the power of mindful learning gives significant effect towards learning achievement. It showed that not only the mindful learning is effective for the learning but also productive in improving the students' creativity and critical thinking.

1 Introduction

Mindfulness is described as paying attention in the presentmoment, on purpose, and without judgment (Kabat-Zinn,1992 [1]). Mindful learning is a learning strategy that is used to raise the awareness of learning, sensitivity of context and novelty, create new definitions and new categories from informations. Mindful learning is very effective to be used as a learning strategy because it drives the students to new perspective, open minded learning, contextual and aware of the self. "When we are mindful, we implicitly or explicitly (1) view a situation from several perspectives, (2) see information presented in the situation as novel, (3) attend to the context in which we perceive the information, and eventually (4) create new categories through which this information may be understood." (Langer, 2000, p.111 [2]). Teasdale et al. (1995[3]) described the nature of this state as''being fully present at the moment, without judging orevaluating it, without reflecting back on memories of thepast, without looking ahead to anticipate the future … andwithout trying to solve problems or otherwise prevent theunpleasant aspects of the current situation''. Other research has beeen conducted by Tarracsh (2015[5]), in which the studysuggests that mindfulness practices can be successfullyintroduced into traditional academic settings and integratedas part of preservice course curricula to enhance professionals'personal mental health and their ability to transmitthe benefits of meditation to their pupils.

Cooper and Boyd (1996[6]) discuss their thoughts about using mindfulness in teaching andlearning. They consider teaching is a process of facilitating and uncovering skills. They state, "Mindfulness takes into account thestudent's inclination to search, to inquire. It considers the student's ability to be aware, toperceive and conceive" (p. 9). Mindfulness involves refining, thinking, formulating meaningand benefiting from learning. In an effective mindful class, the teacher helps his/her studentsbuild meaning, reflect on their learning, and make use of their learning. Other research shows that "Mindfulwriting helped students generate new thoughts and become aware of their thinking;mindfulness facilitated their learning process, cultivated creativity and intelligence; mindfulcooperative learning provided students with an opportunity to discover their awareness, learnfrom others, reflect and think critically (Wang and Liu, 2016[7])."

The context of mindful learning in this research is the professional development course. Professional development course is a course in English Education Department which aims at preparing students to be professionally developed teachers from the professional perspective, pedagogical perspective, personal and interpersonal perspectives. The novelty of this research is that mindful learning applied in professional development course in the context of learning english as a second language in english education especially in university level.

2 Method

Research method used in this study is the grounded theory research, in which the data is used to develop and elaborate the theory. The data is then analysed through the grounded theory research (Corbin and Strauss (1990[8]) in which the steps are notes writing, identification of concepts and development of concept definition or theory elaboration. Participants of this research are 72 students altogether. The treatment of mindful learning was done gradually in two steps. In total, students are given 16 meetings in which the first-half eight meetings are consisted of introduction, initiation and preparation of mindful learning, and the following eight meetings are fully facilitated with mindful learning. The activities in the classroom include lecturing, presentation, discussion, reflection and final projects. The method of data collection is using three kinds of instruments namely, reflective journal, presentation notes, and final projects scoring sheets. These are to measure whether there is a growth in student's awareness of learning, sensitivity to context and new ideas. In reflective journal, students are asked to write down freely on some points, first what they know already about a topic, what they want to know, what they have learned, and what they think about it. In presentation notes, they have to make a note on their presentation, to make a note about their friend's presentation and discussion. In final projects, they wrote their ideas in form of essay about a certain topic that has been given.

3 Results and discussion

The results of research show as follow. The first half of semester with the introduction of profesional development course and the initiation of mindful learning, the students are still in the conventional stage of learning in which they are passive, less creative and less iniative. In lectures section followed by discussion, the students are passive learners in which they listen more and speak less. Very often when they are invited to ask questions, they are silent. Even when they are motivated and supported. From the first two meetings in form of lectures and discussion about a history of national education in Indonesia, it was observed that the student's behaviour were passive. Next was presentation about professional teacher, inspiring and creative teachers. There are six presentations done in six meetings. The students seemed to be more active in speaking, yet they are still unmotivated to raise questions and answer. Here mindful learning is introduced step by step. They are told to ask questions before they present, to reflect after they present and to involve context of their presentation. They were also asked to take notes on what they have presented, and what they have discussed as the part of reflection. Gradually they are used to mindful learning. In the eighth meeting until sixteenth meeting, the mindful learning is fully applied. First they were asked to do mindful meditation to raise their awareness about their own feeling, surrounding context, and open new flexible minds. The students were given a book to analyze, discuss and present. The title of the book is Creativity: Theory, History, and Practice by Rob Pope (2005[9]). The concept of presentation was different from previous presentation. In this phase they are given not only feedback but also criticsm from their friends. If they could not present well, they should repeat the presentation until they get better feedbacks. Presentation was more active, challenging and dynamic.

From reflective journals, presentation notes and final projects scoring sheets, the data has been classified, based on grounded theory research prosedures, and it can be shown that 93,05% of the students have the positive feedback on mindful learning, especially on 3 aspects of mindful learning as the data gained are as follows.

Mindful Learning Raises Self-Awareness of Learning

1. "Be honest and critical thinking. In this learning process, I gained a very basic knowledge for me as a human. We need to be honest, whether I understand or not. I need to be honest not to make friends believe me, but I need to be honest to make me believe myself. Meanwhile the other thing that I gained during this lesson is I know how to start the presentation. It is because I have a mistake while do the presentation then I get a suggestion. I can reflect it to myself, so I can make a development, I feel enjoy to learn this subject because I gained so many new knowledges too such as study about genetics and creativity. As we know we need to be creative as a teacher. So, this subject is useful and helpful for me." (Putu Pramita Dewi)
2. "From what I've got in this course, I have learned about creativity and mindful learning, I tried my best in developing and adapting myself to follow the stream of this course and I have learned some new things, one aspect that I like the most in this course is mindful learning, in which it has some ways to learn peacefully, one of them is by having meditation before studying, it makes me feel better, and become more focused in leaning, mindful learning can also implemented with sharing session in which we can share what we have in mind and gain new things from others. Sharing session also help me in building the critical thinking". (Krisna Sugiarta)

Mindful Learning Raises Contextual Awareness of Learning

1. "Mindful learning is important for education nowadays because it gave contextual learning, since it is about contextual, it will ease students to understand the material. In addition, mindful learning also tells the students why they need to learn the material. So the students will know the benefit of studying the material, and also mindful learning is conducted in a creative way" (Ratih Mas Absari)
2. I like the mindful learning, because the concept is simple and also mindful learning can create comfortable atmosphere for the learners, mindfulness can be practiced through some ways including meditation, KWL chart, writing, found poem, and technology project. Mindful learning in the classroom makes learner become more sensitive about themselves and the environment around them. Mindful learning also can make the students able to absorb the knowledge thoroughly. (Olivia Monty)

Mindful Learning Raises Flexibility to New Ideas

1. In this course, I learn new material and when I present the material I make the mistake. But, I have chance to repair my mistake. I try to make my best when I repair my mistake. But, in this course, I also feel sorry if I should be able to show the best in following this lesson. But, when I cannot show my best. We use our creativity to correct the mistakes we have made. So, we can fix our mistakes. We can be better than ever. (Mia Sapitri)
2. Professional Development is one of many courses that can help us to develop our skill to be professional in doing something especially in teaching as a teacher. From the Professional development courses, we can get many knowledges about how to be the professionalism as a teacher. In this course, we as a teacher must be innovative, creative, and always updating ourselves with knowledge. (Eddy Prayuda)

The results of this study is in line with the study by Flook and Goldberg (2013) that suggest a mindfulnessintervention adapted for educators boosts aspects of teachers'mindfulness and self-compassion, reduces psychologicalsymptoms and burnout, increases effective teaching behavior,and reduces attentional biases. This is also in line with the theory of Langer that says: "When we are mindful, we implicitly or explicitly (1) view a situation from several perspectives, (2) see information presented in the situation as novel, (3) attend to the context in which we perceive the information, and eventually (4) create new categories through which this information may be understood." (Langer,1997, p.111).

According to Yeganeh and Kolb (2009), mindfulness isn't judging, rather accepting the recent condition as a continual experience that is contantly changing. This paradigm is emphasizing the prejudice and overcome the illusion that could take over the mind. Langer (2000) stated that Mindful Learning is the opposite of Mindless Learning.

The strength of Mindful Learning has multi perspective is supported by Jerry E. Fluellen Jrin his article "Mindfulness Learning 2020". Jerry compared Langer's theory to Dewey' s education theory, Piaget's, and Howard Gardner's. In the previous education theories, Langer's has perspective. In Fluellen's article, stated that "From the perspective of Langer"s mindfulness theory, students primarily "created new categories." Some students even "welcomed new information" and "held more than one perspective." This is different than any other education where learner made the category new "knowledge", accepting "new knowledge" and getting more than one "knowledge" in a new perspective. Therefore, it can be stated that Mindful Learning theory is appropiate for 21st century of characteristic learning which is craative, communicative, colaborative and critical.

Therefore it can be stated that the power of Mindful Learning is very significant towards the professional development course. It gives significant effect towards learning achievement. It showed that not only the mindful learning is effective for the learning but also productive in improving the students' creativity and critical thinking.

References

1. Z. Kabat, J. Massion, A. O. Kristeller, J.Peterson, L. G, Fletcher, K. E., Pbert, L., et al. . *American Journal of Psychiatry* **149,** 936–943 (1992)
2. E.Langer, J, Mindful Learning. *Article in Current Directions in Psychological Science*(2000)
3. J.D. Teasdale, Z.V. Segal, J.M.G. Williams, Behavior Research and Therapy, How Does cognitive therapy prevent depressive relapse and why should attentional control (mindfulness) training help? **33**, 25-39 (1995)
4. T. Richard, Mindfulness Meditation Training for Graduate Students in Educational Counseling and Special Education: A Qualitative Analysis (2014)
5. Y. Wang, C. Liu, The IAFOR Journal of Education **4** (2016).

6. C. Cooper, J. Boyd, *Mindful learning* (*Global Learning Communities, Australi,*1996).
7. L. Flook, S. B. Goldberg, L Pinger, K. Bonus, R.J. Davidson, Mind, Brain, and Education **7**,182-197 (2013)
8. C. Juliet, S. Anslem, Qualitative Sociology **13,** 4-21 (1990)
9. P. Rob, *Creativity: Theory, History, Practice* (2005)

The Effectiveness of Learning Instruments of Speaking Containing Character Values toward Speaking Learning

N. M. R. Wisudariani[*], *I N. Sudiana, I. B. Putrayasa,* and *I W. Rasna*

Language Education Study Program, Universitas Pendidikan Ganesha Singaraja, Indonesia

Abstract. This development research aimed at describing the effectiveness of learning instruments of speaking that were contained in character values. The objects of the study were a syllabus, lesson plan, textbook, and assessment. The subjects of the study were the lecturers of Speaking and the 2nd-semester students of IIA, IIB, and IIC in the academic year of 2015/2016. The data was collected through observation, questionnaire, and test. The data was analyzed by administering descriptive qualitative technique and descriptive statistics. The results of the study showed that the learning instruments was effective to be used in the teaching and learning process, to improve students understanding on the materials, to improve students speaking skill, and to shape characters such as religious, polite, communicative, open, democratic, respectful, responsible, curious, discipline, confident, cooperative, appreciate achievement, logic, critic, creative and innovative, and love reading. The results of the study indicated that the learning instruments of speaking contained of character values were effective so that it was feasible to be used by both lecturers and students of Indonesian Language and Literature Department, Language and Art Faculty of Ganesha University of Education.

1. Introduction

The logical consequence of the implementation of a curriculum is right in a learning process. Teaching and learning process become the most important key to achieve learning objectives. To achieve this, learning instruments must be well-prepared in the learning interaction. Preparation of a good learning tool is a strategic step in the management of learning process to achieve the achievement of graduate learning optimally. Lecturer as an educator has a very big responsibility of preparing learning instruments that can optimize the ability of learners to have the readiness to face the demands and needs of the times. Learning instruments consist of at least syllabus, lesson plan, textbook, and assessment.

Learning instruments that are absolutely needed in the teaching and learning process. In teaching language skill courses, the availability of learning instruments is very helpful for lecturers and students. The lecturers do not need to present too many materials in the classroom so that they have more opportunities to guide the students. The students can use the learning instruments as a source of learning and it can lead the students to be independent learning. By using learning instruments, the students will be put up in

[*] Corresponding author: rai.wisudariani85@gmail.com

readiness to join a course and it will enable the students to accelerate their learning [1,2,3,4].

The learning instruments developed by the lecturers should be based on the prediction of future needs. The learning outcomes in multi-purpose speaking course indicate that the students do not have a conceptual understanding of the material. It has affected the students' skill in speaking. The students' behavior in speaking has not yet lead better. The symptom of this character decline is a reflection of the results of education in our country. The achievement educational outcomes seem to indicate that there is something missing which has not yet been actualized in our national education. Moral deterioration gives a strong indication that education in our country faces a dilemma. Some of these facts are proven that character education has not been implemented well at the higher education level.

It is undeniable that news in mass media all the times still highlights the graduates of higher education institutions which are considered "smart and having no character". Mohammad Nasir and Ganjar Pranomo in a separate place stated that higher education is not only responsible for the development of the ratio, but also the character of young generation. The task of higher education institutions do not only produce graduates to improve the quality of education in Indonesia, but they are also responsible for character building of their graduates. Character education in higher education is a stage of character building which is as important as the character building at the school level [5,6,7]. In response to this situation, the lecturer can integrate character values into their all subjects' matters through the development of a multi-purpose speaking learning instruments containing values of character education.

In order to produce good quality learning instruments, learning instruments must have one of the criteria of effectiveness. The effective learning instruments are simply defined that learning instruments are developed appropriately and achieved expected goals as to provide a meaningful experience for users [8,9]. Based on the existing problems, the focus of the study in this study concerns on the effectiveness of learning instruments Speaking is characterized by the value of the characters in the teaching and learning process, increasing the students' understanding of the learning materials, improving students' speaking skills, and the formation of student character.

The results of this study were expected to contribute to the enrichment of science and knowledge, especially for the fulfillment of literature in the field of interactional communication skills. For intuitions, the results of this study can be used as a model for developing the learning instruments which aligned with the graduate competencies, to foster education character in which the teaching and learning atmosphere can truly reflect self-actualization on the basis of the values of character of our nations, and to carry out the concepts of Tri Hita Karana, one of which is to maintain harmonious relations between human and human (*pawongan*) as contained in the vision of our institution. For lecturers, the learning developed instruments can be used in speaking course, so that the teaching and learning process become more effective. In addition, the results of this study can be used as a model for other lecturers to develop the learning instruments as mandated by law. For students, the product developed in this study was expected to facilitate the students in their learning and encourage the students' independence to understand the material, so that the students can master speaking competencies and they will have a good character when communicating orally.

2. Method

This study adapted model 4D and ADDIE. The model consists of five steps, namely define, design, develop, evaluate, and disseminate [10]. In the first step, the researchers conducted a preliminary analysis, ability and character analysis, concept analysis, tasks analysis, and learning objectives analysis, lesson planning analysis, material analysis, and character values analysis relevant to the material. In the second step, the results of the analysis were used to design and develop learning instruments. In the third step, the researchers developed the learning instruments which have been designed. Furthermore, the development included the validation of the learning instruments by a panel of experts and try-out of the learning instruments. Being examined the validity of the learning instruments, then they were tried out.

The try-out of the effectiveness of the learning instruments of Speaking that are contained in character values was conducted at Indonesian Language and Literature Department, Faculty of Language and Art, Ganesha University of Education. The participants were the students of the 2nd-semester students of IIA, IIB, and IIC in academic year 2015/2016. There were totally 54 students involved in this study. In this case, 14 students of class IIA participated in small group try-out, and 19 students of class IIB and 21 students of class IIA participated in field tryout. The instruments used to collect the data in this study were observation sheet, questionnaire, and an interview guide. This study applied two techniques of data analysis, namely the qualitative descriptive and descriptive statistical analysis techniques.

3. Results and Discussion

The results of this study showed that the learning instruments of Speaking have the value of effective characters used in the teaching-learning process, improving students' understanding of learning materials, improving students' speaking skills, and forming the values of the characters in the students themselves. The average understanding of small group students (2C) on learning materials amounted to 79 was good. The average understanding of students of large groups (2B) on learning materials amounted to 81 was good. The average understanding of students of large groups (2A) on learning materials of 76 was also quite good. It indicated that effective learning tools are used in improving students' understanding of learning materials

Based on the result of practice test of speaking to the students of a small group (2C) it was known that the average of students' speaking skill was 84, it was good. The results of midterms and final semester exam showed an excellent achievement improvement. The average score of students' speaking skills in the midterm exam was 87 and the average score of students' speaking skills in the final exam at the end of the semester was 91. The middle and final semester test results were very good. Based on the result of practice test of speaking to the students of a big group (2B) it was known that the average of students' speaking skill was 83, it was good. The results of midterms and final exam results showed an excellent improvement. The average score of students' speaking skills in midterm and final examination of the semester is 85. The middle and final exam results were very good. Based on the result of practice test of speaking to the students of a big group (2A) it was known that the mean value of the practice of students' speaking skill is 78, it is good. The middle and final exam results show a good improvement. The average score of students' speaking skills in the midterm exam and the final exam of the semester of 80 was quite good. The results of this test of speaking skills showed that effective learning tools are used in improving students' speaking skills.

In implementing the value of character education, this learning tool is also effective to use. The value of religious, polite, communicative, open, democratic, respectful, responsibility, curiosity, discipline, self-confidence, cooperation, rewarding achievement, creative and innovative critical logic character, as well as the value of a reading habit which is focused on learning Speaking has begun to develop to students. This suggests that the Speaking learning instruments has consisted with an effective character value used in implanting the student's character values.

Based on the results, there were several findings in this study. Firstly, the use of learning instruments that were structured in the lecture is the application of Behavioristic learning theory. The Behavioristic sees the importance of conditioning environment. The conditioning of the learning environment affects the learning process and outcomes itself. In Connectionism learning theory, the conditioning of learning by using instructional learning instruments as a stimulus is closely related to the three theorems found by Thordike, the law of readiness, the law of exercise and the law of effect. With the existence of learning tools, both students and lecturers become more ready in the teaching-learning process, because lecturers and students understand the direction of the lecture. The more ready the lecturers and students do the teaching-learning process, then the implementation of learning will lead to satisfaction or good results.

The learning of speaking is done through the practice of speaking. The more students are trained to practice speaking, the students' skills in speaking will get better as well. The principle of exercise law states that the ability of students will be more increased and strengthened if being trained repeatedly. In the process of teaching and learning, the lecturer also reflects on the students' performance. The results of this reflection provide a pleasant impact or vice versa for students. When the results of reflection received by students are fun, students tend to improve their speaking skills. If the results of reflection received by students are not very pleasant, the student's tendency will improve the appearance.

Secondly, the use of the learning instruments was able to place the lecturers as a facilitator in the lecture and the students as a subject who owned an independence in the lecture. At the time the lecturer of the course is unable to be present in the classroom, the students can learn independently by answering the questions in the competency test as well as in reflection section. In this case, the learning instruments is an important part of lecture [11]. Through a textbook, the lectures will be easier to teach and the students will be easier to learn. Teaching and learning through learning instruments can be defined as a significantly meaningful strategy for the success of a teaching and learning process. The meaningfulness of the learning instruments can be seen from the ability of the learning instruments especially textbook developed to bring about self-instruction for the students. It means that the developed textbook has the ability to explain clearly how the teaching materials should be learned by the students in the teaching and learning process, both with the help of a lecturer or students independently. These findings more and more strengthen the position of a textbook in the field of education.

Thirdly, the elaboration of the material in the learning instruments containing the values of character education was able to build and instill the character values in the students themselves. The development of this learning instruments containing character values is aimed to increase the students' understanding of the theoretical basis for speaking of various interaction and its applications, and the establishment of character values in the students themselves. The learning instruments do not simply provide an understanding, but it instills values in the students themselves. The three things concern the knowledge of the students of the values, the students' moral feeling, and the students' behavior pertaining to the values. The students' knowledge of the elaboration of the material containing the values of character education has a positive effect on the skill of the students in their learning. It means that the effect of using the learning instruments containing character is not only restricted to knowledge but it also has something to do with skills and attitudes.

The elaboration of the material in the learning instruments serves as a tool to build a competency, character, and a foundation for the appreciation and implementation of the character values. This matter has a connection with the feelings of the students (moral feeling) after using the learning instruments. This study also showed that the understanding and improvement of the students' skills after using the textbook end in the attitudes reflected by the students in the lecture. The characterized behavior come up, develop and strengthen in the students themselves because they know the concepts and natures of characterized behavior, feel and possess a positive attitude towards the concept of a good character, and they are accustomed to practicing it [12-23]. The results of observation and test indicated that the students who can understand the material will be skillful in speaking and they will automatically display mutual respect, appreciate their friend who performs in front of the class.

4. Conclusion

The results of try-out showed that the learning instruments developed can bring about a positive impact on the multi-purpose speaking course and the continuity of character education in this country. In relation to this, there are three important things to be stated here, namely recommendation for the utilization of the product, dissemination of the product, and further development of the product. For the utilization of the product, the learning instruments were developed to meet the needs of the students and lecturers in speaking course at the Department of Indonesian Langauge and Literature Education, Ganesha University of Education. Concerning its relevance to the lecture, these learning instruments need to be reviewed when it is going to be used by the lecturers and the students outside our department. However, this textbook can be used as one of the sources by other institutions in teaching speaking skill. In cases of the utilization and dissemination of the product, the results of this study are limited to own sides. If the product will be used for the interest of the wider community, it needs to be revised by considering the scope of the subject which is not restricted only to the need of the lecturers and the students of language education. For further development of the product, the product developed in this study is simply learning instruments, therefore other researchers are expected to develop suitable media to support the learning instruments.

References

1. Suhadi, *Petunjuk perangkat pembelajaran* (Surakarta, Unmuh, 2007)
2. Hobri, *Metode penelitian pengembangan:aplikasi pada penelitian pendidikan matematika* (Jember: Universitas Jember, 2009)
3. Chodijah, S., *J.* Plt. Pbl Fis **20**, 1-19 (2012)
4. B.G. Davis, *Tools for teaching: perangkat pembelajaran, teknik mempersiapkan dan melaksanakan perkuliahan yang efektif* (Jakarta: Raja Grafindo Persada, 2013)
5. L. Nucci, University of illinois at chicago follows nationwide movement toward character education (ascribe column, Dec 14, 2004).
6. N. Chanifah, *Desain pendidikan karakter di perguruan tinggi umum* (*Prosiding Seminar Nasional Senari III,* 47-51, 2015)
7. Sandra, Graham, D. Marry E., J. Research in Character E **11**, 34 (2015)
8. B. Seel, Richey, R. C., *Teknologi Pembelajaran: Definisi dan Kawasannya* (Washington D.C: Association for Educational Communication and Technology,1994).
9. J.V.D. Akker, *Design Approaches and Tools in Education and Training* (Dortrech: KluwerAcademic Publishers,1999).

10. S.S. Thiagarajan, Semmel, M.L. *Instructional development for training teacher of exceptional children* (Minnesota: Indiana University,1974).
11. Lestari, *Pengembangan bahan ajar berbasis kompetensi: sesuai dengan kurikulum tingkat satuan pendidikan* (Jakarta: Akademia, 2013).
12. M. Elias, J R. Character E **2**, 246-266 (2007)
13. M.B. Adeyemi, T.V. Moumakwa, D. Adeyemi, J In. Stud Home Comm, Sci **3**, 118-128 (2009)
14. T. Lintner, *Soc. Stud. J* **102**, 200-213 (2011)
15. T. Lickona, *Educating For character: mendidik untuk membentuk karakter, bagaimana sekolah dapat mengajarkan sikap hormat dan tanggung jawab* (Jakarta: Bumi Aksara, 2012)
16. Tyra, J Ris Tide **5**, 1-10 (2012)
17. N. Silay, Eur J of E Stud **5**, 131-138 (2013)
18. M. Yaumi, Syahid, *Character education values that work in islamic senior high school* (Seminar Nasional, YPUP Makasar,2013)
19. B.M. Protz, *Character education in print: content analysis of character education in introduction to education textbooks* (Dissertation, University of North Carolina, 2013).
20. A.M. Gina, *R High E J* **26**, 2 (2014)
21. S.A. Khan, J Edu R Inter **3**, 97-106 (2014)
22. B. Tonia, S. Peter, J Character Edu **10**, 155-174 (2014)
23. D.S.S. Mihardi, J Edu Pract **6**,162-171 (2015)

 https://doi.org/10.1051/shsconf/20184200102

Social Entrepreneurship Andragogy-Based for Community Empowerment

Dayat Hidayat

Nonformal Education Department Faculty of Teachers Training and Education University of Singaperbangsa Karawang, hidayat.unsika@gmail.com

Abstract. The essence of social entrepreneurship is the ability to initiate, lead and implement problem-solving strategies, through cooperation with others in all types of social networks. The purpose of this study analyzes: 1) entrepreneurship training, 2) adult education assumptions, and 3) community empowerment. This study is discussed by using a comparative theory analysis method that related to entrepreneurship training, andragogy, and community empowerment from various literary sources. Analysis is conducted to find conclusions that can be justified scientifically. The results of the study conclude that: 1) social entrepreneurship training is an organized learning process from planning, implementation, and evaluation on non-formal education path aimed at enhancing community knowledge, attitude and skills for learning and endeavor, 2) entrepreneurship training based on andragogy conducted through a learning process that implements adults' assumptions of self-conceptual, learning experiences, learning readiness, and learning orientation. 3) Community empowerment is a process of empowerment or strength to community to have the ability of individuals who completely agreed with the community in building social and economic empowerment of the community.

Keywords: Entrepreneurship training, Adult education, Community empowerment.

1. Introduction

Nowadays, almost all institutions, both government and private, are aware of the condition of limited human resources in facing rapidly changing technological changes. Education and training is a learning process aimed at developing human resource capacity, so that it can cope with the changing needs of the community for new technologies.

Education is a process of developing human capabilities in the direction of an organization or institution. While training is part of the educational process that aims to improve the skills or special skills of a person or group of people. In general, training is part of education that describes a learning process. Training is a learning process that aims to improve the knowledge, attitude, skills, and performance of individuals in organizations, institutions or companies. Training programs are designed to improve performance at the individual, group and/or organizational level. Improved performance will show changes that occur within individuals, groups and/or organizations that are measurable in their knowledge, skills, skills, attitudes and social behavior. Various training programs are widely implemented to meet community needs, including social entrepreneurship training.

Entrepreneurship training is the learning process of learning the science of one's values, abilities and behaviors in facing life's challenges and for opportunities with risks that may be faced. In the business field, Zimmerer, T.W. (1996) argued that entrepreneurship is the result of a discipline and a systematic process of applying creativity and innovation in meeting the needs and opportunities in the market.

Training of social entrepreneurship andragogy-based is a learning process that applies adult learning principles. Andragogy or adult education is formulated as a science and art to help adults learning. Andragogy focuses on improving adult life, providing the skills and ability to solve the problems experienced in the life of society.

One of the goals of training of social entrepreneurship andragogy-based is to improve community empowerment. Community empowerment is an effort to increase the ability of the community (poor, marginal, marginalized) to express their opinions and/or needs and actively participate in managing the community institutions for the improvement of their lives.

The problem that is often faced in developing entrepreneurship is the lack of entrepreneurship programs that apply the principles of andragogy learning. Entrepreneurship development is less in line with self-concept, learning experience, learning readiness, and adult learning orientation. In its essence, adults have a self-concept that has been independent. But in the development of entrepreneurship programs are less appropriate in learning atmosphere, diagnosis of learning needs, and adult business planning process.

Adults have a learning experience in accordance with the passage of time to grow and develop toward maturity. But in entrepreneurship practice is not based on adult experience in developing a business. Adults also have learning readiness, so the development of entrepreneurial materials needs to be adjusted between the learning needs and the adult endeavor with its social role.

Adults also have a learning orientation, so have a tendency in developing entrepreneurship centered on solving problems are faced (problem centered orientation). Therefore, the development of entrepreneurship for adults should be practical and can be applied immediately in daily life.

Based on above background, the social entrepreneurship program needs to be developed in accordance with the principles of learning based on andragogy. Through social entrepreneurship based on andragogy; Knowledge, skills and entrepreneurial attitudes of society will increase in accordance with self-concept, learning experience, readiness of study and adult learning orientation to improve community empowerment and economic well-being of his life.

2. Method

This study is discussed by using a comparative theory analysis method that related to entrepreneurship training, andragogy, and community empowerment from various literary sources. Analysis is conducted to find conclusions that can be justified scientifically. This study used analytical research methods that investigate concepts through theoretical analysis. The researchers identified, studied and then cataloged the data to gain an understanding of the concept of social entrepreneurship, andragogy, and community empowerment that can be observed directly or indirectly. The theoretical concept is the main data. The researcher interprets the data and facts to get an explanation of the concepts of social entrepreneurship, andragogy, and the empowerment of collecting societies precisely and clearly to draw conclusions. The technical data analysis used is an interactive model through the stages of data collection, data reduction, display data, and conclusion/verifying.

3. Discussion

3.1. Social Entrepreneurship Training

In general, training is part of education that describes a process. Quite a lot of understanding and limitations of training that we can get from experts with different point of view. Training is a formal procedure facilitated by learning to improve behavioral changes related to improving the goals of the institution or organization. Mayo & DuBois (1987:5) argue that training and education have many similarities. Learning in the educational process is more comprehensive, broad, and builds capabilities of a general nature. While learning in training is more specific, task-oriented, and applicative.

Laird, D. (1985:11) describes training as an activity to improve human performance on working that is currently being worked or is being employed to do. Nedler, L. (1982) defines training as a learning process to improve the performance of people in completing work.

Training programs are expected to increase knowledge, attitudes, and skills of human resources as an important asset in the institution. Increasing knowledge, attitudes, and skills through training programs are expected to improve institutional performance of external change and competition. Robinson, K.P. (1981:12) states about training as a way of learning or experience to develop a person's behavior patterns in the areas of knowledge, skills or competencies to achieve the desired standards.

Entrepreneurship training is an organized learning process from planning, implementation, and evaluation on non-formal education channels aimed at improving community knowledge, attitude and skills for learning and working. Zimmerer, T.W. (1996:51) defines entrepreneurship as applying creativity and innovation to slove the problems and to the exploits of that people face everyday. Entrepreneurship is a combination of creativity, innovation, and courage to face the risks involved by working hard to form and maintain new business. Creativity is 'thinking new things' while innovation is 'doing new things'. So the success of entrepreneurship will be achieved if thinking and doing something new or something old done in a new way (Levitt quoted by Zimmerer, T.W. 1996:51).

Entrepreneurship is an ability in creative thinking and innovative behavior that is used as the basis, resources, driving force, goals, strategies, tips, and processes in the face of life challenges. Meredith, G.G. (1996:9) argues that entrepreneurship means integrating personal character, finances and resources.

Various public entrepreneurship programs have developed very rapidly, including social entrepreneurship. Ontologically, there is no consensus about the precise definition of the definition of social entrepreneurship. According to Borganza et al. (2010) defining social entrepreneurship is a challenge in itself. Nicholls (2006) defines social entrepreneurship in the context of a blend of nonprofit organizational management and commercial firms run by businesspeople, charities, and social drivers.

Epistimologically, Renko et al. (2009) analyzed the effect of entrepreneurial orientation on product innovation and capital investment. According to him entrepreneurship orientation is an important factor that positively influences product innovation and capital investment. While Wu, et al. (2008) analyzed the intellectual impact of capital, including the entrepreneurial orientation, on corporate innovation and found that the moderating effects of entrepreneurial orientation outweigh other intellectual variables of capital.

Axiologically, the study of social entrepreneurial literature contributes to community communities and local governments. Bornstein (2007) argues that the most serious problem of social entrepreneurship is the limited financial resources and the lack of qualified human resources. As a result, many developing country governments have doubts that this entrepreneurial application can work (Borganza, et al., 2010). In its development, the application of social entrepreneurship differs across countries, even among developing countries in Asia (Defourny and Kim, 2011).

Entrepreneurship training has evolved an an important study in the research and development entrepreneurial paradigms related to the study of entrepreneurial academic studies and practical development for new entrepreneurs, but the research is still rare and not much understood well (Deakins, 2000). Entrepreneurship itself is a leaner, who constanly explores the successfull desires of his life journey (Franco and Haase, 2009). The essence of social entrepreneurship involves nonprofit business activities, business activities for social purposes, and combination of both, but for social purposes.

3.2. Andragogy Concepts and Assumptions

Andragogy is often interpreted as an art and science that helps adults to learn (the art and science of helping adult learn). Etymologically, the term andragogy means the study of how adults learn. The definition is in line with the thinking of Knowles (1970) which states that andragogy as the art and science to help adult a learner.

Furthermore, Darkenwald and Merriam (1982) express the notion of adult education is a process where by person of social roles characteristic of adult status undertake systematic and sustained learning activities for the purpose of bringing about chnges in knowledge, attitudes, values, or skills.

Andragogical assumptions consist of self-concept, learning experience, learning readiness, and adult learning orientation. Knowles (1970) in developing the four principal assumptions andragogi concept as follows:

3.2.1. Self Concept

The assumption is that adults have their own self-concept. Because of this independence, adults need the rewards of others as self-determining humans, self-directed. This has implications for the implementation of training programs, particularly those related to climate and learning atmosphere and needs diagnosis and training planning process.

3.2.2. Learning Experience

The assumption is that according to the passage of time adults have grown and progressed towards maturity. Therefore, in adult training or adult technology, there is a decrease in the use of transmittal techniques such as those used in conventional training and to further develop experiential techniques. In this case it is known as "experiential learning cycle". This has implications for the selection and use of adult training methods and techniques. In training practice more use of group discussion, brainstorming, laboratory work, field school, practice and so on, which basically seeks to involve the participation of trainees.

3.2.3. Readiness to Learn

The assumption is that every individual is getting matured according to the passage of time, then learning readiness is not determined by academic or biological necessity or coercion, but more determined by the demands of development and change of task and social role. This brings implications to the learning materials in a particular training. In this case of course learning materials need to be tailored to the needs in accordance with its social role.

3.2.4. Orientation Learning

The assumption is that in adults have a tendency to have a learning orientation centered on solving problems faced. The implications for the nature of learning materials or training for adults, namely that the material should be practical and can be applied immediately in everyday reality.

Based on the above assumptions, it can be concluded that entrepreneurship training based on andragogy considers the principle of learning for adults. Entrepreneurship training is conducted in accordance with the learning needs, experience, readiness, and objectives of adult learning to develop their business.

3.3. Community Empowerment

Word Bank (2001) defines empowerment as an effort to provide opportunities and capabilities to the poor to be able to voice and voice opinions, ideas and ideas, and the ability and courage to choose something (concepts, methods, products, actions and others) is best for the person, the family, and the community. Community empowerment is a process of improving the ability and attitude of community self-reliance.

The main approach in the concept of community empowerment is that the subject of its own development efforts. Schneider (1999) states that: empowerment goes well beyond the narrow realm of polotical power, and differs from classical definition of power by Max Weber.

Rappaport (Weissberg, 1999) suggests that empowerment is viewed as a process: the mechanism by which people, organization and communities gain mastery over their lives. Hacker (1999) mentions empowerment may be understood as a process of transformation. This includes the tranformations of the unequal power relationship, unjust structures of society, and development policies. Empowerment is a way that people, communities, and organizations are directed to be able or powerful over their lives (Rappaport, 1984).

Freire (1972) defines empowerment in the field of education as a practice of self-liberation from ignorance, pressures, and other things that imprison a person and or community to improve his life.

4. Conclusion

Social entrepreneurship training is an organized learning process from planning, implementation, and evaluation on non-formal education channels aimed at improving community knowledge, attitude and skills for learning and working. Andragogy is a science or art that learns adult learning. Andragogical assumptions consist of self-concept, learning experience, learning readiness, and adult learning orientation. Community empowerment is a process of empowerment or strength of community to have the ability in building social and community economic empowerment.

References

1. Bornstein, D. *How to Change the World: Social Entrepreneur and the Power of New Ideas*, Updated Edition, New York, Oxford University Press, (2007)
2. Borganza, C. Depedri, S. dan E, Tortia. *The Growth of Organizational Variety in Market Economies,* The Cs. of Soc. Ent. Eur. Wrk. Pap. J. **03**, 10. (2010)
3. Darkenwald, G D., and Merriam, Sharan B. *Adult Education: Foundations of Practice*, New York, Harper and Row, Publishers, (1982)
4. Deakins, D. and Freel, M. *Entrepreneurial learning and the growth process in SMEs,* The Lrn. Org. **5,** 3. 144-155. (1998).
5. Defourny, J. dan Kim, S. *Emerging. Models of Social Enterprise in Eastern Asia: a Cross-Country Analysis.* Soc. Ent. J. **7,** 1. 86-111. (2011)
6. Franco, M dan Haase, K. *Entrepreneurship: an organisational learning approach*. J. of Smll. Buss. and Ent. Dev. **16,** 4. 628-641. (2009)
7. Freire, P., *The Pedagogy of The Oppressed,* New York, Herder and Herder, (1972)
8. Hacker, H., *Empowerment Projects for and by Women*, Summary of Pilot Study DED Yaounde, (Unpublished), (1999)
9. Knowles, M., *Informal Adult Education: A Guide for Administrator, Leader and Teachers,* New York, Association Press, (1950)
10. Laird, D., *Approach to Training and Development*, New York, Addison-Wesley Publising Company, (1985)
11. Mayo, G.D. & DuBois, P.H., *The Complete Book of Training: Theory, Prinsiples, and Techniques*, San Diego, CA University Associaties, Inc., (1987)
12. Meredith, G., *Kewirausahaan, Teori dan Praktek*, Jakarta, PT. Pustaka Binaman Pressindo, (1996)
13. Nadler, L., *Designing Training Program, The Critical Events Model*, Sidney, Addison Wesley Publishing Company, (1982)
14. Nicholls, A. *The New Social Entrepreneurship. What Awaits Social Entrepreneurial Ventures?* Eql. Opp. Int. **26**, 7. 729-732. (2006)
15. Rappaport, J., *Studies in Empowerment: Introduction to the Issue, Prevention in Human Issue,* USA, (1984)
16. Renko, M., Carsrud, A., dan Brannback, M. *The Effect of a Market Orientation, Enterpreneurial Orientation, and Technological Capability on Innovativeness: A study of Young Biotechnology Ventures in the United States and in Scandinavia*. J. of Smll. Buss. Mgmt. **47**, 3. 331-369. (2009)
17. Robinson, K.P., *A Handbook of Training Management*, London, Kogan Page Ltd, (1981)
18. Schneider, *Participatory Governance: The Missing Link for Poverty Reduction*. OCECD Develompment Center, Policy Brief No 17. Paris, (1999)
19. Srinivasan, L., *Perspectives on Nonformal Adult Learning: Functional Education for Individual, Community and National Development*, Connecticut Prentice Hall, (1977)
20. Wu, W., Chang, M., and Chen, C. *Promoting Innovation through the Accumulation of Intellectual Capital, Social Capital, and Enterpreneurship Orientation*. R&D Mgmt. **38,** 3. 265-277. (2008)
21. Weissberg, R., *The Policies Empowerment*, Praeger, Westport, Connecticut and London, (1999)
22. World Bank, *Sustainable Development in the 21 Century,* (2001)
23. Zimmerer, T. W., & Norman, M. S., *Entrepreneurship And The New Venture Formation*, New Jersey, Prentice-Hall International, Inc., (1996)

The implementation of student team achievement division and picture series on improving the speaking skill of the 3rd grade students of cipta dharma elementary school in academic year 2016/2017

Nengah Dwi Handayani[1,*] and *Dewa Gd. Ag. Gana Kumara*[2,*]

[1,2]English Education Study Program, Mahasaraswati University, Denpasar, Bali, Indonesia

Abstract. The aim of this research was to find out whether or not the Speaking Skill of the 3rd grade students of Cipta Dharma Elementary School could be improved through Student Team Achievement Division (STAD) and Picture Series in Academic Year 2016/2017. This Research used a Classroom Action Research with the total 36 students taken as subject. The result was, in Pre Cycle 52.78% (19 students); in Cycle 1 69.44% (25 students); and in Cycle 2 88.89% (32 students) reached the minimum completeness learning mastery. Based on the result, it could be concluded that the subject's Speaking Skill could be improved the implementation of STAD and Picture Series.

Key words:Speaking skill, STAD, Picture series

1 Introduction

One of the four main skills in English is Speaking Skill. [14] states that speaking is one of the most difficult aspects for students to master. Speaking skills is one aspect that is quite difficult to master because it is associated with several elements of language that must be mastered namely grammar, vocabulary, pronunciation and fluency.

English is currently defined as one of the local content subjects taught in schools. [6] states good speaking activities should be extremely enggaging for the students. It implies that students should be fully engaged in teaching and learning activities that are associated with the improvement of speaking skills. It will also directly challenge teachers to create an interesting and interactive learning process.

One of the techniques that can be used in speaking skill learning is the Student Team Achievement Division (STAD). STAD is one type of cooperative learning (cooperative learning). This STAD is applied by dividing the pupils into groups. What is characteristic of the STAD is that group divisions are based on student achievement in the class itself.

* Corresponding author: ndwihandayani@gmail.com

Besides designing interesting and interactive learning, teachers are also required to be creative and innovative in developing learning media. One of the media that can be used in teaching speech skills is the series image (Picture Series). A series image is a series of related images and forms a story or a cascading explanation.

Based on the initial observations that have been done at Cipta Dharma Denpar Primary School, it was found that the students, especially the 3rd grade, experienced the same thing in expressing something in English. Teachers have been trying to apply some learning techniques and approaches, but speaking skills of grade 3 students of Cipta Dharma Elementary School were still under minimal mastery criteria.

Based on the problems that occur above, Researchers are highly interested in conducting a study on speaking skill. Therefore, this research will focus on Implementation of Student Team Achievement Division and Picture Series as an effort to improve English Speaking Skills of the 3rd Grade Students of Cipta Dharma Elementary School. The purpose of this research are: 1) to know whether or not English Speaking Skills of Grade 3 students of SD Cipta Charma is enhanced by the application of STAD and Picture Series. 2) to know the response of 3rd Grade Students of SD Cipta Dharma Denpasar towards the application of STAD and Picture Series in an effort to improve English Speaking Skills.

2 Method

2.1 Research Design

This research made use of Classroom Action Research. The procedures of this study includedInitial Reflection (IR), Planning (P), Actions (A), Observation (O), Reflection (R) and Revised Planning (RP) performed in cycle form.

2.2 Participants

The subject of the study was the 3rd grade Students of Cipta Dharma Elementary School in academic year 2016/2017. There were 36 students taken as the subject of this study. Based on the initial observations that have been done, it was found that the students experienced difficulties in expressing something in English. Teachers have been trying to apply some learning techniques and approaches, but most of the stuents' speaking skills were still under minimal passing grade.

2.3 Measures

The data gathered from the pre test and post tests were analyzed using formula of :

$$M = \frac{\sum X}{N} \tag{1}$$

In which the total score of the students was divided by the total number of the students, meanwhile, the result of questioanire was analyzed using rating scale 1 to 4 to figure out the students'responses towards the implementation of STAD and Picture Series.

2.4 Procedure and data analysis

IR was conducted in the form of a preliminary test that aimed to determine the initial ability of research subjects. P was a plan that was contained in the form of lesson plan based on the needs of research. A was the action of this study which consisted of the steps of learning. After doing the action, proceed with observing (O) and reflecting (R) by doing post test. The post tests were carried out by the end of each cycle. The test were analysed by using a speaking scoring rubric.

3 Results and Discussion

The Initial Reflection which was in the form of pre test was conducted in order to know the students'pre-existing speaking skill. The result of the questionnaire was 67.03. The test was conducted by asking students to tell about themselves infront of the class. Most of the subject under study were still shy and doubt of what the were saying.

After figuring out the result of the pre test, it was continued by planning the lesson plan. The lesson plan was designed by using STAD and Picture Series. After having a well prepared leeson plan and materials, the research was continued by conducting a teaching learning activites of the cycle 1 which consisted of two sessions. By the end of the last session of this cycle, the post test 1 was administered. The result of the post test was 77.67. There was an improvement of the subject's speaking skill based on the result of the post test. The students were active in doing their learning activities, but there were some of them were still confused and shy of sharing their ideas with the group during the class. The low achievement students were not really active in the group discussion using Pictures Series.

Based on the result of the post test 1, then it continued by planning the lesson followed by applying the plan for cycle 2. This cycle also consisted of two session ended by gathering the post test 2. The result of post test 2 was 82.36. The was a significant improvement from post test 1 to post test 2. The students were very active and really enjoyed the learning activites. Students who were at excellent and good achievement could assisst students who were sufficient. Besides, the questionnaire was also conducted at the end of this cycle.

The data of questionnaire were measured by using scale 1 to 4. The rusult of the questionnaire showed that 85.62% strongly agreed, 10.99% agreed, 03.16% neither, and 00.23% disagreed with the implementation of STAD and Picture Series in improving speaking.

Based on the result above, it could be concluded that: **1)** The Speaking Skill of the 3rd grade students of Cipta Dharma Elementary School could be improved through the implementation of STAD and Picture Series in academic year 2016/2017. **2)** The subject under study gave a positive response on the implementation of STAD and Picture Series in improving Speaking SKill

4 Conclusion

The Result of this research showed the mean score of pre test, post test 1 and post test 2 were 67.03, 77.67, 82.36. And the result of the questionnaire showed 85.62% strongly agreed, 10.99% agreed, 03.16% neither, and 00.23% disagreed with the implementation of STAD and Picture Series in learning Speaking.

References

1. D. Ary, L. C. Jacobs, S. Sorensen, A. Razavieh, *Introduction to Research in Education*, (Wadsworth, California, 2010).
2. H. D. Brown, *Language Assessment*, (Longman, New York, USA, 2004).

3. L. Cameron, *Teaching Language to Young Learners.* (Cambridge University, New York, 2001).
4. R. Carter, *The new grammar teaching. In R Carter (Ed.), Knowledge about language and the curriculum* (Hodder & Stoughton, London, 104-121, 1990).
5. I. K. Ghosn, ELT Journal **6** (2) (2002).
6. J. Harmer, *The Practice of English Language Teaching,* (Longman, Cambridge, 2007).
7. S. Hawanti, International journal of pedagogies and learning **9** (2) (2014).
8. K. Hayriye, The Internet TESL Journal **XII** (11) (2006).
9. W. Jolliffe, *Cooperative Learning in the Classroom* (Paul Chapman Publishing, London, 2007).
10. P. J. Krish, The Internet TESL Journal **VII** (7) (2001).
11. Mantra, et al. International Journal of Linguistics, Language and Culture (IJLLC) **2** (2) (2016).
12. I K. Natasa, The Internet TESL Journal **XII** (11) (2006).
13. L.S. Norton, *Action Research in Teaching and Learning* (Routledge, Oxford, 2011).
14. L. Pollard, *Teaching English : A book to Help You through Your First TwoYears in Teaching* (Cambridge University Press, London, 2008).
15. J. C. Richard, W. Renandya, *Methodology in Language Teaching : An Anthology of Current Practice* (Cambridge University Press, . Cambridge, 2002).
16. G. Shahini, A. M. Riazi, A PBLT Approach to Teaching ESL Speaking, Writing, and Thinking Skills. Retrieved April 15, 2017, from http://eltj.oxfordjournals.org/content/65/2/170.
17. I.A. Md. Sri Widiastuti, The ASIAN EFL Journal **3** (2016).
18. Thornbury, S. *How to Teach Speaking* (Pearson Longman, USA, 2005).

The Preservation of Traditional Knowledge and the Cultural Expression of Craft Rumah Gadang's Walls as the Intellectual Property Of West Sumatera

Armilia Riza[1]*, *and Surya Prahara*[2]

[1]STKIP PGRI SUMBAR, English Department, Padang, Indonesia
[2]STKIP PGRI SUMBAR, Sociology Department, Padang, Indonesia

Abstract. This study discusses the problems of the preservation of Minangkabau craft culture, especially the problem of Intellectual Property Rights in the form of traditional knowledge that is on the craft of Rumah Gadang's walls. The method of this research is qualitative. The informant of this study is determined by using purposive sampling. The chosen informants that are people involved in crafting of Rumah Gadang's walls and the Department of Tourism of West Sumatra Province. This research was conducted at Minangkabau representative area of Luhak Nan Tigo (Agam, Tanah Datar, Lima Puluh Kota). The results of the study shows the extinction of craft of Rumah Gadang's walls was caused by the development factor of modern society's mindset, thus eliminating the value of a local culture of desire to learn old local culture. Moreover, the history and the topic of craft of Rumah Gadang's walls are deleted from Muatan Lokal Subject from the elementary to college level. Besides, the absence of efforts of local governments in trying to preserve the cultural values contained in traditional of craft of Rumah Gadang's walls.

Keywords: Preservation, Craft Rumah Gadang's Walls, Intellectual Property

1 Introduction

The term of World heritage was initiated by UNESCO (United Nations Educational, Scientific and Cultural Organization). UNESCO held a discussion and the first international convention on the world heritage in 1972. In that convention that its theme was The Convention concerning the Protection of World Cultural and Natural Heritage, on November 16, 1972, UNESCO initiated the protection and recognition of the cultural and natural heritage. UNESCO provides protection for a

[1] takuyaeek@gmail.com, armiliariza@stkip-pgri-sumbar.ac.id

world-wide cultural heritage that is so vast and becomes homework for Indonesia. Dealing with that issue, Indonesian Government is aware of the importance of intellectual property value in Indonesian folklore, so that it issued the National Copyright Act 1982 on article 10 which is substituted by article 10 of Law no. 19/2002 on Copyright, and the latest in article 13 of the Copyright Bill of 2010.

Meanwhile, the protection of PTEBT in the National Law is different from what people expect. The people think that the issue of the protection of intellectual property and commercialization of PTEBT are less important than the extinction of PTEBT due to lack of recognition and attention of government[2]. In addition, there is no effort for documenting PTEBT and the custodian. Furthermore, there is culture heritage that is owned collectively and inherited without knowing the first founder.

In its development, at least known several types of intellectual property protection in the form of copyright, brand, geographical indication, industrial product design, patent, integrated circuit layout, and trade secrets (*Article 1 Section 2 of the WTO Agreement on Trade-Related Aspects of Intellectual Property Rights, 1996).*[3] The emergence of such diversification of protection has led to some negative implications, especially in the perspective of IPRs that exist outside of such grouping, one of which is the absence of protection of traditional knowledge[4]. This has concretely hurt some traditional societies. Hence, it takes a great commitment in realizing PTEBT.

The focus of this research is the existing PTEBT on Minangkabau culture that is carving art of Rumah Gadang Minangkabau. Based on the description above, PTEBT should receive attention and protection through IPR system embodied in bentu program that will be used by the Government to realize the program PTEBT. The government should provide protection to those who develop their own local wisdom, ceremony, art, culinary, medicine, and folklore, tailored to the ecosystem environment in which they live.[5] If this program cannot be run, Indonesia will lose intellectual property and wisdom in understandin folklore. According to Darrell Posey, the extinction cultured people affects the world lost thousand years knowledge about life and how to adapt in tropical ecosystems. This valuable

---

[2] Sardjono, Agus, 2005. *Upaya Perlindungan HKI yang terkait dengan GRTKF di Tingkat Nasional dan Internasional (Upaya yang Belum Sebanding),* Jurnal Media HKI vol.II no.2 December 2005, Ditjen HKI. Page. 3

[3] Siagian, Rizaldi, 2006. *Jenis-Jenis Pemanfaatan atas Pengetahuan Tradisional dan Ekspresi Folklor yang perlu dilindungi dan Implikasi Pemanfaatannya, Simposium " Menuju UU Sui Generis Perlindungan terhadap Pemanfaatan Pengetahuan tradisional dan Ekspresi Folklor*", Jakarta 13 November 2006. Hal 5

[4] Citrawinda, Cita, 2006. *Perlindungan terhadap Karya Budaya yang Tidak diketahui Penciptanya*, Jurnal Media HKI vol.III no.1 Feb 2006, Ditjen HKIIASTP AUSAID, 2003. Page. 3

[5] According to LSM, Aliansi Masyarakat Adat Nusantara, *sampai dengan 2010 di Indonesia terdapat 1163 kelompok masyarakat adat dengan populasi antara 50 sampai 70 juta orang yang menempati 5 juta hektar hutan adat*. Read "Membagi Sumber Daya Alam secara Adil", *KOMPAS*, 25 Februari 2010.

information is lost immediately.[6] In addition, the protection of culture will determine the existence and the identity of a nation can be maintained and enhanced and can encourage people to improve the quality of the resulting creation.

The art of carving in Minangkabau is on every side of Rumah Gadang's building. Carving is placed on the whole body of the building. The walls, doors, windows are decorated with carvings consisting of several motifs. Each of the motifs has a deep meaning, Minangkabau traditional carving is divided into three types based on the inspiration of the formation of carvings, the first carvings inspired by the names of plants such as *Aka Barayun, Aka Duo Gagang, Aka Taranang,* and so forth. Secondly, it is inspired by animal names, such as *Ayam Mancotok dalam Kandang, Bada Mudiak,* and so forth. The third is inspired carvings of objects used in everyday life such as *Ambun Dewi, Aie Bapesong, Ati-ati, Carano Kanso, Jalo Taserak* and so forth.

The reason why Rumah Gadang Minangkabau carving is chosen is that the carvings have cultural expression and traditional knowledge values which are known by Minangkabau people only and it is extinct. It is happen because the number of Rumah Gadang that has original motifs of Minangkabau carving is decreased. Another reason is that it is one of the efforts to prevent the stealing of the cultural expression and traditional expression of Rumah Gadang by the foreign parties like 34 cultures that have been stolen[7] such as Reog Ponorogo dance, Pendet dance, Rasa Sayange Song and Rendang Minang.[8]

This research aims to find the causes of the extinction of traditional knowledge that focuses on Rumah Gadang Minangkabau carving. The novelty of this research to produce a program of Traditional Knowledge Protection, Culture and Cultural Expression Preservation that can be used by the government, especially the West Sumatera local government to protect the folklore art of Rumah Gadang carving. The significance of this research is to create the protection law of PTEBT and folklore, so it cannot be claimed by other countries.

2. Methods

There are two different research method used to answer the problem of the real condition of the extinction of the art of wall carvings of Rumah Gadang

---

[6] Posey, D.A., "Indigenous Knowledge and Development: An Ideological Bridge To the Future", dalam Posey, D.A. (K. Plenderleith, ed.), Kayapó Ethnoecology and Culture, Routledge, London and New York, 2002, page. 59, sebagaimana dikutip dalam Dutfield, Graham, *Protecting Traditional Knowledge and Folklore, A Review of Progress in Diplomacy and Policy Formulation*, International Trade & Sustainable Development Series, ICTSD - UNCTAD, Switzerland, 2003, page. 25

[7] Maya Sofia dan Tasya Pramitha, Upaya Pemerintah Agar Budaya Indonesia Tak DIklaim Negara Lain, www.life.viva.co.id, retrieved on December 23, 2016.

[8] www.antaranews.com, Warisan Budaya Indonesia Rawan DIklaim Negara Lain, retrieved on December 23, 2016

Minangkabau in West Sumatra and the stakeholders' problems on the barriers to the preservation and protection of traditional knowledge, especially the art of wall carvings Rumah Gadang. Sociological Research methodology with a qualitative approach is used.[9] To answer the content of IPR protection issues, legal instrument analysis on the protection and preservation of cultural and cultural expression, Law Research methodology with normative legal approach is used.

The methods of collecting the primary data on sociological approach are interviews and observations. Meanwhile, the secondary data is gotten from documents related to this research. Normative legal approach method only relies on the secondary data consisting of primary legal material in the form of international and national law regulation governing folklore and secondary law material in the form of articles and journals related to this research. Informant of this research is conducted by purposive sampling method. The location of the research is in West Sumatera Province, the research area will be represented by the areas that mostly have Rumah Gadang

3. Result and Discussion

Based on the findings of the field, traditional art of Rumah Gadang's carving has faded and lost its identity and role. The picture of Rumah Gadang carving is only a symbol without meaning and becomes a dead object that only has artistic value. Minangkabau society does not know much about the aesthetics value, and the meaning of philosophy contained in it. The influence of globalization and the modern mindset due to the disclosure of information and thought has undermined the Minangkabau customs. This condition can be seen from the social development of Minangkabau people who have abandoned the noble values of birth customs. This is manifested by obsolete, Shrubs and the collapse of Rumah Gadang (baganjong) in Minangkabau hinterland, such as Tanah Datar, Solok, Agam, and Lima Puluh Kota, as well as several other areas.

It is very unfortunate circumstances above occurs when the leader of Minangkabaunese mostly background as *Penghulu* or *Datuak* as the figure holding of *sako jo pusakon lamo*. All these circumstances seem to have occurred systematically symbols of Minangkabau customs are swallowed by the period. The results of this study explain that this happens because of several factors:

1. The development of modern societies' mindset, thereby eroding the value of a culture, forcing the absence of a desire to degrade the skills, knowledge of Rumah Gadang Carving to descendants / custodians and the absence of desires from the descendants as custodians of the engraver to continue the skill.
2. The absence of formal education that gives knowledge of local content to learners in particular.

---

[9] Bungin, Burhan, 2010. *Analisis Data penelitian Kualitatif.* Jakarta: PT Raja Grafindo Persada. Page. 5

3. The existence of a commercial carving industry that produces traditional carvings of Rumah Gadang without knowing the aesthetic value that exists in the traditional carving Rumah Gadang.
4. Less optimal effort of local government in preserving cultural values, especially those contained in the traditional carving art of Rumah Gadang. This is due to the absence of a clear program to address the problem.
5. The absence of legal products at the PERDA level to the PERNA level that focuses on the preservation of regional culture.

From the seven factors that have been described, according to the authors the main cause of the waning beginnings of traditional art of Rumah Gadang is due to the lack of understanding on the aesthetic values and custom meanings contained in the art of carving. The lack of seriousness of governments in preserving the traditional culture of the region is also a factor of extinction of the values and aesthetics that exist in the arts and culture of the region.

In line with the intention and effort to preserve and develop traditional knowledge and culture, it has been agreed upon a charter called Piagam Pelestarian Pustaka Indonesia 2003 which was declared in December 2003 in Ciloto, West Java. The definition of conservation adopted in the charter is the effort of library management through research activities, planning, protection, maintenance, utilization and supervision. The use of the term "protection" has a broad meaning, which means also the effort of preservation and protection of IPR in PTEBT Indonesia. Conservation can also include selective development to maintain continuity, harmony, and carrying capacity in response to the dynamics of the times. (Afrillyanna Purba, 2012: 137).

In the Draft Law on the Protection and Utilization of Intellectual Property of Traditional Knowledge and Traditional Cultural Expression commonly abbreviated as PTEBT explains that Traditional Knowledge and Traditional Culture Protected Expression include cultural elements that are compiled, developed, maintained and transmitted within the tradition and have special characteristics that are integrated with a particular cultural identity of the community that preserves it. Traditional protected cultural expression includes one or a combination of forms one of which is the game. Along with the improvement of technology and cultural transformation towards modern life as well as the influence of globalization, the cultural heritage and traditional values of indigenous peoples face challenges and obstacles to their existence. This needs to be observed because the cultural heritage and traditional values should be preserved, adapted or even developed further.

The question is how the efforts of West Sumatera government in preserving folklore art of wall painting Rumah Gadang as local cultural heritage. Citing the opinion of Koentjaraningrat that the development of Indonesian national culture needs to be oriented to the glorious era of the ancestors of the Indonesian nation that has been past, but also to the present because culture needs to give the ability of the Indonesian nation to face the civilization of today's world. (M.Mandandar Sulaeman, 1998: 43). Efforts of West Sumatera local government in preventing regional culture can be seen through **Conservation concepts contained in West Sumatra Provincial Regulation No. 6 of 2014 on Strengthening of Indigenous Institutions and Minangkabau Culture Preservation**

This regulation was made with the idea that the state recognizes and respects the unity of indigenous and tribal peoples as well as their traditional rights as long as they are alive and in accordance with the development of society and the principle of the Unitary State of the Republic of Indonesia. That due to the decrease of custom and cultural values of Minangkabau due to various negative influences, it is necessary to strengthen the customary institution and preserve the Minangkabau cultural values. Besides that in the framework of the implementation of regional autonomy, the Regional Government has the duty and obligation to strive for the preservation of regional culture as well as to ensure legal certainty towards the effort of preserving the cultural values of the region and based on that consideration it is necessary to stipulate the Regional Regulation on Strengthening of Customary Institution and Preserving Minangkabau Culture Value.

When viewed from the legal instrument, this Regulation is the implementing regulation of several laws and regulations, among others, the Minister of Home Affairs Regulation No. 39 of 2007 on Guidelines for Facilitation of Community Organizations in Culture and Customary Institutions in Strengthening and Preservation of Regional Culture, Regulation of the Minister of Home Affairs Number 52 of 2007 on Guidelines for Strengthening and Preservation of Customs and Social Values of Cultural Society and, Joint Regulation of the Minister of Home Affairs and Minister of Culture and Tourism Number 42 Year 2009 and Number 40 Year 2009 on Guidelines for Cultural Preservation.

Based on Article 4 of West Sumatra Provincial Regulation No. 6 of 2014 on the Strengthening of Indigenous Institutions and Minangkabau Cultural Preservation, the government should be obliged to play a role in strengthening the culture, from the concept that researchers wake up, should the government have obligations in terms of:

1. The regional government should:
   a. facilitate the strengthening of customary institutions;
   b. carry out the preservation of Minangkabau cultural values.
2. Encourage the implementation of strengthening customary institutions including the facilitation, coordination and consultation on strengthening customary institutions.
3. Preserve Minangkabau cultural values including:
   a. realizing, growing, developing and raising awareness and responsibility for the rights and obligations of communities in the management of Minangkabau cultural values;
   b. developing and implementing policies that can ensure the protection and utilization of Minangkabau cultural values;
   c. conducting research and development of Minangkabau cultural values;
   d. providing information on Minangkabau cultural values;
   e. organizing the promotion of Minangkabau cultural values;
   f. preveningt efforts to transfer and shift ownership and control on Minangkabau cultural values;
   g. coordinating district / city governments in order to ensure the implementation of Minangkabau cultural preservation; And / or

h. fostering participation and creativity of the community in preserving the Minangkabau cultural values based on mutual cooperation, independence and justice. The obligation to preserve Minangkabau cultural values as referred to in Article 4 is implemented by the Regional Device Work Unit in charge of culture and tourism.

The strengthening of human resources of adat institutions is carried out by conducting activities to increase pemangku adat's knowledge on the preservation of cultural values of Minangkabau. Meanwhile, strengthening facilities and infrastructure of adat institutions can be implemented by facilitating the provision of facilities and infrastructure supporting the operations of customary institutions in order to preserve the Minangkabau cultural values. Therefore, customary institutions act as partners of local governments in empowering, developing and preserving the customs and cultural values of Minangkabau in the life of society and customary institutions should be involved in:
1. making decisions;
2. formulating decisions relating to the strengthening of customary institutions and the preservation of Minangkabau cultural values; and / or
3. planning and synergizing development programs to conform to Minangkabau cultural values

The development of Minangkabau cultural values is done by:
1. Conducting study and research of Minangkabau cultural values in order to explore the potential of Minangkabau culture;
2. Conducting socialization, seminars, workshops and discussions of Minangkabau cultural values through educational, mass media and other publications.

The development of Minangkabau cultural values as intended is done by maintaining the cultural roots that are owned and not intended to replace existing cultural elements as well as by observing the culture and custom of mutual Nagari.
For the utilization of Minangkabau cultural values is done by:
1. dissemination of information;
2. cultural performances;
3. packaging of teaching materials;
4. packaging of study materials; and
5. developing tourism

Dissemination of information on Minangkabau cultural values is done in the form of:
1. socializing through printed media, electronic media matters relating to the utilization of Minangkabau cultural values, and / or
2. having seminars, discussions, conferences on the theme of Minangkabau culture.

In addition, the dissemination can also be implemented with the performance of Minangkabau cultural values that can be done in the form of:
1. organizing the Minang art festival regularly; And / or
2. displaying the values, norms of customs and traditions of Minangkabau at the exhibition which is followed and organized by the Regional Government.

The five concepts of efforts undertaken by the West Sumatra Regional Government in preserving the culture of the region hve not been fully implemented. This is certainly caused by several important factors. These factors will be discussed through other sub-chapters in this study.

4. Conclusion

The conclusion of this research is the factors causing almost extinction of art of wall carving of Rumah Gadang Minangkabau caused by modern society mindset that is influenced by globalization so that erodes the value of a culture so that force the absence of desire to decrease skill, and knowledge of Rumah Gadang engraving to offspring / custodians and the absence of the desires of the descendants of the tribe as custodians of the engraver to continue the skill. Another factor is the absence of formal education that gives the learners special knowledge about traditional carving art (this should be part of a program that is in the hands of the local government as an effort to preserve local culture). There is no regulation at the level of PERDA to the level of PERNA which is accompanied by the program which is significant to ensure the preservation of regional culture, especially the art of wall carving of Rumah Gadang Minangkabau. Finally, there is no facility to facilitate the engraver (both custodian and not engraver) in the form of an alliance or an association.

A bunch of thank deliver to Directorate General of Research and Technology Research and Development of Ministry of Research, Technology and Higher Education who has appreciated this research by providing funding so that this research can be implemented properly. Moreover, our appreciation goes to all carving artisan, Wali Nagari of Luhak Nan Tigo who have participated as the informant of this research.

References

Bungin, Burhan. *Analisis Data penelitian Kualitatif.* Jakarta: PT Raja Grafindo Persada. (2010)

Citrawinda, Citra. Perlindungan terhadap Karya Budaya yang Tidak diketahui Penciptanya, *Jurnal Media HKI vol.**III** no.1 Feb 2006, Ditjen HKIIASTP AUSAID, 2003.* (2006)

Posey, D.A. Indigenous Knowledge and Development: An Ideological Bridge To the Future", in Posey, D.A. (K. Plenderleith, ed.), *Kayapó Ethnoecology and Culture, Routledge, London and New York, 2002,* page. 59, cited by Dutfield, Graham, *Protecting Traditional Knowledge and Folklore, A Review of Progress in Diplomacy and Policy Formulation*, International Trade & Sustainable Development Series, ICTSD - UNCTAD, Switzerland, 2003, page. 25

Sardjono, Agus. Upaya Perlindungan HKI yang terkait dengan GRTKF di Tingkat Nasional dan Internasional (Upaya yang Belum Sebanding), *Jurnal Media HKI vol.**II** no.2 Desember 2005, Ditjen HKI* (2005)

Siagian, Rizaldi. Jenis-Jenis Pemanfaatan atas Pengetahuan Tradisional dan Ekspresi Folklor yang perlu dilindungi dan Implikasi Pemanfaatannya, Simposium" Menuju UU Sui Generis Perlindungan terhadap Pemanfaatan Pengetahuan tradisional dan Ekspresi Folklor, *Jakarta, 13 November 2006*. page 5 (2006)

Sofia, Maya and Paramitha, Tasya. Upaya Pemerintah Agar Budaya Indonesia Tak DIklaim Negara Lain, www.life.viva.co.id, *retrieved on December 23, 2016*

www.antaranews.com. Warisan Budaya Indonesia Rawan DIklaim Negara Lain, *retrieved on December 2, 2016.*

The Skill in Designing Explorative Learning Tools Of Mathematics Pre-Service Teachers Through Explorative Learning Based On Metacognitive Scaffolding

*Hepsi*Nindiasari[1,*], *IndhiraAsih Vivi*Yandari[1], *Etika*Khaerunnisa[1], *Aan Subhan*Pamungkas[1]

[1]Department of Mathematics Education, University of Sultan AgengTirtayasa, Serang, Indonesia.

Abstract. This study is back grounded by the fact that according to the result of TIMSS survey, Indonesian students' achievement in mathematics is still low. This study aims to see mathematics pre-service teachers' skill in developing learning tools after given explorative learning based on scaffolding metacognitive. The method of study use experimental quasi which is given to one group or class. The population of this study are mathematics pre-service teachers in University of Sultan AgengTirtayasa, Banten Province. The subject of sample taken is one class which is given initial treatment with explorative learning, then students performance in designing learning tools is assessed. Next, they are given explorative learning based on scaffolding metacognitive, and then their performance in designing learning tools is assessed. The instrument used is performance portfolio which is supported by a rubric to assess skill in designing explorative learning tools. The result shows that students in designing learning tools in the form of scenario based on Worksheet and in designing Worksheet after receiving explorative learning based on scaffolding metacognitive (very good category, mean score are 94.20 and 91.27) are better than before (explorative activity without scaffolding metacognitive) with very good category with the mean score are 85.27 and 86.11).

1. INTRODUCTION

Analysis and reasoning ability which is important ability and should be developed in mathematics learning in school still shows its low quality. In fact, students' lack in mathematical ability can be seen from the result of Trends in International Mathematics and Science Study (TIMSS) in 2015 which describe that in mathematics subject, Indonesian students are lack in all aspects of cognitive content. Indonesian students only master the routine problems, simple computation, and measure the knowledge of fact with daily context [1]. Indonesian students need reinforcement of ability in integrating information, drawing conclusion and generalizing the knowledge possessed to anotherthings.

One thing which cause the lack of students' mathematical ability is learning activity in class. Mathematics learning activity in class still give less freedom for students to build their own knowledge. This is confirmed that learning process so far still dominated by teacher and did not give access to students to develop autonomously [2]. This cause the students' knowledge only limited to what had been taught by teacher so their ability to communicate the ideas in solving the problem do not developed optimally. Furthermore, stated that in learning process teacher did not organize students to discuss in heterogeneous group, so interaction and communication among students in learning are not implemented well [3].

Seeing the condition above. then the effort needs to be done to improve this condition. One of them is by developing pre-service teachers in study program they enrolled in lectures. This development direct pre-service teachers to be able to design meaningful learning for students, meaningful learning means that in learning activity, teacher give opportunity to students to investigate mathematical problem in the effort to build their own knowledge. One meaningful learning which can be developed is explorative learning.

Explorative learning offer opportunity to students to broaden their knowledge by using process and skill and connecting prior knowledge with their learning experience to solve the problem by involving students in problem solving process. Through exploration activity, students can find mathematical process in a such way so students experience themselves, able to create a hypothesis (conjecture), and then searching the answer for conjecture students make through observation activity.

In the effort to create explorative learning in class, of course pre-service teachers need to be trained in how to design active learning supported by learning tools among others are explorative Students Worksheet and learning scenario. In order that studentspossesability in designing explorative learning tools, of course they should be habituated by explorative learning activity in their lectures classroom. By experiencing directly explorative learning process, students are expected to be able to create explorative learning ideas for learning in their school in the future.

In the process of explorative learning implementation, students are confronted with a problem. With the problem given in the beginning of learning, students are expected to be able to utilize their knowledge and learning experience they posses before so they are able to generate creative ideas, identify the problem as initial process to solve the problem given. Explorative learning facilitate students to collect data and information to answer the problems posed through invention by investigating, searching relevant information, and making generalization from investigation result. After exploration process is implemented, students are directed to apply the concept which had been obtained by confronting the task to make explorative learning tools.

In exploration process implemented by students, the role of lecturer as facilitator in giving scaffolding to students who find difficulty. Because in initial stage, students need support from more competent person to be able to reach ZPD (zone of proximal development). This scaffolding is intended as help or guidance from lecturer in the initial of learning then giving opportunity to students to take over more responsibility after they can do it by themselves. Lecturer help is very needed to make them become more directed so the process of learning implementation and the achieved goal can be implemented well. This scaffolding is needed because actually students have rich potency but they had not possessed ability to organize information or initial ability. The ability to organize according to [4] is known as one form of metacognition ability.

Metacognition is a part in learning, so metacognition needs to be strived for in the implementation of learning process in class [5]. Someone do metacognitive in two ways:

firstly, individual should aware of their own cognitive process (for example self-monitoring or self-regulation); secondly, individual should be able to apply cognitive process possessed for learning goal or design the solution for problem (for example, using critical thinking or reflective assessment) [6].

The importance of metacognitive in learning not only for mathematics pre-service teachers, but also for students. Metacognitive can motivate performance enhancement and efficiency of problem solving through reflection activation done by students [7]. This is corroborated also by Garciawho said that metacognition in learning can make students to more realize their performance quality, even when mathematical problem is given to students, they can realize how importance to solve that problem not merely for accomplishing it, but also for the quality of task accomplishment [8]. The result study of Yildirim showed that there is positive correlation which is significant between the level of students' metacognitive awareness and the level of their problem solving related to routine and non routineproblem [9]. To involve one's metacognition when he or she is in their ZPD, then it needs help in the form of scaffolding from lecturer or more able person. The help intended here not by giving idea directly related to the task of designing learning tools, but the help in directing students to involve their metacognition in learning. The help in this case are question, direction or command, in which the questions direct to metacognitive skill which is termed by scaffolding metacognitive. The question or help related to idea, knowledge which had been possessed which is related to the present knowledge, planning in developing learning tools, and reflection and evaluate the tools which had been implemented. The question words which had been used are what? why? how?

Seeing explorative learning process based on scaffolding metacognitive is tightly related to investigation activity containing challenging learning tasks and students take bigger role in their learning responsibility. Therefore, it is supposed that explorative learning based on scaffolding metacognitive can contribute well for mathematics pre-service teachers' skill in designing explorative learning tools. So, the problem formulation in this study is *how does the skill mathematics pre-service teachersl in designing explorative learning tools through explorative learning based on scaffolding metacognitive?*

2. RESEARCH METHOD OF STUDY

The aim of this study is to see the skill in designing explorative learning in students before and after they are given explorative learning based on scaffolding metacognitive. So the design of this study use experimental quasi with modified pretest-posttest design.

k	$\mathbf{X_1}$	$\mathbf{O_1}$	$\mathbf{X_2}$	$\mathbf{O_2}$
Experiment	Explorative	Performance	Scaffolding Metakognitive	Performance

The population of this study are mathematics pre-service teachers in University of Sultan AgengTirtayasa. The subject of study sample is one class of mathematics pre-service teachers in University of Sultan Agent Tirtayasa with total of 28 students. In this study, the subject is given explorative learning first, then performance test in designing learning is given. It is followed by explorative learning with scaffolding metacognitive, and in the final experiment, performance test is given anymore.

The instrument used in this study is portfolio of skill in designing explorative learning. The rubric of assessment use analytic Rubric approach. Analytic rubric is a rubric of skill in designing learning scenario comprising assessment of performance in developing explorative worksheet and worksheet (WS) based learning scenario. The assessment rubric of explorative worksheet performance comprise the element of information and question asking with grading score of 4, 3, 2, 1 grading, whereas the assessment of WS based scenario learning comprise the elements of introductory activity, core activity, and closing by using grading score of 3.2. 1. Next. score from this rubric is described and analyzed by using difference test to see whether or not there is difference in students' skill in designing learning before and after treatment given.

Data is analyzed by quantitative and qualitative approach. Data processing use wilcoxon test with related sample. This test aims to see the difference of ability before and after explorative learning based on scaffolding metacognitive is given in the same group.

3. RESULT AND DISCUSSION

The data to see mathematics pre-service teachers' skill is taken from portfolio of learning tools made by students and assessed by using rubric of learning tools performance both in worksheet and learning scenario. Table 1 shows the result of mean score for students' skill in developing learning tools.

Table 1. Mean Score of Students' Skill in Developing Learning Tools

	Worksheet (WS)		WS based scenario	
Mean	Explorative (EP)	Explorative Scaffolding Metacognitive (ESM)	Explorative (EP)	ExplorativeScaffolding Metacognitive (ESM)
	85, 27	94,20	86,11	91,27

From the table 1, mean gain of skill in developing WS based learning tools and learning scenario in EP and ESM learning in very good category. These two learning enable mathematics pre-service teachers to develop learning tools. This is supported by difference test that WS development with ESM learning is better than EP learning. This reality is shown by p-value (sig) 0.004 which is less than α= 0.05 so H_0 is rejected and H_a is accepted.

Similarly with skill in designing WS based learning scenario. H_0 is rejected and Ha is accepted, because p-value (Sig)of 0,004 which is less than$\alpha = 0{,}05$ It shows that skill in designing learning scenario with explorative learning based on scaffolding metacognitive is better than explorative learning.

The result of calculation which is related to difference test can be seen in the following table:

Table 2. Recapitulation Result of Difference Test for Skill in Designing Learning Tools

Learning	Worksheet	WS Based Scenario
Explorative (EP)	P-Value = 0,004 with $\alpha = 0{,}05$ Ho is rejected	P-Value = 0,004 with $\alpha = 0{,}05$ Ho is rejected
Explorative Scaffolding Metacognitive (ESM)		

Based on that result that explorative learning based on scaffolding metacognitive enable mathematics pre-service teachers toposses skill in designing learning tools, then the lecturers are advised to apply that skill. Because this learning not only demand students to conduct investigation, analysis, and discussion activity during learning to train skill in designing learning tools but also lecturer help students to aware of their metacognitive ability by scaffolding metacognitive. The lecturer help students to find idea, to relate the old knowledge to new knowledge, to design learning tools, and to evaluate learning tools which had been designed and implemented. The question asked related to scaffolding metacognitive, for example: what idea will be raised by you in making WS and learning scenario? What media which is appropriate in supporting WS in order that students understand the concept given? How do you feel when implementing scenario and WS based learning? Etc. It is hoped that these questions can make students aware to control their thinking activity so later they are able to reflect on themselves. Explorative learning based on metacognitive is held in group settingso it can facilitate students to interact with their peers and if they find difficulty the can discuss with the others. Lin stated that the creation of supportive social environment can support students' metacognitive development so it can motivate them to do active activity in learning [10]. The study result of Molenaar et al showed that scaffolding influence positively social metacognitive interaction within group and found significant relation between students' participation in social metacognitive interaction within group and students' metacognitive knowledge [11].

The ability to design learning tools given by group setting with good result is one task which is collected and assessed by an assessment rubric.It is included in authentic task criteria. The authentic task which is related to metacognitive instruction in this case scaffolding metacognitive had showed a better result without using metacognitive instruction. This is in accord with (ramarski et al, that students who are taught by learning with cooperative and metacognitive setting have authentic task which is better compared with students without metacognitive instruction [12]. In addition, there is enhancement in this result of study if compared with study of (Nindiasari&Pamungkas) when analyzed students' ability in developing Explorative Worksheet and learning scenario [13]. The study conducted by Nindiasari and Pamungkas in 2016 with students group learning only capable to design learning tools with good category, but in this study both with explorative learning and explorative with scaffolding metacognitive are in very good category.

If it is analyzed, the result of WS design made by students in explorative learning based on scaffolding metacognitive for the element of information is that 82.14% students have been able to arrange the information with not too much and too little, and support problem solving. Similarly with the element of questioning in which 82.14% students have made higher order question which encourage students to conduct investigation and encourage them to discuss. Compared with explorative learning activity without scaffolding metacognitive in WS development for the element of information which is not too little which encourage students to conduct investigation and encourage them to discuss are only 67% and 53.56% respectively. This data shows that there is enhancement from explorative learning activity (EP) to explorative activity with scaffolding metacognitive (ESM). Its enhancement for the element of information is 15.4% and the element of questioningis 28.57%. In fact, the highest enhancement is in questioning. Before explorative scaffolding metacognitive is given, some students only make question which is usual exercise, but after explorative scaffolding metacognitive is given, students are encouraged to reflect and control their thinking activity through questioning encouraged by lecturer so they are aware of what should be done to improve WK they make.

The result of enhancement for skill in designing WK is followed by enhancement in designing learning scenario comprising the elements of introductory activity, core activity and closing activity. Students who design learning scenario after obtaining explorative learning with scaffolding metacognitive for the element of design introductory activity is 100% had designed introductory activity which remind students to prerequisite related to the material which will be delivered (doing apperception), whereas in the element of core activity, 71.4% students had designed activity which reflect WK they make, demand students to do active and explorative learning, and in the element of closing activity, 50% students had done reflection, make conclusion and give task and motivate students to learn. If compared with explorative learning without scaffolding metacognitive, the element of introductory activity is 89.29%, the element of core activity is 60.7% and the element of closing activity is 32.14% respectively. The highest enhancement in designing closing activity is 17.86% compared with another element which only 10.71%. In fact, scaffolding metacognitive which habituate students to control their thinking process through questions given by lecturer can habituate students to do reflection activity and drawing conclusion in their learning scenario. The element of closing activity both in EP and ESM learning needs more attention. Students need to be trained in designing good closing activity particularly reflection activity.

4. CONCLUSION

Students with explorative learning based on scaffolding metacogniveare better in their ability in developing learning tools compare to when they only do explorative activity. Lecturers need to implement explorative activity based on scaffolding metacognitive when designing learning tools.

1

References

1. Kemdikbud. *Diagnosis of TIMSS Result 2015 For Quality and Achievement Enhancement*. (2015).
2. Trianto. *Designing Productive Innovative Learning*. Jakarta: Kencana. (2010).
3. A. Murni. , Y. T. Nurul, and T. Solfitri. *The Application of Active Learning Method with Group to Group Exchange (GGE) Type to Enhance Mathematics Learning Outcome of Students X Social Studies MAN 2 Model Pekanbaru*, JOER , **11**, No. 2, P. 2. (2010).
4. Awi. *The tracing of Scaffolding Metacognitive Types Needed by Students of lass XI Natural Science Senior High School in Solving Mathematical Problem*, The National Seminar Prosiding Seminar of Mathematics and Mathematics,(2010).
5. G. Schraw, K.J. Crippen & K. Hartley. *Promoting Self-Regulation in Science Education: Metacognition as Part of a Broader Perspective on Learning*. JRSE, **36,** pp 111–139 . (2006).
6. Hogan, et all. *Metacognitive Skill Development and Applied Systems Science: A Framework of Metacognitive Skills, Self-regulatory Functions and Real-World Applications*. JML, (2014).
7. B. Hoffman and A. Spatariu. *The influence of self-efficacy and metacognitive prompting on math problem-solving efficiency*. JCEP, **33**, Issue 4, October 2008, Pages 875-893. (2008).
8. T. García, C. Rodríguez, P. González-Castro, et al. *Elementary students' metacognitive processes and post-performance calibration on mathematical problem-solving tasks*. JML, **11,** pp 139–170. (2016).
9. S. Yıldırım & Ersözlü. *The Relationship Between Students' Metacognitive Awareness and their Solutions to Similar Types of Mathematical Problems*. EJOMSATE, **9** (2013).
10. X. Lin, ETR&D. *Designing metacognitive activities*. JETRAD, **49**, pp 23–40. (2001).
11. I. Molenaar, et all . *The effects of scaffolding metacognitive activities in small groups*. JciHB, **26**, Issue 6. (2010).
12. M. Kramarski, et all. *The Effects of Metacognitive Instruction on Solving Mathematical Authentic*. JML. **49,** No. 2 , pp. 225-250 .(2002)
13. H. Nindiasari, & A.Pamungkas, *Analysis of Mathematics Pre-Service Teachers' Ability in Developing Explorative Worksheet along with Learning Scenario* 365–368. (2016).

Self-Regulated Learning in Designing Explorative Learning Tools Among Mathematics Pre-service Teachers through Explorative Module

Indhira Asih Vivi Yandari[1,*], *Hepsi* Nindiasari[1], *Etika* Khaerunnisa[1], *Aan Subhan* Pamungkas[1], *Karso*[2], *Nurjanah*[2]

1) Pendidikan Matematika, Universitas Sultan Ageng Tirtayasa, Serang, Indonesia
2) Pendidikan Matematika, Universitas Pendidikan Indonesia, Bandung, Indonesia

Abstract. This study is back grounded by the importance of self- regulated learning in mathematics pre-service teachers. By possessing self-regulated, students will be able to evaluate learning process, set learning target, arrange the strategy in learning to achieve the goal, search relevant supporting materials for the lectures and show self efficacy. One task which play a role to make students successful to become professional teacher is by creating explorative learning tools. The development of this explorative tools very support learning activity process in class. This study aims to see to what extent self-regulated learning among students in developing explorative learning tools. The method of study is experimental quasi by treatment which is initiated with explorative learning and then explorative module learning is given to one class. The subjects in this study are students of mathematics department with total of 21 students, consist of 4 males and 17 females. Quantitative analysis was performed using the average difference test. The result of study shows that the average of students' self-regulated learning after treatment is in good category and there is significant difference in self regulated learning before and after treatment. It can be concluded that these treatment give good contribution to develop students' self-regulated learning.

Keywords: *Self regulatory learning, explorative, Pre-service teachers, explorative module*

1. INTRODUCTION

Self regulated learning is one important factor in learning. Self regulatory is an ability to direct and control our self in thinking and act, and not depend on the other people emotionally [1]. Self regulated learning contain the meaning of learning based on the ability possessed. Self regulated learning is an ability to monitor own ability and human personality hard effort [2]. Self regulated learning means that students have self consciousness to learn, capable to self determine the steps which should be taken in learning, capable to obtain learning resource by themselves, and doing evaluation activity and reflection on learning activity which had been done. Students who have high self regulated learning tend to learn better, capable to monitor, evaluate and regulate their learning schedule effectively. Self regulated learning is so important to be possessed by students, so development of self regulated learning should be strived for, in order to support the success of lectures process. Self regulated learning has high correlation with the success of their students [3]. Sellf regulated person will show explorative behavior, capable to take decision, self confident and creative [4].

*indhira_1969@untirta.ac.id

Self regulated learning support the process of task accomplishment. Self regulated learning as designing and self monitoring process thoroughly toward cognitive and affective process in accomplishing an academic task [2]. In lectures, of course as students of pre-service teacher need self regulation in solving their academic tasks. One task which play role to make students successfully to become a professional teacher is by making learning tool. Learning tool designed is aimed to be able to develop all potencies in students. One alternative which can be arranged is explorative learning tool.

Explorative learning offer the opportunity to students to broaden their knowledge by using process and skill and connect prior knowledge with their learning experience to solve the problem by involving students in the process of problem solving. Through exploration activity, students can find mathematical process in a such way so they experience themselves, capable to create a hypothesis (conjecture), and then search the answer to conjecture students make through observation activity [1].

Furthermore, Explorative is an approach which aims to delve out the ideas, arguments and the methods which are differ from students through some open-ended questions and commands so it can enable students to understand a concept and solution of problems [5]. Through explorative activity, students are expected can enhance their ability in imagining, intuiting, divergent thinking, producing original work, predicting and conjecturing, trial and error [6].

The essence of that explorative learning is contained in explorative learning stage. According to National Education Department, explorative learning stage comprise: 1) the presentation of exploration problem; 2) data or information collection; 3) data analysis, 4) drawing conclusion [7]. NCTM also give some important things to be given attention to create mathematical explorative activity which is effective, namely: 1) Using real objects (physical); 2) using various topics which is interesting and relevant; 3) question; 4) writing activity; 5) cooperative and collaborative group; 6) listening; 7) assessing [8].

By the strength of that explorative learning, it is hoped that a teacher can design meaningful learning for students by developing explorative learning tool so high order thinking ability can be realized, and students' mathematics learning result is more enhanced. In order that teachers are able to design meaningful learning, the guidance should be given since students study in campus. In this Institution, students need to be trained in how to design active learning supported by learning tools among other are explorative Student Worksheet and learning scenario. To support the arrangement of explorative learning tools, of course students should be accustomed with explorative learning activity in learning classroom. So, by experience directly explorative learning process, students are expected to be able to create explorative learning ideas for learning in school.

In explorative learning implementation, students are confronted by a problem. By the problem given in the initial learning, students are expected to be able to utilize the knowledge and learning experience possessed before so they are able to generate creative ideas, to identify the problem as initial process to solve the problem given. Explorative learning facilitate students to collect data and information to answer the problem posed through discovery by investigating, search relevant information, and make generalization from investigation result. After exploration process is implemented, students are directed to apply the concept which had been obtained by confronted with the task to make explorative learning tools.

Seeing explorative learning process is tightly related to investigation activity, contain challenging learning tasks and students take bigger role in their learning responsibility. Therefore, it is supposed that explorative learning can give good contribution to self regulated learning among mathematics pre-service teachers. So, the

*indhira_1969@untirta.ac.id

 https://doi.org/10.1051/shsconf/20184200106

formulation of problem in this study is how does self regulated learning of mathematics pre-service teachers in designing explorative learning tools through explorative module?

2. METHOD

This study is experimental quasi with modified pretest-posttest group design. In this study, students are given treatment in the form of explorative learning before initial scale of self regulated learning is given, after initial scale, the treatment is given in the form of explorative module and finally students are final scale of self regulated learning. The subjects in this study are students of mathematics department with total of 21 students, consist of 4 males and 17 females.

The instrument used in this study is self regulated learning scale. This scale consist of 40 statements. Based on the response given by students, next the percentage of achievement average of self regulated learning in each aspect is determined.

3. RESULT AND DISCUSSION

Self regulated learning in this study consist of some components, namely initiative, diagnose the needs, set the target, monitor, view the difficulty as challenge, utilize and find relevant learning resource, choose and apply the strategy, evaluate the process and learning outcome, and self ability. The mean score from self regulated learning for each aspect is obtained based on students' response. Recapitulation of self regulated learning scale can be seen in Table 2.

Table 2. Description of Self-Regulated Learning Data

No	Aspect	Average (%)	
		Initial Scale	Final Scale
1	Learning initiative	60,19	72,38
2	Diagnose learning need	60,00	70,95
3	Set learning target	51,19	59,29
4	Monitor and control learning	50,10	55,81
5	See difficulty as challenge	60,76	70,10
6	Utilize and search relevant learning resource	50,71	57,38
7	Choose and apply strategy	55,56	58,41
8	Evaluate the process and learning outcome	47,94	55,87
9	Self Concept	61,22	75,10
	Total Average	56,05	65,26

From Table 2 above, it is known that the initial achievement of students' self regulated learning is 56.05% and fall in enough criteria, and the final achievement of students' self regulated learning is 65.26% and fall in good category. From quality aspect, it is seen that there is enhancement of self regulated learning from enough category to good category.

The following is diagram of comparison between initial and final achievement of students' self regulated learning for each aspect.

*indhira_1969@untirta.ac.id

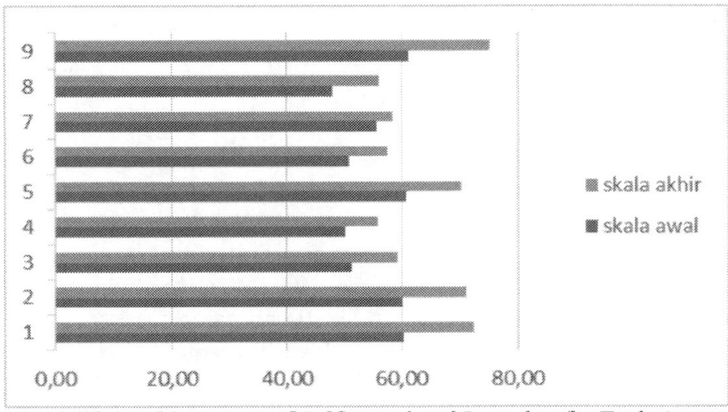

Fig. 1. The Achievement of Self-Regulated Learning for Each Aspect

From the figure above, it can be seen that there are some aspects in self regulated learning which is enhanced quite high, namely learning initiative, diagnose learning needs, see difficulty as challenge and self ability. These four aspects fall in good category. Whereas for achievement in following aspects: set learning target, monitor and control learning, utilize and search relevant supporting materials, choose and apply learning strategy and evaluate process and learning outcome fall in good category. Whereas for achievement in aspects: determine learning target, monitor and control learning, utilize and search relevant supporting material, choose and apply learning strategy and evaluate the process and learning outcome fall in enough category. These four aspects are less developed well even though receive different treatment with initial treatment.

The highest achievement both in initial scale and final scale is achieved in aspect of self ability, that is 61.22% for initial achievement and 75.10 for final achievement, whereas this shows that aspect is not develop well after students receive the treatment, that is, explorative learning with module.

Based on explanation above, it shows that descriptively the final achievement of self regulated learning is higher than initial achievement, but this need to be proved inferentially whether or not the initial and final achievement is differ before and after treatment. The following is the result of paired average difference test.

Table 3. Paired Average Difference Test

t	Df	Sig. (2-tailed)
-8,057	20	0,000

Significance value of 0.000 is less than 0,05. This shows that there is difference in students' self regulated learning before and after given the treatment. This assert that the treatment give big contribution toward development of students' self regulated learning.

There are some findings that explorative learning with or without module cause students have good self regulated learning, among other is that explorative learning is based on problem and students are active participant in their own learning.

In initial learning, the lecturer present exploration problem to students. This presentation stage is the key of whether or not explorative learning is successful. With interesting exploration problem, it is hoped that students are motivated to conduct investigation. Lecturer needs to present problem situation to students as interesting as possible, usually by providing something which can be seen, felt or touched will arouse interest and motivation to investigate, and the problem given can encourage students to initiate learning, diagnose learning needs, set the target/the goal of learning, monitor,

*indhira_1969@untirta.ac.id

regulate and control learning, see difficulty as challenge; utilize and search the relevant supporting material to solve the problem; choose and apply learning strategy; evaluate process and learning outcome; self concept [9]. These abilities are known as self regulated learning.

One example of explorative learning which is implemented in this study is following. ***The First stage is present exploration problem***, in which students are directed to conduct investigation toward a learning media in one of mathematical concepts. With learning media given, it can grow students' learning initiative and they see difficulty as challenge. It is seen when learning which is indicated by students' initiative to do manipulation directly toward media given and all students involvement in manipulating that media. ***The second stage is data and information collection***, in which students are asked to identify whether the media given is included in media hands on activity category based on its characteristics. In this stage, students ability in utilizing learning source is seen, because students are actively searching some information from internet and handbook to finish the task. ***The third stage is students are asked to do analysis toward the existing data*** to support the arrangement of students worksheet based on that hands on activity, and continued by arrangement of learning scenario which can be used by teacher in class. In this stage, indicator of self-regulated learning is seen, in which all students have initiative to share the task with their group members, capable to finish the task with systematical procedure, utilize the relevant source, determine learning strategy to finish the task. ***The fourth stage is present the result report and draw the conclusion***, in which students are asked to present students worksheet and learning scenario which had been arranged, by comparing it with the task in another group, whether or not students worksheet and learning scenario which had been arranged is explorative. This stage can grow students ability in evaluating the process and learning outcome through question answer in group presentation setting.

The importance of learning which is began by the problem is also revealed by Conney which stated that teaching problem solving then students will become more analytical in making decision in their life, autonomous in solving the problem, because students will be skillful about how to collect relevant information, to analyze information and to realize how important to study again the result obtained. Besides, the problem which is given also play role to give good opportunity to broaden thinking because when students try to solve the problem in class, students construct their knowledge actively.

The following is learning which is implemented by lecturer and students in explorative learning through explorative module in concept of designing learning tool. There are some activities which are done by students in explorative learning through module. Students do some series of activities which are suited with explorative module that is in first activity in the form of students activity in distinguishing question and command. The second activity is examine the question of first activity (identify open ended, productive and imaginative question). In the third activity, students make higher order question based on worksheet given. In the forth activity, students examine the result of third activity. In fifth activity, students identify the component of worksheet given. Then it is continued by reflection stage which is guided by lecturer, and student draw conclusion.

Another finding show that the factor which make students' self regulated learning become good is learning process which is done cooperatively or in group. In its implementation, students can enhance their learning initiative because each member have same responsibility for the success of their group, seeing learning difference in each member so they are able to decide the right strategy for themselves. Students in group are triggered to be able to set the working target in group so they can design learning tools.

*indhira_1969@untirta.ac.id

Self regulated learning for mathematics pre-service teachers is important to be possessed, because by this self regulated learning they can develop their knowledge of pedagogical competence more deeply [10]. Self regulated learning possessed by teacher finally can create mathematics learning process which is meaningful for students [10].

4. CONCLUSION

Based on result of study, it can be concluded that students' self regulated learning after given treatment is categorized good. There is difference in the initial and final self regulated learning among students before and after treatment. It means that the treatment given give good contribution in developing students' self-regulated learning. This conclusion is seen from achievement of percentage of self-regulated learning in each aspect measured. Besides, based on result of average difference test, it shows that there is difference in self-regulated learning before and after treatment.

Based on result of study and the importance of students' self-regulated learning, then lecturer needs to develop learning activity which can enhance this aspect. Particularly for the aspect which had not been developed well, that is, in setting learning target, monitoring and controlling learning, utilizing and searching relevant supporting material, choosing and applying learning strategy and evaluating the process and learning outcome.

References

1. Sumarmo, Utari. (2004). (*Self-regulated learning: What, Why and How is it developed in Students.* (online). Available: Kemandirian-belajar-mat-des-06-new.pdf. (13 mei 2012).
2. Sukirwan. (2008). *Explorative Approach Activity to Enhance Mathematical Reasoning and Connection Ability of Elementary School Students:* Quantitative (Thesis in Postgraduate School, Indonesia University of Education).
3. Darr, C dan Fisher, J. (2004). *Self Regulated Learning in Mathematics Class.* (online). Available : www.arb.nzcer.org.nz. (16 Juni 2017).
4. Kramarski, B & Gutman, M. (2006). How can self-regulated learning be supported in mathematical E-learning environments?. *Journal of Computer Assisted Learning.* Online. Volume 22, Issue 1 February 2006 Pages 24–33. Tersedia : http://onlinelibrary.wiley.com/doi/10.1111/j.1365-2729.2006.00157.x/abstract. (15 Juli 2017).
5. Monks, F.J., dkk. (1999). Psychology of Development Introduction in Its Various Parts Yogyakarta: Gadjah Mada University.
6. Rohaeti, E. (2008). *Learning with Exploration Approach to Develop Mathematical Critical and Creative Thinking Ability of Junior High School Students* : Quantitative (Dissertation in Postgraduate School, Indonesia University of Education).
7. Depdiknas. (2003). *S Assessment Standard of Mathematics Textbook* . Jakarta: Pusat Perbukuan.
8. Shadiq, F. (2011).. *Mathematics exploration in Elementary School/MI: Its Example, Definition and Strenght (online). Available on* : *h*ttp://fadjarp3g.files.wordpress.com/2011/03/10eksplorasidisd_fasilitator_.pdf. (16 Juni 2017).
9. Arends, R. (2008). *Learning to Teach* . Yogyakarta: Translation, Pustaka Pelajar.
10. Kramarski, B & Michalsky, Tova. (2010). Preparing preservice teachers for self-regulated learning in the context of technological pedagogical content knowledge. *Journal Learning and Instruction.* Online.Volume 20, Issue 5, October 2010, Pages 434-447. Tersedia : http://dx.doi.org/10.1016/j.learninstruc.2009.05.003. (15 Juli 2017).

*indhira_1969@untirta.ac.id

Cooperatif Learning Models Simulation : From Abstract to Concrete

Ketut Agustini[1,*], and *I Putu Wisna Ariawan*[2]

[1] Teknologi Pembelajaran Pascasarjana, Universitas Pendidikan Ganesha, Jl. Udayana No. 11, Singaraja, Bali.
[2] Pendidikan Matematika, Universitas Pendidikan Ganesha, Jl. Udayana No. 11, Singaraja, Bali.

Abstract. This study aimed to develop a simulation of cooperative learning model that used students as prospective teachers in improving the quality of learning, especially for preparedness in the classroom of the microteaching learning. A wider range of outcomes can be used more widely by teachers and lecturers in order to improve the professionalism as educators. The method used is research and development (R&D), using Dick & Carey development model. To produce as expected, there are several steps that must be done through global research, among others, do steps (a) conduct in-depth theoretical study related to the simulation software that will be generated based on cooperative learning models to be developed , (b) formulate figure simulation software system is based on the results of theoretical study and (c) conduct a formative evaluation is done by content expert, design expert, and media expert to the validity of the simulation media, one to one student evaluation, small group evaluation and field trial evaluation. Simulation results showed that the Cooperative Learning Model can simulated three models by well. Student response through the simulation models is very positive by 60 % and 40% positive. The implication of this research result is that student of teacher candidate can apply cooperative learning model well when teaching real in training school hence student need to be given real simulation example how cooperative learning is implemented in class.

Keywords: Simulation, Cooperative learning Models, formative evaluation, student responses

Introduction

The recognition of teacher status is very high and this is followed by a requirement of a teacher has four major competences, namely pedagogical, personal, social and professional competencies including also the beginning of prospective teachers preparation. Professional competence is the ability that consists of: mastery of the instructional materials, understanding of curriculum and its development, classroom management, use of

* Corresponding author: eghee2006@gmail.com

strategies, media, and learning resources, has an insight into educational innovations, gives help and guidance to the students, etc. [1]

Teacher skills in managing classroom learning need to be improved by applying innovative learning models. Field findings show that not all teachers have implemented innovative learning models so that the learning process is less meaningful. This is also supported by the results of research [2] that ICT teachers tend not to utilize innovative learning models in classroom management so the class seems monotonous. The teacher"s role not just as the one who transfers knowledge to the student, but as mediator and facilitator in learning [3].

Not only ICT teachers but most teachers in the class rarely use the varied learning and multimedia, because the students' success in understanding the basic concepts dependent on the degree of effectiveness of a lecturer in transferring his or her knowledge to the students using multimedia facilities [4].

There are many innovative learning models that have been applied to the current curriculum that require more active learners. One of the learning model is cooperative learning model. The cooperative learning model is one of the lessons developed from the theory of constructivism because it develops the cognitive structure to build its own knowledge through rational thinking [5].

The existence of cooperative learning model in education is not a new one [6]. However, in fact, in the field, there are still many educators who have not really known clearly the correct learning steps on various types of cooperative learning model that exists. In addition, many cooperative learning models confuse educators in distinguishing syntax and are considered still abstract and difficult considering the main characteristics or points that became the concept of each model of cooperative learning.

This is a challenge for the Instructional designer to provide intelligent solutions for the world of education, by visualizing an abstract into a concrete through the development of digital simulations, models of cooperative learning. By concretely it is something abstract that allows students to understand a concept and learning becomes more effective [7,8]. The specific purpose of this study is to develop a simulation system software that can be used to visualize cooperative learning models so as to help facilitate the understanding of educators, especially prospective teachers by looking, listening and certainly more easily understand the techniques and the main points that become Concepts of the various types of cooperative learning models available.

METHOD

The method used is research and development (R&D), using Dick & Carey development model. Software products generated from the development process are validated and tested through two testing stages: (1) Product testing techniques as a software, through SDLC standards with Waterfall-based Model, (2)Testing the product as an instructional system development, through Formative evaluation [9,10].

The development of media begins with the need analysis for three dimensions, namely (1) the characteristics of cooperative learning model identify (2) choose the cooperative learning model that will be visualized; And (3) the syntax of the chosen learning model analysis. This needs analysis takes place continuously following the cycle, because the media will experience adjustments to change, needs, and trends. Then the fourth stage is the development design through standard SDLC with Waterfall Model.

Technical tests include testing software functionality performed in black box testing. Black box testing is the fundamental aspects testing of the system regardless of the internal logic structure of the software. This method is used to determine if the software is working properly. Black box testing is a method of designing test data based on software specification. Test data is generated, executed on the software and then the output of the software is checked whether it has been as expected. Software testing in this study concerns the following:

1. Step-by-step simulation (syntax) in each model;
2. Making figures in the simulation;
3. Description that corresponds to the model step;
4. Displaying simulated videos in the class

Meanwhile, the component of formative evaluation is done by content experts, media experts and design learning experts including the feasibility of content, language, appearance, and graph.

The revisions were made based on the evaluation results of content experts, media experts, and design learning experts. A limited trial was conducted on a small group of students to get a response to the simulation program. Data analysis for student response using questionnaires with Likert scale then analyzed descriptively qualitative.

RESULT AND DISSCUSION

The software functional model provides an overview of the processes that occur in the software without providing details on how these processes are implemented. Functional models also provide an overview of the flow of data that occurs between existing processes and between processes with external entities, such as software users. The flow of data will define the input and output contained in each process that occurs, so the relationship between processes can be seen clearly. The functional model used to describe this software is the Data Flow Diagram (DFD).

The context diagram is the level 0 DFD of the software. In the context, the diagram can be seen the interaction that occurs between users and software. There are three user input menu data required by the software. Next, from level 0 DFD elaborated to level 1 DFD, which explains in more detail the user input menus available in the system software.The design of the developed software interface screen, which consists of three choices of learning model menu, and other options menu, the description, and visualization area of the model in the form of simulation as well as the title and image area. Figure 1 shows the screen interface developed in the simulation model.

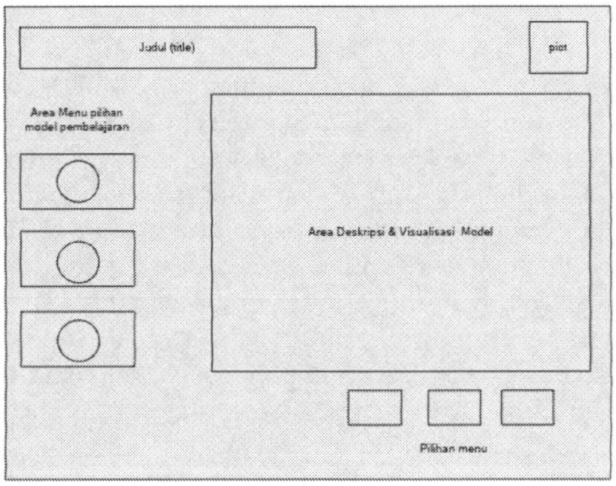

Figure 1. The Software Interface Screen Design.

 https://doi.org/10.1051/shsconf/20184200107

The software interface screen design is implemented using forms in the Macromedia Flash 8 implementation environment. Implementation of the main screen design of this software interface successively can be seen in the Fig. 2, Fig. 3, Fig 5 and Fig. 6 below.

Figure 2. The Software Interface Screen Implementation.

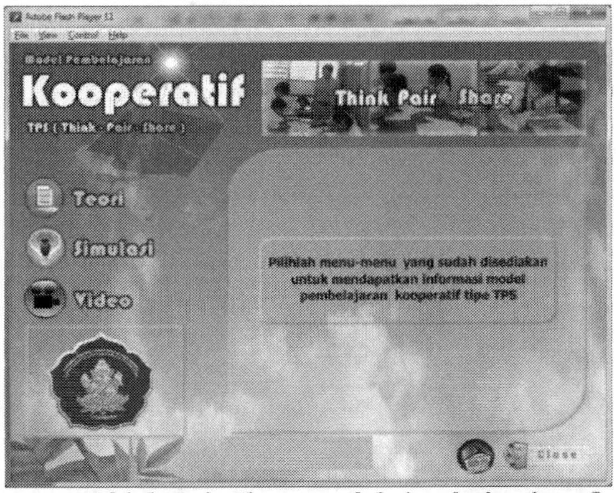

Figure 3. Optional menu on Think Pair Share model simulation interface.

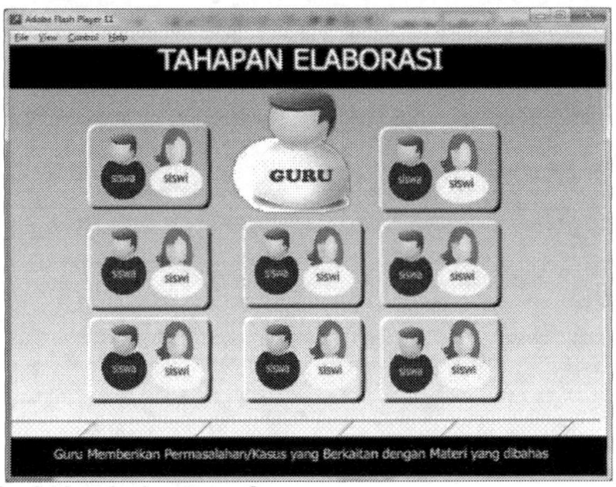

Figure 4. Step model of Simulation Interface.

Figure 5. Video TPS Model Simulation Interface.

User experience is measured after using valid media on students who have already taken real PPL courses. Questionnaire responses consist of 5 points of Likert scale 1-5, so ideal score and ideal lowest score respectively is 25 and 5. The student response rate is 20,50. In general, it shows that the student's response is very positive to the simulation media of cooperative learning model. In detail, it can be seen in the table that in total 60% of the students responded very positively and 40% responded positively to the existence of simulation media cooperative learning model in micro teaching lectures, and no one responded negatively. If visualized in the form of a histogram like a Fig. 6.

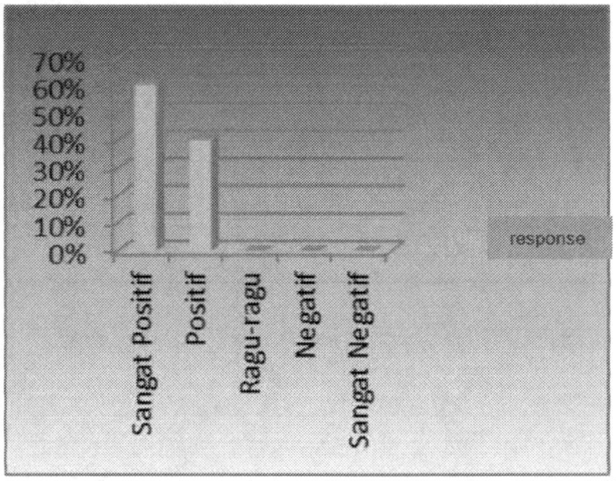

Figure 6. The Histogram of User Experience through Simulation media.

In the development of this simulation media, the accuracy of the use of design is very decisive design. In addition, the type and size of letters and color combinations are also important because a simulation media developed with flash requires a high imagination and ability to analyze to visualize something abstract to be concrete and interesting so that students interested and motivated to learn the material. During the development process has been through several revisions based on inputs provided by content experts, media and instructional design.

With a very positive response from students to the simulation media is expected to be able to improve the professionalism of prospective teachers in preparing before teaching to the real class. The result of this study are consistent with the result of research conducted by the [11] that students provided positive feedback regarding this educational innovation. This year, the development research that has been done is only done for the simulation process of cooperative learning model and limited trial to know the student's response to the developed software. The real experimental stage to further test the effectiveness of the use of simulation software of this cooperative learning model on the learning outcomes in PPL real will be done at the advanced stage of this research.

CONCLUSION

The Cooperative Learning Model Simulation is designed using Data Flow Diagrams (DFD) and implemented using flash 8 and can simulate three models of cooperative learning well, and student response to the development of model simulation Cooperative learning is very positive at 60% and positive by 40%.

Suggestions for further research are: (1) It is necessary to develop more simulation of instructional media based on cooperative learning model to assist students who have abstraction ability that is less in abstract lecture materials, both educational and non-educational field especially in informatics; (2) The utilization of ICT in the learning process should still be implemented within the framework of learning objectives in accordance with the indicators of achievement, with no cause of learning conditions that tend to be considered like playing with the media itself; (3) To know how much influence or effectiveness simulation system that has been developed toward student achievement, it is necessary that the result of this research is followed up by doing further research through experimental research or action research.

ACKNOWLEDGEMENTS

This work was supported by the Daftar Isian Pelaksanaan Anggaran (DIPA) Universitas Pendidikan Ganesha.

References

1 I P. Wisna Ariawan, D.B. Sanjaya, and D.G.H. Divayana, IJARAI, **Vol. 5 No.2** (2016).
2 Agustini. K, *Proceeding,APTEKINDO, ISSN No. 1907-2066, Denpasar* (2010).
3 I M. Ardana, I P.Wisna Ariawan, D.G. Hendra Divayana, IJEME, **Vol. 7 No. 4** (2017).
4 D.G. Hendra Divayana, I M. Ardana, I P. Wisna Ariawan, JATIT, **Vol.95. No. 9** (2017).
5 Rustaman, N., D, S., Yudianto, S.A., Achmad, Y., Subekti, R., Rochintaniawati, D., & Nurjhani, M.. *Common Text Book Strategi Belajar mengajar Biologi*. Edisi Revisi., Bandung: JICA-IMSTEP-UPI (2003).
6 Ina Karlina, *Pembelajaran Kooperatif (Cooperative Learning) sebagai Salah Satu Strategi Membangun Pengetahuan Siswa* (2010).
7 Agustini K., *Proceeding SNATIKA* **Vol 01** (2011).
8 Esther Zaretsky, SCI, **Vol. 4 No. 4** (2006).
9 Dick W, Carey & James O Carey. *The Systematic Design Of Instruction,* New Jersey: Pearson (2009).
10 Santyasa I Wayan. *"Metode Penelitian Tindakan Kelas, Pengembangan, Korelasional, Kausal Komparatif, dan Eksperimen"*, Lembaga Penelitian Universitas Pendidikan Ganesha, Singaraja (2009).
11 M.L. Cason, et al., J Nurs Educ, **Vol. 54 No. 3** (2015).

The Use of Decorative Styles of Woven *Songket* as a Source of Learning in the Course Textile Design and Decoration

I Dewa Ayu Made Budhyani[1,*]

1Department of Primary School Teacher Education, 81111, Singaraja, Indonesia

Abstract. The use of learning sources in the learning process greatly contributes to the achievement of its goal. The use of learning sources greatly affects the teaching and learning process. It is necessary to use different learning sources when creating a textile design in order to produce a quality product. The decorative style of woven *songket* can be used as one of the learning sources. The different decorative types of the woven *songket* contain aesthetic values. The decorative type of woven *songket* is a product of the crafter's feeling, creativity, and intention inspired by the environment. The natural objects such as plants, animals, human beings, and geometric elements are stylized into the decorative forms. The decorative style used on the woven *songket* can be used as the learning source in order to create the decorative style in the course Textile Design and Decoration. There are several aesthetic elements which inspire where the decorative style is placed; they are composition, rhythm, balance, the harmonious layout of the decorative style, and the variation of the decorative style on the textile design. In general, the composition of the placement of the main object of the decorative style is more dominant on the piece of fabric. Rhythm and how the decorative style is composed are shown through the composition of the decorative motive forms such as how big and small they are, how high and low they are and how long and short they are. The ways in which the decorative motives and colors are entirely shown contribute to the harmony and placement of decorative style. The symmetric balance is used to create balance and avoid the impression of being biased. The placement of what the motives contain and how they are composed are used to determine the variation of the decorative style used. The placement of objects and the marginal decoration of the fabric also determine the variation used.

1 Introduction

Indonesia has different types of people's industries, one of which is the woven *songket* industry. It is a cultural heritage which has been developing from generation to generation. The most conspicuous thing in the woven *songket* is the form of its decorative style. The balinese people are well known for their creativity in developing different art industrial

[*] Corresponding author: ayubudhyani@yahoo.com

works. They can be found in different villages and towns with their different art industries. Many develop bamboo handicraft, weaving handicraft, rattan handicraft, wood handicraft, and many others develop other handicrafts with their respective specialty (ardika, 1996: 6).

The decorative style is a composition of lines, forms, colors and figures which are created to contain the values of beauty based on the development of imagination (hery suhersono, 2004: 5). The decorative style is the form of decorative background which is usually used as the repeated pattern of craft works. The decorative style reflects the beauty and aesthetic values causing the art works to be highly interesting and have quality.

The decorative style is based on the human knowledge of the environment which can inspire the creation of the diverse patterns of decoration. The natural objects which are translated into the forms of decorative style are plants, animals, human beings, natural elements, religious values. They are transformed into harmonious beauties (gelebet, 1982: 331). Such objects are transformed in such a way that different decorative styles can be created.

The decorative style on the *songket* woven fabric functions to decorate space which cannot be implicitly separated from the aspects of beauty. As an illustration the decorative style is used to enhance the beauty of a piece of fabric, causing it to look better and more interesting. As a consequence, it is materially and spiritually appreciated (gustmi sp. 1980). The forms of decoration, how they are colored, the way of making and placing them have particular meaning and intention. Decorations are formed in the form of patterns which allow them to be placed in several particular spaces of the elements needing decorating. The cultural factor was the initial factor which had inspired the woven *songket* crafters to produce traditional and modern motives. As we know that the balinese people have been well known for their culture, customs and traditions. They are also well known for their fidelity to hinduism (koentjaraningrat, 1985).

The weaving technique, locally referred to as 'teknik pakan tambahan', is used to create the decorative style of the woven *songket*. The process of making it is that a pinch of lint thread is taken before the golden or silver thread is inserted (cut karmail wardani, 2005:177). The way of raising the lint thread is organized with what is referred to as 'lidi-lidi' (a type of palm leaf rib). The more *lidis* are used the more complicated and richer the decorative style will be.

The decorative style is made up of different types of motives used to decorate an area or object. One, two or more motives are used to decorate an area. Some motives are repeated and some others are stylized. The motives are placed and composed of the elements which can support the beauty of the woven *songket*. The elements include form and structure, harmony, composition, rhythm or motion, balance, and so forth.

The types and forms of the decorative style placed/applied to the fabric of the woven *songket* can be used as the learning source to inspire the development of the source of ideas needed to create the designs of new decorative style. The learning sources include all the sources within and outside someone which allow (ease) the learning process to take place (ahmad rohani, 1997: 102). The educator is not the only source of teaching and learning, although his/her responsibility, role, and function in the teaching and learning process are highly important. The learning sources can be acquired from the learning environment which functions to maximize the learning outcome. Aect describes that the learning sources include: massage, human being, material, tool, technique, and environment (ahmad rohani, 1997: 108). Based on what was described on the decorative style above, the problem of the study can be formulated in the form of a question, namely how the decorative style of the woven *songket* can be used as a learning source viewed from its composition, rhythm, balance, the harmonious layout, and its varieties on the textile design and decoration.

2 Method

This present study is a qualitative one in which the aesthetic approach was used. The object of the study is the decorative style of the woven *songket* as a learning source viewed from its composition, rhythm, balance and harmony. The data were obtained from the woven *songket* crafters. The data were collected through observation, interview, and documentation. All the data obtained from the field are related to composition, rhythm, balance, harmonious layout of the decorative style and the varieties of decorative style. The data were analyzed descriptively and qualitatively.

3 Results and Discussion

Based on the result of the study it can be stated that the elements used as the learning source in the placement of motives on the woven *songket* can be discussed as follows. The composition of the placement of the decorative style of the woven *songket* is adjusted to the type of the decorative style applied to the woven *songket* fabric. In general, not only one type of decorative style which is placed; however, the dominant decorative style is used as the main one, although there are several decorative styles used to complete the area of the woven *songket*. As an illustration, the composition of the placement of the decorative motive of hibiscus; there is space between one and another, causing empty space to be created in between. As a whole, all the decorative motives are arranged in such a way that they all have aesthetic values with respect to form, decorative motive and the type of color used to make the decorative motive on the woven *songket* fabric to vary. In addition to the composition of the main motive placed in the middle of the woven *songket* fabric, the supplementary motive is placed on the marginal part of the woven *songket* fabric. The placement of the decorative style is dependent on the goal and the size of the decorated area. The decorative style can be in the forms of the marginal ornamental style, cornering, regular, composition and repetition (Eko Purnomo, 2013). In general, not only one decorative style which is placed; however, the dominant decorative style is the main one, and the other decorative styles are used to complement the *songket* woven fabric. They can be in the geometric form, stylized plants, and animals.

The rhythm of the decorative style of the woven *songket* is repeatedly and regularly presented. The repeated composition of the decorative style is dependent on the type applied. What is meant by composition in this present study is that the same motives are repeatedly composed; however, they are not monotonously composed; as a result, there seems to be rhythm which can be presented by organizing the form of the decorative motive; it can be big, small, high, low, long, short, and by repeatedly composing different colors. The rhythmic motions can be made by repeating form, changing size, and moving the unbroken line (Diah Angendari, et al., 2014:9). The decorative style which is harmoniously placed can be seen from how the motives are composed or how the colors are presented as a whole on the woven *sangket* fabric. The presented threads used for making the woven fabric, its motives and colors as a whole can be seen from the harmonious aesthetic values. As the motives are placed harmoniously enough the colors composed on several motives with their complementary colors do not contradict one another. The balanced placement of the decorative style on the woven *songket* can be seen from the symmetric balance, which is the composition which is not seriously risky for the reason that it will not lead to the impression of being biased. The decorative motives similarly, repeatedly and entirely placed on the area of the fabric contribute to harmony. The decorative style on the woven *songket* fabric is placed in accordance with the decorative motive made. The reason is that each decorative motive has its own pattern. As an illustration, the pattern of the decorative motive *tirtanadi* is the same regardless of whoever makes it. It is made based on the source and is used as the decorative motive of the woven *songket* fabric. However, the complementary motive composed by the crafter cause the decorative motive to vary. The placement of the objects and the placement of the marginal decoration are also made to vary. When placing the decorative style on the woven *songket*, there are three components which need to be paid attention to; they are a figure as the main decoration, the supporting figures used to support the basic motive, and the complementary ones used to enrich beauty as a whole (Sila, 2013). Where the decorative style is placed is dependent on the goal and the area which is intended to be decorated.

The diversity of motives and the placement of the decorative style on the area of the woven *songket* fabric can be used as the learning source when making the pattern of the textile design and decoration. The knowledge of different types of natural and artificial materials can be used as the learning media (Sumanto et al, 2015). The sources of knowledge from the natural materials can be flora, fauna, geometric and figurative things.

4 Conclusions and Suggestion

The aspects of the decorative style of the woven *songket* which can be used as the learning sources can be concluded as follows: (1) the composition; in general, not only one decorative style is placed on the woven *songket*; however, the dominant decorative style is the main one; and the decorative style is entirely placed on the piece of the fabric. (2) Rhythm; the composition of the decorative style of the woven *songket* is regularly repeatedly presented. The repetition of the composition of the decorative motive depends on the type applied. In this present study, composition means the repeated composition of the same motive; however, the compositionis not monotonous; as a result, there seems to be rhythm. The rhythm which can be presented through the composition of the form of the decorative motive includes the size (big, small), height (high, low), length (long, short), and the repeated composition of different colors. (3) Harmony; the placement of the decorative style of the woven *songket* can be seen from how the decorative motive is composed and how colors are presented on the woven *songket* fabric as a whole. The motives are harmoniously placed enough; several motives are made with complementary colors and the colors used do not contradict one another. (4) Balance; the decorative style of the woven *songket* is placed in such a way that it is symmetrically balanced. The symmetric balance is the composition which is not seriously risky, as it will not give any impression of being biased. In general, the decorative style of the *Jinengdalem* woven *songket* is symmetric for the reason that the decorative style is repeatedly placed on the piece of fabric as a whole. (5) The decorative style of the woven *songket* is placed in accordance with the decorative motive made. The reason is that each decorative motive has a pattern. As an illustration, the decorative motive *tirtanadi* will be the same, whoever makes it. The pattern will be the same and follow the source as the decorative motive of the woven *songket* fabric of *Jinengdalem*. How the crafters compose the complementary motives vary. The placement of objects also varies and so does the placement of the marginal decoration.

References

1. Ahmad Rohani. Media Instruksional Edukatif. Jakarta:Rineka Cipta. (1997).
2. Angendari, Made Diah,dkk. Desain dan Dekorasi Tekstil. Yogyakarta:Graha Ilmu.(2014)
3. Ardika, I wayan. Dinamika kebudayaan Bali. Denpasar: Upada Sastra. (1996).
4. Eko Purnomo, dkk. "Seni Budaya". Jakarta: Kementrian Pendidikan dan Kebudayaan Republik Indonesia. (2013).

5. Glebet, I Nyoman. Arsitektur Tradisional Daerah Bali.Departemen Pendidikan dan Kebudayaan. (1982).
6. Gustami, SP. Pembelajaran Sains Terkini Mendekatkan Siswa dengan Lingkungan Alamiah dan Sosial Budaya. Singaraja: Universitas Pendidikan Ganesha. (2009).
7. Koentjaraningrat. Kebudayaan Jawa. Jakarta : Balai Pustaka Nasution, S. 1988. Metode Penelitian Naturalistik Kualitatif. Bandung : Tersito. (1984)
8. Sanjaya, Wina. Ragam Hias. Denpasar : Departemen Pendidikan dan Kebudayaan.(2006)
9. Sila, I Nyoman dan I Dewa Ayu Made Budhyani. Kajian Estetika Ragam Hias Tenun Songket Jinangdalem Buleleng. *Jurnal Ilmu Sosial dan Humaniora*. **Vol.2 No.1**. 159-165. (2013)
10. Soekarto dan Winardi. Menejemen Pemasaran. Jakarta : Erlangga. (1991).
11. Suhersono, Hery. Desain Bordir Motif Geometris. PT Gremedia Pustaka Utama.(2005).
12. Sujipta. Perubahan Sosial Masyarakat Dalam Hubungannya dengan masyarakat dan pariwisata Bali. Universitas Udanyana, Bali. (1989)
13. Sumanto, dkk. Kerajinan Tangan Di Blitar Sebagai Sumber Belajar Seni Budaya dan Prakarya (SBdP Sekolah Dasar). *Jurnal Sekolah Dasar*. **Tahun 24 Nomor 2**. 111-123. (2015).
14. Susanto, Mikke. Diksi Rupa. Yogyakarta : Penerbit Kanisius (Anggota IKAPI) (2002).
15. Wardani, Cut Karmail dan Ratna Pangabean. Tekstil. Lembaga Seni Nusantara. (2005).
16. Winardi. Menejemen Pemasaran. Jakarta : Erlangga. (1991).
17. Arif Syamsudin. "Antara Pelestarian dan Perlindungan Ekspresi Budaya Tradisional/Pengetahuan Tradisional dan Perlindungan Hak Kekayaan Intelektual". *Media HKI*. **Volume V, No 4**. (2008).

LITERATION MOVEMENT FROM TANAH OMBAK'S COMMUNITY

EkaVidya Putra

Department of Sociology, State University of Padang

Ekavidyaputra@gmail.com

Abstract

Nowadays, Character education becomes the attention of many parties. The key to character education is the availability of a positive character environment. Because character building is not enough just to provide knowledge about which is good attitude and what is bad as taught in school. This misunderstanding of school is the cause of character education and failure. Intended for character building requires a character environment that provides good examples to its members. The problem is that there are many character environments that do not support the building of good characters. Bad character environments can be seen in slum area. Slums, identical to poverty, low levels of education, many criminal acts, promiscuity and other negative behaviors. By using a new institutional approach . Data collection is done by qualitative method, through observation, interview and literature study. The success of the Tanah Ombak Community manipulates the institutional environment can not be separated from two things. First, there are actor who become key figures. Second, there in collectively shared values that drives ideas. How the environment is manipulated can be seen from the three aspects, regulative, normative and cognitive. Regulatively, many emergence of positive rules for character development. Normatively, bad old habits are replaced by positive new habits for character building. Finally, cognitively new knowledge emerges for the new characters building.

Keywords: character environment, manipulation, Komunitas Tanah Ombak

I. INTRODUCTION

Characters associated with mental qualities and unique behavior of one's person. Character is not genetic, but the role model results of the surrounding environment. A good environment could form a positive character, otherwise a bad environment forms a negative character. To be one whose has a good character, it needs a conducive environment, healthy and positive. The environment affects individuals through the interaction's process. The five senses become an important component when one interacts and socializes with the environment. Experience interact with this environment which then affects a person's character.

New Institutional approach besides distinguished on the environment's side. but also on individual and groups as an actor. Actors in the new institutional approach, more free and independent. Besides it, Individuals or groups also have the ability to manipulate the environment. It means that, the environment could affect the behavior, but the actor also has the ability to respond to pressures coming from the environment and at the same time having the ability to influence the environment. The most important thing here is how the environment is produced and reproduced by actors both as individual or collective. How individuals put themselves in social relations, interpret and respond to their institutional context (Walter W. Powell and Patricia Bromley; 2013). The environment is not seen as a fixed and stable reality, but as the construction of individual minds and their actions. Who will manipulate the environment? In order to manipulate the environment, we need creative individuals who have social positions and skills enable them to recognize problems or opportunities and make use of enabling conditions to change the social order (Battilana, Leca and Boxenbaum, 2009). Except to creative's individuals, manipulation can also be done by the authorities (Douglass C. North; 2013).

Lauree Beth Stedje (2010) notes the success of character education never woks itself. School is one of the environment that educuating as long as students are in school environment. In the school environment, teachers are models. But, outside there is a playing environment and family environment. In the family environment, there are family members, especially father and mother who being a model. While the model's playing environment is the community. The playing environment and the larger family environment influence the formation of a person's character, because the interaction takes place longer. During this time many research and attentions was focused on the formation of character in school and family environment. Whereas how the playing environment forms one's characters is rarely as a concern.

However, Not all playing environments are conducive to form the positive characters. In some places, the playing environment provides a negative experience for the formation of one's character. For example in slums. This paper will make a study of how actors - both in individual and collective terms - manipulate the playing environment in slums both in physical and non-physical terms into positive character environments for community, especially children. This paper will take a case in the slums on the coast of Padang.

II. METHODOLOGY

The study was conducted with qualitative approach. Data collection was done through participant observation, interviewing with a number of key actors and collecting documentation data. The study will be conducted on Tanah Ombak's community.

III. RESEARCH FINDINGS

A. PROFILE OF *PASIA* CHILDREN" IN THE VICIOUS GANG

Pasia children literally can be interpreted as a child who lives in the coast. But the term of *Pasia* children" in the local sense is sometimes interpreted not in the same way as literal. *Pasia* Children" has a negative meaning, because there is a stigma in which is attached to a *Pasia* Children. Stigma of *Pasia* Children is close to misbehavior, slums, naughty, and rough talking. While Gang Satan, showing the location, a gang is located at Jalan Purus III, District Padang Barat of Padang, West Sumatera Province. Gang Satan is not the real name of the alley. Gang Satan is a name attached by outsiders to the alley.

Gang Satan is a slum with bad condition and unhealthy living standards. From the economic standards, Gang Satan is a poor area of the city. While from social, cultural and educational, Gang Satan is not feasible for the growth of positive character. When we go to Gang of Satan, we will be familiar hearing harsh words and rough talking *(bacaruik)*. Drinking liquor, narcotics, gambling, motorcycle thieves, are familiar scenery in Gang Satan. The above conditions correlate with low levels of their community education, in which most of them only graduating from high school. The similar condition are also found in the family environment, as scolded, harsh speech, intimidating such a model in educatind the children. Profile of *Pasia* children in Gang Satan, can not be separated from character environment. The character environment provides negative information and then was replicated. Children through their senses, recording and capturing the messages given by their environment.

B. TANAH OMBAK'S COMMUNITY (*Pasia* Children)

Tanah Ombak's Community is an organization which present at Satan Gang and make efforts to manipulate the environment. The institutional community of Tanah Ombak can be described as follows:

First,* the Core *(Core Value) Although not mentioned in a sentence, but what is revealed in the above can be equivalent with the value of the core *(core value)* and objectives core *(core purpose)* of the organization. The core value is the essential lesson and make the organization dynamic. The core value attached to the organization, in this case core inherent in the Tanah Ombak's Community 1 . In this case, the core values are in the two key actors of Tanah Ombak's Founder: Syuhendri Dt. Siri Marajo 2 and Yusrizal KW 3 . Both actors have experience in assisting communities. Both have also worked by founding many communities before.

***Second,* Organizational Structure of Tanah Ombak is simple and sleek.** Formally, organizational structure of Tanah Ombak can be classified simply, consists of two groups, namely founder and board. So there is not much bureaucracy in decision making. The lean organizational structure aims to be more effective for the movement of the Tanah ombak. In this case, the community of Tanah Ombak prefers the functional to the structural position. The main goal of Tanah Ombak's community is the achievement goal in community development.

***Third,* participatory local community.** The management personnel of Tanah Ombak is the Gang Satan community and the community volunteers. Local community participation becomes the solution to the closed structure of Gang Satan society. Participation becomes important for inclusive character education.

***Fourth,* accountability and transparency of financial management.** Financial management is vulnerable parts in organizational management. The source of funding of Tanah Ombak mostly comes from donors. Donors and the amount of money funded is not fixed. Other sources of finance come from the program. Unfortunately, the community does not have a financial SOP yet. Monitoring the use of funds, monitoring the use of funds rests on the principle of transparency and accountability among community members and communities. Because personnel, the relationship between staff is very close and emotionally bound. In addition, the lean organizational structure also facilitates the control function in finance.

***Fifth, Branding* Strategy.** The interesting strategy of Tanah Ombak is in publication. They do not use the conventional media, but through strategy of *branding* that is social media. Through the power of *branding strategy,* At the end of 2016, Tanah Ombak to be the frst regional champion in the Gramedia Reading Community of Sumatra Competition. This accomplishment was followed by many book as prizes from Gramedia. In the same year the Tanah Ombak attract Indonesian Reading Ambassador, Najwa Shihab, to officially release Tanah Ombak"s Vespa4 . In early 2017, KW and Syuhendri were invited to the palace by President Jokowi to receive the award as a literacy activist.

C. MANIPULATING THE PLAYING ENVIRONMENT

As what is in the new institutional approach, actors in terms of individual or collective has the ability to manipulate the environment. The kind of manipulation which is done as follows:

First, Setting up the reading garden. Proverb states *"The book is like an ice breaking ax that could melt the frozen sea that stretches in our souls".* The important thing here is how to attract children to read. The strategy is making reading a book as a *password* to participate in other activities. Like eating together, painting, story telling, singing and more. Until finally reading becomes a habit of children.

***Second,* inviting a lot of good people to Tanah Ombak.** One of the problems in the early years of Tanah Ombak was the exclusive society. Inviting many good people will be a direct example for society. Simultaneously it shows that many people are good and also care about their environment. At last, they encourages desire to imitate and improve themselves and also the environment.

***Third,* keep creativity, innovative and participatory society.** There are three basic concepts that become the key word in implementing the program in Tanah Ombak, that is creativity, innovative and participative. Creativity is the ability to think, develop something new and different in solving problems and finding opportunities.

***Fourth,* open public space for children.** Children are introduced into the public sphere so that could foster their self-confidence. As a first step, they perform in Tanah Ombak and witnessed by Tanah Ombak's community, then watched by outsiders and finally they have a stages outside Tanah Ombak. Even The children of Tanah Ombak are now participating in a number of performances by outside organizers.

Fifth, *Tumpang Sari* Activities. All activities of Tanah Ombak's children titled literation, but in the implementation of activities always directed by efforts to influence the character environment. For example, in instilling courtesy values-children are taught to speak politely not to get dirty words-to get used to greeting every visitor who visit Tanah Ombak, to say thank every one of them to get something. Another example is to instill responsible values. In this case the children always do thoroughly the work given, work without having to wait for orders, and so forth. Thus, character formation does not necessarily have to provide character education in a theoretical sense but rather to the involvement of children in the process. So that process provides a direct experience that characterizes the child.

IV CONCLUSION

Overall the five ways above are effective for manipulating the environment, both the playing environment and the family environment. Effectiveness can be seen from the shift that occurs in the regulation, values and norms and also cognitive of the community. In this case the existence of creative actors became the key to the success of the Tanah Ombak's Community in manipulating the environment. Creative actors succeed in creating shared values and open up more inclusive public spaces

LIST OF LIBRARY

Victor Nee (2003), *New Institutional, Economic and Sociological.* Princeton: Princeton University Press.

Victor Nee (2003), *Sources of the New Institutionalis.* Https://www.russellsage.org

Walter W. Powell & Patricia Bromley (2013) *New institutionalism in the Analysis of Complex Organizations.* Http://www.patriciabromley.com/PowellBromleyEncyclopedia.pdf

Douglass C.North (2013). *Institutional Change: A Framework of Analysis.* *Https://www.iei.liu.se/nek/730A22/filarkiv-2013/del1a.pdf*

Lauree Beth *Stedje,* (2010). *Nuts and Bolts of Character Education.* Character First 13800 N. Benson Road, Suite 206 Edmond, Oklahoma 73013.

Jacques S. Benninga (2003). *The Relationship of Character Education Implementation and Academic Achievenent in Elementary Schools.* Journal of Research in Character Education Vol. 1, No. 1, 2003

William G. Thompson, (2002). *The Effects of Character Education on Student Behavio.* Electronic Theses and Dissertations. East Tennessee State University.

1 Name of the Land of Waves taken from the realities of life of people who live dipingir beach. Tanah Ombak can be understood as the strength of the coastal community in stifling the hard life. Same as the shore survives the strong waves.

2 Syuherdri Dt. Siri Marajo is a civil servant working in Taman Budaya Kota Padang. Graduated from S2 at the Department of Theater at ISI Yogyakarta.

3 Yusrizal KW is a poet and an activist for literacy drive in West Sumatra.

4 Vesva Land Reader Wave is looking for a library moves the reader by driving a Vespa.

DEVELOPMENT OF PRODUCTION UNIT MODEL BASED ON SECONDARY HIGH SCHOOL SCHOOLPRNENEURSHIP

Galeh Nur Indriatno Putra Pratama[1], *Moch. Bruri Triyono*[2]
[1,2] *Faculty of Engineering, Yogyakarta State University, Indonesia*
galeh@uny.ac.id, bruritriyono@uny.ac.id

Abstract. *This study aims to determine: (1) weakness of vocational high school production unit; (2) model of production unit based on schoolpreneurship of vocational high school; and (3) the effectiveness of production unit model based on vocational high school schoolpreneurship. This research uses a R & D approach consisting of three stages: preliminary study, model development and model testing. The sample of this research is the production unit owned by Public Vocational High School (SMKN) 1 Depok, Public Vocational High School (SMKN) 2 Depok, Vocational High School (SMK) Muhammadiyah 3 Yogyakarta, and SMK Muhammadiyah Berbah. Data were collected using in-depth interviews and documentation. Result of research indicate that: (1) weakness of production unit so far is planning still monoton, job descriptions overlap, supervision not running maximal; (2) development of management model of vocational high school production unit as source of learning and alternative funding; (3) the test results indicate that 83.3% agreed that the model component is complete; 91.7% agreed that the structure of the model component was clear; 91.7% agreed that the relationship between model components is clear; 80% agreed that the model was feasible; and 82% agreed that the model is effective if implemented.*
Keywords: schoolpreneurship, vocational high school, production unit.

Introduction

Indonesia is required to be able to compete with other countries in terms of products, services, and the preparation of human resources (HR), which is in line with the objectives of education in XXI century. Efforts to obtain qualified and competitive human resources need to be supported by a national education and training system developed based on the needs of the job market and the dynamics of the accelerated changes occurred in the business and the industrial world. In the context of current developments and changes in Indonesian society, the nation's competitiveness depends on the knowledge and skills of its workforce and producing a knowledgeable and skilled workforce is depend on the quality of education and training, especially on vocational education. Trained and skilled workers will surely be able to increase the added value of products created through characteristics of increased productivity, reduced production costs, high-quality results, and a relatively faster rate of return. In addition, vocational education institutions are required to play their role and ability as an institution which is capable of "supplying" human resources to the needs of the community.

Research Methods

This research uses R & D method, Borg & Gall (1983: 775) argues, research development R & D includes 10 stages: (1) research and information collecting, (2) planning, (3) develop preliminary form of product, (4) preliminary field testing, (5) main product revision, (6) main field testing, (7) operational product revision, (8) operational field testing, (9) final product revision, and (10) dissemination and implementation. The sample of this research is the production unit owned by SMKN 1 Depok, SMKN 2 Depok, SMK Muhammadiyah 3 Yogyakarta, and SMK Muhammadiyah Berbah. The data is collected using in-depth interviews and documentation. Data in this research is analyzed with qualitative technique, so that the data is presented in verbal word form, not in numerical form. In order to make a more meaningful and easy to understand data presentation, the data analysis used in this research is interactive analysis model from Miles and Huberman (1994: 23) which divide the activity of analysis into several parts, including: data collection, data reduction, data presentation, and conclusion.

Result and Discussion

Based on the observation in SMK, some weaknesses are found in production units in general that not all schools have adequate building facilities to be used as production units. The production unit planning is still monotonous. This means that the production unit is only seen for the welfare of the citizens and a place for the practice of students when there is no industry to accommodate. The organization is still overlapping. Many schools have members of production units organization who are still hold a position on other organizations in the school. Thus, the school members are not focused on work. In addition, schools members of the organization in some production units are not experts in the field of competence. This resulted in a lack of production units development because the manager is not competent in their field, the organization of production units in some schools is only a formality for the purposes of school accreditation so that there is no real form of organizational structure.

Implementation of learning in the production unit is not effective in some schools. This happens because the productive subject teacher can not focus on teaching students in the classroom and in the production unit. Nowadays, the lack of teachers who have high loyalty to the management of production units is also a disadvantage. This resulted in the lack of cohesiveness between the personal. Tight effective learning time has resulted in less optimum output of study when students are invited to production units.

Development of UPSMK (Production Unit of Vocational School) management model is a management model of Production Unit that can be done by SMK so that it can maximize the potential of students, teachers, and schools to produce ready to work graduates and income generating. The development of SMK production unit management model has been tested qualitatively and the results show that this model is clear, practical and appropriate for UPSMK development. The explanation and characteristics of Management Model Development of SMK Production Unit are:

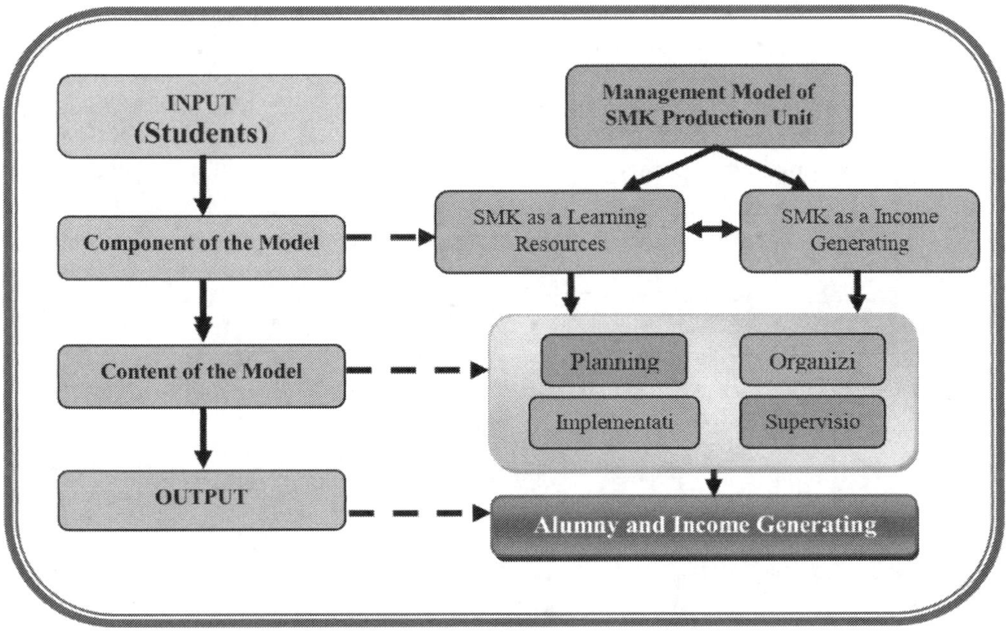

Fig 1. The Management Model of SMK Production Unit

The above pattern can be explained that producing ready to work graduates and increasing income of SMK for improving welfare of its citizen can be done through variable of Production Unit as learning source and Production Unit as funding source. Each stage of the variable will through four stages of management including planning, organizing, implementation and supervision.

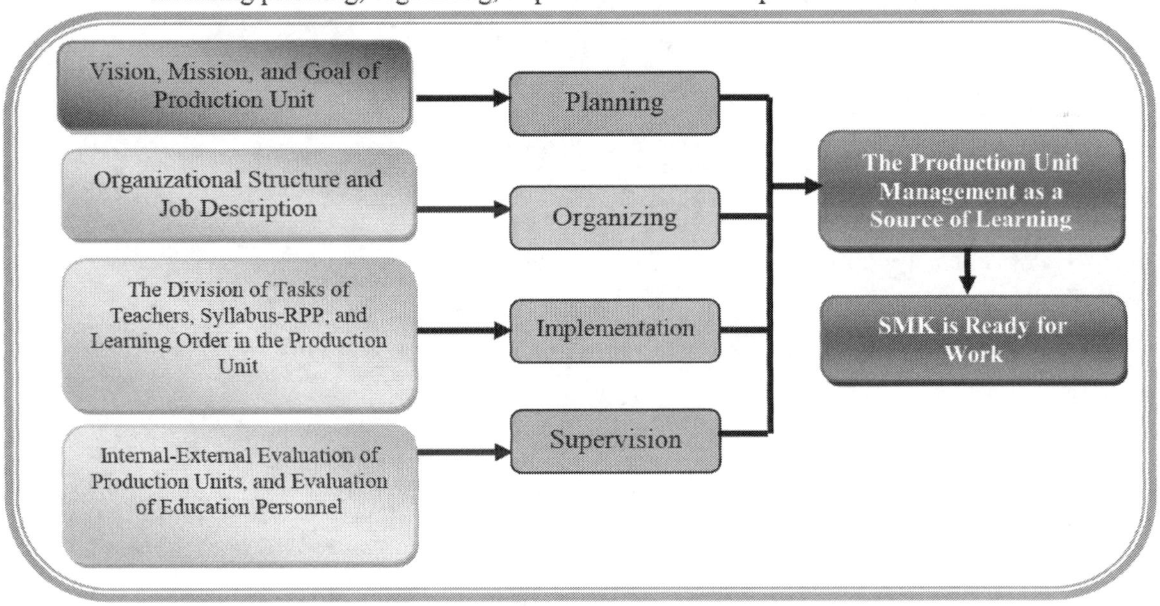

Fig 2. The Management Model as a Source of Learning

In order to produce graduates who are ready to work, can be done through the production unit management as a source of learning. In the production unit students can learn to work like the situation in the industry. To produce a graduate who is ready to work, then the production unit must have good management that supports learning. Management of production units consists of several management functions, namely (1) planning: at this stage it is necessary to declare a clear vision, mission, and objectives as the initial step of establishment of production unit; (2) organizing: the need for guidelines for the preparation of organization, organizational structure, and clear job description to obtain the cohesiveness of the unit manager of production; (3) implementation: it is necessary to divide the task of teachers, the division of tasks of educators, syllabus-RPP, and learning order in the production unit; (4) supervision: for the sustainability of the production unit as a learning resource it is necessary to supervise the students, internal evaluation of production units, and evaluation of education personnel.

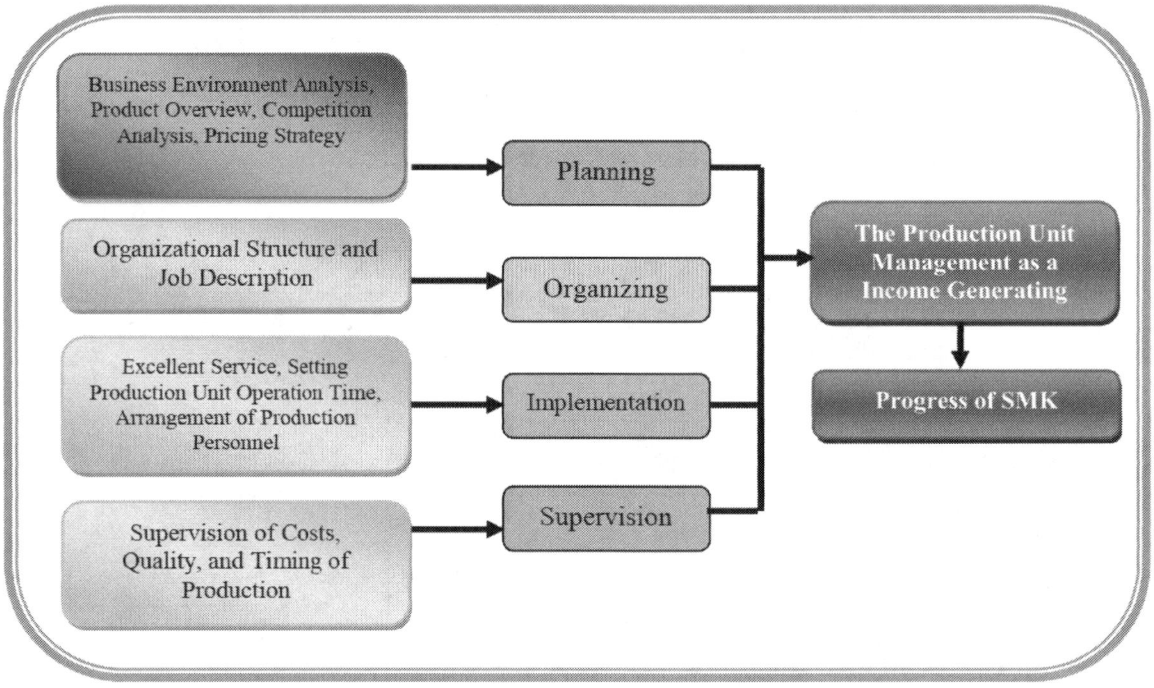

Fig 3. The Management Model as a Source of Funding

The picture above can be explained that the management of production units as a source of funding can be done by: (1) planning includes: business environment analysis, product overview, competition analysis, pricing strategy, description of competitive advantage SMK production unit, description of market segmentation method used, Location description, description of promotion plan, management and personnel identification, unpredictable risk, and identification of raw material suppliers; (2) organizing includes: organizational structure and job description; (3) execution includes: excellent service, setting production unit operation time, arrangement of production personnel, production lay out, production support equipment and marketing management; (4) supervision includes: supervision of costs, quality, and timing of production, personal supervision and filing of claims. The results of internal tests conducted on SMK managers can be described in the table below:

Table 1. The results of Internal Tests

No	Description of Model Variables	Percentage of each Model Component (%)	Management Model of SMK Production Unit	Production Unit as a Learning Source	Production Unit As a source of funding
1	Component Model Completeness	D	16,7%	0%	0%
		A	83,3%	100%	100%
2	Clarity of Model Component Structure	D	8,3%	0%	0%
		A	91,7%	100%	100%
3	Clarity of Model Component Interconnection	D	8,3%	0%	0%
		A	91,7%	100%	100%
4	Readability of Model	D	20%	0%	0%
		A	80%	100%	100%
5	Model Feasibility	D	20%	0%	0%
		A	80%	100%	100%
6	The Model effectiveness if being Implemented	TS	18%	0%	0%
		S	82%	100%	100%

Description: D (Disagree); A (Agree)

Conclusion

Result of research shows that: (1) weakness of production unit is monotonous planning, overlap job descriptions, not optimum supervision; (2) development of management model of SMK production unit as source of learning and alternative funding source; (3) the test results indicate that 83.3% agree that the model component is complete for the management model of SMK production unit and 100% agree the production unit as a source of learning and a source of funding; While 91.7% agree that the model component structure is clear for the management model of SMK production unit and 100% agree that production unit as a source of learning and a source of funding; And 91.7% agree that the relationship between the model components is clear for the management model of SMK production unit and 100% agree that production unit as a source of learning and a source of funding; 80% agree that the model is eligible for the management model of SMK production unit and 100% agree that production unit as a source of learning and a source of funding; And 82% agree that the model is effective if implemented for the management model of SMK Education production unit and 100% agree that production unit as a source of learning and a source of funding.

References

Borg, W.R dan Gall, M.D. *Educational research an introduction.* New York: Longman. (1983).

Klein, S. *Financing vocational education: a state policymaker's guide. Sorting out the byzantine world of state funding formulas, district cost variation, and option for supporting the provision of equitable, quality vocational education in high school.* (Instructional Resource No.30). Athens, GA and College Park, MD: Educational Resources Information Center (ERIC Document Reproduction Service No.ED457329). (2001).

Milles, M.B & A.M. Huberman. *Qualitative Data Analysis.* California: SAGE Publications Inc. (1994).

Pratama, G. N. I. P., & Triyono, M. B. The implementation of CLTSMK method to improve learning quality of SMK. *October 27 th, 2015 Universitas Negeri Jakarta, Indonesia*. (2015).

Preddey, G. An overview of contemporary TVET management practice. Dalam R. Maclean, D. Wilson, & C. Chinien (Ed.), *International Handbook of Education for the Changing World of Work, Bridging Academic and Vocational Learning*. Germany: Springer. (2009).

Rusnani. Keefektifan Pengelolaan Unit Produksi Pada Sekolah Menengah Kejuruan Negeri Kelompok Bisnis dan Manajemen. *Jurnal Pendidikan Vokasi.* Vol. 2 No. 3. Pages 3. (2012).

Sapfo Martoki. *The Contibution Of Vocational Education and Training Preservation and Diffusion of Cultural herigate In Greece: The case of The Specialty " Guardian of museums and Archeological Sites". International Journal of Humanities and Social Science.* Vol 2. No 24. Page 52. (2012).

Schermerhorn, John R., Jr. *Management.* Yogyakarta: ANDI Yogyakarta. (2001).

Timmermann, D. National system of financing TVET. Dalam F. Rauner dan R. Maclean. (2008). *Handbook of Technical and Vocational Education and Training Research*. Bremen: Springer. (2008).

Walther, R. *Financing mechanism and instrument.* Dalam R. Maclean, D. (2009).

Wilson, & C. Chinien (Ed.),International Handbook of Education for the Changing World of Work, Bridging Academic and Vocational Learning. Germany: Springer. (2012).

Making Progress in Content and Language Integrated Learning (CLIL) Lessons: An Indonesian Tertiary Context

Novriani Rabeka Manafe[1*]

[1]Program Studi Pendidikan Bahasa Inggris, FKIP, Universitas Nusa Cendana, Kupang, Indonesia

Jalan Adisucipto, Penfui, Kupang, NTT

Abstract. This paper outlines an attempt to discover students' progress in both content and language skill in a content and language integrated learning (CLIL) lessons at an Indonesia's higher education context. This is a part of a research conducted at Faculty of Science and Technology of Nusa Cendana University in Kupang, East Nusa Tenggara Province. This study employs mixed method approach with 20 participants attending by taking pre-test and post-test as well as joining a focus group interview particularly for 6 students. The tests were aimed at measuring the participants' comprehension of English as the language of CLIL lesson. They were also used as the tool to evaluate students' mastery of Mathematics as the content subject. Based on the post-test results, the findings showed that more students made significant progress in content subject in comparison to their achievement in language proficiency. Regarding the interview, the students admitted that their failure to made progress in both subjects were mainly caused by their inadequate level of English. This, therefore, led to rising anxiety among the students to complete the tests.

1 Introduction

This study was undertaken due to the growing interest of Content and Language Integrated Learning (CLIL) implementation in Asia generally and Indonesia in particular. As the teacher at Nusa Cendana University, the researcher began to explore CLIL as a potential design to teach and learn another language and content subject simultaneously. The understanding of an integrated framework design like CLIL approach and the possibility to succeed in Indonesia has inspired the researcher to conduct CLIL pilot project in Nusa Cendana University context. However, since this was an initial step in the implementation of CLIL, the researcher narrowed the focus of the study on the students' progress in CLIL lessons. Therefore, this paper aims at describing students' progress in CLIL lessons, investigating factors which contributed to the progress, discussing problems preventing the

* Corresponding author: novriani.manafe@gmail.com

students to make progress in CLIL classes and discovering students' expectation from CLIL lessons.

CLIL in higher education

As suggested by Ramsden (1992) as cited in Fortanet-Gomez (2013, pp. 131-132), there are three approaches to the teaching in higher education level. The approaches are as follows:

1. Telling or transmission process causing students to be passive because they only receive information one-sided from the teacher.
2. Student activity process as student-centered learning which position a teacher as a supervisor.
3. A making learning possible process to support students to construct their own understanding of content subject with assistance from a teacher and peer activities.

Obviously, the third approach is best employed in CLIL pedagogy because it offers scaffolding, synergies and social interaction (Coyle, Hood & Marsh, 2010). Adding to this, students working alone with scaffoldings from teachers and peers will make progress. If this happens, teacher can reduce the assistance gradually to enhance students' autonomy (Coyle, Hood & Marsh, 2010).

Furthermore, Rasanen and Klaasen (2004) has pointed out that the major problems in CLIL tertiary classes is the gap between level of language and skills and access to the language used as the vehicular language in the classrooms. Moreover, the language features should be more specific to the subject content. Therefore, the students are expected to have high level of English competence (Dafouz & Guerrini, 2009, p. 104).

2 Methodology

This study employed pragmatic sequential mixed method design to obtain a more sufficient data from both quantitative and qualitative methods (Creswell, 2014). The data quantitative data were the basis for collecting the qualitative data (Mertens, 2010). This research was a part of a larger scale study conducted on piloting CLIL implementation in Kupang, East Nusa Tenggara Province. The research was undertaken in Faculty of Science and Technology, Nusa Cendana University in Kupang, East Nusa Tenggara Province.

Participants

The researcher received 50 applications to participate in the study. However, there were only 20 students (8 male students and 12 female students) recruited to participate as respondents. The researcher used criterion sampling by which she selected the potential 20 participants based on certain criteria (Bryman, 2012). They signed on the informed consent before participating in the lessons. This was done to ensure their rights to withdraw from the research at any stage. The respondents were taken into account based on the following criteria:

1 They were above 18 years old
2 They were enrolled in the Faculty of Science and Technology, Nusa Cendana University
3 The had passed two subjects incorporated in the lessons namely English and Mathematics.

Data collection

The study involved the students in 4 lessons where they were taught Mathematics and English at the same time. The data were collected by triangulation of quantitative and qualitative data collection tools. The primary quantitative data of the study were taken by means of criterion-referenced tests to monitor the students' progress (Salvia, Ysseldyke & Bolt, 2007 as cited in Mertens, 2010, p.356). These tests are divided into pre-test and post-test. Both test consisted of 20 multiple choice questions of English and 10 open-ended Biology questions. Furthermore, the secondary qualitative data were obtained by distributing questionnaires to the 20 students. Preceding the lessons, pre-test was conducted to find out the students proficiency in English as well as the mastery of Mathematics as the content subject. At the end of the lessons, post-test was administered to measure the progress of both the language proficiency and content comprehension. The semi-open ended questionnaires were sent out and returned in sealed and self-addressed envelopes to ensure the anonymity of the respondents from public concern (Mertens, 2010). The questionnaires consist of questions related to problems in making progress in CLIL classes, the contributing factors to the achievement and students' expectation from dual-focused lessons.

Data analysis

The combination of data collection procedure was followed by triangulation of data analysis techniques. The researcher employed descriptive statistics to present the quantitative primary data from both the pre-test and post test results. The descriptive analysis is suitable for describing and presenting data to find the average score of the students' progress without making any inferences or prediction (Cohen, Manion & Morrison, 2011). The statistics were presented in tables and charts to describe not only the initial state of the students' proficiency and mastery but also the progress they made after the lessons. The results of the tests the led to the discussion on the factors contributing to the progress and the problems they encountered. This secondary qualitative data were analyzed by using thematic analysis where data from the questionnaires were transcribed and coded as this process generates new ideas and gathers materials by topics (Creswell, 2013; Richards, 2009).

4 Results and analysis

Students' progress in language and content

Based on data collection and analysis, the findings showed that in general all students did make progress after the four lessons. This raw data is presented in Table 1. However, it appears from the table that the respondents achieved higher scores in content subject than in language proficiency. Adding to this, Chart 1 indicates that there was similar trend occurred in both language and content test scores. This can be illustrated by the fact that the number of students scoring below 50 decreased significantly in both English and Mathematics tests. In terms of English, students scoring below 50 declined from 13 students (65 %) to 4 students (20 %). Similarly, regarding Mathematics, the number of students fell from 10 students (50%) to zero.

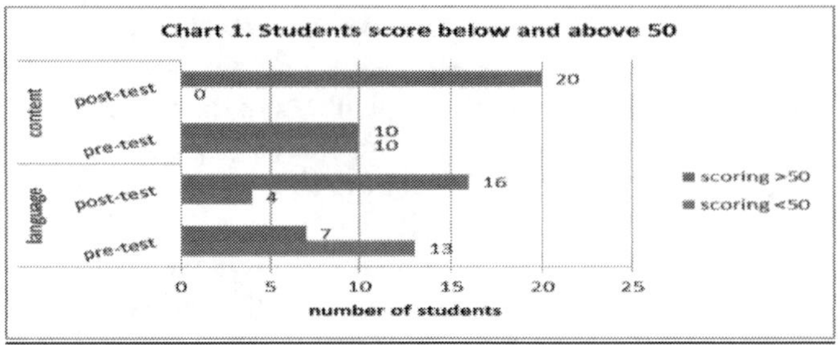

In contrast to the downfall trend in students scoring below 50, there was completely different trend in the number of students scoring above 50. The rising trend was shown by the amount of students in English increased more than double from 7 students (35%) to 16 students (80%). This is followed by similar notion in Mathematics where the number doubled from 10 students (50%) to 20 students (100%).

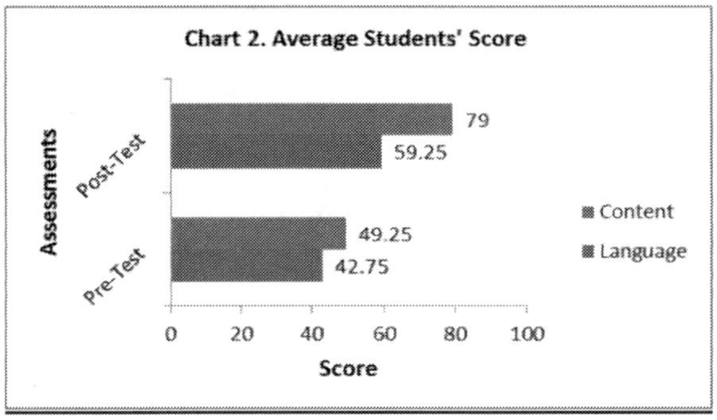

Turning into the average value, Chart 2 clearly points out that there was a growing trend in the average score in language and content subjects. The English average score increased by 16.5% from 42.2 to 59.25. Likewise, the content average score inclined by 29.75 points from 49.25 to 79.

Factors contributing to students' progress

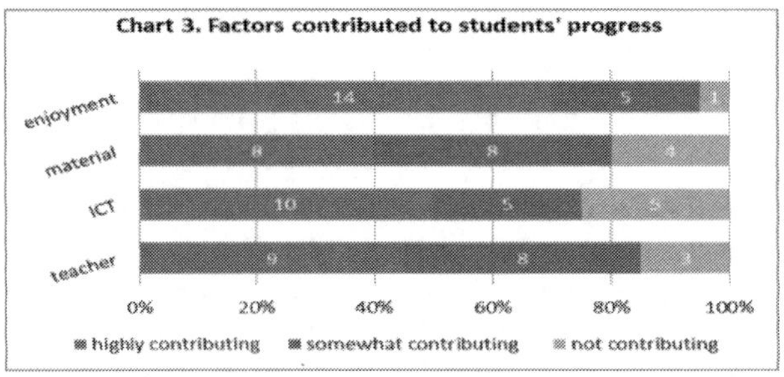

According to the statistics provided in Chart 3, the students revealed the factors contributing to the progress they made in post-tests. The supporting factors relate to enjoyment, teacher's performance, and the use of ICT. In terms of enjoyment, 14 students (70%) claimed that they really enjoyed the new experience. Therefore, they become very much excited to join this newly dual-focused learning experience. Thus, it motivated them to make progress. The second contributing factor is material in which 8 students (40%) stated that the materials used in during the lessons were very helpful to make them understand about the topic. They added that they enjoyed learning with such colorful and updated study resources. Next, 10 students (50%) mentioned the use of ICT improved their scores. The also described the idea of using audiovisual equipment as a very interactive method to support the learning process. Lastly, 9 students (45%) acknowledged the teacher's role in enhancing their progress. These students explained the fact that they liked the teacher's performance in delivering the lessons.

Problems in making progress

Despite the progress they made in the tests, the respondents also pointed out problems that prevented them from making progress in the assessments. The problems mainly arose from language proficiency and little concern from content understanding. First, the language problems lied in the four skills. Based on statistics in Chart 4, problems in writing contributed the highest percentage (90%). Students said that they were not accustomed to writing in English. Following this, 17 students (85%) stated that they had issues in speaking. Again, anxiety in speaking became a major issue. Next, 12 students (60%) admitted that listening also a difficult topic for them. The students admitted that they had problems following the audio recording because they were not familiar with the accents of the speakers. Lastly, 8 students (40%) reported they had difficulties in reading skill. Regarding reading, vocabulary shortage has become a serious problem for the students.

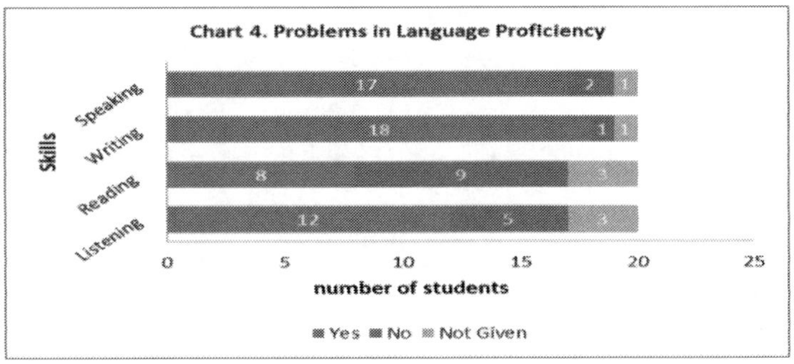

Secondly, regarding content mastery the issues lied in limited vocabulary of the topic being discussed. This can be illustrated by the fact that they had difficulties in understanding content specific vocabularies.

Students' expectation

Concerning the problems that they countered during the dual focus lessons, the students put forward some suggestions for a better teaching and learning process in CLIL classes. The propositions circled around three notions namely time allowance, teacher performance and English language proficiency. First, the respondents stated that learning a content subject delivered in another language rather than their first language is not only exciting but also challenging. Therefore, they suggest longer time spent in CLIL classes. Second, the teacher who conducted the lessons only used English the whole time which caused the students some difficulties to follow the lessons. Thus, they would be very pleased if the teacher could slow down her voice as well as repeating the difficult words. Third, the major issue in attending CLIL classes was the low level of English proficiency. Then the students suggested that those who would like to participate in CLIL lessons need to improve their language skills beforehand.

5 Conclusion

This mixed method study which combined qualitative and quantitative data collection and analysis from 20 participants have brought some evidences into the implementation of tertiary CLIL projects in Kupang context. The findings from the research answered four objectives namely investigating students' progress in both language and content comprehension, addressing the contributing factors to the achievement, delineating the problems preventing them to make progress and discovering students' expectancies from CLIL lessons.

Concerning students' achievements, it was obvious that the learners made progress in both subjects. However, the progress that they made in content knowledge was significantly higher than their language skills. Despite the difference in level of achievement between content and language, there were some factors that contributed to their progress. The supporting variables were enjoyment, learning resources, use of ICT and teacher's delivery method.

Turning into the factors that prevented the students from achieving higher scores in language skills, the main issue was that the proficiency of the students was below average. Therefore, they had difficulties in all the four skills as well as in content comprehension.

In accordance to the contributing factors and the problems stated above, the students exposed positive value towards the research. They brought out some suggestions to improve the quality of CLIL dual-focused lessons. First, they stated that these typical lessons should be conducted in longer time span because they need more opportunity to learn two subjects in one lesson. Next, the students definitely agreed with the idea of improving their English skill to be able to participate in CLIL lessons. Finally, the students proposed repetitions and slow speed from the teacher so they could adhere well to teacher's presentation.

Acknowledgement

This research was a pilot project conducted at Fakultas Sains dan Teknik of Universitas Nusa Cendana in Kupang, East Nusa Tenggara Province. Therefore, I would like to show my gratitude to colleagues at the faculty who provided assistance for the completion of this research. I also thank students at the faculty who participated as respondents in this research.

References

Bryman, A. (2012). Social Research Methods. Oxford: Oxford University Press.

Cohen, L., Manion, L. & Morrison, K. (2011). Research Methods in Education. Oxon: Routledge.

Coyle, D., Hood, P., & Marsh, D. (2010). *CLIL: Content and Language Integrated Learning*. Cambridge: Cambridge University Press.

Creswell, J. W. (2013). Qualitative Inquiry & Research Design: Choosing Among Five Approaches. California: Sage.

Creswell, J. W. (2013). Research Design: qualitative, quantitative and mixed method approaches (4th ed). California: Sage.
Dafouz, E. & Guerrini, M.C. (Eds.), 2009. *CLIL Across Educational Levels*. Madrid: Richmond Publishing.
Fortanet-Gomez, I. (2013). CLIL in Higher Education: Towards a Multilingual Language Policy. Bristol, Buffalo and Toronto: Multilingual Matters.
Gibbons, P. (2002). Scaffolding Language, Scaffolding learning: Teaching Second Language Learners in the Mainstream Classroom. Portsmouth: Heinemann.
Hunt, M. (2011). Learners' perceptions of their experiences of learning through a foreign language. *Educational Review, 63*(3), 365-378.
Mehisto, P., Marsh. D. & Frigols, M. J. (2008). Uncovering CLIL: Content and Language Integrated Learning in Bilingual and multilingual Education. Oxford: Macmillan.
Mertens, D. M. (2010). Research and Evaluation in Education and Psychology: Integrating Diversity with Quantitative, Qualitative and Mixed Methods. California: Sage.
Richards, L. (2009). *Handling Qualitative Data: A Practical Guide*. London: Sage

Elementary school students' mathematical intelligence based on mathematics learning using classical music of the baroque era as the backsound

Karlimah

Primary Teacher Education, Universitas Pendidikan Indonesia Tasikmalaya Campus
Jl. Dadaha No.18 Tasikmalaya, 46115, West Java, Indonesia

Abstract. Many studies suggest that classical music can inccrease the listeners' intelligence, including mathematical intelligence [3, 12, 2, 11]. In this research, we used the classical music of Baroque era as the backsound during math learning. The research method used was quasi experiment with nonequivalent pretest-posttest control group design to grade V SD students in Tasikmalaya city. The results show that the use of classical music of Baroque era during the learning of mathematics gave a high contribution to the mathematical intelligence of fifth grade elementary school students. The student's mathematical intelligence can be seen in the cognitive abilities which were at the high level in the knowledge up to analysis, and at the low level in the synthesis and evaluation. Low mathematical intelligence was shown by students in calculating amount and difference of time, and projecting word problem into the form of mathematical problems. High mathematical intelligence arose in reading and writing integers in words and numbers. Thus, the mathematical intelligence of fifth grade Elementary School students will be better if classical music of Baroque era is used as the backsound in mathematics learning about solving math problems.

KeyWord: Baroque Classical Music, Backsound, Mathematics Intelligence Elementary Students

1. Introduction

The target of learning mathematics for elementary school students is to enable students to understand the concepts of mathematics, being able to explain the interrelationships between concepts and apply concepts or algorithms flexibly, accurately, efficiently, and appropriately in problem-solving; being able to use reasoning on patterns and traits, perform mathematical manipulations in generalizing, compiling evidence, or explaining mathematical ideas and statements; being able solve problems that include the ability to understand problems, design mathematical models, solve models and interpret the solutions obtained; being able to communicate ideas with symbols, tables, diagrams, or other media to clarify circumstances or problems; has an appreciation of the usefulness of mathematics in life that is the curiosity, attention, and interest in learning mathematics, as well as resilience and confidence in problem solving [6].

In 2013, the objective of learning mathematics was focused on the achievements of three domains that require the mastery of mathematical contents according to the objectives of the 2006 curriculum. The achievements of three domains are: first, attitudes that reflect the faithful, noble, confident, and responsible person in interacting effectively with the social environment and nature around his own home, school, and playground. Second, effective and creative thinking and action in the abstract and concrete realm as assigned to him. Third, factual and conceptual knowledge in science, technology, art, culture, humanities, with national, state, and civilization insights related to phenomena and events in the home, school, and playground [7].

The achievement of domain attitudes, skills and knowledge is a concrete form that can be seen and felt when used in the wider community related to learning outcomes. The results of mathematics learning will be seen from the ability to solve problems or math problems as a result of mathematical thinking activities. The quality of mathematical thinking is closely related to mathematical intelligence. Each student has different mathematical thinking skills, so his mathematical intelligence is different. Mathematical intelligence is an intelligence that involves the skills of processing numbers well and/or using reasoning or logic correctly. This intelligence includes sensitivity to logical relationships, causality, and other logic [13, 1, 8]. Thus, mathematical intelligence is a skill or prowess in solving mathematical questions or problems as a result of the process of learning mathematics.

During the process of learning mathematics, all mathematics learning facilities are considered, provided and used to achieve the learning objective of mathematics that most students need as a provision in their daily life that is mathematical intelligence. However, there is a mathematics learning facility that has not been implemented in elementary schools, when in fact it is theoretically recommended for use when students learn math. Mathematics learning facility in question is the using of classical music. Playing music works to improve and foster the development of personal social capabilities [2]. Besides that, it is also useful for entertainment, media in education, relaxation, spiritual interests, discovering hidden aspects of personality, stimulating creativity and imagination, increasing intelligence, and trigger improvements in learning ability [3].

Therefore, music as a medium in learning mathematics is used in an effort to develop cognitive, affective, psychomotor skills, and improve students' intelligence and learning readiness. Studies of brain function showed that music activates the flow of nerve impulses into the corpus callosum, the network of brain fibers that connect the left brain and right brain. When the left and right brain are active, various mental functions will be aligned. This alignment of left and right brain functions will enhance a person's learning readiness [11].

This is supported by several research results which reveal that: 1) music gave a significant positive effect on mathematics learning achievement [9] music enhanced mathematics intelligence of sixth, seventh, and eighth grade students in Saudi Arabia [5] music stimulus could improve students' brain performance when thinking mathematically and linguistically [10] music influenced students' mathematical literacy interest [14] in Rhode Island [2] first grade elementary school students who attended music classes had dramatically increased their reading and math skills. In addition, students who learned about music appreciation earned a score of 46 points higher, and 39 points higher for students who had musical experience. The last result of our research on fifth grade elementary school students in Tasikmalaya City - West Java - Indonesia showed the improvement of students' mathematical intelligence up to high level, through the use of Classical, Romantic and Baroque music during math learning [4].

The facts resulting from our research provide information and understanding to researchers that music can basically improve students' mathematical intelligence. Based on that, researchers are encouraged to reveal the mathematical intelligence shown by students who get the experience of learning mathematics accompanied by Baroque classical music. This caught the attention of the authors because the mathematical intelligence shown by the students who joined the learning process accompanied by Baroque music was at a high level of mathematical intelligence.

2. Experimental Method

This research was conducted by using Quasi Experiments of Nonequivalent Pretest-Posttest Control Group Design to students in four elementary schools (SD) in Tasikmalaya City - West Java - Indonesia. Data were collected using the test instruments of integer operation, time unit, rounding, estimating, and problem-solving questions related to time, distance and speed that have been tested and proven valid and reliable. By using ANOVA's inferential statistical test and SPSS 19, we found the high and the highest level of mathematical intelligence of a series of three-month math learning activities in a sample of elementary schools using the Romantic classical background music. Furthermore, we used descriptive statistics to analyze the data of students work as it was so that we could obtain a complete information about the mathematical intelligence of students who experienced the Baroque classical background music. Descriptive statistical results in question are as follows:

3. Result and Discussion

Based on the empirical treatment, we obtained test results which showed that the mathematical intelligence of students who had experience in learning mathematics with the backsound classical music of Baroque era. The test results showed that students' intelligence in terms of cognitive abilities presented as percentage for knowledge (C1) was 75,64%, comprehension (C2) was 58,65%, application (C3) was 52,78%, analysis (C4) was 53,85%, synthesis (C5) was 38,46%, and evaluation (C6) was 21,15%. Overall, the cognitive abilities of students from highest to lowest sequences ranging from knowledge to evaluation. However, analytical skill palced the third position. In other words, the analysis ability of students who learned mathematical with the backsound classical music of Baroque era was better than their application ability. Judging from the overall level of difficulty, it was found that easy, medium, and difficult questions could be solved correctly by 100,00%; 53,47%; and 32,05% respectively.Recapitulation of math learning result with backsound classical music of Baroque era can be seen in Table 1.

Table 1. Recapitulation of Fifth Grade Elementary School Students' Test Results

TEST PERIOD	COGNITIVE ASPECTS						DIFFICULTY LEVELS		
	C1	**C2**	**C3**	**C4**	**C5**	**C6**	EASY	MODERATE	HARD
PRETEST	56.41	34.62	32.91	44.23	38.46	19.23	92.31	35.61	28.21
POSTEST	75.64	58.65	52.78	53.85	38.46	21.15	100.00	53.47	32.05

Then, the number of students who can answer the problem with the variation of difficulty levels and variation of cognition was also revealed. The student's answers were categorized into low, medium, high, and very high attainment. The ability to correctly answer the questions shown in Table 2 below.

Table 2. Recapitulation of Correct Response Attainment Levels and Various Levels of Difficulty and Cognition

No	Correct Response Attainment Levels	Various Levels of Difficulty and Cognition
1	Low	19.23%
2	Medium	42.31%
3	High	34.62%
4	Very High	3.85%

There was the lowest attainment shown by the students in solving the problem in the evaluation category about projecting word problem into the form of mathematical problems, and the application category about calculating the amount and the difference of time appropriately. The problems were made in the form of multiple choice, and a series of sentences with moderate difficulty. This problem could be solved by 11.54% of students. The questions are as follows.

2 hours + 15 minutes + 1800 seconds = ... minutes
A. 1827 B. 1835 C. 1817 D. 165

Your sister had 5 boxes of marbles. Each box contained 20 marbles. Then your brother bought 10 more marbles. When playing, 2 marbles disappeared. The rest of the marbles would be distributed equally to his two younger siblings and two neighboring children. Which of the following is a mathematical sentence based on the above story?
A. (5x20) + 10 - (8: 4) = 112
B. [(5x20) + 10 - 4]: 4 = 27
C. (5x10) + 10 - (8: 4) = 62
D. [(5x20) + 10 - 2]: 4 = 27

The highest attainment was shown by students in the material of reading and writing integers in words and numbers. Multiple-choice questions, with typical mathematical questions, the cognitive level of knowledge (C1) and easy difficulty. This problem could be solved by 97.5% of students. The problem is as follows.

(-56,487) is read...
A. Fifty six thousand four hundred and eighty seven
B. Negative Fifty six thousand four hundred and eighty seven
C. Negative Fifty six thousand four thousand eight hundred and seven
D. Fifty six hundred four thousand eighty seven

Among the questions with variations of cognitive and difficulty levels, it was found that most students answered incorrectly on each category of the problems, namely, (53.85%) on the knowledge category (C1); 57.69% on the comprehension category (C2); 88.46% on the application category (C3); 69.23% on the analysis category (C4); 88.46% on the evaluation category (C6) all of the problems were all in moderate difficulty; and also 73.08% on the synthesis category (C5) with hard difficulty. Menawhile, another attainment, which showed the most correct answers or good results, existed in the knowledge ability that is 100.00%; 84,62% on understanding ability; 96.15% on application ability; 84.62% on analysis ability; 50.00% on synthesis ability; and also 30.77% om evaluation ability. Therefore, most correct answers at every cognitive level indicated that the higher the cognitive ability, the lower the attainment.

4. Conclusion

The mathematical intelligence of the fifth grade elementary school students who joined the learning by backsound accompaniment of Baroque classical music was proven by the outstanding ability at the level of knowledge about reading and writing integers in words and numbers, and the lowest one was in application and evaluation abilities. This was indicated in answering questions about calculating the exact amount and difference of time and projecting the word problem into a mathematical problem. The students' ability to solve problems based on variations in cognitive and difficulty levels did not provide distinguishing information. The cognitive level of C1 to C6 and the difficulty level from easy to hard did not show the significant difference in students' ability to find the correct answers.

References

1. Campbell, Linda, dkk, *Metode Praktis Pembelajaran Berbasis Multiple Intelligences* (Intuisi Press, Depok, 2006).
2. Djohan, *Psikologi Musik* (Penerbit Buku Baik, Yogyakarta, 2003).
3. S. Gunara, *Manfaat Pendidikan Musik bagi Anak. Ritme,* **6** (1), 19-26. (2008).
4. Karlimah, R. Respati, *Efektivitas Backsound Musik Klasik dalam Meningkatkan Kecerdasan Matematika Siswa Sekolah Dasar* (Laporan Penelitian UPI, tidak diterbitkan, 2016).
5. Khalil, A. *Science of Creativity.* [Online] Tersedia: http://www.adabfalasteeni.org/artical.php. (18 Januari 2015). (2005).
6. --------------- Kurikulum 2006, Standar Isi, Standar Kompetensi, Kompetensi Dasar, (BSNP, Jakarta, 2006).
7. ---------------.. Kurikulum, (Kementerian Pendidikan dan Kebudayaan, Jakarta, 2013)
8. Lwin, May, et al., *How to Multiply Your Child's Intelligence* (Penerbit Indeks, Yogyakarta, 2008).
9. H.B. Lamar, *An Examination Of Congruency Of Musical Aptitude Scores And Mathematics And Reading Achievement Scores Of Elementary Children* (Tesis pada University of Southern Mississippi, Tidak diterbitkan, 1989).
10. C.S. Luiz, *International Symposium on Performance Science* (2007).
11. S. Merrit, *Simfoni Otak: 39 Aktivitas Musik yang Merangsang IQ, EQ, SQ untuk Membangkitkan Kreativitas & Imajinasi* (Penerbit Kaifa, Bandung, 2003).
12. M. Muttaqin, Kustap, *Musik Klasik: Pengantar Musikologi untuk SMK* (Depdiknas, Jakarta, 2003).
13. Saifullah, *Mencerdaskan Anak (Mengoptimalkan Kecerdasan Intelektual, Emosi dan Spiritual Anak* (Lintas Media, Jombang, 2004).
14. Omari, M., dkk, Journal of Science Education and Human **21** (1) (2010).

CHARACTER EDUCATION BASED ON LOCAL WISDOM FOR THE PRISONERS

Buchory Muh Sukemi[1], Okti Purwaningsih[2], Sigit Handoko[3]

buchoryupy@yahoo.com

University of PGRI Yogyakarta

ABSTRACT

This research aims at revealing the existence of character education based on the local wisdom for the prisoners. The subject of this research is the prisoner inhabitant in Wirogunan Prison and Narcotics pakem in Sleman Yogyakarta. The data were gained by interviewing, documenting and demonstrating. The data were analyzed qualitative and quantitative descriptively. The results of this research are: (1) The training of Tembang Maca Pat and Javanese MC are equivalently adapted, (2) The character values being shaped: believe in God, responsibility, respect, fairness, confidence, faithfulness, discipline, careness, spirituality, manners, intelligence, emotion control, character building, the increase of social participants; (3) The competence on Tembang Maca Pat and MC of Javanese can be used as professional earning in society and also can be used as educational model for the prisoners.

Keywords: *Character education, local wisdom, Prison*

1. INTRODUCTION

Since reformation era, the reformation of national education had been conducted, such as Regulation number 20 year 2003 regarding national education system. The function national education attached in Sisdiknas is that developing and characterizing of civilization in order to make citizens smart. The purpose is to develop all youngster potention, both academically and spiritually, social competence, and the whole Indonesian spirit.

The Indonesian human body is being formulated in regulations Sisdiknas have not been realized yet. It is revealed that in daily social life, the attitude of our society is still not match with national heritage perspectives. It is proven by the large amount of corruption, drug abuses, criminalism, and so on. As the result, the inhabitant of prisoners in Yogyakarta increases significantly.

The problems in this research can be formulated as (1) how is the mental condition based on local wisdom of the prisoners of Wirogunan detachment and Narcotics detachment of pakem Sleman in Yogyakarta? (2) What is the approach being implemented in the conducting character education training based on local wisdom for the inhabitant of detachment? (3) How is the effectiveness of conducting character education modelling in detachment of Yogyakarta?

2. THEORETICAL BACKGROUND

The term of 'character' in Kamus Besar Bahasa Indonesia refers to psychological education that differenciates one anothers. It can be also refers to psychological heritance, heart, soul, character values, personalities, temperament. The high quality individual entity refers to someone that does everything good to the God, himself, others, surroundings, nation and the world by optimalizing self potention by realizing, emotion and feelings.

Abdullah Munir (2010:3) states that character is a pattern, including thinking, attitude, acts that attached to someone and can not be ommited. The character is determined genetically, foods, firends, parents and goals. In the character education design (kemendiknas, 2010:9) stated that character configuration in totality context can be grouped into spiritual and emotional development, intelectual development, physical and kinesthetic development and affective and creativity development. Those four psychosocials related holistically and coherently and making characters become valuable values.

The basic principles of character education based on Heri Gunawan (2012:35) are: (1) promoting basic ethic values as the character basis, (2) identifying character comprehensively that includes thinking, feeling and acting, (3) applying effective and efficient approaches to build character, (4) creeating school communities that have cares, (5) giving chances to students to build character and help them to be success, (6) acquiring scope of curriculum that meaningful and challenging and appreciating students, (7) developing motivation of students, (8) functioning all school staff as the moral communities that share responsibility, (9) creating moral leadership and supports character education innitiative and (110 evaluating school character.

To achieve that orders, human should understand, internalize and do the duties as stated in Tri Satya Brata: (1) world prosperity depends on human who has sharp thinking; (2) prime duty of keeping world safety; (3) human's safe depends on the acts. The character values of academical approach are spirituality, well acts, honest, discipline, democratic, fairness, responsibility, work together, creative, appreciate others.

Prison is a acted unit of condemned system that the job is to guide the condemned. In the prison, the prisoners were guided to give and train capabilities in order become a new human after the sentence is over. They are phisically and psychologically different, so that the treats also different compared to normal people.

3. RESEARCH METHOD

The location of the research is limited on the majority of communities exist of Javanese people, especially in DIY, those are prison of IIA Wirogunan in Tamsis Avenue number 6 Yogyakarta and Narcotics Prison IIA in Pakem Sleman yogyakarta. The first step of the first year of research subject is on 2 prisons with 4 administrators and 27 prisoners.

The time of conducting research is around 2,5 months and ended by conducting a competition among trainees with the description as follows:

1. The training of Nembang Maca Pat di Prison of Wiragunan Yogyakarta

   This training were joined by 2 administrators and 14 prisoners and they join from the beginning until the end. The training was for 2,5 months with the total meeting 10 times. The materials are as follows:

Table 1. Time and materials

No	Date	materials
1.	Monday, 23 May 2016	The description
2.	Tuesday, 24 May 2016	Character of Tembang Maca Pat
3.	Monday, 30 May 2016	Practise Nembang Pocung
4.	Tuesday, 31 May 2016	Practice Nembang Gambuh
5.	Monday, 6 June 2016	Practice Nembang Mijil
6.	Tuesday, 7 June 2016	Pratice Nembang Kinanthi
7.	Monday, 18 July 2016	Practise Nembang Sinom & Pangkur
8.	Monday, 25 July 2016	Practise Nembang Dandanggulo
9.	Monday, 1 August 2016	Preparation
10.	Tuesday, 9 August 2016	Competition of Maca Pat of all members

2. The training of MC of Javanese in Prison of Narcotics of Pakem Sleman Yogyakarta

This training was joined by two administrators and 13 followers. The training was conducted for 2,5 months for 10 times. The description is as follows.

Table 2. Time and materials

No	Date	Materials
1.	Friday, 27 May 2016	Explanation
2.	Tuesday, 31 May 2016	Gladen Panata Adicara/MC Basa Jawa
3.	Wednesday, 8 June 2016	Explanation and Practice
4.	Wednesday, 15 June 2016	Explanation & Tulodho MC Pangrupi Laya
5.	Tuesday, 21 June 2016	Explanation & Practice MC pada Rapat RT
6.	tuesday, 28 June 2016	Practice MC Pangrupi Laya
7.	Wednesday, 29 June 2016	Explanation & Tulodho MC Resepsi Temanten
8.	Tuesday, 19 July 2016	Practice MC Resepsi Temanten
9.	Wednesday, 20 July 2016	Preparation
10.	Wednesday, 27 July 2016	Competition

There are some data that were gained during research model development;(1) the data regarding the profile of prisons, identification and need assessment; revealed by using survey method, observation, interview and documentation; (2) the data regarding the capability of instructors in some cases including (a) the ability to absorb the result of research; (b) The capability to apply the result of training in prisons. The data analysis is conducted with the technique of qualitative and quantitative descriptive.

4. RESEARCH RESULT AND DISCUSSION
A. RESEARCH RESULT

Based on the recapitulation of peisoners' idea, it can be concluded that:

1. 83,25% of prisoners agree that these trainings can build somebody's character.
2. 90,50% prisoners agree that local wisdom in these trainings can increase the spirituality.
3. 83,75% of prisoners agree that these trainings increase someone's responsibility.
4. 78,00% of prisoners agree that these trainings increase justice in somebody.
5. 79,50% of prisoners agree that these trainings increase the honesty.
6. 80,00% of prisoners agree that these trainings increase the discipline.
7. 85,50% of prisoners agree that these trainings increase the careness to each others.
8. 84,00% of prisoners agree that these trainings increase someone's trust.
9. 84,25% of prisoners agree that these trainings increase someone's good behavior.
10. 82,50% of prisoners agree that these trainings increase someone intelectuality.
11. 82,75% of prisoners agree that these trainings increase someone's spirituality.
12. 85,00% of prisoners agree that these trainings can control the emotions.
13. 83,25% of prisoners agree that these trainings can shape someone's character.
14. 84,25% of prisoners agree that these trainings can increase someone's participation as social creature.

15. 85,75% of prisoners agree that these trainings can preserve local wisdom values.

16. 87,75% of prisoners agree that these trainings should be conducted continuously.

17. 84,25% of prisoners agree that these trainings will be maintained after the sentence is over.

18. 84,50% of prisoners agree that these trainings will be heritated to the next generation.

19. 83,75% of prisoners agree that it will be applied.

20. 81,25% of prisoners agree that it can be used as additional earnings

21. 86,00% of prisoners agree that these trainings can be used as character education model.

5. CONCLUSION AND SUGGESTION

A. CONCLUSION

1. Most of prisoners (83,25) stated that Maca Pat and Javanese MC can shape somebody's characters.
2. The values of character that can be developed through these kinds of training includes; spirituality, responsibility, respect, justice, honesty, intellectuality. Besides that, it can also be used as emotional control, shaping a good character in daily life.
3. The training of Maca Pat and Javanese MC can be used as a place for preserve the local wisdom and heritated to the next generation. It can be used also as additonal earning after the prisoners end their sentence.
4. The values of local wisdom can be used as character education model.

B. SUGGESTION

Based on the conclusion, it can be suggested that:

1. It is hoped that Maca Pat and Javanese MC can be conducted continuously.
2. It is hoped that these trainings can be used as character education model based on local wisdom.
3. These trainings are also hoped that can be conducted in others formal institution as media of character education in which the teacher of civics can be used as the guide.

6. REFERENCES

1. Ary Ginanjar Agustian. 2002. *Rahasia Sukses Membangun Kecerdasan Emosi dan Spiritual.* Jakarta : Arga.

2. Binsar A. Hutabarat. (2010). *Karakter bangsa, dulu dan kini*, dari file:///G/karakter bangsa dulu dan kini.html. diunduh 27 April 2010.

3. Heri Gunawan. 2012. *Pendidikan Karakter Konsep dan Implementasi.* Bandung : CV. Alfabeta.

4. Kemendiknas. 2010. *Buku Induk Pendidikan Karakter.* Jakarta : Kementerian Pendidikan Nasional.

5. Koesoema, A.D. 2012. *Pendidikan Karakter Utuh dan Menyeluruh.* Yogyakarta : Kanisius.

6. Ki Tyasno Sudarto. (2007). *"Pengembangan nilai-nilai luhur budi pekerti sebagai karakter bangsa"*. Diakses tanggal 12 Maret 2010.

7. Kuntjaraningrat. (2007). *Kebudayaan Mentalitas dan Pembangunan*. Jakarta: PT. Gramedia.

8. Thomas Lickona. 2013. *Educating For Character : Mendidik untuk Membentuk Karakter*. Jakarta : PT. Bumi Aksana.

9. Undang-Undang Republik Indonesia Nomor 20 Tahun2003 Tentang Sistem Pendidikan Nasional. akarta.

APPLICATION OF *TREATMENT TRAIT INTERACTION* TO IMPROVE LEARNING OF CHEMISTRY

Author: Prima Aswirna
Universitas Islam Negeri Imam Bonjol
Jalan Napak Tilas No. 10 Padang
primaaswirna1971@yahoo.com

Abstract. This research leave from the fact that performance study of Majors student of Tadris Natural Sciences Concentration Physics (IPA-FISIKA) State Islamic University Imam Bonjol. The aims of this research were: (1) Searching ability correlation between content of capability in Chemistry and religiosity. (2) Differentiating study performance by using model of *Trait Treatment Interaction* (TTI) between class with capable student group early is high (learner fastern) and class with student group ably early lowering (slower learner) by using method study of autodidact (learning self) and learn habit (teaching regular) added with tutorial. This research use method of *quasy experiment*. Technique data collecting through test and observation, Total of population counted 101 people. But, only 66 students as a sample. The research conclude : (1) The self-learning was suitable to teach for students with high capability of knowledge (faster lesrner). (2) The regular teaching + tutorial was suitable to teach for students with lower capability of knowledge (slower learner). (3) There were correlation between religiosity and capability of content (chemistry of knowledge). (4) Trait treatment interaction can be improve the capability of academic performance of students Chemistry Science.

Keywords: Teaching Models, Trait Treatment Interaction, Chemistry, Students.

INTRODUCTION

Undang-Undang Sisdiknas No. 20 Tahun 2003 had been mention that, "... System education of national have to can guarantee generalization of opportunity of education, make-up of quality and also education management efficiency and relevansi to challenge as according to demand change of local life, national, and global so that require to be done by renewal of education by terencana is, directional, and continual". Pursuant to clarification above earning ism that, government look into important to increase the quality of human resource and education Some effort which is gone through by government to increase the quality of education in this second millennium, one of them is with change of higher education curriculum in Indonesia. This matter is proved to pass change of national curriculum in the year 1994. Curriculum of KBI this is distinguished. for example majoring domination of Science and Technology, do not focus to formulate its ability, consortium specify eye of kuliah is obliged to S-1 counted 100-110 Set Of Credit Semester Then year curriculum 1994 changed to become curriculum of[is core of and is institutional. As for characteristic of Curriculum is core of and Institutional was to: majoring attainment of interest, do not be specified by science definition which need to be studied, study program forum of a kind specify especial interest (60%), supporter interest and is other (40%), existence of educational skills subject, educational skills experterties subject, educational skills developmental.. Here inafter curriculum of[is core of and this institutional more knowledgeable with Curriculum Base On Interest (KBK). Next in the year 2014 Curriculum Base on Interest turned into Curriculum Higher Education (KPT). This Clarification give picture that governmental role and also in effort improve the quality of education have been laboured by as good as possible, as form of social responsibility of government to translate Constitution commendation 1945 namely educate life of nation.

This clarification give picture that governmental role and also in effort improve the quality of education have been laboured by as good as possible, as form of social responsibility of government to translate Constitution commendation 1945 namely educate life of nation. Next is effort the make-up of the quality of human resource through bursary program to educator in college have been run.

Hereinafter in context higher education of System Credit Semester exist is each individual (educative participant) to be able to develop its potency maximally. So that educative participant which pertained capable quickly (learner fastern) in the field of academic can finish its study more or less three half year. Whereas educative participant which have ability of academic which mean of moderate will finish their study during more or less four-five years.

Hereinafter to educative participant group which have tardy ability (lerner slower) will finish its study more than five year.

Problems impact is such as those which elaborated above, can be seen by at storey;level education of base and till higher education shall be as follows : (1) Pursuant to report of United Nationation Development Program year 2014 peringkat of HDI (Human Development Index) Indonesia [among/between] year 2010-2012 HDI Indonesia occupy peringkat 6 from 11 state of Asean. Other Report by Asian QS University Rankings for (2015) discovering in range of year 2014 - 2015 Indonesia only deputized by University Indonesia as one of the 100 best university [in] Asia. Where in the year 2014 University Indonesia occupy peringkat 71 from 100 best university in Asia. While next in the year (2015) University Indonesia occupy peringkat 79 from 100 best university in Asia. This show Indonesia achievement in arena emulation of education world still lower.

Data above indicating Indonesia still have the quality of low education where education in Indonesia is experiencing of big problem. Problem of faced to cover; (1) quality of education which still lower, (2) study system in universitys which not yet is adequate, and (3) moral crisis which knock over our society. Decline of quality education of national is not located to ability of intelegensi all educative by participant of Indonesia, but because of opportunity which do] not flatten in obtaining good education at this nation children. Besides, quality of construction all educator, available opportunity in university environment and society, and also optimal costs which required in education which with quality not yet flattened can enjoy by nation children. Subject metter of Chemistry base on pursuant to majors syllabus of General Science – Physic (Tadris IPA-FISIKA) Faculty of Education (FT) State Islamic University Imam Bonjol Padang (UIN Imam Bonjol Padang). The aim of Chemistry subject is: student can comprehend and apply erudite method, items concept and of energy, stoikiometri, chemical energy, atomic structure, periodic system, chemical bond and also can integrate sentences of Al-Qur'An with concept- chemical concept. Then the target of lecturing of Elementary Chemistry was not yet been reached because pursuant to list assess student academic for the subject metter Chemical base on still many student obtaining mark less than 70 (with category of C). This Matter can be seen by summary assess two the last year at Tables 1.1.

Tabel 1 :Recapitulation Score of 1st Elementary Chemist on General Science - Physic

Semester & Academic Years	Average Score of Students in Gneral Science–Physic										Sum Students
	A	%	B	%	C	%	D	%	E	%	
Semester Anomalous 2013/2014	6	18,2	11	33	12	36	4	12	–	0	33
Semester Anomalous 2013/2014	4	11,1	13	36	15	42	3	8,3	1	2,8	36
Semester Anomalous 2014/2015	6	18,2	11	33	12	36	4	12	–	0	33
Semester Anomalous 2014/2015	4	11,1	13	36	15	42	3	8,3	1	2,8	36

Source: Prima Aswirna (2016:8)

Condition of above happened because process study of chemistry run by educator not to mention paying attention individual difference which owned by educative participant truthfully have interest differ between one with is other. Where there is educative participant which pertained to have high interest in study of chemistry. While on the other hand there is also educative participant which have ability or potency pertained or mean. Otherwise on the other side also there is educative participant which have potency or tardy ability (slower learner). This situation will have an effect on big to knowledge absorpsion had by each educative participant. To educative participant group which have kemampaun quickly (faster learner) hence, they will be able to permeate study items easily. Later;Then they motivated to follow study process swiftly. While educative participant ably they is can follow study of habit (Regular Teaching), its meaning is they can follow study process do not swiftly but, them still can live with in absorbent ability of given by educator. And for the group of tardy capable educative participant (slower lerner) they quickly tire of, because difficult accept items teach better. Educative participant in this group require to obtain;get study addition or remedial (Tutorial) to be able to equip lacking of knowledge absorpsion which they experience of. This matter in line with conducted by research of Kowolik (2009) "show the existence of interaction among cognate ability of educative participant with difference of treatment treatment in course of study". To link problem above hence, researcher consider necessary to use teaching model of *Trait Treatment Interaction* (TTI), where model this study make account of individual difference which owned by educative participant. So that educative participant will give to treat as educative participant group which have kemampaun quickly (learner faster) hence, they will be able to permeate study items easily.

Later, they motivated to follow study process swiftly. While the educative participant ably they can follow study of habit (Regular Teaching), its meaning they can follow study process do not swiftly but them still can live with in absorbent ability of given by study items is educator. And for the group of tardy capable educative participant (learner slower) they quickly tire of, because difficult accept items teach better. Educative participant in this group require to obtain;get study addition or remedial (Tutorial) to be able to equip lacking of knowledge absorpsion which they experience of. This matter in line with conducted by research. Kowolik (2009) where Kowolik show the existence of interaction between cognate ability of educative participant with difference of treatment (treatment) in course of study. To link problem above hence, researcher consider necessary to use model study of *Trait Treatment Interaction* (TTI), where model this study make account of individual difference which owned by educative participant. So that educative participant will give to treat as according to ability or potency which they have each

METHODS

Research design was compiled research structure and plan in such a manner so that researcher obtain; get answer or questions of its research. Research Desain good for providing valid answer, reliable, object and precisely to questions of research and control or control other variant, according to fomentation of Kerlinger (2006). The research used quasy experimental design with quantitative methods. Population in the research 103 students. But, only 66 students joint as a sample of the research. The technique of sampling was used simple random sampling. Because, we had a frame work of reference. Pursuant to the description will be studied: model study of *Trait Treatment Interaction* in study of 1st Elementary Study of chemistry. The data collected with questionaire and using handout. The data analysed with statistical approach with using *Statistical Analysis of Science*.

RESULT AND DISCUSSION

Result of this research discover that: (1) Existence of ability correlation early between ability early interconnected of educative participant chemistry konten with kemampauan early related to aspect of religiusitas educative participant. As for correlation which positive correlation with high hand in glove storey; level between ability early related to educative participant chemistry konten and kemampauan early related to aspect of religiusitas educative participant. (2) With in controlling ability early, performance study of educative participant chemistry which have ability quickly (learner faster) using Self-Supporting Study (Self Learning) compared to higher of Study Of Habit (Regular Teaching) with study addition (Tutorial), effect of effect treatment of study variation of use teaching model of *Trait Treatment Interaction* (TTI). With controlling ability early, performance study of educative participant chemistry which have tardy ability (learner slower) using Study Of Habit (Regular Teaching) with study addition (compared to Higher Tutorial) of Self-Supporting Study (Self Learning), effect of effect treatment of study variation.

The analysis result of hipotetical research: (1) There are difference of performance study of educative participant which [is] signifikan [among/between] self-supporting study (Self Learning) with study of habit (Regular Teaching) with study addition (Tutorial) at Class A. Analyse to yield t = 0,677 db = 64 and sig (2 tailed) or p-value = 0,028 / 2 = 0,014 < 0,05 or H_0 rejected. Thereby raised hypothesis to be tested by data. So that concluded that its meaning of educative by participant of high study performance relative, if using self-supporting study (Self Learning) at capable educative participant group quickly method berbading study of habit (Regular Teaching) with study addition (Tutorial). (2) There are difference of performance result of study of educative participant which is signifikan between self-supporting study (Self Learning) with study of habit (Regular Teaching) with study addition (Tutorial) at Class C. Analyse to yield t = 0,732 db = 64 and sig (2 tailed) or p-value = 0,042 / 2 = 0,021 < 0,05 or H0 refused. Thereby raised hypothesis to be tested by data.

Study conducted Shuell (1986) about "Individually Differences: Changing Conceptions Research Practice and in" discovering that difference of that individual relate to intelegensi and various owned growth duty it. Its meaning of individual difference constituted by the existence of difference of ability had by each educative participant and growth duty which need to be run as according to its age step; child age and adolescent age. Then research conducted by Davis III et al. (1983) about " The Effectiveness Three Methods Teaching Social Studies Concepts Fourth-Grade Students to of of: An Aptitude-Treatment Interaction Study" discovering influence inexistence which is significant in applying of model study of Aptitude Treatment Interaction by using three method learn social science to performance learn educative participant. Method learn this is: (1) critical method concept (esensial). (2) Method giftof example of and do not use example of. (3) Method of practicing through presentation. Later; then the existence of interaction between treatment (applying of method learn) with gender. Its meaning third method learn to have an effect on to gender had by educative participant.

Result of research above giving picture to us that model study of ATI (Aptitude Treatment Interaction) can be combined with study method applied by old fellow at home to be able to improve result of performance study of educative participant. Result of study above indicating that model study of Trait Treatment Interaction can be used to study personality characteristic had by individual.

CONCLUSION

The research conclude : (1) The self-learning was suitable to teach for students with high capability of knowledge (faster lesrner). (2) The regular teaching + tutorial was suitable to teach for students with lower capability of knowledge (slower learner). (3) There were correlation between religiosity and capability of content (chemistry of knowledge). (4) *Trait Treatment Interaction* can be improve the capability of academic performance of students Chemistry Science.

ACKNOWLEDGMENT

I would like to thank for my supervisor; Prof. Dr. Festiyed, MS. , Prof. Dr. Mukhaiyar, M.Pd. , Prof. Jalius Jamma, PhD (Universitas Negeri Padang) Who introduce me with research in Education. I also dedicate my articles to my students who has inspired me to do research in Teaching of Chemistry. Special thanks to Undiksha who want to facilitate for publishing my article.

REFFERENCES

Aswirna, P. 2016. Trait Treatment Interaction Models for Improving Students Achie vemnet in Learning Chemistry. Dissertation. Padang: State University of Padang.

Kowollik, Vanessa. 2009. "The Interaction Between General Cognitive Ability & Training Structure: Using Meta-Analysis to Investigate Aptitude Treatment Interaction (ATI) in The Training Literature". Dissertation. Oklahoma: The Graduate College University of Oklahoma.

Davis III, J.C. McKinney, C.W., Larkins, A.G., Ford, M.J. 1983. The Effectiveness of Three Methods of Teaching Socials Studies Concepts to Fourth-Grade Students: An Aptitude Treatment Interaction (ATI) Study. American Educational Research Journal, Vol. 20, No 4, pp 663-670.

Kerlinger. 2006. Measurement in Psychology. Los Angles: Mc Graw-Hill.

Shuell, Thomas J. 1986. Individual Difference: Changing Conceptions in Research & Practice. American Journal of Education, Vol. 94, No. 3, pp.: 356-377.

Snow, R.E., Federico, P.A., & Montague, William E. (Eds.). 1980. Aptitude, Learning, and Instruction. Vol.: 1. Cognitive Process Analyses of Aptitude. Hillsdale, New Jersey: Lawrence Erlbaum Associates, Publishers.

Snow, R.E., Federico, P.A., & Montague, William E. (Eds.). 1980. "Aptitude, Learning, and Instruction. Vol. 2. Cognitive Process Analyses of Learning and Problem Solving". Hillsdale, New Jersey: Lawrence Erlbaum Associates, Publishers.

Snow, Richard. E. & Farr, Marshall J. (Eds.). 1987. Aptitude, Learning, and Instruction. Vol. 3. Conative & Affective Process Analyses. Hillsdale, New Jersey: Lawrence Erlbaum Associates, Publishers.

Thomas J. Shuell .1986. Individual Differences: Changing Conceptions in Research and Practice. American Journal of Education, Vol. 94, No. 3 (May, 1986), pp. 356-377. Accessed: 04/11/2010 01:59

The Development of Mathematics Comics Media on Linear Equations and Linear Inequalities of One Variable

Nana Sepriyanti[1], Cisil Tapia[2]
[1,2]Tadris Matematika, Fakultas Tarbiyah dan Keguruan, IAIN Imam Bonjol Padang, Indonesia

Abstract. This development research was aimed to produce mathematics comics media which valid and practical. This research was a research and development employing the 4-D model by Thiagarajan & Semmel (1974). The model consisted of four development stages, namely define, design, develop and disseminate. The research instruments were validation sheets and practicality sheets. The data were analyzed by grouping the data based on the product qualification. The calculations were made to obtain mean scores classified into categories in the Likert scale assessment. The validity of product can be seen from the result of experts validation which show that the product was valid (86.25%). The practicality of product can be seen from the result of the teacher and students practicality assessment sheet which show that the product was in a practical category (90.04%). Thus, it can be concluded that the mathematics comics media on linear equations and linear inequalities of one variable was valid and very practical to be used as a medium of learning in mathematics.

1. INTRODUCTION

Mathematics is one of subject with a vital role in education. Mathematics learning consists of skill of knowledge, concept, principle or procedures instructed to the students in steps. It is due to mathematics is not an unbound sequence of knowledge. It implies that, learning mathematics may help students to comprehend other subjects. Nowadays, mathematics becomes a nightmare for many students and comes first among the lessons that are considered difficult to learn. Educators have a great responsibility in this respect. The most important one of these responsibilities is to identify the learning difficulties which are experienced about the lesson to be given and to take necessary precautions in view of these identified difficulties. Although covering a very large scope, learning difficulty in mathematics means a number of inadequacies particular to this field (Durmuş, 2007)

[1] IAIN Imam Bonjol Padang, Indonesia: munzilatululya@yahoo.com
[2] IAIN Imam Bonjol Padang, Indonesia: cisiltapia@yahoo.com

The objective of mathematics education is surely to make students actualize learning in the highest level. However, the fact that the vast majority of the students experience difficulties while a few of them actualize learning is considered a reality of life (Tall & Razali, 1993). To identify and eliminate these experienced difficulties; to assist and guide the students during learning process is not only a requirement of modern education but also among the responsibilities of teachers (Ersoy & Ardahan, 2003). Tall & Razali (1993) stated the reasons for the learning difficulties of students generally as follows: (i)learning the basic concepts in adequately, (ii) inadequacy in formulizing verbal problems mathematically, (iii) insufficiency in algebraic, geometric and trigonometric skills. On scrutinizing the related literature, it has been found out that educators carried out researches on varied learning difficulties students experience in mathematics beginning from the preschool period up to university level and researches that would help eliminate these difficulties (Coşkun, 2008; Durmuş, 2007; Çetinkaya & Ersoy 2009; Harel, 1989; Keşan, Kaya and Güvercin, 2010; Özmantar, Bingölbali & Akkoç, 2008; Rasmussen, 1998; Tall & Razali, 1993; Ural, 2006).

Baştürk (2009) observed that the mistakes made by the students in the concept of absolute value umerous and varied; and the most common mistakes made by them are the ones resulting from he fact that they solved the questions as if there were no absolute value. In their study entitled detection of the difficulties experienced by the elementary school students in absolute value, Yenilmez and Avcu(2009) performed a test composed of open-ended questions on 8thgrade students. As a result of the study, it was detected that the students experienced difficulty in the absolute value of expressions with letters and equation questions that contained absolute value. In the study which was performed on the first-year students who study elementary school mathematics teaching. Today, it is evident that there are various factors that obstruct students' learning and these factors are compiled in three basic sections as sociological, psychological and cognitive. The aim of this study is to detect the learning difficulties of the secondary school students in equation and inequality that contain terms with absolute values in terms of cognitive factors rather than sociological and psychological factors.

Mostly in every level of education; elementary school (known as SD), junior high school (known as SMP/MTsN) senior high school or vocational high school (known as SMA/SMK/MA) and the like, they tend to view mathematics as a terrifying as well as difficult and boring subject. The observation result conducted by the researcher at MTsN Durian Tarung indicates that teacher tend to dominate the class situation and tend to use less media in mathematics learning. The object of mathematics learning is still in abstract form and yet is not supported with an appropriate and attractive media of mathematics learning.

Furthermore, students tend to perform unexpected behaviors during teaching and learning process. The students are obstructing their peers, lots out of class permission, drawing or just doing nothing on their desk. Thus, the low attention display by the students during mathematics and learning process along with the difficulty they face in comprehending the learning material are obstacles experienced by the teacher to achieve the learning goal.

Some students interview result revealed that they find it truly hard to comprehend mathematics since it is identical to numbers, formula and calculation in which according to them is a confusion. These lead them to boredom and less focus in learning mathematics particularly toward material which requires comprehension, analysis and thoroughness.

In addition, teacher also has not yet rely on the use of varying and interesting learning media. Usually, teacher rely heavily on board, marker, Students Activity Sheet (known as LKS) and text book. It is suspected that all these old-fashion media are fail to promote the improvement of students' interest in mathematics learning, so that it may cause the students to think that mathematics is a hard and boring subject.

These problems indicate that most of students accomplishment has not yet achieved minimum criteria of mastery learning (known as KKM. One of the main reason for the students to have a low interest toward mathematics learning is non-varying media utilization. It is against the fact that today the use of media or aids in learning process has improved rapidly. Therefore, the existence of alternative learning media to motivate students to learn mathematics is of vital importance. The selection of appropriate learning media need to fulfill requirements, namely in line with the learning material, the ease to use the media, and attractive for the students, so that optimum learning goal can be achieved.

Mulyasa(2006) stated that "the selection of inappropriate or irrelevant learning media may result in students low comprehension toward the target material as this may also distract the development of students' individual competence". There are three kinds of learning media: graphic media, audio media, and audio visual media. One of the learning media employed is comic media which is part of graphic media. According to Sudjana & Rivai (2013) comics is define as one of the cartoon form which reveal a character and play a plot in a tight sequence and is a picture related which designed to entertain the reader. In this case, the comics is a particular comics with mathematics content. Rohani (1997) stated that comics media is a simple, easy, specified and in personal to comprehend, hence it is informative and educative. Comics is a learning media which facilitates learning core. This kind of media may trigger students motivation in learning since the media being used is interesting and fun.

Thus, may motivate the teachers to search for a more communicative way in teaching mathematics to the student. In doing so, mathematics teacher may actually refer to the research conducted by Richard (2006). Though the research is focus on language teaching, but the tone is actually no different. This research suggested that teacher should be aware about their very own understanding toward certain approach since this may lead to misconception and wrong practice in teaching and learning activity.

In addition, the research conducted by Jacobs & Farrell (2003) also revealed specific points or aspects of communicative language teaching. In case, teacher finds it hard to comprehend this issue or confuse about the point where to begin with, the detail explanation served by these two researchers would ease the teacher. Moreover, Littlewood (2014) hinges on reflection on teachers teaching attitude and perception for both in today situation and the future time. It is expected that these would contribute toward a better result on mathematics learning particularly. Finally, Canagarajah (2014) suggested that learning is long life learning. The keywords are movement and innovation. And so, this is why employing comic as a mathematics learning media is of vital importance.

Pardimin and Widodo (2017) states that Learning devices are tools or equipment to carry out the process that will enable educators and learners perform the learning activities. Learning media is one of the learning tools that support the success of the learning process. Mathematics comic media may help students' comprehension process of learning material since it in the form of comics along with the appropriate story. Mathematics comic media can also support the teaching method employed by the teacher, so that learning process can be more effective and efficient. The purpose of this research is to produce mathematics comic media which fulfil validity and practicality criteria. The result shows that the designed media is valid and practical. Tin Lam Toh proposes the use of cartoons and comics, which are usually enjoyed by school-going children, in teaching algebra. Feedback from teachers showed that their students became more motivated to learn the subject through the use of cartoons and comics in algebra lessons. The teachers were able to engage their students in the process of learning elementary algebra. Mathematics comics media may help students' comprehension process of learning material since it in the form of comics along with the appropriate story.

Mathematics comics media can also support the teaching method employed by the teacher, so that learning process can be more effective and efficient. The purpose of this research is to produce mathematics comics media which fulfill validity and practicality criteria.

2. METHOD

This research was a research and development employing the 4-D model by Thiagarajan and Semmel (1974). The model consisted of four development stages, namely define, design, develop and disseminate. This research development was only in 3 step. Define is to determine and define the requirements need to develop comics media. This stage is also to analyze the goal and limitation of the developed learning material. Next, design step is to pay attention on three product characteristics, namely: content, interface and support. At the develop step, formative evaluation is take place. It is consist of prototyping stage (self evaluation, expert reviews, one-to-one evaluation and Small Group).

The try out subject of this research was six students of class VII MTsN Durian Tarung Padang and two mathematics teachers. The research data is qualitative and quantitative. Qualitative data is obtain from the observation result and interview. Meanwhile, quantitative data is obtain from validation sheet. The research instruments are 1) device validation sheet. 2) practicality test sheet. Data analysis techniques are 1) analysis data which is obtain from observation sheet and interview guidance. 2) data analysis is obtain from validation sheet. Assessment score toward validity is using likert scale.

3. RESULT FINDING AND DISCUSSION

The research product specification are:

1. The comics story is not repeated in the same sub topic

Fig. 1. The Topic of linear equation

2. The title for each sub topic is different from one another. Each of the sub topic is in different theme.

   The story of the comics shows linear equation of one variable that will be learned by the students. The story will be divided into some sub topics:

   - Sub Topic I : Getting to Know Open Sentence and Variable

- Sub Topic II : Solving the Linear Equation of One Variable
- Sub Topic III : Solving the linear inequalities of One Variable
- Sub Topic IV : Create and Solve the Linear Equation Model of One Variable
- Sub Topic V : create and Solve the inequalities of One Variable

3. The mathematics comics comes in color and the language being used is the daily-life language used by the students, so that the students may find it easier to comprehend the learning material.

Fig. 1. Cover of comic

4. Consist of conclusion sheet. This sheet will be fulfilled by the students with the conclusion they obtained after learning the material.

Fig. 2. Student sheet activity

5. Consist of varied exercise to challenge the students to solve.

Exercise

1. Which one of the following is the linear equation of one variable?
   a) 2x – 4 = 8

b) $-4 + 3s = 24$
c) $-8 - d^2 = 32$

2. Find the linear equation of question number 1!

3. Find the compilation solution of this following linear equation.
   a) $7a + 3 = 0$
   b) $5m + 4 = 2m + 16$

4. Find the compilation solution of linear equation
   $2a - 100 = 20$, if:
   a) a is an odd number
   b) a an even number

Good Luck

Fig. 2. Caption of the Figure 1. Below the figure

The designed comics were be validated by mathematics expert which consist of four validator. The validation result of mathematics comics media can be seen at table 1 as follow:

Table 1 Mathematics Comics Media Validation Result

No	Aspect	Score
1	Content appropriateness	83.33%
2	Language	87.50%
3	Display	85.41%
4	Graphics	88.75%
Average		86.25%

The result of mathematics comics media validation displayed on table 1 is valid in criteria. It can be seen from the average score of validity for comics media as 86.25%. every component has validity score between 80% - 89%. It implies that every aspects of mathematics comics is valid. Next, the practicality of mathematics comics media from the students' point of view can be obtained from the practicality questionnaire result that is distributed toward some students as can be seen in table 2.

Table 2.The Result of Mathematics Comic Media Practicality

No	Aspect	Score
1	The ease of the use	86.80%
2	Time effectiveness	89.58%
3	Advantage and fascinate	93.75%
Average		90.04 %

Table 2 shows that according to the students, mathematics comics media is practical in its every aspect with the average score is 90,40%. Mathematics comics media Practicality media according to the teacher is obtain from practicality questionnaire result that can be seen on table 3.

Table 3.The Result of Mathematics Comic Media Practicality

No	Aspect	Score
1	The easy of the use	85.42%
2	Time effectiveness	87.50%
3	Advantage and fascinate	80.00%
Average		84.31 %

Table 3 shows that according to the teacher, mathematics comics media is practical in its every aspect with the average score is 84.31%. Based on the research finding, it can be concluded that the result of mathematics comics media development on linear equation & inequalities of one variable at class VII Yunior High School (SLTP/MTs) Padang is valid and very practical to be used as mathematics learning media. The resulted mathematics comics media obtain 86.25% validity result with valid criteria by validator. In whole, the practicality score given by the students is 90.04% with very practical category. Meanwhile, total practicality score by both teacher and students is 85% with practical category.

4. CONCLUSION

Based on the research finding, it can be concluded that the result of mathematics comic media development on linear equation &inequalities of one variable at class VII Yunior High School (SLTP/MTs) Padang is valid and very practical to be used as mathematics learning media. The resulted mathematics comic media obtain 86.25% validity result with valid criteria by validator. In whole, the practicality score given by the students is 90.04% with very practical category. Meanwhile, total practicality score by both teacher and students is 85% with practical category.

It is recommended for future researcher to dig and explore more on this topic by employing qualitative research method. The future researcher may also invite the participants such as the students and the teacher to work together and participate actively in the research in order to obtain a better understanding toward this particular topic. The future researcher may also refer to the research conducted by Sato & Kleinsasser (1999) who put its participants as the main resources for the research. As the result, a more comprehensive research result is attained. To be precise in doing this kind of recommended research, the future researcher can consult Heigham and Juanita's (2009) book.

REFERENCES

1. Ahmad Rohani. (1997). *Media Instruksional edukatif.* Jakarta: PT Rineka Cipta.
2. Baştürk, S. (2009). Student teachers' approaches to student's mistakes in the case of the absolute value concept. Necatibey Faculty of Education Electronic Journal of Science and Mathematics Education,3(1),174-194.all, D. (1993). Students' difficulties incalculus, proceeding of working group 3 on students' difficulties in calculus. ICME–7, Qoubec, Canada
3. Canagarajah, S. (2014).In search of a new paradigm for teaching English as an international language.*TESOL Journal,* **5**(4), 767-785.
4. Coşkun, O. (2008). Diagnosing learning difficulties in modular arithmetic. Unpublished Master Thesis.Atatürk University Graduate School of Natural and Applied Sciences Departman of Mathematics Education,
5. Durmuş, S. (2007). Matematikte öğrenme güçlüğü gösteren öğrencilere yönelik öğretim yaklaşımları. Mehmet Akif Ersoy Üniversitesi Eğitim Fakültesi Dergisi, 76–83

6. Ersoy, Y. &Ardahan, H. (2003). İlköğretim okullarında kesirlerin öğretimi-II: Tanıya yönelik etkinlikler düzenleme. -Available at http://www.matder.org.tr/index.php?option=com_content&view=article&id=64:ilkogretim-okullarinda-kesirlerin-ogretimi-ii-taniya-yonelik-etkinlikler-duzenleme&catid=8:matematik-kosesi-makaleleri&Itemid=172
7. Harel, G. (1989). Learning and teaching linear algebra: Difficulties and an alternative approach to visualizing concepts and processes. Focus on Learning Problems in Mathematics, 11(2), 139–148.
8. Heigham, Juanita, Dr; Croker, Robert A., Dr (2009). Qualitative Research in Applied Linguistics: A Practical Introduction. Retrieved from http://www.eblib.com
9. Keşan, C., Kaya, D.,& Güvercin S. (2010). The effect of problem posing approach to the gifted student's mathematical abilities, International Online Journal of Educational Sciences, 2 (3), 677-687
10. Jacobs, M. G, & Farrell, T. S. C. (2003).Understanding and implementing CLT (Communicative Language Teaching) Paradigm.*RELC Journal, **34***(1), 5-30.
11. Littlewood, W. (2014). Communicative-oriented language teaching: where are we now? Where do we go from here? *Language Teaching, **47***(3), 349-362.
12. Mulyasa, E. (2006). *Kurikulum yang Disempurnakan Pengembangan Standar Kompetensi dan Kompetensi Dasar.* Bandung: PT. Remaja Rosdakarya..
13. Nana Sudjana dan Ahmad Rivai. (2013). *Media Pengajaran*. Bandung: Sinar Baru Algesindo.
14. Özmantar, M. F., Bingölbali, E. &Akkoç, H. (2008). Matematiksel kavram yanılgıları ve çözüm önerileri.Ankara: Pegem Akademi.
15. Pardimin and Sri Adi Widodo (2017). Development Comic Based Problem Solving in Geometry. IEJME-Mathematics Education Vol.12, No.3, 233-241
16. Rasmussen, C. L. (1998). Reform in differential equations: a case study of students' understandings and difficulties. The Annual Meeting of American Educational Research Association, San Diego, CA. -Available at http://eric.ed.gov/ERICDocs/data/ericdocs2sql/content_stroge_01/0000019b/80/15/8e/cb.pdf
17. Richards, J, C. (2006). Communicative Language Teaching Today. Retrieved from https://learn.uq.edu.au/bbcswebdav/pid-1922751-dt-content-rid-9055584_1/courses/SLAT7826S/Richards-Communicative-Language.pdf
18. Sato, K., &Kleinsasser, R, C. (1999). Communicative Language Teaching (CLT): Practical understanding. The Modern Language Journal, *83 (4)*, 494-517.
19. Tall, D. O. &Razali, M. R. (1993). Diagnosing students' difficulties in learning mathematics. International Journal of Mathematical Education in Science and Technology, 24(2), 209–222
20. Thiagarajan, Semmel & Semmel. (1974). Instructional Development for Training Teachers of Exceptional Children. A sourcebook. Central for Innovation on the Handicapped: Bloomington Indiana
21. Tin Lam Toh . Use of Cartoons and Comics to Teach Algebra in Mathematics Classrooms. National Institute of Education, Nanyang Technological University, Singapura
22. Ural, A. (2006). Conceptual obstacles concerning the learning of the function. Ege Eğitim Dergisi,7(2),75–94.
23. Yenilmez, K. &Avcu, T. (2009). Primary School Students' difficulties in Learning Absolute Value. Dicle Universitesi Ziya Gokalp Egitim Fakultesi Dergisi, 12, 80-88.

THE PROTOTYPE DEVELOPMENT OF CREATIVE THINKING SKILLS MODEL'S BOOK

I Gusti Ayu Tri Agustiana[1], Rudiana Agustini[2], Muslimin Ibrahim[3]
[1] *Elementary School Education of Department Universitas Pendidikan Ganesha Bali Indonesia*
[2] Chemistry Education of Department, State University of Surabaya
[3] Biology Education of Department, State University of Surabaya
Jl. Udayana No.12 Singaraja Bali
Email : igustiayutriagustiana@yahoo.co.id

ABSTRACT

Model of book is an important part of the research process in developing creative thinking skills. Considering the importance of creative thinking skills, it is necessary to attempt developing such skills in the learning process by applying an appropriate and effective model. This research will develop the components of learning model that includes theoretical rationale, syntax, social systems, reaction principle, support systems, and the impact instructional companion. Learning model of science is developed with syntax orientation, excavation creative ideas, research, elaboration, presenting the results of the work, evaluation, and implementation (OPPEMEI) with indicators of creative thinking skills fluency, flexibility, originality, elaboration and evaluation. To achieve the goals, it should need a couple of steps, namely (1) analyzing and providing a rational basis to support the development of learning model for creative thinking skills. (2) analyzing and constructing the theories that support the development of the model. Some synthesis theories for reviews are a constructivism theory, creative thinking theories, a motivation theory, a theory of Vygotsky (the theory of socio-cultural) and cognitive theory, (3) Analyzing the flow of thought decrease hypothetical OPPEMEI Model, (4) providing guidelines for the implementation of learning using OPPEMEI learning model, (5) conducting Focus Group Discussions (FGDs) and (6) Revisions. Validity is assessed by expert judges in terms of content validity and construct validity. The construct validity includes the logic of stage model (syntax) and the support of relevant theory at each stage. Validation of the content reflects the state of the art or the strength of the device and the content materials are assessed based on the conformity with the applicable curriculum. The learning model is stated valid, if the expert judges and practical say that the model is developed based on strong and latest theoretical rationale and there is consistency between the components.

Keywords: creative thinking skills, OPPEMEI Model

I. INTRODUCTION

Creative thinking skills are a factor that will support the children's creativities, and it is believed as a embryo of human resources (Albert, 2016; Munandar, 2009). The superior human resources have the characteristics of high productivity (efficiency, effectiveness, and innovation) (Bono, 2015), Amabile (as cited in Nur, 2014) showed that the key aspects of creative thinking are 1) Happy or entertained when the children have different opinions with the others and trying the deviate solutions from status quo (the social situations nowadays). (2) Combining knowledge that is from many kinds of area that separate before. (3) The students' ability to continue in solving the difficult and not attractive problems. (4) The ability to leave a method and come again with a fresh perspective (incubation). These conditions can prompt the children creating kinds of mindset. Those varieties appear because the areas of creative thinking skills are given to students depend on the appropriate learning model that are applied (Arend, 2012). The learning models that are set depend on the purpose of the learning process. Creativity can be created with scientific approaches (Permendikbud No. 23 tahun 2016). That is why, the use of model is determined by the characteristics of materials that are going to be learned.

Natural sciences (IPA) related with the way how to know about nature systematically by using scientific methods, so IPA is not only about mastering the knowledge, but also as an inquiry process. Therefore, teachers should be able to bring phenomena of nature into the classroom' activities, then their students are expected could more know about themselves and the environment around them. So, the goals of learning IPA is it should teach students in producing a product and the process for it, and also unconsciously it can build creative thinking skills for students. Based on these perspectives, a model that could support the development of creative thinking skills in the natural sciences (IPA) learning will become urgent (Lin & Shih 2016).

All this time, the contextual stimulations that are used to find the creative thinking skills are not optimally implemented in varieties of natural learning (IPA) such as (1) The learning model of problem based learning (Wijnia et al. 2016; Allen et al. 2011; Nuttariya et al. 2012), (2) The Brain Based Learning (BBL) model (Eagleton & Muller 2011), (3) Inquiry model (Hong et al. 2014; Lederman et al. 2014; Lehtinen, 2016).

Thinking creatively is a main model of OPPEMEI in which this model becomes an important model. That condition can be seen from the steps of OPPEMEI Model that consist of (1) Orientation, (2) Extracting creative thinking, (3) Investigation, (4) Elaboration, (5) Developing and presenting results of the work, (6) Evaluating for the process of the product, (7) Implementation. The use of this model comes from the theory of cognitive motivation, this theory is believed that individuals prompted to do some actions based on their mind if the students are given stimulus with varieties of contextual problems, so it can build a deeper understanding.

The purpose of learning model that is able to train the students' skills as stocks that will be used to analyze and evaluate data so further the students are able to explain scientifically the nature phenomena in which are in same steps with OPPEMEI Model. Therefore, it can help students in developing and using the habit of creative thinking skills, understanding the material of IPA, and giving chances for students to take parts during the science process.

Beside that, OPPEMEI model can build students' character. This condition is appropriate with the main purpose of *Undang-undang Nomer 20 Tahun 2003* about National Education System that is expected can create a developing process in which show the quality of students' characters as the next generation for nation in the future, moreover in facing challenges in 21st century known as century of knowledge, knowledge based on society with its principles those are (1) Instruction should be student-centered, (2) Education should be collaborative, (3) Learning should have context, (4) Schools should be integrated with society. In the productivity's demotions, OPPEMEI Model is in the same way with the purpose of Curriculum 2013 in which an education process can produce productive, creative, innovative, and effective society in Indonesia through strengthening attitudes and integrated knowledge (Yasa, 2014).

The development of OPPEMEI model is in the same way with *Permendikbud No. 20 Tahun 2016*, a graduate competent standard for elementary school and junior high school, explain that every graduate of elementary school and junior high school should have competences for three dimensions namely attitudes, knowledge, and skills (Chapter II). For skills' dimension in which students should have skills in thinking and doing, those are creative, productive, critical, independent, collaborative, and communicative through scientific approaches that are appropriate for the development of students in which are relevant with the given tasks.

In the content aspects, the development of OPPEMEI learning model emphasizes the achievement of creative thinking skills based on *Permendikub No. 21 Tahun 2016* which is about the context standard of elementary school and junior high school, ask the students should be able to show the skills of reasoning, processing, and presenting effectively, creatively, productively, critically, independently, collaborating, communicatively, and solution. In the concrete and abstract areas are about the development of what they have learned in the school and enable to use the method based on scientific rules.

In the aspect of process, the development of OPPEMEI learning model emphasize on the achievement of creative thinking skills based on *Permendikub No. 22 Tahun 2016* in which strengthen the scientific approaches (scientific), integrated thematic (thematic between subjects), and thematic (between a subject) that need applying a subject which is provided with attitudes or research (discovery/inquiry learning). In order to prompt the students' ability in producing a contextual product, both individual and groups so then it is really suggested to use an approach that can produce a product based on the problems (project based learning), by the governments' rules in this case students should learn about ideas, solving problems, applying what they have learned, giving opinions, doing a trial, and doing monitoring (Sapurto, 2015). This condition is appropriate with *Permendikbud* in which assessing skills are done through practices, products, projects, portfolios, and/or other techniques based on the assessed competences; *No. 23 tahun 2016* about assessment standard that will measure.

Based on some analysis of learning models that have been implemented in IPA learning in order to increase creative thinking skills, and cognitive understanding about the concept of science learning for students as the candidate teachers of elementary school, so it needs to formulate a hypothetical model that emphasizes some aspects such as (1) focusing orientation in the early stage of learning, (2) There is extracting ideas that can make students feel challenging for their cognitive, (3) Investigative aspects are to solve creative problems, (4) An aspect in which students share their ideas (elaboration), (5) An aspect of developing and producing a product, (6) An aspect of evaluation as reflection and revision, and the last is (7) The aspect of Implementation. Formulating those aspects can be a learning model that should be analyzed more.

Those principles are a new touch in science learning, especially basic natural science (IPA) for students that will become a teacher namely OPPEMEI learning model in which has seven phases namely (1) Orientation, (2) Extracting creative ideas, (3) Investigation (4) Elaboration, (5) Developing and serving results of the work, (6) Evaluating the process of the product, and (7) Implementation. The use of this model comes from the theory of cognitive, this theory believe that individuals prompt to do actions from their mind. The theory of attribution, according to Arend (2012), is the way how the individuals understand and predict the cause of success or failures that can be a motivation. Relating with the efforts in applying OPPMEI Model approach, so the explanation of this article is done by developing a prototype book about learning model of creative thinking skills. By the content validity concept that focus on the existence of a need for the intervention and its design is based on state-of the-art (scientific) knowledge (Nieveen, 2010). In addition, the intervention of this research is OPPEMEI Model.

The purposes of this research are **(1)** Assessing content validity of Orientation Model, extracting the creative ideas, Investigation, Elaboration, Presenting the results, and Evaluating the implementation (OPPEMEI) in terms of (1) rational development model, and (2) The design of this model based on the latest knowledge. **(2)** Measuring the validity in extracting creative thinking model, Investigation, Elaboration, Presenting the results, and Evaluating the implementation (OPPEMEI) in terms of consistency, and the logic of the model's components internally.

II. METHOD

The design of this research is developing research (Development and Research). The validity of learning is done through Focus Group Discussion (FGD) by five expert judges. OPPEMEI Model is valid if the result of validity can fulfill the criteria. The criteria are (1) The expert judges stated that OPPEMEI model is based on theory of learning or strong rational theory (Content Validity), (2) Expert judges stated that the components of OPPEMEI model consistently relate each other (Construct Validity).

The analysis of OPPEMEI Model's validity is done by doing a recapitulation towards the expert judges' statements, finding out the average results of all expert judges, finding out the total of average, the categories of validity are set based on the categorization criteria of quality model in which are adapted from categorizing according (Bloom et al. 1956) as follows: $4{,}5 \leq$ model $\leq 5{,}0$ extremely valid, $3{,}5 \leq$ model $< 4{,}5$ valid, $2{,}5 \leq$ model < 3, valid enough, $1{,}5 \leq$ model $< 2{,}5$ less valid, model $< 1{,}5$ not valid.

The criteria that are used to decide whether the OPPEMEI has sufficient degree of validity or not is if the model of validity is in valid category. If the stage of validity's achievement is lower that the valid category it means that, the model needs revision based on the expert judges' suggestions. Further, the validity is done again. The process continuously as well until the ideal OPPEMEI model is achieved.

III. RESULTS AND DISCUSSION

A. The results of content validity for OPPEMEI Model in terms of some aspects as follows: (a) The rational development of OPPEMEI Model, (b) OPPEMEI Model is designed based on the latest knowledge.

B. The rational development of OPPMEI Model

No	The aspects of the Assessments	Expert judge's assessment					Skor Rerata
		1	2	3	4	5	
1	The development of OPPEMEI Model aims to increase creative thinking skills and the result of students' learning process as a need of primary graduate competences for the elementary school teachers' candidates	4	5	4	4	5	4.4
2	The development of OPPEMEI Model uses scientific and inquiry approaches based on *Permendikbud Nomor 49 Tahun 2014* about the national standard of higher education in which emphasizes that the learning process should give priority to the use of scientific approaches, so it can achieve the academic environment based on system of values, norms, and knowledge' rules.	4	3	4	5	5	4.2
3	The development of OPPEMEI Model bridges the gap between a hope of graduate competences' needs based on standard competencies of Elementary Schools' teachers in which they are able to develop curriculum and learning process creatively and innovatively at Elementary Schools.	5	4	5	4	5	4.6
4	The development of OPPEMEI Model based on a recommendation in improving the innovative learning model that centered in the university students such as PBL, BBL and inquiry.	4	5	5	4	4	4.4
5	The development of OPPEMEI Model really supports the government's programs in order to improve the competence of higher education graduates creatively and responsibly that have been set on *Peraturan Pemerintah RI Nomor 8 Tahun 2012* about KKNI, *Permendikbud Nomor 73 Tahun 2013* about the application of KKNI for higher education and *Permendikbud nomor 49 tahun 2014* about the national standard of higher education. In which it emphasizes that the universities should give a space in developing creativities.	5	5	5	4	5	4.4

C. OPPEMEI Model is designed based on the latest Knowledge

No	The aspects of the Assessment	The expert judge's assessment					Skor Rerata
		1	2	3	4	5	
1	The development of OPPEMEI Model uses the main sources from international journals. By revising the weakness from PBL Model, BBL, and inquiry those are 1) the time that take too long (Lam, 2004), 2) formulating hypothesis was difficult to be done in conditions without cooperation (Soltis, *et al*, 2015), 3) exploration is in PBL usually widened and expanded (Ross & Hurlbert, 2004), without a presentation of coherent data, so the creativity of students are low (Caine, 1995), 4) If the production of report do not provide with facts, data, and a theory, so it makes the ability to keep their ideas weak (Kuby, 2003), 5) There is not a stage that is used to show their products, so it effect their understanding that are still low (Osgood-Campbell 2015), 6) If the incubation process is not verified on the aspect of assessment process and students' understanding product are low (Osgood-Campbell 2015). 7) If the concept is not optimally applied in the everyday life, then the students' creative thinking skills will be low (Caine 1995).	3	4	5	5	4	4.2
2	The development of OPPEMEI Model uses the creative thinking skills theory, motivation theory (ARCS Model), cognitive learning theory (Meta cognitive), the theory of complex cognitive processes (Creativity and problem solving), theory of mind-centered learning and constructivism (observational learning, the reciprocal causation model, constructivism, and scaffolding) from an standard and up-to-date education psychology book namely education psychology book written by (Moreno 2013), (Arends 2012) and (Kauchak & Eggen 2013).	4	5	4	5	5	4.6
3	The development of OPPEMEI Model uses empiric foundation from international and up-to-date journals in which the researches' results by (Hmelo-silver, 2013; Lin & Bruce, 2013; Osgood-Campbell, 2015; Soltis *et al.*, 2015); Talib, 2014)	5	5	5	5	4	4.8
4	The development of OPPEMEI Model more uses references to the latest research journals rather than reference books.	4	5	5	4	4	4.4

5	The development of OPPEMEI Model's planning pays attention on the KKNI competences organize learning that requires the use of ICT media and the use of technology such as laptop, LCD, and the smooth internet network.	3	4	4	4	4	3.8
6	The development of OPPEMEI Model's planning pays attention on the KKNI competences (Model Book) and organize learning that requires the use of ICT media and the use of technology such as laptop, LCD, and smooth internet network.	4	4	4	4	5	4.4
7	The development of OPPEMEI Model' assessment is adjusted with a modified theory of creative thinking skills from Munandar 2009) and (Guilford & Christensen 1973).	4	4	3	4	4	4.35

The results of construct validation of OPPEMEI model are reviewed from some aspects as follows:

A. The overview of OPPEMEI Model

No	**The aspects of the Assessment**	**Expert Judge's Assessment**					**Average scores**
		1	**2**	**3**	**4**	**5**	
1	The suitability between the purposes of the development of OPPEMEI Model in order to increase creative thinking skills with the needs of graduate's competencies.	4	4	4	5	4	4.2
2	The suitability between the purpose of the OPPEMEI Model's development with researchers' recommendation about the improvement of strategies in learning process innovatively that focus on the students such as PBL, BBL, and Inquiry.	4	5	4	4	4	4.2
3	The suitability between The purpose of OPPEMEI Models' development with the governments' programs that have been set on *Peraturan Pemerimtah RI Nomer 8 tahun 2012* about KKNI, *Permendikbud Nomor 73 Tahun 2013* about the application of KKNI for higher education and *Permendikbud Nomor 49 Tahun 2014* about the national standard of higher education	4	4	4	5	4	4.2
4	The meaning and symbols that are on the component of OPPEMEI Model book are not contradictive.	3	4	5	4	4	4

B. The Theoretical support and The empirical of OPPEMEI Model

No	The aspects of the Assessments	Expert Judge's Assessment					Average scores
		1	2	3	4	5	
1	**The consistency of Phase 1: Oriented with:**						
	Theoretical Foundation:	4	4	5	4	4	4.2
	Empirical Foundation:	4	4	4	4	5	4.2
2	**Consistency of Phase 2: Extracting creative ideas with**						
	Theoretical Foundation:	4	3	4	5	4	4
	Empirical Foundation:	4	4	4	4	4	4
3	**Consistency of Phase 3: Investigating with**						
	Theoretical Foundation:	4	4	5	4	5	4.4
	Empirical Foundation:	4	4	4	4	5	4.2
4	**Consistency of Phase 4: Elaborating with**						
	Theoretical Foundation:	4	3	4	3	4	3.6
	Empirical Foundation	4	4	4	4	4	4
5	**Consistency of Phase 5: Presenting the result with**						
	Theoretical Foundation:	3	4	5	4	4	4
	Empirical Foundation:	4	4	3	4	4	3.8
6	**Consistency of Phase 6: Evaluating with**						
	Theoretical Foundation:	4	5	4	5	4	4.4
	Empirical Foundation:	4	3	4	4	3	3.6
7	**Consistency of Phase 7: Implementing with**						
	Theoretical Foundation:	4	4	3	4	4	3.8
	Empirical Foundation:	3	4	3	3	3	3.2

C. The OPPEMEI Model's Planning

No	The aspects of the Assessments	Penilaian Ahli					Average scores
		1	2	3	4	5	
1	The consistency is in determining the basic competencies, indicators, and learning objectives in the field of university.	4	4	4	5	4	4.2
2	The consistency of planning is in making creative questions with an assessment in which measures creative thinking skills on OPPEMEI that has been developed by Hu & Adey (2010).	4	5	4	4	4	4.2

3	The consistency of organizing resources (Understanding the process of skills and the operation of media) and logistic (Students' handbook, students' worksheets, smooth internet network, and LCD availability) with pre-requisite in which will support the success of learning process with OPPEMEI Model.	4	4	4	5	4	4.2

D. The implementation of OPPEMEI Model

No	The Aspects of the Assessment	The expert judge's assessment					Average scores
		1	2	3	4	5	
1	The phase of syntax can give a description about the sequences of learning activities logically, systematically, and consistently. This is to achieve the goals of the OPPEMEI Model's development	4	4	4	5	5	4.4
2	The social systems are able to give a description about the parts of students and lecturers, the interaction between students, the interaction between students and lecturer logically, and consistently achieve the goals of OPPEMEI Model's development	4	5	4	4	4	4.2
3	The reactionary principal gives a description the way the lecturers appreciate and respond the students' behavior during learning process logically and consistently in order to achieve the goals of OPPEMEI Model's development	4	4	4	4	4	4
4	The supportive system (Syllabus, SAP, BAM, LKM, LP) and media (tools and materials of investigation, ICT media, LCD, and smooth internet network logically that are able to support the smooth learning process with OPPEMEI Model	4	4	5	5	5	4.6
5	The impact of instructional consistently aims to develop OPPEMEI Model, and the impact of escort unconsciously effect from the implementation of syntax, social systems, reactionary systems, and supportive systems of OPPEMEI Model	4	5	4	4	5	4.4

E. The management of Learning environment

No	The Aspects of the Assessment	Expert Judge's Assessment					Average scores
		1	2	3	4	5	
1	Learning environments are really important in supporting the achievement for the goals of OPPEMEI Model's development in order to increase the skills of thinking creatively and the result study of students	5	4	4	5	5	4.6
2	The learning environments are arranged based on the theoretical and empirical foundation from latest journals, so it can be implemented logically in KBM in the classroom	5	5	4	5	5	4.8

Based on the result of analysis data by expert judges that learning model which is developed gain score 4,38 for content validity and 4,27 for construct validity. If it is compared with the criteria of validity for a product that gain score $3,5 \leq$ model $< 4,5$, it means that the learning model that has been developed is valid based on the results of content and construct validity. Therefore, the results of this research in which OPPEMEI Model has been developed show that (1) The rational development of OPPEMEI Model are comprehensive, (2) OPPEMEI Model is designed based on the latest knowledge, (3) The overview of OPPEMEI Model is appropriate with theoretical and empirical supports, (4)The syntax's sequences are logical, it can be done a teaching activities, an orientation in which the students practice their thinking skills creatively, (5) The consistency is in determining the basic competencies, indicators, and the goals of learning with KKNI competency (6) The social system can give a description about the parts of lecturers and students, the interaction between students, the interaction between lecturers and students logically, and consistently in achieving the goals of OPPEMEI Model's development, (7) The reactionary principles give a description the way how lecturers appreciate and respond the students' behavior during learning process logically and consistently to achieve the goals of OPPEMEI Model's development, (8) The support system and media logically can support the learning process smoothly by applying OPPEMEI Media, (9) The impact of instructional consistently aims to develop OPPEMEI Model and the impact of escort is unconsciously the effect from the implementation of syntax, social systems, reactionary systems, and supportive systems of OPPEMEI Model, (10) Learning environments are really important in supporting the achievement for the goals of OPPEMEI Model's development in order to increase the skills of thinking creatively and the result of students' study. In other words, relevant aspects are implied by Nieveen (1999) as a requisite of content validity that has been fulfilled with this model.

The selection of cognitive development theory that is correspond with the context of learning nowadays, it seems relevant in solving the learning's problems that happen because the variety of students. Thinking creatively is a authentic and reflective thought and can produce something that is complex and new (original). Gie (2003) gave limits that thinking creatively is a series of actions that are done by using their thought in order to create a new thought that comes from a series of memories that consist of ideas, information, concepts, experiences, and knowledge that exist in their mind as well. Creative thinking is mental activities in order to create connections continuously so it is produced an association product thinking appropriately (Diener et al. 2016; Evans, 1991).

The creative thinking skills identically focus in the strength of the brain. More than two decades ago, the results of researches about brain had brought astonishing results. The new way that are used in order to create and focus to the scientific ability, unconsciously has been found in every research areas of biology and psychology. For some people, the relation between these two researches are invisible, but if it is applied correctly then new scientific views into the utilization of the mind can help a person multiply and even optimize the power of mind. The reinforcement of minds that are meant: the power of studying, remembering, reading, listening and thinking (Stine, 2001).

Amabile and Gardner (as citen in Nur, 2014) emphasize that creative thinking is the key aspect of creative process. Ambile suggested that the key aspects of creative thinking are 1) Happy or entertained when the children have different opinions with the others and trying the deviate solutions from status quo (the social situations nowadays). (2) Combining knowledge that is from many kinds of area that separate before. (3) The students' ability to continue in solving the difficult and not attractive problems, (4) The ability to leave a method and come again with a fresh perspective (Nur, 2014).

Creative thinking according Coleman *et al* (1979) is "thinking which produces new methods, new concepts, new understanding, new inventions, and new work of art" (Rakhmat, 2005). This shows that creative thinking skills emphasize on regenerate toward what have existed. In other words, creative thinking is a kind of thinking that aims to create new ideas that are on the process according Jawwad (2014), it should fulfill four fundamental aspects namely:

the high sensitivity towards kinds of problem, the high productivity in answering one question, the high elasticity in creating varieties of thinking, the high originality in creating varieties of thinking and the high originality in creating unique and new ideas. In the same way with those things, Munandar (2009) and Guilford (1981) explain that creative thinking skill include five aspects namely fluency, flexibility, originality, elaboration and evaluation.

Motivation is a physical and psychology conditions that exist in a person that controls the actions in the certain ways (Gates, et, al, 1954). Greenberg, 1966 (as cited in Djaalin, 2007), says that motivation is a process that are arousing, directing, and stabilizing the behavior in a certain purpose. The mental and human behaviors are a dimension that becomes a reference that is to be discussed into Psychology. According to Miller (1974), psychology is a knowledge that tries to explain, predict, and control the phenomena of mental and human behavior.

Based on thinking concentration theory explain that brain is an organ of human body that is complex and unique. The number of cells contained on the brain varies between 50 to 100 billion brain cells (Jensen, 2011). According to Grisewood (as cited in Syah, 2007), "A brain is the substance of the gray matter that located inside the coconut shell in which commonly has function as a control to the movements and work of the all physical organs as well as thinking tools". On the human brain, there are different parts that served some kinds of functions of mental, thinking, sexuality, memory, defense, emotion, and creativity. Based on constructivism theory explains about a person that understands toward something that has been known (Lorsbach dan Tobin, 1997). All of things that have been learned will be saved in the memories, based on the constructive ability of someone. The theory of socialculture is the last theory about motivation in which social-cultural theory comes from an art of psychologist Rusia Vygotsky (1986), who theories that human' activities happen under cultural control and the influence of the control really depends on what is done and thought.

IV. CONCLUSION

Based on the explanation above then it can be concluded that the first is the content validity of Orientated Model, Extracting creative ideas, Investigation, Elaboration, Presenting Results, and Evaluating the Implementation (OPPEMEI) in terms of: (1) Rational Development Model, and (2) Model's design based on the latest knowledge. The second is measuring the validity of Extracting Creative Ideas Model, Investigation, Elaboration, Presenting the results, and Evaluating the Implementation (OPPEMEI) in terms of consistency aspects and logical components of model internally. Therefore, the results of this research in which OPPEMEI Model has been developed show that (1) The rational development of OPPEMEI Model are comprehensive, (2) OPPEMEI Model is designed based on the latest knowledge, (3) The overview of OPPEMEI Model is appropriate with theoretical and empirical supports, (4) The syntax's sequences are logical, it can be done teaching activities, an orientation in which the students practice their thinking skills creatively, (5) The consistency is in determining the basic competencies, indicators, and the goals of learning with KKNI competency. (6) The social system can give a description about the rules of lecturers and students, the interaction between students, the interaction between lecturers and students logically, and consistently achieving the goals of OPPEMEI Model's development, (7) The reactionary principles give a description the way how lecturers appreciate and respond the students' behavior during learning process logically and consistently to achieve the goals of OPPEMEI Model's development, (8) The support system and media logically can support the learning process smoothly by applying OPPEMEI Media, (9) The impact of instructional consistently aims to develop OPPEMEI Model.

REFERENCES

Albert, D.L., 2016. Teaching Critical and Creative Thinking Skills Through Problem-Solivng in High School Mathematics Classes.

Allen, D.E., Donham, R.S. & Bernhardt, S.A., 2011. Problem-based learning. *New Directions for Teaching and Learning*, (128), pp.21–29.

Arends, R.I., 2012. *Learning to Teach. 9th ed*,

Bloom, B.S. et al., 1956. Taxonomy of Educational Objectives: The Classification of Educational Goals: Handbook I Cognitive Domain. *New York*, 16, p.207.

Bono, E. De, 2015. *Lateral Thinking: Creativity Step by Step*, Available at: http://www.amazon.com/Lateral-Thinking-Creativity-Perennial-Library-ebook/dp/B003V1WTLI/ref=sr_1_1?s=books&ie=UTF8&qid=1403800683&sr=1-1&keywords=Lateral+Thinking:+Creativity+Step+by+Step.

Caine, R.N.C. and G., 1995. Reinventing Schools Through Brain-Based Learning. *Educational Leadership*, pp.43–47.

Diener, M.L. et al., 2016. Socioemotional Correlates of Creative Potential in Preschool Age Children: Thinking Beyond Student Academic Assessments. *Creativity Research Journal*, 28(4), pp.450–457. Available at: https://www.tandfonline.com/doi/full/10.1080/10400419.2016.1229975.

Eagleton, S. & Muller, A., 2011. Development of a model for whole brain learning of physiology. *AJP: Advances in Physiology Education*, 35(4), pp.421–426.

Guilford, J.P. & Christensen, P.R., 1973. The one-way relation between creative potential and IQ. *The Journal of Creative Behavior*, 7, pp.247–252.

Hmelo-silver, C.E., 2013. Creating a Learning Space in Problem-based Learning. *Interdisciplinary Journal of Problem-based Learning*, 7(1), pp.24–39.

Hong, J. et al., 2014. Computers & Education Using a " prediction – observation – explanation " inquiry model to enhance student interest and intention to continue science learning predicted by their Internet cognitive failure. *Computers & Education*, 72, pp.1–11. Available at: http://dx.doi.org/10.1016/j.compedu.2013.10.004.

Kauchak, D. & Eggen, P., 2013. *Education Psychology Windows on Classrooms_Ninth edition* Ninth Edit., United States of America: Peachpit Press.

Lederman, J.S. et al., 2014. Meaningful assessment of learners' understandings about scientific inquiry - The views about scientific inquiry (VASI) questionnaire. *Journal of Research in Science Teaching*, 51(1), pp.65–83.

Lehtinen, A., 2016. Guidance Provided by Teacher and Simulation for Inquiry-Based Learning : a Case Study. *Journal of Science Education and Technology*. Available at: http://dx.doi.org/10.1007/s10956-016-9672-y.

LIN, C.-C. & BRUCE, B.C., 2013. Engaging Youth in Underserved Communities Through Digital-Mediated Arts Learning Experiences for Community Inquiry. *Studies in Art Education*, 54(4), pp.335–348.

Lin, W.-L. & Shih, Y.-L., 2016. Designing EEG Neurofeedback Procedures to Enhance Open-ended versus Closed-ended Creative Potentials. *Creativity Research Journal*, 28(4), pp.458–466. Available at: https://www.tandfonline.com/doi/full/10.1080/10400419.2016.1229979.

Moreno, R., 2013. Educational Psychology. *Journal of Chemical Information and Modeling*, 53(9), pp.1689–1699.

Nuttariya, K.-O. et al., 2012. Designing Framework of Multimedia Learning Environment to Enhance Problem Solving Transfer. *Procedia - Social and Behavioral Sciences*, 46, pp.3421–3425. Available at: http://dx.doi.org/10.1016/j.sbspro.2012.06.077.

Osgood-Campbell, E., 2015. Investigating the educational implications of embodied cognition: A model interdisciplinary inquiry in mind, brain, and education curricula. *Mind, Brain, and Education*, 9(1), pp.3–9.

Soltis, R. et al., 2015. Process-oriented guided inquiry learning strategy enhances students' higher level thinking skills in a pharmaceutical sciences course. *American journal of pharmaceutical education*, 79(1), p.11.

Wijnia, L. et al., 2016. University teacher judgments in problem-based learning: Their accuracy and reasoning. *Teaching and Teacher Education*, 59, pp.203–212.

Identification of Readiness of Developing University to Apply Information and Communication Technology (ICT) in Teaching and Learning

Sholeh Hidayat[1], *Aan* Hendrayana[2], and *Heni* Pujiastuti[2 *]

[1]Instructional Technology Department, Universitas Sultan Ageng Tirtayasa, Indonesia
[2]Mathematics Education Department, Universitas Sultan Ageng Tirtayasa, Indonesia

Abstract. The integration of Information and Communication Technology (ICT) in learning is kind of innovation in education. Several studies have shown that the application of ICT is able to improve students' competence and satisfaction with subject studied and enhancing the ability of both subjects and graduate levels. A developing university must be able to apply ICT in teaching and learning. One of a developing university in Indonesia is Universitas Sultan Ageng Tirtayasa (Untirta). The subject of this research is a student of Faculty of Teacher Training and Education of Untirta. The main purpose of the research is to know the readiness of lecturer and college student in using ICT integrated into the learning process. Based on research it is known that both of lecturers and students have the readiness to adopt integrated ICT in the learning process. The main problem that arises from lecturers is the lack of lecturers' skills in using technology. While the problems that arise from students is the need for a free internet connection. Therefore, the university has to provide the training for the lecturers and provide free connections for the students.
Keywords: Integrated ICT, Readiness, Lecturer, College student, Untirta

1 Introduction

The integration of Information and Communication Technology (ICT) in learning is kind of innovation in education. Several studies have shown that the application of ICT is able to improve students' competence and satisfaction with subject studied [1] and enhancing the ability of both subjects and graduate levels [2]. Although there is a critical view of the use ICT [3] and also needs to be discussed in the technical application of it [4], using ICT remains a priority in learning and teaching. The research involved preservice teacher indicates that there are no differences in pre-service teachers' attitudes and behavioral intentions towards the use of ICT for teaching and learning [5]. Therefore, continuous ICT integration will improve the quality of teaching and learning [6, 7, 8].

In developing integrated ICT should consider several factors that affect the effectiveness of ICT in learning. ICT integration into societies and schools. ICT infrastructure, ICT content into school programs [9]. In other words, the related challenge

* Corresponding author: henipujiastuti@untirta.ac.id

of linking the design of appropriate educational content to the learning process was recognized as one of providing learners with interesting, relevant and transformative contexts to link reflection and practice [10].

In addition to the above survey research, many studies that discuss the advantages that can be a higher education when applied ICT [11]. However, not many are looking at losses obtained if ICT is applied. Some of the findings of the losses derived from the application of ICT in higher education, among others: (1) the lack of lecturers make progress monitoring study and the characteristics of student learning is difficult in monitoring, (2) there is a lot of irrelevant information, (3) feedback from lecturers is limited, and (4) the performance of students in practice are not measured well [12, 13]. The study on these losses is the quickness of the presence of lecturers and their interactions. The point is the use of ICT and the presence of lecturers can simultaneously make a profit to be optimal for higher education. Therefore, research that takes advantage of ICT to find a solution to minimize losses from the application of ICT is wide open. ICT development that accommodates the needs of the above is to develop teaching materials is integrated with the presence of the lecturer. Integrative teaching materials can be video conferencing (interactive video), video, simulations, interactions, chat, text, and materials evaluation.

Thus, the students are expected to have experience in applying ICT in the learning, so that they have competencies to learn. This is in line with purposes of higher education, it should be able to produce graduates who are able to adapt to the needs of the times. Graduates of educational institutions should have the ability; 1) to adapt to changes more quickly than before, 2) adapting the work to be created later, 3) resolve problems that may arise later, and 4) using the latest technology.

In harmony with these needs, a developing university must be able to apply ICT in teaching and learning. One of a developing university in Indonesia is Universitas Sultan Ageng Tirtayasa (Untirta). It has the goal of producing quality graduates, educated, trained, and skilled according to the needs of stakeholders. This means that Untirta always continues to strive to provide graduates according to the needs of stakeholders. Efforts are being made to achieve the objectives set out in any of the missions Untirta, which develops information technology systems that can stimulate the establishment of the college superior, independent, creative, innovative, and competitive. As a manifestation of the mission including by organizing an effective learning process and qualified. Therefore, the mission poured Untirta Research Master Plan (URMP) is one of the strategic plans is there needs to be a research-based instructional innovation of ICT [14].

Base on explanation before, this study aims to identify the readiness of the university to apply ICT in teaching and learning.

2 Research Method

We use research and development (R & D) method to develop an ICT system in teaching and learning for this developing university. The approach of research in education including 10 steps, namely: Preliminary Study, Planning Research, Development Design, Preliminary Field Test, Revised Results of Field Test, Main Field Test, Revised Result of Main Field Test, Feasibility Test, Revised Final Results of Feasibility, and Dissemination and Implementation Final Products [15]. Nevertheless, to identify the readiness of the university to apply ICT in teaching and learning, we use survey method. The instruments used in data collection in this study is in the teacher and student questionnaire and also interview.

This research requires a series of steps that is long enough. From the tenth step, the research step that has been done is step 1. The 9 until 10 steps in the phase of research that is being and will be done. Part of the research development phase is the readiness analysis of research subjects so that will be input on the product to be made. The subject of this research is the sampling of lecturers and students in one faculty. The sample is taken by using cluster random sampling.

3 Results and Discussion

From survey which conducted 100 lecturers at the Faculty of Training and Education and 100 college students, the results are shown in Table 1, Table 2, Table 3, Table 4, and Table 5.

Table 1. Distribution of Subject

Category	Ready	Ready enough	Not ready
Lecturer	83	12	5
College student	61	23	16

Table 2. Gadget Ownership

Category	Handphone	Smartphone	Laptop	Desktop
Lecturer	100	99	100	9
College student	100	86	44	3

Table 3. Gadget Utilization

Category	Telp/SMS Only	Browsing	Chatting	Social Media	Online Learning
Lecturer	0	83	83	62	32
College student	14	72	72	50	44

Table 4. Gadget Selection

Category	Smartphone	Laptop
Lecturer	79	12
College student	57	29

Table 5. Number of Internet Connection

Category	Online	12 hours >	6 hours >	1 hours >	1 hours <
Lecturer	68	9	2	2	1
College student	32	10	9	4	28

From the 100 lecturers given questionnaires, as many as 83 people admitted ready to organize online learning, 12 admitted quite ready, and 5 people said not ready. After being examined, the lecturers who were not ready were those who were in the 50's. Their unfamiliarity is natural because when they took bachelor degree, the computer was not so well known in Indonesia. If time backward, lecturers aged the 50s and pursued a bachelor degree in 1987. At that time computer technology with GUI (Graphic Unit Interface) is not well known.

Lecturers over the age of 50s are more likely to engage in chatting, almost no lecturer found this category active in online learning. Lecturers aged the 50s and over can be active in learning online when universities force them to serve online learning. Meanwhile, lecturers aged less than 50s have a lot of awareness and are active in online learning, even before there is an assignment. Ownership of gadgets is a good indicator for lecturers to be aware of the demands of the online world. Even if the lecturer does not have the gadget it's more on economic factors and just a special case. There are two interesting findings on this data. Firstly, almost all professors have smartphones. Secondly, all lecturers have personal computers in the form of laptops and very few have desktops. Advanced search results, lecturers who have a desktop tend to rarely use it, while active lecturers use the desktop because it still works well. Those with a desktop do not want to replace it with a new one if it breaks. Interestingly, these two findings are the realization that the need for gadgets that have mobility, in short on the packaging, and save energy consumption.

Another interesting finding is that both lecturers and students prefer to use smartphones for internet connection needs. This shows that in the future applications are made applications that are adaptive to smartphone technology or in other words based smartphone operating system or at least accessible by smartphone.

Related to the intensity of internet access, Lecturers dominant online, while students tend to spread. The results of interviews on lecturers obtained the answer that they are required to go online for work needs. Meanwhile, the students' reasons are very diverse and there is a tendency on lifestyle reasons. Apart from that, there are demands on students to get an internet connection for free, especially when on campus.

4 Conclusions

In general, lecturers and students have the readiness to adopt integrated ICT in the learning process. The main problem of the lecturers is the skill in the utilization of technology. The platform to use for e-learning is an app that has a smartphone operating system. The main obstacle of the student is the internet connection for free becomes the main requirement. For that, the effort of the institution for lecturers that is, many provide training and conditioning so that lecturers have demands and awareness in order to use gadgets for online learning needs. In college students, institutions only need to provide free connections because the laboratory is limited and the usage is overwhelming. The university effort by providing alternative connections with the third party is quite helpful but has not solved the problem since the online learning activities are 24 hours.

Acknowledgment

We would like to thank the Ministry of Research, Technology and Higher Education of Indonesia for the research grant (Penelitian Unggulan Perguruan Tinggi). Also similarly to the Institute of Research and Community Service, Universitas Sultan Ageng Tirtayasa who helped the process of research implementation.

References

1. N. Olmedo-Torrea, O. F. Vidala, J. L. Castillob, F. B. Rodríguez, Proc. Soc. and Behav. Sci. **237**, 737-744 (2017)
2. N. Pheeraphan, Proc. Soc. and Behav. Sci. **103**, 365 – 373 (2013)
3. S. Livingstone. Oxf. Rev. of Edu. **38 (1)**, 9-24 (2012)
4. N. Tvenge, K. Martinsen, S. S. V. K. Kolla, Proc. CIRP **54**, 101-106 (2016)
5. T. Valtonen, J. Kukkonen, S. Kontkanen, K. Sormunen, P. Dillon, E. Sointu, Comp. & Edu. (2014)
6. I. N. Umara , A. S. A. Hassan. Proc. Soc. and Behav. Sci. **197**, 2015 – 2021 (2015)
7. M. . Webb. Int. Jour. of Sci. Edu. **27(6)**, 705–735 (2005)
8. K.D. Su, Computers & Education **51**, 1365–1374 (2008)

9. S. Capuk, A. Kara, Proc. Soc. and Behav. Sci. **191**, 56-62 (2015)
10. C. Richards, Tech. Pedag. and Edu, **15 (2)**, 239-255 (2006)
11. U. Toro, M. Joshi, M, Int. Jour. of Innov. Manag. and Tech. , **3(1)**, 20-23 (2012)
12. S. Telebian, H. M. Mohammadi, A. Rezvanfar, Proc. Soc. and Behav. Sci. **152**, 300 – 305 (2014)
13. Y. Perbawaningsih, Proc. Soc. and Behav. Sci. **103**, 717-724 (2013)
14. Universitas Sultan Ageng Tirtayasa [Untirta], *Rencana Induk Penelitian*, Lembaga Penelitian dan Pengabdian Masyarakat Untirta, Serang, (2013)
15. W. R. Borg, M.D. Gall, *Educational Research: An Introduction*. Longman, Inc: London (1987).

EFFECT OF USING PORTFOLIO ASSESSMENT TO IMPROVE SCIENTIFIC ATTITUDE OF STUDENTS LEARNING PHYSICS IN HIGH SCHOOL

Wartawan Putu Gede [1,*]

Guru Fisika SMA Negeri 4 Singaraja Bali Indonesia *putugedewartawan@gmail.com

ABSTRACT. This study aims to investigate the effect of portfolio assessment in teaching physics and scientific attitude. The research was conducted on students of high school in Singaraja. Research was an quasi- experimental study by using "The Posttest-Only Control Group Design". The research involved 152 high school students of class X of science as samples, taken with multistage random sampling technique. Portfolio assessment was integrated with physics learning. The implementation of the portfolio assessment included four key elements such as the students' work folders, clear assessment criteria, and self-assessment, and conference between teacher and students. The data needed in this research was the students' scientific attitude which included the aspect of curiosity, respect for evidence, the willingness to change ideas, and critical reflection. Data needed in this research included scientific attitudes students. A Likert scale instrument was used to measure the scientific attitude students. Data were analyzed using analysis of variance with SPSS 20.0 at significance level $\alpha = 0.05$. The results showed there are differences in the scientific attitude students who take physics learning with assessment portfolios and students who take physics learning with assessment of conventional. The findings of this study indicate that portfolio assessment in learning physics significantly affect the scientific attitude students.

Keywords: portfolio assessment, scientific attitude

INTRODUCTION

The rapid development of science and technology that has a leading position in the exploration of natural phenomena thus could not be separated. Marjorie states that learning science provides the practice of scientific method which develops the scientific attitude, which lead us to think critically and value facts with evidence [1]. Physics as part of science naturally has an integrated concept of thought and understanding in the development of systematic thinking and analysis. Physics essentially includes three things: physics as a process, a product, and a scientific attitude. Therefore, the study of physics in educational institutions particularly in high school should be capable of producing skilled and capable students in developing mental processes to understand nature and its symptoms, developing and applying the concepts of physics and having a scientific attitude that enables them to think and act scientifically.

Pitafi dan Farooq states that a scientific attitude investigates for a certain scientific act or thought [2]. Mukhopadhyay states that a scientific attitude is a characteristic of a scientist [3]. The scientist is someone who solves the problem, who is accustomed to doing research and testing in a planned manner so that a new finding is obtained. Harlen distinguishes two kinds of scientific attitudes: 1) those towards science as an enterprise, 2) those towards objects and events in the environment which are studied in the course of scientific activities [4]. In order for students to have a positive attitude toward science then they must have a correct view of science. This can be done by building a scientific attitude, which includes curiosity, respect for fevidence, a willingness to change ideas, and critical reflection.

However, it should be admitted that the results of education in the field of science and especially physics in Indonesia is still far from expectations. The low quality of Indonesian human resources in global competition, especially in the field of science and physics can be seen from the report of the Program for International Student Assessment (PISA), an international systematic assessment that measures the ability of 15-year-old students in literacy reading, mathematics, and science every three years. The results of the PISA study in 2012 placed Indonesia's position at the 64th position of 65 countries participating in the test. This shows that the ability of Indonesian children aged 15 years in the field of reading, math, and science than other children in the world is still low.

The low scientific attitude of students is certainly a result of the not successful physics learning process is implemented. There are many factors that affect the low scientific attitude of students. One of the predominant factors influencing students' scientific attitude is the assessment used by teachers in assessing physics learning process and outcomes. In the learning cycle, assessment is one of the stages which role is very important and cannot be ignored. Marhaeni said that at least assessment activities can produce two things: first, as feedback on the learning process, and second, can provide information about the quality of acquisition in the subject of students [5]. Suastra states assessment can be used to diagnose the trengths and weakness ot the students, to monitor their progress, to determine their rangking, and to determine the effecveness of instruction done by the teacher [6].

Many efforts have been done to improve the quality of physics learning in Indonesia. Some of the efforts done were providing laboratory facilities and infrastructure & libraries, improving the quality of teachers through education and training, and curriculum improvement. In the 2013 curriculum, there was a revision of the implementation of learning by using scientific approach. Learning in scientific approach is a learning that adopts the steps of scientists in building knowledge through scientific methods. The scientific approach consists of five basic learning experiences: 1) observing, 2) questioning, 3) attempting / collecting data, 4) associating, and 5) communicating.

An assessment model that is compatible with learning with a scientific approach is portfolio assessment. Salvia and Ysseldike define that portfolio is a collection of products used to demonstrate what a person is capable of doing [7]. Dantes, et al quoted Paulson and Paulson who define the portfolio as a purposeful collection of student work the exhibits that the student's efforts, progress and achievement in one ormoreareas. The collection must include student participation in selection, the criteria for selection, the criteria for judging merit and evidence of student self-reflection [8].

Portfolio assessment is an ongoing assessment based on a collection of information that demonstrates the development of learners' abilities within a given period. Through the portfolio assessment teachers will know the progress of learners. For example, the results of their work in preparing or making practical reports. On the basis of the assessment, teachers and learners can make improvements in accordance with the demands of learning.

This study aims to investigate the effect of portfolio assessment in teaching physics and scientific attitude. Research was an quasi- experimental study by using "The Posttest-Only Control Group Design". The results showed there are differences in the scientific attitude students who take physics learning with assessment portfolios and students who take physics learning with assessment of conventional. The findings of this study indicate that portfolio assessment in learning physics significantly affect the scientific attitude students.

METHODS

This research was conducted at students of high cshool in Singaraja City using quasi experiment method with "The Posttest-Only Control Group Design". This study involved 152 high school students of class X of science as research sample, taken with multistage random sampling technique.

Table 1. Distribution of Experiment Group and Control Group Members

No	High School	Experimental Group		Control Group	
		Class	Number	Class	Number
1	SMA Negeri 2 Singaraja	X of science 1	35 students	X of science 3	38 students
2	SMA Negeri 4 Singaraja	X of science 7	40 students	X of science 5	39 students
Number		2	75 students	2	77 students

The treatment variables that were experimented in this study were portfolio assessment. The experimental group follow physics learning with a portfolio assessment, and a control group with a conventional assessment. Portfolio assessment in physics learning that is carried out is the assessment of the process of preparation and product of physics learning tasks in the form of practical reports. After the treatment, data were taken on the students' scientific attitude through the scientific attitude questionnaire.

To measure the scientific attitude variable, the scientific attitude questionnaire was used with the reliability coefficient of 0.960. The scientific questionnaire consisted of 41 items, with ideal minimum score = 1 and ideal maximum score = 205.

Hypothesis testing using variance analysis of SPSS 20.0 program at significance level $\alpha = 0.05$. Preparation test data analysis performed before hypothesis testing includes testing the normality of data distribution and homogeneity of group variance.

RESULTS AND DISCUSSION

Recapitulation of descriptive analysis of scientific attitude scores results is presented in Table 2.

Table 2. Recapitulation of descriptive analysis of scientific attitude

Statistics	Portfolio Assessment	Conventional Assessment
Mean	164,63	148,92
Standard Deviation	9,83	2,67

Scientific attitude data of students who follow physics learning with portfolio assessment have mean = 164,63 and standard deviation 9,83. Referring to the classification of scores, scientific attitude data of students who follow physics learning with portfolio assessment is in the high category. Score data of students' scientific attitude following physics study with conventional assessment had mean score = 148,92 and standard deviation 11,54. Referring to the score classification, students' scientific attitude data following physics learning with conventional assessment is in the high category.

The following table shows the result of analysis of students' scientific attitude in each dimension.

Table 3. Description of Scientific Attitude Dimensions

Dimension	Portfolio Assessment		Conventional Assessment	
	Mean	Category	Mean	Category
Curiosity	42,45	high	41,82	High
Respect for evidence	36,32	high	36,68	High
Willingness to change ideas	40,49	high	34,78	Average
critical reflection	45,49	high	36,65	Average

Further finding on the description of scientific attitude in each dimension for students who learned physic with portfolio assessment are in high category for all dimension (curiosity, respect for evidence, willingness to change ideas, and critical reflection). Whilst, the descriptions of scientific attitude for students' who learned physic with conventional assessment are in high category for two dimensions (curiosity and respect for evidence), and average category for the other two dimensions (willingness to change ideas and critical reflection)

The followings are the results of the data analysis using SPSS 20.0 with significance level $\alpha = 0,05$.

Table 4. Test of Between Subject Effects

Dependen Variabel: Y

Sourch	Type III Sum of Squares	df	Mean Square	F	Sig.
Corrected Model	9370.473[a]	1	9370.473	81.374	.000
Intercept	3735240.184	1	3735240.184	32436.951	.000
A	9370.473	1	9370.473	81.374	.000
Error	17273.079	150	115.154		
Total	3757608.000	152			
Corrected Total	26643.553	151			

a. R Squared = .352 (Adjusted R Squared = .347

The result of analysis of difference of scientific attitude between students who follow physics learning with portfolio assessment and students who follow physics learning with conventional assessment using variance analysis with the SPSS 20.0 program at significance level $\alpha= 0,05$ yield value F = 81,347 and significance number <0, 05. The results of this study indicate that the assessment of portfolio in physics learning has a significant effect on students' scientific attitude.

Table 5. Pairwise Comparisons

Dependen Variabel: Y

(I)A	(J)A	Mean Difference (I – J)	Std. Error	Sig.[b]	95 % Confidence Interval for Difference[b]	
					Lower Bound	Upper Bound
1.00	2.00	15.705*	1.741	.000	12.265	19.145
2.00	1.00	-15.705*	1.741	.000	-19.145	-12.265

Based on estimated marginal means

*The mean difference is significant at the .005 level

[b]Adjustment for multiple comparisons: Least Significant Difference (equivalent to no adjustments)

The result of pairwise comparison shows that the mean difference between the group who learned physic with portfolio assessment and those who learned physic with conventional assessment is 15.705 with level of significance < 0.05. It means that students who learned physic with portfolio assessment has significantly higher scientific attitude than students who learned physic with conventional assessment.

The findings above show that implementing portfolio assessment in physic teaching and learning gives positive effect to the students' scientific attitude. This positive effect is because of the characteristic of portfolio assessment in physic teaching and learning, which are the result of students and teacher collaboration, selection process, self-assessment, and clear criteria of portfolio assessment.

Portfolio assessment through collection of product in physic teaching and learning in the form of experiment report increases the students' scientific attitude effectively. In writing the experiment report the students presents the process of finding a physic product. In this process, students use scientific method. In the process of writing the experiment report by using the given format, which involves self-reflection and self-assessment, discussion and feedback, and assessment by using the agreed rubric, the teacher will be able to recognize the students' mental process. The scientific method that is done by the students will encourage them to develop their scientific attitude which consists of curiosity, respect of evidence, willingness to change ideas, and critical reflection.

The result in this study in line with the research done by Chi-Cheng Chang, Kuo-Hung Tseng, dan Shi-Jer Lou, which found that using portfolio in physic teaching and learning helps students in developing their idea through self reflection and self assessment, and also discussion which gives students feedback so the students gain ideas in writing their experiment report. This process will encourage students to revise their work. It means that students have willingness to change ideas, which is one of the aspects of scientific stain [9]. The result of this study also supports the study done by Mohammad Tawiland which found that the implementation of portfolio assessment in teaching and learning develops higher critical reflection compared to conventional assessment [10]. Further, Pramudya Dwi found that developing and implementing authentic assessment improve students' achievement in cognitive, psychomotor, and affective. These three aspects are the reflection of the scientific thinking ability of students in physic education program [11].

CONCLUTION

Based on the results of hypothesis testing as described, it can be concluded "There is a significant difference in scientific attitude between students who follow physics learning with portfolio assessment and students who follow physics learning with conventional assessment". Further analysis results show that students' scientific attitude following physics learning with portfolio assessment is higher than that of students following physics learning with conventional assessment. This means that portfolio assessment is one of the factors that can be used to improve scientific attitudes.

ACKNOWLEDGEMENTS

We would like to thank the head master of SMA Negeri 2 Singaraja and head master of SMA Negeri 4 Singaraja who given permission for the research, as well as students involved the study. The author thanks, all supporter at completing this article to be published.

REFERENCES

1. M. Lacap, "The Scientific Attitudes of Students Major In Science in the New Teacher Educaton Curriculum," *Asia Pacific Journal of Multidisciplinary Research,* **3**, 7, (2015)
2. A. Pitafi, & M. Farooq, "Measurment of Scientific Attitude of Secondary School Students in Pakistan," *Academic Research International*, **2**, 383 (2012)
3. R. Mukhopadhyay, "Scientific Attitude – Some Psychometric Considerations," *IOSR Journal Of Humanities And Social Science (IOSR-JHSS) OSR-JHSS,* **19**, 98-100 (2014)
4. W. Harlen, *The Teaching of Science*, 40 (1991).
5. A. Marhaeni, "Effect of Portfolio Assessment and Achievement Motivation in Learning English on English Writing Ability," *Dissertation* (not published), 2 (2005)
6. S. Wayan, R. Putu, "Problem Faced by Teacher in Designing and Implementing Authentic Assessment in Science Teaching," *International Research Journal of Enginering, IT & Scientific Research (IRJEIS)*, **3**, 24 – 32 (2017)
7. J. Salvia & J.E. Ysseldike, *Assessment, 6th Edition*, 293 (1996)
8. D. Nyoman, S. Wayan, S. Wayan. *Classroom Based Assessment Guidelines for Cluster of Science Lessons*, 7 (2006)
9. C. Chang, K. Tseng, dan S.Jer Lou, "A Comparative Analysis of the Consistency and Difference Among Teacher-Assesment, Student Self-Assesment and Peer-Assesment in Web-Based Portfolio Assesment Environment for High School Students," *International Journal of Computers & Education,* **58**, 303 – 320 (2012)
10. M. Tawilan, "Portfolio-Based Physics Learning Model to Improve Critical Thinking Skills," *International Journal of Education and Research,* **9**, 1 – 8 (2013)
11. P. Dwi, "Portfolio-Based Physics Learning Model to Improve Critical Thinking Skills," *International Journal of Education and Research,* **15**, 184 (2013)

Analysis of search and rescue emergency evaluation in ship accidents in Indonesia

Arleiny[1,1], *Damoyanto Purba*[2] , Ile Suwondo[3], and *Manungku Trinata P*[4]

[1]POLTEKPEL SBY, Nautica, Indonesia,
[2]POLTEKPEL SBY, Nautica, Indonesia, , [3]POLTEKPEL SBY, Nautica, Indonesia, [4]POLTEKPEL SBY, Nautica, Indonesia

Abstract. The objectives og this research is to describe the factors causing ship accident in Indonesia and know the effectiveness of SAR emergency in ship accident in Indonesia. The research method used in this research is qualitative research. Techniques Collection of literature study data and documents. Data validity method using triangulation. Data analysis uses interactive data analysis. The conclusions of this study are Factors that cause the occurrence of ship accidents in Indonesia, among others, the resources of the crew, the eligibility of ships, supporting facilities for shipping, operators, lack of supervision of apparatus, service users and other factors. The high number of ship accidents in Indonesia shows the ineffective implementation of SAR in ship accident in Indonesia.

1.2 Introduction

The shipping disruption can essentially be a disruption that can be directly addressed, even the need for direct assistance from a particular party, or disruption resulting in the skipper and all the crew having to be involved either to overcome the disruption or to leave the ship. The situation of the shipping disruption according to the situation can be grouped into an emergency based on the type of event itself which, among other things, collision, fire / explosion, rupture, leakage / drowning, people falling into the sea and pollution. Emergency conditions on ships may be detrimental to the Skipper and the crew and the owner of the ship and the link environment may even cause disruption of the 'seabed' ecosystem.

The sinking of the Tampomas II vessel is remembered as a very heartbreaking humanitarian tragedy. The incident on 27 January 1981 was the darkest history for the Indonesian maritime world. This calamity killed 369 people, both passengers and crew. Such negligence, a number of crew members received administrative sanctions by the Ship Court. The sluggish assistance to Motor Vessel (KM) Tampomas II also reflects poor coordination in rescue measures.

The causes of the accident on board are: (1) human error, (2) destruction of ship machinery, (3) external and internal factors, eg fire and impact events, (4) natural or weather factors, (5) a combination of all the causes. In general, the accident that may occur on the ship is the result: (1) collision with other vessels, (2) stranded / grounded, (3) drowning due to bad weather (bedweather), (4) burning), (5) engine damage (engine black out / breakdown), and (6) ship juxtaposed with other vessels.

Research from Jori Nordstrom (2016) with the title Vessel TRIAGE: A method for assessing and communicating the safety status of vessels in maritime distress situation. The result of this research is An evaluation of the proposed method indicates a positive reception among various maritime stakeholders, suggesting that implementing the

1 Corresponding author: arleiny@poltekpel-sby.ac.id/arleinysudardi3@gmail.com

Vessel TRIAGE method in maritime SAR procedures may act as a useful tool to assist in the management of maritime distress situations. As Wan Muhamad at all research (2014) *human error and external factor to be the contributing factor is agreed the most by respondent is the factors that affecting the efficiency of the management of vessels.*

Based on the above exposure the purpose of this study is to describe the factors that cause the occurrence of accidents on the ship in Indonesia and to determine the effectiveness of emergency SAR in ship accident in Indonesia

1.3 Research Method

The method of research conducted in the study this is using case study research methods and descriptive research methods. Study studies case is observing the behavior of that object intensively studied, while descriptive method (descriptive research), that is research method intended to describe phenomena that exists, which lasts as long as the author performs this research. Technique of collecting data in this research is bibliography and document. Data validity techniques use data triangulation and data analysis techniques using interactive analysis techniques

1.3 Factors causing ship accidents in Indonesia

Transportation services with a guarantee of safety will provide a sense of certainty and calm for the traveler or the owner of the goods, so that socio-economic activities of the community can be protected. If the safety aspect of transportation is guaranteed, and the rights of user communities are protected, there will be no unnecessary and counterproductive costs.

The principle of transportation safety has been the concern of the Government for a long time. In 1999, Presidential Decree No. 105/1999 on the establishment of the National Transportation Safety Committee (KNKT) was established. The safety principle, of course, became the government's reference through the Transportation Department when it launched the Road Map to Zero Accident program at the end of the first semester of 2007, in the sense that the government seeks to minimize accidents through accurately programmed activities with continuous control and control.

Based on the results of research that the authors obtained data as follows:

Table 1. Investigation Data of Navy KNKT Year Accident 2010- 2016

NO.	TAHUN	JUMLAH KECELAKAAN	JENIS KECELAKAAN					KORBAN JIWA		REKOMENDASI
			TENGGELAM	TERBAKAR/ MELEDAK	TUBRUKAN	KANDAS	LAIN-LAIN	KORBAN MENINGGAL/ HILANG	KORBAN LUKA-LUKA	
1	2010	5	1	1	3	0	0	15	85	45
2	2011	6	1	3	2	0	0	86	346	82
3	2012	4	0	2	2	0	0	13	10	28
4	2013	6	2	2	2	0	0	65	9	47
5	2014	7	2	3	2	0	0	22	4	25
6	2015	11	3	4	3	1	0	85	2	11
7	2016	15	4	4	3	2	2	51	18	35
TOTAL		54	13	19	17	3	2	337	474	273

Sumber: Database KNKT 30 November 2016

Based on the above table, it can be seen that there is an increase in marine accidents from 2014, 2015 and 2016. Furthermore, it will be clarified on the graph as follows:

Graph I. Percentage of Type of Sea Transport Accident Year 2010 - 2016

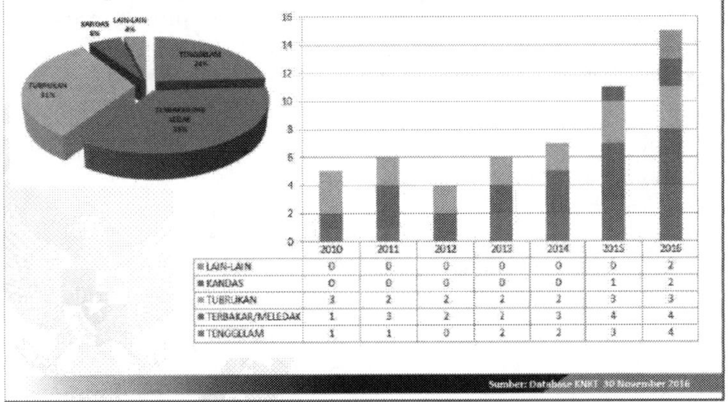

Based on the graph above, it can be seen that the Percentage of Type of Marine Transport Accident in 2010 - 2016 consists of railing, collision, collision / bursting and sinking. In the period 2010 - 2016 the highest accidents are burning / exploding and drowning.

From some data that has been writer collected can be known that the factors causing the occurrence of ship accident in Indonesia among others because (Danny Faturachman, 2012):

1. The resources of the crew

The crew must have the ability to navigate to cross its ships from one port to another, within certain limits of the voyage safely. They are also required to undertake an "economic voyage" that is to travel through the shortest distance safely from navigational hazards so that additional exploitation costs can be avoided.

2. Ship Ability

The ship is in excellent condition, and in accordance with the provisions of the legislation, and declared sea worthy, it would be safer to cross people and goods, on the contrary the ship is doubtful conditions tend to encounter obstacles while in the voyage. If the ship is damaged while on the go will require additional costs such as exploitation costs caused by the delay.

3. Sailing Supporting Facilities

In addition to the technical factors of ships and crew resources, the Sailing Navigation Supporting Facility (SBNP) is also a very important element in shipping safety. This facility consists of marine signs that serve as a guiding tool for sailing vessels, in order to avoid the dangers of navigation. The Coastal Radio Station is also useful as a means of navigating shipping to enable ships to make economical voyages, because without this instrument the vessel must make a "turn" voyage in order to avoid the danger of navigation

4. Operator

Many ships are made traditionally / do not have certificates, shipbuilding does not follow the direction of ship image that has been legalized. Vessel certificate has expired, Communication equipment / navigation ship less functioning. Excessive loading is mainly on-deck. Incremental load / increment placement. Excessive passenger loading. Passenger awareness is still lacking.

5. Lack of supervision of the apparatus

The vessel can exit / enter the place where only The number of locations of supervisory apparatus is limited. Not all shipping places can be monitored. The possibility of examination is less thorough

6. Service User / Community

Lack of public awareness of the importance of shipping safety. Often impose the will regardless of the safety of the voyage. While the causal factors that directly cause the occurrence of marine accidents.

7. Other factors

Human factor is the biggest factor which among others include carelessness in running the ship (Anomin, 1998), lack of ability of the crew in mastering various problems that may arise in ship operations, Based on the above explanation, it can be analyzed that the factors that resulted in ship accidents in Indonesia, among others, due to loading of ships full of excess. Placement of vehicle cargo resulted in the ship having a trim bow (the ship is down). The rampa gate is not closed causing sea water to enter the vehicle deck. The certification process does not refer to the actual condition of the ship. The calculation of the stability of the ship can not be done due to limited time and no vehicle weight data.

1.4 SAR emergency Effectiveness in ship accident in Indonesia

The territory of the Republic of Indonesia consists of territorial waters and islands where as interconnection between islands in order to support the development of the economy is in terms of transportation. Conditions such as this impact transport traffic becomes very crowded. On the other hand public awareness about safety has not been a priority, so that in case of disaster.

A number of incidents of accidents occurring in Indonesian seas need to get attention because the intensity of the incident is still high. In 2014, about 450 incidents occurred in waters. Not only passenger ships, but also freight ships, and so forth. The accident also caused the spill of oil / diesel marine vessels, consequently the fish in the sea dead, sea water became polluted and garbage scattered to the bottom of the sea (Nina Nur Hasanah, 2015).

Based on the authors' analysis, there is an increasing number of ship accidents occurring from 2010 (5 accidents), 2011 (6 accidents), 2012 (4 accidents), 2013 (6 accidents), 2014 (7 accidents), 2015 (11 accidents) Year 2016 (15 crashes) (Secondary Data of KNKT, 2016).

Should be, every ship must have a system and procedures for prevention and prevention of the event of disturbance to the safety (safety) and in its implementation must appoint an officer responsible for supervising the safety of ships and the prevention of pollution from the ship.

In addition, the shipping company must appoint a Manager-level called DPA (Designated Person Ashore). He is responsible and oversees the safety of the Shipping Company. The responsible manager must be responsible and have direct access to the highest Management of the Shipping Company. Management of ships, does not conduct training on the prevention and prevention of safety interruption of shipping activities of the respective Shipping Company periodically to address and prevent safety, the Shipping Company must have facilities and equipment in accordance with the provisions of the ISM Code.

The ineffectiveness of SAR emergency in ship crash in Indonesia is due to several things, most of which captain has low compliance in checking boat with checklist so that the captains are easy to fill out inspection checklist by frequently checking inspection checklist without testing machine and test .

Based on the above explanation, it can be concluded that with the still high number of ship accidents in Indonesia due to lack of education and training for supervisors and crew of the ship this shows the ineffectiveness of the implementation of SAR on ship accident in Indonesia.

1.5 The Acknowledgements

First of all Thanks to GOD, for His mercy and guidance in giving me full strength to complete this paper. Special appreciation to my family for their support. Thanks.

References

Anonim. 1998. ***Improving the Continued Airworthiness of Civil Aircraft: A Strategy for the FAA's Aircraft Certification Service*****. The National Academies of Sciences, Engineering, and Medicine** 500 Fifth St., NW | Washington, DC 20001

Danny Fathurahman, 2012. *Analysis of Indoensian Ship Accident 2005 – 2010*. Conference: International Conference on Computer, Electrical, and Systems Sciences, and Engineering (ICCESSE 2012), At Penang, Malaysia

Jori Nordström.Vessel TRIAGE: A method for assessing and communicating the safety status of vessels in maritime distress situations. ***Safety Science Volume 85, June 2016, Pages 117-129***

Ketua Sub Komite Investigasi. Kecelakaan Pelayaran. Data Investigasi Kecelakaan Pelayaran Tahun 2010-2016. KNKT 2016

Nina Nurhasanah, Asmar Joni, Nur Shabrina. Persepsi Crew dan Manajemen dalam Penerapan ISM Code bagi Keselamatan Pelayaran dan Perlindungan Lingkungan Laut. PROSIDING SEMINAR NASIONAL MULTI DISIPLIN ILMU & *CALL FOR PAPERS* UNISBANK (SENDI_U). *Kajian Multi Disiplin Ilmu untuk Mewujudkan Poros Maritim dalam Pembangunan Ekonomi Berbasis Kesejahteraan Rakyat*. ISBN: 978-979-3649-81-8

Wan Muhamad Amir W Ahmad1, Gobi Krishnan Veluplay2, Kasypi Mokhtar 3, Nurfadhlina Halim4 and Nor Azlida Aleng. Comprehensive analysis of the factors that affecting the efficiency of the management of Vessel Using LRM, RSM and SEM.. International Journal of Engineering and Applied Sciences. July. 2014. Vol. 5. No. 02 ISSN2305-8269

The Complexity integrated-Instruments components media of IPA at Elementary School

Siska Angreni[1*], and *Rona* Taula Sari

[1]FKIP, Univrsitas Bung Hatta, 21571 Padang, Indonesia

Abstract. This research aims at describing the complexity of Integrated Instrument Components media (CII) in learning of science at Elementary schools in District Siulak Mukai and at Elementary schools in District Siulak. The research applied a descriptive method which included survey forms. Instruments used were observation sheets. The result of the research showed Integrated Instrument Components media (CII) natural science that complexity at primary school district Siulak was more complex compared with that at primary school district Siulak Mukai. is better than from primary school district Mukai

1 Introduction

Increasing quality of education is the need of the nation that wants to move forward.With confidence , that quality education can support the development in every sector. Hence, education needs to be most attended so we can catch up in science and technology from another country .One government efforts to improve the quality of education is with a procurement program and infrastructure in education . Provision of infrastructure is intended to support the process and improve learning outcomes of students all levels (primary school , junior high school and senior high school) (Maryandi, 2013). As provision and infrastructure from the government for each school is learning media or the learning tools .Besides junior and senior high school , primary school also received assistance on learning media from the government.

One of the media given by the government is Integrated Instrument Components media (CII) for science. Integrated Instrument Components media (CII) for science provides a pleasant learning on the subjects of science which has perfect adjusted to the curriculum. Every elementary school has Integrated Instrument Components media (CII) provided by the goverment as well as possible in learning. Schools that have Integrated Instrument Components media (CII) should keep and use the media full. With the Integrated Instrument Components media (CII) for science expected to improve achievement and learning outcomes students. Learning of media actually hold a very important role, especially for teaching science. Many the concept of science are very abstract to measure at the primary school, so the concept of this kind of teaching needs to be adequately supported by media. The result of study also found that learning using teaching media will increase the quality of learning. Based on the research done by Maryandi (2013) the utilization of science media kit exerts positive on students study results. While the results of Angreni research shows (2017) that the availability of CII media science in primary schools in Nanggalo allow for learning to use the media but its use is not yet full.

Based on preliminary observations in several public primary schools district Siulak Mukai and public primary schools district Siulak .The elementary school in observation in kecamatan siulak mukai namely , of public elementary school 062 /III Mukai Mudik , of public elementary school 169 /III Mukai Mudik , of public elementary school 122 /III

[*] Corresponding author: siskaangrenisiska@gmail.com

Mukai Tengah .While observation in kecamatan siulak namely of public elementary school 114/III Desa Baru Siulak , of public elementary school 044/III Siulak Gedang , of public elementary school 041 /III koto Beringin, a means of learning at the school is not complete especially for learning science lessons .For example in public primary schools 169 / III Mukai Mudik, amous 10 integrated instrument components media (CII) for science that is observed only 5 integrated instrument components media (CII) were, but not complete. It was because the arrangement integrated instrument components media (CII) for science often chauded .In addition , the school had experienced fire so many facilities and infrastructure burned including integrated instrument components media (CII) for science. Next in public primary schools 122/III Mukai Tengah and public primary schools 062 /III Mukai Mudik , of 2 schools they only found 6 integrated instrument components media (CII) for science but that they are incomplete for example kit electricity there is just cable and the battery missing , then integrated instrument components astronomy only the solar system them just, planets has been no longer .One of the causes of the lack of integrated instrument components media (CII) for science a place arrangement and inadequate care.
School as a recipient integrated instrument components media (CII) for science were supposed to provide a depository properly so as not to broken and eaten by rats .In addition , teachers were supposed to make use of an integrated instrument components media (CII) for science as maximum as possible so that the supply of and the condition of the media an integrated instrument components media (CII) for science always be monitored well .By using an integrated instrument components media (CII) for science learning is fun and students learning outcomes would rise .Based on fact that were found in the field the availability of media kits ipa in public primary schools district Siulak Mukai and public primary schools district Siulak needs to be reconsidered.

As for the purpose of this research is to describe the availability of and complexity of integrated instrument components media (CII) for science in public primary schools Siulak Mukai and Siulak. According to winanto (2011) integrated instrument components media (CII) for science is an apparatus used as an intermediary to explain concepts of science delivered by teachers. Next according to Umra Hi (2013) integrated instrument components media (CII) for science is media used in proving theories ipa associated with the natural environment in which the object get students to be able to develop their potential. According to Smaldino in Anitah (2009) CII is box equipment as a collection of materials that contains more than one kind of props in organize to one particular subject.

2 Method

According to sugiyono in 2008, research methodology is interpreted as a scientific way to get data with purpose and specific use.In accordance with the issues to be examined so method to be used in research is the method descriptive. Method according to Sukmadinata (2011) descriptive research is a form of research describing the facts, good phenomenon which is natural or engineered by man.This study aims to to describe the state of real its about completenessis integrated instrument components media (CII) for science in public primary schools sub district Siulak Mukai and public primary schools sub district Siulak. This research used the kind of research survey. Population in this research is public primary schools which below the uptd Siulak Mukai and uptd Siulak which consisted of 30. Even though they were divided into two district but the uptd Siulak Mukai with Siulak still joined , the uptd office is located in kecamatan siulak .While sample in this research was 12 public primary schools.
Research locations are 12 public primary schools in kecamatan siulak and kecamatan siulak mukai kabupaten kerinci .Research locations in Siulak disric there are 6 public primary schools among other: public primary schools 114/III desa baru siulak , public primary

schools 044/III siulak gedang, public primary schools 123/III Dusun Dalam, public primary schools 075/III Koto Rendah, public primary schools 088/III Siulak Panjang, public primary schools 041/III Koto Beringin. While in Siulak Mukai district there were 6 of public school among other: public primary schools 062/III Mukai Mudik, public primary schools 078/III Mukai Hilir, public primary schools 122/III Mukai Tengah, public primary schools 077/III Mukai Tinggi, public primary schools 169/III Mukai Mudik, public primary schools 216/III Sungai Langkap. The choice of research location was due to the fact that it the observation has been done about integrated instrument components media (CII) in some public in district Siulak and Siulak Mukai.

Techniques used to collect the data in this research were: (1) a direct observation, according to (Hadari Nawawi in 2007), observation directly is the method for data collection done through the observation and recording symptoms looked on objects of research directly in a place where an event, state or situation was going on.An instrument or apparatus to gather the data in this research was sheets of observation, in the implementation of the technique of direct observation, an instrument used to take data is sheets of observation of a register of checks (chek list) about the supply of and completeness of integrated instrument components media (CII) for science in public primary schools district Siulak Mukai and public primary schools sub district Siulak. As for the media in the observation of completeness is media CII neraca, media CII magnet, media CII pesawat sederhana, media CII cahaya, media CII optik, media CII listrik, media CII bunyi, media CII astronomi, media CII air, media CII panas.

Data which have been collected are presented in the form of table and then it would be of the data analysis .Data analysis technique used is a method of non statistics , namely description data analysis which means of the data collected through research are reported as it was and analyzed descriptilvly to get an overview of the facts. For was used to the sheet observation formulas of the percentage of nana sudjana (2011):

$$\mathrm{P} = \frac{\sum X}{N} X100\% \tag{1}$$

Information : P = Percentageg media CII for science

$\sum X$ = the number of the media that was available

N = the maximum number media CII for science

3 Result and discussion

3.1 Result

In accordance with the problems described above, some sort of data collected expressed about complexity integrated instrument components media (CII) for science in primary schools in siulak mukai and primary schools in siulak totaling twelve primary school .The data from the observations are about complexity integrated instrument components media (CII) for science. But before collecting data integrated instrument components media (CII) for science equipment must first be known the availability of integrated instrument components media (CII) for science. The total amount of integrated instrument components media (CII) for science available at elementary school Siulak Mukai sub-district is 42 sets, spread over six elementary schools. Furthermore, the total number of integrated instrument components media (CII) for science in elementary school Siulak subdistrict is 37 sets, spread over six elementary schools. Then the complexity of integrated instrument components media (CII) for science in the two subdistricts. the data obtained about the complexity / completeness is analyzed and then displayed in table form as follows.

Table 1. Description Results of integrated instrument components media (CII) for science completeness.

No	Media name CII	Complete	Quite complete	Incomplete
1	CII neraca		4	1
2	CII magnet			
3	CII pesawat sederhana		1	4
4	CII cahaya		1	3
5	CII optik			3
6	CII bunyi			6
7	CII listrik		2	2
8	CII astronomi		4	1
9	CII Panas		3	2
10	CII Air		2	3
Number			17	25
Average				

Furthermore, the acquisition of data of integrated instrument components media (CII) for science equipment at elementary school Siulak District, the data obtained in the analysis then displayed in the form of tables as follows.

Table 2. Description Results of integrated instrument components media (CII) for science completeness.

No	Media name CII	Complete	Quite complete	Incomplete
1	CII neraca		4	1
2	CII magnet		1	
3	CII pesawat sederhana	1	2	2
4	CII cahaya	1	1	2
5	CII optik	1	1	
6	CII bunyi	1	1	2
7	CII listrik	2		1

8	CII astronomi		1	3
9	CII Panas	1	3	1
10	CII Air	2	2	
Number		9	16	12
Average				

3.2 Discussion

Based on observations about the availability of in integrated instrument components media (CII) for science two district , the availability of integrated instrument components media (CII) for science in public primary schools sub district siulak mukai higher compared with public primary schools sub district siulak . In contrast, for the complexity / completeness of integrated instrument components media (CII) for science elementary school at Siulak sub-district is higher than elementary school at Siulak Mukai sub-district.

10 integrated instrument components media (CII) for science which were observed in 6 primary schools. at Siulak sub-district is available 37 integrated instrument components media (CII) for science t are located in 6 primary schools 9 integrated instrument components media (CII) for science are complete , 16 integrated instrument components media (CII) for science are quite full and 12 integrated instrument components media (CII) for science are incomplete .While at public primary schools sub district Siulak Mukai 42 are not are complete , 17 integrated instrument components media (CII) for science are quite full and 25 integrated instrument components media (CII) for science are incomplete. integrated instrument components media (CII) 17 science are quite complete and 25 integrated instrument components media (CII) for science not complete. So, public elementary school in sub discrict siulak more allows the success of learning by using integrated instrument components media (CII) for science compared with public elementary school sub discrict Siulak Mukai.

Viewed from the availability of integrated instrument components media (CII) for science, elementary sub districts mukai more than elementary school Siulak but when viewed from the complexity, integrated instrument components media (CII) for science in elementary school Siulak subdistrict more complex / complete from elementary school sub district siulak Mukai. This is because from one CII media observed consists of several components, for example CII balance sheet there are several components in it, among other scales, weighing scales, spring balance etc. At elementary school disctrict Siulak, each integrated instrument components media (CII) for science found to have complete components although not in all schools, while elementary school at district Siulak Mukai in general integrated instrument components media (CII) for science is available but the components are not complete.

4 Conclusion

Based on the results of research in obtained , it can be inferred that the availability of integrated instrument components media (CII) for science in public primary schools district Siulak Mukai is higher than that of public elementary school Disrict Siulak but on the whole they allow success of learning using these forms of media . furthermore the complexity / completeness media kits ipa in public primary schools sub discrict siulak is more complete that at public primary schools sub discrict Siulak Mukai.

References

A. Siska, R.T. Sari. J.PDN **2,** 232 (2017)

Anitah, S, *Media Pembelajaran* (LPP UNS, UNS PRESS, 2009)

H. Nawawi, *Metode Penelitian Bidang Sosial* (Gajah Mada University Press, 2007)

M. Riko. J.PP **2,** 6 (2013)

S. Nana, *Penilaian Hasil Proses Belajar Mengajar* (Remaja Rosdakarya, 2009)

Sugiyono, *Metode Penelitian Kuantitatif, Kualitatif dan R & D* (Alfabeta, 2008)

Sukmadinata, *Metode Penelitian Pendidikan. Bandung* (Remaja Rosdakarya, 2011)

Umra Hi. A. Ambai,. I. Said,. Ratman. J.KTO **2,** 78-79 (2013)

W. Adi. J.S **1,** 156 (2011)

The Profile of Adolescent' Moral Intelligence and Practical Solution to Its Improvement Efforts

Hardi Prasetiawan, Muya Barida
Universitas Ahmad Dahlan, Yogyakarta

ABSTRACT: High moral quality is needed by adolescent to be success in their personal and educational life. Those who have high morale quality can be considered as morally intelligent Individuals. This study is aimed to collect information about adolescents' moral intelligent by using moral intelligence Instrument that consists of seven aspects, empathy, self control, conscience, respect, kindness, tolerance dan fairness, furthermore, alternative solution is given to improve their moral intelligence. Subject of this study was selected through simple random sampling. Based on the result of the survey, it was found that students in a State Junior high school in Yogyakarta have 69% moral intelligence percentage. In other words, they have medium moral Intelligence. Adolescents' moral quality develops through continuous process in their life. One of many methods that can be done to improve moral intelligence is guidance and counseling service, in specific, group counseling with self- management technique.
Keywords: Adolescents, Moral Intelligence, Self- Management Technique

1. INTRODUCTION

Moral becomes the primary issue in human life. Its problems exist since the human exist. Even Muhammad SAW was sent to improve human morality, as it is quoted from his words: "I have only been sent to perfect good moral character. Philosophers such as Socrates, Aristotle, Ibn Ruysd, Al Gazhali, up to Imanuel Kant realize the importance of moral factor in human life (Muthahhari, 2008). In psychological study, Piaget and Lawrence Kohlberg focused their study on discussing children until adults' moral development. Those studies indicate that one of primary base of a nations' welfare and advancement substance is moral.

Someone' morality can be seen from their behavior. Data from Koran Rakyat Merdeka (in Azhar & Putri, 2009) about the result of National Child Protection Commission' survey explains that, in 2007 97% of 4500 adolescents in several big cities in Indonesia have watched porn movies, 93% of them have kissed their boyfriend or girlfriend. While 62,7% adolescents in Junior High School level already have sex and 21,2% Senior High School level students have ever done abortion. According to Taher (2007), this condition does not naturally happen, yet it is caused by their environments which have been contaminated by products with outside values. That fact becomes a sign that this nation is experiencing severe moral decadence, and if it is ignored, the existence of this nation and religion will collapse.

Borba (2008) states that many attempts and methods have been done in order to develop positive moral characters in children (Akhlakul Karimah), starts from social approaches, conflict resolution ability, stress management, self confidence that is told by the teachers, until Howard Garder' idea about multiple Intelligence and Daniel Goleman with his idea about Emotional intelligence, yet moral crisis still continues. Therefore, one of effective solutions is directing children and adolescent' ability to steadily understand about what is right and wrong. This ability can be called as Moral Intelligence

Term 'moral' is adopted from Latin "mos" (plural: mores) which means habit. Yusuf (2008: 132) added that moral, asides from habit, also means a rule, values, or procedures in life. Studies about moral that have been widely known are Freud, Piaget, Kohlberg, Gilligan and Turiel (Çam et. al., 2012).

Bull (in Devine, 2006: 4) suggested that moral is All morality consists of relationships between persons; that its three concerns are therefore, self, others and the relationship between them; and that the heart of morality is therefore respect for persons. Furthermore, Lennick & Kiel (2008:31) defined Moral intelligence as the mental capacity to determine how universal human principlesshould be applied to our values, goals, and action.

Borba (in Moghadas & Khalegi, 2013) defined moral intelligence as the capacity to distinguish right from wrong and to act according to a moral certainty in order to provide correct and honorable behavior. This type of intelligence shows the human mental capacity to determine how universal human principles, values and goals related to the actions. Also moral Intelligence represents one's ability to set standards of excellence and passion beyond his interests and even in the midst of such topics as the effectiveness of individual reactions.

Clarken (2009) stated that moral intelligence refers to the ability to apply ethical principles to personal goals, values and actions. Moral Intelligence is considered as Individual' ability in understanding what is right out of what is wrong, having strong ethic beliefs, and act or behave for the right and honorable things (Olusola, Ajayi, & Samson, 2015). Adolescents with high moral intelligence always do good things for themselves, or even other people.

Based on theoretical description about moral intelligence, it can be said that moral intelligence is individual' moral ability that involves emotional and intellectual factor to think, behave, and act based on the society' values which can be applied in life' purposes and actions.

According to Borba (2008: 7-10), there are seven virtues that become the main essence as foundation to act and behave ethically. To describe adolescent' moral intelligence in this study, these virtues become the foundation: 1) Empathy, Ability to feel and understand other people' worries; 2) Conscience, Ability to know and apply how to act correctly; 3) Self Control, Ability to control mind and act in order to resist negative inside or outside encouragement, so they are able to act correctly; 4) Respect, Ability to appreciate other individual and act nicely and politely; 5) Kindness, Ability to show concern to other individual' well-being and feeling; 6) Tolerance, Ability to respect other individual' right and dignity, although they have different faith and behavior; and 7) Fairness, Ability to be open minded and acts honestly and correctly.

Yusuf (2008) revealed that morality concept from childhood until adolescence is dominantly influenced by environment factor, especially family environment, society environment, and school environment. Process of moral concept understanding (right-wrong, good-bad) on children and adolescents is totally different. Children' understanding about moral tends to be *heteronomy* since they understand that adult' rules is a law, this concept belongs to Piaget, which was then developed by Kohlberg at its *pre-conventional* level. While for the adolescents, the concept of moral develops to conventional level and post-conventional level. This adolescents' moral character is important and able to develop through moral character development guidance program so that is able to shape adolescents' personality.

This article discusses the result of adolescent' moral intelligence survey at Junior High School level students in Yogyakarta. This article also gives alternative solution in improving moral intelligence practically through group counseling service with self-management technique.

2. METHODS

This Study used quantitative approach with descriptive survey method (Lodico, Spaulding, Voegtle, 2010: 199). The subjects of this study are chosen through random sampling technique in a State Junior High School of Yogyakarta (Leedy & Ormrod, 2005). The collected data was information about moral intelligence level. It was in form of numerical data.

The used instrument to collect the data of this study was moral intelligence scale, which consists of seven aspects namely empathy, conscience, self-control, respect, kindness, tolerance, and fairness. The scale was consists of eighty eight reliable and valid items. Empathy aspect consists of twenty two items (adapted from Davis, 1980), Conscience aspect consists of eight item (adapted from Lerner, 2005), Self-control aspect consists of ten items (adapted from Tangney, Baumeister, & Boone, 2004), Respect aspect consists of eleven items (adapted from Hendrick & Hendrick, 2006), Kindness aspect consists of thirteen item (adapted from Comunian, 1998), Tolerance aspect consists of twenty two items (adapted Thomae, Birtel, &Wittemann, 2016), and fairness aspect consists of two items (ddapted from Bharadwaj, 2014).

The Instrument used 4 level likert scales, namely very suitable, suitable, unsuitable, and very unsuitable. For favorable items, the score for very suitable answer equals to 4, suitable answer equals to 3, Unsuitable equals to 2, very unsuitable equals to 1. While for unfavorable items, score 1 is given for very suitable answer, 2 for suitable, 3 for unsuitable, and 4 for very unsuitable. Afterwards, the data analyzed by using descriptive statistic analysis technique which is observed from percentage of total score. Percentage of 76% - 100% is categorized as high, 56% - 75% is categorized as medium, while if the percentage is less than 56%, it is categorized as low.

3. RESULTS

3.1 Empathy Aspect Analysis

The average percentage of students' empathy is 68% and it is categorized as Medium.

3.2 Self-Control Aspect Analysis

The average percentage of students' self-control is 66% and it is categorized as Medium.

3.3 Conscience Aspect Analysis

The average percentage of students' conscience is 72% and it is categorized as Medium.

3.4 Respect Aspect Analysis

The average percentage of students' respect is 71% and it is categorized as Medium.

3.5 Kindness Aspect Analysis

The average percentage of students' empathy is 68% and it is categorized as Medium.

3.6 Tolerance Aspect Analysis

The average percentage of students' tolerance is 69% and it is categorized as Medium.

3.7 Fairness Aspect Analysis

The average percentage of students' fairness is 68% and it is categorized as Medium.

3.8 Adolescents' Moral Intelligence level Analysis

The average percentage of students' moral intelligence is 66% and it is categorized as Medium.

4. IMPLICATIONS

Human have different moral reflection and its development and change occurs gradually (Plümer, 2007). Adolescents with medium moral intelligence level need assist when facing their problems since moral intelligence will directly affect them as citizen (Siphai, 2015). Guidance and Counseling services are needed in order to do curative attempt related to personal and social adolescents' problem. There is a guidance and counseling service which is able to be done in solving problem together namely Group Counseling.

One of many abilities that counselors must have is ability to be able to do counseling in group activity. Being seen from the angle of intertwined relationship between counselee and counselors, according to Natawidjaja (2009: 3) group counseling is a guidance technique which is done in form of group. Group guidance and counseling utilizes group atmosphere as treatment and remedial medium or to develop counselee individually. Jacobs & Harvill (2012: 19) stated that group counseling is better because members need the input from others, plus they learn more from listening than talking. In many instances with teenagers, group counseling is better than individual counseling because teenagers often will talk more readily to other teenagers than with adults. For those stuck in the grief process, groups have been found to be very valuable.

Corey (2012: 4) defined group counseling has preventive as well as remedial aims. Generally, the counseling group has a specific focus, which may be educational, career, social, or personal. Counseling groups are often problem oriented, and the members largely determine their content and aims. Group members typically do not require extensive personality reconstruction, and their concerns generally relate to the developmental tasks of the life span. Group counseling tends to be growth oriented in that the emphasis is on discovering internal resources of strength.

Based on some experts' definition about group counseling above, it can be conclude that a service that is done by counselor to a group of individual who experiencing some problems in order to solve them, so their potential will not be hampered.

Gibson & Mitchell (2011: 275) propose the term of group counseling refers to routine adjustment or experience development in group scope. Group counseling is aimed to assist counselee overcome their problems through daily self adjustment and self development.

Group counseling has sequence of process or procedure. Corey (2011:71) defined group process as development steps of a group and characteristic of each step. The followings are process of group counseling: 1) Beginning Stage, 2) Transition Stage, 3) Working Stage, dan 4) Termination Stage. One of many techniques that are able to be applied in group counseling is self management.

Group counseling service is one many guidance and counseling services in school. Group counseling is an attempt given to the students in order to solve their personal and social problems together, so that they can develop dynamic social relation in their group environment. Self management technique can be used in group counseling.

Self Management can be defined as individuals' ability to manage themselves. Nursalim (2013: 149) stated that self management is a process where counselee directs their own behavior through one or more strategies. Counselee should actively move the internal and external variable in order to get the desired changes.

Furthermore, Woolfolk (2009: 409) conveyed that self management is managing our own behavior and being responsible for our own action. In the behavioral theory, behavioral principles are used to change the behavior.

Based on some definition above, it can be concluded that self management is a behavior change strategy which in the process, counselee (adolescent) direct their behavioral change through a technique or therapeutic technique combination.

Self management technique belongs to cognitive behavioral modification and cognitive behaviorism therapy. Woolfolk (2009) said that the root of self management is operant conditioning theory with social cognitive technique. Generally, Self management means technique that makes students include in basic steps in behavior change program. Basic assumption on cognitive behavioral counseling is that the process of group counseling is able to change someone' cognition. It takes important role in self management process and behavioral change that is controlled by internal and external interactions that possibly affect them outside counseling session.

Nursalim (2013) explained that self management consists of self monitoring, stimulus control and positive reinforcement (self reward) Practically, self management has some excellences such as improving individual' understanding about environment and decreasing their addiction to the counselors or others, practical, inexpensive, and easily answered. So in this case, self management technique can improve students or adolescents' moral intelligence.

According to Nursalim (2013: 153), there are three strategies of self management namely Self Monitoring, Stimulus Control and Self Reward. Self monitoring is a process where counselee observes and notes something about themselves and their interactions with environment' situation. Stimulus control is an arrangement or planning of environment' determined conditions that become the reason of why a behavior is done. Stimulus control emphasized the environment rearrangement or modification as particular reason of a certain reaction. Self reward is used to strengthen or improve the desired response (Nursalim 2013: 157-158). The function of self reward is to accelerate the appearance of targeted behavior. Through combination of self monitoring, stimulus control, and self reward, it is expected that adolescents can improve positive moral intelligence.

Self management technique is a strategy of behavioral changing and development that emphasizes the importance of self responsibility and effort in order to change and develop adolescents' behavior. The process of this behavioral change is mostly done by the counselee themselves, not directed by other people (counselor). So in this study it can be said that group counseling model with self management technique can be applied to improve adolescent' moral intelligence.

Group counseling with self management technique focuses on behavioral change and integrates changes that occur between its sessions. In order to improve moral intelligence, identification is needed to find out the level of moral intelligence. Moreover, treatment can't be done just once since evaluation is needed in order to see how far this counseling model can improve adolescents' moral intelligence. Evaluation is done by the counselor together with the counselee to see whether the counseling target is achieved or not.

5. CONCLUSION

Moral intelligence is individual' moral ability that involves emotional and intellectual factor to think, behave, and act based on the society' values which can be applied in life' purposes and actions. Moral intelligence consists of seven aspects namely empathy, conscience, self-control, respect, kindness, tolerance, and fairness. Adolescents can develop those aspects start from the first three aspect, and then when the first three aspects is done, they can develop the last four which are consist of respect, kindness, tolerance, and fairness.

A set of primary virtue that then is called as moral intelligence can be improved through group counseling with self management technique model Self management technique consist of self monitoring, stimulus control dan self reward. Firstly, Adolescents monitors their mind, feeling, and behavior in daily life. Then they control their mind, feeling and behavior to the positive way. If they can positively directing themselves to the positive context, they may give themselves a reward. Through self reward, they will be encouraged to improve their moral intelligence.

6. REFERENCE

Azhar, MH. (2009). Kecerdasan Moral Pada Remaja Yang Mengalami Deviasi Mothering. *Jurnal Psikologi*, 2 (2).

Bharadwaj, P. (2016). Fairness and Social Justice: Distinct Moralities. *Masters These*, 337.

Borba, M. (2008). *Membangun Kecerdasan Moral*. Jakarta: Gramedia Pustaka.

Çam, Z., Çavdar, D., Çok, F., & Seydooğullari, S. (2012). Classical And Contemporary Approaches For Moral Development. *Educational Sciences: Theory & Practice*, 12 (2): 1222-1225.

Clarken, R.H (2009). Moral Intelligence in Schools. *Online Submission*. Paper presented at the Annual Meeting of the Michigan Academy of Sciences, Arts and Letters (Detroit, MI, Mar 20, 2009).

Comunian, A.L. (1998). The Kindness Scale. *Psychological Reports*, 83: 1351-1361.

Corey, G. (2011). *Teori dan Praktek dari Konseling dan Psikoterapi Edisi Ke 4*. Semarang: IKIP Semarang Press.

Corey, G. (2012). *Theory & Practice of Group Counseling Eighth Edition*. Pacific Groove. California: Brooks/Cole.

Davis, M.H. (1980). A Multidimensional Approach to Individual Differences in Empathy. *JSAS Catalog of Selected Documents in Psychology*, 10 (85).

Devine, S. (2006). *What is Moral Education*. Nourthampton: Susan Devine.

Gibson, R. & Mitchell M. (2011). *Bimbingan & Konseling*. Cetakan I. Yogyakarta: Pustaka Pelajar.

Hendrick, S. S., & Hendrick, C. (2006). Measuring Respect in Close Relationships. *Journal of Social and Personal Relationships,* 23: 881-899.

Jacobs, M. & Harvill. (2012). *Group Counseling Strategies And Skills Seventh Edition*. West Virginia University America: Brooks/Cole.

Leedy, P.D. & Ormrod, J.E. (2005). *Practical Research: Planning and Design 6th edition*. Amerika Serikat: Pearson Education.

Lennick, D. & Kiel, F. (2008). *Moral Intelligence, Enhancing Business Perfomance & Leadership Success*. New Jersey: Pearson Education Inc.

Lerner, R.M. (2005). *Social Conscience Scale*. Dikses https://cyfar.org/sites/default/files/PsychometricsFiles/Social%20Conscience%20(middle%20school)_1_0.pdf

Lodico, M.G., Spaulding, D.T., & Voegtle, K.H. (2010). *Methods in Educational Research: From Theory to Practice second edition*. United States of America: John Wiley & Sons, Inc.

Moghadas, M. & Khaleghi, M. (2013). Investigate of Relationship Between Moral Intelligence and Distress Tolerance In Isfahan Staff. *International Journal of Research In Social Sciences*, 2 (2).

Muthahhari, M. (2008). *Falfasatul Akhlaq*. (Terjemahan) *Quantum Akhlak*. Yogyakarta: Arti.

Natawidjaja, R. (2009). *Konseling Kelompok Konsep Dasar & Pendekatan*. Cetakan I. Bandung: Rizqi Press.

Nursalim, M. (2013). *Strategi & Intervensi Konseling*. Cetakan I. Akademia Permata. Jakarta: Indeks.

Olusola, O.I., Ajayi, Samson, O. (2015). Moral Intelligence: An Antidote to Examination Malpractices in Nigerian Schools. *Universal Journal of Educational Research*, 3 (1): 32-38.

Plümer, S.L. (2007). Education in Values and Moral Education in Vocational Colleges. *European Journal Of Vocational Training*, 41 (2): 103-115.

Siphai, S. (2015). Influences of Moral, Emotional and Adversity Quotient on Good Citizenship of Rajabhat University's Students in the Northeast of Thailand. *Universal Journal of Educational Research*, 3 (2): 32-38.

Taher, T. (2007). *Membenahi Moral Bangsa, Meneladani Akhlak Rasul*.

Tangney, J.P., Baumeister, R.F., & Boone, A.L. (2014). High Self-Control Predicts Good Adjusment, Less Pathology, Better Grades, and Interpersonal Success. *Journal of* Personality, 271-324.

Thomae, M., Birtel, M.D., & Wittemann, J. (2016). *The Interpersonal Tolerance Scale (IPTS): Scale Development and Validation*. Paper Presented at the 2016 Annual Meeting of the International Society of Political Psychology, Warsaw, Poland, 13th – 16th July 2016.

Woolfolk, A. (2009). *Educational Psychology Active Learning Edition Tenth Edition*. Cetakan I. Yogyakarta: Pustaka Pelajar

Yusuf, S. (2008). *Psikologi Perkembangan Anak & Remaja*. Bandung: Rosda.

The Effect of Group Investigation Learning Model with Brainstroming Technique on Students Learning Outcomes

Kadek Ayu Astiti

Physics Education Study Program, Mathematics and natural science department
Faculty of Teacher Training and Education, Nusa Cendana University
Kupang, Indonesia

email: kadekayuastiti88@yahoo.com

Abstract

This study aims to determine the effect of group investigation (GI) learning model with brainstorming technique on student physics learning outcomes (PLO) compared to jigsaw learning model with brainstroming technique. The learning outcome in this research are the results of learning in the cognitive domain. The method used in this research is experiment with *Randomised Postest Only Control Group Design.* Population in this research is all students of class XI IPA SMA Negeri 9 Kupang year lesson 2015/2016. The selected sample are 40 students of class XI IPA 1 as the experimental class and 38 students of class XI IPA 2 as the control class using simple random sampling technique. The instrument used is 13 items description test. The first hypothesis was tested by using two tailed t-test. From that, it is obtained that H_0 rejected which means there are differences of students physics learning outcome. The second hypothesis was tested using one tailed t-test. It is obtained that H_0 rejected which means the students PLO in experiment class were higher than control class. Based on the results of this study, researchers recommend the use of GI learning models with brainstorming techniques to improve PLO, especially in the cognitive domain.

Keywords: Group Investigation Learning Model, Brainstroming technique, Students Learning Outcomes

INTRODUCTION

Progress of a nation is determined by the quality of the education system. Aware of this, every nation is competing with various efforts to improve the quality of education in their respective countries. The strategic issues of education in Indonesia continue to roll and are not solved, although various solutions continue to be done. As for the background of the occurrence of these strategic issues is due to discrepancy, diversity, and inequity [7]. These strategic educational issues are reinforced by various research results, like: 1) Research by Suyanto and Hisham in 2000 stated that in the micro scale the learning process in almost all levels of education focuses only on the left brain, while the right brain is rarely developed. This condition causes national education not to produce people who are independent, creative and have self awareness [7], 2) In Education Development Index (EDI) 2011 Indonesia is ranked 69th of 127 countries, 3) According to the Education For All Global Monitoring Report 2012 released by the United Nations Educational, Scientific and Cultural Organization (UNESCO), Indonesia's education is ranked 64th for education worldwide from 120 countries, 4) Based on data from the 2012 Program for International Student Assessment (PISA) 2012, Indonesia ranks 64th out of 65 countries. This shows the quality of education Indonesia is in the second worst position in the world after Peru [16].

Based on these strategic issues, efforts have been made to improve and improve the quality of education in Indonesia. Efforts to improve the quality of education is marked by the renewal of the education system from time to time, one of which is through curriculum renewal. The learning process that develops in the classroom is generally determined by the role of teachers and students as individuals directly involved in the process. Students' ability depends on the way the teacher teaches. According to Law no. 20 in 2003, learning is the process of interaction of learners with educators and learning resources in a learning environment.

In science lessons, particularly physics, active learning methods are very important to thoroughly understand. Physics is one of the science lessons, and is a discipline based on qualitative and quantitative measurements for understanding natural phenomena around us. Students have difficulty learning physics because physics consists of more abstract concepts. Many studies were conducted by researchers about physics, especially static fluid, to better understand them. Cooperative learning is a method in which students are assigned to small groups in the classroom, as well as other environments, where the students help each other to learn together. Students achieve more and increase their self-confidence as individuals, develop communication skills and participate actively in this method [11]. It has been concluded from a great deal of scientific research related to the cooperative learning method, that it is more effective for student achievement [11].

The Group Investigation (GI) technique was developed by Sharan and Sharan in 1989. In this technique, the class is divided into several groups that study in a different phase of a general issue. The study issue is then divided into working sections among the members of the groups. Students pair up the information, arrangement, analysis, planning and integrate the data with the students in other groups. In this process, the teacher must be the leader of the class and ensure that students comprehend the explanations. This technique is suitable in science lessons because it encourages students to learn and attracts them to scientific research [17].

Brainstorming is method is oriented towards students' ability to express ideas, further stimulating students to be good, courageous and active thinkers in learners who have to worry about criticism or mistakes. Based on these the authors are interested in conducting research by title "The effect of Group Investigation Learning Model with Brainstroming technique on Students Learning Outcomes".

METHODE

This research is experiment in SMA N 9 Kupang with *Randomised Postest Only Control Group Design* which are illustrated as in the Table below

Sample	Treatment	Post test
Experiment	X_1	T_1
Control	X_2	T_2

Table 1. Research Design

Population in this research is all student of class XI IPA SMA Negeri 9 Kupang year lesson 2015/2016. The sample consisted of students from ages 15 to17. The selected sample is 40 students of class XI IPA 1 as the experimental class and 38 students of class XI IPA 2 as the control class using Simple Random Sampling technique. All classes were given the same task and allotted time. The instrument used is 13 item description tests. Data collection and analysis techniques are carried out as follows.

Instrument	Instrument validation	Test prerequisite analysis	Hypothesis testing
description tests	Validity test, reliability test	normality test, homogeneity test	two tiled t-test, one (right) tailed t-test

Table 2. Data collection and analysis techniques

Instruments that have been made in the trial then tested the validity and reliability first so as to obtain the appropriate instrument used. Before performing the hypothesis test, then first test prerequisite analysis is tested normality and homogeneity test. The first hypothesis test was performed using a two tailed t-test. The statistical equations used for the first and second hypothesis are:

$$t = \frac{\overline{X_1} - \overline{X_2}}{S\sqrt{\frac{1}{n_1} + \frac{1}{n_2}}} \quad \text{.......(1)}$$

with $S^2 = \frac{(n_1 - 1)S_1^{\ 2} + (n_2 - 1)S_2^{\ 2}}{n_1 + n_2 - 2}$

(Sudjana, 2002)

testing criteria: accept H_0 if $-t_{1-1/2\alpha} < t_{acount} < t_{1-1/2\alpha}$ when t_{tabel} with opportunities $(1-1/2\alpha)$, dk 76 For the other t value H_0 is rejected.

The second hypothesis test was performed using a one tailed (right sided) t-test. Testing criteria: accept H_0 if $t_{acount} < t_{(1-\alpha)}$ when t_{tabel} with opportunities $(1-\alpha)$, dk 76 For the other t value H_0 is rejected.

RESULT & DISCUSSION

The results obtained in this study are average value of student learning outcomes of experiment class 72.05 and control class 63.31. This data is shown as the following diagram.

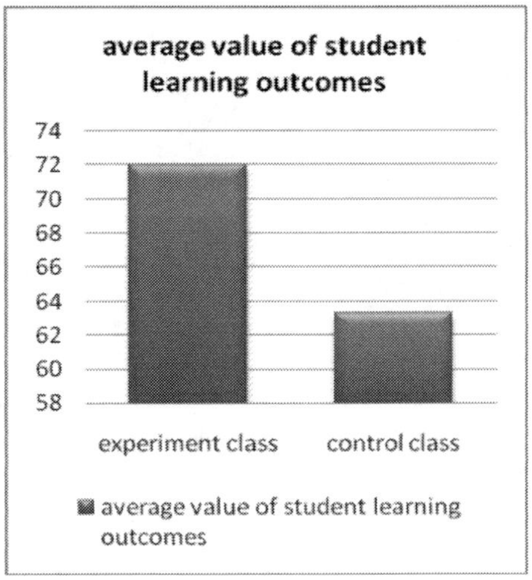

Image1.1 Average value of student learning outcomes.

Based on these data it can be seen that the students learning outcomes in the cognitive domain differ between the experimental class and the control class. It appears that the experimental class learning outcomes are higher than the control class. Statistical tests were performed using t-test. The first hypothesis test using two tailed t-test is obtained t_{count} 7.27 and t_{tabel} 1.99 with opportunities (1–1/2α), dk 76. Because $t_{count} > t_{tabel}$ so H_0 rejected and Ha accepted which means there are differences in learning outcomes in the cognitive domain. The second hypothesis test using one tailed t-test is obtained t_{count} 7.27 and t_{tabel} 1.66 with opportunities (1–α), dk 76. Because $t_{count} > t_{tabel}$ so H_0 rejected and Ha accepted which means the students physics learning outcome who were taught using brainstorming techniques on the GI learning model were higher than students who were taught using the brainstorming technique on the Jigsaw Learning model.

Based on the data obtained is seen GI learning model with brainstroming technique provide better influence on student learning outcomes, especially in the cognitive domain. Use of this model makes students more active in following the lessons, because students are involved since the selection of topics to be discussed and studied further. Students feel free to contribute to the topics covered. In addition students feel more enthusiastic because they will study the topics learned in accordance with the interest so that students do not feel burdened. At the group division stage, students are also given responsibilities in groups. At the stage of investigation after students have been in their respective groups, students will be free in creating and imagining so that self-confidence will also increase. Students working in groups will practice to work together through the discussion phase. At the discussion stage each student is required to express his opinion, so that students will get used to speaking systematically.

In the control class, students will be divided into groups of origin and expert groups. The sub topics students learned were obtained by means of students counting 1 to 2 in the original group. Students who get number 1 will study subtopic number 1 and students who get number 2 will discuss sub topic 2. Once all the students in the original group get their own sub topics, they will join students from other groups of origin who get the same sub topic. They will form a group of experts to discuss and study the material together. Each group of experts consists of 16 students. They will form a small group of 4 students to experiment. Based on the observations during the research, there are students who are less enthusiastic in studying the material discussed. This is because the sub topic obtained less favorable. In addition, some groups convey inappropriate information. It causes misconception. Some of these causes may be factors so that the learning outcomes in the control class are lower than the experimental class.

This study clearly supports the research that showed a relationship between Group Investigation learning model and student achievement, which identifies Group investigation learning model as a means to increase students' academic performance. Relevant research is a study that has previously been done by Garonia L. Parchment which examined the effectiveness of jigsaw and group investigation cooperative learning methods in a 9th grade living environment (LE) class. This study was carried out in four different LE classes during the 2008-2009 academic year. This comparison showed the GI method rendering a slightly higher assessment average and lower standard deviation value (M=79, SD=12, N=45) than the jigsaw assessments (M=78, SD=24 & N=46). Percentages of Assessment Scores 65 or Higher in Jigsaw, Group Investigation, and Traditional Instruction Methods are shown on the following table:

Instruction	Mean
Jigsaw	79
GI	90
Traditional	68

Table 3. Percentages of Assessment Scores 65 or Higher in Jigsaw, Group Investigation, and Traditional Instruction Methods (G. L. Parchment, 2009)

Based on the data of table, GI seemed to be the most effective on student academic performance.

Based on the results of research conducted by David W. Johnson, Roger T. Johnson, and Mary Beth Stanne [2] it was found that GI learning model is superior to jigsaw learning model. The cooperative learning methods may be ranked by the size of the effect they have on achievement and by the number of comparisons available. When the impact of cooperative lessons is compared with competitive learning, when the impact of cooperative lessons is compared with individualistic learning. It is seen that the investigative group learning model is more awaited than the jigsaw learning model. In aditional, methods of cooperative learning were evaluated on five dimensions: (a) ease of learning the method, (b) ease of initial use in the classroom, (c) ease of long-term maintenance of use of the method, (d) robustness of the method (applicability to a wide variety of subject areas and grade levels), and (e) ease of method's adapting to changing conditions. Based on the evaluation, obtained the total score of group investigation learning model higher than jigsaw learning model.

Some of the results are in line with the results of this study. The updated result of this research is using brainstroming method in each model.

CONCLUSION

Based on the results of research and data analysis, it can be concluded that there are differences in student physics learning outcomes in the cognitive domain between students taught by using Group Investigation (GI) learning model with brainstorming technique and students taught using Jigsaw learning model with brainstorming technique. The result of student physics learning outcomes that was taught by using Group Investigation (GI) learning model with brainstorming technique was higher than students taught using Jigsaw learning model with brainstorming technique.

The results of this study can be used as a reference for teachers to apply the Group Investigation (GI) learning model with brainstorming technique in the learning process in the classroom so that it can help overcome the problems faced, especially in the learning outcomes in the cognitive domain. For other researchers need to do research on other factors that can give a direct impact or influence on student physics learning outcomes such as learning methods, how to learn students, the quality of interpretation and student responses so that the learning process that occurs in the classroom more effective and efficient.

The preparation of this research and article supported by headmaster and science teachers of SMA N 9 Kupang, all those who have contributed comments and insights.

REFERENCES

1. A. Pitoyo, H. J. Waluyo, S. Suwandi, Adnyani, JEP, **5(1)**, 21-29 (2014)
2. D. W. Johnson, R. T. Johnson, M. B. Stanne, *Cooperative Learning Methods: A Meta Analysis*, (University of Minnesota, 2000)
3. E. A. Al-ziadat, M. H. Alsaaideh, D. R. Al-Elaimat, IJE, **5(3),** 184-199 (2013)
4. F. Ahsanah, TELL J., **3(1)**, 57-69 (2015)
5. G. L. Parchment, *A Study Comparing Cooperative Learning Methods: Jigsaw & Group Investigation* (2009)
6. H. C. Chen. AJKN, 69-82 (1999)
7. H. Nanang, S. Cucu, *Konsep Strategi Pembelajaran*, (Bandung: Refika Aditama, 2010)
8. I M. Astra, C. Wahyuni, H. Nasbey, JEP, **6(11),** 75-79 (2015)
9. J. F. D. Sitinjak, Sahyar, J. Inpafi, **3(2)**, 60-66 (2015)
10. J. S. Mari, S. A. Gumel, IJIET, **5(3)**, 196-199 (2015)
11. N. Okur akçay, K. doymuş, EBAD-JESR, **2**,109-123 (2012)
12. K. Smith, *Startegies for Effective Teaching a Handbook for Teaching Assistants.* University of Minnesota : E-book (2009)
13. M. N. Adora, IJHMS, **2**,146-147 (2014)
14. S. Jongsermtrakoon and J. Nasongkhla, IJIAET, **5(10)**, 783-788 (2015)
15. S. Ying Chu, IJMSR, **2**. 166-172 (2014)
16. Setiawan, Hakim, *Indeks Pembangunan Manusia*, (Universitas Islam Indonesia, 2013)
17. S. J. Sherman, *Cooperative learning and science* (Westport, CT: Greenwood Press. pp. 226–244, 1994)
18. S. Mehta, A. K. Kulshrestha, *Implementation of Cooperative Learning in science: A-Developmental-cum-Experimental Study,* (Hindawi Publishing corporation, 2014)
19. Sudjana, Metode Statistika, (Bandung: Tarsito, 2002)
20. V. D. Tran, IJHE, **3(2)**, 131-140 (2014)

Exploring the values of the local wisdom of kediri through harinjing and ceker inscription

Ita Kurnia[1] and Alfi Laila

Primary School Teacher Education, Teacher Training and Education Department, Nusantara PGRI University of Kediri, East Java, Indonesia

Abstract. Indonesia needs human resources in sufficient quantity and quality as a major supporter of development. To fulfill these human resources, education has a very important role is character development. One effort that can be done by the people of Kediri as an effort to build Indonesia is to use the values of local wisdom as a force. This values can be explored from Harinjing and Ceker Inscription which is located in Kediri, East Java. This research uses qualitative approach with descriptive method. The research purpose is to explore the potentials of local wisdom of Kediri people through (1) the history of Harinjing Inscription and (2) the history of Ceker Inscription. The results of this research that (1) The construction of embankments and holy buildings on the Serinjing River and the making of irrigation systems in Siman, Kepung, Kediri by Bhagawanta Bari proves that since the time of the kingdom, the people of Kediri have a caring to environmental and social, mutual cooperation, and religious character; (2) The people of Kediri also has the character of love of the homeland and faithful. This is evidenced through the inscriptions in Ceker found in the Sukoanyar, Mojo, Kediri.
Key Words: local wisdom, Kediri, Harinjing Inscription, Ceker Inscription.

1. Introduction

Local Wisdom consists of two words of wisdom and local (KBBI). In general, local wisdom can be interpreted as local ideas that are wise, full of wisdom, good value, which is embedded and followed by members of the community. In the discipline of Anthropology local wisdom known by the term local genius. Local genius is the term first introduced by Quaritch Wales. Many anthropologists are discussing about the mean of local genius, such as Haryati Soebadio who said that local genius is also local identity, identity or cultural personality of the nation that causes the nation is able to absorb and cultivate foreign culture according to the character and ability itself (K.A, Soekarto, 1985:18). While Moendardjito (K.A, Soekarto, 1985:40) said that elements of potential regional culture as local genius because it has proven its ability to survive until now.

Characteristic features local genius is 1) can to survive against the outside culture; 2) has the ability to accommodate elements of external culture; 3) have the ability to integrate elements of external culture into indigenous cultures; 4) have the ability to control; And 5) able to give direction to the development of culture. Local wisdom is basically tested naturally and undoubtedly is good because it is a repetitive social action reinforcement. If an action is not considered good by the public, it will automatically not get a continuous reinforcement. The movement naturally occurs voluntarily because it is considered good and contains goodness according to the local community.

Philosophically, local wisdom is a reflection of the cultural behavior of people who have background from the values of life held by the community for generations. For example, in the people of Kediri believes Javanese philosophy such as Urip Iku Urup, Sura Dira Jayaningrat Lebur Dening Pangastuti, Aja Kuminter Mundak Keblinger Aja Cidra Mundak Cilaka, Alon-alon Waton Klakon, Nrimo Ing Pandum, and others will be manifested in living the life of Javanese society. In addition, the philosophical values and order of daily life practices will also affect social life. Examples of life cooperative practiced in the life of the community for generations will inspire the next generation to do the same thing. Potential culture is the potential of local wisdom that needs to be explored and developed in modern society, especially the people of Kediri.

Disclosure of local wisdom values of the people of Kediri is limited to two things. The first is the Harinjing Inscription which is a testament to the success of Bhagawanta Bari in running the farming and irrigation system, thus being awarded tax-exempt land. The second is Ceker Inscription which is a proof of the loyalty of the people of Kediri against the king, so it is rewarded with tax-free land.

[1]itakurnia@unpkediri.ac.id

Local wisdom of the people of Kediri must be preserved because culture is not enough just to be told but must be inherited so as not to destroy. The values contained in the local wisdom of the people of Kediri can certainly be integrated into everyday life as one way to preserve local culture and introduce local wisdom to the next generation.

All of this can not be separated from the history of Kediri Regency which is had famous agricultural products since years a go. As one of the evidence is the village of Siman. It is the oldest village located in Kepung, Kediri, East Java. In Siman village there is Harinjing Inscription which is a monument which aims to give appreciation to Bhagawanta Bari who has been instrumental in establishing dawuhan or irrigation system consisting of channel and weir or embankment in Harinjing River (the people of Kediri called Serinjing River).

2. Method

This research uses qualitative approach with descriptive method. Qualitative research discloses comprehensively some cases relevant to the purpose of the study (S.E, Robert, 1995) with an eight-month research period beginning in August 2016 until March 2017.

The object of research is Harinjing Inscription and Ceker Inscription. The research was conducted in Siman Village, Kepung, Kediri Regency and Sukoanyar Village, Mojo, Kediri Regency.

Data collection techniques used by conducting participant interviews and documentation, as for the procedure of data analysis techniques using interpretive data analysis by re-checking by participants/member check (F.L, David, 1993).

3. Results And Discussion

3.1. Values of Local Wisdom in Harinjing Inscription

The people of Kediri knowledge about the environment is wide enough. The interaction between nature and humans can not be separated from the view that humans must be able to adapt to the natural environment, able to manage the natural wealth, and utilize natural wealth wisely. A number of empirical facts mention that some people live and take advantage of the natural wealth it has. The experiences and knowledge people get from the environment make them aware of the kinds of resources they have.

Based from the knowledge, the Siman community, Kepung, Kediri has a tradition of managing a fertile farm. The Siman community has the utmost responsibility to continue the tradition for future generations. The success of agriculture the people of Siman began since the time of Kediri Kingdom. This is evidenced by the existence of Harinjing Inscription.

Inscriptions harpings are carved in andesite stone. According to a report of PV van Stein Callenfels (1334) in De inscriptie van Soekaboemi, this inscription is found in Sukabumi plantations, Kepung, Kediri. It is written using ancient Java characters and languages. The inscription has a length of 75 cm, width 26 cm, and height 117 cm.

According to the sources, a research had once been done by experts of Javanologists such as Prof. MM. Soekarto Karto Atmojo. There is mention of Kediri on three Harinjing Inscriptions which are in Siman, Kepung, Kediri. The inscription consists of three charter written with ancient Javanese script and language on both sides. Here is the transcription of Harinjing Inscription.

3.1.1. *Harinjing A Inscription (25 March 804 AD)*

Front Side Inscription (Recto)

a. Swasti sakawarṣ ātīta 726 Caitra(Selamat tahun saka telah berlalu 726 Bulan Caitra [Maret-April])
b. māsa tithi ekādaśī śuklapaksa wāra ha. wa. so tatkā (tanggal sebelas. paro terang. harinya [Ha]riyang, [Wa]gai, [So]ma/Senin (diekuivalenkan ke masehi menjadi 25 Maret 804 Masehi). Ketika
c. la Bhagawanta Bārī i Culanggi sumaksyakan sīmaniran mula ḍawu(seorang pendeta agung [Bhagawanta] bernama Bārī dari Culanggi menyaksikan [mengudang orang untuk menjadi saksi] penetapan tanah sīmanya [tanah bebas pajak/perdikan] diperoleh dari pembuatan tanggul [ḍawuhan])
d. han gawainira kali i Hariñjing hana ta ḷ maḥ ḍapu bhī sang apatih a(dibuatlah sungai di Hariñjing adapun tanah milik Ḍapu Bhī seorang patih tua/sepuh)
e. tuha kāmbah deni kali hinêlyan ḷ maḥ satamwah de Bhagawanta Bārī pa(tergenang aliran [air] sungai yang mengalir [seluas] satu tamwah oleh Bhagawanta Bārī)
f. mêgêt sang panggumulan parttaya sang kamalagyan anu hinanākan rika kāla wa (Pemutus perkara bernama Sang Panggumulan, orang kepercayaan bernama Sang Kamalagyan [datang sebagai saksi], serta yang [juga] didatangkan pada waktu itu seorang pejabat)

g. huta sang labihan pitungtung ḍaman waca panigran ḍaman gundu parwuwus sang apatih ḍa(bernama Sang Labihan, tokoh yang ditinggikan bernama Ḍaman Waca, pembuat kesepakatan [?] bernama Ḍaman Gundu, Juru Bicara bernama Sang Apatih Ḍaman)
h. man gundu winêkas ḍaman gandhang parwuwus ḍaman gamêl lawan ḍaman cali gusti(Gundu, Pemberi Perintah bernama Ḍaman Gandhang, Juru Bicara bernama Ḍaman Gamêl serta Ḍaman Cali, Bangsawan)
i. ḍapu landap tuhaḍ ihulu ḍapu gundu …. watan. acur hulêr ḍapu …. kan apkan ḍ(bernama Ḍapu Landap, Kepala pimpinan para petugas bernama Ḍapu Gundu ……watan[?]. petugas pengatur pengairan bernama Ḍapu ….. kan[?], Petugas pengurus pasar bernama)
j. pu undun hulu kuwu ḍapu santan wuang atuha sang wulawan lawan ḍapu patêt tuha wêrêh ḍaman(Ḍapu Undun, Pimpinan Permukiman bernama Ḍapu Santan, orang yang dituakan bernama Sang Wulawan dan Ḍapu Patêt, Pimpinan para pemuda bernama Ḍaman Er)
k. er wariga ḍaman udaya citralekha ḍapunta waca nāhan sira rāma i paraḍah anu hinanākan(Ahli pertanggalan bernama Ḍaman Udaya, Penulis prasasti bernama Ḍapunta Waca, demikian [juga] para rāma [pimpinan daerah yang terhormat] dari Paradah yang didatangka)
l. rika kāla // rāma i bagu winkas ḍapu tahani gusti ḍapu natu tuhaḍ ihulu ḍapu garbu wka wuang(pada waktu itu // rāma dari Bagu [di antaranya] Penyampai pesan bernama Ḍapu Tahani, Bangsawan bernama Ḍapu Natu, Kepala pimpinan para petugas bernama Ḍapu Garbu, anak-anak muda)
m. matuha saka sang er lāwan ḍapu tiwul awatas ḍaman wanyaga apêkan ḍaman wahang …(pengatur hubungan pertemanan bernama Sang Er dan Ḍapu Tiwul, Petugas penentu batas tanah bernama Ḍaman Wanyaga, Petugas Pasar bernama Ḍaman Wahang …..)
n. tu ḍaman sampanna parwuwus sang majalu bsi lāwan ḍaman bangun wariga ḍapu taji tuha wêrêh daman dahara (tu [?] bernama Ḍaman Sampanna, Juru bicara bernama Sang Majalu Bsi dan Ḍaman Bangun, Ahli pertanggalan bernama Ḍapu Taji, Pimpinan para pemuda bernama Daman Dahara)
o. ni tuha buru ḍaman kuñja tuhalas ḍaman wacana lāwan ḍaman wihar tuha kalang ḍaman wanu(Pengawas perburuan bernama Ḍaman Kuñja, Pengawas hutan bernama Ḍaman Wacana dan Ḍaman Wihar, Pemimpin golongan kalang [mungkin tukang kayu atau pembuat bangunan] bernama Ḍaman Wanua)
p. parwuwus ḍaman lampuyang wadahuma ḍaman waca tuha padahi si bunta lāwan si kāryya kulapati i(Juru Bicara bernama Ḍaman Lampuyang, Pengatur masalah rumah tangga bernama Ḍaman Waca, Pimpinan pemain gendang bernama Si Bunta dan Si Karyya, tokoh masyarakat dari)
q. Hariñjing ḍapunta manḍi kulapati i lalatêng ḍapu aman kulapati i bubul dapunta karugnan tp siring anung (Hariñjing bernama Ḍapunta Manḍi, tokoh masyarakat dari Lalatêng bernama Ḍapu Aman, tokoh masyarakat dari Bubul bernama Dapunta Karugnan, dari daerah batas wilayah yang menjadi)
r. sākṣi ḍaman guntar i garaga ḍaman damên nāhan sira. aligrama sakṣinira sakṣyakên ikana sīmanira mu(saksi bernama Ḍaman Guntar, dari daerah Garaga bernama Ḍaman Damên itulah semuanya. Orang-orang yang telah datang menjadi saksi untuk melihat secara langsung penetapan status sīma-nya [Bhagawanta Bārī] yang diperoleh dari pembuatan)
s. ḍawuhan // kunang tinaḍah bhagawanta i ramanta gawai sang aswa syamah (?) warangan pitu, kawêrêhan inuman (tanggul [ḍawuhan] // Adapun yang diterima oleh Bhagawanta dari para pimpinan wilayah yang paling terhormat membuat selera makan menjadi nikmat, hidangan sayur-sayuran syamah [?] untuk berpesta ria tujuh minuman yang bisa membuat awet muda [kuat])
t. amunuhatasamada (?) awaknira sira kunang yan gawai i sabya …. salwiranira // niyanta wṛ ddhinira mu. an. sang pa (menaruh perhatian terhadap tubuh beliau apabila mengerjakan sesuatu yang pantas dilakukan ….. segala macam // benar-benar sebuah kemajuan [daerah] yang dikerjakan beliau [Bhagawanta Bārī] beserta Sang pemutus perkara)
u. mgat lasun deni …. bhagawanta ḍapu lanḍap …. anta ... an …(berasal dari Lasun oleh ... Bhagawanta, Ḍapu Lanḍap …. anta ….. an …)
v. ḍapu hadyan ḍapu sūra ḍaman kuñja panigran sang taruk ramānta ri tajam wina … sangka. i. sang arawisa(Ḍapu Hadyan, Ḍapu Sūra, Ḍaman Kuñja, pembuat kesepakatan [?] bernama Sang Taruk, Pimpinan wilayah yang terhormat dari Tajam wina[?]…. sangka[?]. dari Sang Arawisa)
w. i sangkawu i sangkāra i sang … i sangsala i sangngang nāhan kawṛ ddhini gawe Bhagawanta i(di Sangkawu, di Sangkāra, di Sang ... di Sangsala, di Sangngang, itulah sebuah kemajuan [daerah] yang dilakukan oleh Bhagawanta dari)
x. Culanggi //-// … i bagu i kukap i watu walu (Culanggi //-// … di Bagu, di Kukap, di Watu Walu)
y. …. ralan tpisiringnya watêk (… daerah perbatasan wilayah)

Right Side Inscription

a. Rangga manggêhan tangkilsugih ranḍa(Rangga, Manggêhan, Tangkilsugih, Raṇdamman)
b. mman nahan sakweh nikang wanua ka (itulah semua desa yang diikutsertakan)
c. samwahan deni gawai Bhagawanta Bārī (dalam pekerjaan [yang dilakukan oleh] Bhagawanta Bārī)

3.1.2. *Harinjing B Inscription (19 September 921 AD)*

a. // Swasti śakawarṣ ātīta 843(Selamat tahun saka telah berlalu 843)
b. aśujimāsa tithi pañcadasi śuklapaksa wāra ha. u. (Bulan Asuji [september-oktober]. tanggal limabelas. Paro terang. Harinya [Ha]riyang, [U]manis/legi)
c. bu nakṣ atra uttarabhadrawāda ahnibudhna dewāta wṛ sayoga ta([Bu]dha/Rabu (diekuivalenkan ke masehi menjadi 19 September 921 Masehi). Pada perbintangan terletak pada uttarabhadrawāda. Penguasa waktu terletak pada Waktu selama gerak bersamaan antara bulan dan matahari dalam posisi 13°20' terletak padawṛ sa)
d. tkala ajña Śrī Mahārāja Rake Layang Dyah Tuloḍ ong tinadah rakryan(ketika perintah dari Śrī Mahārāja Rake Layang Dyah Tuloḍ ong diterima oleh Rakryan)
e. mapatih i hino mahāmantri śrī ketudharamanimantaprabhāprabhusaktitri (Mapatih i Hino Mahāmantri Śrī Ketudharamanimantaprabhāprabhusaktitri)
f. wikrama umingsor ing rakryan mapatih halu wka sirikan kalungwarak tiruan muang (wirama kemudian diturunkan kembali kepada [pejabat tinggi istana yaitu] Rakryan, Mapatih dari daerah Halu, Wka, Sirikan, Kalungwarak, Tiruan dan)
g. pamgat bawang tiruan halaran kumonnakan saṣ ana sang dewata lumah i kwak ka(pemberi keputusan [terdiri dari] Bawang, Tiruan, Halaran memerintahkan surat perintah dari almarhum yang dicandikan di Kwak)
h. pagĕhakna ni wka Bhagawanta Bārī sang magawai kali i Hariñjing ikanang tan kolāhala (untuk diperkuat kembali oleh anak Bhagawanta Bārī seorang tokoh yang telah membuat sungai di Hariñjing [agar] tidak diganggu gugat)
i. deni patih wahuta rāma muang tan katamāna dening saprakāra ni mangilāla dṛ byahaji(oleh [pejabat istana terdiri dari] Patih, Wahuta, Rāma serta tidak didatangi oleh para abdi dalem istana [mangilāla dṛ byahaji])
j. kring paḍ ammapuy tapahaji erhaji tuhān tuha dagang tuhā huñjamman manghuri kutak ka([di antaranya] petugas pemadam kebakaran, penjaga tempat pertapaan milik raja [?], Penjaga air milik raja, para pimpinan, ketua perdagangan, ketua kelompok masyarakat, juru tulis di istana, Pemukul)
k. payungngan pakalangkang pamanikan limus galuh pengaruhan manimpiki malañjang lba kalungwa (Petugas Pembawa Payung [?], Pekerja Bangunan Kayu [?], Pembuat Manik-manik, Pembuat Perhiasan Emas, Tukang Emas, Tukang Ukir, Pengawas Perjudian, Asisten Pengawas Perjudian, Kalungwarak [?],)
l. rak wungkal tajam paranakan kḍ i walyan widu mangidung singgah mamṛ si hulun haji wata(Tukang Mempertajam Batu, para anak buah dari dukun wanita, dukun laki-laki, Biduan, Singgah [?], Pejabat Keagamaan, Pelayan Raja, [golongan] yang tinggal di)
m. k i jro ityewamādi saprakāra ni mangilala dṛ byahaji miśra paramiśra tan tamātah i(dalam [keraton] dan selanjutnya seluruh abdi dalem istana [mangilala dṛ byahaji] [khususnya] pemungut pajak usaha kerajinan tidak ingkar [memasuki])
n. sthā ni wka Bhagawanta Bārī i Culanggi sang girigil muang ikāng kabikuan mula sa .... mā ma (daerah tempat tinggal para anak Bhagawanta Bārī di Culanggi, Sang Girigil dan sebuah bangunan suci untuk para wiku [bhiksu] ...mā ma [?])
o. kangaran i wulahya ing waruk ing śambung ing wilang kewala pamasangnya mās su 1 maparah. (bernama di Wulahya, di Waruk, di Śambung, di Wilang hanya memasang [membayar] emas 1 ṣuwarṇ a diberikan)
... until
w. praśasti sāṣ ana śrī mahārāja //–//(dalam prasasti perintah śrī mahārāja)

3.1.3. *Harinjing C Inscription (7 March 927 AD)*

a. muwah śaka 849 caitramasa tithī pratipāda śukla (serta tahun saka 849 Bulan Caitra [Maret-April] Tanggal pertama paro terang)
b. wara wā. u. bu. Tatkala ni wka Bhagawanta Bārī mintonakan anugraha sang lumāh ri kwa (Harinya [wā]s, [u]manis, [bu]dha/rabu(diekuivalenkan ke masehi menjadi 7 Maret 927 Masehi). Ketika para anak [keturunan] Bhagawanta Bārī memperlihatkan sebuah anugerah dari almarhum yang dicandikan di Kwak)
c. k tinaḍ ah rake hino mpu ketu makasambandha pakon sang pamgat momahhumah kaka (diterima oleh Rakai Hino Mpu Ketu sebagai alasannya yaitu perintah dari Sang Pemutus Perkara tentang urusan perumahan kakak)
d. n rake sumba sang pamgat anggĕhan sang parpāt sira ta kumonnakan ikang prasasti umnahhakna ya wung (dari Rakai Sumba, Sang Pemutus Perkara tentang hubungan seseorang bernama Sang Parpāt, mereka memerintahkan agar prasasti dipahatkan pada batu)
e. kal kweh ni wka Bhagawanta Bārī i Culanggi sang giwil sang trayi sang saśrā sang pulas ri wilang sang banat (semua anak keturunan Bhagawanta Bārī dari Culanggi [di antaranya] Sang Giwil, Sang Trayi, Sang Saśrā, Sang Pulas di Wilang, Sang Banat)

f. sang durak ring sambung sang ngrawit sang uwir ri wulahya sang wasêh sang bayakā ri waruk sang wadhi sang kinang nāhan (Sang Durak di Sambung, Sang Ngrawit, Sang Uwir di Wulahya, Sang Wasêh, Sang Bayakā di Waruk, Sang Wadhi, Sang Kinang. Itulah [nama-nama]).

Left Side Inscription

a. kwehnira wka Bhagawanta Bārī umanhhakan nikang (semua anak keturunan Bhagawanta Bārī yang [meminta untuk] menuliskan)
b. prasasti ri wungkal kunang ta kweh (?) sang lumah ri kwa (prasasti di batu. Demikian pula sang tokoh yang dicandikan di Kwak)
c. k asing umulahhulaha ikang anugraha sangyang su (?) (siapapunn yang berani mengganggu gugat anugerah sang hyang su [?])
d. kram (?) upadrawa yan ….. kuha …. jnā….. (kram [?] akan mendapat bencana apabila…… kuha [?]…. jnā…….)
e. ………………… //// (………………………..).

(Casparis (1978) dalam Annisa, 2011; K.A, Soekarto, 1985:49-63; and J.Z, Petrus, 1982)

Harinjing A Inscription dated March 25, 804 AD, tells of a prominent priest (Bhagawanta) named Bari who built the embankment along with the river in Harinjing by bringing in some official figures as witnesses. The year is the reign of Rakai Warak Dyah Wanara who ascended the throne of Mḍang Kingdom (Ancient Mataram) in 803 AD (K.A, Soekarto, 1985:78).

Harinjing B Inscription Dated September 19, 921 AD, recounts the reinforcement of Bhagawanta Bari's land grant status (Perdikan) by King Rakai Layang Dyah Tulodong and given to the offspring of Bhagawanta Bari. The Harinjing Inscription C dates from March 7, 927 AD, tells the inscription to be a legitimate decree of the king and presumably when it is still written on a rontal (Ripta Inscription) or metal plate (Tamra Inscription) to be written on a stone. In the year 921 and 297 AD was the reign of Rakai Layang Dyah Tulodong in the Kingdom of Mḍang (Ancient Mataram) (about 919-927 AD) (S.H, Endang, et al, 2010:178 &182).

On lines 1 to 23 the Harinjing B Inscription states that Sri Maharaja Rake Layang Dyah Tulodong on 15 Suklapaksa month of Asuji in 843 Saka (19 September 921 AD) acknowledged the rights of pastors in Culanggi because they still have to maintain Harinjing channel.

Starting the next line, line 24 to last, this inscription states that a similar right is recognized at 1 Suklapaksa Caitra month of 849 Saka (June 7, 1015 AD). The contents of the three Harinjing Inscriptions are figures from the village of Culanggi named Bhagawanta Bari who received a land grant from Sri Maharaja Rake Layang Dyah Tulodong. The success of Bagawanta Bhari at that time, the effort to save the environment from the raging annual floods that threaten the region.

Based on the information in Harijing C Inscription can be seen that slate Harbing Inscription which is currently a collection of National Museum-Jakarta (inventory number D-173) was written on March 7, 927 AD.

Since then Kediri began to be mentioned as a country or kingdom because it is an independent region with a lot of farming. In Harinjing Inscription also mentioned the construction of irrigation system (which consists of channels and embankments) called dawuhan in Konto River, namely Harinjing River. Soil fertility around Brantas river caused by volcanic material deposits from some active volcano in the upstream of the river, namely Mount Kelud and Mount Semeru.

The story of Bhagawanta Bari as it is written in the Harinjing Inscription is full of positive values that should be exemplified by the whole society, especially Kediri people. Efforts to construct embankments to cope with floods are exemplary in order to foster environmental and social concerns.

In addition Bhagawanta Bari's attitude of building a holy building (dharma) around the Harinjing River which was later reinforced by his offspring, is highly relevant to religious values, ie obedient behavior of religious teachings adopted, tolerant of the implementation of other religious services, and living in harmony with followers of other religions (H.H, Said, et al, 2010:9-10). The values of local wisdom like this are very important to be passed on to the young generation today.

Bhagawanta Bari figure is a human figure who has a caring soul, high mutual cooperation, and religious. Three things that are reflected and easily understood and recognized to date.

a. Caring, both care about the environment and social. This attitude was shown by Bhagawanta Bari through building a embankment (irrigation canal) to overcome the flood disaster that occurred at the time.
b. Mutual cooperation, can be called working together is a value that is rooted in Javanese culture. The value of mutual cooperation is a basic value that can create a social network that serves to solve daily problems faced, namely working together to create a embankment (irrigation canal).
c. Religious. Local wisdom of Javanese is born, lived, and developed. Javanese culture manifested in various forms of representation either in the form of ideas, thoughts, moral/spiritual teachings, philosophical teachings as well as ideas and concepts of leadership and society. Religious values can be seen from the establishment of a holy building around the Harinjing river by Bhagawanta Bari.

3.2. Values of Local Wisdom in Ceker Inscription

Unlike the Harinjing Inscription, not much can be showed in the Inscription of Ceker. Ceker Inscription originating from Sukoanyar Village, Mojo, Kediri Regency from 1107 Saka (September 11, 1185 AD).

The inscription was made by Paduka Sri Maharaja Sri Kameswara Triwikramawatara Aniwaryyawiryya Parakrama Digjayotunggadewanama. The inscription was once declared missing (K.A, Soekarto, 1985:72). But later the Ceker Inscription known stored in the Museum Airlangga Kediri.

The inscription contains the story of the people of Ceker (duwan i Ckêr) came to the king and told him that he had received the grace from the previous ruler of the perdikan land. They asked that the inscription be asked to be strengthened or written on a stone. Furthermore, it is also told about the story that Sri Maharaja who returned to his position in Bhumi Kadiri (mantuk ri sīmanira ring Bhūmi Kadiri), thus raising the allegation that the event was related to the enemy's attack, so the king had to leave his palace (S.H, Endang, et al, 2010:292). Attitudes made by residents of Ceker (now Sukoanyar village) is relevant to the value of love of the homeland in the concept of cultural education and character of the nation.

Regarding the figure of Sri Maharaja Sri Kameswara once alleged by Prof. Dr. R.M.Ng. Poerbatjaraka as the character of Panji Inu Kertapati in the script of Panji Story. If there is a Panji Inu Kertapati name that has the wife of Goddess Candrakirana, while in Kakawin Smaradahana composed by Mpu Dharmaja during the reign Sri Maharaja Sri Kameswara, the king has a wife named Sri Kiranaratu (Poerbatjaraka, 1986:XI).

But this opinion is still causing debate among experts. More interesting again in Kakawin Smaradahana, it turns out Mpu Dharmaja in the manggala (opening) states that Sri Maharaja Sri Kameswara also as Panji. The complete quote is as follows, "yan riṅ praṅ kita siŋha wīra taruṇ ārjāpañji śūreṅraṇ a",meaning that in the battlefield you are like a brave lion, young and handsome named Panji Surengraṇ a (Manu, 1986).

The value of local wisdom that can be exemplary in Ceker Inscription is the love of the homeland and faithful. Because of the love and loyalty to the king, the people of Ceker get a land grant. This value is exemplary by the young generation in the middle of the moral gradation of children of the nation that is currently happening.

4. Cover

4.1. Conclusion

Referring to the findings of research and discussion on the values of local wisdom Kediri as an effort to preserve the culture of the nation, it can be concluded as follows:

a. The construction of embankments and holy buildings on the Serinjing River and the making of irrigation systems in Siman, Kepung, Kediri by Bhagawanta Bari proves that since the time of the kingdom, the people of Kediri have a caring to environmental and social, mutual cooperation, and religious.
b. The people of Kediri also has the character of love of the homeland and faithful. This is evidenced through the inscriptions in Ceker found in the Sukoanyar, Mojo, Kediri.

4.2. Acknowledgement

After exploring the potentials of local wisdom of the people of Kediri through the history of Harinjing Inscription and the history of Ceker Inscription located in Kepung and Mojo district, it is found that since the time of Kediri Kingdom have some values of character that should be emulated by the young generation today. These character values are among the environmental and social attitudes, the mutual cooperation, and the religious as they are written in the Harinjing Inscription.

Ceker Inscription also have the value of the love character of the homeland and faithful. But it is unfortunate considering the people of Kediri today not many people know the existence as well as exemplary values of the inscription. It is a duty for local governments and researchers and culturalists to work together to uncover this "treasure" in order to create a society of character as it has been abandoned by its predecessors.

References

Annisa, *Penggunaan kronogram di Indonesia,Vietnam, dan Kamboja abad VII-XIV M: pendekatan arkeologis dan epigrafis* (Tesis tidak diterbitkan, Fakultas Ilmu Budaya UI, Depok, 2011)

F.L, David, *Qualitative research in education, an introduction to the major tradition (Longman, New York, 1993)*

H.H, Said, et al, *Bahan pelatihan: penguatan metodologi pembelajaran berdasarkan nilai-nilai budaya untuk membentuk daya saing dan karakter bangsa (Badan Penelitian dan Pengembangan Pusat Kurikulum, Jakarta, 2010)*

J.Z, Petrus, *Kalangwan: sastra Jawa kuno selayang panjang*. Terjemahan Dick Handoko S.J. 1985 (Djambatan, Jakarta, 1974)

J.Z, Petrus, *Kamus Jawa Kuna Indonesia*. Terjemahan Darusuprapta & Sumarti S. 1994 (Gramedia Pustaka Utama, Jakarta, 1982)

K.A, Soekarto, *Hari jadi Kediri* (Lembaga Javanologi, Yogyakarta, 1985)

M, Pidarta, *Landasan kependidikan: stimulus ilmu pendidikan bercorak Indonesia* (Rineka Cipta, Jakarta, 2007)

Manu, *Kakawin smaradahana: sebuah studi filologi dalam rangka ilmu sastra* (Skripsi tidak diterbitkan, Fakultas Sastra Universitas Gajah Mada, Yogyakarta, 1985)

N.W, Sartini, Jurnal filsafat **37 (2),** 111—120 (2004). Dapat diunduh di: http://scholar.google.co.id/citations?user=nzQpUlcAAAAJ&hl=id

Poerbatjaraka, *Tjerita pandji dalam perbandingan* (Gunung Agung, Djakarta, 1986)

S, Muchlas dan Hariyanto, *Konsep dan model pendidikan karakter* (Remaja Rosdakarya, Bandung, 2012)

S.E, Robert, *The art of case study research* (Sage Publications, Thousand Oaks, CA,1995)

S.H, Endang, et al, *Zaman kuno. Dalam sejarah nasional Indonesia II* (Balai Pustaka, Jakarta, 2010)

S.S, Nana, *Metode penelitian pendidikan* (Remaja Rosdakarya, Bandung, 2005)

Suhartini, *Prosiding seminar nasional penelitian, pendidikan, dan penerapan MIPA* (Yogyakarta, 2009). Dapat diunduh di: http://eprints.uny.ac.id/12149/1/Bio_Suhartini2%20UNY.pdf

Tim Penyusun, *Kamus Besar Bahasa Indonesia* (Pusat Bahasa, Jakarta, 2008)

---. *Kebudayaan Jawa* (Gajah Mada University Press, Yogyakarta, 1984)

---. *Penduduk Kabupaten Kediri* (BPS, Jakarta, 2015)

---. *Buku kearifan lokal di tengah modernisasi. pusat penelitian dan pengembangan kebudayaan badan pengembangan sumber daya kebudayaan dan pariwisata Kementerian Kebudayaan dan Pariwisata Republik Indonesia* (Kementerian Kebudayaan dan Pariwisata Republik Indonesia, Jakarta, 2011)

AUTHOR INDEX

AUTHOR INDEX

AUTHOR INDEX

EDP Sciences – Web of Conferences
17, Avenue du Hoggar
Parc d'Activité de Courtabouf
BP 112
F-91944 Les Ulis Cedex A
France

ISBN 978-1-5108-5661-5

9 781510 856615

CURRAN ASSOCIATES INC.
proceedings.com